T0092797

Next Generation Wireless Terahertz Communication Networks

Next Generation Wireless Terahertz Communication Networks

Edited by

Saim Ghafoor, Mubashir Husain Rehmani
and Alan Davy

CRC Press
Taylor & Francis Group
Boca Raton London New York

CRC Press is an imprint of the
Taylor & Francis Group, an informa business

First edition published 2021
by CRC Press
6000 Broken Sound Parkway NW, Suite 300, Boca Raton, FL 33487-2742

and by CRC Press
2 Park Square, Milton Park, Abingdon, Oxon OX14 4RN

CRC Press is an imprint of Taylor & Francis Group, LLC

Library of Congress Cataloging-in-Publication Data
A catalog record has been requested for this book

ISBN: 9780367430726 (hbk)
ISBN: 9780367770426 (pbk)
ISBN: 9781003001140 (ebk)

Typeset in Sabon
by Newgen Publishing UK

Contents

Acknowledgements

First, we are thankful to Allah Subhanahu Wa-ta'ala that by His grace and bounty we were able to complete the editing of this book.

This book focuses on the next generation wireless terahertz networks and presents different wireless communication aspects for the terahertz frequency band communication.

We would like to thank all the chapter authors for their hard work and involvement from the beginning till the finishing of this book. We would also like to thank our reviewers for their valuable comments. They were of immense help in improving the overall quality of the book.

We would like to thank the TERAPOD project, which funded us throughout the book writing phase. We would also like to thank all the TERAPOD project partners for their appreciation and involvement.

Mubashir wishes to express gratitude to Sheikh Hazrat Mufti Mohammad Naeem Memon of Hyderabad, Pakistan. He could not have edited this book without his prayers, spiritual guidance, and moral support.

Last but not the least, we would like to thank our parents and families for their continuous love, motivation, and prayers.

Saim Ghafoor
Mubashir Husain Rehmani
Alan Davy

Editor Biographies

Saim Ghafoor

Saim Ghafoor received a Ph.D. in computer science from University College Cork, Ireland, in 2018. He completed his M.S. in computer science and engineering from Hanyang University, South Korea, in 2010 and B.E. in computer systems engineering from Mehran University of Engineering and Technology, Pakistan, in 2005. Currently, he is working as an assistant lecturer in Letterkenny Institute of Technology (LYIT), Ireland. He worked at Telecommunications Software and Systems Group (TSSG), Waterford Institute of Technology (WIT), Waterford, Ireland as a post-doctoral researcher from June 2018 to August 2020. He is currently an associate editor of the *Computer and Electrical Engineering Journal* (Elsevier) and serving as a reviewer for many reputable journals and conferences. He co-organized a workshop on terahertz wireless communication in IEEE INFOCOM 2019. He also received a best student paper award in International Conference on Disaster Management in 2017. He received best reviewer of the year award from *Computer and Electrical Engineering Journal* in 2014 and 2015, and many reviewer recognition certificates. His research interests are Terahertz and millimetre communication networks, autonomous and intelligent communication networks, cognitive radio networks and wireless sensor networks.

Mubashir Husain Rehmani

Mubashir Husain Rehmani received B.Eng. degree in computer systems engineering from Mehran University of Engineering and Technology, Jamshoro, Pakistan, in 2004, M.S. from the University of Paris XI, Paris, France, in 2008, and Ph.D. from the University Pierre and Marie Curie, Paris, in 2011. He is currently working as an assistant lecturer at Cork Institute of Technology (CIT), Ireland. He worked at Telecommunications Software and Systems Group (TSSG), Waterford Institute of Technology (WIT), Waterford, Ireland, as a post-doctoral researcher from September 2017 to October 2018. He served for five years as an assistant professor at COMSATS Institute of Information Technology, Wah Cantt., Pakistan. He is currently an area editor of *IEEE Communications Surveys and Tutorials*. He served for three years (from 2015 to 2017) as an associate editor of *IEEE Communications Surveys and Tutorials*. He is also serving as a column editor for book reviews in *IEEE Communications Magazine*. Currently, he serves as an associate editor of the *Journal of Network and Computer Applications* (JNCA) and the *Journal of Communications and Networks* (JCN). He is also serving as a guest editor of *Ad Hoc Networks* journal, *Future Generation Computer Systems* journal, the *IEEE Transactions on Industrial Informatics* and *Pervasive and Mobile Computing* journal. He has authored and edited two books published by IGI Global, USA, one book published by CRC Press, USA, and one book with Wiley, UK He received the 'Best Researcher of the Year 2015 of COMSATS Wah' award in 2015. He received the certificate of appreciation, 'Exemplary Editor of the IEEE Communications Surveys and Tutorials for the year 2015' from the IEEE Communications Society. He received Best Paper Award from IEEE ComSoc Technical Committee on Communications Systems Integration and Modeling (CSIM), in IEEE ICC 2017. He consecutively received research productivity award in 2016–17 and also ranked # 1 in all Engineering disciplines from Pakistan Council for Science and Technology (PCST), Government of Pakistan. He also received Best Paper Award in 2017 from Higher Education Commission (HEC), Government of Pakistan. He is the recipient of Best Paper Award in 2018 from Elsevier Journal of Network and Computer Applications. He received Best Paper Award in 2017 from Higher Education Commission (HEC), Government of Pakistan. He is the recipient of Best Paper Award in 2018 from Elsevier Journal of Network and Computer Applications. He has been selected for inclusion on the annual Highly Cited Researchers™ 2020 list from Clarivate. His performance in this context features in the top 1% in the field of Computer Science. The highly anticipated annual list identifies researchers who demonstrated significant influence in their chosen field, or fields, through the publication of multiple highly cited papers during the last decade. Their names are drawn from the publications that rank in the top 1% by citations for field and publication year in the Web of Science™ citation index.

Alan Davy

Alan Davy was awarded B.Sc. (with honours) in applied computing and Ph.D. from the Waterford Institute of Technology, Waterford, Ireland, in 2002 and 2008 respectively. He has worked at TSSG originally as a student since 2002 and became a postdoctoral researcher in 2008. In 2010 he worked at IIT Madras in India as an assistant professor lecturing on network management systems. He received a Marie Curie international Mobility fellowship in 2010, which brought him to work at the Universitat Politècnica de Catalunya for 2 years. Following that, he returned to the TSSG and was appointed a senior research fellow. In 2014 he became Research Unit Manager of the Emerging Networks Laboratory at TSSG. In 2019 he was appointed the head of Department of Computing and Mathematics at the Waterford Institute of Technology.

Contributor Biographies

Alexandros-Apostolos A. Boulogeorgos

Alexandros-Apostolos A. Boulogeorgos was born in Trikala, Greece in 1988. He received the Electrical and Computer Engineering (ECE) diploma degree and Ph.D. degree in wireless communications from the Aristotle University of Thessaloniki (AUTh) in 2012 and 2016, respectively.

From November 2012, he has been a member of the wireless communications system group of AUTh, working as a research assistant/project engineer in various telecommunication and networks projects. In 2017, he joined the Information Technologies Institute, while from November 2017, he has joined the Department of Digital Systems, University of Piraeus, where he conducts research in the area of wireless communications. Moreover, from October 2012 until September 2016, he was a teaching assistant at the department of ECE of AUTh. But from February 2017, he serves as an adjunct lecturer at the Department of Informatics and Telecommunications Engineering of the University of Western Macedonia and as visiting lecturer at the Department of Computer Science and Biomedical Informatics of the University of Thessaly.

Boulogeorgos has authored and co-authored more than 55 technical papers, which were published in scientific journals and presented at prestigious international conferences. Furthermore, he has submitted two (one national and one European) patents. Likewise, he has been involved as member of Technical Program Committees in several IEEE and non-IEEE conferences and served as a reviewer in various IEEE journals and conferences. Moreover, he is an associate editor of *Frontier in Communications and Networks*.

Boulogeorgos was awarded with the "Distinction Scholarship Award" of the Research Committee of AUTh for the year 2014 and was recognized as an exemplary reviewer for *IEEE Communication Letters* for 2016 (top 3% of reviewers). Moreover, he was named a top peer reviewer (top 1% of reviewers) in Cross-Field and Computer Science in the Global Peer Review Awards 2019, which was presented by the Web of Science and Publons. His current research interests spans in the area of wireless communications and networks with emphasis in high-frequency communications, optical wireless communications and communications for biomedical applications.

He is a senior member of the IEEE and a member of the Technical Chamber of Greece.

Amalendu Patnaik

Amalendu Patnaik received Ph.D. in electronics from Berhampur University, Berhampur, India, in 2003. From 2004 to 2005, he was with the University of New Mexico, Albuquerque, New Mexico, USA, as a visiting scientist. He is currently a professor with the Department of Electronics and Communication Engineering, Indian Institute of Technology Roorkee, Roorkee, India. He has authored/co-authored over 100 peer-reviewed papers in international journals and conferences, and co-authored one book on *Compact Antennas for High Data Rate Communication: Ultra-wideband (UWB) and Multiple-Input Multiple-Output (MIMO) Technology* (Springer) and one book chapter for "Neural Networks for Antennas" in *Modern Antenna Handbook* (Wiley). His current research interests include plasmonic antennas, antenna array signal processing, application of soft-computing techniques in electromagnetics, CAD for patch antennas, EMI and EMC. He is a recipient of the IETE Sir J. C. Bose Award in 2000, the URSI Young Scientist Award in 2005, and the BOYSCAST Fellowship from the Department of Science and Technology, Government of India, from 2004 to 2005. He was the Chairperson of the IEEE Roorkee Subsection from 2015 to 2017. He was an IEEE AP-S Region-10 Distinguished Speaker from 2015 to 2016. He serves as an editorial board member of the *International Journal of RF and Microwave Computer-Aided Engineering* (Wiley).

Ana Carolina Borges Monteiro

Ana Carolina Borges Monteiro is a Ph.D. student at the Faculty of Electrical and Computer Engineering (FEEC) at the State University of Campinas – UNICAMP, where she develops research projects regarding health software with emphasis on the development of algorithms for the detection and counting of blood cells through processing techniques and digital images. These projects led in 2019 to a computer program registration issued by the INPI (National Institute of Industrial Property). She holds a master's degree in electrical engineering from the State University of Campinas – UNICAMP (2019) and graduated in biomedicine from the University Center Amparense (UNIFIA) with a degree in clinical pathology – clinical analysis (2015). In 2019, she acquired a Master's degree in health informatics. She has experience in the areas of molecular biology and management with research animals. Since 2017, she has been a researcher at the FEEC/UNICAMP Visual Communications Laboratory (LCV) and has worked at the Brazilian Technology Symposium (BTSym) as a member of the Organizational and Executive Committee and as a member of the Technical Reviewers Committee. In addition, she works as a reviewer of the health magazines of the Federal University of Santa Maria (UFSM Brazil), *Medical Technology Journal* (MTJ, Algeria) and *Production Planning and Control* (Taylor & Francis). She is interested in digital image processing, hematology, clinical analysis, cell biology, medical informatics, Matlab, and teaching.

Angeliki Alexiou

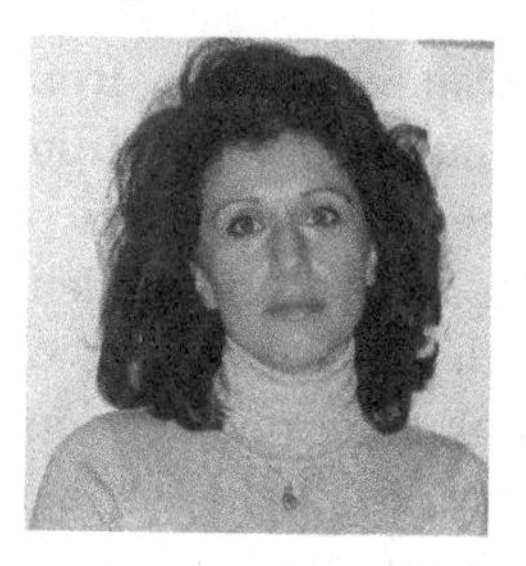

Angeliki Alexiou is a professor at the department of Digital Systems, ICT School, University of Piraeus. She received the Diploma in Electrical and Computer Engineering from the National Technical University of Athens in 1994 and the Ph.D. in Electrical Engineering from Imperial College of Science, Technology and Medicine, University of London in 2000. Since May 2009 she has been a faculty member at the Department of Digital Systems, where she conducts research and teaches undergraduate and postgraduate courses in Broadband Communications and Advanced Wireless Technologies. Prior to this appointment she was with Bell Laboratories, Wireless Research, Lucent Technologies, (later Alcatel-Lucent, now NOKIA), in Swindon, UK, first as a member of technical staff (January 1999-February 2006) and later as a Technical Manager (March 2006-April 2009). Professor Alexiou is a co-recipient of Bell Labs President's Gold Award in 2002 for contributions to Bell Labs Layered Space-Time (BLAST) project and the Central Bell Labs Teamwork Award in 2004 for role model teamwork and technical achievements in the IST FITNESS project. Professor Alexiou is the Chair of the Working Group on Radio Communication Technologies and of the Working Group on High Frequencies Radio Technologies of the Wireless World Research Forum. She is a member of the IEEE and the Technical Chamber of Greece. Her current research interests include radio interface for systems beyond 5G, MIMO, THz wireless technologies and Reconfigurable Intelligent Surfaces, efficient resource management for Ultra Dense wireless networks, machine-to-machine communications and Artificial Intelligence and Machine Learning for future wireless systems. She is the project coordinator of the Horizon 2020 TERRANOVA project (ict-terranova.eu) and the technical manager of Horizon 2020 ARIADNE project (ict-ariadne.eu).

Antonio Ramírez

Antonio Ramírez (Valencia, Spain, 1980) is a Telecommunication Engineer since 2004 by the Universitat Politecnica de Valencia (Spain). In 2005 he joined the Nanophotonics Technology Center as junior research engineer on radio over fibre systems and network technologies. In 2007 he moved to DAS PHOTONICS S. L. as R&D engineer involved in projects funded by the European Commission. In 2010 he joined FIBERNOVA SYSTEMS S. L. as Project Manager. He has accumulated over 15 years of experience coordinating tasks in different Horizon 2020 (H2020) projects related to wireless communications infrastructures in millimeter wave (mm-wave) and Terahertz (THz) bands and managing wired and wireless deployments in different technologies for the company's customers.

Bilal Hussain

Bilal Hussain has received his Masters degree in 2011 from Chalmers University of Technology, Sweden, and Ph.D. degree in 2019 from Scuola Superiore Sant'Anna, Pisa, Italy in the field of microwave photonics. He has been the recipient of two Marie-Curie fellowships awarded by the European Union. Currently, He is currently working as a Post-Doctoral fellow at INESC TEC, Porto, Portugal. His research interests include CMOS RFIC design, millimeter-wave antenna design, and Photonic Integrated Circuits (PICs) for microwave applications. He has authored/co-authored 20+ publications in international conferences and peer-reviewed journals.

Chahé Nerguizian

Chahé Nerguizian received B.Eng., M.Eng., and Ph.D. in electrical engineering from École Polytechnique Montréal, McGill University and Université du Québec (INRS-EMT) in 1981, 1983, and 2003, respectively. After two years of industrial experience as a microwave engineer with the Canadian Marconi Canada (avionic division), he joined the Department of Electrical and Computer Engineering at Polytechnique Montreal in 1986. Currently, he is a full professor in the telecommunication group. His research interests are in wireless radio propagation (narrow band, wide band and ultrawide band channel characterization), indoor–outdoor radiolocation (RFID and WLAN) and advanced wireless communication (OFDM, MIMO and UWB). Nerguizian is a member of the Institute of Electrical and Electronic Engineering (IEEE) and the Order of Engineers of Quebec (OIQ). He acts, since June 2007, as the head of the Telecommunication and Microwave section within the Electrical Engineering Department.

Claudio Paoloni

Claudio Paoloni received laurea degree cum laude in electronic engineering from the University of Rome "Sapienza", Italy. Since 2012, he has been Cockcroft Chair with the Engineering Department, Lancaster University, U.K. Since 2015, he has been the head of Engineering Department. He is member at large of the Board of Governors of IEEE Electron Devices Society. He is a senior member of IEEE and senior fellow of the Higher Education Academy. He was chair of the IEEE Electron Devices Society Vacuum Electronics Technical Committee (2017–2020). He was guest editor for the Special Issue of *Transactions on Electron Devices on Vacuum Electronics* (June 2014). He organised numerous international conferences and workshops, on vacuum electronics, millimetre wave and terahertz communications and technology. He has been coordinator of two European Commission Horizon 2020 projects, TWEETHER and ULTRAWAVE.

He is the author of more than 240 articles in international journals and conferences in the field of high-frequency electronics, millimetre waves and THz vacuum electronics devices, wireless communications.

Daniel M. Mittleman

Daniel M. Mittleman received his B.S. in physics from the Massachusetts Institute of Technology in 1988, and his M.S. in 1990 and Ph.D. in 1994, both in physics from the University of California, Berkeley. He then spent two years at the AT&T Bell Laboratories as a post-doctoral member of the technical staff, before joining the Electrical and Computer Engineering Department at Rice University in 1996. In 2015, he moved to the School of Engineering at Brown University. His research interests involve the science and technology of terahertz radiation. He is a Fellow of the Optical Society (OSA), the American Physical Society (APS), and the IEEE, and is a 2018 recipient of the Humboldt Research Award.

Davide Cimbri

Davide Cimbri received B.Sc. in engineering physics and the M.Sc. in nanotechnology from Politecnico di Torino, Torino, Italy. He worked on the M.Sc. thesis at the Boston University, developing a quantum mechanical-based code to simulate and analyse the electronic band structure of antimonide (Sb)-based type-II superlattice Mid-Infrared (MIR) photodetectors.

He is currently with the High-Frequency Electronics Group, Department of Electronics and Nanoscale Engineering, University of Glasgow, Glasgow, U.K., working towards a Ph.D. in electronics and electrical engineering sponsored by a Marie Skłodowska-Curie fellowship. His research focuses on resonant tunneling diode (RTD) terahertz (THz) oscillators for imaging, radar and communication applications.

Deepak Mishra

Deepak Mishra (Member, IEEE) received B.Tech in electronics and communication engineering from the Guru Gobind Singh Indraprastha University, New Delhi, India, in 2012, and Ph.D. in electrical engineering from IIT Delhi, New Delhi, India, in 2017. He has been a senior research associate with the School of Electrical Engineering and Telecommunications, University of New South Wales (UNSW) Sydney, Australia, since August 2019. Before that, he was a postdoctoral researcher at the Department of Electrical Engineering (ISY), Linköping University (LiU), Linköping, Sweden, from August 2017 to July 2019. He has also been a Visiting Researcher at Northeastern University,

Boston, MA, USA, University of Rochester, Rochester, NY, USA, Huawei Technologies, France, and Southwest Jiaotong University, China. His current research interests include energy-harvesting cooperative communication networks, massive MIMO, backscattering, physical layer security, as well as signal processing and energy optimization schemes for uninterrupted operation of wireless networks. He was also a TPC member of the IEEE International Conference on Communications (ICC) in 2018, 2019, and 2020 and for WPSN in 2019. He was a recipient of the IBM Ph.D. Fellowship Award in 2016, the Raman Charpak Fellowship Award in 2017 and the Endeavour Research Fellowship Award in 2018. He was selected as an Exemplary Reviewer of *IEEE Transactions on Wireless Communications* in 2017 and 2018, and *IEEE Transactions on Communications* in 2019.

Douglas Sicker

Douglas Sicker a professor of Computer Science and Senior Associate Dean of Computing at the University of Colorado, Denver. Doug is also the Executive Director of BITAG and Co-Director of the Institute for Regulatory Law and Economics (IRLE). Previously, Doug served as the Lord Endowed Chair in Engineering, Department Head, interim Director of CyLab, and Professor of Computer Science at Carnegie Mellon University. He was a founder and continues on the Board of CMMB Vision, a high-power L-band satellite company. Previously, Doug was the DBC Endowed Professor in the Department of Computer Science at the University of Colorado at Boulder with a joint appointment as Director of the Interdisciplinary Telecommunications Program. Doug recently served as the Chief Technology Officer and Senior Advisor for Spectrum at the National Telecommunications and Information Administration (NTIA). Doug also served as the CTO of the Federal Communications Commission (FCC) and prior to this he served as a senior advisor on the FCC National Broadband Plan. Earlier he was Director of Global Architecture at Level 3 Communications, Inc. In the late 1990s, Doug served as Chief of the Network Technology Division at the FCC. Doug has published extensively in the fields of networking, wireless systems and network security.

Edward Wasige

Edward Wasige received B.Sc. (Eng.) in electrical engineering from the University of Nairobi, Nairobi, Kenya, in 1988, M.Sc. (Eng.) in microelectronic systems and telecommunications from the University of Liverpool, Liverpool, UK, in 1990, and the Dr.-Ing. degree in electrical engineering from Kassel University, Kassel, Germany, in 1999.

Between 1990–1993 and 1999–2001, he was a lecturer at Moi University in Kenya. In 2001–2002, he was a UNESCO postdoctoral fellow at the Technion – Israel Institute of Technology. He has been a lecturer at the University of Glasgow since 2002, where he is now a professor in high-frequency electronics. His research interests include compound semiconductor micro-/nanoelectronics and applications with focus on gallium nitride (GaN) electronics and resonant tunnelling diode based terahertz electronics.

Elmustafa Sayed Ali Ahmed

Elmustafa Sayed Ali Ahmed is a Ph.D. candidate, received his M.Sc. in electronics and communication engineering from the Sudan University of Science and Technology in 2012 and B.Sc. in 2008. He worked as a senior engineer in Sudan Sea Port Corporation for five years where he was a team leader of new projects in wireless networks includes (Tetra system, Wi-Fi, Wi-Max, and CCTV). He is a senior lecturer in Electrical and Electronics Engineering Department in the Red Sea University. Currently he is the head of marine systems department in Sudan marine industries. Elmustafa published papers in peer-reviewed academic international journals and book chapters on wireless communications, computer and networking. His areas of research interest include routing protocols, wireless networks, low-power wide-area network (LPWAN) and Internet of Things. He is a member of IEEE Communication Society (ComSoc), International Association of Engineers (IAENG), Six Sigma Yellow Belt (SSYB), and Scrum Fundamentals certified (SFC).

François Magne

François Magne has 37 years of experience in detection and communication R&D. Since 2006, he founded two small and medium enterprises and has been previously part of big companies such as Thales and Philips as given below:

2014: Founder and CEO of WHEN-AB – research in microwave communication and detection systems.

2006: Founder and Chief scientist of BLUWAN – SME of microwave backhaul communication networks

2003: CTO deputy of Thales Group

1994: CTO of Thales Communication

1989: CTO of TH-CSF CNI

1983: Chief Scientist of TRT-Défense (Philips Group)

Five patents in communication systems

Education: Master's degrees, ENST and USC

François Magne began is carrier as a Military Engineer in the French Ministry of Defence and then in the Centre National de la Recherche Scientifique (CNRS) before joining industry in 1983.

Guofu Xu

Guofu Xu received the B.Sc. degree in Optical Information Science and Technology from the Army Engineering University of PLA, China, in 2012, and the M.Eng. degree in Electrical Theory and Advanced Technology from the Hefei University of Technology, China, in 2019. He is currently pursuing the Ph.D. degree under the supervision of Prof. Maksim Skorobogatiy at Polytechnique Montréal, where he is involved in the development of optical waveguide components for THz applications.

Hadi Sarieddeen

Hadi Sarieddeen received B.E. (summa cum laude) in computer and communications engineering from Notre Dame University – Louaize (NDU), Zouk Mosbeh, Lebanon, in 2013, and Ph.D. in electrical and computer engineering from the American University of Beirut (AUB), Beirut, Lebanon, in 2018. He is currently a post-doctoral research fellow with the Computer, Electrical and Mathematical Sciences and Engineering (CEMSE) Division, King Abdullah University of Science and Technology (KAUST), Thuwal, Makkah, Saudi Arabia. His research interests are in the areas of communication theory and signal processing for wireless communications with an emphasis on large, massive and ultra-massive MIMO systems, and terahertz communications.

Hang Yuan

Hang Yuan received B.S. in information engineering from Beijing Institute of Technology, Beijing, China, in 2016, where he is currently pursuing Ph.D. in the School of Information and Electronics. From 2017 to 2019, he was a visiting student with the Research School of Electrical, Energy and Materials Engineering at the Australian National University. His research interests include millimeter wave and terahertz communications.

Henrique M. Salgado

Henrique M. Salgado Graduated in Applied Physics (Optics and Electronics) from the University of Porto in 1985 and received the PhD degree in Electronic Engineering and Computer Systems from University of Wales in 1993. Presently he is Associate Professor at the Department of Electrical Engineering of the University of Porto, member of scientific council of INESC TEC and the scientific committee of the MAP-tele doctoral programme. He was also the founder of the research area in Optical and Electronics Technologies at INESC TEC. Formerly he was Research Fellow at the Department of Electrical and Electronic Engineering of the University College London, UK and Research Assistant at the School of Electronic Engineering and Computer Systems, Bangor, UK.

He has successfully managed and coordinated several national and international R&D projects and is the author and co-author of over 200 publications, including scientific journal and international conferences, in the field of optical fibre communications and microwaves. His research interests include: radio-over-fiber technology and microwave photonics, digital equalization in coherent optical systems, all-optical networks, modelling of nonlinearities and design of compact multiband antennas. He is a member of the Photonics and Communications societies of the IEEE.

Hichem Guerboukha

Hichem Guerboukha is a postdoctoral research fellow at Brown University, School of Engineering. He received the B.Sc. degree in engineering physics, M.Sc. in applied science and the Ph.D. in engineering physics from Polytechnique Montreal in 2014, 2015 and 2019, respectively. His previous research included THz instrumentation and waveguides, THz computational imaging and THz communications. Guerboukha was the recipient of the 2015 Releve Étoile Louis-Berlinguet from Fonds de recherche – Nature et technologies. He obtained the Best M.Sc. Thesis Award and the Best Ph.D. Thesis Award from Polytechnique Montreal in 2015 and 2019 respectively. He is currently a FRQNT postdoctoral research fellow and working on THz communications in the group of Prof. Daniel Mittleman at Brown University.

Janne Lehtomäki

Janne Lehtomäki obtained his doctorate from the University of Oulu in 2005. Currently, he is an adjunct professor (docent) at the University of Oulu, Centre for Wireless Communications. He spent the fall 2013 semester at Georgia Tech, Atlanta, as a visiting scholar. Currently, he is focusing on terahertz band wireless communication. He co-authored a paper that received the Best Paper Award at IEEE WCNC 2012. He is an associate editor of *Elsevier, Physical Communication.*

Jianping An

Jianping An received Ph.D. degree from the Beijing Institute of Technology, China, in 1996. He joined the School of Information and Electronics, Beijing Institute of Technology in 1995, where he is now a full professor. He is currently the Dean of the School of Information and Electronics, Beijing Institute of Technology. His research interests are in the field of digital signal processing, cognitive radio, wireless networks, and high-dynamic broadband wireless transmission technology.

Joonas Kokkoniemi

Joonas Kokkoniemi received Dr. Sc. in communications engineering from the University of Oulu, Finland, in 2017. He is a postdoctoral researcher with the Centre for Wireless Communications, University of Oulu. From September to November 2013, he was a visiting researcher at Tokyo University of Agriculture and Technology, Japan. From March to October 2017, he was a visiting postdoctoral researcher with the University at Buffalo, New York. His research interests include THz band channel modeling.

Kai Yang

Kai Yang received the B.E. and Ph.D. from the National University of Defense Technology and Beijing Institute of Technology, China, in 2005 and 2010, respectively, both in communications engineering. From January to July 2010, he was with the Department of Electronic and Information Engineering, Hong Kong Polytechnic University. From 2010 to 2013, he was with Alcatel-Lucent Shanghai Bell, Shanghai, China. In 2013, he joined the Laboratoire de Recherche en Informatique, University Paris Sud 11, Orsay, France. Now, he is with the School of Information and Electronics, Beijing Institute of Technology, Beijing, China. His current research interests include convex optimization, massive MIMO, mmWave systems, resource allocation, and interference mitigation.

Kathirvel Nallappan

Kathirvel Nallappan received B.E. in electronics and communication engineering and M. Tech. in laser and electro-optical engineering during 2010 and 2013 respectively from Anna University, Tamil Nadu, India. After working as a senior project fellow in the Structural Engineering Research Center (CSIR), India, for a period of one year, he joined as a research intern in the research group of Prof. Maksim Skorobogatiy at Polytechnique Montréal, Canada, until 2015. He then joined as a Ph.D candidate under the supervision of Prof. Maksim Skorobogatiy and co-supervision of Prof. Chahé Nerguizian at the department of Electrical Engineering, Polytechnique Montréal and graduated in 2020. Currently, he is working as a postdoctoral research fellow in the area of Terahertz communications at Polytechnique Montréal, in the research group of Prof. Maksim Skorobogatiy. His research interest includes millimeter and Terahertz wireless communications, Waveguide modelling, Terahertz spectroscopy, Terahertz imaging and optical free space communication.

Luís Pessoa

Luís Pessoa received the "Licenciatura" degree in 2006 and PhD degree in 2011, both in Electrical and Computer Engineering at the Faculty of Engineering of the University of Porto. Currently he is Senior Researcher and Manager of the Optical and Electronic Technologies (OET) group from the Centre of Telecommunications and Multimedia at INESC TEC. He is responsible for the overall coordination of the OET group, including the management of the team, the conception and management of R&D projects, coordination of research students and fostering the valorisation of research results through new contracts with industry. He has been involved in teaching of RF/microwave engineering and optical communications as an invited assistant professor at University of Porto. He has authored/co-authored 60+ publications in international conferences and journals with peer-review (including 10 journal papers and 4 book chapters) and 2 patents. He has coordinated several national and international research projects and participated in several European Union funded research projects. His main research interests include coherent optical systems, radio-over-fibre, RF/microwave devices and antennas, underwater radio/optical communications and underwater wireless power transfer.

Maksim Skorobogatiy

Maksim Skorobogatiy graduated in 2001 from the Massachusetts Institute of Technology (MIT) with Ph.D. in physics and MSc in electrical engineering and computer science. He then worked at the MIT spin-off Omniguide Inc. on the development of hollow-core fibers for guidance of high-power mid-IR laser beams. He was hired by Polytechnique Montréal, Canada, in 2003 and was awarded a Tier 2 Canada Research Chair (CRC) in Micro and Nano Photonics and then a Tier 1 CRC in Ubiquitous THz Photonics in 2016. Thanks to the support of the CRC program, he could pursue many high-risk exploratory projects in guided optics, and recently THz photonics, which have resulted in significant contributions in these two booming research fields. In 2012, he was awarded the rank of Professional Engineer by the Order of Engineers of Québec, Canada. In 2017, Skorobogatiy was promoted to be the Fellow of the Optical Society of America for his pioneering contributions to the development of microstructured and photonic crystal multi-material fibers and their applications to light delivery, sensing, smart textiles and arts. Additionally, in 2017 he was promoted as the senior member of the IEEE for his contribution to engineering and applied research. Skorobogatiy is also an author of two books on the subjects of photonic crystals and guided optics, as well as ten book chapter on materials science, smart textiles, biosensors and terahertz technologies. He is also an author of more than 150 articles in the peer-reviewed journals, has made over 200 contributions to conference proceedings.

Markku Juntti

Markku Juntti received his doctorate in 1997. He was a visiting scholar at Rice University, Houston, Texas, 1994–1995, and a senior specialist with Nokia Networks, 1999–2000. Since 2000 he has been a professor of communications engineering at the University of Oulu, Centre for Wireless Communications (CWC), where he serves as head of the CWC – Radio Technologies (RT) Research Unit. His research interests include signal processing for wireless networks, and communication and information theory. He is a fellow of IEEE.

Meltem Civas

Meltem Civas received B.Sc. and M.Sc. in electrical and electronics engineering from the Bilkent University, Ankara, Turkey, and Koç University, Istanbul, Turkey in January 2016, and September 2018, respectively. She is currently a research assistant at the Next-generation and Wireless Communications Laboratory and pursuing her Ph.D. in electrical and electronics engineering, Koç University, under the supervision of Prof. Ozgur B. Akan.

Mira Naftaly

Mira Naftaly is a laser physicist and spectroscopist, and is now working in the area of terahertz measurements. Her main research interests are terahertz metrology, spectroscopy, nondestructive testing, and characterisation of devices and systems. She is a Senior Research Scientist at the National Physical Laboratory, UK.

Mohamed-Slim Alouini

Mohamed-Slim Alouini received Ph.D. in electrical engineering from Caltech, California, in 1998. He served as a faculty member at the University of Minnesota, Minneapolis, then at Texas A&M University at Qatar, Doha, before joining KAUST as a professor of electrical engineering in 2009. His current research interests include the modeling, design, and performance analysis of wireless communication systems.

Mona Bakri Hassan

Mona Bakri Hassan received her M.Sc. in electronics engineering and telecommunication in 2020 and B.Sc. (Honors) in electronics engineering and telecommunication in 2013. She has completed Cisco Certified Network Professional (CCNP) routing and switching course and passed the Cisco Certified Network Administrator (CCNA) routing and switching. She has gained experience in network operations center as a network engineer in back office department in Sudan Telecom Company (SudaTel) and in vision valley as well. She has also worked as a teacher assistant in Sudan University of Science and Technology, which has given her a great opportunity to gain teaching experience. She published book chapters, and papers on wireless communications and networking in peer-reviewed academic international journals. Her areas of research interest include cellular LPWAN, IoT and wireless communication.

Nan Yang

Nan Yang received the Ph.D. degree in electronic engineering from the Beijing Institute of Technology in 2011. He has been with the Research School of Electrical, Energy and Materials Engineering at the Australian National University since July 2014, where he currently works as a Associate Professor. He received the IEEE ComSoc Asia-Pacific Outstanding Young Researcher Award in 2014 and the Best Paper Awards from the IEEE GlobeCOM 2016 and the IEEE VTC 2013-Spring. He also received the Top Editor Award from the Transactions on Emerging Telecommunications Technologies, the Exemplary Reviewer Award from the *IEEE Transactions on Communications*, *IEEE Wireless Communications Letters* and *IEEE Communications Letters*, and the Top Reviewer Award from the *IEEE Transactions on Vehicular Technology* from 2012 to 2018. He is currently serving in the editorial board of the *IEEE Transactions on Wireless Communications*, *IEEE Transactions on Molecular, Biological, and Multi-Scale Communications*, *IEEE Transactions on Vehicular Technology* and *Transactions on Emerging Telecommunications Technologies*. His general research interests include ultra-reliable low latency communications, millimeter wave and terahertz communications, massive multi-antenna systems, cyber-physical security, and molecular communications.

Nasir Saeed

Nasir Saeed received B.E. in telecommunication engineering from the University of Engineering and Technology, Peshawar, Pakistan, in 2009 and M.S. in satellite navigation systems from Polito di Torino, Italy, in 2012. He received Ph.D. in Electronics and Communication Engineering from Hanyang University, Seoul, South Korea in 2015. He was an Assistant Professor at the Department of Electrical Engineering, Gandhara Institute of Science and IT, Peshawar, Pakistan from August 2015 to September 2016. Saeed worked as an assistant professor at IQRA National University, Peshawar, Pakistan from October 2017 to July 2017. He is currently a postdoctoral research fellow in the King Abdullah University of Science and Technology (KAUST). Saeed is also a member of technical program committee for various IEEE conferences, such ICCC, GLOBECOM and VTC. He is also serving as an associate editor for *Frontiers in Communications and Networks*. He has participated in various projects in KAUST relevant to communications, localization and networking in extreme environments. His current areas of interest include cognitive radio networks, underwater and underground wireless communications, satellite communications, dimensionality reduction, and localization.

Ozgur B. Akan

Ozgur B. Akan completed his Ph.D. in ECE at the Georgia Institute of Technology, Atlanta, Georgia, in 2004. He is the head of the Internet of Everything (IoE) Group at the Department of Engineering, University of Cambridge, UK, and also a Professor of Electrical and Electronics Engineering, Koç University, Istanbul, Turkey. He conducts highly advanced theoretical and experimental research on nanoscale, molecular and neural communications, IoE, 5G and THz wireless mobile networks. He is a fellow of IEEE.

Quang Trung Le

Quang Trung Le received the Diploma of Engineering in optronics and the Ph.D. in physics from the University of Rennes 1, France, in 2006 and 2010, respectively. From 2010 to 2015, he was with Orange Labs, France and then with Technical University of Darmstadt, Germany, working on opto-electronics devices and systems for optical networks and radio-over-fiber. Since 2015, he has been with HF Systems Engineering GmbH, a part of Hübner Photonics, as development engineer and served as Technical Team Leader since 2019. His research interests include the design and integration of mm-wave and terahertz components and systems for sensing and communication applications. He has contributed to several research projects, including European Commission Horizon 2020 TWEETHER and ULTRAWAVE projects for millimeter-wave high-capacity wireless networks.

Rangel Arthur

Rangel Arthur holds a degree in electrical engineering from the Paulista State University Júlio de Mesquita Filho (1999), a master's degree in electrical engineering (2002) and a Ph.D. in electrical engineering (2007) from the State University of Campinas. Over the years from 2011 to 2014, he was the Coordinator and Associate Coordinator of Technology Courses in Telecommunication Systems and Telecommunication Engineering of FT, which was created in its management. From 2015 to 2016, he was a associate director of the technology (FT) of Unicamp. He is currently a lecturer and advisor to the Innovation Agency (Inova) of Unicamp. He has experience in the area of electrical engineering, with emphasis on telecommunications systems, working mainly on the following topics: computer vision, embedded systems and control systems.

Rashid A. Saeed

Rashid A. Saeed received Ph.D. in communications and network engineering from the Universiti Putra Malaysia (UPM). Currently he is a professor in Computer Engineering Department, Taif University. He is also working in Electronics Department, Sudan University of Science and Technology (SUST). He was a senior researcher in Telekom Malaysia™ Research and Development (TMRND) and MIMOS. Rashid published more than 150 research papers, books and book chapters on wireless communications and networking in peer-reviewed academic journals and conferences. His areas of research interest include computer network, cognitive computing, computer engineering, wireless broadband and WiMAX Femtocell. He was successfully awarded three U.S patents in these areas. He supervised more 50 M.Sc./Ph.D. students. Rashid is a senior member of IEEE, member of The Institute of Engineers, Malaysia (IEM), SigmaXi, and Sudanese Engineering Council (SEC).

Ravikant Saini

Ravikant Saini received B.Tech in Electronics and Communication Engineering and M.Tech in Communication Systems from Indian Institute of Technology Roorkee, India in 2001 and 2005, respectively. After completing his Ph.D. from Indian Institute of Technology Delhi, India in 2016, he worked as an assistant professor in Shiv Nadar University, Greater Noida, India till December 2017. Currently, he is working as an assistant professor in the Indian Institute of Technology Jammu, India.

Reinaldo Padilha França

Reinaldo Padilha França is a Ph.D. candidate at the Faculty of Electrical and Computer Engineering (FEEC) at the State University of Campinas (UNICAMP), Master in Electrical Engineering from UNICAMP (2018), Bachelor's degree in Computer Engineer (2014), and Logistics Technology Management (2008) from the Regional University Center Espírito Santo of Pinhal. He has scientific knowledge in programming and development in C/C++, Java, and .NET languages, from projects and activities. He is a natural, dynamic, proactive, communicative, and creative self-taught person. Since 2016 he has developed research projects regarding precoding and bit coding in AWGN channels, telecommunications channels, and digital image processing with an emphasis on blood cells resulting in two registrations of computer programs with the INPI (National Institute of Industrial Property) (2019). Since 2017 he has worked as a researcher at the Laboratory of Visual Communications (LCV) at FEEC/UNICAMP. He has served, in 2019, as Assistant Editor in Chief at the journal SET IJBE. He was the editor of the published Scopus Springer book entitled "Proceedings of the 5th Brazilian Technology Symposium (BTSym19) – Emerging Trends and Challenges, vol. 1". He has interest and affinity in the areas of technological and scientific research, teaching, digital image processing, and Matlab.

Roberto Llorente

Roberto Llorente (IEEE M'89, OSA SM'17) is full professor at the Universitat Politècnica de València (UPV), Spain, since 2017. He is with the Nanophotonics Technology Center (NTC) (www.ntc.upv.es) of this university, a designated singular scientific and technical institution (ICTS) by the Government of Spain, and has been appointed its deputy director in 2015. He has been leading the Optical Networks and Systems Research Unit in the NTC as a coordinator or principal researcher of several projects in the framework of European Union's Research and Innovation programmes FP6, FP7 and Horizon 2020. He has also led more than 60 national-level research and technology-transfer projects bringing together highly specialised companies and academia in Spain. He has authored or co-authored more than 200 papers in leading international journals and conferences, and has three patents to his credit on photonic processing technology, some of them being licensed in the telecommunications and defence sectors. His current research interests include electro-optic processing techniques applied to high-performance photonic applications as optical beamforming, photonic sensing and advanced optical network architectures.

Rohit Singh

Rohit Singh is currently a post-doctoral researcher at the University of Colorado Denver. He received Ph.D. in engineering and public policy from Carnegie Mellon University, Pittsburgh, and an M.S. degree in Computer Science from the Illinois Institute of Technology, Chicago. He has published and reviewed in top IEEE and ACM conferences and journals. His current research interests are THz Communication, 5G and Beyond, radio resource management, spectrum management, security and privacy.

Sasmita Dash

Sasmita Dash received Ph.D. from the Department of Electronics and Communication Engineering, Indian Institute of Technology (IIT) Roorkee, India, in 2019. Prior to Ph.D., she received M.Tech in electronics from the Department of Electronic Science, M.Phil in Physics and M.Sc. in Physics from the Department of Physics, Berhampur University, India. She worked towards post-doctoral research as a Special Scientist in a European Union FET OPEN project at the Department of Computer Science, University of Cyprus, Cyprus, from April 2019 to July 2020. She is currently a Post-Doc Researcher at IRIDA Research Centre for Communication Technologies, Department of Electrical and Computer Engineering, University of Cyprus, Cyprus. She has more than 30 peer-reviewed research paper publications in international journals and conferences. Her research interest is on interdisciplinary topics at the frontier of Electromagnetics, Electronics, Material science, and Physics, including Graphene, CNT and Nanomaterial based structures, Antennas and Absorbers, Electromagnetics, Plasmonics, Metamaterial & Metasurface, Reconfigurable surface, Modelling and Simulation, Terahertz, Communication, etc.

Sean Ahearne

Sean Ahearne is a senior research scientist for Dell Technologies, Ireland. His current research areas include 5G technologies, THz wireless communication, Software Defined Network (SDN)/Network Functions Virtualization (NFV), Graphical Processing Unit (GPU) virtualization and acceleration, and edge computing. Sean is contributing to or leading research efforts in several current European Horizon 2020 research projects and has contributed to several past projects also. In particular, he is determining the integration of THz wireless links into datacenter communications as part of the TERAPOD project.

Tareq Y. Al-Naffouri

Tareq Y. Al-Naffouri received B.S. in mathematics and electrical engineering (with first honors) from KFUPM, Saudi Arabia, M.S. in electrical engineering from Georgia Tech, Atlanta, in 1998 and Ph.D. in electrical engineering from Stanford University, California, in 2004. He is currently a professor in the Electrical Engineering Department at KAUST. His research interests lie in the areas of sparse, adaptive and statistical signal processing and their applications, localisation and machine learning.

Turker Yilmaz

Turker Yilmaz received B.S., M.Sc. and Ph.D. in electrical and electronics engineering from the Boğaziçi University, University College London and Koç University in 2008, 2009 and 2018, respectively. He is currently a research engineer at the Communication Systems Department, EURECOM, France. His current research interests include mobile, millimeter wave and terahertz communications, and Internet of Things.

Yang Cao

Yang Cao received B.Sc. (2014) in electronic science and technology and M.Sc. (2017) in optical engineering from TianJin Univ. He is currently a Ph.D. candidate at Polytechnique Montréal, under the supervision of Prof. Maksim Skorobogatiy. His research interests include THz waveguide components design and fabrication, integrated circuits for THz communication, and THz sensing.

Yuzo Iano

Yuzo Iano received BS (1972), master's (1974) and a Ph.D. degree (1986) in electrical engineering from the State University of Campinas, Brazil. Since then, he has been working in the technological production field, with one patent granted, eight patent applications filed and 36 projects completed with research and development agencies. He successfully supervised 29 doctoral theses, 49 master's dissertations, 74 undergraduate and 48 scientific initiation work. He has participated in 100 master's examination boards, 50 doctoral degrees, author of two books and more than 250 published articles. He is a currently professor at the State University of Campinas, Brazil, editor-in-chief of the *SET International Journal of Broadcast Engineering* and general chair of the Brazilian Symposium on Technology (BTSym). He has experience in electrical engineering, with knowledge in telecommunications, electronics and information technology, mainly in the field of audio-visual communications and data.

Preface

The advancement in mobile technology and heavy dependability on data access by users in daily life demands search for new and sophisticated technologies to meet growing demand. At one end, the service providers require high-capacity backbone networks and at the other, the users are demanding high-speed connectivity to access heavy data in the form of video and data. Although the existing technologies using advanced physical layer techniques and improved hardware components like sources, detectors, and antennas can be used to achieve high data rates, they still cannot achieve the Gigabits per second (Gbps) and Terabits per second (Tbps) speed, which is due to the limited bandwidth availability.

New frequency bands are being explored for high-speed links and to fulfill the requirements for Beyond 5th Generation (B5G) networks like ultra-high throughput and ultra-low latency. One of the alternatives is millimeter wave bands. Still it is not enough to satisfy the requirements of future B5G wireless communication networks. For example, applications like wireless local area networks, virtual reality devices, data centres, and autonomous vehicles require a data rate up to at least 100 Gbps.

The Terahertz frequency bands (0.1 GHz to 10 THz) is therefore envisioned as a promising candidate for B5G networks. These bands are being used in different applications like imaging, spectroscopy, and wireless communication. Among them, wireless communication is the least explored area, which is currently in heavy demand for B5G communication networks. Currently, the wireless communication research and trials for above 100 GHz links are underway for both indoor and outdoor channels for different transmitter and receiver antenna configurations and polarizations at multiple frequencies. The main experiments for channel modeling are focused on propagation measurements, directional path losses, and penetration losses.

There are many books available on the imaging and spectroscopy applications using the Terahertz band. The most relevant are those discussing the device and materials aspects of Terahertz technology. However, none of the books discusses Terahertz wireless communication and its suitability for B5G communication networks. So the wireless communication aspects using Terahertz bands like channel, transceivers, antenna, beam management, and communication layer aspects with deployments are presented for the first time in this book. This book also highlights the relevant challenges for bridging the gap between the existing technologies and the Terahertz wireless communication for B5G communication networks.

The contents of this book show the relevancy of the book for researchers at different levels, postgraduate students, service providers looking to enhance their backbone and frontend network capacity, and data centre operators for reshaping the future data centre geometry.

Terahertz Communication Networks

The global traffic is expected to reach 400 exabytes by 2022,[1] which requires high bandwidth and capacity links to cater to future traffic demands. The lower frequency bands cannot meet these requirements due to limited bandwidth availability. Therefore, the interest is shifting toward more potential bands like the Terahertz band, which can provide speeds up to Tbps with huge bandwidth availability. The Terahertz band communication appears as a promising technology for future wireless communication networks and can be used to fulfill the future traffic demands for many indoor and outdoor communication networks. Using the Terahertz band, a file in gigabytes can be downloaded in a few seconds, which would take several minutes previously, using lower frequency bands including Industrial, Scientific, and Medical (ISM) and cellular bands.

Besides the ultra-high bandwidth availability, the Terahertz band suffers from high path loss and molecular absorption loss, which affects the achievable communication distance and throughput. To enhance the communication distance, directional antennas are required with multiple antennas and beam management techniques. Due to these unique band features and requirements like path and molecular absorption loss, scattering, and reflection phenomenon, novel techniques are required for Terahertz communication networks at the nano and macro scale for indoor and outdoor applications including channel and propagation models; antenna technology and beam management; physical layer techniques including modulation and coding, frame error control and bit error rate; medium access control (MAC) layer techniques including link establishment, interference mitigation, resource management, blockage mitigation, and addressing; and network layer techniques for routing and end-to-end network efficiency.

The existing literature on Terahertz communication covers device technology including resonant tunnelling diodes, Uni-Traveling-Carrier Photodiodes (UTC-PDs), and Schottky diodes (STDs) with material like graphene and antenna design. However, only a few studies are available on communication aspects to enable high-bandwidth Terahertz communication networks. Due to the potential of the Terahertz band, it can be used in applications like small cell deployment, backhaul communication, inter-satellite communication, indoor wireless access networks like local area networks, personal area networks, data center networks, and nano-communication networks. For each scenario, different challenges need to be addressed. Due to huge bandwidth availability, it can fulfill the requirements for 5G and B5G networks like high throughput, low latency, and massive connectivity.

The continuous evolution of wireless networks from the first generation to fourth generation provides additional capacity. The B5G networks in this regard are also being engineered to provide high-capacity links with high throughput and low latency to satisfy user demands. Therefore, the prime objectives for Terahertz communication for B5G communications are ultra-high data rate up to Tbps for sufficient transmission distance, ultra-low latency for massively connected devices, reliability for mobility environments, and high energy efficiency for various mobile networks and terminals.

The 5G and B5G have opened the floodgates for higher bandwidth. After the wider adoption of millimeter-wave bands, the Terahertz bands using more than 100 GHz bandwidth will be inevitable to realize as sixth generation and beyond networks.

Intended Audience

This book would be of interest to multiple groups of researchers, postgraduate students, service provides, and network operators, who are working on B5G development and deployment.

- *Academics:* B5G has already been a major area of research and study for many educational institutions all over the world. As Terahertz frequency is used in B5G wireless communication networks, there is no book available that can be used for teaching and Terahertz components understanding this interesting field.
- *Network operators:* The network operators will be looking to adopt B5G technology to offer new services to their customers. This book will provide guidelines and techniques to deploy and develop Terahertz technology.
- *Data centre operators:* The data centre operators will look for new strategies and technologies to satisfy the future traffic demands. This book will help the operators in reshaping the existing data centre geometry and to go for an alternative to achieve above 100 Gbps backbone networks.
- *Telecommunication researchers:* B5G forms one of the key interest areas for telecommunication researchers with regard to high-capacity and flexible Terahertz backhaul links. This book will offer a single source of major Terahertz technology components including channel, antenna, and beam management.
- *Vehicular industry:* The vehicular industry is continuously attempting to design autonomous driving vehicles and smart cars. This book will give insights into the use of appropriate channel models and antenna technology to exchange above 100 Gbps data to advance smart car technology.

Organization of the Book

The book covers various parts in Terahertz wireless communication networks and is divided into four parts: Terahertz transceivers and devices, Terahertz channel characteristics and modeling, Terahertz antenna design, and Terahertz links, application, and deployment. First, a perspective of future Terahertz systems is presented in Chapter 1, in which the interaction between the Terahertz sensing, imaging, and localization is presented.

Part I presents Terahertz transceivers and devices with an overview of Terahertz networks. Chapter 2 describes the basic working principle of resonant tunnelling diode (RTD) driven oscillator and current state-of-the-art RTD-based transmitters, receivers, and transceivers technology, with performance and possible future solutions. Chapter 3 provides an overview of main characterization techniques for emitter (transmitter) and detector (receiver) devices used in Terahertz wireless links, with emitter frequency spectrum, power, and beam profile, and detector responsivity and spatial acceptance.

Part II deeply looks into the channel characteristics, measurements, and modeling with line/non-line of site propagation for different applications with interference issues. Chapter 4 gives basics of stochastic modeling of the interference in above 100 GHz wireless systems with directional antennas and various channel impairments. Chapter 5 introduces key aspects of Terahertz channel characteristics and measurements. It also describes the theoretical concepts of Terahertz channel characterization, limitations, requirements, and methodologies for measurements and analysis. Chapter 6 provides initially a brief overview of Terahertz communication networks and then different channel propagation categories like line/non-line of site propagation with their background and importance.

Part III describes the antenna design and technology for Terahertz communication networks and focuses on antenna advancements, beamforming, and blockage mitigation. In Chapter 7, an overview of different types of Terahertz antenna like planar, reflectarray, horn, and lens antenna are given. The latest trend of designing the Terahertz antenna using graphene and carbon nanotubes is describe and their performance against the traditional copper-based Terahertz antenna compared. A discussion on Terahertz's power sources is also presented along with different fabrication techniques and processes of Terahertz components. The use of directional antennas in Terahertz communication requires tight alignment and faces blockage issues. Chapter 8 discusses the antenna misalignment and blockage issues for Terahertz links and presents suitable models for their characteristics. The impact of antenna misalignment is quantified in terms of total directional gain and blockage in terms of blocking probability. Further, different antenna misalignment and blockage mitigation approaches are presented with their performance, advantages, and disadvantages. In Chapter 9, an overview of hybrid beamforming techniques for wireless Terahertz communication systems is presented. The basic concepts and capacity performance analysis for narrow-band and frequency selective hybrid beamforming is presented. The multi-user scenario and distance-aware multicarrier modulation are also discussed with performance ana1.lysis, design, and implementation. In Chapter 10, the fundamental aspects of the ultra-massive multiple input multiple output (MIMO) antenna system and plasmonic nano-antenna array are introduced with their design aspects, model, applications, and implementation challenges. In Chapter 11, different techniques for the fabrication of passive components like antennas and waveguides are discussed with an overview of planar antennas and phased array. Some technical challenges of millimeter-wave signal generation and the current state of the art using microwave photonics is also presented.

Part IV presents the use of Terahertz technology and links for different applications and their deployments. Chapter 12 presents an overview of inter-satellite links and also proposes a Terahertz band-based inter-satellite link system. The state-of-the-art on the suitability of inter-satellite links adoption using size, power, and data rate, is presented. In Chapter 13, the integration and capability of Terahertz components and systems above 100 GHz are described to support the 5G radio access network and B5G networks with their challenges. In Chapter 14, the recent advancement in Terahertz waveguides is presented with specific applications and design challenges. Besides, various waveguide designs of active and passive components for signal processing are also explained. In Chapter 15, a solution framework for resource management in the future Terahertz system is highlighted. Different types of resources such as fixed, variable, and imposed are presented with their system performance and models. In Chapter 16, the Terahertz communication zones are discussed to tackle connectivity issues in Terahertz communication networks. The issues like channel modeling and designing suitable modulation and coding schemes are discussed followed by signal processing, energy-efficient, and cooperative communication challenges. Finally, in Chapter 17, possible types of data centre network architectures are discussed for Terahertz link utilization, which is based on the software-defined networking principles to promote automated configuration. The network function virtualization is also described to utilize software-defined network-enabled Terahertz links.

Note

1 Cisco per month forecast. https://davidellis.ca/wp-content/uploads/2019/05/cisco-vni-feb 2019.pdf.

Chapter 1

The Meeting Point of Terahertz Communications, Sensing, and Localization

Hadi Sarieddeen, Nasir Saeed, Tareq Y. Al-Naffouri, and Mohamed-Slim Alouini

Contents

1.1 Introduction

As the demand for bandwidth continues to increase, wireless communication carrier frequencies continue to expand. Recently, efficient communication paradigms have been demonstrated at the millimeter (mmWave) band [1,2], as well as at the optical band in free space optics (FSO) and visible light communications (VLC) [3,4]. In between, the terahertz (THz) band stands as an unexamined part of the radio frequency (RF) spectrum. RF engineers mark any system operating beyond 100 GHz as a THz system, below which popular mmWave applications are placed. On the other hand, optical engineers observe all frequencies beneath the far-infrared (10 THz threshold) as THz frequencies. Nevertheless, as defined by *IEEE Transactions on Terahertz Science and Technology*, the THz band ranges between 300 GHz and 10 THz.

Today, researchers are exploring technologies from both neighboring bands to advance THz communications and close the so-called THz gap that existed because of the absence of efficient and compact THz devices. Contemporary THz transceiver design research is employed mostly in electronics and photonics [5–8]. Even though a significant data rate gain is observed in photonic technologies, electronic solutions continue to be superior in generating higher power. Electronic solutions [5] are principally based on silicon complementary metal-oxide-semiconductor (CMOS) technology, high electron mobility transistors (HEMTs), and III–V-based semiconductors in heterojunction bipolar transistors (HBTs). Photonic solutions [6], on the other hand, are based on photoconductive antennas, uni-traveling carrier photodiodes, quantum cascade lasers, and optical down-conversion systems. Besides, integrated hybrid electronic-photonic solutions [7] are gaining popularity as they can achieve a good trade-off between reconfigurability and performance. Similarly, plasmonic solutions based on novel materials are emerging, graphene-based solutions [9], in particular.

Traditional THz-band use cases have been in imaging and sensing [10–14]. Recently, the progress in signal generation and modulation techniques at the THz band is paving the way toward THz communication-based use cases [15–20]. THz communications promise to enable ultra-low latency and ultra-high bandwidth communication schemes, to support mobile wireless medium-range communications at both the access and device levels in the context of indoor and outdoor communications (Figure 1.1). By merging THz communications, sensing, imaging, and localization, THz technology can realize 6G ubiquitous wireless intelligence [21–23]. This chapter advocates the merging of these applications by detailing the corresponding system models and illustrating proof-of-concept results.

1.2 THz Communications

1.2.1 Use Cases for THz Communications

THz communications are expected to be realized in the future sixth-generation (6G) of wireless mobile communications [24–27] and beyond, enabling ultra-low latency and ultra-high bandwidth communication models. Consequently, several research groups have drawn substantial funds to carry THz research, and standardization attempts have started [28–30]. THz communications promise a terabit/second data rate, which opens the door for applications that cannot be accomplished in mmWave systems. Compared to mmWave communications, THz communications sustain higher directionality, maintain greater resilience to eavesdropping, and are less sensitive to inter-antenna interference and free-space diffraction. This is mainly due to the inherently shorter wavelengths at THz frequencies, further resulting in THz systems being realized in much smaller footprints. Furthermore, as opposed to VLC/FSO, THz signals are less influenced by factors such as cloud dust, scintillation, ambient light, atmospheric turbulence, and others. Nevertheless, by simultaneously using mmWave, THz, and optical communications in a heterogeneous plan, availability can be enhanced.

In a particular use case, THz-band communications can enhance future vehicular networks, both in terms of reliability and latency [31]. Reliable and high-speed communications are critical demands of future vehicular networks, where the bird's-eye view for a vehicle necessitates 50 ms latency and 50 Mbps data rate [32]. Correspondingly, automatic over-take requires less than 10 ms latency for 99.999% reliability. Therefore, researchers envision that using the THz band will improve safety solutions and enable various other applications such as remote driving and vehicle platooning.

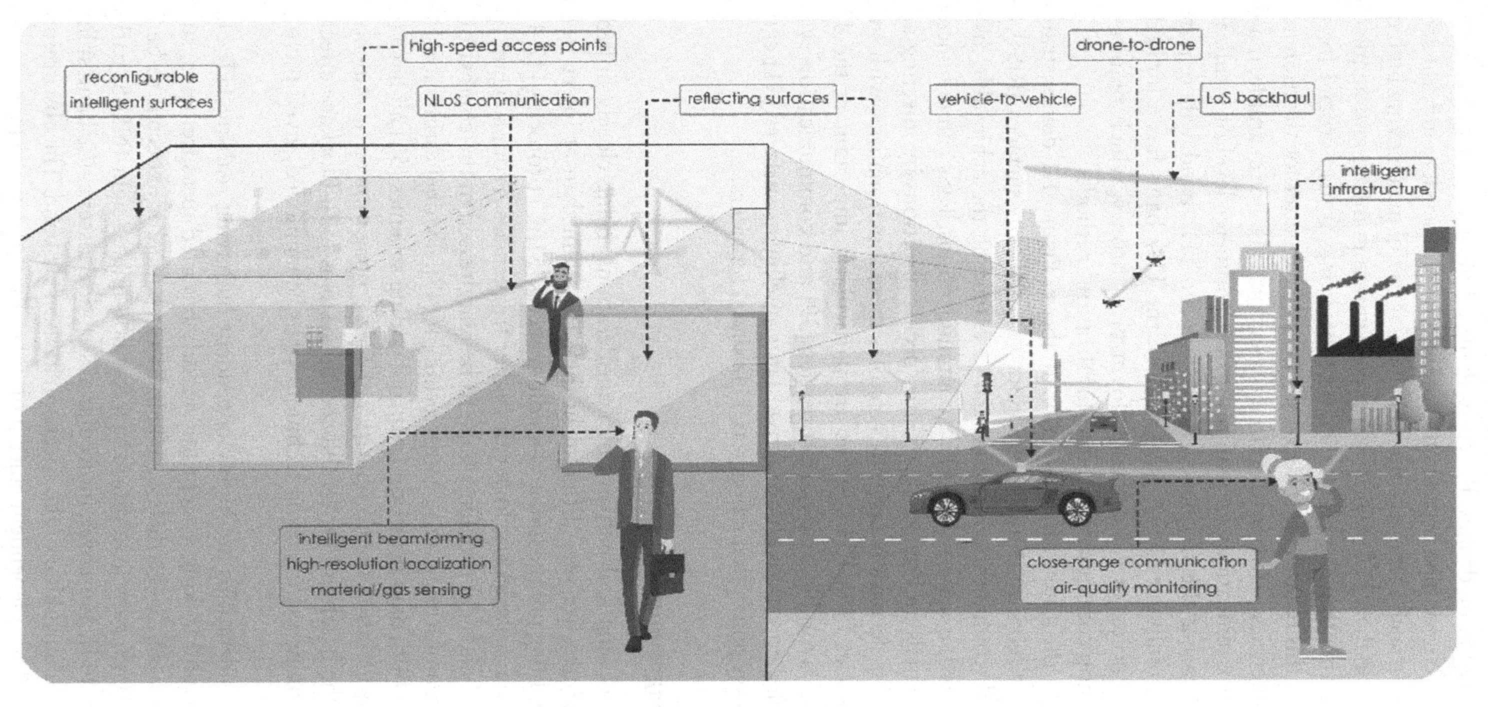

Figure 1.1 Prospective outdoor and indoor THz-band applications in communications and sensing.

In another plausible use case, the THz band can accommodate high-speed communications among drones. Recently, flying ad-hoc networks (FANETs) consisting of several drones have enabled broadband communication services in rural areas or disaster-affected places. In such scenarios, the THz band is useful in achieving high capacity and is also more flexible than FSO, which requires accurate pointing-and-acquisition methods. Furthermore, estimating the precise location of drones is also crucial for path planning and tracking. In this regard, the THz band can achieve better localization accuracy due to the higher frequencies [33] (more on that in Section 1.4). For example, if a drone is using 30 GHz for communications, it requires a sub-centimeter level of localization accuracy. Therefore, increasing the frequency further to the THz band will require localization accuracy at the millimeter level. Nevertheless, the THz band has a limited transmission range that may necessitate a dense deployment of drones, which requires the development of multi-hop communication-and-localization methods.

In addition to the applications mentioned above, Figure 1.1 demonstrates the importance of specular reflections and reconfigurable intelligent surfaces (RISs) [34,35] at the THz band. Both configurations are crucial to extending the achievable distances of THz communications and supporting NLoS communications. RISs can enhance the power of received THz signals and reflect them into specific directions by injecting arbitrary phase shifts [36,37]. An electrically large RIS enables these functionalities with small footprints at THz frequencies [38]. RIS systems are particularly important at the THz band due to THz-specific favorable material properties [39] and communication system considerations [40]. Furthermore, regular (active) large intelligent surfaces [41] can also provide distributed access points that can act as THz signal repeaters.

1.2.2 Challenges and Solutions

The signal processing and networking considerations, being closely linked to the transceiver architectures, are very different in the THz realm than at lower frequencies. This necessitates enabling unique link and medium access (MAC) protocols [42,43] that cannot be considered for networks at lower or higher frequencies. From a signal processing perspective, several classical problems have to be readdressed in the context of THz [44,45]. For instance, beamforming and beamsteering [46,47], channel estimation [48], precoding and combining [49], data detection [50], and coding schemes [51] all have to be revisited, taking into consideration the peculiarities of THz propagation environments as well as the limitations in the baseband for terabit/second systems [52]. Blockage effects also need to be modeled. Compressive sensing techniques can be employed, given the high sparsity in THz channels [53].

All these considerations are highly dependent on having realistic THz channel models [54], which are still lacking, except for recently reported measurements, such as those in Ref. [22] (140 GHz). Nevertheless, a THz-band channel is expected to be dominated by a line-of-sight (LoS) path in addition to a few non-line-of-sight (NLoS) paths due to large reflection losses [44]. Moreover, molecular absorption losses cause band splitting, which in turn results in the shrinking of the spectrum at the THz band. The absorption-free spectral windows are distance-dependent since molecular absorptions are more profound at larger communication distances. To combat spectrum shrinkage, distance-adaptive solutions [55,56], which optimized resource allocation and antenna array designs, are required.

Note that novel multicarrier (or single-carrier [57]) waveform designs are also needed as an alternative to the complex orthogonal frequency-division multiplexing (OFDM).

In another regard, generating continuous modulations at the THz band is challenging. For example, only a few milli-watts short pulses can be generated at room temperature with graphene. In order to combat transceiver limitations, pulse-based on–off keying modulations (femtosecond pulses) [58] can be used. However, because the pulse-based systems' frequency response comprises a wide range of THz frequencies, it is impossible to avoid absorption spectra. Re-transmissions resulting from molecular absorption generate an additional colored noise factor [9,59]. Note, however, that this colored noise model has not yet been verified in measurements. Therefore, accurate noise modeling is still a remaining challenge.

Even though THz communications possess quasi-optical traits, they retain some microwave propagation properties, where THz signals can still make use of antenna array processing techniques and reflections. In fact, power limitations and severe propagation losses result in limited communication distances. To overcome this limitation, ultra-massive multiple input multiple output (UM-MIMO) antenna arrangements [45,60] are crucial. Very dense UM-MIMO array-of-subarrays (AoSA) configurations can accommodate the needed array gains to combat communication distance limitations. At THz frequencies, many antenna elements (AEs) can be set in a footprint of a few square millimeters. Furthermore, subarrays (SAs) facilitate hybrid beamforming techniques, reducing power consumption and hardware costs, and providing adaptability to exchange beamforming with multiplexing gains for enhanced spectral efficiency. Configurable AoSA designs can further support multicarrier paradigms and varieties of spatial/index modulation schemes [44]. Such adaptive transmission schemes can be efficiently complemented with blind parameter estimation techniques at the receiver side [61,62].

1.2.3 A Model of the THz Communications System

A standard system model for THz communications is depicted in Figure 1.2, in which multiple transmitting adaptive AoSAs serve multiple devices. Each SA's operation can be tuned to a particular frequency and can be attributed to a distinct modulation mode. Moreover,

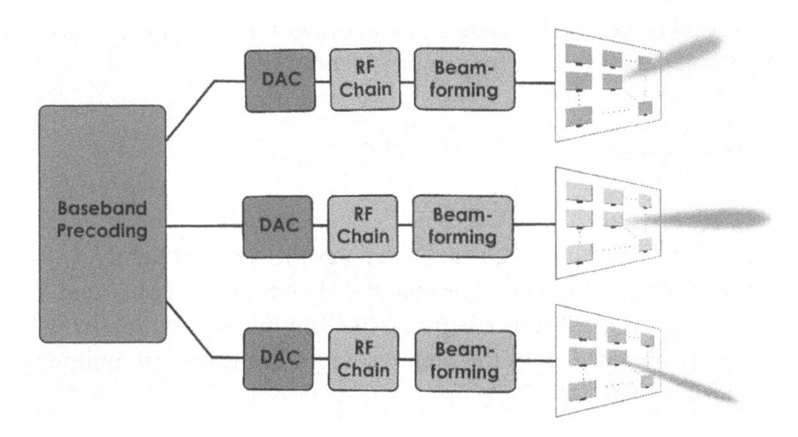

Figure 1.2 THz communication baseband and front-end system model.

following digital-to-analog conversion (DAC), the SAs are supplied by dedicated RF chains. The "pencil" beams caused by the high directivity of THz signals result in each SA being disconnected from its adjacent SAs, where the purpose of baseband precoding gets diminished to merely determining the utilization of SAs (turning them on and off). Considering that AEs can be contiguously installed over a 3D structure, SAs can be virtually allocated and interleaved frequency allocations for AEs can be supported. For a required communication distance, the necessary number of AEs per SA is assigned. The diversity/multiplexing gain is then dictated by the number of potential SA allocations, which is limited by the number of RF chains and array dimensions.

We hereby adopt a three-dimensional AoSA UM-MIMO system model [44], [63]. The effective channels are point-to-point channels, consisting of AoSAs [64] at both the transmitter and receiver, having $M_t \times N_t$ and $M_r \times N_r$ SAs, respectively. Each SA is considered to be formed of $Q \times Q$ nano-AEs. The overall configuration is a large MIMO system [50]. Denote by bold upper case, bold lower case, and lower case letters matrices, vectors, and scalars, respectively. The end-to-end system model at the baseband for a specific frequency is expressed as

$$y = \mathbf{W}_r^H \boldsymbol{H} \mathbf{W}_t^H \boldsymbol{x} + \mathbf{W}_r^H \boldsymbol{n}$$

where $\boldsymbol{x} = [x_1 x_2 \cdots x_{Ns}]\ T \in \mathcal{X}_{Ns} \times 1$ is the modulated symbol vector that holds information, $\boldsymbol{y} \in \mathcal{C}^{N_s \times 1}$ is the received symbol vector, $\boldsymbol{H} = \left[\boldsymbol{h}_1 \boldsymbol{h}_2 \cdots \boldsymbol{h}_{M_t N_t}\right] \in \mathcal{C}^{M_r N_r \times M_t N_t}$ is the channel matrix, $\mathbf{W}_t \in \mathcal{R}^{N_s \times M_t N_t}$ and $\mathbf{W}_r \in \mathcal{R}^{M_r N_r \times N_s}$ are the baseband precoder and combiner matrices, and $\boldsymbol{n} \in \mathcal{C}^{M_r N_r \times 1}$ is the additive white Gaussian noise (AWGN) vector with a power of σ^2. Note that $(\cdot)^T$ and $(\cdot)^H$ denote transpose and conjugate transpose, respectively.

An element of $\boldsymbol{H}$, $h_{m_r n_r, m_t n_t}$, the frequency response between the (m_t, n_t) and (m_t, n_t) SAs, is thus defined as

$$h_{m_r n_r, m_t n_t} = \boldsymbol{a}_r^H(\phi_r, \theta_r) \dot{G}_r \alpha_{m_r n_r, m_t n_t} \dot{G}_t\, \boldsymbol{a}_t(\phi_t, \theta_t)$$

for $m_r = 1, \cdots, M_r, n_r = 1, \cdots, N_r, m_t = 1, \cdots, M_t$, and $n_t = 1, \cdots, N_t$, where α is the path gain, a_t and a_r are the transmit and receive steering vectors per SA, ϕ_t, θ_t and ϕ_r, θ_r are the transmit and receive angles of departure and arrival (ϕ's are the azimuth angles and θ's are the elevation angles), and $\dot{\mathrm{G}}_t$ $\dot{\mathrm{G}}_t$ are the transmit and receive antenna gains, respectively.

The LoS path gain at a specific frequency is defined as

$$\alpha^{LoS}_{m_r n_r, m_t n_t} = \frac{c}{4\pi f d_{m_r n_r, m_t n_t}} e^{-\frac{1}{2}\mathcal{K}(f) d_{m_r n_r, m_t n_t}} e^{-j\frac{2\pi f}{c} d_{m_r n_r, m_t n_t}}$$

where f denotes the operating frequency, c denotes the speed of light, $\mathcal{K}(f) = \sum_{g=1}^{G} \mathcal{K}^g(f)$ is the absorption coefficient that is accumulated over all G gazes that constitute the medium, and $d_{m_r n_r, m_t n_t}$ is the distance between the transmitting and receiving SAs. All these parameters can be obtained from the high-resolution transmission molecular absorption database (HITRAN) [65].

1.3 THz Sensing and Imaging

THz technology has long been utilized for wireless sensing and imaging [10–14] in many applications, such as quality control and security. Numerous propagation properties allow THz signals to be suitable for gas and material sensing. For instance, several chemical and biological elements present unique spectral fingerprints at the THz band. In particular, THz signals can penetrate various dielectric, amorphous, and non-conducting materials (wood, plastic, glass). Furthermore, metals fully reflect THz signals, facilitating inspecting the presence of metallic components. Similarly, THz radiation can be utilized to observe water dynamics and detect gaseous compositions [23].

Time-domain spectroscopy (THz-TDS) [66] is the most popular procedure for THz sensing. THz-TDS starts by probing a specimen with broadband THz signals using short pulses. The pulses' temporal profiles are then recorded, with and without the specimen, to determine the sample material's optical properties. The spectral behavior of these profiles yields amplitude and phase information that allows direct estimation of the frequency-specific index of refraction, absorption coefficient, and sample thickness. THz-TDS can also be used for imaging, which can be accomplished by extracting the required spectral information to discover the material's type and shape. Because of limited scattering, THz-TDS imaging produces images of high contrast. Furthermore, THz imaging is more powerful than infrared-based imaging because of the more limited effect of ambient light and weather on THz channel conditions. The much broader THz channel bandwidths also encourage applications that require generating images of large fields of view. Moreover, high-gain directional THz antennas support very fine spatial resolution (sub-millimeter spatial differentiation [22]) for highly directional sensing and imaging.

Two types of spectroscopic measurements are distinguished: transmission-based and reflection-based. Transmission spectroscopy investigates the quantity of light absorbed by a specimen material, and reflection spectroscopy analyzes the scattered and reflected light. THz sensing and imaging are principally carried in transmission mode because absorbance THz spectroscopy has a more convenient setup, and it provides greater signal contrast. Nonetheless, various attributes encourage using reflection mode. For example, reflection spectroscopy permits identifying objects on non-transparent substrates. Furthermore, reflection geometry is the solution of choice in open-field applications. In the particular case of joint THz communication and sensing, the handheld user equipment can enable imaging and sensing in the reflection mode without demanding infrastructure support.

Following recent progress in THz technology, carrier-based sensing and imaging operations are no longer far-fetched. Carrier-based systems allow higher adaptability and fine-tuning capacities across the THz frequency range. In fact, carrier-based sensing is a particular form of frequency-domain spectroscopy [67]. We demonstrate a proof-of-concept simulation of wireless THz gas sensing (also known as electronic smelling [5]). Carrier-based THz gas sensing can be achieved using a source that transmits multiple specific single-carrier high-frequency signals into a medium and a receiver that estimates the corresponding channel response to detect distinct absorption spikes (rotational spectroscopy). The calculated channel responses are compared to a database of molecular absorption spectra of reference gases (HITRAN), and a conclusion is constructed on the components of the studied medium.

To demonstrate that a sensing system can be piggybacked over a communication system in a seamless manner, we examine a MIMO setup as described in Section 1.2. In Ref. [44], it is demonstrated that the maximum spatial degrees of freedom can be guaranteed in a LoS THz environment by tuning the SA spacings such that spatially uncorrelated channel matrices are generated. Hence, we assume the channel to be orthogonal by design. We further consider each SA to operate at a specific frequency in a perfectly symmetric setup that matches the transmitting and receiving arrays. Therefore, the MIMO channel can be reduced into multiple single-input single-output (SISO) streams (a diagonal channel), where each stream corresponds to the channel response between a specific transmitting SA operating at a specific frequency and its matching SA at the receiver.

In the context of pure sensing applications, $\boldsymbol{x}$ can be designed to have a specific structure. But, in the context of joint communications and sensing, the elements of $\boldsymbol{x}$ are expected to be drawn from a specific constellation of a specific modulation type. Such prior knowledge of the distribution of $\boldsymbol{x}$ can be used to further improve the sensing performance. By detecting $\boldsymbol{x}$, we can detect the gases comprising a medium and estimate their corresponding concentrations in this medium, where a value of 0 for a specific absorption coefficient would indicate the absence of the corresponding gas. Several methods can be used to solve for $\boldsymbol{x}$, such as optimal and exhaustive maximum likelihood detectors and varieties of compressive sensing and machine learning techniques. For example, we could first detect the presence of specific spikes and then build decision trees for classification [67].

In the particular cases of measuring the percentage of water vapor and oxygen, Figures 1.3 and 1.4 illustrate the error rate performance versus signal-to-noise ratio (SNR). The measurements are conducted over a distance of 5 m, assuming a 396 K temperature and a pressure of 0.1 atm. The carrier frequencies are either selected uniformly between 1 and 2 THz, or set to coincide with resonant frequencies corresponding to the target molecule. We perform an exhaustive, yet not optimal, heuristic search to identify the concentrations with a resolution of 0.01. Performance enhancement is seen when

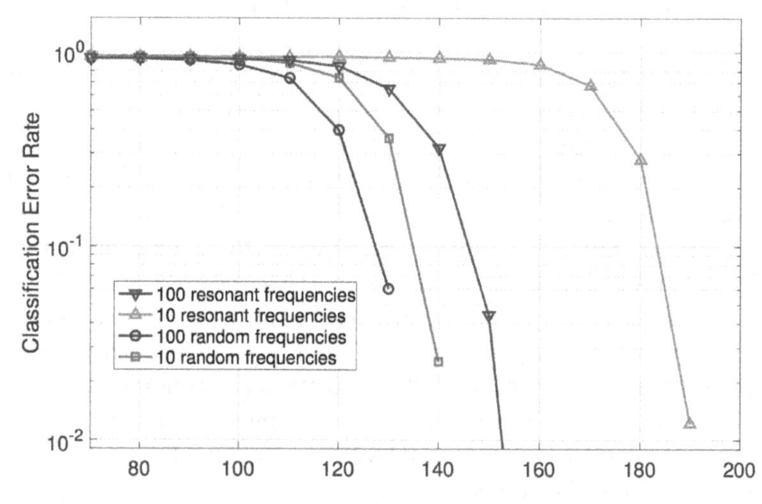

Figure 1.3 Classification error rates of THz carrier-based sensing of water vapor concentration as a function of the number of carriers and SNR.

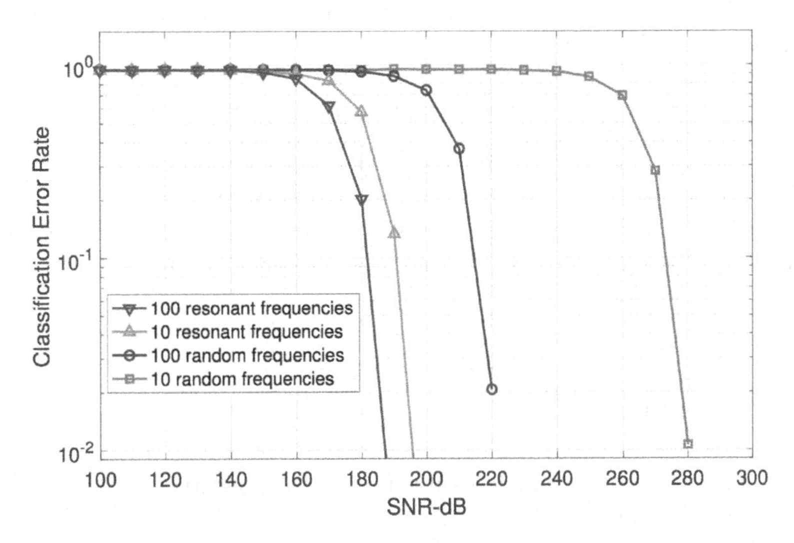

Figure 1.4 Classification error rates of THz carrier-based sensing of oxygen concentration as a function of the number of carriers and SNR.

expanding the set of carriers from 10 to 100. Yet, SNRs higher than 100 dB (assuming 0 dB antenna gains) are still needed to overcome the losses over a 5 m range. This gap can be lessened by adding antenna gains and beamforming gains. Furthermore, it is noted that a random choice of frequencies is favored when detecting water vapor, whereas testing resonant frequencies is a better option in the case of oxygen. This might be caused by the fact that absorption spikes are orders of magnitude higher for water vapor. Lastly, we note that the spikes' linewidths appear pressure-broadened to a few GHz at standard pressure, posing a serious challenge for sensing. However, at low pressures, such absorption lines confine to widths of single-megahertz.

1.4 THz Localization

Over the past few years, the demand for services that depend on precise localization of objects and people has increased. This growing demand has led to the evolution of various localization systems. This section briefly introduces the main ranging techniques based on time of arrival (ToA), time difference of arrival (TDoA), angle of arrival (AoA), and received signal strength (RSS). Further, it introduces the localization methods for THz systems.

1.4.1 Time of Arrival Ranging

ToA is a well-known ranging method that uses the time of flight of the signal. The ToA-based systems estimate the distance from the travel time of the propagating signal. For instance, the beacon nodes broadcast the ranging signal that arrives at the receiving node. Since the electromagnetic (EM)/optical signals travel with the speed of light, the distance is estimated from the signal's propagation speed and time of flight. The receiving node collects the ranging signals from multiple beacon nodes and then uses multi-lateration to estimate its location. The noisy ToA measurement between the l-th beacon and i-th receiving node is written as

$$r_{\text{ToA}} = d_{il} + \eta_{\text{ToA}} = \sqrt{(x_{\text{i}} - x_{\text{l}})^2 - (y_{\text{i}} - y_{\text{l}})^2} + \eta_{\text{ToA}},$$

where $\{x_i, y_i\}$ is the unknown location of the i-th receiving node, $\{x_l, y_l\}$ is the location of the l-th anchor node, and η_{ToA} represents the ranging error. Considering that there are L beacon nodes, the location of i-th node is estimated by minimizing the following cost function

$$f_{\text{ToA}}(p_{\text{i}}^*) = \sum_{l=1}^{L} \left(r_{\text{ToA},l} - \sqrt{(x_{\text{i}}^* - x_{\text{l}})^2 - (y_{\text{i}}^* - y_{\text{l}})^2} \right)^2, \tag{1.1}$$

where $\{x_i^*, y_i^*\}$ represents the smallest value for $f_{\text{ToA}}(p_{\text{i}}^*)$, resulting in the estimated location of the i-th receiving node, i.e.,

$$\hat{p}_i = \arg\min_{p_{\text{i}}^*} f_{\text{ToA}}(p_{\text{i}}^*),$$

where $\hat{p}_{\text{i}} = \{\hat{x}_{\text{i}}, \hat{y}_{\text{i}}\}$. Solving (1.1) is not an easy task since there exist local minima. Both global and local optimization techniques can be used to solve this optimization function. A major problem with the ToA method is that it suffers from multipath, leading to different signal propagation times. Hence, most of the time, it is assumed that the LoS is available and pre-dominant [68]. The range of error in multipath channels can be much greater than that caused by noise. The error can be produced by an early arriving path, which spoils the exact arrival time of the LoS path. Another possible situation is that the LoS is highly attenuated, leading to an error of presuming another component as LoS. The accuracy of the ToA-based scheme is strongly dependent on the receiver to determine the propagation time of the LoS signal [69].

1.4.2 Time Difference of Arrival (TDoA) Ranging

Unlike ToA, the TDoA-based systems estimate the distance by collecting two distinct signals from the same beacon or alike signals from separate beacons. The Cricket system [70] is among the famous indoor positioning system based on TDoA ranging that uses both RF and ultrasound signals. In TDoA systems, the time difference between two distinct signals is calculated at the receiving node, resulting in a hyperbola. The point of intersection of these hyperbola yields the position of the receiving node. One major advantage of TDoA systems is that they do not need synchronization among the beacon and the receiving node. Mathematically, the TDoA measurement at the i-th node for L beacons can be written as

$$r_{\text{TDoA}} = d_{\text{i,l}} + \eta_{\text{TDoA}} = \sqrt{(x_{\text{i}} - x_{\text{l}})^2 - (y_{\text{i}} - y_{\text{l}})^2} + \eta_{\text{TDoA}},$$

where $l=1,2,3,\ldots L$, $d_{i,l}=\sqrt{(x_i-x_{l+1})^2-(y_i-y_{l+1})^2}-\sqrt{(x_i-x_l)^2-(y_i-y_l)^2}$, and η_{TDoA} is the ranging error. Similarly to the ToA error function, the cost function for the TDoA location estimation is given as

$$f_{\mathrm{TDoA}}(p_i^*)=\sum_{l=1}^{L}\left(r_{\mathrm{TDoA},l}-\sqrt{(x_i^*-x_l)^2-(y_i^*-y_l)^2}\right)^2, \tag{1.2}$$

This cost function can also be solved using various global and local optimization methods. In (1.2), $\{x_i^*, y_i^*\}$ corresponds to the smallest value of $f_{\mathrm{TDoA}}(p_i^*)$. Minimization of (1.2) yields the estimated location of the i-th receiving node, i.e.,

$$\hat{p}_i=\arg\min_{p_i^*} f_{\mathrm{TDoA}}(p_i^*).$$

1.4.3 Received Signal Strength (RSS) Ranging

RSS measurements are ubiquitous in wireless communication systems. The localization systems based on RSS are less accurate compared to the time-based systems; however, they do not require any modification to the available infrastructure. Mainly, RSS measurements are readily obtainable in most of the wireless communication systems without any supplementary hardware. Indeed, the RSS values are necessary for defining various wireless standards for basic radio propagation, such as link quality estimation, channel assessment, and radio resource management. In RSS-based positioning systems, the receiving node localizes itself by collecting signals from multiple beacons. However, unlike in the ToA systems, the receiving node measures the strength of the arriving signal [71]. The received signal's strength indicates the distance between the beacon and the receiving node that can be estimated from the free-space path loss model. The path loss (in dB's) between l-th beacon node and i-th receiving node is given as

$$PL_{i,l}=PL_0-10\alpha\log_{10}\left(\frac{\|p_i-p_l\|}{d_0}\right)+w_{i,l}, \tag{1.3}$$

where PL_0 is the power loss at reference distance d_0 and α represents the path loss exponent. $p_i=\{x_i,y_i\}$ and $p_l=\{x_l,y_l\}$ represent the location of i-th ordinary node and l-th beacon node, respectively. The term $w_{i,l}\sim\mathcal{N}(0,\sigma_{i,l}^2)$ denotes zero-mean log-normal shadowing effect with variance $\sigma_{i,l}^2$. Based on the path loss expression in (1.3), maximum likelihood estimation can be used to find the location of i-th node, i.e.,

$$\hat{p}_i=\arg\min_{p_i}\sum_{l=1}^{L}\frac{1}{\sigma_{i,l}^2}\left[(PL_{i,l}-PL_0)-10\alpha\log_{10}\left(\frac{\|p_i-p_l\|}{d_0}\right)\right]. \tag{1.4}$$

This cost function can be solved using various recursive optimization techniques such as Newton's method or gradient descent method. However, these methods can result in a locally optimal solution. Hence, convex relaxation methods can be used to solve the cost function in (1.4) for a global optimal solution [72].

1.4.4 Angle of Arrival (AoA) Ranging

AoA ranging is computed from the direction of the received signal. The AoA-based methods are more straightforward than the time-based mechanisms because only two AoA estimations are needed to locate the receiving node in a two-dimensional space. Despite that, acquiring accurate AoA estimation is a complicated task, especially with NLoS situations. Moreover, in the case of indoor environments where the LoS signal is difficult to acquire, AoA estimations are significantly erroneous. In comparison to the other ranging techniques, AoA calculation requires an array of antennas. Further, each of the antenna elements has to be equipped with an RF component, which is one of the expensive parts of a radio system. Besides, the RF components consume reasonably high power; thus, it is anticipated that the AoA estimation methods require more energy and are more complicated compared to their counterparts. Like ToA and RSS ranging, in AoA, the beacon position is known in prior. However, unlike ToA and RSS, in AoA, only two beacon nodes are required for the receiving node to locate itself. To find the AoA, the main lobe of the antenna array is directed toward the receiving node.

1.4.5 Localization Using THz signals

5G-and-beyond wireless communication networks are expected to enable accurate location-based services. Unlike the existing GPS and cellular positioning systems, 5G-and-beyond networks should support centimeter-level accuracy using higher frequency bands, including mmWaves and THz frequencies [22]. Localization systems operating in these frequency bands mainly use network visualization tools such as simultaneous localization and mapping (SLAM). In SLAM-based methods, the localization accuracy improves with an increase in the resolution of the collected images. Since high-frequency bands can capture good quality images of the environment, they can provide better localization accuracy. There are three main steps involved in SLAM-based methods: (i) taking images of the surrounding environment, (ii) estimating the ranges to the user, and (iii) fusing the

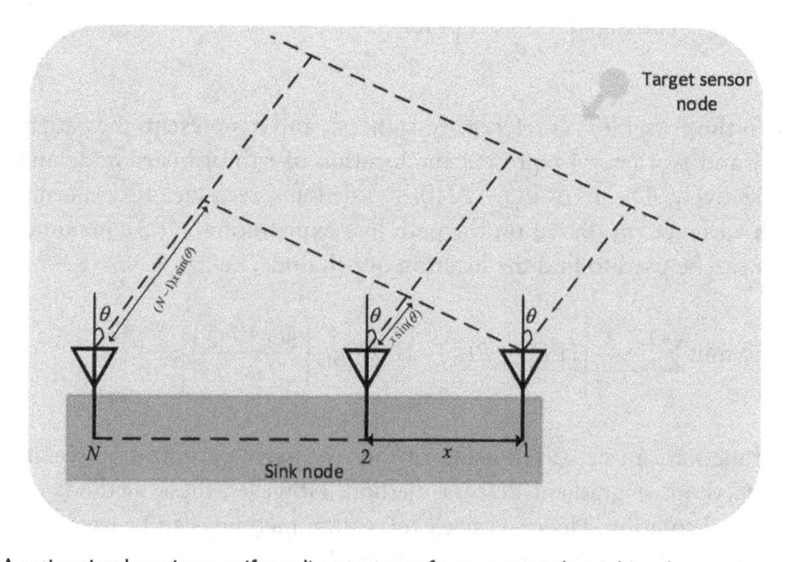

Figure 1.5 AoA estimation by using a uniform linear array of antennas at the sink node.

ranging information with the images. For instance, using the frequency band between 200 and 300 GHz for collecting the 3D images of the environment and fusing them with the ToA and AoA can result in a sub-centimeter level of accuracy. Using only ToA or AoA does not achieve higher accuracy, especially in indoor environments. Therefore, it is crucial to fuse ranges from multiple high-frequency bands to achieve the sub-centimeter level of accuracy [73].

Besides the use of SLAM-based methods, few other localization techniques exist in the literature for THz communications. For example, the location of a transceiver operating at the THz band can be estimated using the well-known weighted linear least square (WLLS) estimator [74]. In this method, first, the anchors (tags with dielectric resonators) transmit the beacon signal. Then, the receiver calculates the round-trip time-of-flight (RToF) from various anchors. The time-based ranging methods require synchronization among the anchors and the receivers; however, RToF can avoid this challenging issue. Finally, the receiver applies the WLLS estimator to the RToF measurements for estimating its location. This approach achieves a millimeter-level of accuracy.

The sensing and detection capabilities of nano-senor networks (NSNs) makes them an interesting area of research, especially for body-centric networks [75]. The nano-devices are supposed to communicate using the THz band, where these devices' location estimation is of paramount importance. For instance, the nano-devices in an NSN operating at THz frequencies can be localized using AoA ranging along with the multiple-signal-classification (MUSIC) technique [76]. In an NSN, the sink node consists of a uniform and linear array with a number of antennas, which finds the angle of incoming THz signals (see Figure 1.5). The target sensor node (nano-device) employs the MUSIC technique for estimating the AoA, resulting in a sub-degree accuracy for a distance of 6 m [76].

The above studies focus mainly on ranging for THz networks while ignoring the final localization process. Accordingly, the development of accurate and robust localization methods for THz networks remains an open research problem at the moment. Therefore, we propose a well-known dimensionality-reduction approach, multidimensional scaling (MDS), for localization in THz networks. The MDS method visualizes higher-dimensional data into a lower-dimensional space using the correlation among the data's different variables [77]. MDS has many applications, including psychology, data analysis, sports, ecology, environmental monitoring, and localization. In the context of localization, the MDS method takes the estimated ranges among the nodes as an input, where the ranges are estimated from the time, angle, or RSS-based techniques [78]. In reality, the ranges are corrupted by distance-dependent additive Gaussian noise, leading to noisy pairwise distances. Based on these noisy distances, the MDS method finds the location of the nodes in the network. In the following, we present the major steps required for MDS-based localization in THz networks:

1. First, we need to compute the shortest path pairwise distances for all the nodes in the network. Note that the single-hop distances can be calculated by using any available ranging method.
2. The MDS method is then applied to the complete pairwise distance matrix, where a two- or three-dimensional map of the network can be created from the most significant eigenvectors and eigenvalues. In the case of two-dimensional (or three-dimensional) maps, the first two (or three) largest eigenvectors and eigenvalues are chosen.

3. Finally, with the help of available beacon nodes, the relative map created in the second step can be converted into actual geographical locations using linear transformation methods.

To elaborate more, consider that there are 20 sensor nodes (circles) randomly distributed in an area of 15×15 m^2, and a single anchor node (squares) placed at each corner. Figure 1.6 displays the actual locations of the anchor and sensor nodes. We also consider that the ranging error variance is 8 cm.

Based on the available noisy range measurements, the MDS creates a network map, mainly showing all the nodes' proximity information, as shown in Figure 1.7. Note that the output map from the MDS only indicates the proximity information without any actual coordinates. To estimate the sensor nodes' actual coordinates, the map in Figure 1.7 needs a linear transformation. This can be achieved with anchors and linear transformation methods, such as Helmert's transform or Procrustes analysis. Figure 1.8 shows the final estimated location of all sensor nodes, where accuracy of 2.4 cm is achieved.

1.5 Implementation Aspects

As research on THz devices advances, it is becoming clearer that signal processing, networking, and communications considerations are very different in the THz realm [79]. In fact, optimizing signal processing algorithms and networking protocols requires an understanding of the devices, and optimizing devices requires knowledge of what these algorithms and protocols are trying to accomplish. Nevertheless, joint hardware optimization and algorithm design for joint THz communications, sensing, imaging, and localization will soon be a hot research topic [80].

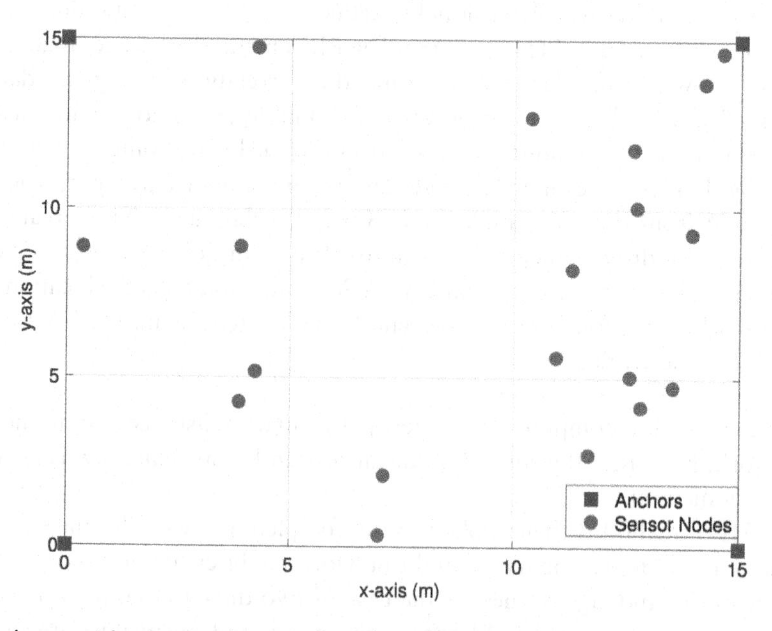

Figure 1.6 Actual setup with sensor nodes and anchors placed in 15×15 m^2 area.

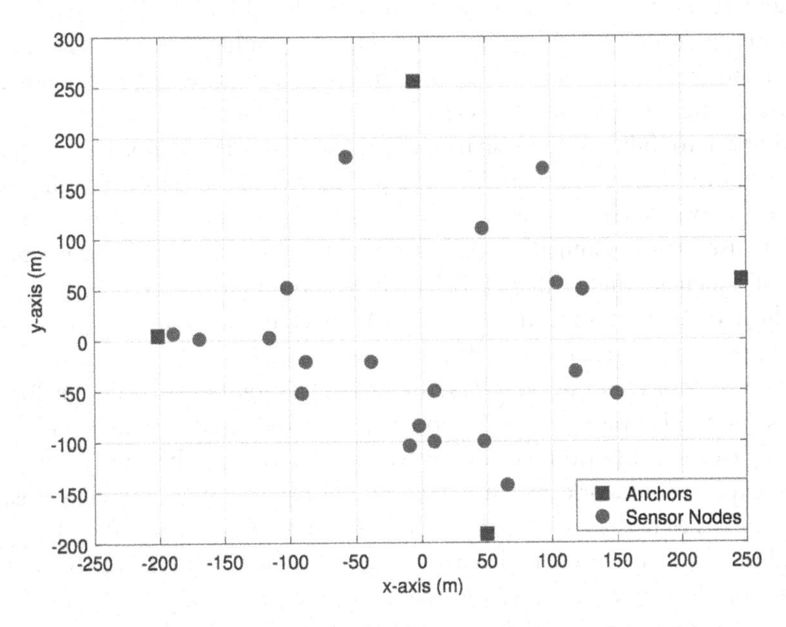

Figure 1.7 Relative MDS map with sensor nodes and anchors centered around the origin.

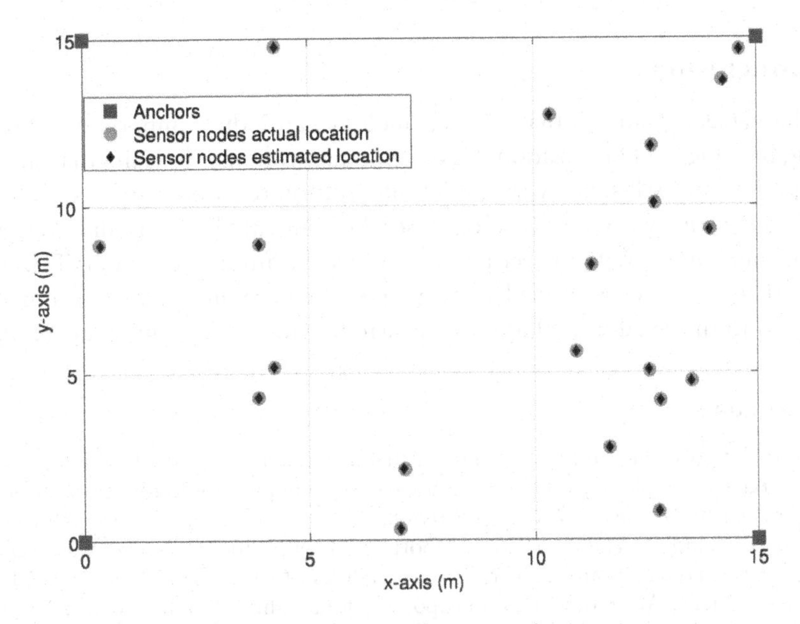

Figure 1.8 Final map with actual and estimated location of sensor nodes.

As THz sensing and localization applications are expected to coexist, it is natural to strive to have them piggybacked onto THz communications. Toward this end, resources can be shared in time, frequency, and space. For example, gas sensing or reflection-mode imaging can be executed by identifying changes in the channel responses over specific information-bearing carrier signals, irrespective of the exact data that is modulated over these carriers. Furthermore, sharing resources is feasible in pulse-based systems, where pulses can be utilized to conduct pulse-based communications, TDS-based sensing/imaging, and localization. However, the efficiency of pulse-based systems is questionable due to the limited power of bulky THz-TDS spectrometers.

In addition to full resource sharing, dedicated resource allocation schemes remain important in multiple scenarios. For example, carrier-based sensing might require tuning carriers over absorption spectra; such spectra are dominated by absorption lines and hence cannot be used to transmit information. In general, sensing applications perform much better at frequencies higher than 1 THz, where communication remains very challenging. In a dedicated resource allocation setup, each RF chain in Figure 1.2 can be assigned to a specific application, perhaps on dedicated time slots as well.

Finally, it is worth noting that THz band applications raise a few health and privacy concerns. Since THz radiation is not ionizing, it can be assumed that only heating could potentially cause health problems. However, as THz sources become more available, and with the expected network densification and anticipated high antenna gains, it is crucial to carry studies to understand further THz radiation's biological and molecular impact on health [22]. A privacy concern also arises from bringing THz signals to handheld devices and enabling them to cover large distances. For instance, a malicious device can conduct remote sensing and see-through imaging (that sees through clothing but not the body). For smartphone use, limited-range (10 cm) low-power pulse-based TDS should be sufficient to carry near-field imaging to capture gestures and other useful applications.

1.6 Conclusion

This chapter demonstrates that THz technology can induce important improvements in sensing, imaging, and localization besides communications. We claim that this merging of THz applications will shape wireless communication research activities in the future. We analyze the distinctiveness of each of these THz band applications and provide proof-of-concept simulations. With proper configurations, we attest that various THz applications can be seamlessly accomplished in real time. We recommend that serious health and privacy concerns be addressed in future research before THz technology matures further.

References

[1] Xiao, Ming, Shahid Mumtaz, Yongming Huang, Linglong Dai, Yonghui Li, Michail Matthaiou, George K. Karagiannidis et al. "Millimeter wave communications for future mobile networks." *IEEE Journal on Selected Areas in Communications* 35, no. 9 (2017): 1909–1935.

[2] Rangan, Sundeep, Theodore S. Rappaport, and Elza Erkip. "Millimeter-wave cellular wireless networks: Potentials and challenges." *Proceedings of the IEEE* 102, no. 3 (2014): 366–385.

[3] Grobe, Liane, Anagnostis Paraskevopoulos, Jonas Hilt, Dominic Schulz, Friedrich Lassak, Florian Hartlieb, Christoph Kottke, Volker Jungnickel, and Klaus-Dieter Langer. "High-speed visible light communication systems." *IEEE Communications Magazine* 51, no. 12 (2013): 60–66.

[4] Pathak, Parth H., Xiaotao Feng, Pengfei Hu, and Prasant Mohapatra. "Visible light communication, networking, and sensing: A survey, potential and challenges." *IEEE Communications Surveys & Tutorials* 17, no. 4 (2015): 2047–2077.

[5] Choi, Wooyeol, Qian Zhong, Navneet Sharma, Yaming Zhang, Ruonan Han, Z. Ahmad, Dae-Yeon Kim et al. "Opening terahertz for everyday applications." *IEEE Communications Magazine* 57, no. 8 (2019): 70–76.

[6] Nagatsuma, Tadao, Guillaume Ducournau, and Cyril C. Renaud. "Advances in terahertz communications accelerated by photonics." *Nature Photonics* 10, no. 6 (2016): 371.

[7] Sengupta, Kaushik, Tadao Nagatsuma, and Daniel M. Mittleman. "Terahertz integrated electronic and hybrid electronic–photonic systems." *Nature Electronics* 1, no. 12 (2018): 622–635.

[8] Koenig, Swen, Daniel Lopez-Diaz, Jochen Antes, Florian Boes, Ralf Henneberger, Arnulf Leuther, Axel Tessmann et al. "Wireless sub-THz communication system with high data rate." *Nature photonics* 7, no. 12 (2013): 977.

[9] Jornet, Josep Miquel, and Ian F. Akyildiz. "Channel modeling and capacity analysis for electromagnetic wireless nanonetworks in the terahertz band." *IEEE Transactions on Wireless Communications* 10, no. 10 (2011): 3211–3221.

[10] Woolard, Dwight L., R. Brown, Michael Pepper, and Michael Kemp. "Terahertz frequency sensing and imaging: A time of reckoning future applications?" *Proceedings of the IEEE* 93, no. 10 (2005): 1722–1743.

[11] Choi, Min Ki, Alan D. Bettermann, and Daniel W. van der Weide. "Biological and chemical sensing with electronic THz techniques." In *Chemical and Biological Standoff Detection*, vol. 5268, pp. 27–35. International Society for Optics and Photonics, 2004.

[12] Fischer, B., M. Hoffmann, H. Helm, G. Modjesch, and P. Uhd Jepsen. "Chemical recognition in terahertz time-domain spectroscopy and imaging." *Semiconductor Science and Technology* 20, no. 7 (2005): S246.

[13] Pickwell, E., and V. P. Wallace. "Biomedical applications of terahertz technology." *Journal of Physics D: Applied Physics* 39, no. 17 (2006): R301.

[14] Ho, Louise, Michael Pepper, and Philip Taday. "Terahertz spectroscopy: Signatures and fingerprints." *Nature Photonics* 2, no. 9 (2008): 541.

[15] Akyildiz, Ian F., Josep Miquel Jornet, and Chong Han. "TeraNets: Ultra-broadband communication networks in the terahertz band." *IEEE Wireless Communications* 21, no. 4 (2014): 130–135.

[16] Akyildiz, Ian F., Josep Miquel Jornet, and Chong Han. "Terahertz band: Next frontier for wireless communications." *Physical Communication* 12 (2014): 16–32.

[17] Kleine-Ostmann, Thomas, and Tadao Nagatsuma. "A review on terahertz communications research." *Journal of Infrared, Millimeter, and Terahertz Waves* 32, no. 2 (2011): 143–171.

[18] Chen, Zhi, Xinying Ma, Bo Zhang, Yaxin Zhang, Zhongqian Niu, Ningyuan Kuang, Wenjie Chen, Lingxiang Li, and Shaoqian Li. "A survey on terahertz communications." *China Communications* 16, no. 2 (2019): 1–35.

[19] Huang, Kao-Cheng, and Zhaocheng Wang. "Terahertz terabit wireless communication." *IEEE Microwave Magazine* 12, no. 4 (2011): 108–116.

[20] Song, Ho-Jin, and Tadao Nagatsuma. "Present and future of terahertz communications." *IEEE transactions on Terahertz Science and Technology* 1, no. 1 (2011): 256–263.

[21] Latva-aho, Matti, and Kari Leppänen. "Key drivers and research challenges for 6G ubiquitous wireless intelligence." *6G Research Visions* 1 (2019). http://jultika.oulu.fi/files/isbn9789526223544.pdf

[22] Rappaport, Theodore S., Yunchou Xing, Ojas Kanhere, Shihao Ju, Arjuna Madanayake, Soumyajit Mandal, Ahmed Alkhateeb, and Georgios C. Trichopoulos. "Wireless communications and applications above 100 GHz: Opportunities and challenges for 6G and beyond." *IEEE Access* 7 (2019): 78729–78757.

[23] Sarieddeen, Hadi, Nasir Saeed, Tareq Y. Al-Naffouri, and Mohamed-Slim Alouini. "Next generation terahertz communications: A rendezvous of sensing, imaging, and localization." *IEEE Communications Magazine* 58, no. 5 (2020): 69–75.

[24] Huq, Kazi Mohammed Saidul, Sherif Adeshina Busari, Jonathan Rodriguez, Valerio Frascolla, Wael Bazzi, and Douglas C. Sicker. "Terahertz-enabled wireless system for beyond-5G ultra-fast networks: A brief survey." *IEEE Network* 33, no. 4 (2019): 89–95.

[25] Rajatheva, Nandana, Italo Atzeni, Emil Bjornson, Andre Bourdoux, Stefano Buzzi, Jean-Baptiste Dore, Serhat Erkucuk et al. "White paper on broadband connectivity in 6G." arXiv preprint arXiv:2004.14247 (2020).

[26] Rajatheva, Nandana, Italo Atzeni, Simon Bicais, Emil Bjornson, Andre Bourdoux, Stefano Buzzi, Carmen D'Andrea et al. "Scoring the terabit/s goal: Broadband connectivity in 6G." arXiv preprint arXiv:2008.07220 (2020).

[27] Saad, Walid, Mehdi Bennis, and Mingzhe Chen. "A vision of 6G wireless systems: Applications, trends, technologies, and open research problems." *IEEE Network* 34, no. 3 (2020): 134–142, May/June, doi: 10.1109/MNET.001.1900287.

[28] Kürner, Thomas, and Sebastian Priebe. "Towards THz communications-status in research, standardization and regulation." *Journal of Infrared, Millimeter, and Terahertz Waves* 35, no. 1 (2014): 53–62.

[29] Tekbıyık, Kürşat, Ali Rıza Ekti, Güneş Karabulut Kurt, and Ali Görçin. "Terahertz band communication systems: Challenges, novelties and standardization efforts." *Physical Communication* 35 (2019): 100700.

[30] Elayan, Hadeel, Osama Amin, Basem Shihada, Raed M. Shubair, and Mohamed-Slim Alouini. "Terahertz band: The last piece of RF spectrum puzzle for communication systems." *IEEE Open Journal of the Communications Society* 1 (2019): 1–32.

[31] Choi, Junil, Vutha Va, Nuria Gonzalez-Prelcic, Robert Daniels, Chandra R. Bhat, and Robert W. Heath. "Millimeter-wave vehicular communication to support massive automotive sensing." *IEEE Communications Magazine* 54, no. 12 (2016): 160–167.

[32] Mumtaz, Shahid, Josep Miquel Jornet, Jocelyn Aulin, Wolfgang H. Gerstacker, Xiaodai Dong, and Bo Ai. "Terahertz communication for vehicular networks." *IEEE Transactions on Vehicular Technology* 66, no. 7 (2017): 5617–5625.

[33] Mendrzik, Rico, Danijela Cabric, and Gerhard Bauch. "Error bounds for terahertz MIMO positioning of swarm UAVs for distributed sensing." In *2018 IEEE International Conference on Communications Workshops (ICC Workshops)*, pp. 1–6. IEEE, 2018.

[34] Ma, Qian, Guo Dong Bai, Hong Bo Jing, Cheng Yang, Lianlin Li, and Tie Jun Cui. "Smart metasurface with self-adaptively reprogrammable functions." *Light: Science & Applications* 8, no. 1 (2019): 1–12.

[35] Ashraf, Nouman, Taqwa Saeed, Rafay Iqbal Ansari, Marios Lestas, Christos Liaskos, and Andreas Pitsillides. "Feedback based beam steering for intelligent metasurfaces." In *2019 2nd IEEE Middle East and North Africa COMMunications Conference (MENACOMM)*, pp. 1–6. IEEE, 2019.

[36] Nadeem, Qurrat-Ul-Ain, Abla Kammoun, Anas Chaaban, Merouane Debbah, and Mohamed-Slim Alouini. "Intelligent reflecting surface assisted wireless communication: Modeling and channel estimation." arXiv preprint arXiv:1906.02360 (2019).

[37] Di Renzo, Marco, Alessio Zappone, Merouane Debbah, Mohamed-Slim Alouini, Chau Yuen, Julien de Rosny, and Sergei Tretyakov. "Smart radio environments empowered by reconfigurable intelligent surfaces: How it works, state of research, and road ahead." arXiv preprint arXiv:2004.09352 (2020).

[38] Ntontin, K., M. Di Renzo, J. Song, F. Lazarakis, J. de Rosny, D-T. Phan-Huy, O. Simeone et al. "Reconfigurable intelligent surfaces vs. relaying: Differences, similarities, and performance comparison." arXiv preprint arXiv:1908.08747 (2019).

[39] La Spada, Luigi, Valeria Loscrí, and Anna Maria Vegni. "MetaSurface structure design and channel modelling for THz band communications." In *IEEE INFOCOM 2019-IEEE Conference on Computer Communications Workshops (INFOCOM WKSHPS)*, pp. 708–713. IEEE, 2019.

[40] Han, Chong, and Ian F. Akyildiz. "Three-dimensional end-to-end modeling and analysis for graphene-enabled terahertz band communications." *IEEE Transactions on Vehicular Technology* 66, no. 7 (2016): 5626–5634.

[41] Hu, Sha, Fredrik Rusek, and Ove Edfors. "Beyond massive MIMO: The potential of data transmission with large intelligent surfaces." *IEEE Transactions on Signal Processing* 66, no. 10 (2018): 2746–2758.

[42] Ghafoor, Saim, Noureddine Boujnah, Mubashir Husain Rehmani, and Alan Davy. "MAC protocols for terahertz communication: A comprehensive survey." arXiv preprint arXiv:1904.11441 (2019).

[43] Lemic, Filip, Sergi Abadal, Wouter Tavernier, Pieter Stroobant, Didier Colle, Eduard Alarcón, Johann Marquez-Barja, and Jeroen Famaey. “Survey on terahertz nanocommunication and networking: A top-down perspective.” arXiv preprint arXiv:1909.05703 (2019).

[44] Sarieddeen, Hadi, Mohamed-Slim Alouini, and Tareq Y. Al-Naffouri. “Terahertz-band ultra-massive spatial modulation MIMO.” *IEEE Journal on Selected Areas in Communications* 37, no. 9 (2019): 2040–2052.

[45] Faisal, Alice, Hadi Sarieddeen, Hayssam Dahrouj, Tareq Y. Al-Naffouri, and Mohamed-Slim Alouini. “Ultra-massive MIMO systems at terahertz bands: Prospects and challenges.” *IEEE Vehicular Technology Magazine* 15, no. 4 (2020): 33–42, doi10.1109/MVT.2020.3022998

[46] Ghasempour, Yasaman, Rabi Shrestha, Aaron Charous, Edward Knightly, and Daniel M. Mittleman. “Single-shot link discovery for terahertz wireless networks.” *Nature Communications* 11, no. 1 (2020): 1–6.

[47] Wang, Junyi, Zhou Lan, Chang-Woo Pyo, Tuncer Baykas, Chin-Sean Sum, Mohammad Azizur Rahman, Jing Gao et al. “Beam codebook based beamforming protocol for multi-Gbps millimeter-wave WPAN systems.” *IEEE Journal on Selected Areas in Communications* 27, no. 8 (2009): 1390–1399.

[48] Gao, Xinyu, Linglong Dai, Yuan Zhang, Tian Xie, Xiaoming Dai, and Zhaocheng Wang. “Fast channel tracking for terahertz beamspace massive MIMO systems.” *IEEE Transactions on Vehicular Technology* 66, no. 7 (2016): 5689–5696.

[49] Li, Dan, Deli Qiao, Lei Zhang, and Geoffrey Ye Li. “Performance analysis of indoor THz communications with one-bit precoding.” In *2018 IEEE Global Communications Conference (GLOBECOM)*, pp. 1–7. IEEE, 2018.

[50] Sarieddeen, Hadi, Mohammad M. Mansour, and Ali Chehab. “Large MIMO detection schemes based on channel puncturing: Performance and complexity analysis.” *IEEE Transactions on Communications* 66, no. 6 (2017): 2421–2436.

[51] Süral, Altuğ, E. Göksu Sezer, Yiğit Ertuğrul, Orhan Arikan, and Erdal Arikan. “Terabits-per-second throughput for polar codes.” In *2019 IEEE 30th International Symposium on Personal, Indoor and Mobile Radio Communications (PIMRC Workshops)*, pp. 1–7. IEEE, 2019.

[52] Weithoffer, Stefan, Matthias Herrmann, Claus Kestel, and Norbert Wehn. “Advanced wireless digital baseband signal processing beyond 100 Gbit/s.” In *2017 IEEE International Workshop on Signal Processing Systems (SiPS)*, pp. 1–6. IEEE, 2017.

[53] Eltayeb, Mohammed E., Tareq Y. Al-Naffouri, and Robert W. Heath. “Compressive sensing for blockage detection in vehicular millimeter wave antenna arrays.” In *2016 IEEE Global Communications Conference (GLOBECOM)*, pp. 1–6. IEEE, 2016.

[54] Han, Chong, A. Ozan Bicen, and Ian F. Akyildiz. “Multi-ray channel modeling and wideband characterization for wireless communications in the terahertz band.” *IEEE Transactions on Wireless Communications* 14, no. 5 (2014): 2402–2412.

[55] Zakrajsek, Luke M., Dimitris A. Pados, and Josep M. Jornet. “Design and performance analysis of ultra-massive multi-carrier multiple input multiple output communications in the terahertz band.” In *Image Sensing Technologies: Materials, Devices, Systems, and Applications IV*, vol. 10209, p. 102090A. International Society for Optics and Photonics, 2017.

[56] Han, Chong, A. Ozan Bicen, and Ian F. Akyildiz. “Multi-wideband waveform design for distance-adaptive wireless communications in the terahertz band.” *IEEE Transactions on Signal Processing* 64, no. 4 (2015): 910–922.

[57] Doré, Jean-Baptiste, Yoann Corre, Simon Bicais, Jacques Palicot, Emmanuel Faussurier, Dimitri Kténas, and Faouzi Bader. “Above-90GHz spectrum and single-carrier waveform as enablers for efficient Tbit/s wireless communications.” In *2018 25th International Conference on Telecommunications (ICT)*, pp. 274–278. IEEE, 2018.

[58] Jornet, Josep Miquel, and Ian F. Akyildiz. “Femtosecond-long pulse-based modulation for terahertz band communication in nanonetworks.” *IEEE Transactions on Communications* 62, no. 5 (2014): 1742–1754.

[59] Kokkoniemi, Joonas, Janne Lehtomäki, and Markku Juntti. “A discussion on molecular absorption noise in the terahertz band.” *Nano Communication Networks* 8 (2016): 35–45.

[60] Akyildiz, Ian F., and Josep Miquel Jornet. "Realizing ultra-massive MIMO (1024× 1024) communication in the (0.06–10) terahertz band." *Nano Communication Networks* 8 (2016): 46–54.

[61] Loukil, Mohamed Habib, Hadi Sarieddeen, Mohamed-Slim Alouini, and Tareq Y. Al-Naffouri. "Terahertz-band MIMO systems: Adaptive transmission and blind parameter estimation." *IEEE Communications Letters* 25, no. 2 (2021): 641–645, doi=10.1109/LCOMM.2020.3029632

[62] Sarieddeen, Hadi, Mohammad M. Mansour, Louay Jalloul, and Ali Chehab. "High order multi-user MIMO subspace detection." *Journal of Signal Processing Systems* 90, no. 3 (2018): 305–321.

[63] Han, Chong, Josep Miquel Jornet, and Ian Akyildiz. "Ultra-massive MIMO channel modeling for graphene-enabled terahertz-band communications." In *2018 IEEE 87th Vehicular Technology Conference (VTC Spring)*, pp. 1–5. IEEE, 2018.

[64] Lin, Cen, and Geoffrey Ye Li Li. "Terahertz communications: An array-of-subarrays solution." *IEEE Communications Magazine* 54, no. 12 (2016): 124–131.

[65] Gordon, Iouli E., Laurence S. Rothman, Christian Hill, Roman V. Kochanov, Y. Tan, Peter F. Bernath, Manfred Birk et al. "The HITRAN2016 molecular spectroscopic database." *Journal of Quantitative Spectroscopy and Radiative Transfer* 203 (2017): 3–69.

[66] Jepsen, P. Uhd, David G. Cooke, and Martin Koch. "Terahertz spectroscopy and imaging–Modern techniques and applications." *Laser & Photonics Reviews* 5, no. 1 (2011): 124–166.

[67] Ryniec, R., P. Zagrajek, and N. Palka. "Terahertz frequency domain spectroscopy identification system based on decision trees." *Acta Physica Polonica—Series A General Physics* 122, no. 5 (2012): 891.

[68] Gu, Yanying, Anthony Lo, and Ignas Niemegeers. "A survey of indoor positioning systems for wireless personal networks." *IEEE Communications surveys & tutorials* 11, no. 1 (2009): 13–32.

[69] Guvenc, Ismail, and Chia-Chin Chong. "A survey on TOA based wireless localization and NLOS mitigation techniques." *IEEE Communications Surveys & Tutorials* 11, no. 3 (2009): 107–124.

[70] Priyantha, Nissanka B., Anit Chakraborty, and Hari Balakrishnan. "The cricket location-support system." In *Proceedings of the 6th Annual International Conference on Mobile Computing and Networking*, pp. 32–43. 2000.

[71] Wang, Gang, and Kehu Yang. "A new approach to sensor node localization using RSS measurements in wireless sensor networks." *IEEE Transactions on Wireless Communications* 10, no. 5 (2011): 1389–1395.

[72] Tomic, Slavisa, Marko Beko, and Rui Dinis. "RSS-based localization in wireless sensor networks using convex relaxation: Noncooperative and cooperative schemes." *IEEE Transactions on Vehicular Technology* 64, no. 5 (2014): 2037–2050.

[73] Kanhere, Ojas, and Theodore S. Rappaport. "Position locationing for millimeter wave systems." In *2018 IEEE Global Communications Conference (GLOBECOM)*, pp. 206–212. IEEE, 2018.

[74] El-Absi, Mohammed, Ali Alhaj Abbas, Ashraf Abuelhaija, Feng Zheng, Klaus Solbach, and Thomas Kaiser. "High-accuracy indoor localization based on chipless RFID systems at THz band." *IEEE Access* 6 (2018): 54355–54368.

[75] Saeed, Nasir, Mohamed Habib Loukil, Hadi Sarieddeen, Tareq Y. Al-Naffouri, and Mohamed-Slim Alouini. "Body-centric terahertz networks: Prospects and challenges." (2020). TechRxiv. Preprint. https://doi.org/10.36227/techrxiv.12923498.v1

[76] Prasad, M. Shree, Trilochan Panigrahi, and Mahbub Hassan. "Direction of arrival estimation for nanoscale sensor networks." In *Proceedings of the 5th ACM International Conference on Nanoscale Computing and Communication*, p. 21. ACM, 2018.

[77] Saeed, Nasir, Haewoon Nam, Mian Imtiaz Ul Haq, and Dost Bhatti Muhammad Saqib. "A survey on multidimensional scaling." *ACM Computing Surveys (CSUR)* 51, no. 3 (2018): 1–25.

[78] Saeed, Nasir, Haewoon Nam, Tareq Y. Al-Naffouri, and Mohamed-Slim Alouini. "A state-of-the-art survey on multidimensional scaling-based localization techniques." *IEEE Communications Surveys & Tutorials* 21, no. 4 (2019): 3565–3583.

[79] Sarieddeen, Hadi, Mohamed-Slim Alouini, and Tareq Y. Al-Naffouri. "An overview of signal processing techniques for terahertz communications." arXiv preprint arXiv:2005.13176 (2020).

[80] Bourdoux, Andre, Andre Noll Barreto, Barend van Liempd, Carlos de Lima, Davide Dardari, Didier Belot, Elana-Simona Lohan et al. "6G White Paper on Localization and Sensing." arXiv preprint arXiv:2006.01779 (2020).

Part I

Terahertz Transceiver and Devices

Chapter 2

Terahertz Communications with Resonant Tunnelling Diodes

Status and Perspectives

Davide Cimbri and Edward Wasige

Contents

Acronyms and Symbols

5G	Fifth generation	MOVPE	Metal organic vapour phase epitaxy
ASK	Amplitude shift keying	MIMO	Multiple-input multiple-output
AlAs	Aluminium arsenide	mW	Milli-Watt
AlGaAs	Aluminium gallium arsenide	NDC	Negative differential conductance
AlSb	Aluminium antimonide	NDR	Negative differential resistance
AlN	Aluminium nitride	NLOS	Non-line of sight
BER	Bit error rate	OOK	On-off keying
CMOS	Complementary metal-oxide-semiconductor	PAM	Pulse amplitude modulation
CPS	Coplanar stripline	PVCR	Peak-to-valley current ratio
CPW	Coplanar waveguide	QAM	Quadrature amplitude modulation
CW	Continuous wave	QPSK	Quadrature phase shift keying
DBQW	Double barrier quantum well	QCL	Quantum cascade laser
FEC	Forward error correction	QW	Quantum well
f_{max}	Maximum frequency of oscillation	PRBS	Pseudo-random binary sequence
GaAs	Gallium arsenide	RF	Radio frequency
GaInAs	Gallium indium arsenide	RTD	Resonant tunnelling diode
GaN	Gallium nitride	Rx	Receiver
Gb/s	Gigabits per second	SBD	Schottky barrier diode
GHz	Gigahertz	Si	Silicon
HBT	Heterojunction bipolar transistor	Tb/s	Terabits per second
HEMT	High electron mobility transistor	THz	Terahertz
InAs	Indium arsenide	TRx	Transceiver
InGaAs	Indium gallium arsenide	Tx	Transmitter
InP	Indium phosphide	TMIC	Terahertz monolithic integrated circuit
I_p	Peak current	UTC-PD	Uni-travelling carrier photodiode
I_v	Valley current	VCO	Voltage controlled oscillator
IoT	Internet of things	V_p	Peak voltage
I–V	Current–voltage	V_v	Valley voltage
J_p	Peak current density	ΔI	Peak-to-valley current difference
LOS	Line of sight	ΔJ	Peak-to-valley current density difference
LNA	Low-noise amplifier	ΔV	Peak to valley voltage difference
MBE	Molecular beam epitaxy	μW	Micro-Watt
MIM	Metal–insulator–metal		

2.1 Introduction

2.1.1 Need for High-Speed Wireless Connectivity

For the last two decades, we have witnessed an extraordinarily fast evolution of mobile cellular networks from first generation (1G) to fourth generation (4G) with the fifth generation (5G) wireless communication networks now being deployed [1]. Figure 2.1 shows the data-rate demand versus time trend for wireline, nomadic and wireless communication technologies between 1970 and 2020 [2]. It shows that wireless networks data transfer capability has been continuously improved at a speed much greater than wirelines over the last four decades. Indeed, the tremendous increase of mobile data traffic and widespread diffusion of wireless networks generate an unceasing demand for ultra-broadband multi-gigabit wireless communication technology, capable of extremely large channel bandwidths and ultra-high data rates required by modern multimedia services [3], including the Internet of Things (IoT) [4,5]. This is in line with Edholm's law, which states that the demand for bandwidth performance in wireless short-range communications has doubled every 18 months since 1980 [6], and so data rates of tens of gigabits per second (Gb/s) have to be accommodated starting from 2020 onwards [7] while hundreds of Gb/s and even terabits per second (Tb/s) wireless communication links are expected within the next ten years [2].

Therefore, it becomes clear that 5G technology will bring us into an era of ubiquitous high-capacity radio links with ever-lower levels of latency and extensive Gb/s capacity. For the path beyond 5G or upcoming sixth generation (6G) mobile communication systems [8, 9], the mobile network is envisioned as becoming more advanced since it can self-adapt based on users' experience thanks to intelligent learning mechanisms, allowing flexible and fast spectrum reallocation with resulting large bitrates available to users (over 100 Gb/s single links speeds [10]), ultralow latency (not exceeding 1 millisecond (ms)) and ultra-reliability (package error rate of 10^{-5} or less). In addition, 3D/holographic type communication will lead to an improvement of the tele-interaction quality. By supplying mobile edge computing and edge caching capabilities, together with Artificial Intelligence (AI), to the proximity of end users, the beyond 5G network will allow for

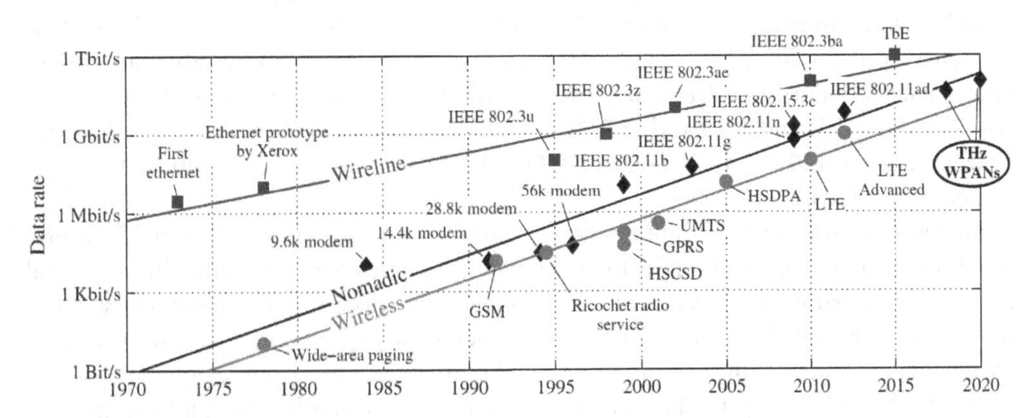

Figure 2.1 Data-rate demand versus time trend for wireline, nomadic and wireless communication technologies between 1970 and 2020, pointing out the faster exponential increase associated with wireless systems compared to wireline technologies (reprinted from [2] with permission).

powerful computational processing and massive data acquisition locally at edge networks to support emerging applications such as self-driving cars, remote surgery, intelligent transport systems, Industry 4.0, smart energy, e-health, and augmented reality and virtual reality (AR/VR) services. In addition, new realities such as the COVID-19 pandemic has brought the need for more bandwidth and network flexibility into sharp focus. The use of online meetings will be extended to all kinds of human interactions, from medical visits and client interactions to real-time remote/tele-control applications, which will drastically increase the need for more capacity or bandwidth.

2.1.2 Increasing the Speed of Wireless Transmission

Future wireless communication technologies will have to feature ultra-high bandwidths in excess of tens of Gb/s and beyond [2,11,12]. In order to respond to this demand, there has been a significant amount of research and development (R&D) effort in the digital signal processing (DSP), complex modulation scheme and multiple-input multiple-output (MIMO) technology fields at the microwave frequency spectrum region, mainly in the 1–6 GHz range and reaching up to the 60 GHz band. The aim has been to improve the data capacity of radio links through advanced transmission methods, such as orthogonal frequency-division multiplexing (OFDM) [13] and very large-scale MIMO [14,15].

Despite numerous efforts, the current technology operating in the lower microwave range (<6 GHz) is inherently limited by the narrow bandwidth, which typically gives bit rates ranging from several hundreds of megabits per second (Mb/s) up to at most few Gb/s in both long- and short-range scenarios [17]. Therefore, it is becoming increasingly difficult to accommodate, year after year, the performance requirements empirically predicted by Edholm's law, where spectral efficiencies of up to at least 14 bit/s/Hz would be required to attain bit rates of the order of 100 Gb/s [2]. This can be considered extremely challenging and practically unrealistic since it would inevitably involve extremely complex transmission approaches [18].

Currently, the largest globally available frequency band allocated for mobile telecommunication services is set in V-band (40–75 GHz) [19], around 60 GHz (57–64 GHz) and with up to 7 GHz of continuous bandwidth [2,11,20,21]. In order to meet the requirements imposed by the upcoming 5G technology, the United States has already started to extend the bandwidth up to 14 GHz (57–71 GHz) by including the unlicensed portion 56–71 GHz [22,23]. This adds to the total bandwidth belonging to the licensed standard low-frequency bands (<6 GHz) [24] and high-frequency K-band (18–27 GHz) [19] and Ka-band (27–40 GHz) [19] portions [24] (such as 24.25–27.5 GHz in Europe [24] and 27.5–28.35 GHz, 37–38.6 GHz and 38.6–40 GHz in the United States [22]).

However, in order to significantly increase transmission data rates and to enable future-proof scenarios and applications as foreseen in 6G, an increase in the bandwidth by several tens of GHz is required. Based on the current spectrum allocation, it is clear that wireless communications will have to operate in frequency bands where more spectral resources are available. Therefore, the exploitation of higher frequency spectrum regions, specifically the terahertz (THz) band, to alleviate the spectrum scarcity and bandwidth limitations of current microwave systems is inevitable [25,26] and it represents an attractive candidate for ultra-high-speed wireless communication applications [27].

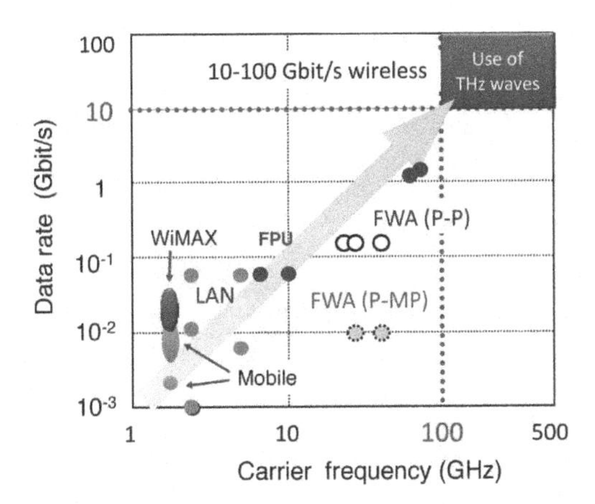

Figure 2.2 Data rate versus carrier frequency trend (reprinted from [29] with permission). THz carriers allow for data rates beyond 10 Gb/s.

Moving up in frequency to the THz band becomes crucial, since higher carrier frequencies allow, in principle, wider modulation bandwidths and higher data rates according to Shannon–Hartley theorem [28]. This concept can be clearly seen in Figure 2.2, which shows the relationship between data rate and carrier frequency [29], revealing that THz carriers allow for data rates beyond 10 Gb/s. Although candidate bands are several, particularly interesting is the still unallocated yet unregulated THz gap, which is attracting great interest due to recent technological advances [12,30,31]. Actually, different frequency windows in this range have already been employed for several decades in radio astronomy, since they are in correspondence with specific vibrational and rotational resonant modes of molecules under study [27]. Therefore, a careful approach will have to be adopted to avoid any kind of possible conflict and interference between mobile telecommunication and radio astronomy systems [32]. Another advantage in choosing THz carrier waves is the sub-millimetre dimension of the antennas, which become cheaper and more integrable [33,34], even though integrated antenna solutions suffer from low gain (< 6 dBi).

Further, high-bandwidth systems have a tendency to decrease the end-to-end latency due to high rate data packet transfer, a reason why high frequencies are seen as opportunistic bands for haptic applications requiring seemingly 'zero-perceptible' latency. However, due to the lack of devices capable of generating and detecting THz waves, its usage in the scope of wireless communication links has been limited, but this situation is changing fast. This chapter describes the key developments in solid-state THz electronics in this regard with special focus on resonant tunnelling diode (RTD) technology.

Typical ultrafast THz communications applications rely on both long-range and short-range line-of-sight (LOS) and non-line-of-sight (NLOS) operation [2,35]. Figure 2.3 shows some potential applications including wireless backhaul/fronthaul [36], wireless local area networks (WLAN) and wireless personal area networks (WPAN) [37], kiosk downloading [23,38] and nano-networks [39]. THz technology is also appealing in the context of other potential scenarios such as wireless data centres [40,41], intra-chip connectivity [42,43] and even space communications [32]. Thus, one can distinguish three categories of application scenarios: (i) short range (a few centimetres) like in a case of

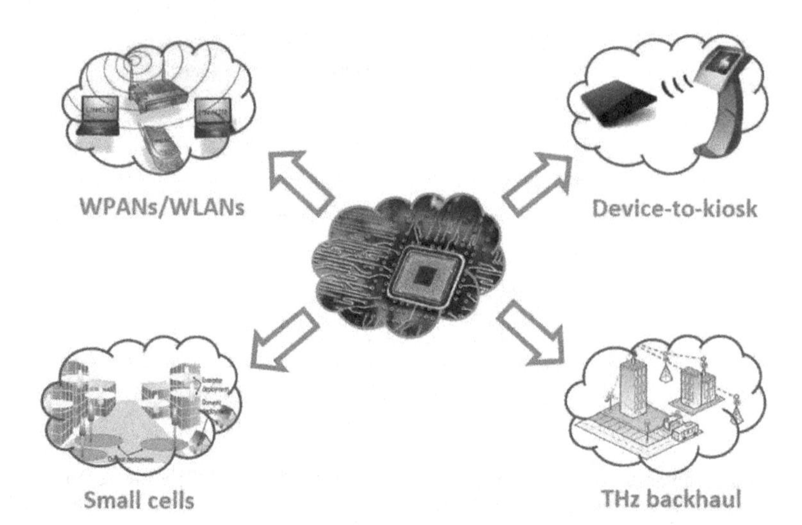

Figure 2.3 Potential Applications of millimetre-wave (mm-wave) and terahertz communications including wireless backhaul/fronthaul, wireless local area networks (WLAN) and wireless personal area networks (WPAN), kiosk downloading and nano-networks.

intra-device communication on the electronic boards or kiosk downloading, (ii) medium (a few metres) like in the data centres or computers in the same office space and (iii) long-range like links (distance up to 100 m). It is worth noting that these long-range links can reach kilometres or more in the case of wireless communication in space (no attenuation). It is evident that these communication scenarios impose very different requirements on the sources, antennas, optics and detectors from the point of view of power consumption, dimensions, weight, potential market volume and price.

2.1.3 Challenges to Realising THz Communications

Despite the clear advantages, THz wireless communications is still quite challenging to realise from a technological standpoint [12,30]. There are huge signal losses caused by the strong absorption by oxygen (O_2), water (H_2O) and other gas molecules in the atmosphere, where the relative humidity can reach 100% in outdoor rainy conditions [11]. Therefore, both the signal-to-noise ratio (SNR) and the link data capacity are strongly deteriorated by the ambient humidity. The THz range is therefore mostly promising, in the short-term, for short-range indoor applications, where the link distance is relatively short (metre range) and the environmental relative humidity is a less stringent constrain (typically ~50%) [11].

Figure 2.4 shows the atmospheric specific attenuation versus frequency in (standard) indoor conditions up to 1 THz for dry (relative humidity ~0%) and moist (relative humidity ~50%) conditions [11]. Of particular interest is the transmission window around the 300 GHz band (275–330 GHz) [11,44], since it features 55 GHz of continuous bandwidth with low attenuation (water absorption gives 0.04 dB/m) [11]. This is insignificant compared to free-space losses and also there are no resonant absorption peaks, making short-range applications feasible. Therefore, standardisation efforts for

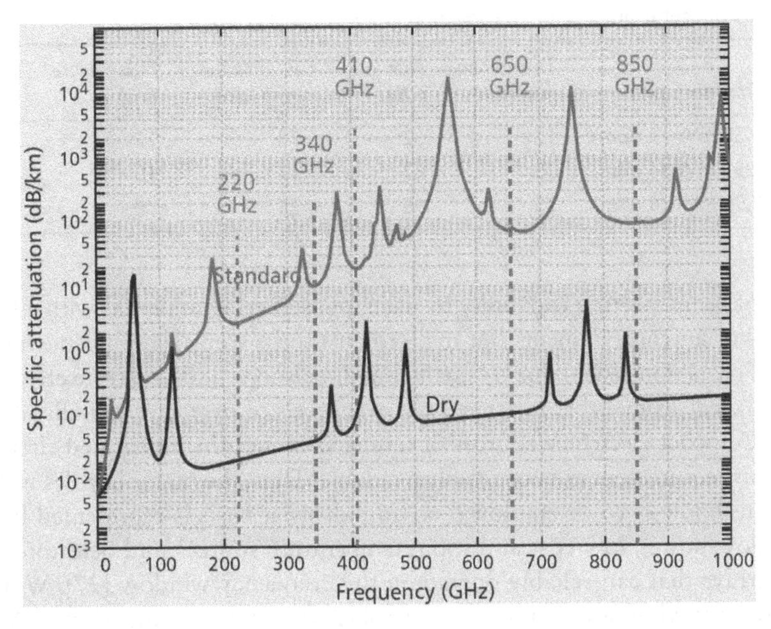

Figure 2.4 Atmospheric specific attenuation versus frequency in (standard) indoor (relative humidity ~50%) and dry (relative humidity ~0%) conditions up to 1 THz (reprinted from [11] with permission).

this band have recently been started through the IEEE standard 802.15.3d in the definition of physical layers for the next generation ultra-high-speed wireless personal area networks (WPAN) [45].

For THz links, good alignment between the transmitter (Tx) and receiver (Rx) antennas is required [46] and care has to be taken in the design and realisation of building structures in order to reduce signals attenuation through walls, ceilings, etc. [47]. Long distance outdoor applications (kilometre range) are feasible but will rely on future technological advances of both sources (output power) and detectors (sensitivity), together with antennas (gain, so both directivity and radiation efficiency) due to the higher attenuation at relative humidities of up to 100%.

2.1.4 Link Budget and Antennas

To get a sense of the basic transceiver requirements, we consider the link budget at 300 GHz in a transmission bandwidth of 55 GHz for the short-range line of sight (LOS) scenario, which is of prime interest for the first feasible products. Table 2.1 [11] lists some realistic link budget examples. Calculations are based on the Friss formula for free space attenuation. Furthermore, the signal-to-noise ratio is assumed to be about 10 dB, while a margin for implementation losses of 10 dB is also included. Good alignment between the Tx and Rx is assumed and therefore misalignment losses are neglected.

From Table 2.1 we learn that antenna gain in the range of 20–25 dBi is required. This is currently provided either through the use of horn antennas, which require the integration of Tx/Rx chips within a rectangular waveguide [48] or by mounting the chips on Si lenses [49]. Future systems might follow the traditional approach to achieve this with

Table 2.1 Link budget examples for transmission at 300 GHz [11]

Minimum range	*Transmit power*	*Tx antenna gain [dBi]*	*Path loss [dB]*	*Rx antenna gain [dBi]*	*Noise figure [dB]*	*SNR + margin [dB]*
10 cm	1.58 mW (2 dBm)	12	62	12	12	18
1 m	2.51 mW (4 dBm)	25	82	17	10	20
5 m	3.98 mW (6 dBm)	25	96	25	8	18

inclusion of beam steering capability by using phased array antennas with a large number (hundreds) of antenna elements but that remains to be seen. The implementation of such arrays at THz is non-trivial due to small chip dimensions and substrate effects.

It is also clear from Table 2.1 that Tx power levels of over several milliwatts (mW) are required. Compact electronic sources or terahertz monolithic integrated circuits (TMICs) meeting this requirement are not readily available. Thus, the major hurdles which still prevent THz technology to be employed on an industrial basis is represented by the lack of compact, low-power, low-cost and room temperature sources and detectors, and lack of antenna arrays that can reliably operate in this frequency window [37]. With regards to THz sources for wireless communications, coherent and continuous-wave (CW) operation are required. Although several solid-state semiconductor-based competitor technologies are being developed [50], none of them has made a real technological breakthrough. The operation frequency of electronic devices is typically limited by the carrier transit time in the active region [51], while photonic devices face the tough requirement of band-to-band optical transitions towards long emission wavelengths [52].

In order to aid sources and detectors to meet link budget requirements, antenna gain has to be enhanced, for instance, through innovative beam-steering solutions [53,54] based on micro/nano-electro-mechanical systems (MEMS/NEMS) [55,56] or metamaterials [57,58].

2.1.5 Enabling Technologies for THz Communications: Electronic and Photonic

Recent technological innovations regarding THz system components, where demonstrations of both electronic and photonic approaches have been shown, indicate the viability of THz wireless communications. Regarding transmitters, there are two main approaches: (1) photonic techniques for THz signal generation and (2) all-electronic devices. Approach (1) has proven to be effective to achieve higher data rates of >10 Gb/s, since telecom-based high-frequency components such as lasers, modulators and photodiodes are available together with low-loss optical fibre cables. Generally, this approach consists in photo-mixing two optical wavelengths on a high-speed photodiode such as a uni-travelling-carrier photodiode (UTC-PD), to generate a THz carrier wave signal, and therefore remains too complex to bring the THz wireless communications technology to the consumer marketplace. Regarding approach (2), there are various candidates for semiconductor electronic THz emitters or oscillators operating at room temperature such as tunnel transit-time (TUNNET) diodes, impact ionisation avalanche transit-time (IMPATT) diodes, Gunn diodes, resonant tunnelling diodes (RTDs) and transistor-based monolithic microwave integrated circuits (MMICs) or terahertz monolithic integrated circuits (TMICs), which utilise frequency multiplication.

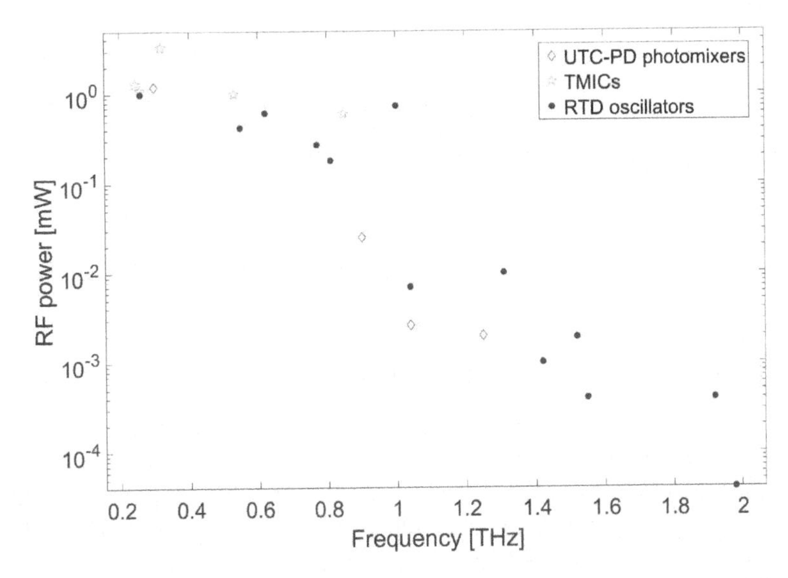

Figure 2.5 RF power versus operation frequency for UTC-PD-based photomixer, transistor-based TMIC and RTD-based oscillator sources in the 0.2–2 THz range from the most relevant results reported in the literature.

UTC-PD-based photo-mixing, transistor-based TMICs and RTD-based oscillators represent the most promising candidates for practical THz wireless communication technology. Figure 2.5 shows the output RF power versus frequency for UTC-PD, transistor TMIC and RTD-based oscillator sources in the 0.2–2 THz range. It can be seen that all the three technologies can provide above an mW threshold in the 300 GHz band, while both UTC-PDs and RTDs have also demonstrated to work above 1 THz with associated output power of few microwatts (μW). Therefore, an overview of the UTC-PD and TMIC technologies for THz communications is first provided, followed by a detailed description of the RTD technology for THz communications including developments to date, the current status and future perspectives.

2.1.6 Photonic-based THz Sources

Photonic-based sources used for generation on THz signals under 1 THz are generally based on the optical heterodyne down-conversion, usually referred as to photo-mixing [59,60]. In this technique, two optically selected laser emission lines interact with a photoconductive device (such as a UTC-PD), which optically mixes the input signals and generates THz radiation, whose frequency (known as beat frequency) is the result of the difference of the input lines frequencies [61]. The photonic input signals are typically provided through a pulsed mode-locked laser source [62] or, alternatively, employing an optical frequency comb generator (OFCG), whose main elements typically include a CW infrared laser source (such as highly tuneable distributed feedback (DFB) lasers [63–65], vertical-cavity surface-emitting lasers (VCSELs) [66] or quantum cascade lasers (QCLs) [67]), phase modulators (PMs) and a low-frequency local oscillator (LO) [68,69].

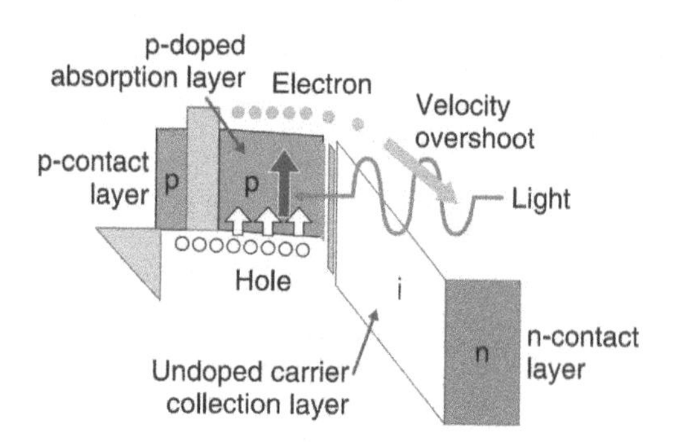

Figure 2.6 UTC-PD band diagram (reprinted from [73] with permission).

Indium phosphide (InP) based UTC-PDs are room temperature devices and have been shown to work above 1.2 THz [70] and to be able to provide radio frequency (RF) powers of up to 1.2 mW around 300 GHz in CW regime [71]. The UTC-PD was proposed by Ishibashi et al. in 1997 [72]. Figure 2.6 shows its band diagram [73] and the working principle [70,74]. The active region consists of a neutral (p-type) narrow-bandgap light absorption layer (typically gallium indium arsenide (GaInAs)-based) and an undoped or lightly n-type doped wide-bandgap carrier collection layer (usually InP-based). Upon light illumination and related electron–hole pair generation, the photo-generated majority holes in the quasi-neutral absorption layer relax extremely quickly and diffuse very fast, with a timescale of the order of the dielectric relaxation time, towards the anode while they are blocked by the carrier collection layer. The photo-generated minority electrons diffuse into the depleted collection layer while they are blocked by a wide-bandgap layer placed in between the absorption layer and the anode and then drift towards the cathode layer by means of the intrinsic electric field and the applied DC bias, taking advantage of velocity overshoot. Therefore, this device configuration allows for electrons operation only, where they act as active carriers running through the depleted carrier collection layer. Consequently, the photo-response of a UTC-PD device is uniquely determined by electron transport, which is typically much faster than hole transport. Indeed, the electron drift velocity in the carrier collection depletion layer is an order of magnitude higher than that of the holes. Thus, the device response time is mainly limited by the electron diffusion time across the absorption layer. However, electron diffusion velocity in the employed semiconductor material (GaInAs) is typically very large due to the high minority mobility and transit time can be minimised by an appropriate design of the absorption layer.

The optical-to-electrical photo-mixing heterodyne down-conversion process used to generate the THz beat note is obtained from the difference frequency generation (DFG) based on second-order non-linear optical processes in semiconductors at relatively high input optical fields [73,75].

Table 2.2 shows a summary of UTC-PD based THz wireless communication systems and shows that they have been demonstrated with bit-rate capabilities beyond 100 Gb/s in the 300–500 GHz band. Using multi-channel data transmission, up to 160 Gb/s through QPSK over a 50 cm long link [76,77] and up to 260 Gb/s through 16-QAM over a 50 cm long link [78] have been reported, both with bit error rate (BER) below the forward error correction (FEC) limit.

Table 2.2 UTC-PD THz wireless communication links: some recent results from the literature

Ref.	*UTC-PD Tx Technology*	*Rx*	f_c *[GHz]*	*Modulation*	*Distance*	*Data rate (Gb/s)*	*BER*
[62]	InP/GaAs	35 nm HEMT	225–250	8/16-QAM	20 m	100	$< 4.5 \times 10^{-3}$
[68]	InP/–	–	320	OOK	100 m	50	9.5×10^{-4}
[69]	InP/–	–	320	QPSK	5–10 cm	100	–
[69]	InP/–	–	320	QPSK	5–10 cm	90	1.7×10^{-3}
[79]	InP/InP	PC	287–325	QPSK	58 m	30	< FEC
[80]	InP/–	–	280	16-QAM	50 cm	100	$< 4 \times 10^{-3}$
[81]	InP/–	SBD	400	16-QAM	50 cm	106	$< 2 \times 10^{-2}$
[77]	InP/–	SBD	300–500	QPSK	50 cm	160	7×10^{-3}
[77]	InP/–	SBD	300–500	QPSK	50 cm	260	< FEC

Notes:
f_c, carrier frequency; BER, bit error rate; HEMT, high electron mobility transistor; SBD, Schottky barrier diode; PC, photoconductor.

A block diagram of a UTC-PD transmitter is shown in Figure 2.7. This fully analogue 300 GHz-band wireless communication system employed a CW coherent photonic UTC-PD-based transmitter and an electronic sub-harmonic mixer (SHM)-based coherent receiver. The THz carrier signal was generated photonically through OFCG by employing an infrared laser source, electro-optic phase modulators (EOPM) and a local oscillator (LO). After that two of the comb lines were optically selected, modulation took place employing an electro-optic amplitude modulator (EOAM) or a quadrature phase shift keying modulator (QPSKM) and the lines were then combined and sent to the UTC-PD through an optical fibre, where the modulated THz carrier signal was generated by photo-mixing and then irradiated though an antenna. This system demonstrated single-channel wireless data transmission of up to 50 Gb/s through on-off keying (OOK) over a 100 m long link [70] and up to 100 Gb/s through quadrature phase shift keying (QPSK) in the 5–10 cm range [71], both with BER below the FEC limit.

Despite the impressive performances, the photonic THz sources are usually complex and expensive, in certain cases bulky including optical discrete components, such as laser infrared sources, modulators, filters, lenses and cumbersome optical setups, and some need proper cryogenic cooling such as p-type Ge lasers and QCLs, which increases cost, complexity, reduces integrability and increases reliability issues.

2.1.7 Electronic Transmitters

Transistor-based terahertz monolithic integrated circuits (TMICs) are being developed for wireless communication applications. Silicon (Si)-based complementary metal-oxide-semiconductor (CMOS) field-effect transistor (FET) [82], silicon germanium (SiGe) heterojunction bipolar transistor (HBT) [83] technologies already offer maximum unity power gain cut-off frequency (f_{max}) of up to 450 GHz [82] and 720 GHz [83], respectively. On the other hand, InP-based heterojunction bipolar transistors (HBTs) [84], double heterostructure HBTs (DHBTs) [84] and high electron mobility transistors (HEMTs) [85] including pseudomorphic HEMTs (pHEMTs) [86] and metamorphic HEMTs (mHEMTs) [87]) have been demonstrated with f_{max} in excess of 1 THz [88] and of up to 1.5 THz [85],

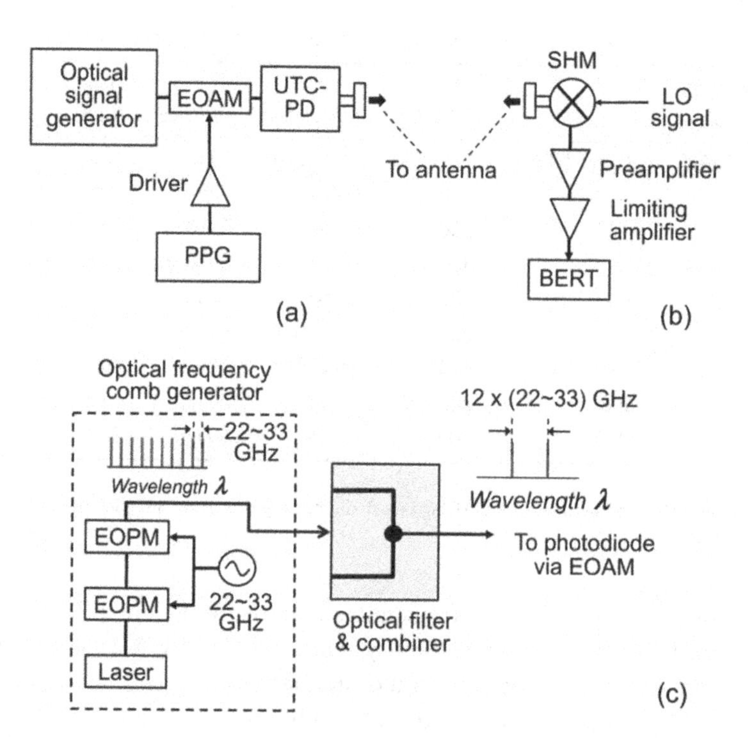

Figure 2.7 Block diagram of 300-GHz-band coherent wireless communication system (adapted and reprinted from [68] with permission). (a) Transmitter. (b) Receiver. (c) Optical signal generator. EOAM: electro-optic amplitude modulator. EOPM: electro-optic phase modulator. PPG: pulse pattern generator. BERT: bit error rate tester. UTC-PD: uni-travelling-carrier photodiode. SHM: sub-harmonic mixer.

respectively. The unity power gain defines the frequency range over which good circuit performance is achievable, which is up to around a third of this value. Therefore, available InP transistor technologies can underpin circuit realisations in the 275–330 GHz transmission window. In general, TMIC sources are usually designed including frequency multipliers [89], up-conversion mixers [48], push-push [90,91] or triple-push [92] harmonic topologies, and RF power amplifiers stages [85,93–97]. The extremely complex and inefficient circuitry lowers the DC-RF conversion efficiency to well below 1% above 300 GHz.

TMIC sources have been reported to work at up to 850 GHz [98,99] and to provide output powers in the mW range [100–102]. Table 2.3 shows a summary of various TMIC-based THz wireless links. Data rates of over 100 Gb/s in the 70–300 GHz band have been demonstrated in single-channel links without employing multiplexing techniques [103] or MIMO methods [104].

Figure 2.8 shows a block diagram of an InP HEMT-based THz wireless communication system that was reported in [105] and was used to carry out high-speed data transmission experiments, while Figure 2.9 shows a photograph of the fabricated mixer MMIC chip and packaged mixer module. At the CW Tx side, an intermediate frequency (IF) signal with central frequency of 20 GHz was generated through an arbitrary waveform generator (AWG) and modulated through 16-QAM. The signal was then mixed up with a LO of frequency of 15 GHz, which was then upconverted by a 18× frequency extender and amplified, providing an output signal of frequency of 270 GHz. The modulated upconverted

Table 2.3 TMIC THz wireless communication links: some recent results from the literature

Ref.	*Technology*	*Tx*	*Rx*	*f_c [GHz]*	*Modulation*	*Distance*	*Data Rate [Gb/s]*	*BER*
[48]	InP	80 nm HEMT	80 nm HEMT	272–302	16-QAM	2.22 m	100	$< 10^{-3}$
[105]	InP	80 nm HEMT	80 nm HEMT	296	16-QAM	9.8 m	120	10^{-3}
[106]	Si	40 nm CMOS	40 nm CMOS	296	16-QAM	1 cm	28	10^{-3}
[107]	Si	40 nm CMOS	40 nm CMOS	300	16-QAM	5 cm	28	$< 10^{-3}$
[108]	Si	65 nm CMOS	65 nm CMOS	70–105	16-QAM	20 cm	120	$< 10^{-3}$
[109]	SiGe	130 nm HBT	130 nm HBT	220–255	16-QAM	1 m	100	4×10^{-3}
[110]	SiGe	130 nm HBT	130 nm HBT	225–255	QPSK	1 m	65	$< 10^{-3}$
[111]	SiGe	130 nm HBT	130 nm HBT	230	16-QAM	1 m	90	10^{-3}
[112]	GaAs	35 nm HEMT	35 nm HEMT	240	QPSK	850 m	64	7.9×10^{-5}

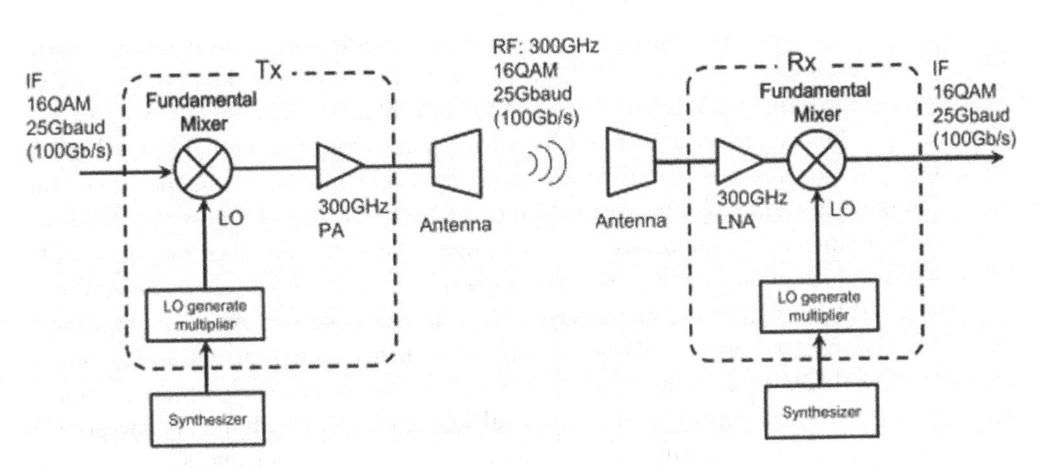

Figure 2.8 Conceptual schematic of a TMIC 300-GHz transceiver (reprinted from [48] with permission).

THz signal was then sent to a high-pass filter (HPF) to suppress the LO leak and the image signal, amplified by a 300 GHz power amplifier (PA) and then sent to the antenna and radiated. The PA, the low-noise amplifier (LNA) at the Rx side and both the LO amplifiers and fundamental mixer stages at both the transmitter and receiver sides were based on 80 nm InP HEMT technology. This wireless communication system enabled single-channel data transmission up to 120 Gb/s at a carrier frequency of 296 GHz using 16-QAM over a 9.8 m long link with BER below the FEC limit.

Diode-based TMIC sources are also being developed for wireless communications, in particular resonant tunnelling diodes (RTDs) [113–116]. RTDs are the fastest demonstrated solid-state semiconductor-based electron devices operating at room temperature. They

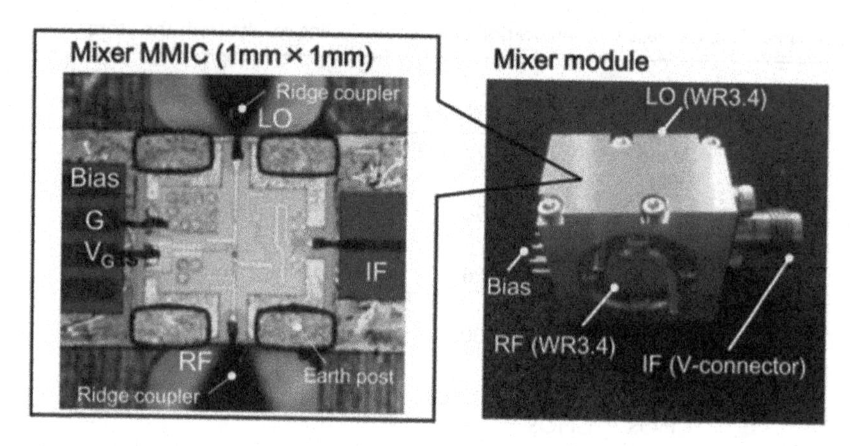

Figure 2.9 Photograph of fabricated mixer MMIC and module for the electronic InP HEMT-based THz wireless communication TRx (adapted and reprinted from [48] with permission).

have enabled fundamental oscillators of up to 1.98 THz [117] and output powers of up to 1 mW in J-band (220–325 GHz [118]) [119,120]. RTD-based 300 GHz-band wireless communication has already reached single-channel data transmission up to 30 Gb/s (error-free) and data rates of up to 56 Gb/s with BER below FEC limit using amplitude shift keying (ASK) over a 7 cm long link [121]. Baseband data is superimposed on the bias line to directly modulate the oscillator and requires no mixers.

THz RTD oscillators are characterised by interesting advantages with respect to the other two technologies. Compared to UTC-PD-based photo-mixers, RTD oscillators do not require any external input laser source, LO and optical component, such as modulators, filters and combiners, reducing cost and increasing compactness and reliability, where all the circuit components are monolithically fabricated and integrated on the same chip [122]. On the other hand, if compared to transistor-based TMICs, RTD oscillators do not require any frequency multiplication and RF power amplification stage but only a DC power supply [120], which drastically reduces complexity, cost and increases integrability. Moreover, RTD oscillators are characterised by relatively uncomplex circuit topologies compared to transistor-based oscillators, which makes them easy to fabricate and extremely compact [120, 122]. At the same time, RTD oscillators have relatively moderate phase-noise [123,124] (even though narrow spectral linewidths and high-quality factors (Q) have been reported by integrating varactor diodes [125] or by employing photonic crystal cavities [126,127]) and do not typically suffer from thermal stability issues. RTDs can also be employed to design high-sensitive THz detectors as well [128], opening up the possibility to build up full RTD-based transceivers [121,129]. For these reasons, RTD oscillators are considered a promising compact and low-cost solution in order to bring ultra-broadband THz wireless communications technology to a widespread consumer marketplace. The rest of the chapter is therefore dedicated to describing this technology in detail.

2.2 Resonant Tunnelling Diode Technology

2.2.1 RTD Device Technology

A resonant tunnelling diode (RTD) is a one-dimensional (1D) vertical transport unipolar two terminal semiconductor device, which is characterised by a highly non-linear

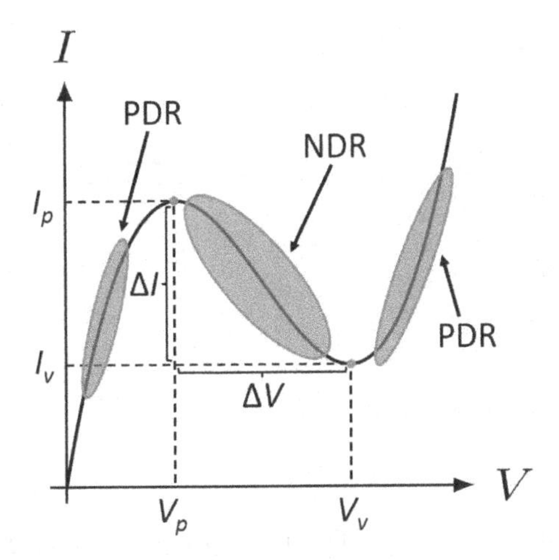

Figure 2.10 Schematic illustration of the current–voltage (*I–V*) characteristic of an RTD.

current–voltage (*I–V*) characteristic comprising a negative differential resistance (NDR) region and positive differential resistance (PDR) regions, as illustrated in Figures 2.10 and 2.20 [130]. This non-linearity is a result of the quantum mechanical resonant tunnelling of electrons through the device [131]. The precise shape of the *I–V* characteristic depends on different elements such as the device size, material composition and epitaxial structure and the temperature [116, 132].

The NDR region is characterised by I_p and I_v, which are the peak current, valley current and corresponding voltages V_p and V_v, respectively. Further, it can be described by $\Delta I = I_p - I_v$, which is the peak-to-valley current difference, $\Delta V = V_p - V_v$ the peak-to-valley voltage difference and $PVCR = I_p/I_v$, the peak-to-valley current ratio (PVCR). The span of the NDR region determines the maximum radio frequency (RF) power the diode can deliver to a load and this can be approximated by $(3/16)\Delta I \Delta V$ [133]. The actual RF output power would depend on different elements such as the operation frequency, device and circuit parasitic elements and impedance matching considerations [134,135].

Figure 2.11 shows the typical layer structure of an RTD in the InP-based material system. The core of the device comprises a low-bandgap semiconductor, in the case the indium gallium arsenide (InGaAs) quantum well layer, sandwiched by high bandgap semiconductor layers, in this case the aluminium arsenide AlAs layers, forming the so-called double barrier quantum well (DBQW) structure. The device has a spacer layer on either side of the DBQW (unmarked in the figure) and completed by highly doped *n*+ InGaAs contact layers. The DBQW is nanometric in dimensions, typically under 10 nm, and therefore thin enough to allow quantum mechanical tunnelling of the electrons through the structure. It also means that energy states in the quantum well are quantised, that is, only certain energy levels are allowed [136]. The conduction band diagram of the RTD is shown alongside for the unbiased and biased cases. In the illustration, the lowest allowed energy state in the quantum well is shown. The term 'resonant' in the name of resonant tunnelling diode refers to the behaviour of electrons with kinetic energy lower than the barrier potential but that still are able to travel though the double barriers. This is a consequence of the wave–particle duality, where the electron can be described through a wavefunction ψ. The possibility of

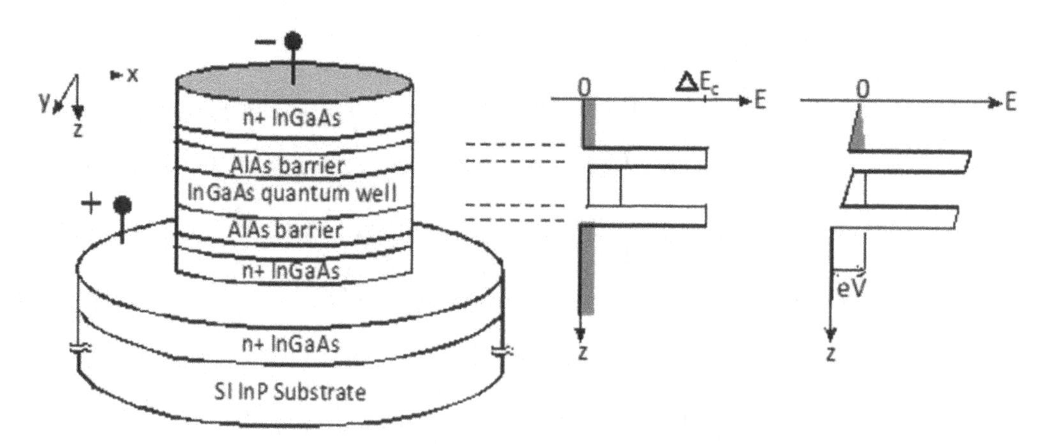

Figure 2.11 Typical layer structure of an RTD which is usually grown by epitaxy and the corresponding conduction band diagram at zero bias and higher bias. The energy states in the quantum well are quantised.

electrons tunnelling through the barriers is defined by the transmission coefficient. At the resonant state, the transmission coefficient is close to unity. This corresponds to the electrons entering the DBQW having the same energy level as the allowed energy state in the quantum well. Thus, as the transmission coefficient of electrons tunnelling through the DBQW changes with the bias voltage, the *I–V* characteristic of resonant tunnelling devices exhibits negative differential resistance (NDR) [137]. The details of how this happens are provided next.

The general working principle of a DBQW intraband RTD is illustrated in Figure 2.12 [130]. In the description which follows, E_F^e, E_F^c, E_C^e, E_C^c and E_0 are the emitter and collector electronic quasi-Fermi levels (electrochemical potentials), emitter and collector conduction band edges and fundamental quantum well quasi-bound energy state, respectively. In this description, one state only is supposed to exist in the quantum well. When the contact on the right hand side is forward biased (Figure 2.12a), resonant current starts flowing as soon as $E_0 = E_F^e$ (while a negligible thermionic non-resonant current component occurs for $E_0 > E_F^e$) and grows as E_0 moves towards E_F^e (first PDR region), up to the point where $E_0 = E_C^e$ (Figure 2.12b), where the current has a peak. When the bias is increased further, E_0 moves below E_C^e (Figure 2.12c), where no electronic states are filled on the emitter side, and the current dramatically drops with increasing bias, that is, the device exhibits a negative differential resistance (NDR). If the bias is increased even further (Figure 2.12d), the second barrier moves below E_F^e and thermionic emission across the first barrier takes place, where the current dramatically increases (second PDR region).

Since resonant tunnelling in these 1D vertical transport semiconductor-based nanostructures is a very fast process (how fast and the concept of 'tunnelling time' itself are still subjects of debate [138–140]), the NDR is characterised by an extremely wide bandwidth [141], which can extend up to the THz range [142,143]. Therefore, RTDs can be embedded in resonators to build up sub-millimetre-wave or THz continuous wave sources and highly sensitive detectors.

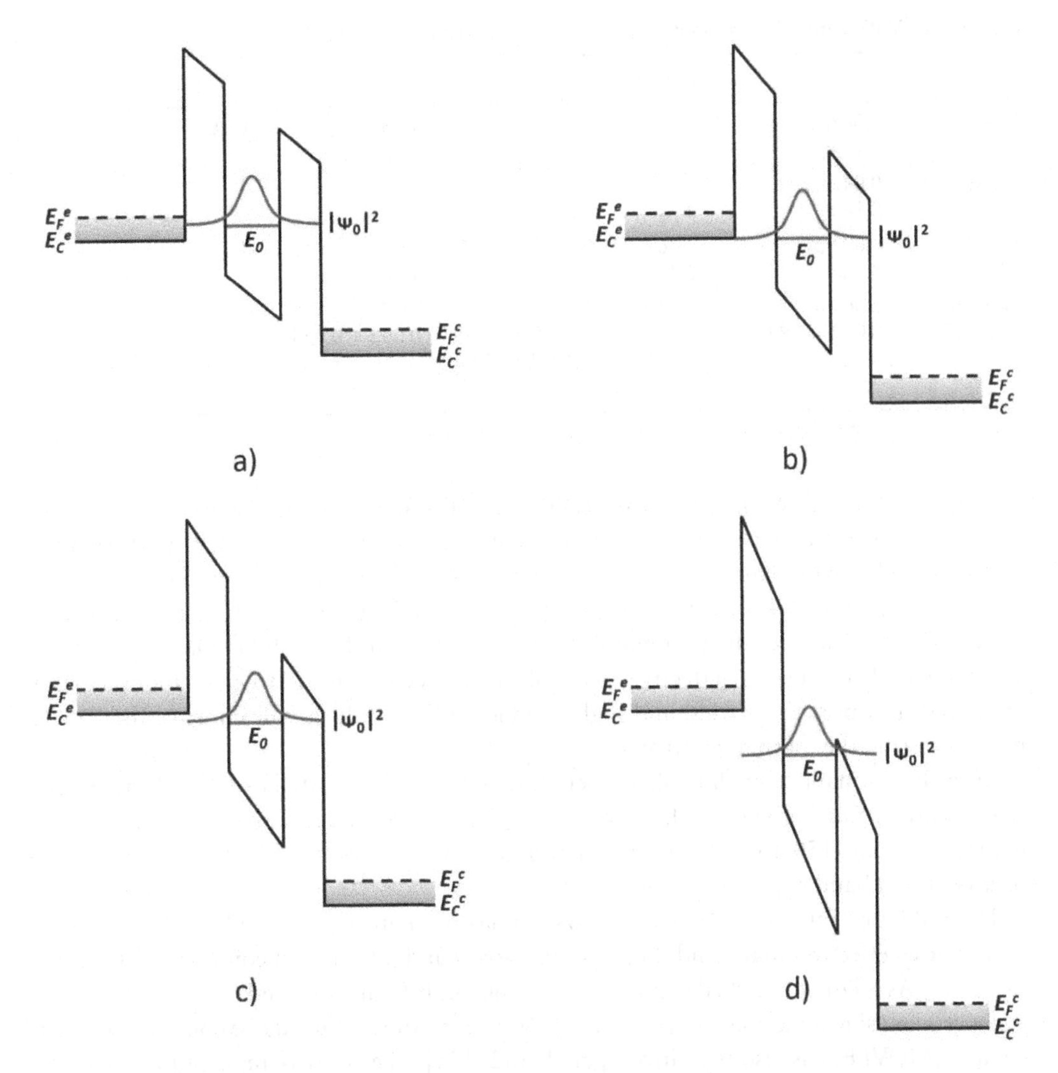

Figure 2.12 Intraband RTD conduction band diagrams at different bias points. This approximated representation assumes the right contact to be forward biased with respect to the left one, where E_F^e, E_F^c, E_C^e, E_C^c and E_0 are the emitter and collector quasi-Fermi levels (electrochemical potentials), conduction band edges and ground state resonant energy level, respectively, the latter assumed to be sole in the well. A schematic representation of the quasi-bound electronic wavefunction (evanescent envelope) square modulus $|\psi_k|^2 = |\psi_0|^2$ associated to $E_k = E_0$ is provided. The operation principle can be divided into four regions: (a) first PDR region, (b) peak current, (c) NDR region and (d) second PDR region.

Resonant tunnelling theory was first proposed by Tsu and Esaki in 1973 [144] and the *I–V* characteristics with peak tunnelling current at resonant energy were observed experimentally on double barrier GaAs/$Al_{0.7}Ga_{0.3}As$ heterostructure in 1974 [145]. The development of epitaxial crystal growth techniques such as molecular beam epitaxy (MBE) in the 1970s led to a great improvement in high-quality RTD heterostructure materials growth. Since then extensive research has been devoted to resonant tunnelling devices.

Table 2.4 III–V RTD material parameters (at room temperature (300 K)) [149]

Material	m^*	E_g (eV)	ε_r	ΔE_c (eV)
GaAs	$0.067m_0$	1.42	12.9	0.28
AlAs	$0.15m_0$	3	10.1	
$In_{0.53}Ga_{0.47}As$	$0.042m_0$	0.74	13.9	1
AlAs	$0.15m_0$	3	10.1	
InAs	$0.027m_0$	0.36	15.1	1.35
AlSb	$0.12m_0$	2.4	12.04	
GaAs	$0.067m_0$	1.42	12.9	
$Al_xGa_{1-x}As$	0.063+0.083x	1.424 + 1.247 if $0<x<0.45$ 1.9 + 0.125x + 0.143x^2 if $0.45<x<1$	0.063 + 0.083x	0.79x if $0<x<0.41$ 0.475 − 0.335 + 0.143x^2 if $0.41<x<1$

Note: $m_0 = 9.11 \times 10^{-31}$ kg is the electron rest mass.

RTD devices realised in III–V materials show attractive characteristics such as THz intrinsic cut-off frequency, high peak current density and high PVCR performance [146,147]. The parameters of commonly used III–V semiconductor materials for RTDs are shown in Table 2.4, in which the effective mass (m^*), band gap (E_g), relative dielectric constant (ε_r) and conduction band offset (ΔE_c) of the different RTD material systems are compared. In general, effective mass of small electron leads to high mobility and improved transport properties, and high conduction band offset will improve the PVCR by suppressing the thermal electron current [148].

After the resonant tunnelling phenomenon was first demonstrated in 1974 [145], extensive research contributed to the first prototype RTD device based on the GaAs/Al_xGa_{1-x} As material system [150,151]. GaAs (low band gap) was sandwiched between Al_xGa_{1-x} As barriers (high band gap).

The InP-based InGaAs/AlAs material system has been attractive for THz RTDs because of the low effective mass and high conduction band offset semiconductor materials. $In_{0.53}Ga_{0.47}As$ is lattice matched to a semi-insulating (SI) InP substrate and the structure is typically grown using molecular beam epitaxy (MBE), even though metal organic vapour phase epitaxy (MOVPE) has also been investigated [152–154]. The effective mass of $In_{0.53}Ga_{0.47}As$ is $0.044m_0$, which is much smaller than $0.067m_0$ for GaAs and the conduction band offset $\Delta E_C = 0.65$ eV higher than GaAs/AlGaAs system [150]. A low specific contact resistance (ρ_c) of the order of 10^{-8} Ωcm² (10^{-6} Ωcm² for GaAs) with a saturation velocity more than 1.5×10^7 cm/s (10^7 cm/s for GaAs) is attainable in InGaAs/AlAs system [155,156].

A generic example is depicted in Figure 2.13, while variations in the quantum well design and doping level and position of spacers/contacts are found among the different RTDs reported in the literature. The DBQW comprises the low- bandgap GaInAs layer sandwiched between two widebandgap AlAs barriers. Often, an In- rich GaInAs well or an indium arsenide (InAs) subwell is also used. Metal contacts typically employ a titanium/palladium/gold (Ti/Pd/Au) stack scheme. In this representation, the nomenclature of layers assumes the top metal contact to be forward biased with respect to the bottom one. The thicknesses of the DBQW layers and the indium (In) molar fraction reflect on the diode I–V characteristic and impedance [157,158], affecting speed and RF power performances. The small electron effective mass of GaInAs and the high AlAs/GaInAs conduction band offset (which decreases the valley current associated with thermionic emission [159]) lead

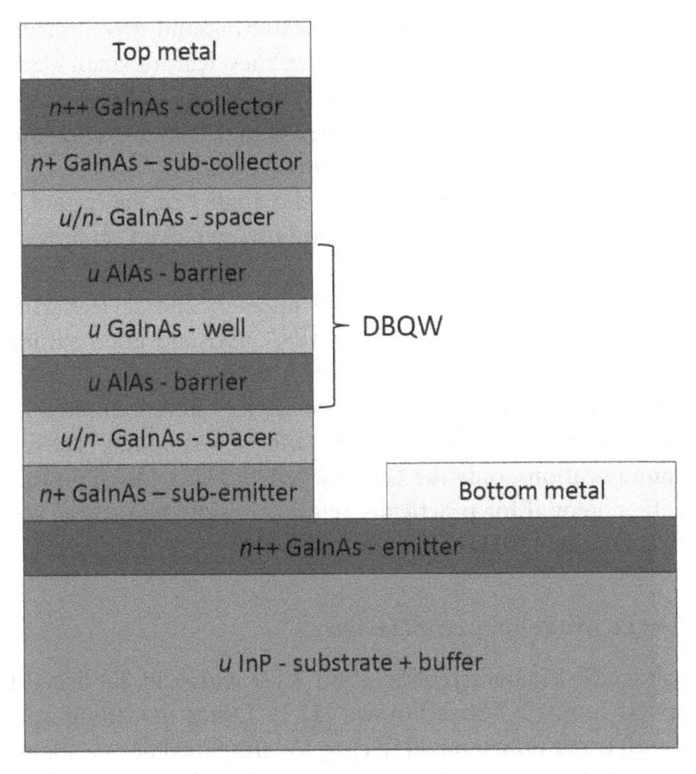

Figure 2.13 Schematic representation of a generic AlAs/GaInAs-based DBQW RTD epitaxial structure for THz applications grown on a semi-insulating (SI) InP substrate (variations in well design and spacers/ contacts doping level and position are found among the different RTDs reported in the literature). The layers nomenclature assumes the top metal contact to be forward biased with respect to the bottom one. Labels u, n-, n+ and n++ mean undoped, lightly doped, highly doped and heavily doped, respectively. Metal contacts typically employ a Ti/Pd/Au stack scheme [120].

to high-available current density and PVCR [160]. Outside of the undoped active region, spacers and heavily doped (typically employing Si as a dopant) electron reservoirs complete the epitaxial structure of the diode. Spacers are chosen to avoid dopant diffusion into the active region, to reduce the self-capacitance of the device and to maximise speed and power performances [161,162–165], while doping level of contact layers and In molar fraction are optimised to enhance current density and reduce the associated contact resistance [49,122].

Indium arsenide/aluminium antimonide (InAs/AlSb) heterostructures [166,167–171] including those based on gallium-antimonide (GaSb) well [172–176]) have also been investigated as THz sources since they are characterised by superior speed and power performance. The InAs/AlSb system has several advantages over GaAs/AlGaAs and InGaAs/AlAs system such as low effective mass and high conduction band offset. The effective mass of InAs is $0.023m_0$ and the conduction band offset is about 1.35 eV [177]. The low specific contact resistance (10^{-9} Ωcm^2) and high saturation velocity of electrons (5×10^7 cm/s) benefit the InAs/AlSb material system over other materials [166,178,179]. Indeed, they feature high-available current density due to the staggered type-II conduction band offset [180], which allows electrons to tunnel through the barriers close to the valence band edge of AlSb, reducing the associated attenuation coefficient [181–183]), low effective mass of InAs, absence of ternary/quaternary alloy scattering and low electron-to-longitudinal optical phonon scattering rate [178]. Moreover, they feature small electron transit time across the device depletion regions [178], which lowers the diode self-capacitance. In addition, a nearly ideal Ohmic contact can be formed at the metal–InAs junction since the Fermi level pinned in the conduction band [179], removing the Schottky barrier and lowering the associated contact resistance [184]. However, the quality of the epitaxial growth is poor due to the lattice mismatch arising with commonly employed semi-insulating substrates [168], such as GaAs and GaSb.

Recently, III-nitrides have gained interest for intersubband devices. They feature a large conduction band discontinuity (~2.1 eV in AlN/GaN) and good saturation velocity of 2.5×10^7 cm/s, but have large electron effective mass ($m^* \sim 0.2–0.3m_0$) and the Ohmic contacts are poor ($\sim 10^{-6}$ Ωcm^2). Demonstrated devices have exhibited f_{max} of under 200 GHz [185] and so the jury is still out for this material system for THz sources.

For THz communications, only the GaInAs/AlAs RTDs realised on InP substrates have demonstrated the potential for practically relevant results and so the rest of the chapter will focus on this family of RTDs.

2.2.2 *RTD Device Modelling and Design*

The current voltage (*I–V*) characteristic of an RTD is usually modelled by an analytical expression derived from its device physics [113]. Other modelling approaches include fitting with an *n*-th order polynomial [131] or for simplified device analysis, a third order polynomial in which the origin (reference point) is shifted to the mid-point of the negative differential resistance (NDR) region [133].

At any specific bias point of the *I–V* characteristic, the RTD can be modelled using a linear model, usually a lumped element equivalent circuit as shown in Figure 2.14. In this case, the device is represented by its self-capacitance $C_{rtd} = C_0 + C_{qw}$ where C_0 is the device geometrical capacitance and C_{qw} is the quantum well capacitance in parallel with the series of the negative differential conductance (NDC) $-G_{rtd}$ and the negative quantum well inductance $-L_{qw}$ with both G_{rtd} and L_{qw} being positive values. C_{qw} is associated to the quantum well charge distribution change as a function of the voltage [186,187] and this is related to the electron escape rate from the well towards the collector and with the transit delay of depletion regions [188], while G_{rtd} and L_{qw} model the RF gain capability of the device and the lag associated with the quantum well charging and discharging as a function of the voltage (which is related to the electron quasi-bound state lifetime in the well) [187, 189], respectively. In the PDR regions of the *I–V* characteristic, both

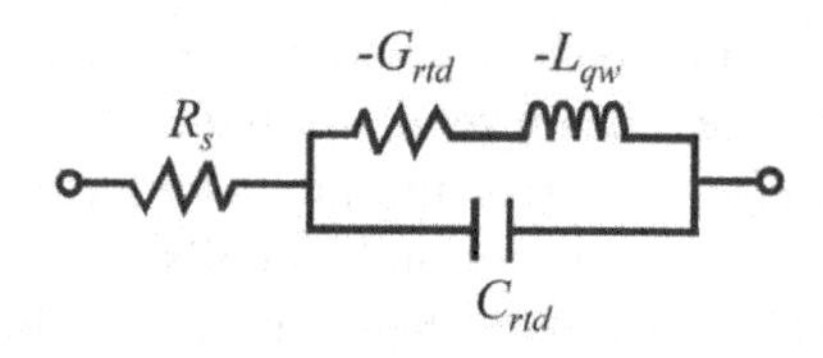

Figure 2.14 Intrinsic small-signal equivalent circuit of an RTD in the NDR region. R_S is the contact and access resistance, $-G_{rtd}$ the device conductance, C_{rtd} the device self-capacitance and $-L_{qw}$ the quantum-well inductance.

the differential conductance and quantum well inductance are positive. The model is completed by a series resistance R_S, which models the contact and access resistance.

Both the device *I–V* characteristics and small signal model provide feedback required for technology optimisation and can, for instance, be used in the estimation of the maximum operation frequency f_{max} [190] and the gain roll-off [191] of the RTD device. The device cut-off frequency f_{max} is given by [116,120]:

$$f_{max} = \frac{G_{rtd}}{2\pi C_{rtd}} \sqrt{\frac{1}{R_S G_{rtd}} - 1} \tag{2.1}$$

Here, $G_{rtd} = \frac{3}{2}\frac{\Delta I}{\Delta V} = \frac{3\Delta J A}{2\Delta V}$ and $R_S = \frac{\rho_c}{A}$, with ρ_c being the specific contact resistance; ΔI and ΔV are the peak-to-valley current and voltage differences, respectively, and are found from the measured *I–V* characteristics; and $\Delta J = \Delta I/A$ is the peak to valley current density. Therefore (1) can be rewritten as [120]:

$$f_{max} = \frac{d}{2\pi\epsilon_0\epsilon_r} \frac{2\Delta J}{3\Delta V} \sqrt{\frac{2\Delta V}{3\Delta J \rho_c} - 1} \tag{2.2}$$

For a given RTD device, it can be deduced from (2.2) that the cut-off frequency is independent of device sizing and is only related to ΔJ and ρ_c, which are mostly determined by the layer design and the fabrication process, respectively. This means that the largest possible RTD devices can be used to maximise oscillator output power. However, this has still to be proved experimentally and work in this regard is already underway. Here, the RTD and load (G_L) conductance must also be matched to the negative differential conductance, that is, it should satisfy the relationship $G_{rtd} = 2G_L$ [133].

Linear modelling of RTDs is based on S-parameter characterisation of the device. However, the existence of the NDR means that RTD devices are prone to oscillations when biased in this region, and so to avoid these bias oscillations, a shunt resistance R_{st} is usually connected across the device. It must satisfy $R_{st} > \frac{1}{G_{rtd}}$ which for a given R_{st} establishes the maximum device size, A_{max}, as [120]:

$$A_{max} R_{st} < \frac{2\Delta V}{3\Delta J} \tag{2.3}$$

R_{st} is usually chosen to be in the 10–20 Ω range. From (2.3), it is clear that the use of RTD design with large ΔV and moderate ΔJ is key to realising large devices which can provide high power at THz frequencies. For the small-signal device modelling, the stabilising resistance, has to be de-embedded from the measurements [192]. A review of the small signal modelling of RTDs can be found in Refs. [116,186–191, 193–203] for the interested reader.

A unified or non-linear large-signal model is needed to carry out large signal analysis of an RTD oscillator, for example, estimate the diode RF output power capabilities, or in the design of a direct detector [116,204]. Such a model should be capable of reproducing both the device I–V and bias-dependant small-signal characteristics. To date, however, such a model is yet to be developed but efforts are underway in this regard [205]. Therefore, RTD oscillator design is typically carried out by relying on an empirical approach [135,189,199]. Based on the device's I–V characteristics and small-signal parameters, basic oscillator design is done and then variations of these designs are manufactured and then characterised to identify the best or desired designs.

2.2.3 RTD Oscillator Design

RTD-based negative differential resistance (NDR) oscillators can produce either sinusoidal or non-sinusoidal/relaxation oscillations. The latter are usually at low frequency (MHz range) and set by the inductance of the bias line [178] and so not discussed here. State-of-the-art THz RTD transmitters exclusively employ sinusoidal oscillators and their principle of operation conforms to classical electronic oscillators. When the RTD is biased in the NDR region, electronic noise in the circuit is amplified by the NDR and the system filters out all frequency components except those defined by the resonator passband. The shape of the spectrum defines its quality factor (Q) and the corresponding relative bandwidth which depend upon the LC resonator and losses in the circuit. If the large-signal NDR can compensate for circuit, device and antenna losses within the resonator bandwidth, a stable oscillating signal is obtained across the load [206]. An RTD oscillator can be therefore considered a DC-to-RF power converter, where the energy provided by the bias supply is transformed into an RF output signal with a certain efficiency and then radiated through an antenna [122].

A schematic representation of a general RTD-based sinusoidal oscillator lumped equivalent circuit topology is shown in Figure 2.15(a) [207]. The DC part of the circuit is composed of the DC supply V_b, the bias line that is modelled by its parasitic resistance R_b and line inductance L_b, the decoupling capacitor C_{dc} and the stabilising shunt resistor R_{st}, while the RF part of the circuit comprises the resonating inductance L, the RTD device and the load resistance R_L. The resonating inductance is usually realised from a short section of a transmission line such as a coplanar waveguide (CPW) [208], a coplanar stripline (CPS) or a microstrip line, while the resonating capacitance is provided by the intrinsic self-capacitance of the diode. In this case where the resonating inductance L is realised

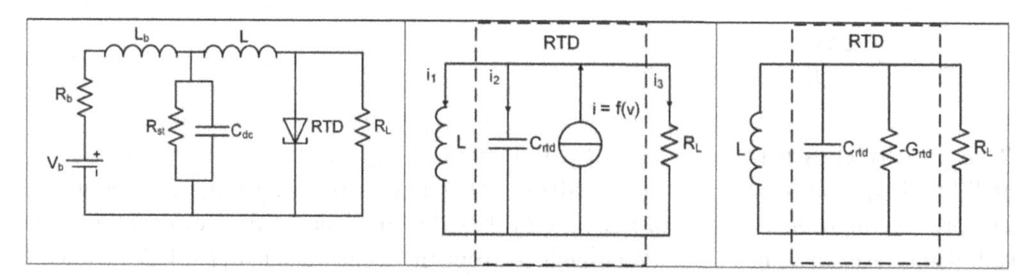

Figure 2.15 (a) Single RTD oscillator topology with shunt resistor R_{st} and decoupling capacitor C_{dc}. R_b and L_b denote the bias cable resistance and inductance. R_L is the load resistance and L the resonating inductance. (b) Large signal model. RTD is represented by its self-capacitance C_{rtd} in parallel with voltage controlled current source $i(v)$. (c) Small signal equivalent circuit. RTD is represented by its self-capacitance C_{rtd} in parallel with the negative conductance $-G_{rtd}$. The parasitic contact resistance is neglected.

from a transmission line, the relation $\omega_0 L = Z_0 \tan\beta l$ is used, where ω_0 is the oscillation frequency, β the phase constant and l the length of the stub. This is often the case for RTD oscillators operating below 300 GHz. Also, for such oscillators, an external load is often used. For higher frequencies, an on-chip integrated antenna is usually employed in which case it works both as the resonating inductance and radiator, i.e. load resistance. The shunt resistor R_{st} suppresses low-frequency bias oscillations. And since the DC bias is fed via the resonating inductance, the decoupling capacitor C_{dc} is used to ground the inductor and also to short-circuit the stabilising resistor at the oscillation frequency, thereby decoupling the oscillator circuit from the bias supply.

The decoupling capacitor is chosen using equation (2.4), that is, it should act as a short circuit at the desired frequency f_{osc}:

$$\left(2\pi f_{osc} C_{dc}\right)^{-1} < 0.1 \tag{2.4}$$

while the value of the stabilising resistance should satisfy [209,210]:

$$R_{st} < \frac{1}{G_{rtd}} = R_{rtd} = \frac{2\Delta V}{3\Delta J A} \tag{2.5}$$

Equation (2.5) indicates important information for RTD oscillator design: for a large-size RTD device, the absolute value of negative resistance (R_{rtd}) is small, which requires a small shunt resistor R_{st} to suppress the bias oscillations. For example, for an RTD device with $\Delta V = 0.4$ V, $\Delta J = 9$ mA/μm^2, if the device size $A = 10$ μm^2, the calculated $R_{st} = 2.9\ \Omega$, which is an impractically small resistor. Also, a large portion of DC power would be dissipated by such a small value shunt resistor. Instead, from Equation (2.5), it can be noted that for a given shunt resistor value R_{st}, the RTD device size can be maximised for high-output power capability while the low-frequency bias oscillations can also be suppressed [209,210].

The RF equivalent circuit of the circuit shown in Figure 2.15(a) is represented in Figure 2.15(b), where the RTD is replaced by its large single model, which consists of a self-capacitance (C_{rtd}) in parallel with voltage controlled current source. By approximating he RTD I–V characteristic with a third order polynomial of the form $i = f(v) = -av + bv^3$ for which the origin is shifted to the mid-point of the NDR region, the performance of the RTD oscillator can be evaluated analytically. Specifically, it can be shown that the maximum power delivered to the load $G_L = 1/R_L$, that is, the maximum power generated by the diode, is given by [133,211]:

$$P_{max} = \frac{3}{16}\Delta V \Delta I \tag{2.6}$$

Finally, the small signal equivalent circuit of the single RTD oscillator is shown in Figure 2.15(c) where the RTD is represented by its small signal model, with the parasitic series resistance R_S neglected. The frequency of the oscillation (f_{osc}) can be derived from the susceptance of the circuit being set to zero, that is [207]:

$$2\pi f_{osc} C_{rtd} - \frac{1}{2\pi f_0 L} = 0 \tag{2.7}$$

and therefore

$$f_{osc} = \frac{1}{2\pi\sqrt{LC_{rtd}}} \tag{2.8}$$

Further, it can be shown that a non-negligible contact resistance R_S of the RTD causes the variation of the RTD oscillator output power with frequency [135]. The resonant frequency f_{osc} can be determined to be [209]:

$$f_{osc} = \frac{\sqrt{(L - C_{rtd} R_S^2)}}{2\pi L\sqrt{C_{rtd}}\,(1 + R_S G_L)} \tag{2.9}$$

while the power delivered to the actual load resistance R_L is given by [209]:

$$P_L = \frac{R_L''}{R_S + R_L''} \frac{2(G_{rtd} - G_L')G_L'}{3b} \tag{2.10}$$

where

$$R_L'' = \frac{\omega^2 L^2 G_L}{1 + \omega^2 L^2 G_L^2} \tag{2.11}$$

and G_L is the apparent load, which changes with frequency and so does not present an ideal load for maximum output power, that is, output power drops with increasing frequency. At any given frequency, an optimum value of the oscillator load G_L may be found. For the simplified RTD equivalent circuit with only one parasitic component (R_S), the maximum output power predicted by (2.6) can be considered as an upper limit.

To illustrate the importance of reducing R_S, the calculated/expected output power as a function of frequency for a single 4 μm × 4 μm RTD device oscillator is shown in Figure 2.16 (solid trace). Here, the device parameters ΔV = 0.6 V and ΔI = 24 mA and R_S = 2.8 Ω have been used. As seen in Figure 2.16, the cut-off frequency is around 340 GHz. The expected output power is about 0.5 mW at 300 GHz. Figure 2.16 (dashed trace) also shows the expected oscillator performance if a lower contact resistance of, say 1.4 Ω, was used – higher output power and higher bandwidth become possible. The expected output power becomes 2.5 mW at 300 GHz.

2.3 THz RTD Oscillators

2.3.1 Overview

Different approaches have been adopted to develop THz RTD oscillators featuring different device epitaxial structures and circuit designs. Figure 2.17 shows the output power and corresponding oscillation frequency of these oscillators to date. Clearly, THz operation has been demonstrated, but the output power is still under a 1 mW beyond 300 GHz, which is perhaps the main limitation of this technology. The oscillators are almost all exclusively in InP technology.

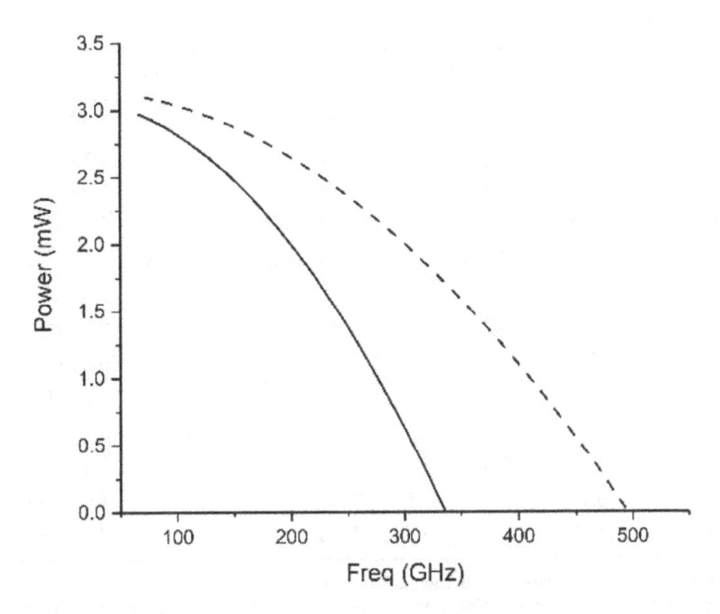

Figure 2.16 Variation of output power with oscillation frequency for an RTD oscillator employing a 4 μm × 4 μm device (C_{rtd} = 60 *fF*) with a R_S = 2.8 Ω (solid line, f_{max} ≈ 340 GHz & P_{max} ≈ 0.5 mW at 300 GHz; while for R_S = 1.4 Ω (dashed line, f_{max} ≈ 500 GHz & P_{max} ≈ 2.5 mW at 300 GHz).

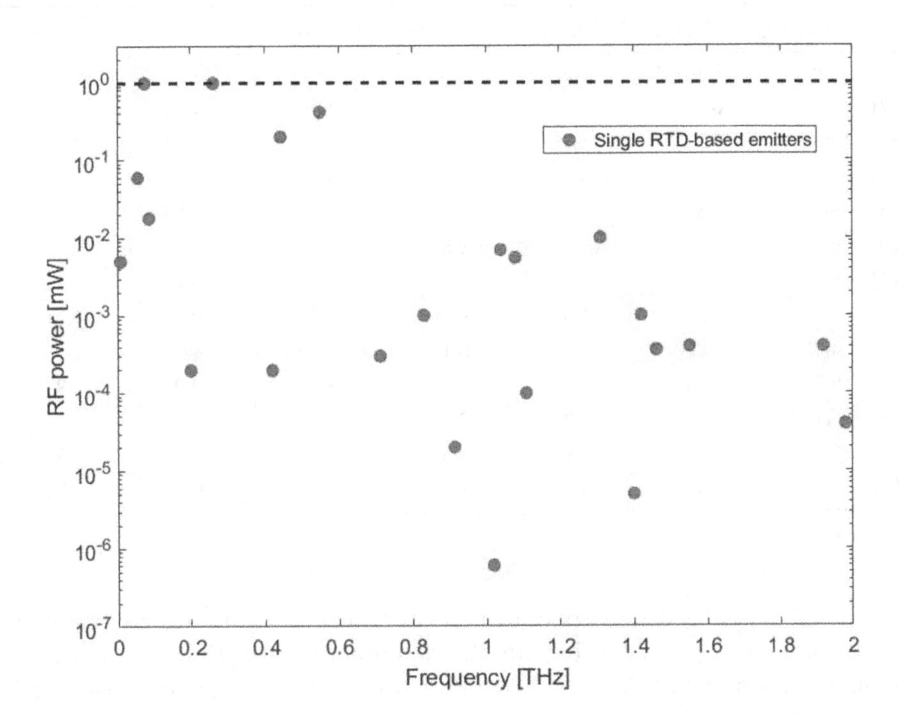

Figure 2.17 Output and fundamental frequency of oscillation for reported single-chip RTD oscillators.

Table 2.5 State of the THz RTD oscillators

Ref.	*Technology*	*ΔI [mA]*	*ΔV [V]*	*PVCR*	*P_{max} [mW]*	*Antenna*	*f_osc*	*RF power*
[213]	InP	18–23	0.36	2	1.21–1.55	slot	548 GHz	0.42 mW
[122]	InP	2.1	0.4	1.7	0.16	slot	1.92 THz	0.4 μW
[117]	InP	2.8	0.5	1.8	0.26	slot	1.98 THz	40 nW
[214]	InP	19.6	0.36	–	1.32 (single)	slot	620 GHz	0.61 mW
[215]	InP	7.2	0.34	2	0.46 (single)	dipole	1 THz	0.73 mW
[120]	InP	25	0.7	3	3.28	/	260 GHz	1 mW
[212]	InP	16.6	0.8	2.5	2.49 (single)	/	84 GHz	2 mW
[216]	InP	29.6	1.3	12	7.21	/	62.5 GHz	3.12 mW
[217]	InP	1.74	0.22	1.4	0.07	*patch*	1.52 THz	1.9 μW
[218]	InP	3.1	0.2	1.4	0.11	*dipole*	675 GHz	47 μW
[219]	GaN	2.5	0.8	1.05	0.37	/	0.98 GHz	3 μW
[194]	GaAs	6.25	0.4	1.5	0.47	/	420 GHz	0.2 μW
[166]	GaAs*	4.95	0.3	3.4	0.28	/	712 GHz	0.3 μW

Notes: f_0 stands for oscillation frequency, J_p is the peak current density while ΔI, ΔV, PVCR and P_{max} are the peak-to-valley current difference, peak-to-valley voltage difference, peak-to-valley current ratio, and maximum possible power, respectively. *GaAs substrate but InAs/AlSb RTD structure.

Table 2.6 Electrical characteristics of RTDs in Table 2.5

Ref.	*Technology*	*Mesa [μm²]*	*ρ_c [Ω μm²]*	*C_{rtd} [fF/μm²]*	*J_p [mA/μm²]*	*J_v [mA/μm²]*	*ΔI [mA]*	*V_p [V]*
[213]	InP	1.5–1.9	9.5–12	–	24	12	24	–
[122]	InP	0.1	–	–	50	29.4	20.6	0.4
[117]	InP	0.2	-	–	31	17.1	13.9	–
[214]	InP	1.4	3	–	–	–	–	–
[215]	InP	0.5	–	–	29	14.5	14.5	–
[120]	InP	16	50	3.8	3	1	2	1
[212]	InP	16	83	2.5	1.9	0.75	1.12	0.9
[216]	InP	26.4	106	4.5	1.2	0.1	1.21	1.4
[217]	InP	0.88	4.6	8.1	6.9	4.9	2	0.67
[218]	InP	1.65	–	–	6.7	4.8	1.9	0.65
[219]	GaN	25	100	2.4	2.2	2.1	0.1	9.2
[194]	GaAs	12.5	10	1.5	1.5	1	0.5	0.6
[166]	GaAs*	2.5	5	1.1	2.8	0.82	1.98	1.2

Note: *InAs/AlSb RTD.

Table 2.5 provides details some of oscillators including the highest reported fundamental oscillations at 1.98 THz with 40 nW output power [117] and highest reported output powers of 1 mW at 0.26 THz [120], while Table 2.6 provides electrical details on the RTD devices that were used. The corresponding epitaxial layer structures of these devices are provided in Appendix A (Tables 2.8 and 2.9) for reference.

Note that for frequencies beyond around 300 GHz, the oscillator designs employ integrated antennas and the devices are characterised by high peak current density J_p. The device sizes are small and the output power is typically low in the μW range. On the other hand, for lower frequencies, low J_p, large ΔV and large mesa area devices are used as well as external loads/antennas. The output power are in the mW range. ΔI is higher for devices operating below 600 GHz, while ΔV is very modest, under 0.5 V for the InP-based RTDs. The discrepancy between the maximum available power and actual measured power is also significant.

From Table 2.6, it can be seen that the typical self-capacitances in the range 3–8 fF/μm², while device sizes are under 2 μm² for oscillators operating over 100 GHz apart from those reported in Refs. [120, 212]. The contact resistances are also very varied, ranging from 3 to 106 Ωμm². The peak current density J_p varies over a large range, 1–50 mA/μm², which in some ways shows the relative immaturity of the epitaxial design approaches and associated oscillator circuit realisation.

The achieved results show large variations in output power with frequency, and so a coordinated approach using best practise may result in THz sources with output powers in the mW range. Clearly, effective THz power combining techniques suited to RTDs also need to be developed.

2.3.2 RTD Oscillators Up To 300 GHz

We first describe in some detail RTD oscillators operating below 300 GHz. As noted above, the device design is characterised by low J_p, large ΔV and large RTD mesa area. Due to the large device sizes, photolithography and wet etching can be adopted, which make fabrication relatively easy and inexpensive. The load is usually either an external antenna or the input impedance of the characterisation equipment such as the spectrum analyser or power meter, which is typically 50 Ω. Figure 2.18 shows the circuit topology of W-band RTD oscillator [212], while a micrograph of the fabricated oscillator is shown in Figure 2.19. The characterisation setup for these oscillators involved on-wafer probing and down-conversion mixers for display of the signal on a 50 GHz spectrum analyser [212].

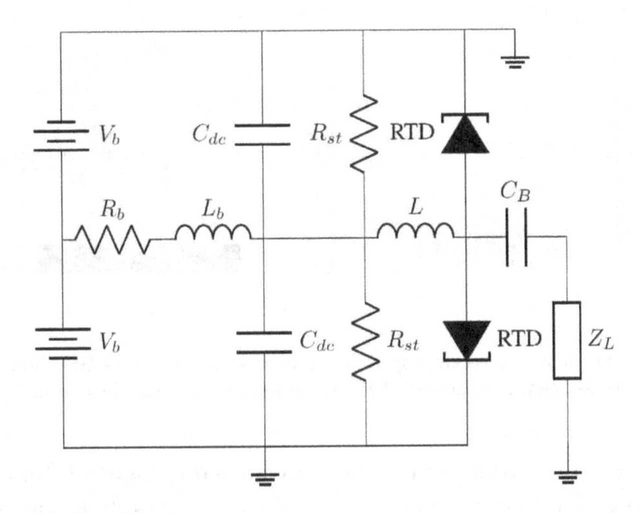

Figure 2.18 Double-device RTD oscillator lumped equivalent circuit topology, where each device is biased and stabilised individually. The circuit elements are the same as in Figure 2.15(a), where a DC block capacitor is added.

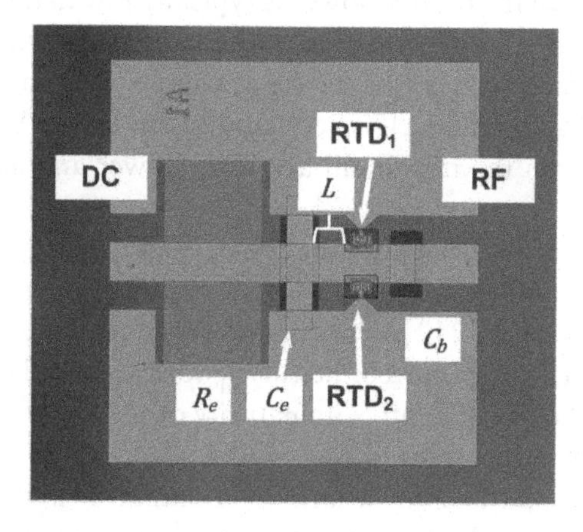

Figure 2.19 Micrograph of the fabricated 84 GHz double-RTD oscillator [212].

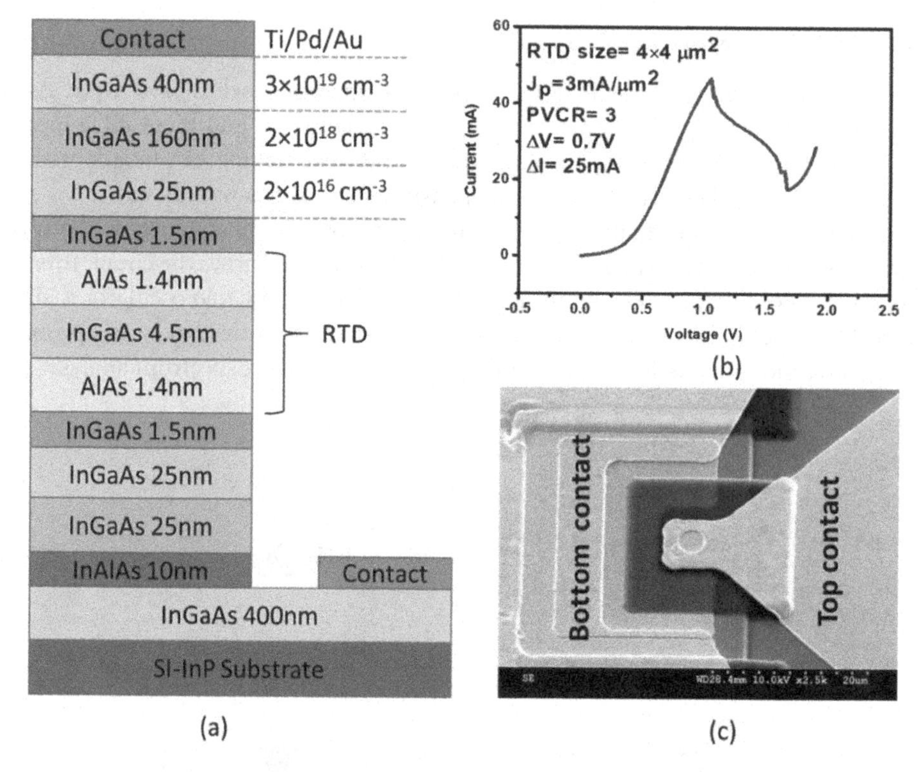

Figure 2.20 RTD epitaxial structure employed in the realisation of the 84 GHz and 260 GHz oscillators, measured *I–V* characteristics and a micrograph of a 4×4 μm² device [222].

The epitaxial layer structure for the RTD is depicted in Figure 2.20, which includes the *I–V* characteristic of a 4 μm × 4 μm device and its micrograph. It was characterised by a GaInAs/AlAs DBQW heterostructure with ~1.4 nm thick barriers, ~4.5 nm thick well and ~26 nm thick lightly doped spacers [220]. It was grown onto a SI InP substrate by MBE.

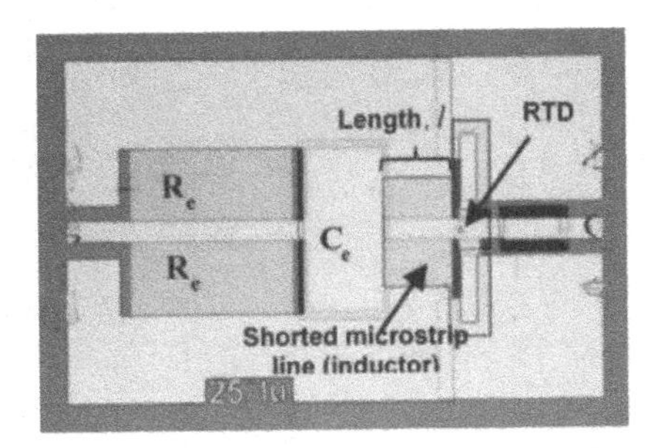

Figure 2.21 Micrograph of the fabricated microstrip 260 GHz RTD oscillator [120].

Two RTDs were connected in parallel for high-output power operation and each one was biased and stabilised individually, but they shared the same RF load [208]. The RTDs mesa was ~16 μm² with the following electrical characteristics: J_p = 1.87 mA/μm², PVCR = 2.5, ΔI = 16.6 mA and ΔV = 0.8 V. The self-capacitance of the RTDs resonated with a common inductance provided by a 42 μm long CPW line of characteristic impedance Z_0 = 50 Ω. Stabilising resistors (R_{st} = 10 Ω) were fabricated using nichrome (NiCr) films, while the MIM decoupling capacitors (~2 pF) and the DC block (~1.5 pF) were realised using silicon nitride (SiN_x) as the dielectric layer. Ohmic contacts were fabricated through a Ti/Pd/Au metal stack, where the contact resistivity ρ_c was ~50 Ω-μm². The maximum oscillator output power was 2 mW.

By scaling (reducing) the resonating inductance, a J-band RTD oscillator was demonstrated using the same epitaxial wafer structure and RTD device size [208,221]. The oscillator design featured a single RTD device, a 88 μm long microstrip based resonating inductance with characteristic impedance Z_0 = 10.4 Ω with which Polyimide PI-2545 was used as the dielectric. Reducing Z_0 made the inductance per unit of length to decrease, which allowed the use of a long length of resonating inductance. Fabrication was done solely with photolithography. Ohmic contacts, a 0.1 pF SiN_x MIM decoupling capacitor, the DC block (~1.3 pF) and the stabilising resistor (R_{st} = 20 Ω) were fabricated as in Ref. [212]. A micrograph of fabricated oscillator is shown in Figure 2.21. The reported RTD electrical parameters were J_p = 3 mA/ μm², PVCR = 3, ΔI = 25 mA and ΔV = 0.7 V. Figure 2.22 shows the spectrum of the fundamental oscillations for the oscillator at 260 GHz. The corresponding output power was 1 mW, with a DC-to-RF efficiency of ~0.7% [120].

The modulation bandwidth of this oscillator was measured using 2-port S-parameters of an unbiased oscillator and then deduced from the forward transmission coefficient, S_{21}, to be ~110 GHz [222].

2.3.3 RTD Oscillators above 300 GHz

For RTD oscillators operating beyond 300 GHz, different antenna types, such as slot antennas [122], Vivaldi antennas [223], radial line slot antenna [224], patch antennas [225–227], dipole antennas [218,215,228] and bow-tie antennas [229,230], have been employed. The most common on-chip antenna is the slot antenna, which is used in conjunction with a hemispherical Si lens [122,231,232]. This is so because of the high dielectric permittivity of InP (ε_0 ~ 12.5 [233], ε_∞ ~ 9.6 [234]), and so most of the output power

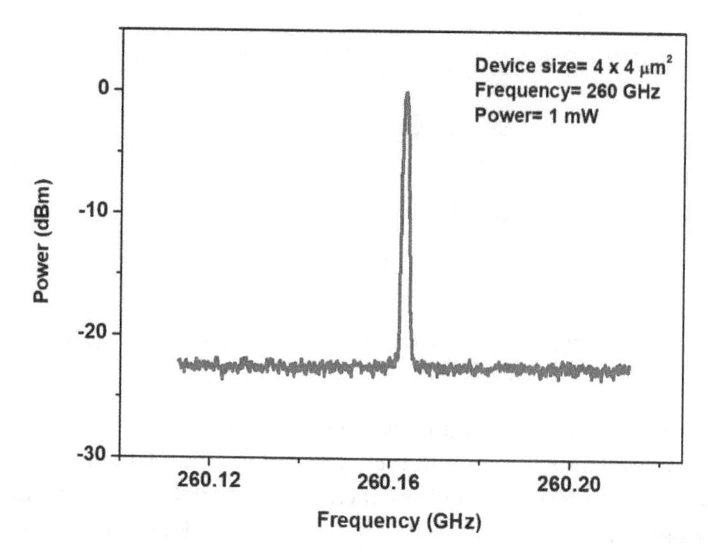

Figure 2.22 Measured spectrum of a 1 mW 260 GHz microstrip single RTD oscillator [120].

is radiated into the substrate. Thus, the oscillator chip is mounted onto a hemispherical Si lens to extract and collimate the radiated power [235], with the thickness of the oscillator chip and the lens designed to maximise the power extraction efficiency.

Figure 2.23(a) and 2.23(b) show an illustration of the fabricated structure of an RTD oscillator integrated with slot antenna and its equivalent circuit, respectively [134,195]. The RTD is located at the centre of the slot. The electrodes of the RTD are connected to the left and right electrodes of the antenna. At both edges of the antenna, the electrodes are overlapped with a silicon dioxide (SiO_2) layer between them forming a metal–insulator–metal (MIM) capacitor across which a stabilising resistor is connected. The equivalent circuit of this oscillator is identical to that of Figure 2.15. The oscillation frequency is determined by the parallel resonance of L and C_{rtd}, where the inductance L is produced by the antenna and the capacitance C_{rtd} is the intrinsic self-capacitance of the RTD. The SiO_2 MIM capacitor plays the same role as C_{dc} in Figure 2.15. The antenna inductance L could also be viewed as the inductance of the metallisation (connection) between C_{dc} and the RTD, and looking at it this way the circuit is identical to that shown in Figure 2.15.

Since the input impedance of the slot antenna is infinity at the centre and zero at the edges of the slot, for good impedance matching between the RTD and the antenna, the RTD should be located away from the centre [236]. Actual location can be determined from 3D electromagnetic simulations. For this realisation, the dimension of the shorter part determines the antenna susceptance which is mainly inductive and so defines the oscillation frequency. On the other hand, the dimensions of the longer part determines the radiation conductance which defines the output power of the oscillator. Using this approach led to output powers of up to 0.42 mW at a fundamental oscillation frequency of 548 GHz [213].

The layout of a 1.92 THz RTD slot antenna oscillator is shown in Figure 2.24. The upper RTD electrode was connected to the top antenna electrode along the back side of the slot through an air-bridge structure as shown in Figure 2.25a, while the RTD bottom contact was connected to the bottom antenna electrode on the front side of the slot, integrating

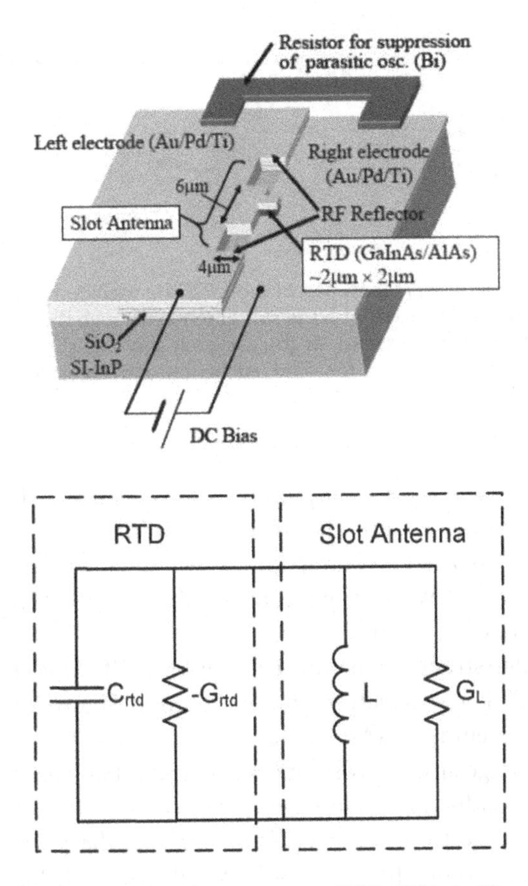

Figure 2.23 (a) A planar resistor-stabilised millimetre-wave and THz RTD oscillator integrated with a slot antenna load (reprinted from [134] with permission). (b) RF equivalent circuit of the oscillator in (a) (reprinted from [134] with permission).

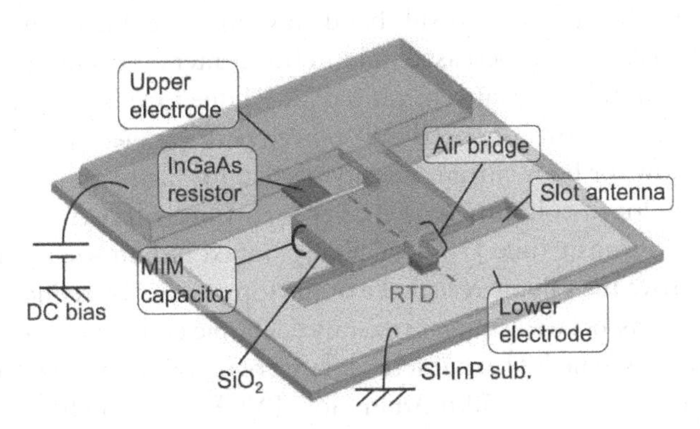

Figure 2.24 Schematic of the circuit layout employed in the realisation of the 1.92 THz RTD-based oscillator (reprinted from [122] with permission).

the RTD device with the antenna. The decoupling MIM capacitor was placed on the back side and it was fabricated through a thin SiO_2 layer sandwiched between the upper and lower Au-based antenna electrodes. The shunt resistor was realised from the highly doped GaInAs underlayer and connected between the upper and bottom electrodes of the MIM

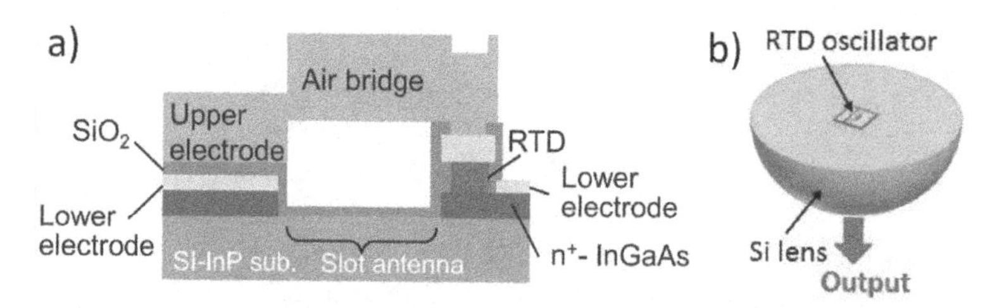

Figure 2.25 In (a), a schematic representation of the 1.92 THz oscillator cross section, showing the integration of the RTD device with the slot-antenna through the air-bridge structure (reprinted from [122] with permission). In (b), a sketch showing the oscillator chip mounted onto the hemispherical Si lens for field collection and focusing (reprinted from [235] with permission).

capacitor and RTD. The measured output power was 0.4 μW at 1.92 THz [122]. The RTD device had a mesa area of ~0.1 μm^2. By further optimising the antenna electrode thickness (3 μm) to reduce the associated conduction losses, fundamental oscillations at 1.98 THz with output power of 40 nW [117] were achieved. The 1.98 THz oscillator employed an RTD with a mesa size of ~0.2 μm^2.

The epitaxial wafer structure for the 1.92 and 1.98 THz RTD oscillator comprises a GaInAs/AlAs DBQW heterostructure and is shown in Figure 2.26. It features a gallium aluminium indium arsenide (GaAlInAs) graded emitter layer [237], which allows for a reduction of the voltage drop across the RTD when it is biased in its NDC region by moving the emitter conduction band edge closer to the well resonant level and therefore shifting the peak voltage to lower bias. This increases the DC-to-RF efficiency, reduces thermal heating and prevents the generation of high electric fields in the depletion regions [237]. An indium-rich (90%) and ~2.5 nm thick GaInAs quantum well was employed to allow further DC bias reduction by depressing the quantum well ground state sub-band. Employing a thin well layer reduces the electron transit time and increases the operation frequency [157], while the second sub-band stays in place due to the low well electronic effective mass caused by the increased indium (In) content. This increases both the current density ΔJ and peak-to-valley current ratio (PVCR) due to valley current reduction, which benefits the device in terms of speed and power performances [238]. Thin ~1 nm thick AlAs barriers allow for high-current density needed for high-speed operation [235], while a 12-nm-thick undoped collector spacer layer for a trade-off between geometrical capacitance and carriers transit time [161]. A highly doped indium-rich graded cap layer was designed to improve the Ohmic resistance of the top contact by reducing the Schottky barrier height and the associated contact resistance. Ohmic contacts were fabricated using the Ti/Pd/Au metal stack. The RTD characteristics were reported in [122] and are as follows: a high peak current density J_p = 50 mA/μm^2, low PVCR = 1.7, high ΔJ = 20.6 mA/μm^2 and ΔV = 0.4 V.

For the measurement of the oscillation frequency, the RTD oscillator die was mounted are on a Si hemispherical lens and an FTIR (Fourier-transform infrared) spectrometer with a liquid helium-cooled bolometer as a THz receiver was used. For the output power measurement, the radiated THz wave was focused by a parabolic mirror and THz lens and received by a power meter via a horn antenna.

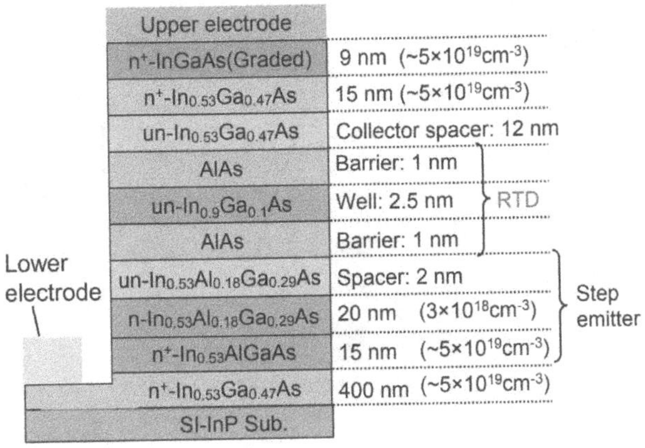

Figure 2.26 Schematic of the RTD epitaxial structure design employed in the realisation of the 1.92 THz [122] and 1.98 THz [117] oscillators (reprinted from [122] with permission).

2.3.4 Other THz RTD Oscillators

THz RTD oscillator concepts other than the aforementioned ones have also been reported in the literature. They include a 675 GHz differential double-RTD oscillator; a 165 Hz push-push [239] and triple-push [217,240] (1.52 THz) based on second and third harmonics, respectively.

A varactor diode can be integrated with the RTD to increase the frequency tunability range, for example, up to 40% [241]. This is a way to overcome the weak bias dependence of the device self-capacitance and so realise a voltage controlled oscillator (VCO).

For spectral narrowing, the use of phase locked loops (PLLs) with RTDs have also been demonstrated [125]. In this regard, free running very low-phase noise microwave RTD transmitter with –95 dBc/Hz @ 100 kHz and –114 dBc/Hz @1 MHz have been demonstrated in Ref. [242]. These meet the requirements for applications in wireless communications where very low-phase noise oscillators (< –90 dBc/Hz at 100 kHz offset and < –110 dBc/Hz at 1 MHz offset) are needed.

As discussed in the Chapter 3 ("Characterisation of Emitters and Detectors"), the coherence length of the 300 GHz RTD sources is around 100 m as determined from the measured spectrum. Therefore, RTD sources can be highly coherent. Also, from a broader view of the frequency peak, its lineshape or spectral profile, the RTDs sources may be judged to be low-noise emitters.

2.4 THz RTD detectors

2.4.1 Overview and General Working Principles

DBQW RTDs can be used to realise high-sensitivity THz detectors as well and can operate both as direct detectors [243,244] or coherent detectors [121,245]. In the first case, the RTD is typically biased at the peak current and can perform standard envelope detection of amplitude modulated signals according to the square-law scheme by exploiting the non-linearity associated with this region [245]. This is illustrated in Figure 2.27. It is worth to make it clear that, in DBQW RTDs, this non-linearity is not associated with thermionic

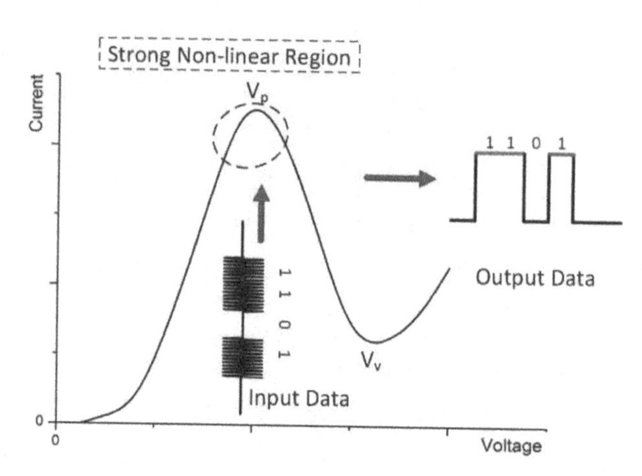

Figure 2.27 Operation principle of an RTD envelope detector for OOK demodulation.

emission but rather with thermal broadening of the quantum well sub-band associated with the resonant level. This makes the sensitivity of an RTD detector to be typically higher than of a Schottky barrier diode (SBD) even at room temperature [116,128].

In the second case, an RTD oscillator can operate as a coherent detector [246]. Here, the RTD is biased within its NDR region and the circuit acts as a local oscillator (LO). If the incoming carrier frequency is close enough to the LO one, injection locking [247,248] takes place and the two signals synchronise [249] performing coherent homodyne detection [245]. At the same time, the incoming signal is demodulated through the non-linear mixing properties of the NDR region [121], where the RTD acts as a self-oscillating mixer [246]. Since the RTD works, at the same time, as a LO and radio frequency (RF) mixer, this approach allows to realise ultra-compact and high-sensitivity receiver chips. The principle of operation of the RTD coherent detector is illustrated in Figure 2.28.

The coherent scheme typically provides better performance in comparison with direct detection. Indeed, envelope detectors are usually limited by Johnson–Nyquist noise [250], which lowers the signal-to-noise ratio (SNR). They rely on the incoming amplitude information of the signal only, which reduces spectral efficiency and sensitivity, limiting transmission distance or the corresponding bit-rate according to the specific link budget [246]. On the other hand, the coherent scheme allows to increase the sensitivity due to the RF gain provided by the device NDR and it is able to provide better spectral efficiency, where the phase, frequency and polarisation information of the received signal can be retrieved through injection locking. Despite all the positive attributes, the use of a coherent RTD receiver is still very new and appropriate design methodologies to guarantee injection locking and maximise sensitivity need to be developed.

For completeness, we mention here InP-based RTDs employing triple-barriers, which have been used in the design of highly sensitive zero-bias THz detectors [116,251–254]. Here, a high degree of freedom is achieved in the design of the asymmetry associated to the first non-linear PDR region. These triple barrier RTDs have also been designed for both high-power sources [255,256], typically featuring large PVCR due to thermionic current reduction but limited peak current density. The interested reader can refer to the provided reference, and references therein.

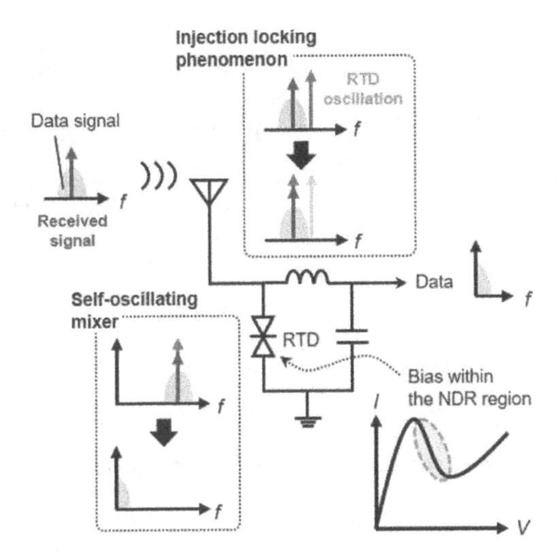

Figure 2.28 Operation principle of an RTD coherent detector (reprinted from [245] with permission).

2.5 RTD-based THz Wireless Communication

2.5.1 Overview

THz RTD wireless transmitters employ amplitude (direct) modulation schemes, such as amplitude shift keying (ASK) [121,257] and on-off keying (OOK) [129] and they transmit data over single-channel links, where the RF power is modulated through the DC bias line [258–260]. Recently, multi-channel wireless data transmission employing polarisation and frequency-division multiplexing (PDM/FDM) schemes have also been reported [261]. The links are mainly line of sight and so accurate alignment between the transmitter and receiver is important. Specifically, the antenna must be aligned in the correct way so that the polarisation of transmitted and received signal is the same at both antennas.

An RTD oscillator is usually employed as the transmitter. Both ASK and OOK modulation are applicable to the RTD transmitter depending on the bias level and the amplitude of data as illustrated in Figure 2.29. ASK modulation has been widely used because of the advantages of being a simple, low-cost, high-bandwidth and high-efficiency technique [262]. OOK as a special case of ASK represents the data modulation by switching on and off the carrier. For OOK modulation, the RTD is biased near the peak voltage (V_p) position for input NRZ (non-return to zero) data to switch on-off the oscillator while for ASK modulation, RTD device is biased in the middle of NDR region. In contrast of OOK, the data amplitude requirement for ASK is low within NDR region as illustrated in Figure 2.29.

2.5.2 Wireless System Architecture

The block diagram of the wireless system is illustrated in Figure 2.30. The transmitter (Tx) consists of a fundamental frequency voltage controlled RTD oscillator (RTD-VCO) and an external antenna or integrated on-chip antenna mounted on a hemispherical Si lens. The data is superimposed over DC bias through a bias tee. Power amplifiers are not available at THz and therefore none is employed in this stage. On the receiver (Rx) side,

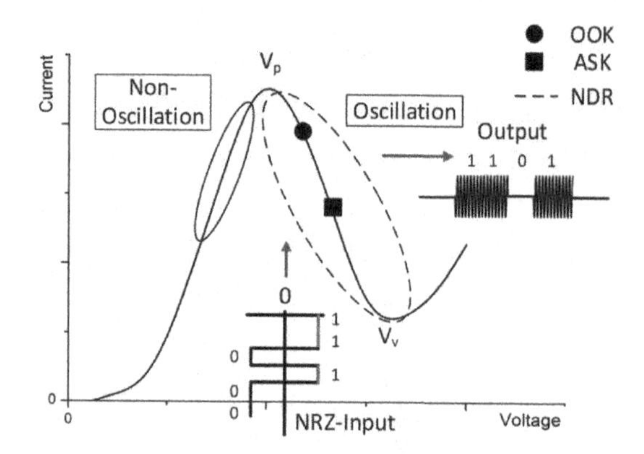

Figure. 2.29 Typical RTD device IV characteristics and illustration of ASK /OOK modulation [212].

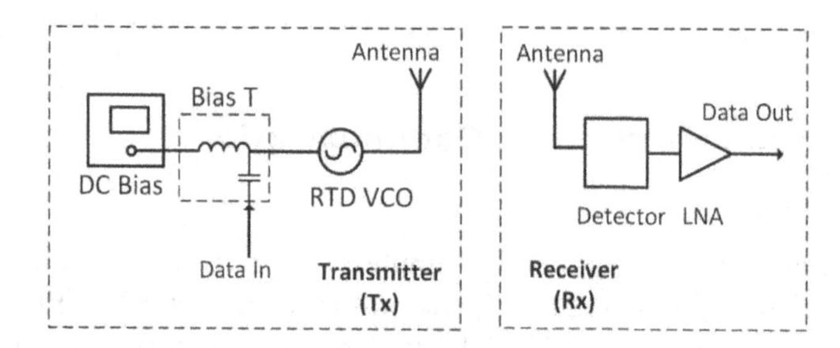

Figure 2.30 Block diagram of the wireless system architecture [212].

a receive antenna, which may be external or integrated, is used. The received signal is demodulated using either a square-law-based direct detection approach using a Schottky barrier diode (SBD) or RTD direct detector, or a coherent scheme where the RTD device acts as a self-oscillating mixer and performs homodyne detection. The received signal is amplified by an LNA.

Table 2.7 summarises some of the recent wireless communication link results reported in literature involving RTDs, either as a Tx or Rx or both. The longest link distances of 50 cm [212] and 150 cm [263] feature transmitter oscillators operating under 0.1 THz. For higher frequency transmitters, all link distances are under 30 cm. There is a clear correlation between the link distance and the transmitter power, in the mW range for the sub-100 GHz transmitters and μW range for the higher frequencies. Data rates from around 10 to 60 Gb/s have been demonstrated. An all RTD Tx and Rx enabled error-free 30 Gb/s transmission over a 7 cm link [121].

2.5.3 THz RTD Transmitters

Here, we describe some to the wireless transmission experiments in detail, the first in which a 490 GHz oscillator was used as the transmitter [257]. Direct modulation was carried out by superimposing a modulation signal onto the bias voltage through ASK. The experimental setup is shown in Figure 2.31, where a pulse pattern generator (PPG) was

Table 2.7 THz RTD wireless communications links

Ref.	*RTD Technology*	*Tx*	*Rx*	f_{osc} *[GHz]*	*Modulation*	*Distance [cm]*	*Date rate [Gb/s]*	*BER*
[257]	InP	RTD	SBD	490	ASK	20–30	22	Error free
[257]	InP	RTD	SBD	490	ASK	20–30	34	1.9×10^{-3}
[264]	InP	RTD	SBD	650	ASK	20–30	25	Error free
[264]	InP	RTD	SBD	650	ASK	20–30	44	5×10^{-4}
[263]	InP	RTD	SBD	62.5	OOK	30	10	Error free
[263]	InP	RTD	SBD	62.5	OOK	150	15	10^{-5}
[265]	InP	UTC-PD	RTD	297	OOK	3	17	Error free
[245]	InP	UTC-PD	RTD	322	OOK	2	27	Error free
[245]	InP	UTC-PD	RTD	322	OOK	2	32	2.7×10^{-2}
[243]	InP	UTC-PD	RTD	350	OOK	3	32	Error free
[243]	InP	UTC-PD	RTD	350	OOK	3	36	9×10^{-2}
[266]	InP	RTD	RTD	286	OOK	10	9	Error free
[266]	InP	RTD	RTD	286	OOK	10	12	4×10^{-3}
[129]	InP	RTD	RTD	345	OOK	7.5	13	Error free
[129]	InP	RTD	RTD	345	OOK	7.5	20	2.1×10^{-3}
[121]	InP	RTD	RTD	343	ASK	7	30	Error free
[121]	InP	RTD	RTD	343	ASK	7	56	1.39×10^{-5}
[214]	InP	RTD	SBD	500	ASK	20	56	2.3×10^{-4}
[244]	InP	UTC-PD	RTD	324	16-QAM	1	60	2×10^{-3}
[212]	InP	RTD	SBD	84	ASK	50	15	4.1×10^{-3}
[119]	InP	RTD	SBD	260	ASK	1	13	-

Note: They all use a RTD either as a transmitter (Tx), receiver (Rx) or both as a Tx and Rx. Some links employ the UTC-PD as the Tx.

used to impress digital data over the carrier. The modulated THz signal was then received through a horn antenna, demodulated by an SBD direct detector and amplified through an LNA. The link distance between the Tx and the Rx was set to 20–30 cm. The demodulated signal was measured through an error detector and an oscilloscope. The results are shown in Figure 2.32. Clear eye opening and error-free transmission were obtained at a data rate of 22 Gb/s, and data rates of up to 34 Gb/s with correctable bit error rate (BER) of 1.9×10^{-3}. The rapid increase in BER from 22 to 34 Gb/s was attributed to the limitations associated with the system direct modulation bandwidth, which was determined by the external components of the circuit around the RTD. In particular, the MIM shunt capacitor (~0.1 pF) and the parasitic inductance of bonding ribbons used to package the oscillator, which limited the modulation bandwidth to 15 GHz.

For multi-channel data transmission, ASK and PDM/FDM schemes have been investigated [261]. Figure 2.33 shows a schematic representation of the Tx chip. Four RTD oscillators were integrated together: two of them oscillated at 500 GHz and the other two at 800 GHz for FDM, while the two oscillators at each of those two frequencies had polarisations that were orthogonal to each other realised through the orientations of their antennas layouts which were perpendicular to each other. The RTDs epitaxial structure and circuit design (employing offset-fed slot antennas) were similar to the in

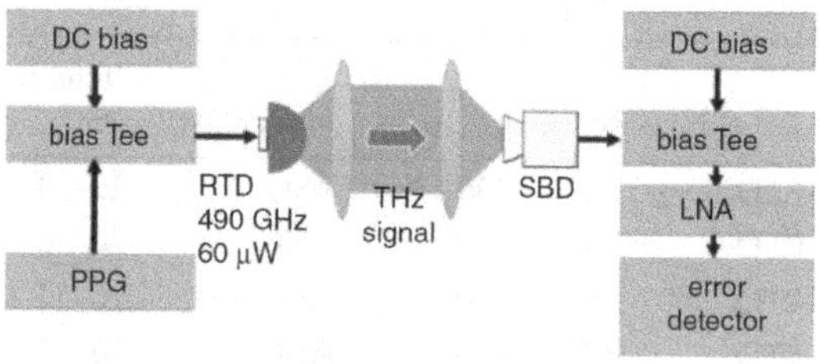

Figure 2.31 Block diagram of the THz wireless data transmission experimental setup employed in Ref. [257] (reprinted from [257] with permission).

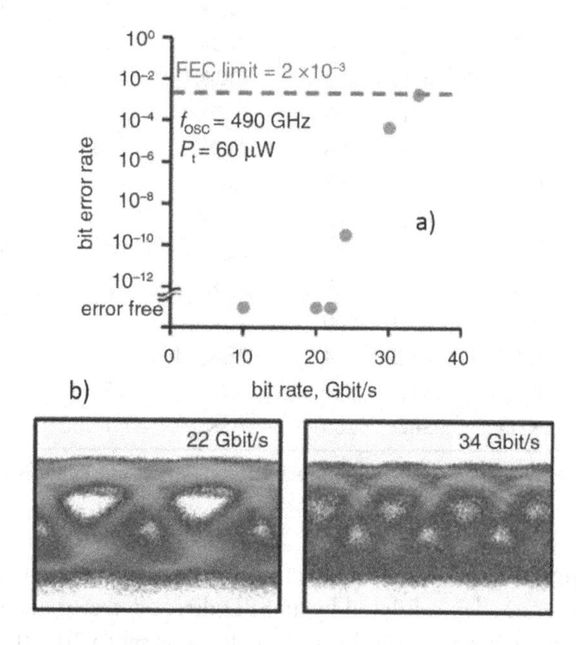

Figure 2.32 (a) The measured bit error rate (BER) as a function of the bit-rate, showing error-free (EF) data transmission up to 22 Gb/s and bit rates of up to 34 Gb/s with BER below the FEC limit. (b) The corresponding measured eye-diagrams (reprinted from [257] with permission).

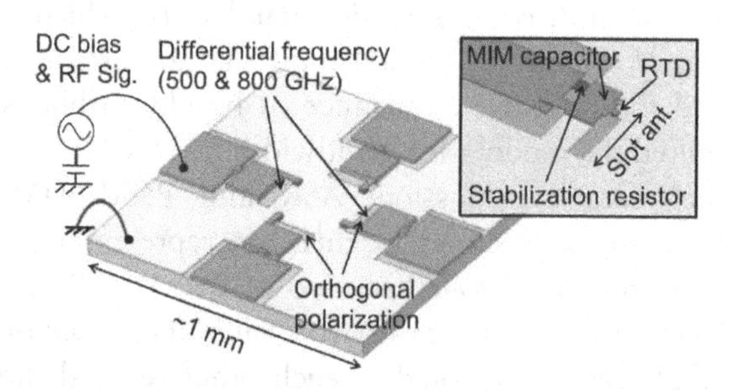

Figure 2.33 Schematic representation of the oscillator source circuit layout employed at the Tx side for wireless data transmission through FDM and PDM (reprinted from [261] with permission).

Figure 2.26. Data rates of up to 56 Gbps (28 Gbps per channel) were obtained using FDM in the 500 and 800 GHz channels with BERs of 2.3×10^{-4} and 1.5×10^{-3}, respectively, and through PDM at 500 GHz with BERs of 1.5×10^{-3} and 1.4×10^{-4} for the vertical and horizontal polarisation channels, respectively.

2.5.4 THz RTD Receivers

As already described, THz RTD receivers can be based on both the direct and coherent detection approach. An example of an RTD-based 300 GHz-band receiver (Rx) was reported in [243]. The detector was integrated with a Si-based photonic crystal waveguide platform [267] and had a measured responsivity around 4 kV/W. Figure 2.34a shows a photograph of the Rx module showing the metal-based tapered-slot mode converter and the photonic crystal waveguide. This approach has the advantage of integrating micron-scale electronic devices and photonic components on the same planar substrate, and therefore omitting bulky optics and realising ultra-high compact THz systems needed for practical applications.

Figure 2.34b shows the wireless experimental setup. The transmitter was a UTC-PD based source providing a carrier frequency of 350 GHz modulated through OOK. Figure 2.35 shows the experimental results, where error-free data transmission up to 32 Gb/s and data rates of up to 36 Gb/s with BER of 9×10^{-2}. Using a similar setup with a 16

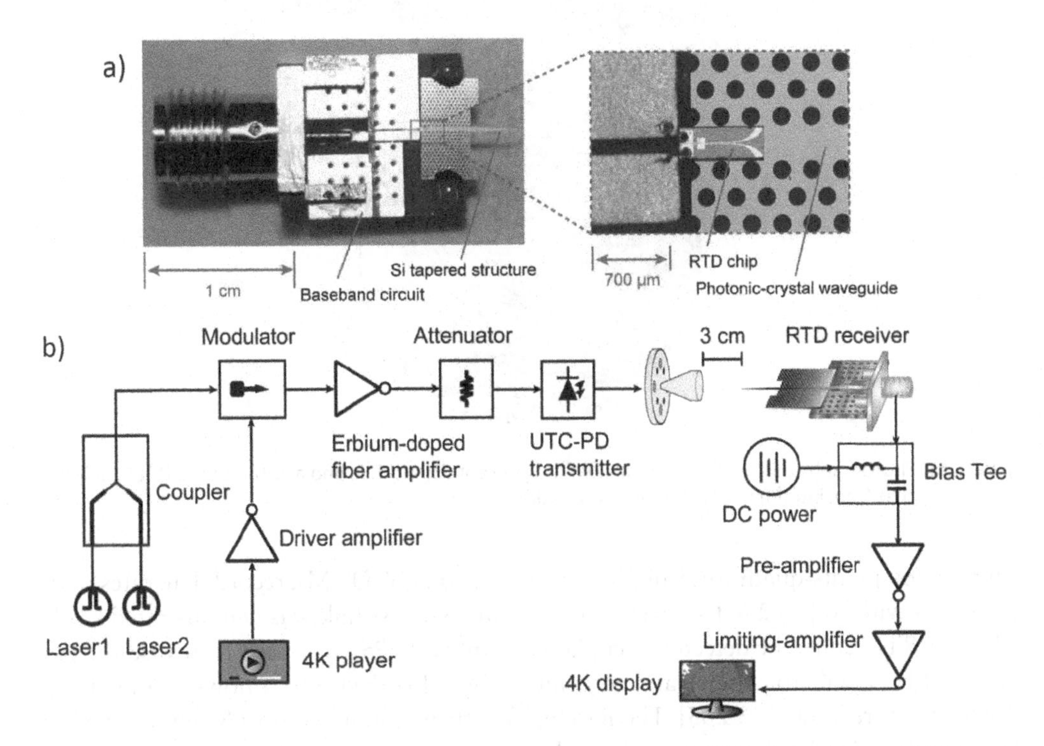

Figure 2.34 The 300 GHz-band wireless communication link employing a RTD direct detector. (a) A photograph of the RTD-based Rx module with a zoom in of the RTD integrated with a Si-based photonic crystal waveguide through a tapered-slot mode coupler. (b) The wireless experimental setup used to transmit a high-resolution 4k video employing a UTC-PD Tx and the RTD direct Rx (reprinted from [243] with permission).

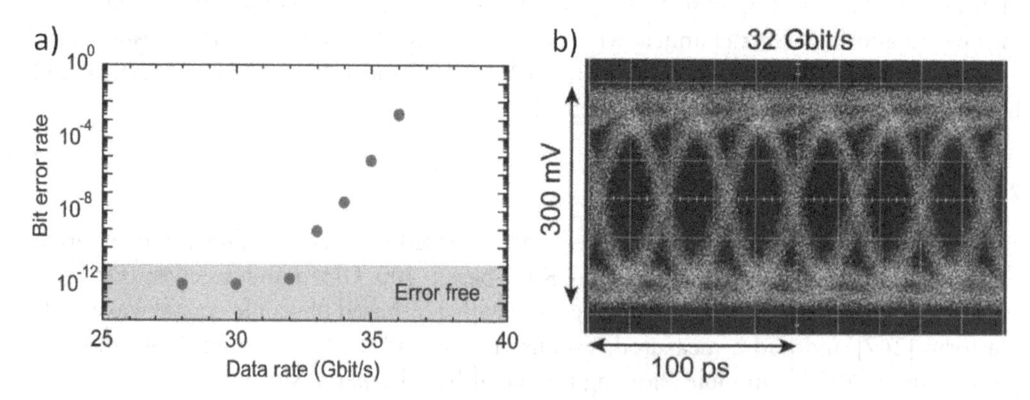

Figure 2.35 (a) The BER as a function of the data-rate. (b) The eye diagram at 32 Gb/s. Error free communication and a clear eye opening were achieved up to 32 Gb/s, while bit rates of up to 36 Gb/s were obtained with BER < FEC limit (reprinted from [243] with permission).

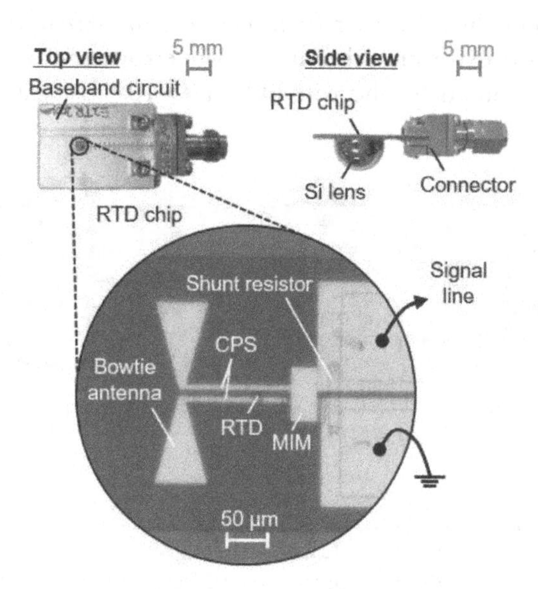

Figure 2.36 Photographs of the 300 GHz-band RTD coherent Rx module and a zoom of the RTD oscillator chip (reprinted from [245] with permission).

constellation points-quadrature amplitude modulation (16-QAM), record date rates of up to 60 Gb/s with BER = 2×10^{-3} over a 10 mm long wireless link was demonstrated [244].

For an RTD coherent detector, a coplanar stripline (CPS) was used to realise the resonator of the oscillator circuit and to connect the RTD device to a bow-tie antenna as shown in Figure 2.36 [243,265]. The Rx chip was then mounted onto a Si lens to increase the antenna gain. The experimental wireless communication setup was similar to that shown in Figure 2.34b. The link distance was set to 2 cm, while an oscilloscope and an error detector were used to measure the demodulated signal. Error-free single-channel data transmission of up to 27 Gb/s and data rates of up to 32 Gb/s with BER of ~10^{-6} were reported [245], with a sensitivity enhancement of around 10 dB with respect to the direct detection counterpart.

2.5.5 *All-RTD THz Transceivers*

All-RTD-based THz transceivers were reported in Refs. [121,129,266,268]. In these systems, a RTD was employed at both the Tx and Rx sides by exploiting both the direct [129,266,268] as well as the coherent detection scheme [121]. In the setup using the coherent detector described in Ref. [121], error-free ASK data transmission up to 30 Gb/s and data-rates of up to 56 Gb/s with BER of 1.39×10^{-5} were reported in the 300 GHz-band over a link distance of 7 cm. A sensitivity enhancement of up to 40 dB with respect to the direct detection approach was observed. The measured BER versus data rate is shown in Figure 2.37.

Using a direct RTD detectors, error free single-channel OOK wireless data transmission up to 9 Gb/s and data rates of up to 12 Gb/s with BER of $\sim 10^{-3}$ was reported in the 300 GHz-band over a distance of 10 cm [266]. Using this setup, real-time error free transmission of an uncompressed high-resolution 4k video was demonstrated at a data rate of 6 Gb/s.

In the setup described in Ref. [268], data in the optical domain was converted into an electrical signal and used to modulate the RTD. In this work, OOK data was modulated using an intensity modulator and after passing through a 1 km long fibre, it was converted to an electrical signal using a photodiode. The output was added to a DC biasing voltage using a bias-T and the resultant signal modulated the RTD. Using multi-chip code division multiple access (CDMA), data rates of up to 13 Gb/s error-free and of up to 20 Gb/s with BER of 2.1×10^{-3} are possible to be transmitted.

2.6 Challenges and Future Perspectives

The development of THz transmitter and receiver systems based on compact solid-state electronic devices is urgently required to meet the data rate and bandwidth requirements of 5G and futures systems. Figure 2.38 shows the state-of-the-art THz wireless

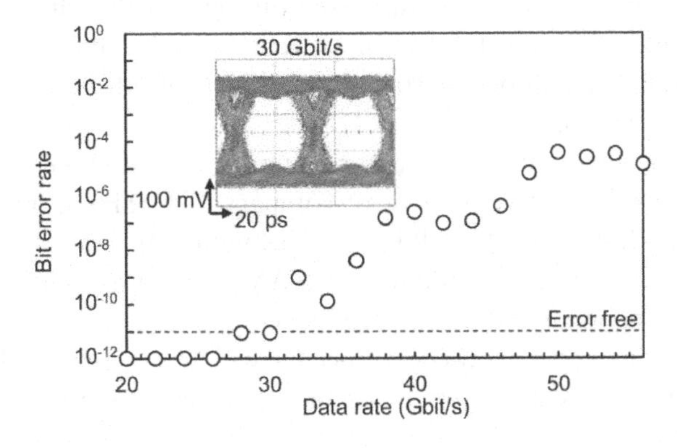

Figure 2.37 Measured BER versus data-rate for the wireless data transmission experiment employing an all-RTD TRx and a direct detection approach. A clear eye opening and EF transmission up to 30 Gbps were reported, while bit rates of up to 56 Gb/s were achieved with BER below the FEC limit (reprinted from [121] with permission).

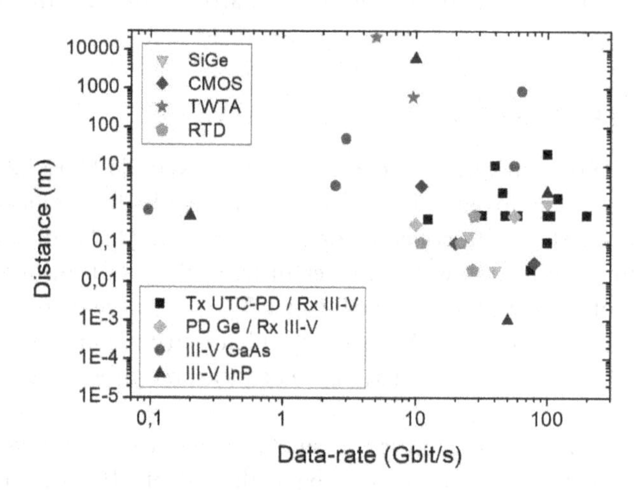

Figure 2.38 **The state-of-the-art THz wireless technologies with data rates of up to 100 Gb/s. SiGe: silicon germanium, CMOS: Complementary metal-oxide-semiconductor (silicon), TWTA: Travelling wave tube amplifier, RTD: Resonant tunnelling diode.**

technologies with data rates of up to 100 Gb/s. This has been achieved by several nodes (III–V, Si, SiGe, ...); however, the link distance is still very limited to under 10 m, with the longer ranges largely enabled by extremely high-gain antennas (>50 dBi) and coherent detectors. It is clear that we must increase the link distance to realise practical THz wireless communications [7,269]. This means that we must significantly improve the performance of the compact semiconductor solutions. In addition, most of these systems use complex signalling (QPSK, QAM) so far. While featuring high spectral density, signal processing and synchronisation is mandatory, which makes it further challenging for QAM-16 and higher, and therefore the designed systems are energy-hungry, especially at the receiver side.

RTD technology seems to offer the simplest and highest performance technology option for THz transceivers. The main weakness of the RTD technology is represented by the low RF output power of the sources. As discussed throughout this chapter, the main reasons for this include underdeveloped device and circuit design techniques, including epitaxial designs, circuit implementation approaches, and lack of effective design techniques for arrays of RTD oscillators.

Current research trends suggest that increasing the output power of THz RTDs is feasible and would enable to development of compact RTD THz ultra-high-speed wireless transceivers [270]. Individual THz RTD sources with mW output powers would facilitate the use of pulse amplitude modulation (PAM) signalling. Such an approach would offer a way to dramatically reduce the receiver complexity, as no carrier synchronisation is required. In that case, the total energy per bit will be far improved and will make the system more energy efficient. We are aware of unpublished but promising research efforts in this direction.

Even though arrays of RTD oscillators have been reported, they have fallen short of delivering the expected high output power. Output powers of up to 0.61 mW at 620

GHz, 0.27 mW at 770 GHz and 0.18 mW at 810 GHz were reported by employing a two-element frequency locked oscillator array for CW coherent emission [214]. Most realisation are, however, unsynchronised, for instance, a 16-element and a 64-element arrays provided output powers of up to 28 μW at fundamental frequencies of 290 GHz and up to 650 GHz, respectively [271–273]. Here, each array element, consisting in a single oscillator, was coupled to the other elements and radiated independently. Power combining was based on the spatial interference of the corresponding radiation pattern. Earlier efforts in this regard included quasi-optical resonators for oscillators stabilisation [274–276] and for power combining [277]. And recently, pulsed emission with RF powers of up to 0.73 mW at around 1 THz was reported by employing an unsynchronised 89-element large-scale array [215], but output powers have remained below the 1 mW level.

It remains a challenge to achieve mutual coupling with large-scale arrays. Spatial power combining in any desired direction is difficulty to realise by acting on the array geometry without compromising mutual locking between the oscillator elements [207]. Thus, new approaches for this challenge are required.

At the device level, accurate characterisation and modelling of the RTD, especially of the key NDR region, is still non-trivial due to device instability. It would be advantageous to develop robust characterisation techniques that enable the development of a complete non-linear model [201]. A calibrated physics based numerical modelling approach may also be useful in this regard. The availability of such a model would enable the full non-linear dynamic analysis of the device in its entire operation frequency range [116].

And at the circuit level, despite the low DC power consumption (typically of the order of tens/few hundreds of mW [120,218], with values as low as of hundreds of μW have also been reported [278]), which makes RTDs low-power devices, the DC-to-RF conversion efficiency of RTD oscillators is typically below 1% [120,218,278]. Recently, however, improved circuit implementation has led to high efficiencies of up to 10% for mm-wave RTD oscillators [279]. This and similar approaches need to be adopted for THz RTD oscillators.

Clearly, a seamless interface between wireless transceivers and the optical fibre backbone is important. In this regard, the idea of radio-over-fibre optoelectronic transducers based on RTDs has also been considered [280–285], with, for instance, optical direct intensity modulation of a RTD photodetector optoelectronic oscillator at 79 GHz has been reported [286]. Also, as noted in the previous section, such an interface via a high-speed photodiode for conversion of the optical signal into an electrical one, has also been demonstrated [268]. This approach used discrete components but a monolithic integration of the RTD Tx and photodiode seems feasible.

A further challenge to the RTD technology is created by the fact that it is realised on high permittivity InP substrate. On-chip antennas have low gain, typically under 6 dBi, and the radiation is directed into the substrate as dictated by the physics of the system. Therefore, the semiconductor dies are usually mounted on hemispherical lenses to collimate and focus the radiation, but these are bulky and this is a cumbersome procedure. As such, efforts to design sources with airside or upward radiation from the chips are underway. In this regard, a radial line slot antenna array for upward emission has recently been reported [224], though further innovations to this challenge are needed.

There is also the requirement on good alignment between Tx and Rx for THz wireless links which is not easy to meet. To simplify this task, RTD oscillators with circularly polarised radiation are being developed. The circularly polarised radiation from the radial slot antenna array [224] is one such an option. This approach would benefit from higher power sources than currently available.

The coherent scheme, moreover, has the potential for multi-channel division multiplexing, longer transmission distances and higher bit rates [246]. Moreover, duplex capability is, in principle, feasible upon specific circuit design since the same oscillator is employed at both the transmitter and receiver front ends [246].

Even though today THz RTDs are dominated by those based on InP, the future may belong to antimonide based RTDs. Beyond about 300 GHz, the advantages of semi-insulating high-permittivity substrates such InP become less important and often we need to isolate the substrate from the THz circuitry [120,217]. In earlier efforts, antimonide RTDs were grown on either GaAs or GaSb substrates and so suffered from poor material quality due to the lattice mismatch between the substrates and the device epitaxial layers. At THz, there is no apparent need for such substrates since we can electrically isolate them from the passive THz circuitry and so growth on the conductive but lattice matched InAs substrates of high-quality InAs/AlSb RTDs can be undertaken. These RTDs may underpin future THz transceivers.

Besides wireless communications, it is important to mention that THz RTD oscillators are also being developed for other applications such as compact imaging systems [287], spectroscopy [288], radar [289–295,296,297] and even high-speed brain-inspired optical memories and neuromorphic systems [298]. It would therefore seem that it is just a question of time before we see RTD technology embedded in practical applications.

Acknowledgements

The authors would like to thank members of the High Frequency Electronics Group at the James Watt School of Engineering, University of Glasgow, both past and present, including Dr Abdullah Al-Khalidi, Dr Jue Wang, Dr Liquan Wang, Dr Khalid Alharbi, Dr Andrei Cornescu and Dr Razvan Morariu for their contributions to some of the results and perspectives presented in this chapter. They would also like to thank Professors Safumi Suzuki and Masahiro Asada of Tokyo University of Technology and Professors Tadao Nagatsuma, Masayuki Fujita and Julian Webber of Osaka University for providing details and insight to their RTD work. In addition, thanks are also due to partners on European Commission Horizon 2020 projects in which a significant part of our RTD has been developed, including the iBROW, TERAPOD and TeraApps projects.

Appendix A

Table 2.8 Epitaxial structure details of THz RTDs – Part 1

Ref.	*Emitter*	*Spacer*	*Barrier*	*Well*	*Barrier*	*Spacer*	*Collector*
[213]	$Ga_{0.47}In_{0.53}As$ 400 nm 2×10^{19} cm^{-3} $Ga_{0.47}In_{0.53}As$ 20 nm 3×10^{18} cm^{-3} $Ga_{0.49}In_{0.51}As$ 2.5 nm 3×10^{18} cm^{-3} $Ga_{0.51}In_{0.49}As$ 2.5 nm 3×10^{18} cm^{-3}	$Ga_{0.47}In_{0.53}As$ 2 nm undoped	AlAs 1.2 nm undoped	$Ga_{0.2}In_{0.8}As$ 4.5 nm undoped	AlAs 1.2 nm undoped	$Ga_{0.47}In_{0.53}As$ 25 nm undoped	$Ga_{0.47}In_{0.53}As$ 15 nm 2×10^{19} cm^{-3} $Ga_{0.3}In_{0.7}As$ 8 nm 2×10^{19} cm^{-3}
[122]	$Ga_{0.47}In_{0.53}As$ 400 nm 5×10^{19} cm^{-3} $GaIn_{0.53}AlAs$ 15 nm (graded) 5×10^{19} cm^{-3} $Ga_{0.29}In_{0.53}Al_{0.18}As$ 20 nm 3×10^{18} cm^{-3}	$Ga_{0.29}In_{0.53}Al_{0.18}As$ 2 nm undoped	AlAs 1 nm undoped	$Ga_{0.1}In_{0.9}As$ 2.5 nm undoped	AlAs 1 nm undoped	$Ga_{0.47}In_{0.53}As$ 12 nm undoped	$Ga_{0.47}In_{0.53}As$ 15 nm 5×10^{19} cm^{-3} GaInAs 9 nm (graded) 5×10^{19} cm^{-3}
[117]	$Ga_{0.47}In_{0.53}As$ 400 nm 5×10^{19} cm^{-3} $GaIn_{0.53}AlAs$ 15 nm (graded) 5×10^{19} cm^{-3} $Ga_{0.29}In_{0.53}Al_{0.18}As$ 20 nm 3×10^{18} cm^{-3}	$Ga_{0.29}In_{0.53}Al_{0.18}As$ 2 nm undoped	AlAs 1nm undoped	$Ga_{0.1}In_{0.9}As$ 2.5 nm undoped	AlAs 1 nm undoped	$Ga_{0.47}In_{0.53}As$ 12 nm undoped	$Ga_{0.47}In_{0.53}As$ 15 nm 5×10^{19} cm^{-3} GaInAs 9 nm (graded) 5×10^{19} cm^{-3}
[214]	$Ga_{0.47}In_{0.53}As$ 400 nm 2×10^{19} cm^{-3} $Ga_{0.47}In_{0.53}As$ 20 nm 3×10^{18} cm^{-3} $Ga_{0.49}In_{0.51}As$ 2.5 nm 3×10^{18} cm^{-3} $Ga_{0.51}In_{0.49}As$ 2.5 nm 3×10^{18} cm^{-3}	$Ga_{0.47}In_{0.53}As$ 2 nm undoped	AlAs 1.2 nm undoped	$Ga_{0.2}In_{0.8}As$ 4.5 nm undoped	AlAs 1.2 nm undoped	$Ga_{0.47}In_{0.53}As$ 25 nm undoped	$Ga_{0.47}In_{0.53}As$ 15 nm 2×10^{19} cm^{-3} $Ga_{0.3}In_{0.7}As$ 8 nm 2×10^{19} cm^{-3}

(*continued*)

Table 2.8 Cont.

Ref.	*Emitter*	*Spacer*	*Barrier*	*Well*	*Barrier*	*Spacer*	*Collector*
[215]	$Ga_{0.47}In_{0.53}As$ 400 nm 2×10^{19} cm^{-3}	$Ga_{0.47}In_{0.53}As$ 2 nm undoped	AlAs 1.2 nm undoped	$Ga_{0.2}In_{0.8}As$ 4.5 nm undoped	AlAs 1.2 nm undoped	$Ga_{0.47}In_{0.53}As$ 25 nm undoped	$Ga_{0.47}In_{0.53}As$ 15 nm 2×10^{19} cm^{-3}
	$Ga_{0.47}In_{0.53}As$ 20 nm 3×10^{18} cm^{-3}						$Ga_{0.3}In_{0.7}As$ 8 nm 2×10^{19} cm^{-3}
	$Ga_{0.49}In_{0.51}As$ 2.5 nm 3×10^{18} cm^{-3}						
	$Ga_{0.51}In_{0.49}As$ 2.5 nm 3×10^{18} cm^{-3}						

Note: The emitter/collector/spacer layers doping is *n*-type everywhere and made with silicon (Si) donors.

Table 2.9 Epitaxial structure details of THz RTDs – Part 2

Ref.	*Emitter*	*Spacer*	*Barrier*	*Well*	*Barrier*	*Spacer*	*Collector*
[120]	$Ga_{0.47}In_{0.53}As$ 400 nm 3×10^{19} cm^{-3}	$Ga_{0.47}In_{0.53}As$ 2.5 nm 2×10^{16} cm^{-3}	AlAs 1.4 nm undoped	$Ga_{0.47}In_{0.53}As$ 4.5 nm undoped	AlAs 1.4 nm undoped	$Ga_{0.47}In_{0.53}As$ 1.5nm undoped	$Ga_{0.47}In_{0.53}As$ 160 nm 2×10^{18} cm^{-3}
	Al0.48In0.52As 10 nm 3×10^{19} cm^{-3}	$Ga_{0.47}In_{0.53}As$ 1.5 nm undoped				$Ga_{0.47}In_{0.53}As$ 25 nm 2×10^{16} cm^{-3}	$Ga_{0.47}In_{0.53}As$ 40 nm 3×10^{19} cm^{-3}
[212]	$Ga_{0.47}In_{0.53}As$ 400 nm 3×10^{19} cm^{-3}	$Ga_{0.47}In_{0.53}As$ 2.6 nm 2×10^{16} cm^{-3}	AlAs 1.4 nm undoped	$Ga_{0.47}In_{0.53}As$ 4.5 nm undoped	AlAs 1.4 nm undoped	$Ga_{0.47}In_{0.53}As$ 1.5nm undoped	$Ga_{0.47}In_{0.53}As$ 160 nm 2×10^{18} cm^{-3}
	$Al_{0.48}In_{0.52}As$ 10 nm 3×10^{19} cm^{-3}	Ga0.47In 0.53As 1.5 nm undoped				$Ga_{0.47}In_{0.53}As$ 25nm 2×10^{16} cm^{-3}	$Ga_{0.47}In_{0.53}As$ 40 nm 3×10^{19} cm^{-3}
[216]	$Ga_{0.47}In_{0.53}As$ – 4×10^{18} cm^{-3}	$Ga_{0.47}In_{0.53}As$ 3 nm undoped	AlAs 1.4 nm undoped	$Ga_{0.47}In_{0.53}As$ 1.9 nm undoped	AlAs 1.4 nm undoped	$Ga_{0.47}In_{0.53}As$ 3 nm undoped	$Ga_{0.4}In_{0.6}As$* 30 nm 6×10^{19} cm^{-3}
				InAs 2 nm undoped		$Ga_{0.47}In_{0.53}As$ 100 nm 2×10^{17} cm^{-3}	
				$Ga_{0.47}In_{0.53}As$ 1.9 nm undoped			
[217]	GaInAs 500 nm –	GaInAs 1 nm undoped	AlAs 1.2 nm undoped	GaInAs 1 nm undoped	AlAs 1.2 nm undoped	GaInAs 1 nm undoped	GaInAs 20 nm –
	GaInAs 60 nm –			InAs 1 nm undoped			GaInAs 20 nm –

Table 2.9 Cont.

Ref.	*Emitter*	*Spacer*	*Barrier*	*Well*	*Barrier*	*Spacer*	*Collector*
				GaInAs 1 nm undoped			
[218]	GaInAs 500 nm –	GaInAs 1 nm undoped	AlAs 1.2 nm undoped	GaInAs 1 nm undoped	AlAs 1.2 nm undoped	GaInAs 1 nm undoped	GaInAs 20 nm –
	GaInAs 60 nm –			InAs 1 nm undoped			GaInAs 20 nm –
				GaInAs 1 nm undoped			
[219]	GaN 100 nm 1×10^{19} cm^{-3}	GaN 10 nm undoped	AlN 2 nm undoped	GaN 1.75 nm undoped	AlN 2 nm undoped	GaN 6 nm undoped	GaN 100 nm 1×10^{19} cm^{-3}
[194]	GaAs 200 nm (graded) 2×10^{18} cm^{-3}	GaAs 100 nm 2×10^{17} cm^{-3}	AlAs 1.1 nm undoped	GaAs 4.5 nm undoped	AlAs 1.1 nm undoped	GaAs 6 nm undoped	GaAs 50 nm (graded) 2×10^{18} cm^{-3}
[166]	InAs 1 μm 5×10^{18} cm^{-3}	InAs 75 nm 2×10^{17} cm^{-3}	AlSb 1.5 nm undoped	InAs 6.4 nm undoped	AlSb 1.5 nm undoped	InAs 20 nm 2×10^{16} cm^{-3}	InAs 100 nm 2×10^{18} cm^{-3}
	InAs 200 nm 2×10^{18} cm^{-3}						

Note: The emitter/collector/spacer layers doping is *n*-type everywhere and made with silicon (Si) donors (*tin (Sn) atoms were used for the collector layer in Ref. [216].

References

[1] M. Shafi, A. F. Molisch, P. J. Smith, T. Haustein, P. Zhu, P. De Silva, F. Tufvesson, A. Benjebbour and G. Wunder, "5G: A Tutorial Overview of Standards, Trials, Challenges, Deployment, and Practice", *IEEE Journal on Selected Areas in Communications*, vol. 35, no. 6, pp. 1201–1221, 2019.

[2] T. Kürner and S. Priebe, "Towards THz Communications – Status in Research, Standardization and Regulation", *Journal of Infrared, Millimeter, and Terahertz Waves*, vol. 35, no. 1, pp. 53–62, 2014.

[3] Z. Chen, X. Ma, B. Zhang, Y. Zhang, Z. Niu, N. Kuang, W. Chen, L. Li and S. Li, "A Survey on Terahertz Communications", *China Communications*, vol. 16, no. 2, pp. 1–35, 2019.

[4] L. Atzori, A. Iera and G. Morabito, "The Internet of Things: A Survey", *Computer Networks*, vol. 54, no. 15, pp. 2787–2805, 2010.

[5] A. Al-Fuqaha, M. Guizani, M. Mohammadi, M. Aledhari and M. Ayyash, "Internet of Things: A Survey on Enabling Technologies, Protocols, and Applications", *IEEE Communications Surveys & Tutorials*, vol. 17, no. 4, pp. 2347–2376, 2015.

[6] S. Cherry, "Edholm's Law of Bandwidth", *IEEE Spectrum*, vol. 41, no. 7, pp. 58–60, 2004.

[7] M. Koch, "Terahertz Communications: A 2020 vision, in Terahertz Frequency Detection and Identification of Materials and Objects", chapter 4, *Springer*, pp. 325–338, 2007.

[8] S. Dang, O. Amin, B. Shihada and M. -S. Alouini, "What Should 6G Be?", *Nature Electronics*, vol. 3, no. 1, pp. 20–29, 2020.

[9] P. Yang, Y. Xiao, M. Xiao and S. Li, "6G Wireless Communications: Vision and Potential Techniques", *IEEE Network*, vol. 33, no. 4, pp. 70–75, 2019.
[10] M. Latva-aho and K. Leppänen (eds.), "Key Drivers and Research Challenges for 6G Ubiquitous Wireless Intelligence (White Paper)", 6G Flagship, University of Oulu, Finland.
[11] P. Smulders, "The Road to 100 Gb/s Wireless and Beyond: Basic Issues and Key Directions", *IEEE Communications Magazine*, vol. 51, no. 12, pp. 86–91, 2013.
[12] J. Federici and L. Moeller, "Review of Terahertz and Subterahertz Wireless Communications", *Journal of Applied Physics*, vol. 107, no. 11, pp. 111101, 2010.
[13] Y. Li and G. L. Stüber, "Orthogonal Frequency Division Multiplexing for Wireless Communications", *Springer*, 2006.
[14] F. Rusek, D. Persson, B. K. Lau, E. G. Larsson, T. L. Marzetta, O. Edfors and F. Tufvesson, "Scaling Up MIMO: Opportunities and Challenges with Very Large Arrays", *IEEE Signal Processing Magazine*, vol. 30, no. 1, pp. 40–60, 2013.
[15] S. Yang and L. Hanzo, "Fifty Years of MIMO Detection: The Road to Large-Scale MIMOs", *IEEE Communications Surveys & Tutorials*, vol. 17, no. 4, pp. 1941–1988, 2015.
[16] A. J. Paulraj, D. A. Gore, R. U. Nabar and H. Bolcsket, "An Overview of MIMO Communications – A Key to Gigabit Wireless", *Proceedings of the IEEE*, vol. 92, no. 2, pp. 198–218, 2004.
[17] U. Pfeiffer, "Integrated Circuit Design for Terahertz Applications", 6G Wireless Summit, 2020.
[18] T. Omiya, M. Yoshida and M. Nakazawa, "400 Gbit/s 256 QAM-OFDM Transmission over 720 km with a 14 bit/s/Hz Spectral Efficiency by Using High-Resolution FDE", *Optics Express*, vol. 21, no. 3, pp. 2632–2641, 2013.
[19] "IEEE Standard Letter Designations for Radar-Frequency Bands", *IEEE Std 521–1984*, pp. 1–8, 1984.
[20] T. Baykas, C. -S. Sum, Z. Lan, J. Wang, M. A. Rahman, H. Harada and S. Kato, "IEEE 802.15.3c: The First IEEE Wireless Standard for Data Rates over 1 Gb/s", *IEEE Communications Magazine*, vol. 49, no. 7, pp. 114–121, 2011.
[21] T. S. Rappaport, J. N. Murdock and F. Gutierrez, "State of the Art in 60-GHz Integrated Circuits and Systems for Wireless Communications", *Proceedings of the IEEE*, vol. 99, no. 8, pp. 1390–1436, 2011.
[22] "Fact Sheet: Spectrum Frontiers Rules Identify, Open up Vast amounts of New High-Band Spectrum for Next Generation (5G) Wireless Broadband", transition.fcc.gov, 2019.
[23] F. J. O'Hara, S. Ekin, W. Choi and I. Song, "A Perspective on Terahertz Next-Generation Wireless Communications", *Technologies*, vol. 7, no. 2, 2019.
[24] Y. Wang, J. Li, L. Huang, Y. Jing, A. Georgakopoulos and P. Demestichas, "5G Mobile: Spectrum Broadening to Higher-Frequency Bands to Support High Data Rates", *IEEE Vehicular Technology Magazine*, vol. 9, no. 3, pp. 39–46, 2014.
[25] I. Hosako, N. Sekine, M. Patrashin, S. Saito, K. Fukunaga, Y. Kasai, P. Baron, T. Seta, J. Mendrok, S. Ochiai and H. Yasuda, "At the Dawn of a New Era in Terahertz Technology", *Proceedings of the IEEE*, vol. 95, no. 8, pp. 1611–1623, 2007.
[26] M. Hangyo, "Development and Future Prospects of Terahertz Technology", *Japanese Journal of Applied Physics*, vol. 54, no. 12, pp. 120101, 2015.
[27] H. -J. Song and T. Nagatsuma, "Present and Future of Terahertz Communications", *IEEE Transactions on Terahertz Science and Technology*, vol. 1, no. 1, pp. 256–263, 2011.
[28] C. E. Shannon, "Communication in the Presence of Noise", *Proceedings of the IRE*, vol. 37, no. 1, pp. 10–21, 1949.
[29] T. Nagatsuma, "Terahertz Technologies: Present and Future", *IEICE Electronics Express*, vol. 8, no. 14, pp. 1127–1142, 2011.
[30] T. Kleine-Ostmann and T. Nagatsuma, "A Review on Terahertz Communications Research", *Journal of Infrared, Millimeter, and Terahertz Waves*, vol. 32, no. 2, pp. 143–171, 2011.
[31] G. Chattopadhyay, Technology, "Capabilities, and Performance of Low Power Terahertz Sources", *IEEE Transactions on Terahertz Science and Technology*, vol. 1, no. 1, pp. 33–53, 2011.

[32] T. Nagatsuma and A. Kasamatsu, "Terahertz Communications for Space Applications", Asia-Pacific Microwave Conference (APMC), pp. 73–75, 2018.

[33] T. Nagatsuma, "Antenna Technologies for Terahertz Communications", *International Symposium on Antennas and Propagation (ISAP)*, pp. 1–2, 2018.

[34] C. H. Chan, "THz Antennas - Design, Fabrication and Testing", 13th European Conference on Antennas and Propagation (EuCAP), pp. 1–4, 2019.

[35] I. F. Akyildiz, J. M. Jornet and C. Han, "Terahertz Band: Next Frontier for Wireless Communications", *Physical Communication*, vol. 12, pp. 16–32, 2014.

[36] K. M. S. Huq, S. A. Busari, J. Rodriguez, V. Frascolla, W. Bazzi and D. C. Sicker, "Terahertz-Enabled Wireless System for Beyond-5G Ultra-Fast Networks: A Brief Survey", *IEEE Network*, vol. 33, no. 4, pp. 89–95, 2019.

[37] H. -J. Song and T. Nagatsuma, *Handbook of Terahertz Technologies: Devices and Applications*, Jenny Stanford Publishing, 2015.

[38] D. He, K. Guan, A. Fricke, B. Ai, R. He, Z. Zhong, A. Kasamatsu, I. Hosako and T. Kürner, "Stochastic Channel Modeling for Kiosk Applications in the Terahertz Band", *IEEE Transactions on Terahertz Science and Technology*, vol. 7, no. 5, pp. 502–513, 2017.

[39] A. Galal and X. Hesselbach, "Nano-networks Communication Architecture: Modeling and functions", *Nano Communication Networks*, vol. 17, pp. 45–62, 2018.

[40] A. Al-Khalidi, K. Alharbi, J. Wang and E. Wasige, "THz Electronics for Data Centre Wireless Links – The TERAPOD Project", 9th International Congress on Ultra Modern Telecommunications and Control Systems and Workshops (ICUMT), pp. 445–448, 2017.

[41] S. Mollahasani and E. Onur, "Evaluation of Terahertz Channel in Data Centers", IEEE/IFIP Network Operations and Management Symposium (NOMS), pp. 727–730, 2016.

[42] S. Moghadami, F. Hajilou, P. Agrawal and S. Ardalan, "A 210 GHz Fully-Integrated OOK Transceiver for Short-Range Wireless Chip-to-Chip Communication in 40 nm CMOS Technology", *IEEE Transactions on Terahertz Science and Technology*, vol. 5, no. 5, pp. 737–741, 2015.

[43] V. Petrov, D. Moltchanov, M. Komar, A. Antonov, P. Kustarev, S. Rakheja and Y. Koucheryavy, "Terahertz Band Intra-Chip Communications: Can Wireless Links Scale Modern x86 CPUs?", *IEEE Access*, vol. 5, pp. 6095–6109, 2017.

[44] I. Kallfass, I. Dan, S. Rey, P. Harati, J. Antes, A. Tessmann, S. Wagner, M. Kuri, R. Weber, H. Massler, A. Leuther, T. Merkle and T. K¨urner, "Towards MMIC-Based 300GHz Indoor Wireless Communication Systems", *IEICE Transactions on Electronics*, vol. E98.C, no. 12, pp. 1081–1090, 2015.

[45] "802.15.3d-2017 - IEEE Standard for High Data Rate Wireless Multi-Media Networks – Amendment 2: 100 Gb/s Wireless Switched Point-to-Point Physical Layer", IEEE Std 802.15.3d-2017 (Amendment to IEEE Std 802.15.3–2016 as amended by IEEE Std 802.15.3e-2017), pp. 1–55, 2017.

[46] T. Nagatsuma, K. Oogimoto, Y. Inubushi and J. Hirokawa, "Practical Considerations of Terahertz Communications for Short Distance Applications", *Nano Communication Networks*, vol. 10, pp. 1–12, 2016.

[47] C. M. Armstrong, "The Truth about Terahertz", *IEEE Spectrum*, vol. 49, no. 9, pp. 36–41, 2012.

[48] H. Hamada, T. Fujimura, I. Abdo, K. Okada, H. -J. Song, H. Sugiyama, H. Matsuzaki and H. Nosaka, "300-GHz. 100-Gb/s InP-HEMT Wireless Transceiver Using a 300-GHz Fundamental Mixer", *IEEE/MTT-S International Microwave Symposium-IMS*, pp. 1480–1483, 2018.

[49] M. Asada and S. Suzuki, "Room-Temperature Oscillation of Resonant Tunneling Diodes close to 2 THz and Their Functions for Various Applications", *Journal of Infrared, Millimeter, and Terahertz Waves*, vol. 37, no. 12, pp. 1185–1198, 2016.

[50] M. Shur, "Terahertz Technology: Devices and Applications", *Proceedings of 35th European Solid-State Device Research Conference*, pp. 13–21, 2005.

[51] J. M. Chamberlain, R. E. Miles, C. E. Collins and D. P. Steenson, "Introduction to Terahertz Solid-State Devices," in *New Directions in Terahertz Technology*, Springer, chapter 1, pp. 3–27, 1997.

[52] M. S. Vitiello, "Terahertz Photonic Devices," in *THz and Security Applications: Detectors, Sources and Associated Electronics for THz Applications*, Springer, chapter 5, pp. 91–111, 2014.

[53] K. -I. Maki, T. Shibuya, C. Otani, K. Suizu and K. Kawase, "Terahertz Beam Steering via Tilted-Phase Difference-Frequency Mixing", *Applied Physics Express*, vol. 2, no. 2, pp. 022301, 2009.

[54] K. Maki and C. Otani, "Terahertz Beam Steering and Frequency Tuning by Using the Spatial Dispersion of Ultrafast Laser Pulses", *Optics Express*, vol. 16, no. 14, pp. 10158–10169, 2008.

[55] Y. Monnai, V. Viereck, H. Hillmer, K. Altmann, C. Jansen, M. Koch and H. Shinoda, "Terahertz Beam Steering Using Structured MEMS Surfaces for Networked Wireless Sensing", *Ninth International Conference on Networked Sensing (INSS)*, pp. 1–3, 2012.

[56] S. B. Celik, K. Demir, A. B. Sahin and M. Unlu, "Design of a Terahertz Beam-Steering Photomixer Array", IEEE International Symposium on Antennas and Propagation & USNC/URSI National Radio Science Meeting, pp. 303–304, 2017.

[57] Z. Tan, F. Fan, X. Dong, J. Cheng and S. Chang, "Nonreciprocal Terahertz Beam Steering based on Magneto-optic Metagratings", *Scientific Reports*, vol. 9, no. 1, pp. 20210, 2019.

[58] X. Fu, F. Yang, C. Liu, X. Wu and T. J. Cui, "Terahertz Beam Steering Technologies: From Phased Arrays to Field-Programmable Metasurfaces", *Advanced Optical Materials*, vol. 8, no. 3, pp. 1900628, 2020.

[59] S. Preu, G. H. Döhler, S. Malzer, L. J. Wang and A. C. Gossard, "Tunable, Continuous-wave Terahertz Photomixer Sources and Applications", *Journal of Applied Physics*, vol. 109, no. 6, pp. 061301, 2011.

[60] R. Safian, G. Ghazi and N. Mohammadian, "Review of Photomixing Continuous-wave Terahertz Systems and Current Application Trends in Terahertz Domain", *Optical Engineering*, vol. 58, no. 11, pp. 1–28, 2019.

[61] C. C. Renaud, M. Natrella, C. Graham, J. Seddon, F. Van Dijk and A. J. Seeds, "Antenna Integrated THz Uni-Traveling Carrier Photodiodes", *IEEE Journal of Selected Topics in Quantum Electronics*, vol. 24, no. 2, pp. 1–11, 2018.

[62] S. Koenig, D. Lopez-Diaz, J. Antes, F. Boes, R. Henneberger, A. Leuther, A. Tessmann, R. Schmogrow, D. Hillerkuss, R. Palmer, T. Zwick, C. Koos, W. Freude, O. Ambacher, J. Leuthold and I. Kallfass, "Wireless Sub-THz Communication System with High Data Rate", *Nature Photonics*, vol. 7, no. 12, pp. 977–981, 2013.

[63] A. J. Deninger, A. Roggenbuck, S. Schindler and S. Preu, "2.75 THz Tuning with a Triple-DFB Laser System at 1550 nm and InGaAs Photomixers", *Journal of Infrared, Millimeter, and Terahertz Waves*, vol. 36, no. 3, pp. 269–277, 2015.

[64] S. -H. Yang, R. Watts, X. Li, N. Wang, V. Cojocaru, J. O'Gorman, L. P. Barry and M. Jarrahi, "Tunable Terahertz Wave Generation Through a Bimodal Laser Diode and Plasmonic Photomixer", *Optics Express*, vol. 23, no. 24, pp. 31206–31215, 2015.

[65] N. Kim, S. -P. Han, H. -C. Ryu, H. Ko, J. -W. Park, D. Lee, M. Y. Jeon and K. H. Park, "Distributed Feedback Laser Diode Integrated with Distributed Bragg Reflector for Continuous-Wave Terahertz Generation", *Optics Express*, vol. 20, no. 16, pp. 17496–17502, 2012.

[66] M. T. Haidar, S. Preu, J. Cesar, S. Paul, A. S. Hajo, C. Neumeyr, H. Maune and F. Küppers, "Systematic Characterization of a 1550 nm Microelectromechanical (MEMS)-Tunable Vertical-Cavity Surface Emitting Laser (VCSEL) with 7.92 THz Tuning-Range for Terahertz Photomixing Systems", *Journal of Applied Physics*, vol. 123, no. 2, pp. 023106, 2018.

[67] M. Ravaro, C. Manquest, C. Sirtori, S. Barbieri, G. Santarelli, K. Blary, J. -F. Lampin, S. P. Khanna and E. H. Linfield, "Phase-Locking of a 2.5 THz Quantum Cascade Laser to a Frequency Comb Using a GaAs Photomixer", *Optics Letters*, vol. 36, no. 20, pp. 3969–3971, 2011.

[68] T. Nagatsuma, K. Oogimoto, Y. Yasuda, Y. Fujita, Y. Inubushi, S. Hisatake, A. M. Agoues and G. C. Lopez, "300-GHz-Band Wireless Transmission at 50 Gbit/s over 100 Meters", 41st International Conference on Infrared, Millimeter, and Terahertz waves (IRMMW-THz), pp. 1–2, 2016.

[69] T. Nagatsuma, Y. Fujita, Y. Yasuda, Y. Kanai, S. Hisatake, M. Fujiwara and J. Kani, "Real-Time 100-Gbit/s QPSK Transmission Using Photonics-Based 300-GHz-Band Wireless Link", IEEE International Topical Meeting on Microwave Photonics (MWP), pp. 27–30, 2016.

[70] T. Nagatsuma and H. Ito, "High-Power RF Uni-Traveling-Carrier Photodiodes (UTC-PDs) and Their Applications", in *Advances in Photodiodes*, IntechOpen, chapter 14, pp. 291–314, 2011.

[71] H. -J. Song, K. Ajito, Y. Muramoto, A. Wakatsuki, T. Nagatsuma and N. Kukutsu, "Uni-Travelling-Carrier Photodiode Module Generating 300 GHz Power Greater Than 1 mW", *IEEE Microwave and Wireless Components Letters*, vol. 22, no. 7, pp. 363–365, 2012.

[72] T. Ishibashi, N. Shimizu, S. Kodama, H. Ito, T. Nagatsuma and T. Furuta, "Uni-Traveling-Carrier Photodiodes, Ultrafast Electronics and Optoelectronics", *OSA Trends in Optics and Photonics Series*, vol. 13, 1997.

[73] A. Wakatsuki, Y. Muramoto and T. Ishibashi, "Development of Terahertz-wave Photomixer Module Using a Uni-Traveling-Carrier Photodiode", *NTT Technical Review*, vol. 10, no. 2, 2012.

[74] H. Ito, S. Kodama, Y. Muramoto, T. Furuta, T. Nagatsuma and T. Ishibashi, "High-Speed and High-Output InP-InGaAs Unitraveling-Carrier Photodiodes", *IEEE Journal of Selected Topics in Quantum Electronics*, vol. 10, no. 4, pp. 709–727, 2004.

[75] R. W. Boyd, "The Nonlinear Optical Susceptibility", in *Nonlinear Optics*, Academic Press, chapter 1, pp. 1–67, 2008.

[76] O. Graydon, "Ultrafast Wireless Link", *Nature Photonics*, vol. 10, no. 11, pp. 691, 2016.

[77] X. Yu, S. Jia, H. Hu, M. Galili, T. Morioka, P. U. Jepsen and L. K. Oxenløwe, "160 Gbit/s Photonics Wireless Transmission in the 300–500 GHz Band", *APL Photonics*, vol. 1, no. 8, pp. 081301, 2016.

[78] X. Pang, S. Jia, O. Ozolins, X. Yu, H. Hu, L. Marcon, P. Guan, F. D. Ros, S. Popov, G. Jacobsen, M. Galili, T. Morioka, D. Zibar and L. K. Oxenkwe, "260 Gbit/s Photonic-Wireless Link in the THz Band", IEEE Photonics Conference (IPC), pp. 1–2, 2016.

[79] T. Harter, S. Ummethala, M. Blaicher, S. Muehlbrandt, S. Wolf, M. Weber, M. M. H. Adib, J. N. Kemal, M. Merboldt, F. Boes, S. Nellen, A. Tessmann, M. Walther, B. Globisch, T. Zwick, W. Freude, S. Randel and C. Koos, "Wireless THz Link with Optoelectronic Transmitter and Receiver", *Optica*, vol. 6, no. 8, pp. 1063–1070, 2019.

[80] V. K. Chinni, P. Latzel, M. Z´egaoui, C. Coinon, X. Wallart, E. Peytavit, J. F. Lampin, K. Engenhardt, P. Szriftgiser, M. Zaknoune and G. Ducournau, "Single-Channel 100 Gbit/s Transmission using III-V UTC-PDs for Future IEEE 802.15.3d Wireless Links in the 300 GHz Band", *Electronics Letters*, vol. 54, no. 10, pp. 638–640, 2018.

[81] S. Jia, X. Pang, O. Ozolins, X. Yu, H. Hu, J. Yu, P. Guan, F. Da Ros, S. Popov, G. Jacobsen, M. Galili, T. Morioka, D. Zibar and L. K. Oxenløwe, "0.4 THz Photonic-Wireless Link with 106 Gb/s Single Channel Bitrate", *Journal of Lightwave Technology*, vol. 36, no. 2, pp. 610–616, 2018.

[82] H. -J. Lee, S. Rami, S. Ravikumar, V. Neeli, K. Phoa, B. Sell and Y. Zhang, "Intel 22nm FinFET (22FFL) Process Technology for RF and mm Wave Applications and Circuit Design Optimization for FinFET Technology", IEEE International Electron Devices Meeting (IEDM), pp. 14.1.1–14.1.4, 2018.

[83] B. Heinemann, H. Rücker, R. Barth, F. Bärwolf, J. Drews, G. G. Fischer, A. Fox, O. Fursenko, T. Grabolla, F. Herzel, J. Katzer, J. Korn, A. Krüger, P. Kulse, T. Lenke, M. Lisker, S. Marschmeyer, A. Scheit, D. Schmidt, J. Schmidt, M. A. Schubert, A. Trusch, C. Wipf and D. Wolansky, "SiGe HBT with f_T/f_{max} of 505 GHz/720 GHz", IEEE International Electron Devices Meeting (IEDM), pp. 3.1.1–3.1.4, 2016.

[84] J. C. Rode, H. -W. Chiang, P. Choudhary, V. Jain, B. J. Thibeault, W. J. Mitchell, M. J. W. Rodwell, M. Urteaga, D. Loubychev, A. Snyder, Y. Wu, J. M. Fastenau and A. W. K. Liu, "Indium Phosphide Heterobipolar Transistor Technology Beyond 1-THz Bandwidth", *IEEE Transactions on Electron Devices*, vol. 62, no. 9, pp. 2779–2785, 2015.

[85] X. Mei, W. Yoshida, M. Lange, J. Lee, J. Zhou, P. Liu, K. Leong, A. Zamora, J. Padilla, S. Sarkozy, R. Lai and W. R. Deal, "First Demonstration of Amplification at 1 THz Using 25-nm InP High Electron Mobility Transistor Process", *IEEE Electron Device Letters*, vol. 36, no. 4, pp. 327–329, 2015.

[86] D. -H. Kim, J. A. del Alamo, P. Chen, W. Ha, M. Urteaga and B. Brar, "50-nm E-Mode In0.7Ga0.3As PHEMTs on 100-mm InP Substrate with f_{max} > 1 THz", International Electron Devices Meeting, pp. 30.6.1–30.6.4, 2010.

[87] D. -H. Kim, B. Brar and J. A. del Alamo, "f_T = 688 GHz and f_{max} = 800 GHz in Lg = 40 nm In0.7Ga0.3As MHEMTs with gm_{max} > 2.7 mS/μm", International Electron Devices Meeting, pp. 13.6.1–13.6.4, 2011.

[88] M. Urteaga, R. Pierson, P. Rowell, V. Jain, E. Lobisser and M. J. W. Rodwell, "130nm InP DHBTs with f_t >0.52THz and f_{max} >1.1THz", 69th Device Research Conference, pp. 281–282, 2011.

[89] A. Nikpaik, A. H. M. Shirazi, A. Nabavi, S. Mirabbasi and S. Shekhar, ""A 219-to-231 GHz Frequency-Multiplier-Based VCO With ~3% Peak DC-to-RF Efficiency in 65-nm CMOS", *IEEE Journal of Solid-State Circuits*, vol. 53, no. 2, pp. 389–403, 2018.

[90] E. Seok, C. Cao, D. Shim, D. J. Arenas, D. B. Tanner, C. -M. Hung and K. O. Kenneth, "A 410GHz CMOS Push-Push Oscillator with an On-Chip Patch Antenna", *IEEE International Solid-State Circuits Conference – Digest of Technical Papers*, pp. 472–629, 2008.

[91] F. Ahmed, M. Furqan, B. Heinemann and A. Stelzer, "0.3-THz SiGe-Based High-Efficiency Push-Push VCOs With > 1-mW Peak Output Power Employing Common-Mode Impedance Enhancement", *IEEE Transactions on Microwave Theory and Techniques*, vol. 66, no. 3, pp. 1384–1398, 2018.

[92] A. H. M. Shirazi, A. Nikpaik, S. Mirabbasi and S. Shekhar, "A Quadcore-Coupled Triple-Push 295-to-301 GHz Source with 1.25 mW Peak Output Power in 65nm CMOS Using Slow-Wave Effect", IEEE Radio Frequency Integrated Circuits Symposium (RFIC), pp. 190–193, 2016.

[93] L. A. Samoska, "An Overview of Solid-State Integrated Circuit Amplifiers in the Submillimeter-Wave and THz Regime", *IEEE Transactions on Terahertz Science and Technology*, vol. 1, no. 1, pp. 9–24, 2011.

[94] W. R. Deal, "Solid-State Amplifiers For Terahertz Electronics," IEEE MTTS International Microwave Symposium, pp. 1122–1125, 2010.

[95] W. R. Deal, K. Leong, V. Radisic, S. Sarkozy, B. Gorospe, J. Lee, P. H. Liu, W. Yoshida, J. Zhou, M. Lange, R. Lai and X. B. Mei, "Low Noise Amplification at 0.67 THz Using 30 nm InP HEMTs", *IEEE Microwave and Wireless Components Letters*, vol. 21, no. 7, pp. 368–370, 2011.

[96] J. Hacker, M. Urteaga, M. Seo, A. Skalare and R. Lin, "InP HBT Amplifier MMICs Operating to 0.67 THz", *IEEE MTT-S International Microwave Symposium Digest (MTT)*, pp. 1–3, 2013.

[97] V. Radisic, K. M. K. H. Leong, X. Mei, S. Sarkozy, W. Yoshida and W. R. Deal, "Power Amplification at 0.65 THz Using InP HEMTs", *IEEE Transactions on Microwave Theory and Techniques*, vol. 60, no. 3, pp. 724–729, 2012.

[98] K. M. K. H. Leong, X. Mei, W. H. Yoshida, A. Zamora, J. G. Padilla, B. S. Gorospe, K. Nguyen and W. R. Deal, "850 GHz Receiver and Transmitter Front-Ends Using InP HEMT", *IEEE Transactions on Terahertz Science and Technology*, vol. 7, no. 4, pp. 466–475, 2017.

[99] K. M. K. H. Leong, X. Mei, W. Yoshida, P. -H. Liu, Z. Zhou, M. Lange, L. -S. Lee, J. G. Padilla, A. Zamora, B. S. Gorospe, K. Nguyen and W. R. Deal, "A 0.85 THz Low Noise Amplifier Using InP HEMT Transistors", *IEEE Microwave and Wireless Components Letters*, vol. 25, no. 6, pp. 397–399, 2015.

[100] U. R. Pfeiffer, Y. Zhao, J. Grzyb, R. A. Hadi, N. Sarmah, W. Förster, H. Rücker and B. Heinemann, "A 0.53THz Reconfigurable Source Array with up to 1mW Radiated Power for Terahertz Imaging Applications in 0.13μm SiGe BiCMOS", IEEE International Solid-State Circuits Conference Digest of Technical Papers (ISSCC), pp. 256–257, 2014.

[101] R. Han, C. Jiang, A. Mostajeran, M. Emadi, H. Aghasi, H. Sherry, A. Cathelin and E. Afshari, "A SiGe Terahertz Heterodyne Imaging Transmitter with 3.3 mW Radiated Power and Fully-Integrated Phase-Locked Loop", *IEEE Journal of Solid-State Circuits*, vol. 50, no. 12, pp. 2935–2947, 2015.

[102] K. Schmalz, R. Wang, J. Borngräber, W. Debski, W. Winkler and C. Meliani, "245 GHz SiGe Transmitter with Integrated Antenna and External PLL", *IEEE MTT-S International Microwave Symposium Digest (MTT)*, pp. 1–3, 2013.

[103] J. Yu, X. Li, J. Zhang and J. Xiao, "432-Gb/s PDM-16QAM Signal Wireless Delivery at W-Band Using Optical And Antenna Polarization Multiplexing", The European Conference on Optical Communication (ECOC), pp. 1–3, 2014.

[104] R. Puerta, J. Yu, X. Li, Y. Xu, J. J. V. Olmos and I. T. Monroy, "Demonstration of 352 Gbit/s Photonically-Enabled D-Band Wireless Delivery in One 2×2 MIMO System", Optical Fiber Communications Conference and Exhibition (OFC), pp. 1–3, 2017.

[105] H. Hamada, T. Tsutsumi, G. Itami, H. Sugiyama, H. Matsuzaki, K. Okada and H. Nosaka, "300-GHz 120-Gb/s Wireless Transceiver with High-Output-Power and High-Gain Power Amplifier Based on 80-nm InP-HEMT Technology", IEEE BiCMOS and Compound Semiconductor Integrated Circuits and Technology Symposium (BCICTS), pp. 1–4, 2019.

[106] S. Lee, R. Dong, S. Hara, K. Takano, S. Amakawa, T. Yoshida and M. Fujishima, "A 6-mW-DC-Power 300-GHz CMOS Receiver for Near-Field Wireless Communications", IEEE MTT-S International Microwave Symposium (IMS), pp. 504–507, 2019.

[107] K. Takano, K. Katayama, S. Amakawa, T. Yoshida and M. Fujishima, "Wireless Digital Data Transmission from a 300 GHz CMOS Transmitter", *Electronics Letters*, vol. 52, no. 15, pp. 1353–1355, 2016.

[108] K. K. Tokgoz, S. Maki, J. Pang, N. Nagashima, I. Abdo, S. Kawai, T. Fujimura, Y. Kawano, T. Suzuki, T. Iwai, K. Okada and A. Matsuzawa, "A 120Gb/s 16QAM CMOS Millimeter-Wave Wireless Transceiver", *IEEE International Solid-State Circuits Conference (ISSCC)*, pp. 168–170, 2018.

[109] P. Rodrìguez-Vázquez, J. Grzyb, B. Heinemann and U. R. Pfeiffer, "A 16-QAM 100-Gb/s 1-M Wireless Link With an EVM of 17% at 230 GHz in an SiGe Technology", *IEEE Microwave and Wireless Components Letters*, vol. 29, no. 4, pp. 297–299, 2019.

[110] P. Rodrìguez-Vázquez, J. Grzyb, N. Sarmah, B. Heinemann and U. R. Pfeiffer, "A 65 Gbps QPSK One Meter Wireless Link Operating at a 225–255 GHz Tunable Carrier in a SiGe HBT Technology", *IEEE Radio and Wireless Symposium (RWS)*, pp. 146–149, 2018.

[111] P. Rodrìguez-Vázquez, J. Grzyb, N. Sarmah, B. Heinemann and U. R. Pfeiffer, "Towards 100 Gbps: A Fully Electronic 90 Gbps One Meter Wireless Link at 230 GHz", 48th European Microwave Conference (EuMC), pp. 1389–1392, 2018.

[112] I. Kallfass, F. Boes, T. Messinger, J. Antes, A. Inam, U. Lewark, A. Tessmann and R. Henneberger, "64 Gbit/s Transmission over 850 m Fixed Wireless Link at 240 GHz Carrier Frequency", *Journal of Infrared, Millimeter, and Terahertz Waves*, vol. 36, no. 2, pp. 221–233, 2015.

[113] J. P. Sun, G. I. Haddad, P. Mazumder and J. N. Schulman, "Resonant Tunneling Diodes: Models and Properties", *Proceedings of the IEEE*, vol. 86, no. 4, pp. 641–660, 1998.

[114] E. T. Croke and J. N. Schulman, "Resonant Tunneling Diodes", *Encyclopedia of Materials: Science and Technology*, 2nd edition, Pergamon, pp. 8185–8192, 2001.

[115] M. Asada and S. Suzuki, "Resonant-Tunneling-Diode Terahertz Oscillators and Applications", IEEE International Electron Devices Meeting (IEDM), pp. 29.3.1–29.3.4, 2016.

[116] M. Feiginov, "Frequency Limitations of Resonant-Tunnelling Diodes in Sub-THz and THz Oscillators and Detectors", *Journal of Infrared, Millimeter, and Terahertz Waves*, vol. 40, no. 4, pp. 365–394, 2019.

[117] R. Izumi, S. Suzuki and M. Asada, "1.98 THz Resonant-Tunneling-Diode Oscillator with Reduced Conduction Loss by Thick Antenna Electrode", 42nd International Conference on Infrared, Millimeter, and Terahertz Waves (IRMMW-THz), pp. 1–2, 2017.

[118] H. Ito, T. Furuta, Y. Muramoto, T. Ito and T. Ishibashi, "Photonic Millimetre- and Sub-Millimetre Wave Generation Using J-Band Rectangular Waveguide-Output Uni-Travelling-Carrier Photodiode Module", *Electronics Letters*, vol. 42, no. 24, pp. 1424–1425, 2006.

[119] A. Al-Khalidi, J. Wang and E. Wasige, "Compact J-band Oscillators With 1 mW RF Output Power and Over 110 GHz Modulation Bandwidth", 43rd International Conference on Infrared, Millimeter, and Terahertz Waves (IRMMW-THz), pp. 1–2, 2018.

[120] A. Al-Khalidi, K. H. Alharbi, J. Wang, R. Morariu, L. Wang, A. Khalid, J. M. L. Figueiredo and E. Wasige, "Resonant Tunneling Diode Terahertz Sources With up to 1 mW Output

Power in the J-Band", *IEEE Transactions on Terahertz Science and Technology*, vol. 10, no. 2, pp. 150–157, 2020.

[121] Y. Nishida, N. Nishigami, S. Diebold, J. Kim, M. Fujita and T. Nagatsuma, "Terahertz Coherent Receiver Using A Single Resonant Tunnelling Diode", *Scientific Reports*, vol. 9, no. 1, pp. 18125, 2019.

[122] T. Maekawa, H. Kanaya, S. Suzuki and M. Asada, "Oscillation up to 1.92 THz in Resonant Tunneling Diode By Reduced Conduction Loss", *Applied Physics Express*, vol. 9, no. 2, pp. 024101, 2016.

[123] J. Wang, K. Alharbi, A. Ofiare, H. Zhou, A. Khalid, D. Cumming and E. Wasige, "High Performance Resonant Tunneling Diode Oscillators for THz Applications", IEEE Compound Semiconductor Integrated Circuit Symposium (CSICS), pp. 1–4, 2015.

[124] K. Karashima, R. Yokoyama, M. Shiraishi, S. Suzuki, S. Aoki and M. Asada, "Measurement of Oscillation Frequency and Spectral Linewidth of Sub-Terahertz InP-Based Resonant Tunneling Diode Oscillators Using Ni-InP Schottky Barrier Diode", *Japanese Journal of Applied Physics*, vol. 49, no. 2R, pp. 020208, 2010.

[125] K. Ogino, S. Suzuki and M. Asada, "Spectral Narrowing of a Varactor-Integrated Resonant-Tunneling-Diode Terahertz Oscillator by Phase-Locked Loop", *Journal of Infrared, Millimeter, and Terahertz Waves*, vol. 38, no. 12, pp. 1477–1486, 2017.

[126] K. Okamoto, K. Tsuruda, S. Diebold, S. Hisatake, M. Fujita and T. Nagatsuma, "Terahertz Sensor Using Photonic Crystal Cavity and Resonant Tunneling Diodes", *Journal of Infrared, Millimeter, and Terahertz Waves*, vol. 38, no. 9, pp. 1085–1097, 2017.

[127] X. Yu, J. -Y. Kim, M. Fujita and T. Nagatsuma, "Highly Stable Terahertz Resonant Tunneling Diode Oscillator Coupled to Photonic-Crystal Cavity", *Asia-Pacific Microwave Conference (APMC)*, pp. 114–116, 2018.

[128] Y. Takida, S. Suzuki, M. Asada and H. Minamide, "Sensitivity Measurement of Resonant-Tunneling-Diode Terahertz Detectors", 44th International Conference on Infrared, Millimeter, and Terahertz Waves (IRMMW-THz), pp. 1–2, 2019.

[129] J. Webber, N. Nishigami, J. -Y. Kim, M. Fujita and T. Nagatsuma, "Terahertz Wireless CDMA Communication Using Resonant Tunneling Diodes", *IEEE Globecom Workshops (GC Wkshps)*, pp. 1–6, 2019.

130. H. Mizuta and T. Tanoue, *The Physics and Applications of Resonant Tunnelling Diodes*, Cambridge University Press, 1995.

[131] B. Ricco and M. Y. Azbel, "Physics of Resonant Tunneling. The One-Dimensional Double-Barrier Case", *Physical Review B*, vol. 29, no. 4, pp. 1970–1981, 1984.

[132] M. Asada, S. Suzuki and T. Fukuma, "Measurements of Temperature Characteristics and Estimation of Terahertz Negative Differential Conductance in Resonant-Tunneling-Diode Oscillators", *AIP Advances*, vol. 7, no. 11, pp. 115226, 2017.

[133] C. Kim and A. Brandli, "High-Frequency High-Power Operation of Tunnel Diodes", *IRE Transactions on Circuit Theory*, vol. 8, no. 4, pp. 416–425, 1961.

[134] L. Wang, "Reliable Design of Tunnel Diode and Resonant Tunnelling Diode Based Microwave Sources", Ph.D. Thesis, chapter 5, University of Glasgow, 2012.

[135] L. Wang, "Output Power Analysis and Simulations of Resonant Tunneling Diode Based Oscillators", in *International Computer Science Conference, System Simulation and Scientific Computing*, Springer, pp. 47–55, 2012.

[136] J. Davies, *The Physics of Low-Dimensional Semiconductors*, Cambridge University Press, 1998.

[137] H. Mizuta and T. Tanoue, *The Physics and Applications of Resonant Tunnelling Diodes*, Cambridge University Press, 2006.

[138] M. Büttiker and R. Landauer, "Traversal Time for Tunneling", *Physical Review Letters*, vol. 49, no. 23, pp. 1739–1742, 1982.

[139] R. Landauer, "Barrier Traversal Time", *Nature*, vol. 341, no. 6243, 567–568, 1989.

[140] R. Landauer and T. Martin, "Barrier Interaction Time in Tunneling", *Reviews of Modern Physics*, vol. 66, no. 1, pp. 217–228, 1994.

[141] W. R. Frensley, "Quantum Transport Calculation of the Small-Signal Response of a Resonant Tunneling Diode", Applied Physics Letters, vol. 51, no. 6, pp. 448–450, 1987.

[142] T. C. L. G. Sollner, W. D. Goodhue, P. E. Tannenwald, C. D. Parker and D. D. Peck, "Resonant Tunneling Through Quantum Wells at Frequencies Up To 2.5 THz", *Applied Physics Letters*, vol. 43, no. 6, pp. 588–590, 1983.

[143] J. S. Scott, J. P. Kaminski, M. Wanke, S. J. Allen, D. H. Chow, M. Lui and T. Y. Liu, "Terahertz Frequency Response of an $In_{0.53}Ga_{0.47}As$/Al as Resonant-Tunneling Diode", *Applied Physics Letters*, vol. 64, no. 15, pp. 1995–1997, 1994.

[144] R. Tsu and L. Esaki, "Tunneling in a Finite Superlattice," *Applied Physics Letters*, vol. 22, no. 11, pp. 562–564, 1973.

[145] L. L. Chang, L. Esaki, and R. Tsu, "Resonant Tunneling In Semiconductor Double Barriers," *Applied Physics Letters*, pp. 24, no. 12, pp. 593–595, 1974.

[146] M. Feiginov, C. Sydlo, O. Cojocari, and P. Meissner, "Resonant-Tunnelling Diode Oscillators Operating at Frequencies above 1.1 THz," *Applied Physics Letters*, vol. 99, no. 26, pp. 233 506–233 509, 2011.

[147] E. R. Brown, J. R. Soderstrom, C. D. Parker, L. J. Mahoney, K. M. Molvar, and T. C. McGill, "Oscillations up to 712 GHz in InAS/AlSb Resonant Tunnelling Diodes," *Applied Physics Letters*, vol. 58, no. 20, pp. 2291–2293, 1991.

[148] T. Inata, S. Muto, Y. Nakata, T. Fujii, H. Ohnishi, and S. Hiyamizu, "Excellent Negative Differential Resistance of InAlAs/InGaAs Resonant Tunnelling Barrier Structures Grown by MBE," *Japanese Journal of Applied Physics*, vol. 25, no. 12, pp. 983–985, 1986.

[149] J. Singh, *Semiconductor Optoelectronics: Physics and Technology*, McGraw-Hill Ryerson, 1995.

[150] C. I. Huang, M. J. Paulus, C. A. Bozada, S. C. Dudley, K. R. Evans, C. E. Stutz, R. L. Jones, and M. E. Cheney, "AlGaAs/GaAs Double Barrier Diodes with High Peak-To-Valley Current Ratio," *Applied Physics Letters*, vol. 51, no. 2, pp. 121–123, 1987.

[151] T. J. Shewchuk, P. C. Chapin, P. D. Coleman, W. Kopp, R. Fischer, and H. Morkoc, "Resonant Tunneling Oscillations in a $_{Ga}As$-$Al_xGa_{1-x}As$ Heterostructure At Room Temperature," *Applied Physics Letters*, vol. 46, no. 5, pp. 508–510, 1985.

[152] H. Sugiyama, H. Yokoyama, A. Teranishi, S. Suzuki and M. Asada, "Extremely High Peak Current Densities of over 1×10^6 A/cm^2 in InP-Based InGaAs/AlAs Resonant Tunneling Diodes Grown by Metal-Organic Vapor-Phase Epitaxy", *Japanese Journal of Applied Physics*, vol. 49, no. 5R, pp. 051201, 2010.

[153] H. Sugiyama, H. Yokoyama, A. Teranishi, S. Suzuki and M. Asada, "High-Uniformity InP-Based Resonant Tunneling Diode Wafers with Peak Current Density of Over 6×10^5 A/Cm^2 Grown by Metal-Organic Vapour-Phase Epitaxy", *Journal of Crystal Growth*, vol. 336, no. 1, pp. 24–28, 2011.

[154] H. Sugiyama, A. Teranishi, S. Suzuki and M. Asada, "Structural and Electrical Transport Properties of MOVPE-Grown Pseudomorphic AlAs/InGaAs/InAs Resonant Tunneling Diodes on InP Substrates", *Japanese Journal of Applied Physics*, vol. 53, no. 3, pp. 031202, 2014.

[155] A. K. Baraskar, M. A. Wistey, V. Jain, U. Singisetti, G. Burek, B. J. Thibeault, Y. Ju Lee, A. C. Gossards and M. J. W. Rodwell, "Ultralow Resistance, Nonalloyed Ohmic Contacts to n-InGaAs", *Journal of Vacuum Science & Technology B: Microelectronics and Nanometer Structures Processing, Measurement, and Phenomena*, vol. 27, no. 4, pp. 2036–2039, 2009.

[156] S. Masudy-Panah, Y. Wu, D. Lei, A. Kumar, Y. -C. Yeo and X. Gonga, "Nanoscale Metal–InGaAs Contacts with Ultra-Low Specific Contact Resistivity: Improved Interfacial Quality And Extraction Methodology", *Journal of Applied Physics*, vol. 123, no. 2, pp. 024508, 2018.

[157] H. Kanaya, T. Maekawa, S. Suzuki and M. Asada, "Structure Dependence of Oscillation Characteristics of Resonant-Tunneling-Diode Terahertz Oscillators Associated with Intrinsic and Extrinsic Delay Times", *Japan Journal of Applied Physics*, vol. 54, no. 9, pp. 094103, 2015.

[158] H. Kanaya, "Study of Resonant-Tunneling-Diode Terahertz Oscillators for High Frequency", Ph.D. thesis, Tokyo Institute of Technology, 2015.

[159] K. J. P. Jacobs, B. J. Stevens, R. Baba, O. Wada, T. Mukai and R. A. Hogg, "Valley Current Characterization of High Current Density Resonant Tunnelling Diodes for Terahertz-Wave Applications", *AIP Advances*, vol. 7, no. 10, pp. 105316, 2017.
[160] E. R. Brown, "High-Speed Resonant-Tunneling Diodes, in Heterostructures and Quantum Devices", *VLSI Electronics Microstructure Science*, vol. 24, chapter 10, Elsevier, pp. 305–350, 1994.
[161] H. Kanaya, R. Sogabe, T. Maekawa, S. Suzuki and M. Asada, "Fundamental Oscillation up to 1.42 THz in Resonant Tunneling Diodes by Optimized Collector Spacer Thickness", *Journal of Infrared, Millimeter, and Terahertz Waves*, vol. 35, no. 5, pp. 425–431, 2014.
[162] N. Kishimoto, S. Suzuki, A. Teranishi and M. Asada, "Frequency Increase of Resonant Tunneling Diode Oscillators in Sub-THz and THz Range Using Thick Spacer Layers", *Applied Physics Express*, vol. 1, no. 4, pp. 042003, 2008.
[163] H. M. Yoo, S. M. Goodnick and J. R. Arthur, "Influence of Spacer Layer Thickness on the Current–Voltage Characteristics of AlGaAs/GaAs and AlGaAs/InGaAs Resonant Tunneling Diodes", *Applied Physics Letters*, vol. 56, no. 1, pp. 84–86, 1990.
[164] M. A. M. Zawawi and M. Missous, "Dependencies of Peak Current Density on Barrier and Spacer Thickness in InGaAs/AlAs Resonant Tunneling Diode", UK Semiconductor, 2014.
[165] S. Javalagi, V. Reddy, K. Gullapalli and D. Neikirk, "High Efficiency Microwave Diode Oscillators", *Electronics Letters*, vol. 28, no. 18, pp. 1699–1701, 1992.
[166] E. R. Brown, J. R. Söderström, C. D. Parker, L. J. Mahoney, K. M. Molvar and T. C. McGill, "Oscillations up to 712 GHz in InAs/AlSb Resonant-Tunneling Diodes", *Applied Physics Letters*, vol. 58, no. 20, pp. 2291–2293, 1991.
[167] E. R. Brown, S. J. Eglash, G. W. Turner, C. D. Parker, J. V. Pantano and D. R. Calawa, "Effect of Lattice-Mismatched Growth on InAs/AlSb Resonant Tunneling Diodes", *IEEE Transactions on Electron Devices*, vol. 41, no. 6, pp. 879–882, 1994.
[168] J. R. Söderström, E. R. Brown, C. D. Parker, L. J. Mahoney, J. Y. Yao, T. G. Andersson and T. C. McGill, "Growth and Characterization of High Current Density High-Speed InAs/AlSb Resonant Tunneling Diodes", *Applied Physics Letters*, vol. 58, no. 3, pp. 275–277, 1991.
[169] J. R. Söderström, D. H. Chow and T. C. McGill, "InAs/AlSb Double-Barrier Structure with Large Peak-To-Valley Current Ratio: A Candidate for High-Frequency Microwave Devices", *IEEE Electron Device Letters*, vol. 11, no. 1, pp. 27–29, 1990.
[170] L. F. Luo, R. Beresford and W. I. Wang, "Resonant Tunneling in AlSb/InAs/AlSb Double-Barrier Heterostructures", *Applied Physics Letters*, vol. 53, no. 23, pp. 2320–2322, 1988.
[171] E. Ozbay, D. M. Bloom, D. H. Chow and J. N. Schulman, "1.7-ps, Microwave, Integrated-Circuit-Compatible InAs/AlSb Resonant Tunneling Diodes", *IEEE Electron Device Letters*, vol. 14, no. 8, pp. 400–402, 1993.
[172] J. N. Schulman, D. H. Chow and T. C. Hasenberg, "InAs/Antimonide-Based Resonant Tunneling Structures with Ternary Alloy Layers", *Solid-State Electronics*, vol. 37, no. 4–6, pp. 981–985, 1994.
[173] P. Fay, J. Lu, Y. Xu, N. Dame, D. H. Chow and J. N. Schulman, "Microwave Performance and Modeling of InAs/AlSb/GaSb Resonant Interband Tunneling Diodes", *IEEE Transactions on Electron Devices*, vol. 49, no. 1, pp. 19–24, 2002.
[174] J. Shen, G. Kramer, S. Tehrani and H. Goronkin, "Static Random Access Memories Based on Resonant Interband Tunneling Diodes in the InAs/GaSb/AlSb Material System", *IEEE Electron Device Letters*, vol. 16, no. 5, pp. 178–180, 1995.
[175] E. D. Guarin Castro, F. Rothmayr, S. Krüger, G. Knebl, A. Schade, J. Koeth, L. Worschech, V. Lopez-Richard, G. E. Marques, F. Hartmann, A. Pfenning and S. Höfling, "Resonant Tunneling of Electrons in AlSb/GaInAsSb Double Barrier Quantum Wells", *AIP Advances*, vol. 10, no. 5, pp. 055024, 2020.
[176] J. R. Söderström, D. H. Cho and T. C. McGill, "New Negative Differential Resistance Device Based on Resonant Interband Tunneling", *Applied Physics Letters*, vol. 55, no. 11, pp. 1094–1096, 1989.
[177] H. Kroemer, "The 6.1 Å Family (InAs, GaSb, AlSb) and Its Heterostructures: A Selective Review", *Physica E: Low-dimensional Systems and Nanostructures*, vol. 20, no. 3–4, pp. 196–203, 2004.

[178] E. R. Brown, C. D. Parker, "Resonant Tunnel Diodes as Submillimetre-Wave Sources", *Philosophical Transactions of the Royal Society of London. Series A: Mathematical, Physical and Engineering Sciences*, vol. 354, no. 1717, pp. 2365–2381, 1996.

[179] C. A. Mead and W. G. Spitzer, "Fermi Level Position at Metal-Semiconductor Interfaces", *Physical Review*, vol. 134, no. 3A, pp. A713-A716, 1964.

[180] G. Ghione, "Semiconductors, Alloys, Heterostructures", in *Semiconductor Devices for High-Speed Optoelectronics*, Cambridge University Press, chapter 1, pp. 1–51, 2009.

[181] J. R. Söderström, E. T. Yu, M. K. Jackson, Y. Rajakarunanayake and T. C. McGill, "Two-Band Modeling of Narrow Band Gap and Interband Tunneling Devices", *Journal of Applied Physics*, vol. 68, no. 3, pp. 1372–1375, 1990.

[182] G. Bastard, "Theoretical Investigations of Superlattice Band Structure in the Envelope-Function Approximation", *Physical Review B*, vol. 25, no. 12, pp. 7584–7597, 1982.

[183] M. A. Davidovich, E. V. Anda, C. Tejedor and G. Platero, "Interband Resonant Tunneling and Transport in InAs/AlSb/GaSb Heterostructures", *Physical Review B*, vol. 47, no. 8, pp. 4475–4484, 1993.

[184] T. Nittono, H. Ito, O. Nakajima and T. Ishibashi, "Non-Alloyed Ohmic Contacts to n-GaAs Using Compositionally Graded $In_xGa_{1-x}As$ Layers", *Japanese Journal of Applied Physics*, vol. 27, part 1, no. 9, pp. 1718–1722, 1988.

[185] W. D. Zhang, T. A. Growden, E. R. Brown, P. R. Berger, D. F. Storm and D. J. Meyer, "Fabrication and Characterization of GaN/AlN Resonant Tunneling Diodes", in *High-Frequency GaN Electronic Devices*, Springer, chapter 9, pp. 249–281, 2020.

[186] R. Lake and J. Yang, "A Physics Based Model for the RTD Quantum Capacitance", *IEEE Transactions on Electron Devices*, vol. 50, no. 3, pp. 785–789, 2003.

[187] Q. Liu, A. Seabaugh, P. Chahal and F. J. Morris, "Unified AC Model for the Resonant Tunneling Diode", *IEEE Transactions on Electron Devices*, vol. 51, no. 5, pp. 653–657, 2004.

[188] M. Asada, S. Suzuki and N. Kishimoto, "Resonant Tunneling Diodes for Sub-Terahertz and Terahertz Oscillators", *Japanese Journal of Applied Physics*, vol. 47, no. 6R, pp. 4375–4384, 2008.

[189] E. R. Brown, C. D. Parker, and T. C. L. G. Sollner, "Effect of Quasibound-State Lifetime on the Oscillation Power of Resonant Tunneling Diodes", *Applied Physics Letters*, vol. 54, no. 10, pp. 934–936, 1989.

[190] M. N. Feiginov, "Does the Quasibound-State Lifetime Restrict the High-Frequency Operation of Resonant-Tunnelling Diodes?", *Nanotechnology*, vol. 11, no. 4, pp. 359–364, 2000.

[191] M. N. Feiginov and D. R. Chowdhury, "Operation of Resonant-Tunneling Diodes beyond Resonant-State-Lifetime Limit", *Applied Physics Letters*, vol. 91, no. 20, pp. 203501, 2007.

[192] R. Morariu, J. Wang, A. C. Cornescu, A. Al-Khalidi, A. Ofiare, J. M. L. Figueiredo and E. Wasige, "Accurate Small-Signal Equivalent Circuit Modeling of Resonant Tunneling Diodes to 110 GHz", *IEEE Transactions on Microwave Theory and Techniques*, vol. 67, no. 11, pp. 4332–4340, 2019.

[193] E. R. Brown, W. D. Goodhue and T. C. L. G. Sollner, "Fundamental Oscillations up to 200 GHz in Resonant Tunneling Diodes and New Estimates of Their Maximum Oscillation Frequency from Stationary-State Tunneling Theory", *Journal of Applied Physics*, vol. 64, no. 3, pp. 1519–1529, 1988.

[194] E. R. Brown, T. C. L. G. Sollner, C. D. Parker, W. D. Goodhue and C. L. Chen, "Oscillations up to 420 GHz in GaAs/AlAs Resonant Tunneling Diodes", *Applied Physics Letters*, vol. 55, no. 17, pp. 1777–1779, 1989.

[195] N. Orihashi, S. Hattori, S. Suzuki and M. Asada, "Experimental and Theoretical Characteristics of Sub-Terahertz and Terahertz Oscillations of Resonant Tunneling Diodes Integrated with Slot Antennas", *Japanese Journal of Applied Physics*, vol. 44, part 1, no. 11, pp. 7809–7815, 2005.

[196] D. D. Coon and H. C. Liu, "Frequency Limit of Double Barrier Resonant Tunneling Oscillators", *Applied Physics Letters*, vol. 49, no. 2, pp. 94–96, 1986.

[197] S. Luryi, "Frequency Limit of Double-Barrier Resonant-Tunneling Oscillators", *Applied Physics Letters*, vol. 47, no. 5, pp. 490–492, 1985.

[198] J. P. Mattia, A. L. McWhorter, R. J. Aggarwal, F. Rana, E. R. Brown and P. Maki, "Comparison of a Rate-Equation Model With Experiment for the Resonant Tunneling Diode in the Scattering-Dominated Regime", *Journal of Applied Physics*, vol. 84, no. 2, pp. 1140–1148, 1998.

[199] H. P. Joosten, H. J. M. F. Noteborn, K. Kaski and D. Lenstra, "The Stability of the Self-Consistently Determined Current of a Double-Barrier Resonant-Tunneling Diode", *Journal of Applied Physics*, vol. 70, no. 6, pp. 3141–3147, 1991.

[200] M. N. Feiginov, "Effect of the Coulomb Interaction on the Response Time and Impedance of the Resonant-Tunneling Diodes", *Applied Physics Letters*, vol. 76, no. 20, pp. 2904–2906, 2000.

[201] M. Feiginov, C. Sydlo, O. Cojocari and P. Meissner, "High-Frequency Nonlinear Characteristics Of Resonant-Tunnelling Diodes", *Applied Physics Letters*, vol. 99, no. 13, pp. 133501, 2011.

[202] M. N. Feiginov, "Displacement Currents and the Real Part of High-Frequency Conductance of the Resonant-Tunneling Diode", *Applied Physics Letters*, vol. 78, no. 21, pp. 3301–3303, 2001.

[203] S. Clochiatti, K. Aikawa, K. Arzi, E. Mutlu, M. Suhara, N. Weimann and W. Prost, "Large-Signal Modelling of sub-THz InP Triple Barrier Resonant Tunneling Diodes", *Third International Workshop on Mobile Terahertz Systems (IWMTS)*, 2020.

[204] C. Kidner, I. Mehdi, J. R. East and G. I. Haddad, "Power and Stability Limitations of Resonant Tunneling Diodes", *IEEE Transactions on Microwave Theory and Techniques*, vol. 38, no. 7, pp. 864–872, 1990.

[205] S. Diebold, S. Nakai, K. Nishio, J. Kim, K. Tsuruda, T. Mukai, M. Fujita and T. Nagatsuma, "Modeling and Simulation of Terahertz Resonant Tunneling Diode-Based Circuits", *IEEE Transactions on Terahertz Science and Technology*, vol. 6, no. 5, pp. 716–723, 2016.

[206] M. E. Hines, "High-Frequency Negative-Resistance Circuit Principles for Esaki Diode Applications", *The Bell System Technical Journal*, vol. 39, no. 3, pp. 477–513, 1960.

[207] L. Wang, "Reliable Design of Tunnel Diode and Resonant Tunnelling Diode Based Microwave Sources", Ph.D. Thesis, chapter 2, University of Glasgow, 2012.

[208] J. Wang, A. Al-Khalidi, R. Morariu, I. A. Ofiare, L. Wang and E. Wasige, "15 Gbps Wireless Link Using W-Band Resonant Tunnelling Diode Transmitter", 48th European Microwave Conference (EuMC), pp. 1405–1408, 2018.

[209] L. Wang, "Reliable Design of Tunnel Diode and Resonant Tunnelling Diode Based Microwave Sources", Ph.D. Thesis, chapter 8, University of Glasgow, 2012.

[210] C. Kidner, I. Mehdi, J. East and G. Haddad, "Bias Circuit Instabilities and Their Effect on the D.C. Current–Voltage Characteristics of Double-Barrier Resonant Tunneling Diodes", *Solid-State Electronics*, vol. 34, no. 2, pp. 149–156, 1991.

[211] W. F. Chow, *Principles of Tunnel Diode Circuits*", Wiley, 1964.

[212] J. Wang, A. Al-Khalidi, L. Wang, R. Morariu, A. Ofiare and E. Wasige, "15-Gb/s 50-cm Wireless Link Using a High-Power Compact III–V 84-GHz Transmitter", *IEEE Transactions on Microwave Theory and Techniques*, vol. 66, no. 11, pp. 4698–4705, 2018.

[213] M. Shiraishi, H. Shibayama, K. Ishigaki, S. Suzuki, M. Asada, H. Sugiyama and H. Yokoyama, "High Output Power (~400 μW) Oscillators at around 550 GHz Using Resonant Tunneling Diodes with Graded Emitter and Thin Barriers", *Applied Physics Express*, vol. 4, no. 6, pp. 064101, 2011.

[214] S. Suzuki, M. Shiraishi, H. Shibayama and M. Asada, "High-Power Operation of Terahertz Oscillators With Resonant Tunneling Diodes Using Impedance-Matched Antennas and Array Configuration", *IEEE Journal of Selected Topics in Quantum Electronics*, vol. 19, no. 1, pp. 8500108–8500108, 2013.

[215] K. Kasagi, S. Suzuki and M. Asada, "Large-Scale Array of Resonant Tunneling- Diode Terahertz Oscillators for High Output Power at 1 THz", *Journal of Applied Physics*, vol. 125, no. 15, pp. 151601, 2019.

[216] M. Egard, M. Arlelid, L. Ohlsson, B. M. Borg, E. Lind and L. Wernersson, "$In_{0.53}Ga_{0.47}As$ RTD-MOSFET Millimeter-Wave Wavelet Generator", *IEEE Electron Device Letters*, vol. 33, no. 7, pp. 970–972, 2012.

[217] J. Lee, M. Kim and K. Yang, "A 1.52 THz RTD Triple-Push Oscillator with a μW-Level Output Power", *IEEE Transactions on Terahertz Science and Technology*, vol. 6, no. 2, pp. 336–340, 2016.

[218] M. Kim, J. Lee, J. Lee and K. Yang, "A 675 GHz Differential Oscillator Based on a Resonant Tunneling Diode", *IEEE Transactions on Terahertz Science and Technology*, vol. 6, no. 3, pp. 510–512, 2016.

[219] J. Encomendero, R. Yan, A. Verma, S. M. Islam, V. Protasenko, S. Rouvimov, P. Fay, D. Jena and H. G. Xing, "Room Temperature Microwave Oscillations in Gan/Aln Resonant Tunneling Diodes With Peak Current Densities Up To 220 kA/cm^2", *Applied Physics Letters*, vol. 112, no. 10, pp. 103101, 2018.

[220] J. Wang, A. Ofiare, K. Alharbi, R. Brown, A. Khalid, D. Cumming and E. Wasige, "MMIC Resonant Tunneling Diode Oscillators for THz Applications", *11th Conference on Ph.D. Research in Microelectronics and Electronics (PRIME)*, pp. 262–265, 2015.

[221] E. Wasige, A. Al-Khalidi, K. Alharbi and J. Wang, "High Performance Microstrip Resonant Tunneling Diode Oscillators as Terahertz Sources", 9th UK-Europe-China Workshop on Millimetre Waves and Terahertz Technologies (UCMMT), pp. 25–28, 2016.

[222] A. Al-Khalidi, J. Wang and E. Wasige, "Compact J-band Oscillators with 1 mW RF Output Power and over 110 GHz Modulation Bandwidth", 43rd International Conference on Infrared, Millimeter, and Terahertz Waves (IRMMW-THz), pp. 1–2, 2018.

[223] M. Feiginov, C. Sydlo, O. Cojocari and P. Meissner, "Resonant Tunnelling-Diode Oscillators Operating at Frequencies above 1.1 THz", *Applied Physics Letters*, vol. 99, no. 23, pp. 233506, 2011.

[224] D. Horikawa, Y. Chen, T. Koike, S. Suzuki and M. Asada, "Resonant Tunneling-Diode Terahertz Oscillator Integrated with a Radial Line Slot Antenna for Circularly Polarized Wave Radiation", *Semiconductor Science and Technology*, vol. 33, no. 11, pp. 114005, 2018.

[225] M. Asada, K. Osada and W. Saitoh, "Theoretical Analysis and Fabrication of Small Area Metal/Insulator Resonant Tunneling Diode Integrated with Patch Antenna for Terahertz Photon Assisted Tunneling", *Solid-State Electronics*, vol. 42, no. 7–8, pp. 1543–1546, 1998.

[226] K. Okada, K. Kasagi, N. Oshima, S. Suzuki and M. Asada, "Resonant-Tunneling-Diode Terahertz Oscillator Using Patch Antenna Integrated on Slot Resonator for Power Radiation", *IEEE Transactions on Terahertz Science and Technology*, vol. 5, no. 4, pp. 613–618, 2015.

[227] K. Urayama, S. Aoki, S. Suzuki, M. Asada, H. Sugiyama and H. Yokoyama, "Sub-Terahertz Resonant Tunneling Diode Oscillators Integrated with Tapered Slot Antennas for Horizontal Radiation", *Applied Physics Express*, vol. 2, no. 4, pp. 044501, 2009.

[228] K. Kasagi, S. Fukuma, S. Suzuki and M. Asada, "Proposal and Fabrication of Dipole Array Antenna Structure in Resonant-Tunneling-Diode Terahertz Oscillator Array", 41st International Conference on Infrared, Millimeter, and Terahertz waves (IRMMW-THz), pp. 1–2, 2016.

[229] H. Yamakura and M. Suhara, "Proposal of Bow-Tie Antenna-Integrated Resonant Tunneling Diode Transmitter Utilizing Relaxation Oscillations and Its Application to Short-Distance Wireless Communications", *Journal of Infrared, Millimeter, and Terahertz Waves*, vol. 39, no. 11, pp. 1087–1111, 2018.

[230] K. H. Alharbi, A. Ofiare, M. Kgwadi, A. Khalid and E. Wasige, "Bowtie Antenna for Terahertz Resonant Tunnelling Diode Based Oscillators on High Dielectric Constant Substrate", 11th Conference on Ph.D. Research in Microelectronics and Electronics (PRIME), pp. 168–171, 2015.

[231] M. Reddy, S. C. Martin, A. C. Molnar, R. E. Muller, R. P. Smith, P. H. Siegel, M. J. Mondry, M. J. W. Rodwell, H. Kroemer and S. J. Allen, "Monolithic Schottky-Collector Resonant Tunnel Diode Oscillator Arrays to 650 GHz", *IEEE Electron Device Letters*, vol. 18, no. 5, pp. 218–221, 1997.

[232] N. Orihashi, S. Suzuki and M. Asada, "One THz Harmonic Oscillation of Resonant Tunneling Diodes", *Applied Physics Letters*, vol. 87, no. 23, pp. 233501, 2005.

[233] R. E. Neidert, S. C. Binari and T. Weng, "Dielectric Constant of Semi-Insulating Indium Phosphide", *Electronics Letters*, vol. 18, no. 23, pp. 987–988, 1982.

[234] M. Fox, *Optical Properties of Solids*, Oxford University Press, 2010.

[235] M. Asada and S. Suzuki, "Room-Temperature Oscillation of Resonant Tunneling Diodes Close to 2 THz and Their Functions for Various Applications", *Journal of Infrared, Millimeter, and Terahertz Waves*, vol. 37, no. 12, pp. 1185–1198, 2016.
[236] K. Hinata, M. Shiraishi, S. Suzuki, M. Asada, H. Sugiyama and H. Yokoyama, "Sub-Terahertz Resonant Tunneling Diode Oscillators with High Output Power (~ 200 μW) Using Offset-Fed Slot Antenna and High Current Density," *Applied Physics Express*, vol. 3, no. 1, pp. 014001, 2009.
[237] S. Suzuki, M. Asada, A. Teranishi, H. Sugiyama and H. Yokoyama, "Fundamental Oscillation Of Resonant Tunneling Diodes above 1 THz at Room Temperature", *Applied Physics Letters*, vol. 97, no. 24, pp. 242102, 2010.
[238] H. Kanaya, S. Suzuki and M. Asada, "Terahertz Oscillation of Resonant Tunneling Diodes with Deep and Thin Quantum Wells", *IEICE Electronics Express*, vol. 10, no. 18, pp. 20130501, 2013.
[239] J. Lee, M. Kim, J. Lee and K. Yang, "A Sub-mW D-band 2nd Harmonic Oscillator Using InP-Based Quantum-Effect Tunneling Devices", 26th International Conference on Indium Phosphide and Related Materials (IPRM), pp. 1–2, 2014.
[240] J. Lee, M. Kim, and K. Yang, "An InP-based RTD Triple-Push Oscillator Operating at over 1 THz", IEEE International Conference of Indium Phosphide and Related Materials, 2015.
[241] S. Kitagawa, S. Suzuki and M. Asada, "Wide Frequency-Tunable Resonant Tunnelling Diode Terahertz Oscillators Using Varactor Diodes", *Electronics Letters*, vol. 52, no. 6, pp. 479–481, 2016.
[242] J. Wang, L. Wang, C. Li, B. Romeira, and E. Wasige, "28 GHz MMIC Resonant Tunnelling Diode Oscillator of around 1 mW Output Power", *Electronics Letters*, vol. 49, no. 13, pp. 816–818, Jun. 2013.
[243] X. Yu, J. -Y. Kim, M. Fujita and T. Nagatsuma, "Efficient Mode Converter to Deep-Subwavelength Region with Photonic-Crystal Waveguide Platform for Terahertz Applications", *Optics Express*, vol. 27, no. 20, pp. 28707–28721, 2019.
[244] T. Yamamoto, N. Nishigami, Y. Nishida, M. Fujita and T. Nagatsuma, "16-QAM Wireless Communications Using Resonant Tunneling Diodes at 60 Gbit/s", IEICE General Conference, 2020.
[245] N. Nishigami, Y. Nishida, S. Diebold, J. Kim, M. Fujita and T. Nagatsuma, "Resonant Tunneling Diode Receiver for Coherent Terahertz Wireless Communication", Asia-Pacific Microwave Conference (APMC), pp. 726–728, 2018.
[246] D. Pozar, *Microwave and RF Design of Wireless Systems*, Wiley, 2000.
[247] M. Asada and S. Suzuki, "Theoretical Analysis of Coupled Oscillator Array Using Resonant Tunneling Diodes in Sub-Terahertz and Terahertz Range", *Journal of Applied Physics*, vol. 103, no. 12, pp. 124514, 2008.
[248] K. Arzi, S. Suzuki, A. Rennings, D. Erni, N. Weimann, M. Asada and W. Prost, "Subharmonic Injection Locking for Phase and Frequency Control of RTD-Based THz Oscillator", *IEEE Transactions on Terahertz Science and Technology*, vol. 10, no. 2, pp. 221–224, 2020.
[249] Y. Nishida, S. Diebold, K. Tsuruda, T. Mukai, J. -Y. Kim, M. Fujita and T. Nagatsuma, "Injection-Locked Resonant Tunneling Diode Receiver for Terahertz Communications", 4th International Symposium on Microwave/Terahertz Science and Applications, 2017.
[250] A. Van Der Ziel and E. R. Chenette, "Noise in Solid State Devices", *Advances in Electronics and Electron Physics*, vol. 46, pp. 313–383, 1978.
[251] G. Keller, A. Tchegho, B. M¨unstermann, W. Prost, F. Tegude and M. Suhara, "Triple Barrier Resonant Tunneling Diodes for Microwave Signal Generation and Detection", European Microwave Integrated Circuit Conference, pp. 228–231, 2013.
[252] K. Arzi, S. Clochiatti, S. Suzuki, A. Rennings, D. Erni, N. Weimann, M. Asada and W. Prost, "Triple-Barrier Resonant-Tunnelling Diode THz Detectors with on-chip antenna", 12th German Microwave Conference (GeMiC), pp. 17–19, 2019.
[253] K. Arzi, A. Rennings, D. Erni, N. Weimann, W. Prost, S. Suzuki and M. Asada, "Millimeter-wave Signal Generation and Detection via the same Triple Barrier RTD and on-chip Antenna", First International Workshop on Mobile Terahertz Systems (IWMTS), pp. 1–4, 2018.

[254] K. Arzi, S. Clochiatti, E. Mutlu, A. Kowaljow, B. Sievert, D. Erni, N. Weimann, W. Prost, K. Arzi, A. Rennings, D. Erni, N. Weimann, W. Prost, S. Suzuki and M. Asada, "Broadband Detection capability of a Triple Barrier Resonant Tunneling Diode", Second International Workshop on Mobile Terahertz Systems (IWMTS), pp. 1–4, 2019.

[255] Y. Koyama, R. Sekiguchi and T. Ouchi, "Oscillations up to 1.40 THz from Resonant-Tunneling-Diode-Based Oscillators with Integrated Patch Antennas", *Applied Physics Express*, vol. 6, no. 6, pp. 064102, 2013.

[256] R. Sekiguchia, Y. Koyama and T. Ouchi, "Subterahertz Oscillations from Triple-Barrier Resonant Tunneling Diodes with Integrated Patch Antennas", *Applied Physics Letters*, vol. 96, no. 6, pp. 062115, 2010.

[257] N. Oshima, K. Hashimoto, S. Suzuki and M. Asada, "Wireless Data Transmission of 34 Gbit/S At A 500-Ghz Range Using Resonant-Tunnelling Diode Terahertz Oscillator", *Electronics Letters*, 52, no. 22, pp. 1897–1898, 2016.

[258] K. Ishigaki, M. Shiraishi, S. Suzuki, M. Asada, N. Nishiyama and S. Arai, "Direct intensity Modulation and Wireless Data Transmission Characteristics of Terahertz-Oscillating Resonant Tunnelling Diodes", *Electronics Letters*, vol. 48 no. 10, pp. 582–583, 2012.

[259] K. Ishigaki, K. Karashima, M. Shiraishi, H. Shibayama, S. Suzuki and M. Asada, "Direct Modulation of THz-Oscillating Resonant Tunneling Diodes", International Conference on Infrared, Millimeter, and Terahertz Waves, pp. 1–2, 2011.

[260] S. Bhattacharjee, J. H. Booske, C. L. Kory, D. W. van der Weide, S. Limbach, S. Gallagher, J. D. Welter, M. R. Lopez, R. M. Gilgenbach, R. L. Ives, M. E. Read, R. Divan and D. C. Mancini, "Folded Waveguide Traveling-Wave Tube Sources for Terahertz Radiation", *IEEE Transactions on Plasma Science*, vol. 32, no. 3, pp. 1002–1014, 2004.

[261] N. Oshima, K. Hashimoto, S. Suzuki and M. Asada, "Terahertz Wireless Data Transmission With Frequency and Polarization Division Multiplexing Using Resonant-Tunneling-Diode Oscillators", *IEEE Transactions on Terahertz Science and Technology*, vol. 7, no. 5, pp. 593–598, 2017.

[262] B. B. Purkayastha and K. K. Sarma, "A Digital Phase Locked Loop based Signal and Symbol Recovery System for Wireless Channel", *Springer*, 2015.

[263] L. Ohlsson and L. -E. Wernersson, "A 15-Gb/s Wireless ON-OFF Keying Link", *IEEE Access*, vol. 2, pp. 1307–1313, 2014.

[264] M. Asada and S. Suzuki, "Terahertz Oscillators Using Resonant Tunneling Diodes", *Asia-Pacific Microwave Conference (APMC)*, pp. 521–523, 2018.

[265] K. Nishio, S. Diebold, S. Nakai, K. Tsuruda, T. Mukai, J. -Y. Kim, M. Fujita and T. Nagatsuma, "Resonant Tunneling Diode Receivers for 300-GHz-band Wireless Communications", *URSI-Japan Radio Science Meeting*, 2015.

[266] S. Diebold, K. Nishio, Y. Nishida, J. -Y. Kim, K. Tsuruda, T. Mukai, M. Fujita and T. Nagatsuma, "High-Speed Error-Free Wireless Data Transmission Using a Terahertz Resonant Tunnelling Diode Transmitter and Receiver", *Electronics Letters*, vol. 52, no. 24, pp. 1999–2001, 2016.

[267] X. Yu, R. Yamada, J. Kim, M. Fujita and T. Nagatsuma, "Integrated Circuits Using Photonic-Crystal Slab Waveguides and Resonant Tunneling Diodes for Terahertz Communication", *Progress in Electromagnetics Research Symposium (PIERS-Toyama)*, pp. 599–605, 2018.

[268] J. Webber, N. Nishigami, X. Yu, J. -Y. Kim, M. Fujita and T. Nagatsuma, "Terahertz Wireless Communication using Resonant Tunneling Diodes and Practical Radio-over-Fiber Technology", *IEEE International Conference on Communications Workshops (ICC Workshops)*, pp. 1–5, 2019.

[269] R. Piesiewicz, T. Kleine-Ostmann, N. Krumbholz, D. Mittleman, M. Koch, J. Schoebel and T. Kurner, "Short-Range Ultra-Broadband Terahertz Communications: Concepts and Perspectives", *IEEE Antennas and Propagation Magazine*, vol. 49, no. 6, pp. 24–39, 2007.

[270] T. Nagatsuma, M. Fujita, A. Kaku, D. Tsuji, S. Nakai, K. Tsuruda and T. Mukai, "Terahertz Wireless Communications Using Resonant Tunneling Diodes as Transmitters and Receivers", Proceedings of the Third International Conference on Telecommunications and Remote Sensing – Volume 1: ICTRS, pp. 41–46, 2014.

[271] R. P. Smith, S. T. Alien, M. Reddy, S. C. Martin, J. Liu, R. E. Muller and M. J. W. Rodwell, "0.1 µm Schottky-Collector AlAs/GaAs Resonant Tunneling Diodes", *IEEE Electron Device Letters*, vol. 15, no. 8, pp. 295–297, 1994.

[272] M. Reddy, M. J. Mondry, M. J. W. Rodwell, S. C. Martin, R. E. Muller, R. P. Smith, D. H. Chow and J. N. Schulman, "Fabrication and dc, Microwave Characteristics of Submicron Schottky-Collector AlAs/$In_{0.53}Ga_{0.47}As$/InP Resonant Tunneling Diodes", *Journal of Applied Physics*, vol. 77, no. 9, pp. 4819–4821, 1995.

[273] R. E. Muller, S. C. Martin, R. P. Smith, S. A. Allen, M. Reddy, U. Bhattacharya and M. J. W. Rodwell, "Electron-Beam Lithography for the Fabrication of Air-Bridged Submicron Schottky Collectors", *Journal of Vacuum Science & Technology B: Microelectronics and Nanometer Structures Processing, Measurement, and Phenomena*, vol. 12, no. 6, pp. 3668–3672, 1994.

[274] K. D. Stephan, E. R. Brown, C. D. Parker, W. D. Goodhue, C. L. Chen and T. C. L. G. Sollner, "Resonant-Tunnelling Diode Oscillator Using a Slot-Coupled Quasioptical Open Resonator", *Electronics Letters*, vol. 27, no. 8, pp. 647–649, 1991.

[275] E. R. Brown, C. D. Parker, K. M. Molvar and K. D. Stephan, "A Quasi-Optically Stabilized Resonant-Tunneling-Diode Oscillator for the Millimeter- and Sub-Millimeter-Wave Regions", *IEEE Transactions on Microwave Theory and Techniques*, vol. 40, no. 5, pp. 846–850, 1992.

[276] E. R. Brown, C. D. Parker, A. R. Calawa, M. J. Manfra and K. M. Molvar, "A Quasi-Optical Resonant-Tunneling-Diode Oscillator Operating above 200 GHz", *IEEE Transactions on Microwave Theory and Techniques*, vol. 41, no. 4, pp. 720–722, 1993.

[277] T. Fujii, H. Mazaki, F. Takei, J. Bae, M. Narihiro, T. Noda, H. Sakaki, K. Mizuno, "Coherent Power Combining of Millimeter Wave Resonant Tunneling Diodes in a Quasi-Optical Resonator", *IEEE MTT-S International Microwave Symposium Digest*, vol. 2, pp. 919–922, 1996.

[278] J. Lee, M. Kim, J. Park and J. Lee, "225 GHz Triple-Push RTD Oscillator with 0.5 mW dc-Power Consumption", *IET Circuits, Devices & Systems*, vol. 14, no. 2, pp. 209–215, 2020.

[279] A. C. Cornescu, R. Morariu, A. Ofiare, A. Al-Khalidi, J. Wang, J. M. L. Figueiredo and E. Wasige, "High-Efficiency Bias Stabilization for Resonant Tunneling Diode Oscillators", *IEEE Transactions on Microwave Theory and Techniques*, vol. 67, no. 8, pp. 3449–3454, 2019.

[280] H. I. Cantu, B. Romeira, A. E. Kelly, C. N. Ironside and J. M. L. Figueiredo, "Resonant Tunneling Diode Optoelectronic Circuits Applications in Radio-Over-Fiber Networks", *IEEE Transactions on Microwave Theory and Techniques*, vol. 60, no. 9, pp. 2903–2912, 2012.

[281] J. Wang, A. Al-Khalidi, C. Zhang, A. Ofiare, L. Wang, E. Wasige and J. M. L. Figueiredo, "Resonant Tunneling Diode as High Speed Optical/Electronic Transmitter", 10th UK-Europe-China Workshop on Millimetre Waves and Terahertz Technologies (UCMMT), pp. 1–4, 2017.

[282] S. Watson, W. Zhang, J. Tavares, J. Figueiredo, H. Cantu, J. Wang, E. Wasige, H. Salgado, L. Pessoa and A. Kelly, "Resonant Tunneling Diode Photodetectors for Optical Communications", *Microwave and Optical Technology Letters*, vol. 61, no. 4, pp. 1121–1125, 2019.

[283] B. Romeira, L. M. Pessoa, H. M. Salgado, C. N. Ironside, and J. M. L. Figueiredo, "Photo-Detectors Integrated with Resonant Tunneling Diodes", *Sensors*, vol. 13, no. 7, pp. 9464–9482, 2013.

[284] S. Watson, W. Zhang, J. Wang, A. Al-Khalidi, H. Cantu, J. Figueiredo, E. Wasige and A. E. Kelly, "Resonant Tunneling Diode Oscillators for Optical Communications", Proceedings of the SPIE, Third International Conference on Applications of Optics and Photonics, vol. 10453, 2017.

[285] B. Romeira, J. M. L. Figueiredo, T. J. Slight, L. Wang, E. Wasige and C. N. Ironside, "Wireless/Photonics Interfaces Based on Resonant Tunneling Diode Optoelectronic Oscillators", *Conference on Lasers and Electro-Optics and International Quantum Electronics Conference*, pp. 1–2, 2009.

[286] W. Zhang, S. Watson, J. Figueiredo, J. Wang, H. I. Cant´u, J. Tavares, L. Pessoa, A. Al-Khalidi, H. Salgado, E. Wasige and A. E. Kelly, "Optical Direct Intensity Modulation of a 79GHz Resonant Tunneling Diode-Photodetector Oscillator", *Optics Express*, vol. 27, no. 12, pp. 16791–16797, 2019.

[287] T. Miyamoto, A. Yamaguchi and T. Mukai, "Terahertz Imaging System with Resonant Tunneling Diodes", *Japanese Journal of Applied Physics*, vol. 55, no. 3, pp. 032201, 2016.

[288] S. Kitagawa, M. Mizuno, S. Saito, K. Ogino, S. Suzuki and M. Asada, "Frequency-Tunable Resonant-Tunneling-Diode Terahertz Oscillators Applied to Absorbance Measurement", *Japanese Journal of Applied Physics*, vol. 56, no. 5, pp. 058002, 2017.

[289] A. Dobroiu, R. Wakasugi, S. Suzuki and M. Asada, "Toward a Solid-State, Compact, Terahertz-Wave Radar", *AIP Conference Proceedings*, vol. 2067, no. 1, pp. 020004, 2019.

[290] A. Dobroiu, R. Wakasugi, Y. Shirakawa, S. Suzuki and M. Asada, "Amplitude-Modulated Continuous-Wave Radar in the Terahertz Band Using a Resonant-Tunneling-Diode Oscillator", 44th International Conference on Infrared, Millimeter, and Terahertz Waves (IRMMW-THz), pp. 1–2, 2019.

[291] A. Dobroiu, R. Wakasugi, Y. Shirakawa, S. Suzuki and M. Asada, "Absolute and Precise Terahertz-Wave Radar Based on an Amplitude-Modulated Resonant-Tunneling-Diode Oscillator", *Photonics*, vol. 5, no. 4, 2018.

[292] J. Hu, R. Wakasugi, S. Suzuki and M. Asada, "Amplitude-Modulated Continuous-Wave Ranging System with Resonant-Tunneling-Diode Terahertz Oscillator", 43rd International Conference on Infrared, Millimeter, and Terahertz Waves (IRMMW-THz), pp. 1–2, 2018.

[293] A. Dobroiu, S. Suzuki and M. Asada, "Terahertz-Wave Radars Based on Resonant-Tunneling-Diode Oscillators", *Proceedings of the SPIE, Terahertz Emitters, Receivers, and Applications X*, vol. 11124, 2019.

[294] A. Dobroiu, R. Wakasugi, Y. Shirakawa, S. Suzuki and M. Asada, "Amplitude-Modulated Continuous-Wave Radar in the Terahertz Range Using Lock-In Phase Measurement", *Measurement Science and Technology*, 2020.

[295] Y. Shirakawa, A. Dobroiu, S. Suzuki, M. Asada and H. Ito, "Principle of a Subcarrier Frequency-modulated Continuous-Wave Radar in the Terahertz Band Using a Resonant-Tunneling-Diode Oscillator", 44th International Conference on Infrared, Millimeter, and Terahertz Waves (IRMMW-THz), pp. 1–2, 2019.

[296] O. K. Nilsen and W. D. Boyer, "Amplitude Modulated CW Radar", *IRE Transactions on Aerospace and Navigational Electronics*, vol. ANE-9, no. 4, pp. 250–254, 1962.

[297] T. Jaeschke, C. Bredendiek and N. Pohl, "A 240 GHz Ultra-Wideband FMCW Radar System With On-Chip Antennas For High Resolution Radar Imaging", *IEEE MTT-S International Microwave Symposium Digest (MTT)*, pp. 1–4, 2013.

[298] B. Romeira, J. M. L. Figueiredo and J. Javaloyes, "Delay Dynamics Of Neuromorphic Optoelectronic Nanoscale Resonators: Perspectives and Applications", *Chaos: An Interdisciplinary Journal of Nonlinear Science*, vol. 27, no. 11, pp. 114323, 2017.

Chapter 3

Characterisation of Emitters and Detectors

Mira Naftaly

Contents

3.1 Introduction

This chapter presents an overview of main characterisation techniques for emitter (transmitter) and detector (receiver) devices used in THz wireless links. As such, it is focused on devices operating in free space, an on optical (free space) measurement techniques and instrumentation.

As the demand for wireless bandwidth escalated over the last decade, the terahertz (THz) frequency range has attracted a growing interest for the next generation of wireless communications [1–5]. To date the majority of effort in developing devices and components for the physical layer has been primarily concerned with frequencies between 100 and 500 GHz [2]. The measurement methods and instruments surveyed in this chapter are particularly suited to such devices, and in all cases are applicable to frequencies up to 1.5 THz.

Device characterisation is important and necessary for two reasons. First, accurate and detailed knowledge of device performance is essential in order to be able to design, simulate and build well-functioning wireless links, and to deploy them effectively.

Second, the ability to specify device performance in accordance with a standardised methodology makes it possible to compare devices reliably. This sets benchmarks for device development and optimisation. It also makes it possible to select the most suitable components from different manufacturers.

THz technologies, including THz wireless communications, are currently transitioning from academic research to industrial roll-out, with widespread uptake being envisaged in the course of the next decade. Robust metrological underpinning [6] is necessary to support both research and development (R&D) and industrial uptake. As the telecom industry prepares for 6G and beyond, both equipment manufacturers and network operator end-users demand component performance standards and calibration services.

Standardisation bodies are at work on developing standards for THz wireless links and components; and regulatory bodies require such standards be in place before granting approval. The ultimate goal is to establish a robust framework of metrological traceability and calibration services, and to equip all users for reliable implementation of basic required device characterisation measurements.

3.2 Metrology Definitions and Parameters

In order to quantify system performance, it is necessary to define the performance parameters and to determine their values and the uncertainties in those values [6–8]. Metrological performance parameters include sensitivity and resolution, precision and accuracy, dynamic range (DR) and signal-to-noise ratio (SNR).

Sensitivity and resolution are related, in that sensitivity refers to the minimum measurable value, whereas resolution refers to the minimum measurable difference between two values. Precision is the numerical value of measurement reproducibility. These three quantities – sensitivity, resolution and precision – define the measurement uncertainty.

Experimental values are determined with associated uncertainty, as $\bar{x}\pm\Delta x$, where $\bar{x}$ represents the mean value

$$\bar{x} = \frac{1}{N}\sum_{i=1}^{N} x_i \tag{3.1}$$

and Δx is the measurement uncertainty given by the standard deviation of the measured value x, calculated as:

$$\Delta x = \sigma = \sqrt{\frac{1}{N}\sum_{i=1}^{N}\left(x_i - \bar{x}\right)^2} \tag{3.2}$$

where N is the number of measurements.

This is so-called Type A uncertainty, derived directly from the measured data. If the distribution of measured values is normal (Gaussian), then 68.2% of data points will lie in the range $\bar{x}\pm\Delta x$. Type B uncertainty applies to derived values and is calculated using standard methods of error propagation from known uncertainties in the data and system operation [8].

In some types of measurement, it is also customary to define the 'noise floor', which is the standard deviation when the mean signal equals zero, that is, $NF = (\bar{x} = 0)$.

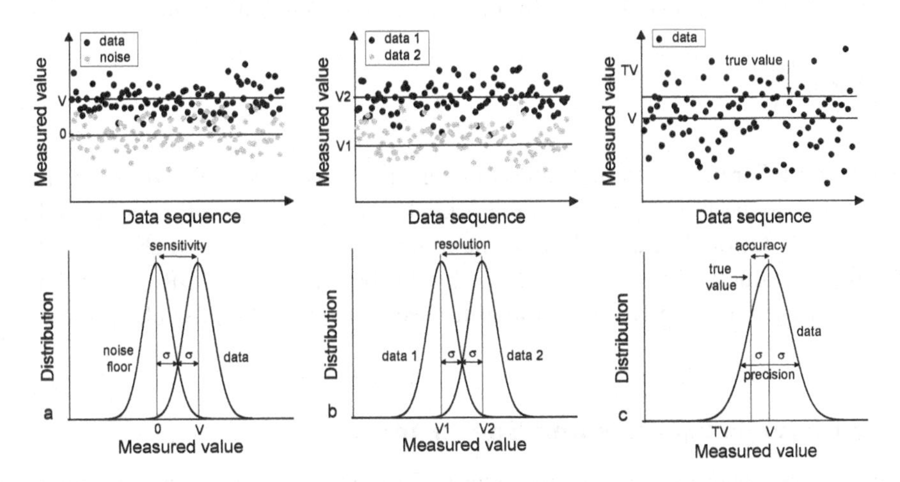

Figure 3.1 Schematic depiction of metrological performance parameters: sensitivity, noise floor, resolution, precision and accuracy. Top: sequentially measured data points. Bottom: data distribution.

Sensitivity, or the minimum measurable value, can be understood as being roughly equal to twice the noise floor, or twice the uncertainty for a small signal (Figure 3.1a). Similarly, resolution, as the minimum measurable difference between two values, is roughly equal to twice the uncertainty of those values (Figure 3.1b).

Precision is the reproducibility of the measurement, and as such can also be interpreted as being represented by the uncertainty (Figure 3.1c). That is because reproducibility can be viewed as the probability that a given data point will lie within one standard deviation (i.e. uncertainty) of the mean value, which can be calculated for any distribution of the measured data.

Sensitivity, resolution and precision are all determined internally to the experimental system, by performing repeated measurements and analysing the data to determine its mean, standard deviation and form of distribution.

In contrast, accuracy is evaluated by reference to an externally known quantity. Accuracy is defined as the difference between the 'true' value and the value measured by the experimental system. As such, it must be determined by calibrating the system against a known standard or calibration artefact (Figure 3.1c).

For a measurement to be considered correct it is necessary that the measured and true values 'agree within error'; in other words, that the true value lie within one standard deviation of the mean data, that is, $\bar{x}_{meas} - \sigma < x_{true} < \bar{x}_{meas} + \sigma$. This requires that accuracy have a lower value than precision (i.e. accuracy must be encompassed within precision, as seen in Figure 3.1c).

The DR and SNR of a system are also defined internally to the system, based on the analysis of the measured data. However, unlike the parameters discussed above, DR and SNR are generally used to refer to the directly measured signal, and not to any derived quantities. DR is the ratio of the maximum absolute value of the measurable signal to the noise floor:

$$DR = \frac{|x_{max}|}{NF} \tag{3.3}$$

whereas SNR is the ratio of the absolute mean signal to its uncertainty:

$$SNR = \frac{|\bar{x}|}{\Delta x} \tag{3.4}$$

SNR and DR reflect different and complementary aspects of the system performance. SNR indicates the minimum detectable signal change, and therefore is related to sensitivity and resolution. DR describes the maximum quantifiable signal change and indicates the range of values that can be measured by the system. Clearly, DR must always be significantly larger than SNR for meaningful measurement to be possible; preferably DR > 10 SNR.

3.3 Emitter Frequency and Spectrum

Time and frequency standards are some of the most important internationally agreed standards, underpinning essential areas of everyday life such as communications, computing, the internet, global positioning and navigation, time-stamping of financial transactions and many more. Accurate and precise knowledge of frequency is also essential for all types of telecommunication links. Time and frequency are of course interchangeable, being reciprocal quantities. Many frequency standards exist in different areas of technology and in all frequency ranges. All, however, are ultimately traceable to the International Atomic Time Standard based on atomic clocks [9,10]. Currently, the precision of these clocks is of the order of 10^{-16} [11], making time one of the highest-precision quantities measurable by human endeavour.

3.3.1 Heterodyne Frequency Measurements

Frequency standards in the form of high-precision oscillators also underpin frequency measurements of THz emitters. Heterodyne detection has long been established as the most widely used, accurate, high-resolution, high-precision, traceably calibrated technique for measuring signal frequencies and spectral profiles. It is employed across the whole of the electromagnetic spectrum, from the optical [12] down to millimetre-wave [13] and RF.

Heterodyne measurements use a local oscillator (LO) of a known frequency, amplitude and phase. The LO signal is fed together with the source signal into a 'mixer', which is a non-linear device whose output is proportional to the product of the two input signals (see Figure 3.2):

$$E_S \cos \omega_S t \times E_{LO} \cos \omega_{LO} t = E_S E_{LO} \left[\frac{1}{2} \sin(\omega_S - \omega_{LO})t - \frac{1}{2} \sin(\omega_S + \omega_{LO})t \right] \tag{3.5}$$

where E_S and ω_S respectively are the amplitude and angular frequency of the source, and E_{LO} and ω_{LO} are those of the LO. The LO frequency is chosen such that $(\omega_S - \omega_{LO}) \ll \omega_S$. The difference-frequency signal $E_S E_{LO} \sin(\omega_S - \omega_{LO})t$ is then easily detected and analysed using low-frequency circuits (whereas the sum-difference signal generally lies outside the detector bandwidth). An IF spectrum analyser is commonly used, whose readout is produced as $E_S(\omega_S)$.

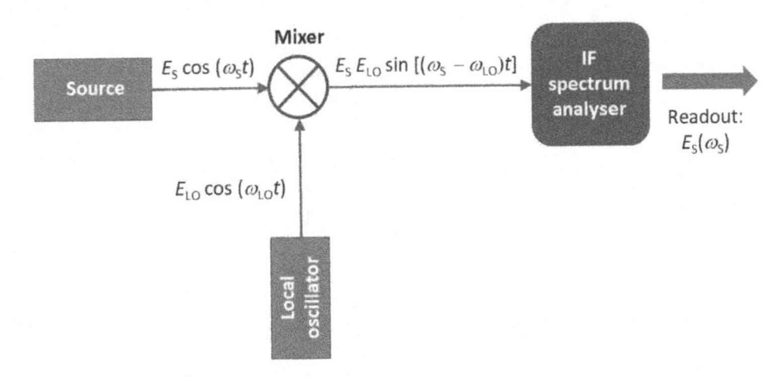

Figure 3.2 Schematic depiction of heterodyne measurement.

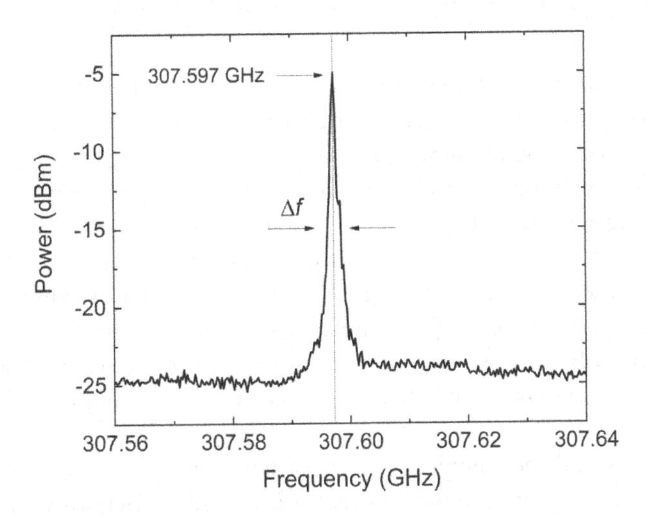

Figure 3.3 Frequency of a resonant tunnelling diode measured using the heterodyne technique. Courtesy of Abdullah Al-Khalidi, University of Glasgow, UK [14].

An example of heterodyne measurement is presented in Figure 3.3, which shows the emission spectrum of a resonant tunnelling diode (RTD) [14]. The linewidth of the frequency peak reflects the coherence length (L_c) of the source, which can be calculated from:

$$L_c = \sqrt{\frac{2\ln 2}{\pi}}\frac{c}{\Delta f} \tag{3.6}$$

where c is the speed of light and Δf is the FWHM linewidth of the source (full width at half-maximum). The coherence length therefore increases inversely with the linewidth. In communication transmitters long coherence length helps reduce the BER. For the RTD shown in Figure 3.3, the centre frequency is 307.597 GHz and the coherence length is 100 m. It is therefore highly coherent.

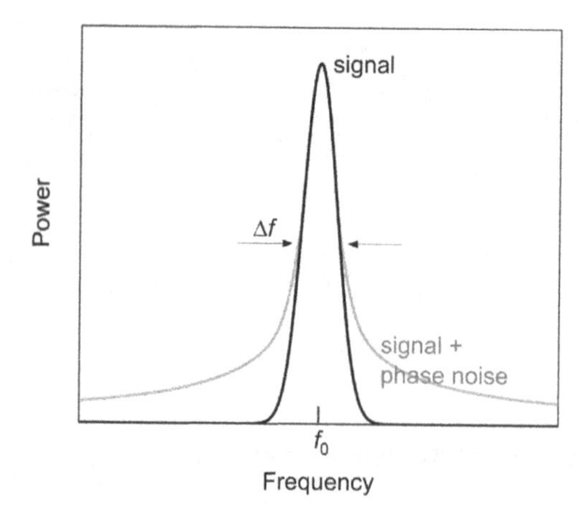

Figure 3.4 The effect of phase noise on the source lineshape (model data).

In measuring the emitter linewidth, it is important to take into account the frequency resolution of the spectrum analyser and the shape factor of its filters, both of which may broaden the observed signal linewidth.

The broader lineshape or spectral profile of the signal, and especially its low wings, indicates the phase noise of the source [15]. Phase noise is measured as the noise power relative to the signal, and varies with the frequency offset from the peak centre. It is typically symmetric around the signal peak, and its major component is $1/f$ noise. Figure 3.4 depicts the effect of phase noise using model data: it increases the linewidth and reduces the coherence length. In communication transmitters phase noise is highly detrimental, as it increases the BER. The RTD depicted in Figure 3.3 is seen to have a low phase noise.

The techniques and metrological aspects of heterodyne signal analysis are well understood, and the instrumentation industry provides a wide variety of well-characterised and traceably calibrated instruments, including vector network analysers VNAs and signal (or spectrum) analysers. In heterodyne measurements of THz sources the main technological challenges are: (i) a THz frequency LO and (ii) a mixer device operating adequately at THz frequencies. A state of the art spectrum analyser can measure frequencies up to 1.5 THz with the accuracy of around 10^{-7} [16].

3.3.2 *Interferometric Spectral Measurements*

The centre frequencies and linewidths of narrow-band THz emitters, such as those used for THz wireless links, can thus be measured with great accuracy and resolution by heterodyne-based techniques and instruments. However, these are unsuitable for characterising the spectra of sources that produce extended frequency features such as side-lobes or harmonics, or have a broad spectral profile. In such cases, the full-source spectrum can be measured by employing a broadband optical interferometer.

Although there are several commonly used interferometer designs, they all employ the same basic principle of homodyne detection, which may also be termed autocorrelation or self-interference. As its name suggests, in contrast to heterodyne, homodyne does not

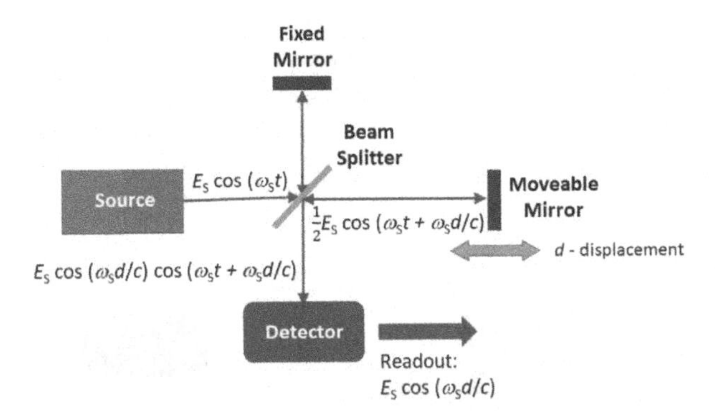

Figure 3.5 A schematic depiction of operation of a Michelson interferometer with a narrow-frequency source.

employ an external LO. Instead, the source signal is split into two parts which are made to interfere, while one of them is phase-shifted relative to the other.

Figure 3.5 depicts the operation of a Michelson interferometer, which is the simplest type of interferometric design. The beam from the source to be analysed is split into two parts and directed to two mirrors, one of which is fixed while the other is moveable. The reflected beams are then recombined and interfere on a detector. Translation of the moveable mirror introduces a variable phase difference in one of the beams, and a corresponding interference pattern on the detector. The source spectrum is calculated from the interferogram by applying Fourier transform. An ideal narrow-frequency source will produce a pure sinusoidal interference pattern as a function of mirror displacement d:

$$\frac{1}{2}E_S \cos\omega_S t + \frac{1}{2}E_S\cos\left(\omega_S t + \omega_S \frac{d}{c}\right) = E_S \cos\left(\omega_S \frac{d}{c}\right)\cos\left(\omega_S t + \omega_S \frac{d}{c}\right) \tag{3.7}$$

A type of interferometer that is particularly suited for spectroscopy at THz frequencies is a Michelson interferometer with a lamellar mirror, shown in Figure 3.6, where a split mirror acts as both a beam-splitter and a moveable mirror [17], thus avoiding the need for a transmission beam-splitter which is problematic at THz frequencies. The lamellar mirror consists of two parts, each comprising several lamellae or 'fingers', with one part being fixed and the other moveable. The operational bandwidth of the interferometer is determined by the number (N) and height (h) of the lamellae. The ratio of maximum to minimum frequency is roughly equal to the number of lamellae in each half of the mirror: $f_{max}/f_{min}{\approx}N$. The minimum frequency is set by the lamellae height: $f_{min}>2c/h$. However, the fall-off at both high and low band edges is shallow, allowing operation over a much larger range, albeit with reduced contrast. However, it should be noted that in practice the bandwidth of an interferometer is often limited by the power of the emitter and the sensitivity of the detector, together with the requirement to maintain sufficient SNR for a valid measurement.

The frequency resolution of a Michelson interferometer is determined by the scanning length of the moveable mirror D, and is given by $\Delta f{=}c/2D$, where the factor 2 arises due to double-pass. Typical achievable resolutions are of the order of 1 GHz, which

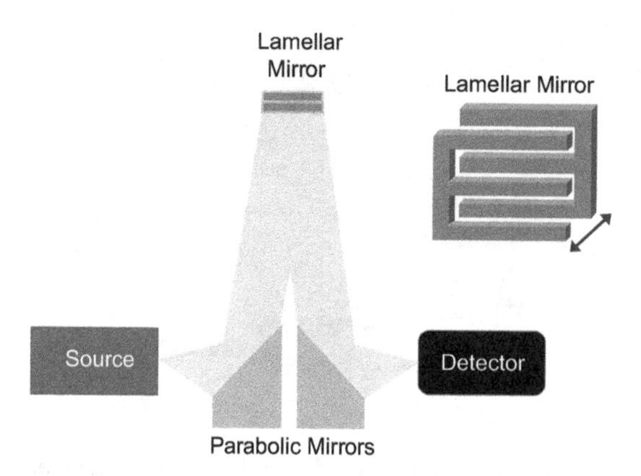

Figure 3.6 Schematic drawing of a lamellar interferometer and a split mirror.

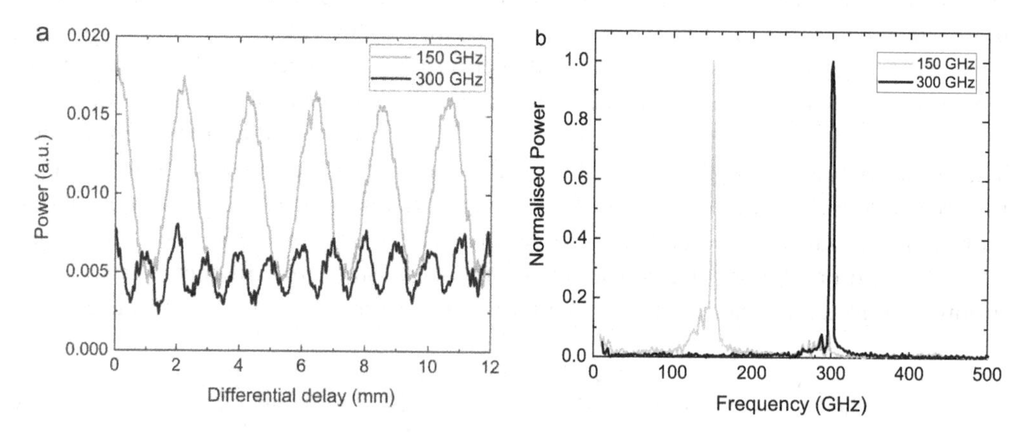

Figure 3.7 (a) Lamellar interferometer signal from a PIN diode emitter at nominal 150 and 300 GHz. (b) Calculated emitter spectra.

is much coarser than that of a signal analyser. Unlike heterodyne-based instruments, interferometers do not employ a LO; therefore careful attention must be given to their testing and calibration.

Figure 3.7 shows the interferograms and spectra of a PIN diode source operating at the nominal frequencies of 150 and 300 GHz. The instrument used for these measurements had a mirror with 10 lamellae in each half, with a height of 5 mm. The reduced amplitude at 300 GHz is due to lower emitter power. A significant shoulder is seen on the low-frequency side of the 150 GHz spectrum.

A lamellar interferometer has an additional advantage in that it can be used to determine the spectral performance of both emitters and detectors. To characterise an emitter, a calibrated power detector is required, having a flat response over the bandwidth of interest. Conversely, to characterise a detector, a calibrated broadband source with a known spectrum is necessary.

In order to fully characterise a THz emitter, its spectrum must be measured using both a narrow-band heterodyne signal analyser and a broadband optical interferometer. Of particular importance is the possible presence of features outside the fundamental oscillation line, such as harmonics and side-lobes, and their dependence on the operating conditions.

3.4 Emitter Power

Accurate knowledge of emitter power is essential first and foremost for the design and operation of wireless communication links. It is also required for health and safety and electromagnetic compatibility (EMC) certification. Beyond that, power measurements form the basis of other types of characterisation measurements, such as interferometry and beam profiling. However, unlike frequency, high precision is not required for power measurements: precision of 10^{-3} is sufficient for most purposes.

All the THz power meters described in this section are thermal devices. In a thermal detector, absorbed radiation heats the sensing element and the output signal is proportional to the temperature change. Since the detector response is determined by the absorbed power, it is frequency-independent, and therefore thermal detectors are typically highly broadband. However, thermal response is by its nature slow, and therefore thermal detectors have slow response times of the order of 10 Hz. In contrast, electronic and photonic detectors are fast, with response times ranging from kHz to MHz, but with responsivity that is strongly frequency-dependent. As a result, they are generally not suitable for accurate power measurements.

3.4.1 Types of Power Measurement Devices, Their Operation and Properties

Calorimeters have long been used for power measurement of RF, microwave and millimetre-wave radiation. An isothermal calorimeter has two identical radiation inputs terminated with absorbers; radiation is coupled into one of these, while the other is electrically heated to maintain identical temperature. The electrically dissipated power should then be equal to the absorbed radiation power. The operational frequency range is determined by the absorber. Calorimeters can be traceably calibrated, and can have high sensitivity and accuracy. The main disadvantage is their slow response time, especially for low incident power. Calorimeters measure average input power or continuous-wave (unmodulated) power. In recent years, a THz calorimeter has been developed [18] that has become a de facto industry standard [19]. Unlike the other THz power meter devices described below, which are all free space, the VDI-Erickson calorimeter is waveguide-coupled (Figure 3.8).

Pyroelectric detectors are the most widely used THz detectors / power meters. The sensing element is a pyroelectric material: absorbed radiation causes its temperature to rise, producing a change in its permanent electric dipole moment, which can be monitored in either current or voltage mode. The pyroelectric element is often coated with absorber in order to increase responsivity and the frequency range. Because signal is produced by temperature changes, only modulated (chopped) power can be detected. For THz and sub-THz power meters, the modulation frequency is typically 10–100 Hz. Pyroelectric detectors are robust, compact, inexpensive and convenient to use (Figure 3.9). They can also be traceably calibrated. In recent years PTB (Physikalisch-Technische Bundesanstalt) together with SLT (Sensor und Lasertechnik) have developed a calibrated THz power

Figure 3.8 The VDI-Erickson THz power meter PM5B [19].

Figure 3.9 Calibrated pyroelectric detectors from SLT [21].

meter with spectrally flat response from 0.3 to 10 THz and sensitivity down to ~1 μW [20,21].

Thermopiles are a series array of thermocouples which use the thermoelectric (Peltier) effect to convert a thermal gradient created by the absorption of radiation into an electrical signal. They have sensitivity approaching that of pyroelectric detectors; the frequency

Figure 3.10 A Golay cell detector from Tydex [23].

bandwidth is similarly determined by the radiation absorbing layer. They are generally larger than pyroelectric detectors, and equally robust. Unlike pyroelectric detectors, they produce output signal in response to continuous-wave illumination. The main disadvantage is slow response time. Calibrated thermopile detectors are commercially available [21].

The Golay cell is a type of acousto-optic sensor that has had a distinguished history in the development of THz measurements [22,23]. THz radiation entering the detector is absorbed by a thin metal layer, which then heats the gas in a cell and raises its pressure. One wall of the cell is a flexible mirror whose deformation is optically detected, producing the readout signal. Golay cells have the highest sensitivities of all room-temperature THz detectors (Figure 3.10). However, they are extremely fragile and easily damaged by overexposure to illumination. Note that damage occurs even when the device is switched off; therefore it should always be capped with an opaque lid when not in use. Their responsivity drops with ageing and exposure to radiation due to the loss of elasticity in the flexible mirror; and their response becomes sub-linear at higher power levels. For that reason, they also cannot be reliably calibrated. They are slower than pyroelectric detectors; and similarly detect only modulated power. Because the absorber is a thin metal film, they operate over a broad frequency range, 0.1–10 THz.

Cryogenic bolometers have by far the highest sensitivity of all THz detectors. They exploit free-carrier absorption by electrons in a doped semiconductor. Many different designs and materials are available. The advantages are high sensitivity, fast response time and low noise. The great disadvantage is that these are cryogenic devices, bulky and inconvenient, expensive and requiring support facilities. No calibrated bolometers are available; they are not suitable for accurate power measurements.

Table 3.1 summarises the properties of commercially available THz detectors/power meters.

Table 3.1 Commercially available THz detectors/power meters

Detector Type	*Minimum detectable power*	*NEP (W/√Hz)*	*Frequency range (THz)*	*Modulation bandwidth*	*Calibrated?*	*Ease of use*
Calorimeter	1 μW	10^{-8}	0.1–3	CW	Yes	Very slow, robust, waveguide-coupled. De facto industry standard.
Pyroelectric	1 μW	10^{-9}–10^{-10}	0.1–10	5–200 Hz	Yes	Robust and compact, convenient, inexpensive
Thermopile	10 μW	10^{-9}	0.1–10	CW – 1 Hz	Yes	Slow, robust and compact, convenient, inexpensive.
Golay Cell	<0.1 μW	10^{-10}	0.1–10	5–50 Hz	No	Slow, very fragile, responsivity drops with age, non-linear at higher powers.
Composite bolometer	0.1 nW	10^{-12}–10^{-14}	0.1–30	10^3 Hz	No	Cryogenic, expensive, large and inconvenient; requires support facilities.

3.4.2 Issues in Power Measurements

In performing THz power measurements, it is necessary to be aware of several measurement issues relating to detector operation and measurement procedures.

Detector calibration and linearity. Detector responsivity is provided by the manufacturer and should be calibrated using a well-characterised source. Preferably, calibration should be traceable to a standard power source, such as a black body [24] (see also Section 3.6). Detector responsivity should vary minimally throughout its lifetime. However, it is a good practice to check its power calibration at regular intervals, since responsivity may be reduced by wear or damage. This is especially the case for Golay cells, whose responsivity tends to decline with age. Most power detectors have a highly linear response within their range of operation, that is, their output signal is proportional to the incident power. The maximum measurable power is usually limited by the saturation level of the amplifier circuit. Golay cells are an exception, showing a gradually falling response at higher power levels.

Spectral responsivity. The responsivity of the power metre may vary with the frequency of the incident radiation. All the power detectors described above are activated thermally, that is, by the absorbed radiation heating the sensor element. Therefore in principle their responsivity can be independent of frequency. However, in all these devices absorption of incident radiation is mediated by a special layer whose absorptivity is frequency-dependent. Therefore the spectral absorption properties of the radiation absorber in a detector are a crucial factor in determining its frequency range and the flatness of its spectral response. The 'THz range' is commonly defined as encompassing frequencies from 0.1 to 30 THz, that is, two and half decades. Consequently, designing absorbers with flat spectral response over the largest possible frequency range is a major component of device engineering for THz power meters [25]. The spectral responsivity data for a THz power meter should be provided by the manufacturer.

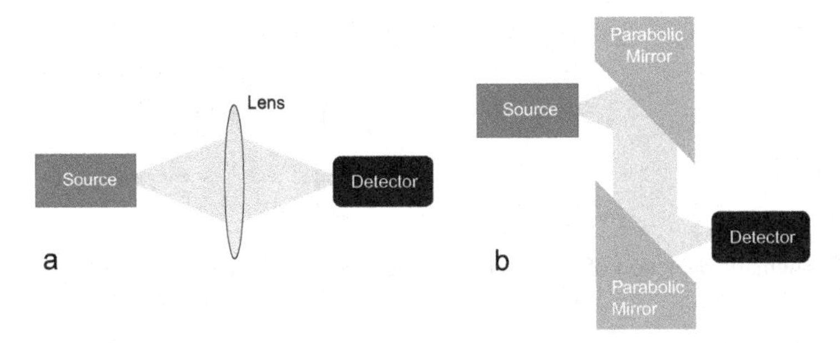

Figure 3.11 Examples of optics for maximising beam collection using (a) a lens and (b) two parabolic mirrors.

Dependence on chopping frequency. Of the detectors in Table 3.1, the calorimeter and the thermopile accept a CW (i.e. DC) input. The pyroelectric and Golay detectors require modulated (AC) input. Both have a slow response time; therefore the chopping frequency must be low. Detector responsivity decreases rapidly when the chopping frequency exceeds the optimum for the specific detector. For THz pyroelectric detectors, the optimum chopping frequency is 20–50 Hz; For the SLT detector mentioned above, it is 25 Hz. For Golay cells the optimum chopping frequency is around 10 Hz.

Beam collection efficiency. THz emitters typically produce strongly diverging beams. In contrast, THz power meters typically have apertures that are smaller than the beam diameter even in close proximity (~10 mm) to the emitter. In consequence, in order to measure accurately the emitter power, it is necessary to quantify and maximise the fraction of the beam collected by the detector. When using a waveguide-coupled detector, the equivalent is to quantify the coupling losses. It is advisable therefore to measure beam divergence first (see Section 3.5), and then to use collection optics to maximise the collection efficiency. Figure 3.11 depicts two examples of such optical setups using (a) a lens and (b) two parabolic mirrors. Care must be taken to use optical elements of sufficient aperture diameter.

Standing waves. Emitters for wireless links are highly coherent. As a consequence, standing waves can form between the emitter and detector, that affect the detected power and introduce large errors in power data. A standing wave arises when the surfaces of the emitter and detector are partially reflective and form a (low-Q) cavity, causing some proportion of the beam to undergo multiple reflections [26]. A standing wave manifests as a sinusoidal variation of the detected power with the distance between the emitter and detector; the distance between two power maxima is $c/2f$ (Figure 3.12). In the visible or near-infrared this distance is <1 μm, which is unresolvable and is also smaller than the commonly present mechanical jitter; therefore standing waves do not affect optical power measurements. In contrast, at THz and sub-THz frequencies, the separation between power maxima is of the order of 0.1–1 mm, causing large positional variations in the observed power.

Standing waves are possibly the most severe problem in power measurements, because they are ubiquitous and extremely difficult to eliminate completely. Figure 3.13 shows the effect on power measurement. The effect can be greatly reduced by tilting the detector at 45° to the beam axis, as seen in Figure 3.13. Note, however, that this reduces the effective

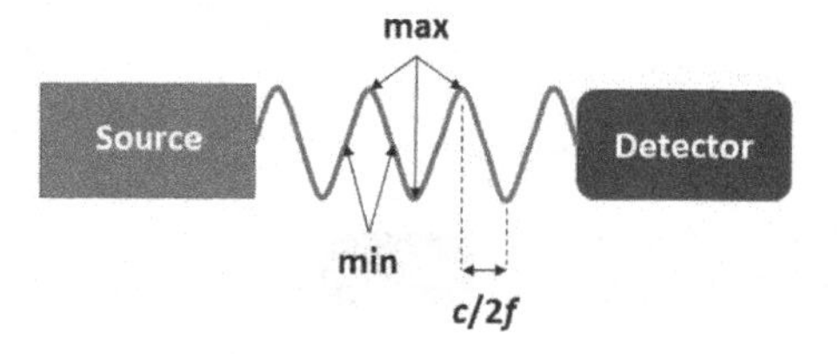

Figure 3.12 Schematic depiction of a standing wave between source and detector.

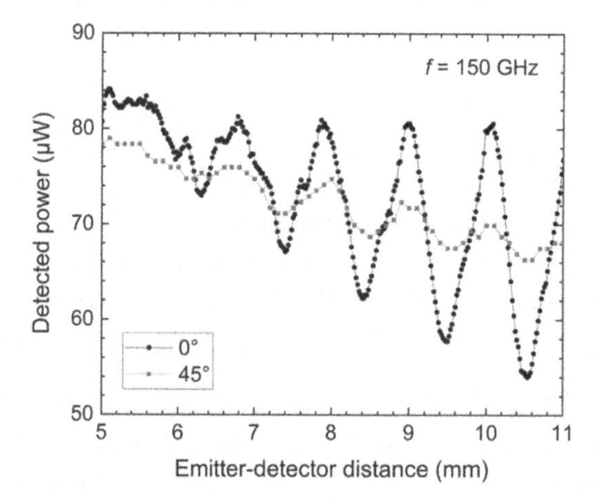

Figure 3.13 Standing waves in power measurements of a PIN diode at 150 GHz. The drop in the average power with distance is caused by the reduced collection efficiency due to beam divergence. Tilting the detector at 45° to the beam axis decreases the standing wave amplitude (the tilt data is adjusted for reduced collection efficiency). Courtesy of Jess Smith, University of Surrey, UK.

detector area, and therefore may reduce the collection efficiency; the tilt may also reduce detector absorptivity and thus its responsivity.

3.5 Emitter Beam Profile

Knowledge of the spatial profile and divergence of emitter beams is necessary for the design of wireless links and the requisite optical components. It is therefore highly desirable to spatially map both the field amplitude and the phase of propagating beams. Such characterisation of transmitter beam profiles has long since been accepted as an essential tool in designing microwave and millimetre wave communication systems. There are extensively developed and well understood techniques for antenna characterisation [27,28], and specialised facilities are available to perform the required measurements. None of these as yet exist for THz devices. Unlike microwave sources, THz emitters produce relatively low powers (commonly <100 μW), have short wavelengths (<1 mm) and there is a lack of compact, high-sensitivity detectors, in particular phase-sensitive detectors. These factors combine to make spatial characterisation of THz beams challenging.

3.5.1 Characterisation of Emitter Beam Profiles

In characterising beam profiles of THz emitters, it is important to ensure that measurements take place in the far field, that is, at the distance from the emitter where the angular variation of the beam remains constant with distance. The far-field distance L_{far} is determined by the radiation wavelength λ and the emitter aperture D, and is defined broadly as [28]:

$$\text{For } D < \frac{\lambda}{2} \quad L_{far} > 2\lambda = \frac{2c}{f} \tag{3.8a}$$

$$\text{For } D > \frac{\lambda}{2} \quad L_{far} > \frac{2D^2}{\lambda} = \frac{2D^2 f}{c} \tag{3.8b}$$

Measurements of RF and microwave antennas are often carried out in the near field due to spatial constraints, as their far field distance may be many metres long; and far field profiles are then calculated from near field data using tools developed by the antenna theory. However, at THz or sub-THz frequencies, where the far-field distance is commonly below 100 mm, it is preferable to measure beam profiles in the far field, thus yielding directly highly accurate beam profile information.

There are two approaches to describing beam profiles in the far field, depicted in Figure 3.14: as a function of lateral distance from the beam axis; and as a function of angular inclination relative to the beam axis. Lateral mapping is appropriate for collimated beams because they preserve their lateral dimensions and profile with distance from the source. For this reason, that is the method commonly adopted for optical beams, especially laser beams. Angular mapping is appropriate for diverging beams because they preserve their angular spread and variation with distance from the source. Therefore this is the method employed for mapping microwave sources, and which is similarly appropriate for THz emitters.

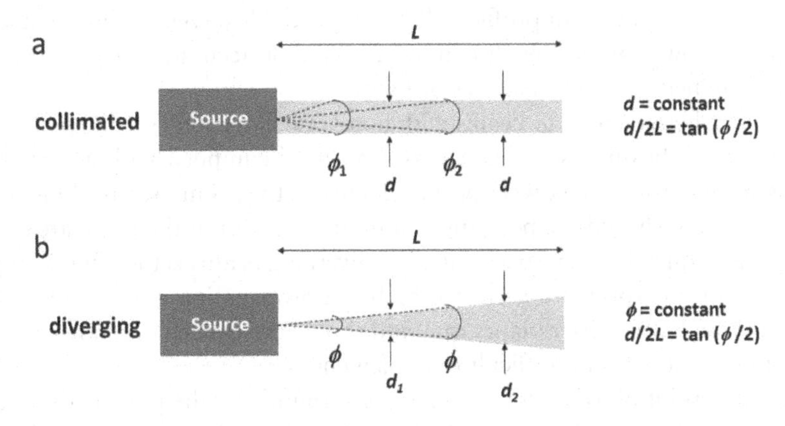

Figure 3.14 Simplified depiction of beam geometry of (a) a collimated and (b) a diverging beam.

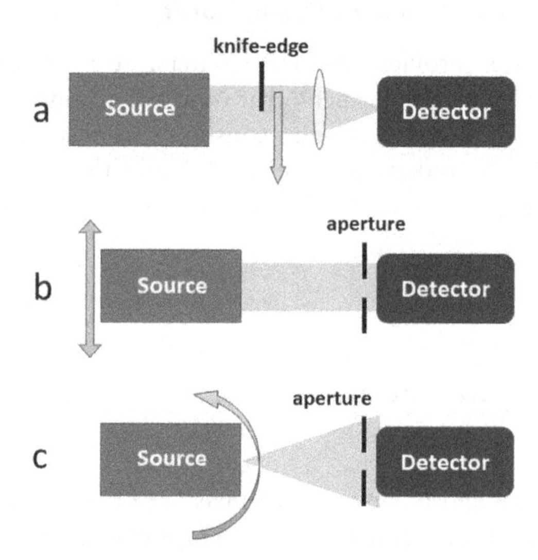

Figure 3.15 Measurement setups for beam intensity profile: (a) knife-edge scan; (b) pinhole line scan; (c) pinhole angular scan.

3.5.2 Intensity Profile Mapping

The intensity profile of a beam can be mapped using a power detector, such as described in Section 3.4. Figure 3.15 depicts setups commonly employed for measuring spatial beam profiles.

A technique frequently used in optics for collimated beams and to examine beam waist in the focal plane is the 'knife-edge', where an opaque edge (usually a thin metallic plate) is traversed across the beam area while a detector records the total beam power arriving (Figure 3.15a). A collecting lens is included in the setup because it is important to ensure that all of the beam is received by the detector. For a perfect Gaussian beam the dependence of power on the edge position is the error function, which can be differentiated to give the beam profile. The beam waist diameter can thus be determined, and broad deviations from the Gaussian profile will be revealed. However, as this method measures the integrated power along the chosen axis, it cannot accurately depict beams that are distorted, asymmetric or have features or lobes.

An approach better suited to beams with non-Gaussian profiles is the pinhole line scan, where the emitter beam is scanned across a detector equipped with a narrow aperture, producing an intensity (or power) profile (Figure 3.15b). Line scans along two orthogonal axes can (and should) be performed; a full raster scan of the beam area may also be carried out. An equivalent approach suited to diverging beams is the pinhole angular scan, where the emitter is rotated normal to the beam propagation axis (Figure 3.15c). Here also, rotations in both emitter planes can (and should) be carried out; however, an angular equivalent to a raster scan is difficult to implement.

The spatial resolution of a pinhole scan is determined by the pinhole diameter and the emitter–detector distance. This in turn is limited by the emitter power and detector sensitivity because sufficient radiation must reach the detector to provide adequate SNR.

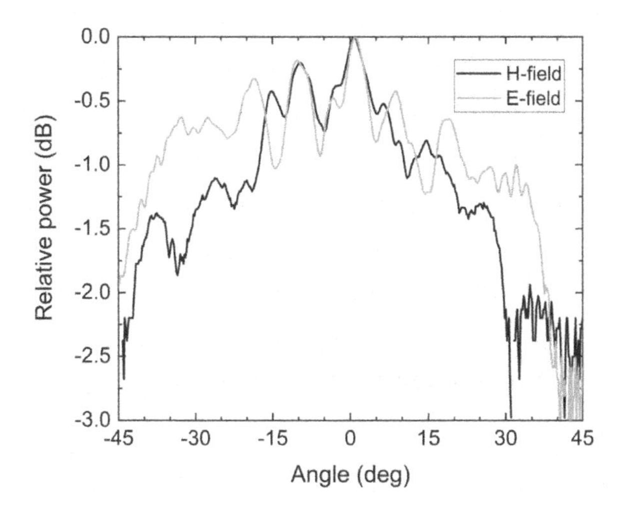

Figure 3.16 Angular pinhole scan of a PIN diode at 300 GHz. E-field is parallel and H-field is normal to beam polarisation.

An example of a pinhole angular scan measurement is shown in Figure 3.16. The emitter was a PIN diode operating at 300 GHz; E-field is parallel and H-field is normal to beam polarisation. It is seen that the beam profile has multiple lobes and irregular features, emphasising the importance of such measurements.

Instead of raster-scanning a detector, a THz camera can be employed to produce an image of the whole beam area. Several THz cameras are available, reviewed in a buyer's guide [29]. They employ different THz technologies, including pyro-electric, microbolometer, CMOS (complementary metal–oxide–semiconductor) and FET (field-effect transistor); and their performance specifications and prices vary widely. Sensitivity, pixel size and aperture size are particularly important criteria when choosing a camera suitable for imaging beam profiles of THz wireless transmitters. Sensitivity limits the minimum power required to obtain a beam image; it can range from around 20 pW for a microbolometer to a few μW for a pyro-electric. Pixel size determines the spatial resolution of the beam profile; it can vary from less than 50 μm to over 1 mm. Aperture limits the diameter of the beam that can be captured; these are typically 30–40 mm. Clearly, angular mapping of the beam is not possible using a camera.

3.5.3 Field Profile Mapping

The main disadvantage of pinhole scans using a power detector or a camera, as above, is that they reveal only the intensity profile of the beam, but are incapable of providing phase information. For coherent beams, such as produced by THz transmitters for wireless links, knowledge of the phase variation across the wavefront is crucial for calculations of beam propagation. Electro-optic (EO) detection offers the most promising route to THz beam imaging capable of determining both the field amplitude and phase across the beam profile [30,31].

The detection principle is depicted schematically in Figure 3.17. A THz beam and a colinear optical probe beam from a laser are incident on an EO crystal. Both beams must

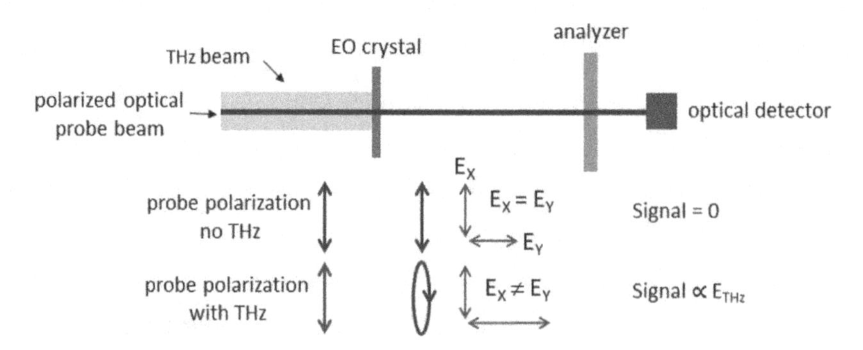

Figure 3.17 Schematic explanation of electro-optic detection of THz field.

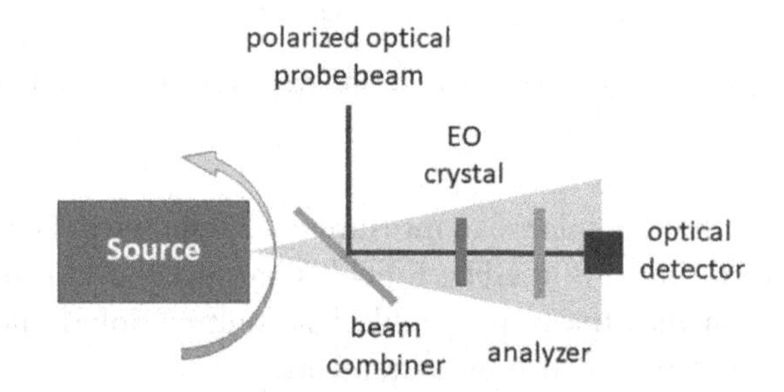

Figure 3.18 Measurement setup for beam imaging using EO detection.

be linearly polarised (either parallel or orthogonal to each other). In the presence of the THz field, the polarisation of the probe beam is rotated by the EO crystal, with the degree of rotation being proportional to the THz field amplitude. The probe beam then passes through an analyser, and an optical detector receives a signal that is proportional to the THz field amplitude. Unlike power measurements, EO detection is phase sensitive, because the signal is proportional to field amplitude, rather than intensity.

Beam profiling using EO detection can be implemented in a manner similar to that using a power detector, as depicted in Figure 3.18. The spatial resolution in this case is limited by the diameter of the probe beam. The main limitation of EO sensing is its relatively low sensitivity, which means that high emitter power is required to generate detectable signal. Although few implementations of EO beam imaging have been reported for continuous-wave THz or sub-THz emitters, a very detailed and thorough description of the technique is presented in Ref. [32].

As with beam intensity mapping described above, EO detection can be implemented using an optical camera to image the beam wavefront. The beam profile image can then be acquired at video rates, limited only by the camera speed and data analysis. However, the emitter power required to produce an image is considerably greater than for single-point detection. EO beam imaging using a camera has been reported for pulsed THz beams [33] whose peak power is three to four orders of magnitude higher than that of continuous-wave sources.

3.6 Detector Responsivity

Responsivity of power detectors is their fundamental and most important property. Responsivity is the conversion factor between the radiative power incident on the detector and the electrical signal at the detector output; it is given in units of W/V or W/A. Spectral responsivity refers to the variation of responsivity with the frequency of incident radiation. Knowing spectral responsivity of a detector therefore makes it possible to measure accurately the incident power of known frequency.

Detector responsivity is calibrated using a standard radiation source. An ideal radiation standard is a perfect 'black body', which is a body whose absorptivity and emissivity equal 1 at all frequencies. Although perfect black bodies do not exist, close approximations are available for various frequency ranges. A black body at a constant temperature radiates a spectrum that is determined only by its temperature, described by Plank's law [34]. THz frequencies correspond to cryogenic temperatures: for example, a black body whose peak emission is at 1 THz has a temperature of 9.7 K; while a peak at 300 GHz corresponds to 2.9 K. Therefore THz detectors have to be calibrated using the low wing of black body radiation [24,35]. Figure 3.19 shows examples of black body spectra at room temperature, below it, and above.

As an alternative to a black body, THz detectors may also be calibrated using a stabilised laser source, which has itself been calibrated against a black body. Established procedures exist for calibrating THz detectors using both black body sources and lasers [36–38]. However, all of these require highly specialised equipment and meticulous attention to all factors that may affect measurement.

A practical approach to determining the responsivity of a detector is to compare its response to that of a known calibrated power meter.

1. Measure the spectral power of a source using a calibrated power meter.
2. Measure the same source using the detector being tested. Ensure that the beam collection efficiency is the same for both detectors, or else factor in the difference.
3. Calculate the responsivity of the tested detector from the known power of the source.

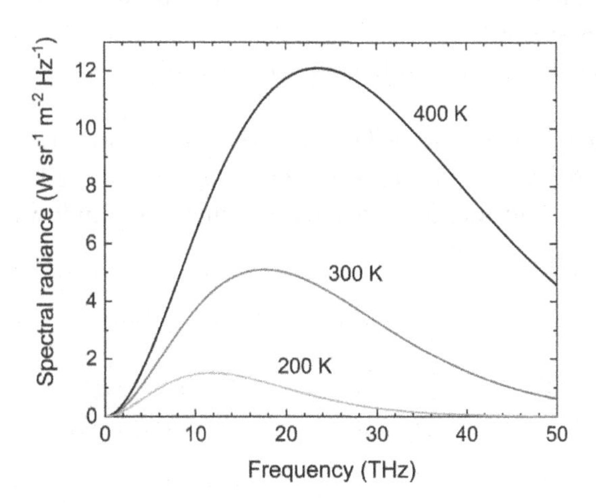

Figure 3.19 Examples of black body spectra at 200, 300 and 400 K, as given by Plank's law.

It is also important to determine the DR of a detector and to verify its power linearity. This can be done by varying the power of the source in a series of steps, increasing the power by a known factor at each step, and noting whether the detected signal varies by the same factor.

Although high responsivity is clearly an advantage in measuring weak signals, the detector sensitivity is of equal importance because it determines the lowest measurable signal. In everyday practice, sensitivity of a measurement may be defined as the signal level where SNR drops below 2.

Noise-equivalent power (NEP) of a detector is a measure of its lowest detectable signal. NEP is commonly defined as the input signal that produces SNR=1 in a 1 Hz output bandwidth (BW); its units are W/√Hz [39,40]. The minimum detectable signal is then calculated as NEP × √BW. A detector with a lower NEP is more sensitive. Note that NEP increases with the measurement bandwidth, therefore it benefits from bandwidth-narrowing techniques. The NEP of a detector can be evaluated by measuring its noise spectrum in the absence of input signal and then calculating the 'equivalent' signal using known detector responsivity.

3.7 Detector Acceptance

Similarly to emitter beam profiles, knowledge of the spatial acceptance properties of receivers is also important for designing and modelling wireless links. The acceptance cone of a detector refers to the variation of the detector response with the angle of incidence of the incoming beam.

The acceptance cone may be measured in a straightforward manner by rotating the detector normal to the beam axis, as depicted in Figure 3.20. In order to obtain accurate measurements with good angular resolution, the incident beam must be well collimated and its dimensions restricted by an aperture. To determine the cone in both detection planes, the detector is rotated along both axes.

Angular variation in the detected signal may be due to three factors:

- Angular dependence of absorptivity of the radiation absorption layer
- Angular dependence of reflectivity of entrance window or lens, in detectors equipped with such components
- Geometric factors that restrict beam collection efficiency.

It is not possible to separate these factors in the angular variation data. Power detectors such as described in Section 3.4 are commonly designed to have low sensitivity of absorptivity on the angle of incidence, and have no entrance windows or lenses, so in such

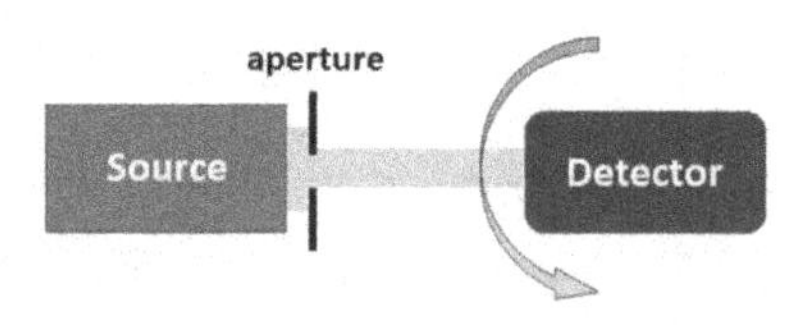

Figure 3.20 Measurement setup for determining the acceptance cone of a detector.

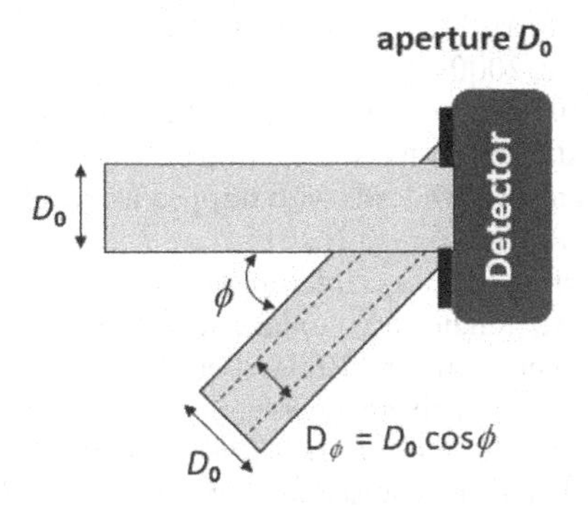

Figure 3.21 Angular dependence of beam collection efficiency imposed by the entrance aperture.

detectors these effects are negligible. However, in other types of detectors they may be significant.

Geometric factors restricting beam collection efficiency of a detector with the angle of incidence can be straightforward to identify and quantify in some cases. An example is shown in Figure 3.21, where an entrance aperture imposes a well-defined angular dependence. However, many types of detectors, especially those designed for wireless links, have much more complex geometries. In particular, many such detectors are equipped with collection lenses or receiver antennas whose beam collection behaviour is strongly angle-dependent.

THz sources are commonly polarised. Therefore it may be advisable to determine also the polarisation sensitivity of a detector. Power detectors described in Section 3.4 are polarisation insensitive; however, other types of detectors may have significant polarisation dependence. The simplest method of determining polarisation sensitivity of a detector is to employ a polarised source, and to compare the detected signal with the source polarisation set to be either horizontal or vertical.

References

[1] Key drivers and research challenges for 6G ubiquitous wireless intelligence (white paper), Editors: Matti Latva-aho, Kari Leppänen, Publisher: 6G Flagship, University of Oulu, Oulu, ISBN: 978-952-62-2353-7, September 2019, http://jultika.oulu.fi/files/isbn9789526223544.pdf

[2] Elayan, Hadeel, Osama Amin, Basem Shihada, Raed M. Shubair, and Mohamed-Slim Alouini. "Terahertz band: The last piece of RF spectrum puzzle for communication systems." *IEEE Open Journal of the Communications Society* 1 (2019): 1–32.

[3] Ghafoor, Saim, Noureddine Boujnah, Mubashir Husain Rehmani, and Alan Davy. "MAC protocols for terahertz communication: A comprehensive survey." *IEEE Communications Surveys & Tutorials* 22, no. 4 (2020): 2236–2282.

[4] Tekbıyık, Kürşat, Ali Rıza Ekti, Güneş Karabulut Kurt, and Ali Görçin. "Terahertz band communication systems: Challenges, novelties and standardization efforts." *Physical Communication* 35 (2019): 100700.

[5] Rappaport, Theodore S., Yunchou Xing, Ojas Kanhere, Shihao Ju, Arjuna Madanayake, Soumyajit Mandal, Ahmed Alkhateeb, and Georgios C. Trichopoulos. "Wireless communications and applications above 100 GHz: Opportunities and challenges for 6G and beyond." *IEEE Access* 7 (2019): 78729–78757.

[6] Naftaly, Mira. *Terahertz Metrology*. Artech House, 2015.
[7] Bendat, J. S. and S. G. Piersol, *Random Data Analysis and Measurement Procedures*, 3rd ed., John Wiley & Sons, Inc, 2000.
[8] Taylor, John. *Introduction to Error Analysis, the Study of Uncertainties in Physical Measurements*, 2nd ed., University Science Books, 1997.
[9] Margolis, H. S. "Frequency standards with trapped ions." In *Proceedings International School of Physics Enrico Fermi*, pp. 139–152. 2015.
[10] Katori, Hidetoshi. "Tricks for ticks." *Nature Physics* 13, no. 4 (2017): 414.
[11] Baynham, Charles F.A., Rachel M. Godun, Jonathan M. Jones, Steven A. King, Peter BR Nisbet-Jones, Fred Baynes, Antoine Rolland et al. "Absolute frequency measurement of the optical clock transition in with an uncertainty of using a frequency link to international atomic time." *Journal of Modern Optics* 65, no. 5–6 (2018): 585–591.
[12] Hirano, M., "Optical heterodyne measurement method", in *Handbook of Optical Metrology*, T. Yoshizawa, ed., CRC Press, 2015.
[13] N. B. Carvalho, D. Schreurs, *Microwave and Wireless Measurement Techniques*, Ch. 2, Cambridge University Press, 2013.
[14] Al-Khalidi, Abdullah, Khalid Hamed Alharbi, Jue Wang, Razvan Morariu, Liquan Wang, Ata Khalid, José Figueiredo, and Edward Wasige. "Resonant tunnelling diode terahertz sources with up to 1 mW output power in the J-band." *IEEE Transactions on Terahertz Science and Technology* (2019).
[15] Demir, Alper, Amit Mehrotra, and Jaijeet Roychowdhury. "Phase noise in oscillators: A unifying theory and numerical methods for characterization." *IEEE Transactions on Circuits and Systems I: Fundamental Theory and Applications* 47, no. 5 (2000): 655–674.
[16] www.keysight.com/gb/en/assets/7018-05092/brochures/5992-1316.pdf
[17] Naftaly, Mira, Paul Dean, Robert E. Miles, John R. Fletcher, and Andrei Malcoci. "A simple interferometer for the analysis of terahertz sources and detectors." *IEEE Journal of Selected Topics in Quantum Electronics* 14, no. 2 (2008): 443–448.
[18] Erickson, Neal. "A fast, very sensitive calorimetric power meter for millimeter to submillimeter wavelengths." In Thirteenth International Symposium on Space Terahertz Technology, Harvard University. 2002.
[19] http://vadiodes.com/en/products/power-meters-erickson
[20] Müller, Ralf, Berndt Gutschwager, Jörg Hollandt, Mathias Kehrt, Christian Monte, Ralph Müller, and Andreas Steiger. "Characterization of a large-area pyroelectric detector from 300 GHz to 30 THz." *Journal of Infrared, Millimeter, and Terahertz Waves* 36, no. 7 (2015): 654–661.
[21] www.pyrosensor.de/Home-924526.html
[22] Golay, Marcel JE. "A pneumatic infra-red detector." *Review of Scientific Instruments* 18, no. 5 (1947): 357–362.
[23] www.tydexoptics.com/products/thz_devices/golay_cell/
[24] Monte, C., B. Gutschwager, A. Adibekyan, and J. Hollandt. "A Terahertz blackbody radiation standard based on emissivity measurements and a Monte-Carlo simulation." *Journal of Infrared, Millimeter, and Terahertz Waves* 35, no. 8 (2014): 649–658.
[25] Müller, Ralf, Werner Bohmeyer, Mathias Kehrt, Karsten Lange, Christian Monte, and Andreas Steiger. "Novel detectors for traceable THz power measurements." *Journal of Infrared, Millimeter, and Terahertz Waves* 35, no. 8 (2014): 659–670.
[26] Rao, R. Srinivasa. *Microwave Engineering*. PHI Learning Pvt. Ltd., 2015.
[27] C. A. Balanis, *Modern Antenna Handbook*, Part IV, Wiley, 2011.
[28] Balanis, Constantine A. *Antenna Theory: Analysis and Design*. Wiley, 2016.
[29] Simoens, François. "Buyer's guide for a terahertz (THz) camera". *Photoniques* Special EOS Issue 2 (2018): 58–62. www.photoniques.com/articles/photon/pdf/2018/02/photon2018S3p58.pdf
[30] van der Valk, Nick C.J., Tom Wenckebach, and Paul C.M. Planken. "Full mathematical description of electro-optic detection in optically isotropic crystals." *JOSA B* 21, no. 3 (2004): 622–631.

[31] Planken, Paul C.M., Han-Kwang Nienhuys, Huib J. Bakker, and Tom Wenckebach. "Measurement and calculation of the orientation dependence of terahertz pulse detection in ZnTe." *JOSA B* 18, no. 3 (2001): 313–317.

[32] Deng, Yuqiang, Heiko Füser, and Mark Bieler. "Absolute intensity measurements of CW GHz and THz radiation using electro-optic sampling." *IEEE Transactions on Instrumentation and Measurement* 64, no. 6 (2014): 1734–1740.

[33] E. Abraham, H. Cahyadi, M. Brossard, J. Degert, E. Freysz, T. Yasui, "Development of a wavefront sensor for terahertz pulses", *Optics Express* 24 (2016) 5203–5211.

[34] Csele, Mark. *Fundamentals of Light Sources and Lasers*. Wiley, 2004.

[35] Gutschwager, Berndt, Christian Monte, Hossein Delsim-Hashemi, Oliver Grimm, and Jörg Hollandt. "Calculable blackbody radiation as a source for the determination of the spectral responsivity of THz detectors." *Metrologia* 46, no. 4 (2009): S165.

[36] Steiger, Andreas, Berndt Gutschwager, Mathias Kehrt, Christian Monte, Ralf Müller, and Jörg Hollandt. "Optical methods for power measurement of terahertz radiation." *Optics Express* 18, no. 21 (2010): 21804–21814.

[37] Müller, Ralf, Berndt Gutschwager, Jörg Hollandt, Mathias Kehrt, Christian Monte, Ralph Müller, and Andreas Steiger. "Characterization of a large-area pyroelectric detector from 300 GHz to 30 THz." *Journal of Infrared, Millimeter, and Terahertz Waves* 36, no. 7 (2015): 654–661.

[38] Deng, Yuqiang, Jing Li, and Qing Sun. "Traceable measurements of CW and pulse terahertz power with terahertz radiometer." *IEEE Journal of Selected Topics in Quantum Electronics* 23, no. 4 (2016): 1–6.

[39] Mackowiak, Verena, Jens Peupelmann, Yi Ma, and Anthony Gorges. "NEP–noise equivalent power." Thorlabs, Inc., Newton, NJ, USA, White Paper (2015).

[40] Richards, P. L. "Bolometers for infrared and millimeter waves." *Journal of Applied Physics* 76, no. 1 (1994): 1–24.

Part II

Terahertz Channel Characteristics and Modelling

Chapter 4

Fundamentals of Interference Modelling by Stochastic Geometry in THz Networks

Joonas Kokkoniemi, Janne Lehtomäki, and Markku Juntti

Contents

4.1 Introduction

The high-frequency wireless communication links and networks are seen as promising platforms to satisfy the ever-increasing amount of wireless and mobile data. Millimeter-wave band (30–300 GHz), and especially the above 100 GHz bands, such as D band (110–170 GHz) and THz band (300 GHz–10 THz), offer vast frequency resources for future +100 Gbps communication links. These frequencies have large numbers of use cases from long range backhauls, small cell base station fronthauls, all the way to device-to-device (D2D) communications. Especially above 100 GHz frequencies are seen the next big thing for achieving promises and goals set for the upcoming sixth generation (6G) mobile communication systems [1]. The biggest challenges for conquering these frequencies lie in the still prototype phase transceivers. Although some backhaul applications are not very sensitive to the device sizes due to stationary nature of those, the small base station fronthaul applications and user devices will require small and efficient radios. However, a lot of research is ongoing to overcome the hardware problems in the hope to see compatible consumer ready products on the markets in the near future.

The high-frequency bands suffer from high channel losses that will require directional communications. High-gain antennas and larger channel losses pose their own challenges for initial access and link maintenance in mobile applications. Extremely high bandwidths also bring some challenges on signal processing and hardware imperfections, such as the phase noise, limiting the performance even further. However, regardless of the unsolved problems, these frequencies show great potential to achieve very high capacities and very low latencies that enable various applications from high-speed connectivity to virtual and augment reality (VR/AR), sensor networks, and tactile applications, to mention a few.

As with all the networks at any frequency band, large numbers of users sharing the resources also unavoidably cause interference to coexisting links. High antenna gains effectively mean lower probability for the interference, but it also means high interference level when it occurs. To what extent the interference is a problem depends on the system and environment. For instance, a stationary backhaul connection would most likely be safe from interference due to very high gains and designed nature of such links. The mobile applications and especially D2D connections are far more likely to suffer from the interference. The interference and its impact on the network performance are traditionally studied by means of computer simulation models. However, there are also better tools available for certain network realizations.

The stochastic geometry is a very powerful tool to solve spatial network problems. Interference modeling is one very important application for the stochastic modeling due to the unwanted signals tend to arrive to Rx from spatially independent and random locations and directions. The stochastic geometry offers tools to solve network-level interference in a closed form without a need for time- and effort-consuming simulation models. The stochastic geometry has been studied widely on various network settings on different frequency bands [2–13]. The resultant stochastic models often give more accurate results due to closed form expression rather than looping over large simulation sets. However, the stochastic geometry is mostly usable in the case of random networks. Thus, regular simulation models do remain relevant in the case of well-organized networks with complex resource allocation and interference management schemes. Where applicable, the stochastic models have a clear advantage over the network simulators.

The THz frequencies suffer from molecular absorption loss and very high path loss in general. On the other hand, the channels tend to be simpler than the lower frequency systems due to highly directional links and the high losses attenuating the non-line-of-sight (NLOS) paths. Due to high penetration losses and directional antennas at transmitter (Tx) and receiver (Rx) ends, the THz frequencies in general do not support multipath propagation very efficiently, at least to the extent that some below 3 GHz low gain systems do. Thus, the channel model is inherently different from the traditional low frequency wireless communication systems. When utilizing the stochastic geometry for THz networks, the directionality and in general line-of-sight (LOS) type communications have to be taken into account in the interference modeling.

This chapter presents the basics of the stochastic geometry and stochastic interference modeling for the THz networks. Those offer a starting point to utilize and expand the stochastic geometry models for a variety of different network configurations in various usage scenarios. The next section gives a brief background of the stochastic geometry. Section 4.3 goes through the fundamental THz channel propagation features. Finally, the stochastic model for the interference is derived in Section 4.4.

4.2 Brief Background on Stochastic Geometry

The stochastic geometry relies on modeling the nodes and transmission in the network as point processes. Several types of point processes can be utilized, such as Poisson points processes (PPPs), binary point processes (BPPs), and many others [9]. By modeling the network as random node pairs having certain stochastic density, the stochastic geometry gives tools to mathematically characterize the interference, and for instance the moments of it. Thus, closed form expressions can be derived, for example, for the mean interference in the network. The stochastic geometry has been studied and proven by simulations to be very accurate in a large numbers of works, such as in Refs. [2–13]. Particularly good papers to get detailed fundamentals on the stochastic geometry are Refs. [3,9].

The basic setting is to model the network as randomly dropped Tx–Rx node pairs. In an infinite network, any pair of nodes experiences average behavior of the surrounding network of node pairs. The idea of the stochastic geometry is to utilize mathematics to describe the surround network as homogeneous (or heterogeneous) density field of interfering transmitters. By integrating over the space, the total aggregated interference can be obtained. How this aggregate interference is calculated exactly will be described later in this chapter. The starting point to solve the aggregate interference is the familiar expression for a summed interference at the Rx:

$$I=\sum_{i\in\Phi} P^{i}_{\mathrm{Tx}}\, h(r_{\mathrm{i}}), \tag{4.1}$$

where Φ is the set of active interfering nodes, P^{i}_{Tx} is the transmit power of the ith node, and $h(r_{\mathrm{i}})$ is the channel coefficient, where r_i is the distance. The latter is often taken as a simple power law $h(r_{\mathrm{i}})= \delta_{\mathrm{f}}\, r_{\mathrm{i}}^{-\alpha}$, where r_{i} is the Tx–Rx distance, α is the path loss exponent, and δ_{f} is the fading coefficient, for example, a Rayleigh fading coefficient. In an absorption dominated environment, this power law can be changed to $h(r_{\mathrm{i}}) = \delta_{\mathrm{f}}\exp(-\kappa r_{\mathrm{i}})$, where κ is the absorption coefficient [3]. The absorption loss model is covered in more detail below, but as a rule of thumb, in the THz band communications, the dominance of the loss mechanism strongly depends on the distance and frequency. Free space path loss (FSPL) tends to dominate at short distances and the molecular absorption loss dominates over long distances. However, this is also dependent on a particular frequency band of interest. Thus, in the following, we utilize the power law and the exponential absorption law make sure the derived interference is truly applicable to an arbitrary frequency band in general and within the THz frequencies.

There are many papers on the stochastic geometry for high-frequency networks, such as Refs. [14–20]. Those closely follow the general stochastic models that have been used in large numbers of papers in the past. What is different in the THz band is the usually lack of or low fading and molecular absorption loss. As a whole, the THz-specific papers focus on the propagation features that are typical to high-frequency communications. Those include high-gain antennas and large penetration loss causing blocking to become an issue. Some common channel specific problems are covered in the next section before going into the stochastic modeling of the interfering signals.

4.3 THz Band Propagation

In general, the channel losses tend to rise to a square of frequency when moving to higher frequencies. This is caused by the antenna effective area being proportional to the square of the wavelength. The path loss at the THz regime is very high and the communications always require directional antennas. This section looks into some basic phenomena that have impact on the propagation modeling of even simple THz systems.

4.3.1 Free Space Path Loss

The general free space channel is described by the Friis transmission equation. The channel gain with free space path loss becomes

$$h_{\mathrm{FSPL}}(r) = \frac{A_{\mathrm{e}}^{\mathrm{Rx}}}{4\pi r^2} G_{\mathrm{Tx}} = \frac{c^2}{(4\pi f r)^2} G_{\mathrm{Tx}} G_{\mathrm{Rx}}, \tag{4.2}$$

where r is the distance between Tx and Rx, G_{Tx} and G_{Rx} are the Tx and Rx antenna gains, respectively, c is the speed of light, f is the frequency, $4\pi r^2$ accounts for the free space expansion of the transmitted wave, and $A_{\mathrm{e}}^{\mathrm{Rx}}$ is the effective aperture of the Rx antenna, which is equivalent to

$$A_{\mathrm{e}}^{\mathrm{Rx}} = \frac{\lambda^2}{4\pi} G_{\mathrm{Rx}}, \tag{4.3}$$

where λ is the wavelength. Notice the importance of the effective antenna aperture above. When moving to high-gain systems, large aperture antennas may have considerable gain due to short wavelength. However, the last term in (4.2) also shows that the FSPL can be converted to the regular form by inserting the effective area into the equation. It is still good to remember what this expression resembles and that the physical and effective aperture antennas have significant impact on the achievable gain. This is certainly a good news for many THz band applications that require large antenna gains and can afford large aperture antennas. For instance, a parabolic reflector antenna has an effective gain described by $G = e_{\mathrm{a}} \left(\frac{\pi d}{\lambda} \right)^2$, where e_{a} is the aperture efficiency and d is the diameter of the parabolic reflector. We can see that the gain can be very large given the wavelength is small compared to the diameter of the antenna. As we can intuitively guess, the antenna effective aperture is multiplication between its physical aperture and the aperture efficiency. In the following we consider the last term of (4.2) as the FSPL since we do not apply any particular antenna structure herein, and ultimately the FSPL can be represented by the last term of (4.2) even if we did.

Notice the difference to a path loss model given above as $h(r_i) = \delta_f \, r_i^{-\alpha}$. Lower frequency propagation is often described by fading coefficient δ_{f} and a path loss exponent α. The latter is used to model different environments and is usually higher than two given by the pure spherical expansion of wave in the FSPL model. The reason is because the lower frequencies tend to have wide beam antennas and the radiation penetrates objects more easily. This leads to high numbers of signal propagation paths that are often difficult to characterize in closed from. Thus, the path loss exponent plus fading gives easier way to model complex environments. In the THz frequencies, there usually are not many paths available and different propagation phenomena are easier to handle in closed form. Therefore we use the path loss exponent two in (4.2).

4.3.2 *Molecular Absorption Loss*

The molecular absorption loss is caused by the signal energy exciting the molecules in the air to higher energy states. This leads to loss of the signal energy. The molecular absorption causes discrete deterministic fading to the signals and it is usually modeled by the Beer-Lambert law [21,22]:

$$\tau(r) = \mathrm{e}^{-\kappa_a(f)r}, \tag{4.4}$$

where $\kappa_a(f)$ is the absorption coefficient. This coefficient is a function of frequency, molecular composition of the channel, temperature, and pressure of the channel. There are databases [23], detailed physical models [21], software [22], and simplified models [24] that give tools and models to calculate this coefficient at different channel conditions and frequencies. As it can be seen in the above equation, the molecular absorption loss increases exponentially with distance compared to square law in (4.2) for the free space propagation. This causes the free space propagation to dominate the low-distance communications and the molecular absorption to take over at higher distances. However, this is also dependent on the absorption coefficient $\kappa_a(f)$, which is a function of frequency.

Figure 4.1 shows the FSPL and FSPL plus molecular absorption losses from 100 to 1000 GHz at 22°C and 50% relative humidity at three different distances; 10, 100, and 1,000 meters. This figure shows that the severity of the molecular absorption loss depends strongly on the link distance as well as the frequency. Below 300 GHz, frequencies have very wide low loss regions for all link distances. The same applies to the high frequencies, but the link distance plays an important role on how the wing absorption attenuates the signals. The wing absorption is caused by the spectral broadening of the absorption lines. On surface the level of Earth, this is mostly caused by the pressure of the atmosphere via pressure broadening. This causes severe losses on the signals by the molecular absorption loss even on the low loss regions. The shorter distances, on the other hand, offer very opportunistic spectral windows at very wide spectral region, albeit the FSPL is always high and operation on the THz band requires high-gain antennas.

In addition, but not covered herein, the THz band communications close to the molecular absorption lines causes the utilizable bandwidth of the transmissions to narrow as a function distance. This follows from the increasing wing absorption around the centers of the molecular absorption lines. In some system specific studies, this has to be included into the modelling if, for example, the utilized modulation takes into account the distance from the access point [25,26]. In the general case, such as here, the impact of the molecular absorption only has impact on the channel quality as presented above.

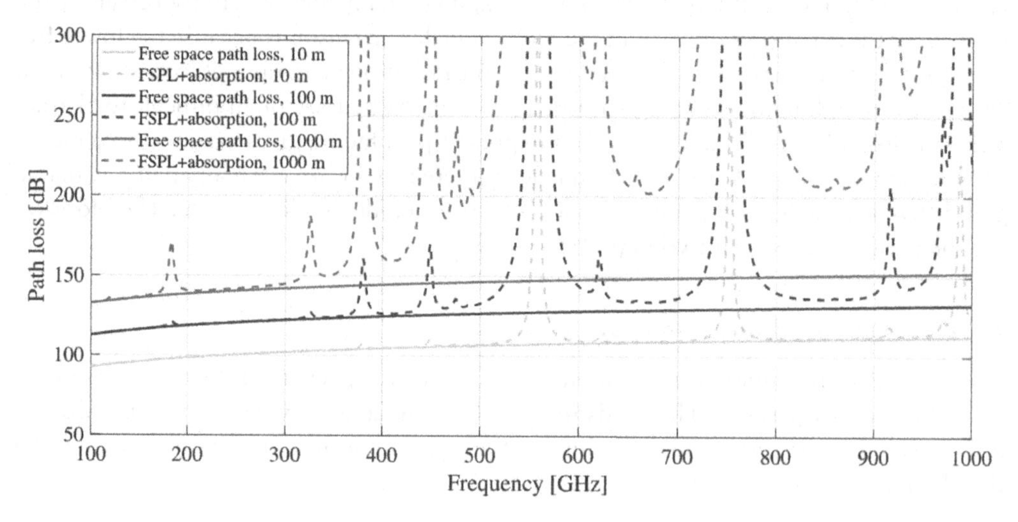

Figure 4.1 Illustration of the FSPL and FSPL plus molecular absorption losses at three distance: 10, 100, and 1,000 m.

4.3.3 Multipath Propagation

The multipath propagation primarily requires two or more simultaneously active communication paths between Tx and Rx. The Rx response to multiple copies of the signal shows as constructive and destructive summation of the signals propagating different path lengths and experiencing different channel losses. At lower frequencies with wide antenna beams we often can use Rayleigh, Rician, or similar fading mechanisms to model the multipath propagation. The THz channels are very sensitive to blockage by objects due to high penetration loss. High antenna gains, on the other hand, limit the number of possible paths simultaneously visible to the Rx and Tx. Furthermore, in general higher path loss on LOS paths alone limits the number of possible multipath components by attenuating the weak NLOS components. The large free space does not allow very high additional loss on NLOS paths to take the signal level below the detection threshold. These issues lead to vastly simplified channels for the THz communications. Instead, the problem usually is how to find the good propagation channels and where to point the Rx and Tx antennas. Thus, the fading is usually omitted in the THz band channels and link budget calculations. This also applies to most of the stochastic models. Because of this, also herein we do not consider fading for the THz links, but rely on single path propagation models. Regardless of this, the fading can be included by utilizing a fading channel model as it was shown above for a generic path loss model.

4.3.4 Signal Blockage by Objects

The THz frequencies are very vulnerable to blocking of the signal by objects, walls, humans, and other possible obstructions in the propagation path. In real environments, the NLOS propagation tends to be common. This is partially true at the lower frequencies the large cell sizes. On the THz band operation, the density of the access point needs to be sufficiently high to mitigate the large path loss and allow high probability for LOS and possible low loss NLOS paths. Regardless of this, there is always a possibility that an object or human blocks the signal path. This is true for the desired link and it is also true for the interference. There are several papers that have modeled the stochastic blocking.

The stochastic geometry relies on transmit probabilities of the interfering nodes. These will be detailed below, but in the presence of blocking objects, the probability to receive interference decreases. Although the blocking also means that the desired signal can be blocked. Herein we do not take that into account since the focus is on the interference modelling. One way to calculate the blocking probability is to assume that the blockers (e.g., humans) are simple cylinders of radius, r_B. This gives a good way to mathematically characterize the blocking probability by assuming Poisson distributed blockers as [17]

$$p_B(r_i) = 1 - e^{-\lambda_B (r_i - r_B) r_B}, \tag{4.5}$$

where $p_B(r_i)$ is the probability that the path between desired Rx and the interferer at distance r_i is blocked, λ_B is density of the blockers (number of blockers/m^2), and r_B is the diameter of the blocker. From this, we get the probability that the path is not blocked as $p_N(r_i) = 1 - p_B(r_i)$. It will be shown later how to incorporate this non-blocking probability into the stochastic model for the aggregate interference. There also exist models that take into account the height of the blockers [18,27]. It should be noticed that there can be many types of blocking, such as from the walls or car. Respective models should be utilized or derived based on the environment that is modeled.

4.3.5 Beam Misalignment

In the Friis transmission equation for the FSPL, we have the antenna gains assuming maximum possible gain. In the ideal case, this is true when the antennas are stationary and the antennas are perfectly aligned. It is possible that the motion causes the average gain of the antennas to be lower than the maximum due to random motion. In this case, we need to employ expected antenna gain based on the antenna motion statistics [28]:

$$E_\beta\left[G(\Theta)\right] = \int_\theta \beta(\theta) G(\theta) \mathrm{d}\theta, \tag{4.6}$$

where $\beta(\theta)$ is function describing the antenna motion and $G(\theta)$ is the antenna gain pattern. It should be noticed that the motion in general case does not have influence on the received interference, since it is arriving from the random directions in any case. Therefore, some random motion the interferers experience does not increase the randomness of the interference. However, there are situations where extra interference would be introduces by the interferer motion. For instance, if the interfering signal is sent by the desired Tx to some node spatially close to the desired Rx, it is possible that the source of the interference is the motion itself. However, in the basic case herein, we assume that the interference is coming from the surrounding random nodes. Therefore, the antenna motion does not have impact on the interference, albeit it may influence the experienced signal-to-interference-plus-noise ratio (SINR) level by corrupting the desired links antenna gains.

4.3.6 Received Power and SINR

Taking into account the loss mechanisms described above, we adopt the following channel model for the THz band propagation:

$$h(r)=h_{\text{FSPL}}(r)\tau(r)=\frac{\left(c^2e^{-\kappa_a r}\right)}{(4\pi f r)^2}E\left[G_{\text{Tx}}(\Theta)\right]E\left[G_{\text{Rx}}(\Theta)\right]. \tag{4.7}$$

With perfectly aligned antennas, the expected antenna gains are given by the maximum gains. However, those can be deteriorated, for example, by antenna movement as discussed above. Given this channel model, the expected received desired signal becomes

$$E[S]=P_{\text{Tx}}h(r)+n_{\text{g}}, \tag{4.8}$$

where n_{g} is additive white Gaussian noise (AWGN). It is possible to expand this channel model to take into account the fading mechanisms, NLOS propagation features (in the case the main communication path is one possible NLOS path), and any possible antenna gain model and possible hardware imperfections. Utilizing the above channel model, the receiver SINR becomes

$$\text{SINR}=\frac{E[S]}{E[I]+n_{\text{g}}}=\frac{P_{\text{Tx}}h(r)}{\sum_{i\in\Phi}P_{\text{Tx}}^{i}h_{\text{I}}(r_{\text{i}})+n_{\text{g}}}, \tag{4.9}$$

where $h_{\text{I}}(r_i)$ is the expected channel gain for the interference. This component is fundamentally similar to the desired signal's expected channel gain with the difference coming from the random interference directions. Those are discussed in the next section where the stochastic interference level for the THz band networks is obtained. Knowing the stochastic interference level opens doors to stochastic SINR, and subsequently capacity and other network statistics that can be derived from expected signal levels.

4.4 Interference in THz Networks

The interference modeling by the stochastic geometry will be covered in this section. First, we make the basic assumptions of the interference modeling herein and then give the corresponding system model based on the channel phenomena discussed above. Then we go through the derivation of the aggregate interference in the considered system. Throughout this derivation, we assume carrier-based transmissions. Pulse-based transmission techniques would require additional modeling of the collision probabilities of the pulses. Especially in the case that the pulses would be spread in time to avoid interference. However, this is true also for time slotted transmission techniques. Here we focus on simple carrier-based approach where users use the channel randomly.

4.4.1 Interference Modeling by Stochastic Geometry

The stochastic modeling relies on utilizing stochastic average values, that is, expected values of various phenomena, such as antenna gains and blockage probabilities. The stochastic geometry furthermore relies on spatial properties of the nodes in the network. For instance, the random interference can be modeled as a shot noise process leading to models that we utilize below [3]. When we know the spatial properties of the network, we can obtain the statistics of, for instance, the interference from the Laplace transform of the spatial model. In the other words, we model the total network interference as a spatial process that depends on the path losses, transmit powers, etc. Assuming that the interference is a spatial shot noise process, we can utilize the Laplace transform to characterize interference. As a consequence, we can calculate the statistics of this interference. As stated above, several other point processes could be utilized as well [9], but herein we utilize the PPP due to straightforward interference modeling.

In this chapter, we assume homogeneous PPP, in which the nodes are homogeneously distributed in the network. The interference generated by the nodes is following a Poisson distribution in this setting, that is, the nodes cause interference by a shot noise process on the studied Rx, or the desired Rx, the typical node of the network in average similar to any other node. The desired Rx is usually placed in the origin of the coordinate system in the Euclidean space. This gives an advantage that we can use simple radial integrals to sum all the interference in the network. We also assume that the PPP is isotropic, that is, the process is independent of the rotations (homogeneous interference space with studied node at the origin). We assume that the PPP is simple, that is, no two nodes can be in the same position. To say this in the other words, we assume that we model random N node network to produce interference from N nodes.

The PPPs further have a couple of interesting simplifying properties that simplify the network modeling. An independent thinning of a PPP is a PPP. As it will be shown below, the density of the network is modeled by Poisson density parameter λ. This gives the average number of nodes per unit are (number of nodes/m^2). Each node has a certain probability to transmit (denoted as $p_t p_t$). This follows from a usual assumption of ALOHA channel access protocol due to simple handling of the transmissions in the network. Thus, the transmit probability thins the PPP by the transmit probability. The resultant network is still a PPP and has a density of $p_t\lambda$. Furthermore, we can add other phenomena to this thinning process. For instance, the above probability that the path between interfering node and the desired node is not blocked ($p_N(r_i)$) results in a PPP network of density $p_N(r_i)p_t\lambda$. With the distance-dependent blocking probability, the network will not be homogeneous anymore, but it can still be handled by adding the blocking probability to the stochastic spatial interference model. This will be further discussed below where the stochastic model for the aggregate interference is derived. Similarly, other possible thinning processes could be added in without breaking the Poisson distribution.

The second important feature of the PPPs is that the superimposition of multiple PPPs is a PPP. This means that we can model multiple tiers of nodes. In the below, we assume K different tiers of nodes. Each tier of nodes can have independent transmit probabilities, antennas patterns, transmit power, and other parameters as well. As a result, sum of the interference over all the tiers of nodes does not break the PPP. This gives tools to model complex networks with different types of nodes, for example, regular users, sensor nodes, and base stations, to mention few possibilities. Below we give an example of such network model and derive the mean interference for a generic multi-tier THz network.

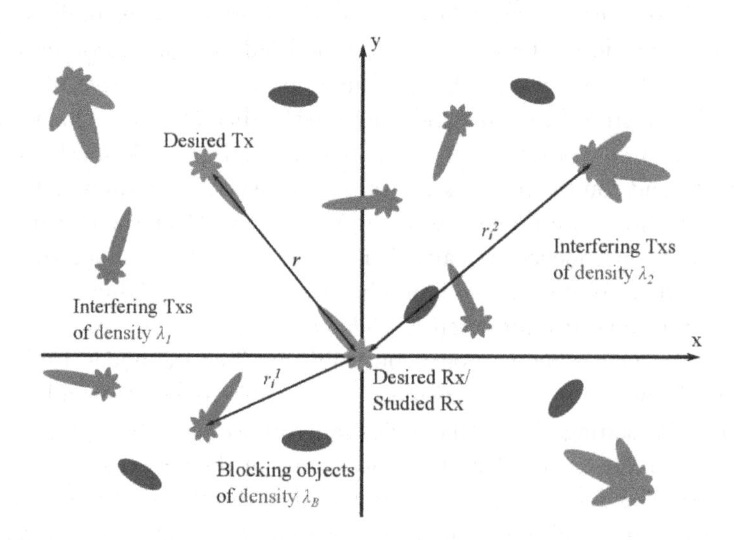

Figure 4.2 An illustration of the simple system model considered in the stochastic model herein.

4.4.2 System Model

The considered system model is given in Figure 4.2. The considered model consists of the desired link, or the studied link that is utilized to characterize the system behavior. As stated above, we can have multiple tiers of different types of nodes with different antenna gains, node densities, and transmit powers. Signal blockage by objects on the channel may decrease the probability to receive interference. As discussed above, the desired Rx is usually set into the origin in order to make it easy to integrate the interference power over the space around the origin by simple radial integration. However, in the THz band due to large channel losses, it is also possible to limit the space size since the far away interferers are too weak to contribute to the total interference. Thus, it is also possible to approach to interference modeling in Cartesian coordinate in order to easily model, for instance, indoor locations [29]. However, herein we focus on general infinite network model as detailed in the next section.

4.4.3 The Aggregate Interference

In this derivation, we focus on potential practical system geometry in which the nodes are dropped on two-dimensional area, whereas the channel losses follow three-dimensional expressions as shown in (4.7). It is also possible to utilize arbitrary dimensions for the network and the channel losses. More information on those can be found, for example, in Refs. [14,15]. Having these assumptions, the aggregate interference of the network is described as it was given above:

$$I = \sum_{i \in \Phi} P_{\mathrm{Tx}}^{i} h_{\mathrm{I}}\left(r_{\mathrm{i}}\right), \tag{4.10}$$

that is, as sum over the active interfering nodes. As shown above in the system model, we can have multiple tiers of nodes with different parameters. Then the aggregate interference is a superimposition over all the active interfering nodes of all the tiers K as

$$I = \sum_{k=1}^{\mathrm{K}} \sum_{\mathrm{i} \in \Phi_k} P_{\mathrm{Tx,k}}^{i} h_{\mathrm{I}}^{k}(r_{\mathrm{i}}), \tag{4.11}$$

where Φ_k is the set of active interfering nodes in tier k. Each tier can have its own set of parameters for the node densities, transmit powers, transmit probabilities, and blocking probabilities. The channel coefficient in the above equation is

$$h_{\mathrm{I}}^{k}(r_{\mathrm{i}}) = \frac{c^2 e^{-\kappa_{\mathrm{a}} r_{\mathrm{i}}}}{(4\pi f r_{\mathrm{i}})^2} E\left[G_{\mathrm{Tx}}(\Theta)\right] E\left[G_{\mathrm{Rx}}(\Theta)\right]. \tag{4.12}$$

This is the same expression as in above for the desired link. The main difference is in the expected antenna gains. In the case of random networks, the expected antenna gains equals unity. This follows from the conservation of energy as the transmitted is only capable, regardless of the antenna patters, to emit at most certain maximum amount of energy to the environment. High gain antennas emit more energy to certain directions, but far less on the other directions. Thus, the net radiated energy remains constant and the expected antenna gain equals unity. However, there are some notable exceptions to the expected antenna gains. For instance, in a single access point downlink scenario, the desired Rx could be pointed toward the access point and have a full gain toward the interfering source, that is, the access point itself. Or one could calculate the interference experienced by the access point in the uplink scenario where all the nodes are pointed toward the access point. These types of scenarios, the modeling of the system can be handled via the expected antenna gains by utilizing appropriate models to take into account the system peculiarities. Herein, we consider the classical fully random mesh network, where the interference comes from all the possible directions. As a consequence, the antenna gains disappear (they equal unity), but due to general representation, those are kept in the below derivations.

We characterize the interference by the Laplace transform of the aggregate interference in order to later calculate the statistics of the interference. The Laplace transform of the aggregate interference is

$$L_{\mathrm{I_{aggr}}}(s) = E_{\Phi}\left[\exp\left(-s \sum_{k=1}^{K} \sum_{i \in \Phi_k} h_{\mathrm{I}}^{k}(r_i),\right)\right] \tag{4.13}$$

where $E_{\Phi}[\cdot]$ is the expectation over all the active sets of nodes over the K tiers. This can be rearranged as

$$L_{\mathrm{I_{aggr}}}(s) = \prod_{K} E_{\Phi_k}\left[\prod_{\mathrm{i} \in \Phi_k} \exp\left(-s h_{\mathrm{I}}^{k}(r_{\mathrm{i}})\right)\right], \tag{4.14}$$

where

$$E_{\Phi_k}\left[\prod_{\mathrm{i} \in \Phi_k} \exp\left(-s h_{I}^{k}(r_{\mathrm{i}})\right)\right] = E_{\Phi_k}\left[\prod_{x \in \Phi_k} v(x)\right] = \Omega(v)$$

is the probability generating functional (PGFL) [2] with $v(x)=\exp\left(-s h_I^k(r_x)\right)$. This PGFL for a PPP can be solved as [2]

$$\Omega(v)=\exp\left(-\int_{R^2}(1-v(x))\Lambda_k(x)dx\right), \tag{4.15}$$

where the nodes are dropped in R^2 and $\Lambda_k(x)$ is the intensity function of the PPP of tier k. With homogeneous PPP and two-dimensional drop space

$$\Lambda_k(x)=2\pi p_k\lambda_k x, \tag{4.16}$$

where λ_k is an intensity parameter, that is, the density of the nodes of tier k per unit area as discussed above, and p_k is the probability of a node in tier k to transmit. As it was shown above, the possible blocking probability and other phenomena can be added as a thinning operation to the network density λ via thinning the density by the non-blocking probability, similarly as with the ALOHA transmit probability. Then the effective density of the kth tier would become $p_k(x)=p_N(x)p_k\lambda_k$. Notice that the blocking probability is a function of the distance x between the interferer and the desired Rx. Thus, the blocking probability is integrated over the space to account for the distance x in the above spatial integral. As discussed above, the blocking probability will cause the interference to become non-homogeneous. However, the integration over the space will take this into account, but it will also make the integration more difficult. Notice also that the density of the nodes could be non-homogeneous by adopting an appropriate, for example, a distance-dependent, model for the density of the users. In the below numerical examples we assume no blockage and perfectly homogeneous environment.

Due to assuming that the desired Rx is in the origin, the radial/spherical integral is an integration over the distance r. After some manipulations, the Laplace transform becomes

$$L_{I_{aggr}}(s)=\prod_K \exp\left[-2\pi p_k\lambda_k\int_0^\infty r\left(1-\exp\left(-s h_I^k(r)\right)\right)dr\right]. \tag{4.17}$$

This is the final form of the Laplace transform with our assumptions. From this expression, we can calculate the statistics of the interference, but first, a few words on the directional antennas and the interference statistics.

4.4.4 Directional Antennas

The high-frequency systems always require directional antennas due to high channel losses. We have to take this into account in the stochastic modeling of the antenna gains. In the case of fully random network, the antenna gain does not have an impact on the mean interference. The reason is that due to conservation of energy, the average energy emitted by the Tx is constant. The antenna pattern emphasizes some directions at the expense of the others. As a consequence, the average energy in the system is not dependent on the antennas, but only on the energies emitted in average by the nodes. However, the directional antennas do have an impact on the high moments, like variance. This follows from the higher received energy when the antenna pattern is pointed at the Rx. We can take this into account by introducing factor

$$f_G(n) = \left(\frac{\max_\Theta \left[G_{\mathrm{Tx,k}}(\Theta) \right] \max_\Theta \left[G_{\mathrm{Rx,k}}(\Theta) \right]}{E_\Theta \left[G_{\mathrm{Tx,k}}(\Theta) \right] E_\Theta \left[G_{\mathrm{Rx,k}}(\Theta) \right]} \right)^{n-1}, \tag{4.18}$$

where n is the order of the moment, that is, below we derive the moments of the interference. The nth moment of the interference is the nth derivative of the above Laplace transform of the aggregate interference. We can see that this factor is directly proportional to the maximum antenna gains. The expected antenna gains in the denominator are in the general case equal to unity due to the above-mentioned conservation of the energy. However, the expected antenna gain also depends on the assumptions made on the antennas. For instance, the expected antenna gain in the general case is the gain of isotropic antenna's gain. This isotropic gain could also be below unity if one assumes that the total integrated gain equals unity.

4.4.5 Moments of the Interference

The moments of the interference can be calculated from the zero points of the derivatives of the Laplace transform. Then the nth moment of the interference is the zero point of nth derivative with respect to the transform variable s as

$$E\left[I^n\right] = (-1)^n \frac{d^n}{ds^n} L_{I_{\mathrm{aggr}}}(s), \tag{4.19}$$

which is evaluated at $s = 0$. Let us mark the above Laplace transform in simpler terms to make it easier to extract the closed form expression for the derivatives

$$L_{I_{\mathrm{aggr}}}(s) = \exp(-L(s)), \tag{4.20}$$

where

$$L(s) = \sum_{K} 2\pi p_k \lambda_k \int_0^\infty r\left(1 - \exp\left(-s h_I^k(r)\right)\right) dr.$$

Based on this expression, calculating the derivatives, we obtain the first two moments of the interference as

$$E\left[I_{\mathrm{aggr}}\right] = L', \tag{4.21}$$

$$E\left[I_{\mathrm{aggr}}^2\right] = -f_G(n=2) L'' + (L')^2, \tag{4.22}$$

where the factor $f_g(n)$ was given above with discussion on utilizing the directional antennas, and

$$L'(s=0)=\sum_{K}\frac{c^2}{8\pi}p_k\lambda_k E_\Theta\left[G_{\mathrm{Tx,k}}(\Theta)\right]E_\Theta\left[G_{\mathrm{Rx,k}}(\Theta)\right]\int_0^\infty r^{-1}\int_W\frac{P_{\mathrm{Tx,k}}}{Wf^2}\exp\left(-\kappa_a(f)r\right)df dr,$$

$$L''(s=0)=-\sum_{K}\frac{c^4}{128\pi^3}p_k\lambda_k\left(E_\Theta\left[G_{\mathrm{Tx,k}}(\Theta)\right]E_\Theta\left[G_{\mathrm{Rx,k}}(\Theta)\right]\right)^2\int_0^\infty r^{-3}\int_W\frac{P^2_{\mathrm{Tx,k}}}{W^2f^4}\exp\left(-2\kappa_a(f)r\right)df dr.$$

In this expression, we included the integration over the frequency band W to obtain mean interference over some particular band. The THz band is characterized by frequency selective fading due to molecular absorption. This may in the case of very wide band require taking into account the variations in the channel loss. The division of the transmit power by bandwidth mainly means that we send the full power on this band, and this power is equally distributed on this band. In the case of, for instance, some water filling algorithms or otherwise uneven power distribution, the transmit power should be a function of the frequency. This particular model was first derived in Ref. [16]. The mean aggregate interference is obtained directly from (4.21) and the variance is given by

$$\mathrm{var}\left(I_{\mathrm{aggr}}\right)=-f_G(n=2)L''(s=0),$$

since $\mathrm{var}(X)=E[X^2]-(E[X])^2$, which is easy to solve from (4.21) and (4.22). The higher moments can be calculated similarly, although usually the mean and variance of the interference are enough to analyze the system performance. The above calculations were given for a two-dimensional node drop and three-dimensional path loss model. One could also calculate the interference statistics for the any spatial dimensions as it was shown in Refs. [14,15].

4.4.6 Numerical Examples

As an example result on THz band peculiarities, Figures 4.3 and 4.4 show the signal-to-interference ratios (SIRs) for the same interference statistics imposed on two different desired signal link distances. The SIR level is equivalent to SINR in the interference limited networks. The interference limited network was chosen here for simplicity and the noise could be added by summing the noise level to the interference. All the sources have the same transmit power, $P_{\mathrm{Tx}}=1$ W. The desired signal in Figure 4.3 is at 10 cm distance and at 100 cm distance in Figure 4.4. The interfering nodes are formed of two tiers, forming independent aggregated interference levels I_1 and I_2. Tier I_1 has a node density of $\lambda_1 = 1/\mathrm{m}^2$ and an antenna gain $G_{\mathrm{Tx},1} = 20$ (linear scale), and tier I_2 has a node density of $\lambda_2=2/\mathrm{m}^2$ and an antenna gain $G_{\mathrm{Tx},2} = 10$. That is, the first tier has twice the lower density, but it also has twice the higher antenna gain. From these figures, we can see the impact of the molecular absorption versus the FSPL. The very short, desired link distance does not suffer from the molecular absorption loss and the molecular absorption actually increases the SIR level by rejecting the far away interference. On the other hand, when the desired link distance is increased, the desired link itself starts to suffer from the increased absorption loss. However, at these link distances, the bulk of the loss comes from the FSPL and the molecular absorption loss mostly has impact on the regions with intense absorption lines.

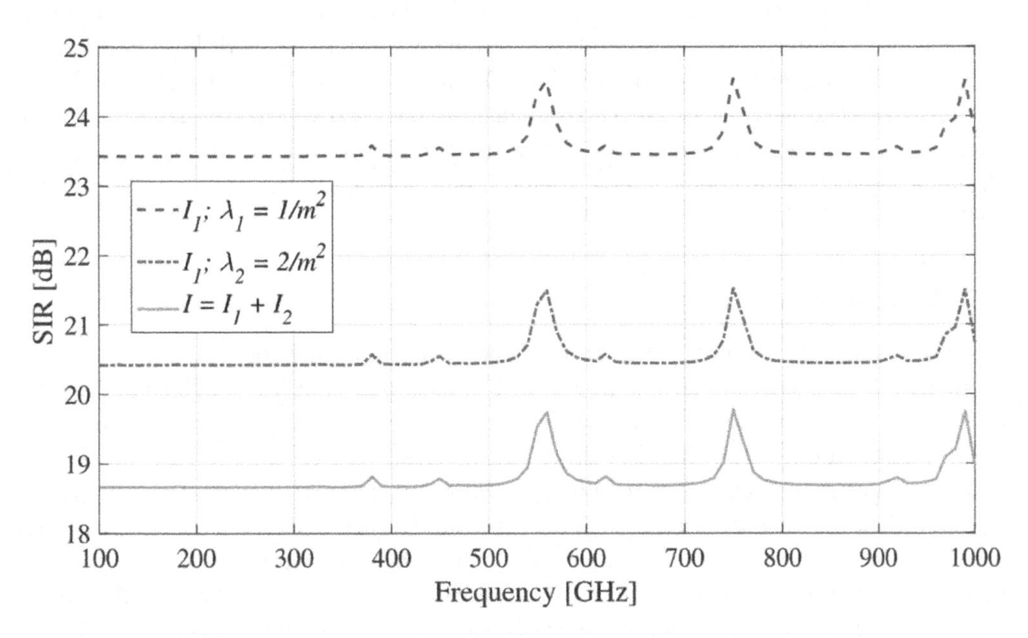

Figure 4.3 The SIR levels for two tiers of nodes of densities $\lambda_1 = 1/m^2$, $\lambda_2 = 2/m^2$, and their joint impact. The desired Rx is at 10 cm away from the desired Tx.

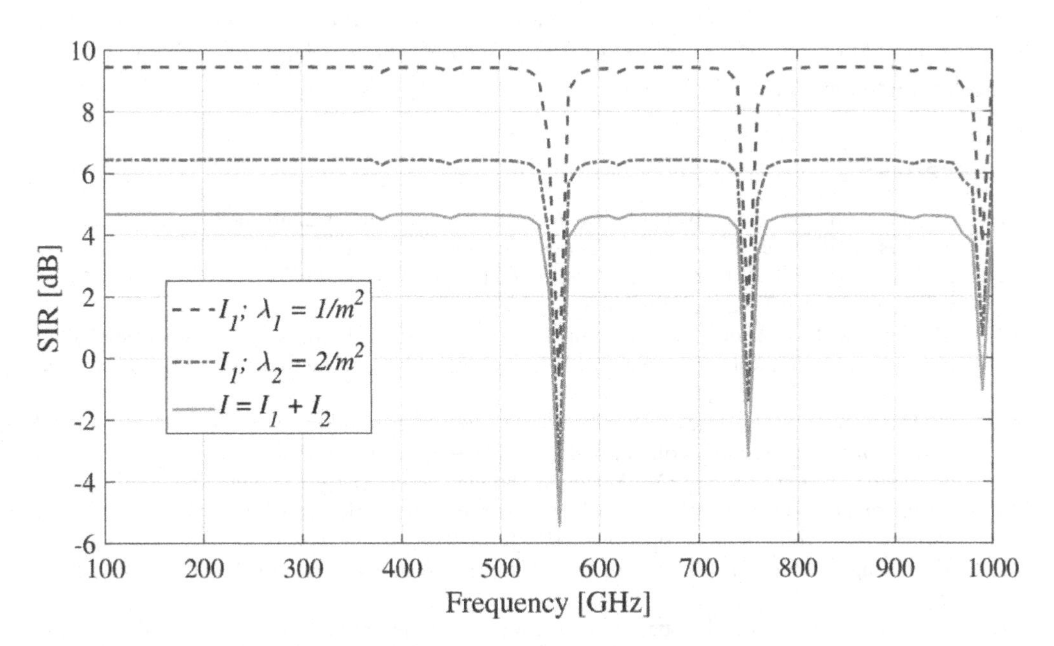

Figure 4.4 The SIR levels for two tiers of nodes of densities $\lambda_1 = 1/m^2$, $\lambda_2 = 2/m^2$, and their joint impact. The desired Rx is at 100 cm away from the desired Tx.

This simple example shows that the THz band has many interesting aspects to study. The stochastic geometry is a very good tool to quickly produce network-level interference on top of the in general trivial desired link. By manipulating the stochastic models, we can easily predict behavior of the THz signals in many types of environments and usage scenarios. The closed form expressions offer fast "simulation" platform. As such, the stochastic geometry should be found in the toolbox of every researcher doing system level analysis.

4.5 Conclusions

This chapter gave some fundamentals on THz band interference modeling by the stochastic geometry. As such, the models herein represent the starting point for the THz band system level interference modeling. The stochastic models can be applied to various different network configurations by adding and modifying different components in the model, such as antenna pattern, blockage probabilities, and node properties. In general, the stochastic geometry is a very handy tool to estimate the interference and other statistics without a need for complicated system simulators. In the problems where stochastic tools can be utilized, they do not only provide faster results but also more accurate results due to effective handling of all the possible network realizations simultaneously in closed form.

Acknowledgments

The work of the authors is supported by ARIADNE project under Horizon 2020, European Union's Framework Programme for Research and Innovation, under grant agreement no. 871464. This work was also supported in part by the Academy of Finland 6Genesis Flagship under grant no. 318927.

References

[1] M. Latva-Aho and K. Leppänen, Eds., "Key drivers and research challenges for 6G ubiquitous wireless intelligence", 6G research visions. University of Oulu, September 2019, no. 1.

[2] M. Haenggi and R. K. Ganti, "Interference in large wireless networks", *Found. Trends Netw.*, vol. 3, no. 2, pp. 127–248, Nov. 2008.

[3] M. Haenggi, J. G. Andrews, F. Baccelli, O. Dousse, and M. Franceschetti, "Stochastic geometry and random graphs for the analysis and design of wireless networks", *IEEE J. Sel. Areas Commun.*, vol. 27, no. 7, pp. 1029–1046, Sep. 2009.

[4] M. Haenggi, "Outage, local throughput, and capacity of random wireless networks", *IEEE Trans. Wireless Commun.*, vol. 8, no. 8, pp. 4350–4359, Aug. 2009.

[5] H. ElSawy, E. Hossain, and M. Haenggi, "Stochastic geometry for modeling, analysis, and design of multi-tier and cognitive cellular wireless networks: A survey", *IEEE Commun. Surveys Tuts.*, vol. 15, no. 3, pp. 996–1019, Jun. 2013.

[6] M. Haenggi, *Stochastic Geometry for Wireless Networks*, Cambridge University Press, 2013.

[7] P. Nardelli, "Analysis of the Spatial Throughput in Interference Networks", Ph.D. dissertation, University of Oulu, Finland, 2013.

[8] J. Venkataraman, M. Haenggi, and O. Collins, "Shot noise models for outage and throughput analyses in wireless ad hoc networks", in Proc. IEEE Mil. Commun. Conf., 2006, pp. 1–7.

[9] F. Baccelli and B. Blaszczyszyn, "Stochastic geometry and wireless networks, volume I – Theory", *Found Trends Netw.*, vol. 3, no. 3–4, pp. 249–449, Dec. 2009.

[10] P. Cardieri, "Modeling interference in wireless ad hoc networks", *IEEE Commun. Surveys Tuts.*, vol. 12, no. 4, pp. 551–572, May 2010.
[11] M. Z. Win, P. C. Pinto, and L. A. Shepp, "A mathematical theory of network interference and its applications", *Proc. IEEE*, vol. 97, no. 2, pp. 205–230, Feb. 2009.
[12] J. Wildman, P. Nardelli, M. Latva-aho, and S. Weber, "On the joint impact of beamwidth and orientation error on throughput in directional wireless Poisson networks", *IEEE Trans. Wireless Commun.*, vol. 13, no. 12, pp. 7072–7085, Dec. 2014.
[13] O. Georgiou, S. Wang, M. Z. Bocus, C. P. Dettmann, and J. P. Coon, "Directional antennas improve the link-connectivity of interference limited ad hoc networks", in Proc. Int. Symb. Personal Indoor Mobile Radio Commun., 2015, pp. 1311–1316.
[14] J. Kokkoniemi, J. Lehtomäki, and M. Juntti, "Stochastic geometry analysis for mean interference power and outage probability in THz networks", *IEEE Trans. Wireless Commun.*, vol. 16, no. 5, pp. 3017–3028, May 2017.
[15] J. Kokkoniemi, J. Lehtomäki, and M. Juntti, "Stochastic analysis of multi-tier nanonetworks in THz band", in Proc. ACM Int. Conf. Nanoscale Comput. Commun., 2017, pp. 1–6.
[16] J. Kokkoniemi, J. Lehtomäki, and M. Juntti, "Stochastic geometry analysis for band-limited terahertz band communications", in IEEE Vehic. Techol. Conf. (spring), 2018, pp. 1–5.
[17] V. Petrov, M. Komarov, D. Moltchanov, J. M. Jornet, and Y. Koucheryavy, "Interference and SINR in millimeter wave and terahertz communication systems with blocking and directional antennas", *IEEE Trans. Wireless Commun.*, vol. 16, no. 3, pp. 1791–1808, Mar. 2017.
[18] Y. Wu and C. Han, "Interference and coverage analysis for indoor terahertz wireless local area networks", in Proc. IEEE Globecom Workshops, 2019, pp. 1–6.
[19] D. Moltchanov, P. Kustarev, and Y. Koucheryavy, "Analytical approximations for interference and SIR densities in terahertz systems with atmospheric absorption, directional antennas and blocking," *Phys. Commun.*, vol. 26, pp. 21–30, 2018.
[20] M. Di Renzo, "Stochastic geometry modeling and analysis of multi-tier millimeter wave cellular networks," *IEEE Trans. Wireless Commun.*, vol. 14, no. 9, pp. 5038–5057, Sep. 2015.
[21] J. M. Jornet and I. F. Akyildiz, "Channel modeling and capacity analysis for electromagnetic nanonetworks in the terahertz band", *IEEE Trans. Wireless Commun.*, vol. 10, no. 10, pp. 3211–3221, Oct. 2011.
[22] S. Paine, "The am atmospheric model", *Smithsonian Astrophysical Observatory, Tech. Rep.* 152, 2012.
[23] L. S. Rothman et al., "The HITRAN 2012 molecular spectroscopic database", *J. Quant. Spectrosc. Radiat. Transfer*, vol. 130, no. 1, pp. 4–50, Nov. 2013.
[24] J. Kokkoniemi, J. Lehtomäki, and M. Juntti, "A line-of-sight channel model for the 100–450 gigahertz frequency band," *EURASIP J. Wireless Commun. Netw.*, vol. 88, pp. 1–15, Apr. 2021.
[25] C. Han and I. F. Akyildiz, "Distance-aware multi-carrier (DAMC) modulation in Terahertz Band communication", 2014 IEEE International Conference on Communications (ICC), Sydney, NSW, 2014, pp. 5461–5467.
[26] H. Sarieddeen, M. Alouini and T. Y. Al-Naffouri, "Terahertz-Band Ultra-Massive Spatial Modulation MIMO", *IEEE J. Sel. Areas Commun.*, vol. 37, no. 9, pp. 2040–2052, Sept. 2019.
[27] R. Kovalchukov, D. Moltchanov, A. Samuylov, A. Ometov, S. Andreev, Y. Koucheryavy, and K. Samouylov, "Evaluating SIR in 3D millimeter-wave deployments: Direct modeling and feasible approximations", *IEEE Trans. Wireless Commun.*, vol. 18, no. 2, pp. 879–896, Feb. 2019.
[28] J. Kokkoniemi, A.-A. A. Boulogeorgos, M. Aminu, J. Lehtomäki, A. Alexiou, and M. Juntti, "Impact of beam misalignment on THz wireless systems," *Elsevier Nano Communication Networks*, vol. 24, pp. 1–9, May 2020.
[29] J. Kokkoniemi, A. Boulogeorgos, M. Aminu, J. Lehtomäki, A. Alexiou, and M. Juntti, "Stochastic analysis of indoor THz uplink with co-channel interference and phase noise", in Proc. ICC Workshops, June 2020, pp. 1–6.

Chapter 5

Terahertz Communication Channel Characteristics and Measurements

Elmustafa Sayed Ali, Mona Bakri Hassan, and Rashid A. Saeed

Contents

5.1 An Overview of THz Communication Channels

Sub-microwaves and millimeter waves for a radio system operate in gigahertz frequency and support point-to-point digital links that operate in a duplex mode. Radio channel waves consist of a frequency pair for both transmitting and receiving directions. GHz bandwidth, which is determined by link capacity and modulation scheme, is quite symmetric and wide [1]. The widespread use of various electronic devices, that is, laptops, mobile phones, and tablets, have dramatically increased connectivity across the world. This increase in connectivity results in an extensive media sharing of videos, music, and social networks, which requires a massive enhancement in data rate between devices over a large bandwidth capacity. Gigahertz has been an exciting band for modern technologies and one of the most important co-factors of rapidly expanding information age [2].

The recent interest in THz band has led to studies and development of new models for high-performance communications systems, which enable widespread networks with high-quality definition video applications such as radar imaging. THz is expected to be a key spectrum of wireless technologies that would be capable of meeting the demand of 5G communications for high bandwidth and data rates. THz wireless band communication suffers from severe transmission path loss and attenuation, which lead to constraints in distance [3]. In this chapter, the characterizations and measurements of THz communication channels are presented to develop a new transceiver with high frequencies close to THz for many applications, such as indoor wireless, vehicular networks, drone communications, and device-to-device (D2D) and nano communications.

THz band provides high data rates for future communications and large bandwidths, but many challenges and issues regarding its physical layer efficiency remain. In THz communications, the large bandwidth that can be achieved is disrupted by signal propagation loss and a high-gain directional antenna is required to communicate over a distance of a few meters [4]. MIMO techniques can slightly reduce the loss in THz bands and can be used to design smaller THz system antenna embedded in a small footprint. Other technical difficulties with higher carrier frequencies in THz communication are related to effect of molecular or atmospheric absorption, which lead to THz channels facing the problem of very high attenuations [5]. The attenuation problem needs to be evaluated by characterizing THz channels and testing different methodologies according to communication environment scenarios.

In this chapter, an overview of terahertz channel characteristics and measurements is presented. The chapter is organized as follows. A brief introduction and overview of terahertz communication channel is presented in this section. Related works and a background description of mm-Wave versus THz communication channel environments are provided in Sections 5.2 and 5.3 respectively. The THz channelization and measurements in the THz band are discussed in Section 5.4. In Section 5.5, different terahertz channelization scenarios, such as large-scale THz statistics and small-scale THz statistics, as well as structures are presented. Section 5.6 provides THz channel measurement metrics such as antenna, distance measurements, spectrum capacity, dynamic range, and Doppler frequency characteristics. An extensive discussion of THz channel measurement methodologies is presented in Section 5.7. A general review of THz channel measurement and channel analysis approaches and differences are presented in Section 5.8. Finally, conclusions and a summary for the chapter are given in Section 5.9.

5.2 Related Works

Several works since the early 2000s have covered THz channelization and measurement. Some of them designated and precisely modeled the effect of oxygen and water vapor [1]. Work done in THz based on temporal domain spectroscopy (TDS) [2] has shown that open space can be exactly demonstrated by utilizing high-resolution transmission molecular absorption (HITRAN) database [3]. Significantly, this contribution showed how field assimilation is often handled incorrectly, which is due to inconsistencies in total assimilation. Recently, another study proved that THz measurements adopting the recommended International Telecommunication Union-Radio Communication Sector (ITU-R) for absorption experiments was not precise [4].

Several THz measurements were made around fading, absorption, and attenuation [5,6]. The precise stochastic analysis for THz wireless communication needs certain experimental setting with realistic channel sounds, that is, realistic environmental for rain [7], fog [8], and dust [9]. The vector network analyzer (VNA) stages scanned over array of detached, narrowband frequency of the bandwidth of interest is used to determine channel size and phase response in addition to channel Inverse Discrete Fourier transform (IDFT), which leads to imaginary channel impulse response (CIR) [13].

VNA utilized for actual specific channel quantities have been employed in stationary environments using an anticipated frequency determination [12] and channel coherence time [13]. For short-range indoor laboratory experiment, a suitable wired connection is needed for synchronization [14]. In addition, the test needs high signal to interference ratio (SINR) and good dynamic range (DR) [15]. Direct signal spread spectrum (DSSS) is one of well-known transmission methods for narrowband channel sounding and measurements, where a train of pulses are used rather than a single pulse.

DSSS signal has great immunity for in-band interference and acceptable dynamic range (DR) [16]. For bandwidth enhancement, several solutions have been proposed such as introducing time delay in sliding window [17]. This method introduces a cheaper storing and processing method for CIR measurements [18]. Another solution is by correlating a delayed version of transmitted signal with received signal and filter results. This method is called temporal dilation for CIR [19]. The parameters that can be measured by using enhanced sliding window are (a) multipath components, (b) received power, (c) antenna patterns, and (d) time delays [20].

5.3 mm-Wave Versus THz Channels

The millimeter waves occupy frequencies of 30–300 GHz in the electromagnetic spectrum and wavelength of 1–10 mm, while THz waves occupy a range of 300 GHz to 10 THz with a wavelength of 100 microns to 1 mm. The mm-Wave channels can provide up to 100 Gbps data rates, which is promising for 5G networks and wireless applications. THz channels may exceed expectations by even providing data rates higher than 100 Gbps, opening a new opportunity for biological and nano communications, in addition to high spatial resolution in imaging applications. Both millimeter and terahertz channels suffer from severe attenuations that can occur due to rain or resonant absorption in oxygen and water molecules [9]. The characterization of mm-wave channels show that it depends on multiple components, in addition to other parameters such as time delay, angle of departure (AoD), Doppler shift, path weights polarimetry matrix, and angle of arrival (AoA). The

Table 5.1 A comparison of conventional, mm-waves and THz communication channels

	Conventional (2.4 or 5 GHz)	*Mm Waves (60 GHz)*	*THz (More Than 300 GHz)*
Bandwidth	40 MHz	2 GHz	100 GHz
Path loss (10 m)	60dB	88 dB	101dB
Data rates	600 Mbps	4 Gbps	100 Gbps and beyond
Antennas	Omnidirectional, 3dBi	Medium Directivities Max 25 dBi	High Directivities Max 40 dBi
Output power	Limited by regulations, 22dBm	Limited by technology & regulations, 10dBm	Limited by technology, <10dBm

bandwidth of the mm-waves channels is less than 10 GHz and can be affected by these parameters. In THz channels, parameters that affect the most are molecular absorption and transmission distances, which may actually degrade THz channel bandwidth but still can offer a few tens of Gbps data rates with acceptable bandwidth [9]. In addition, THz frequency bandwidth can be classified as a narrow spectrum, but it is still a broadband divided into sub-windows. The channel characteristics for mm-waves and THz have some common and distinguishing characteristics, differing from other traditional communication channels. Table 5.1 shows a comparison between conventional, mm waves and THz communication channels.

Both mm-waves and THz channels follow two kinds of propagation paths: (i) light of sight (LoS) and (ii) non-line of sight (NLoS). These result from reflection and scattering phenomena. Any extra atmospheric attenuation and free space attenuation due to water vapor absorption can impact the THz channel propagation. The major absorption in mm-wave channels is from oxygen molecules. In mm-wave channels, attenuation ranges lie in a few tens of dB/km, but it can increase to hundreds of dB/km with higher frequencies in THz channels. The reflection and scattering loss would be greater for THz channels rather than in mm-wave channels because of differences in wavelengths [10]. Consequently, higher mm-waves and THz band channels are considered only up to second order reflections. Generally, in THz channel, increase in reflection loss decreases the number of rays and the coherent bandwidth increases up to 5 GHz, but the overall angular spread becomes smaller. For mm-wave channels in an indoor scenario, researchers report angular spread up to 119° at 60 GHz, but for THz channels in indoor environments, the angular spread observed is up to 40°. THz channels are different from mm-wave channels and highly dependent on distance and frequency [10]. Change in communication distance would not only result in a variation of path loss but also of available transmission channel bandwidth when compared to mm-wave channels. The THz channel bandwidth would shrink if the communication distance increases.

5.4 Motivation for THz Channelization and Measurements

The THz band has been researched extensively in the last three decades for its excellent wave characteristics and massive potential for possible applications due to these properties. These characteristics can be broadly elaborated in four main properties [12]: (a) the properties of low scattering and high penetration rate compared with the conventional waves; (b) clear-cut resolution for three-dimensional imaging; (c) low ionization properties, making it safer than other conventional frequency bands; (d) finally, spectral lines in THz band, which are useful to identify atoms that can be utilized for deep space communications.

Although THz channels have many attractive properties, there are several issues that need to be characterized and measured due to high absorption loss in free space. The main problem is that THz band has a lack of competent, comprehensible, and solid THz facilities and transceivers, i.e. transistors and antennas. Such components have frequencies rolling over a square frequency due to the properties of reactive resistance and large transfer delays. The THz channel has to be analyzed and measured for spectral efficiency and power reduction to compensate for high losses. Another motivation for THz channelization and measurement is that there are many proposals for using THz bands in aerospace, deep space, imaging, and short range. All these applications need proof-of-concepts based on real-time experiments and test. Addressing issues like high losses due to water vapor, rain drops, and dust are vital to these technologies. Hence, THz band and applications have very ingenious instruments for essential soundings in several study fields such as chemistry and physics.

Recently, a number of studies have been conducted to analyze the frequency characteristics of microwave channels in order to develop 5G systems. This has led to development of models for channels operating on millimeter waves, where frequencies are limited to less than 100 GHz [3]. THz band communication requires many evaluations of channel characterizations to assess statistical propagation, transceiver designs, reconfigurable platforms, in addition to several important statistical channel parameter measurements.

The importance of THz channel characterization and measurements is that they help to study the behavior of THz signal propagation in different scenarios related to nature of the region in which THz signals would be transmitted. Moreover, such characterization would help in modeling THz channels and gaining knowledge about different influences that can hinder or reduce speed of THz signals. The transmission and cohesion of THz channels are studied to gather sufficient information about changes that can be made for the propagation of THz waves, while measurements would help in assessing the performance of THz communication systems.

5.5 The Terahertz Channelization Scenarios

THz band is known as sub-millimeter band with a range between 100 GHz and 10 THz, having wavelengths between 3 mm and 30 μm. Generally, to describe properties of any communication channel, it is important to adopt a specific model for typical channel in order to evaluate its behavior. In THz communications, the channel might suffer from different constraints such as atmospheric environment or molecular absorption, in addition to free-space path loss due to atmospheric attenuation. These constraints are related to transmission distance. Accordingly, a suitable carrier frequency is selected based on the intended application. The THz channel model can be described through THz bands ranging between 275 and 325 GHz. A statistical parameter for D2D channels can also be described for THz bands [4]. These models simulate the reflectance patterns and scattering in a sub-THz D2D environment. There are three types of THz channels considered: indoor, outdoor, and intra-device channels. Figure 5.1 shows printed circuit board (PCB) model for intra-device THz channels.

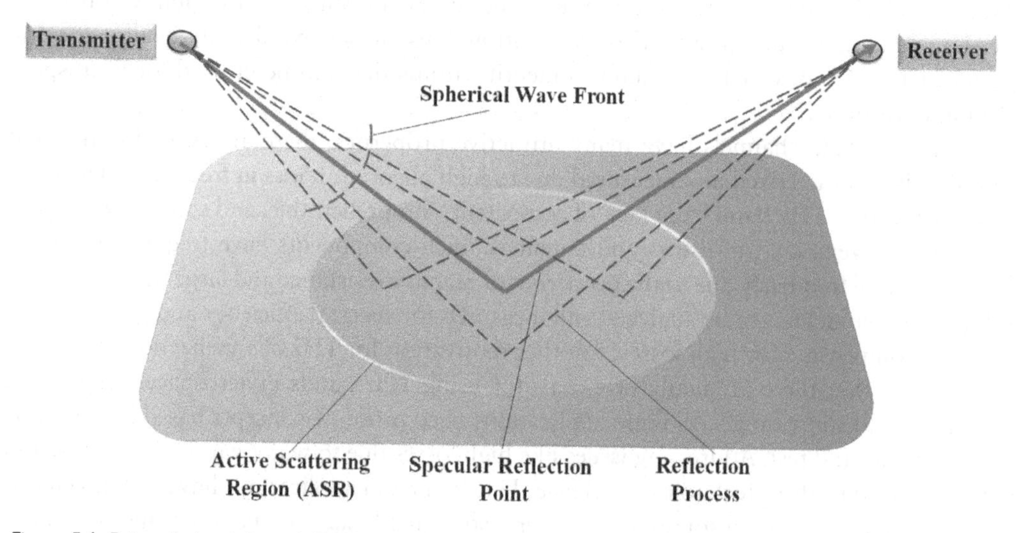

Figure 5.1 Printed circuit board (PCB) model for intra-devices THz channels.

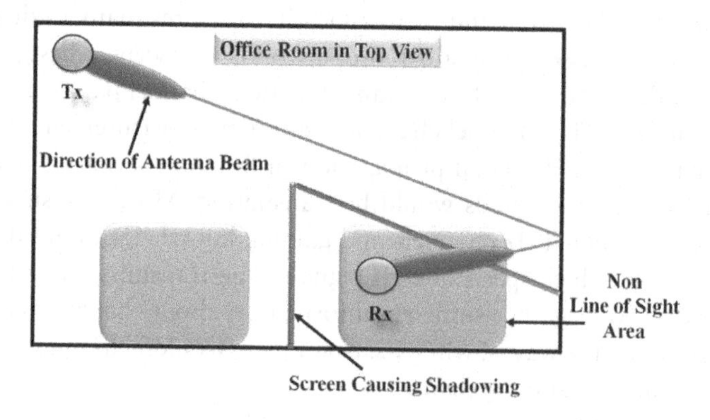

Figure 5.2 Shadowing in indoor communication channel environment scenario for THz Channels using directional antenna.

The most important channels are in indoor and outdoor environments that have much more complex propagation conditions. Figure 5.2 shows an indoor environment communication channel scenario for THz channels. THz channel properties related to propagation and path loss are categorized by special and molecular absorption loss. The free space loss for omnidirectional antennas and molecular absorption are categorized as a special loss. These kind of losses occur due to internally vibrating molecules on the frequencies of the THz band and as derived from coefficients from HITRAN database. The database of HITRAN has combined spectroscopic parameters, which are used by different computer scripts in order to simulate and predict THz transmission in the atmosphere. This process will help to describe the high-resolution transmission of molecular absorption [5]. Figure 5.3 presents an outdoor communication channel environment scenario for THz channels.

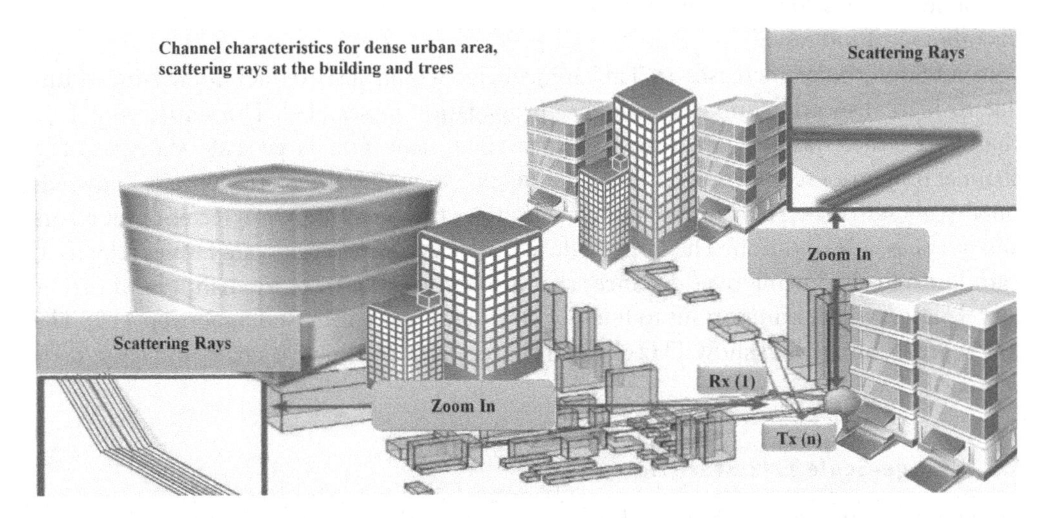

Figure 5.3 Non-line of-sight (NLOS) outdoor with scattering ray communication channel environment scenario for THz channels.

In fact, applications on terahertz frequencies are still under analysis where studies focus on THz wave propagation in various environments, whether in an outdoor environment with short distances or an indoor/in-chip environment such as nano-scale applications. A number of laboratory experiments have been carried out for study of THz bands and their effects.

Theoretically, characterizations of THz channels would allow to capture important parameters related to THz channels. These parameters are gain, delay, and angular spread. Also the THz characterizations will allow to utilization of important channel characteristics, such as degrees of freedom and capacity. THz channel characterization also includes several factors related to the channel and propagation such as directivity, geometry, and motion, in addition to radiation patterns in humidity, vapor, or dust environmental conditions [11]. The impact of spreading and molecular absorption losses should be considered with the abovementioned channel characterization factors. Moreover, the considerations should include the line-of-sight (LoS), non-line-of-sight (NLoS), scattered and reflected paths, in addition to static and time-variant environments. THz signals in receiving systems at pass band are affected by the propagation environment. When considering a constant general spread, the received signal is modeled as an overlay of multiple copies of the transmitted THz signals with different levels of attenuation and delays. So then THz baseband channel includes an impulse response to a number of resolvable multipath components represent delay and attenuation factors corresponding to the number of multipaths [12].

For LOS, two static distributions can be used to describe first-order attenuation factor: Rice and Nakagami distributions [13]. These static distributions are useful to describe the special case of fading mechanisms that are an intrinsic characteristic of THz channels. Water vapor can affect THz response. Molecular absorption losses are predominant depending on distance range, which contributes to dividing the available spectrum into smaller windows depending upon the distance. The water vapor will reduce the available transmission channel bandwidth and increase the path loss due to any increase in communication range.

In addition, more increases in THz frequencies would also reduce total bandwidths due to high absorption spikes and higher propagation losses [14]. The statistics of THz channel characterization is categorized into small-scale and large-scale statistics. The channel properties related to small-scale fading due to multipath propagation is known as small-scale statistics. This kind of fading determines the performance in terms of bit errors and average fade duration. The large-scale statistics evaluate large-scale fading related to path loss and shadowing over distances that are of the order of more than tens of carrier wavelengths, which is important to find out coverage area of cells and handovers [15]. The following section would show THz channel characterization in both small- and large-scale statistics.

5.5.1 Large-Scale THz Statistics

In wireless communications, several different propagations models like Okumura-Hatta and Wolfish-Ikegami can be proposed for path loss in many environments, that is, in urban, rural, and indoor environments [21,24]. Diffraction, reflection, and absorption are constrains affecting LoS and NLoS propagation. Obstacles such as buildings and trees result in a path loss through distance as random variations of received power, which is known as shadow fading that can be distributed according to following equation [14]:

$$F_{\Omega}(x) = \frac{10}{X\sigma_{\Omega}\sqrt{2\pi}ln10}\exp\left[-\frac{\left(10log_{10}x - \mu_{\Omega}(dBm)\right)^{2}}{2\sigma_{\Omega}^{2}}\right] \tag{5.1}$$

where Ω denotes the mean squared envelope level and σ the shadow fading standard deviation. μ represents the mean of area in dBm.

The large-scale statistics of a THz channel are characterized by two parameters: path loss exponent that can be expressed by γ and standard deviation of shadow fading σ, which is obtained from the distance domain of path loss due to displacement. These two parameters determine the increase in path loss as the transmitter / receiver THz separation distance increases, in addition to variation in the measured path loss. The equation of path loss calculation in dB at distance d for a single slope is shown as follows [15]:

$$\text{PL}(d) = 10\gamma log_{10}\left(\frac{d}{d_{o}}\right) + \text{PL} + X_{\sigma} \tag{5.2}$$

where PL(d) represents path loss at distance d, PL(d_0) denoted for free space path loss at distance d_0 and X_σ represents the shadow fading.

For a large -scale THz indoor scenario, studies report that delay spread does not exceed more than few tens of nanoseconds. But the spread decreases because of weak multipath components disappearing below the noise floor, resulting in delay spread decreases at large distances [16]. The large distances in between the transmitter and receiver system and the indoor environment walls lead to higher propagation delay.

5.5.2 Small-Scale THz Statistics

For small-Scale statistics, it found that THz channels can be affected by scattering in direct path between THz transmitter and receiver system and generate an NLoS path. Due to scattering, THz signals take many different directions with different delay in its path and at THz receiver with multipath fading. This means that THz components arise under NLOS propagation due to the scattering environment in a form of Rayleigh distribution fading. The fading at the receiver is described by Rician distribution [17]. THz channel can be characterized at the receiver as a low pass impulse response by the following equation:

$$h(t-\tau) = \sum_{l=1}^{L} g_l(\tau)\ \delta(t-\ \tau_l) \tag{5.3}$$

where L represents the total number of multipath components, $g_l(\tau)$ denotes time-varying envelope associated with l^{th} resolvable multipath component, with average time delay τ_l. g_l is either Rayleigh or Rician fading.

THz channel impulse response can be affected by the motion of the transmitter or receiver described by time selectivity, which represents the characteristic of the impulse response channel that changes over time. The time sensitivity in frequency domain appears with Doppler shifts in the transmitted signal, causing the spectrum of the transmitted signal to expand. The other consideration is related to the impulse response channel changing with frequency [17]. The frequency selectivity is based on the multipath components that arrive with different time delays. This leads to fact that as more multipaths are detected, more variation occurs in frequency response. Therefore, small-scale statistics of THz channel characterization is associated with RMS delay spread that enables us to estimate the multipath-rich channel, which is directly related to coherence bandwidth [18]. In the case of transmission over the narrow band compared to the channel coherence bandwidth, the channel can be considered as a flat frequency. Otherwise, the channel becomes frequency selective.

5.5.2.1 Outdoor THz Channel Characterization

It is known that the THz band provides a high bandwidth that reaches several terahertz and depends on the propagation distance, but there are a number of challenges of communication in these bands. One of these challenges is an attribute of high propagation loss due to the communication distance limitations. In designing the transceiver systems, consideration must be given to the calculation accuracy for system sensitivity and noise parameters to make THz systems capable of reducing high path loss at the THz band frequencies. In addition, to overcome the problem of high path loss in THz bands, the use of very large antenna array will help to reduce the loss. Other limitation to THz channel is related to THz free space propagation radiation. The THz traveling waves show the path loss as absorbing losses [26]. Attenuation in outdoor scenario is considered one of main impediments affecting the THz channels due to absorption loss that are atmospheric and molecular in nature.

Attenuation is considered as one of the most disruptive factors in THz applications. Attenuation due to water molecules cause intense rotational transitions in ambient atmosphere. In outdoor THz environment, water vapor causes atmospheric absorption

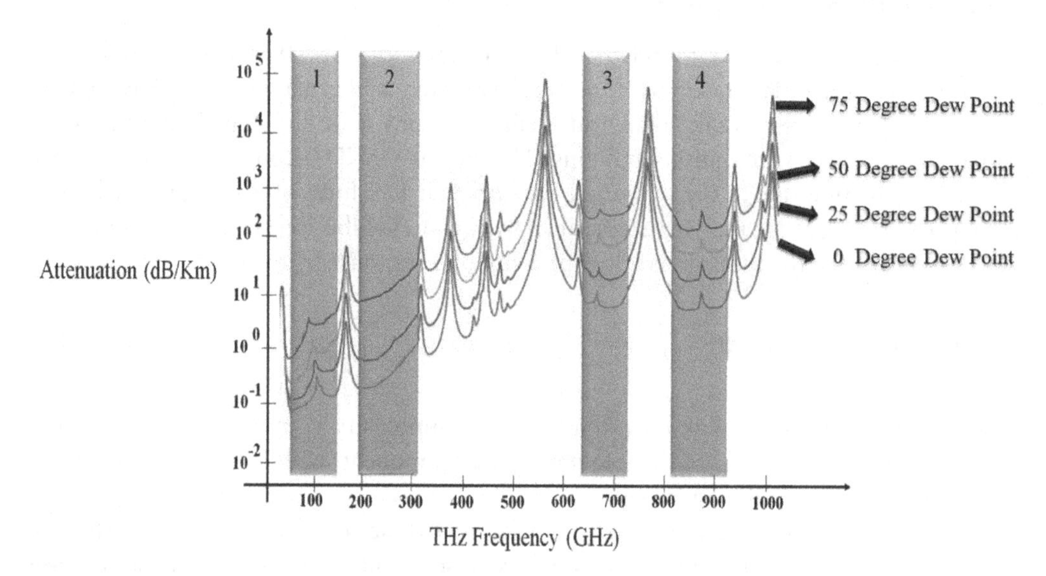

Figure 5.4 Four band windows at 100, 250, 680, 870 GHz atmospheric attenuation of THz propagation with four (0, 25, 50, 75) degree dew points.

attenuation. This kind of attenuation is due to parameters of spectral absorption line from atmospheric transmitter. High-resolution transmission molecular absorption database (HITRAN) is used to describe the limitations of THz propagation through water depending on temperature, pressure, and humidity. HITRAN is a sensitive database to ambient air pressure and relative pressure. It is used to calculate the absorption lines widths according to the air and self-broadened width [27]. In short-distance THz applications, the selection of operating frequencies is very important to avoid extreme degradation of system performance under adverse environmental conditions, especially in the situation of high water vapor content. The impact of atmospheric water vapor with scattering loss would attenuate THz pulses according to the atmosphere conditions if it were a heavy fog or light especially in THz up to 2 THz.

In outdoor THz environment, the THz electromagnetic propagation pulses in atmosphere are controlled by the water vapor parameters. These parameters are frequency-dependent absorption and dispersive index of refraction represented by α (ω) and $n(\omega)$ respectively [28]. Terahertz radiation faces severe limitations due to atmospheric attenuation. These severity of these limitations depend on water vapor present in the atmosphere, which can restrict the THz applications. Figure 5.4 shows the THz propagation atmospheric attenuation in four atmospheric windows dew points.

5.5.2.1.1 Water Vapor Absorption

Gaseous attenuation occurs due to the absorption of water vapor, but there are spectral regions in which oxygen lines predominate that do not allow absorption to occur. The water vapor absorption combines the water vapor lines and continuum component, which is contribute to gaseous attenuation in the frequencies far from the absorption line peaks. This attenuation in kilometers by an accurate estimation of scattering and

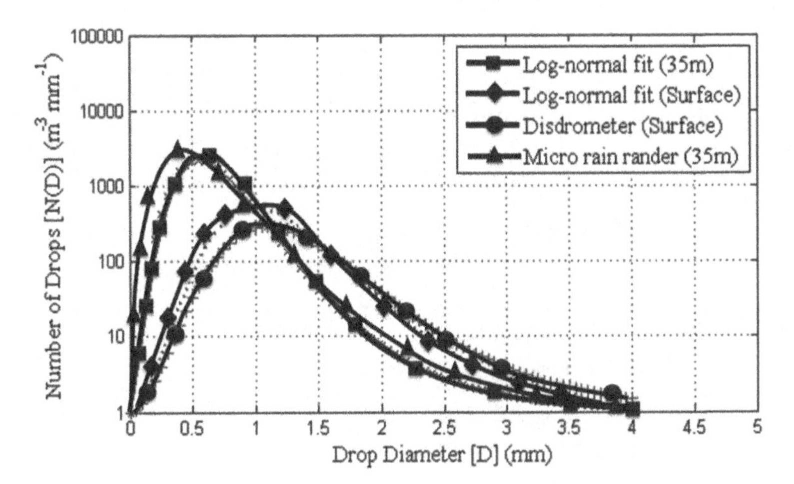

Figure 5.5 Comparison of four rain drop size distribution (DSD) by using the parameters number of drops vs. drop diameters.

absorption coefficients as an extinction cross-section σ_e can be calculated according to of water spheres with different radii r in mm by the following equation [29]:

$$\gamma_c = 4.34 * 10.3 * \int \sigma_e(r) N(r) dr \tag{5.4}$$

where $N(r)$ represents the particle size distribution of clouds. $N(r)dr$ determines the water drop size distribution (DSD) density [29]. Figure 5.5 presents a comparison for rain drop size distribution (DSD), which plays an important role in the evaluation of the effects of rain on THz communication channels. DSDs are considered as a key parameter that limits the predictions of rain attenuation effects on THz bands.

5.5.2.1.2 Molecular Gas Absorption

Molecular absorption is very important issue limiting the applications of using THz bands. It contributes to signal loss due to the partial conversion of electromagnetic energy into internal energy of particles. In general, molecular absorption is considered similar to sky noise in outdoor environment. The molecular absorption occurs due to absorption energy transfer in the atmosphere. There is a difference between the molecular absorption noise and sky noises as they vary by the cause of occurrence. The atmospheric temperature accompanies the sky noise while the molecular absorption noise occurs due to transmission in the radio channel [30]. In addition to this reason, and more precisely, molecular absorption occurs due to the transformation of the electromagnetic wave particles in the medium to higher energy states, which affects the absorption frequency. Figure 5.6 shows different kinds of molecular absorption reactions in the range of THz frequencies.

In atmospheric propagation, gas absorption attenuation absorbs more THz waves. This phenomenon depends on the spectral line intensity, wave numbers, and atmospheric molecular density. Moreover, the parameters like temperature and pressure influence the

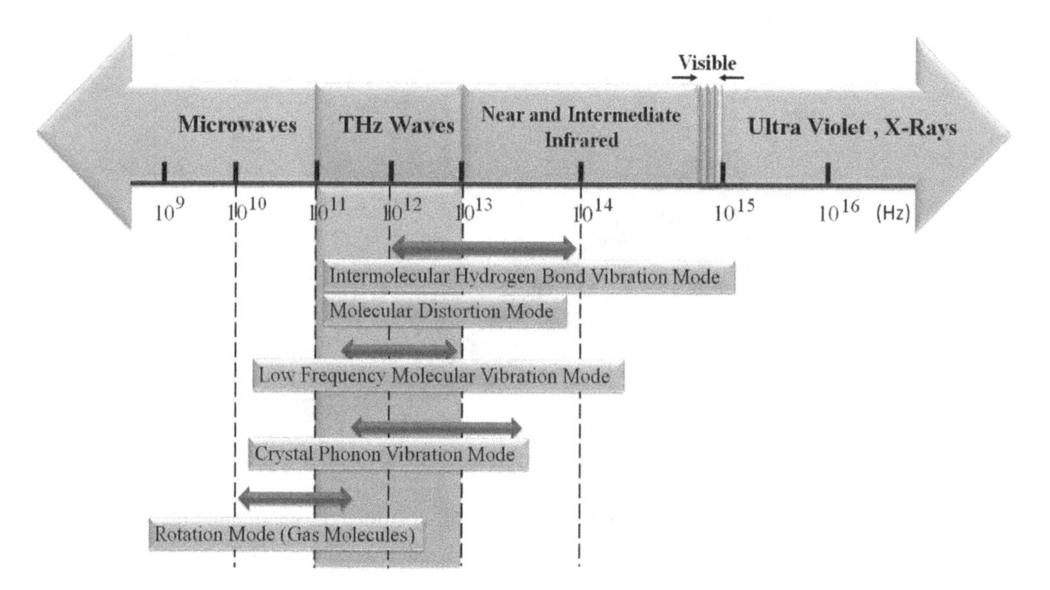

Figure 5.6 Types of molecular absorption throughout the frequencies from microwave up to x-ray.

environmental conditions of atmospheric propagation [30]. The molecular absorption coefficient variation with respect to wavelength is fluctuates according to the change in wavelength. The molecular absorption coefficient can be calculated as follows [31]:

$$K_{\mathrm{a}}^{i}(f) = \mu_{\mathrm{i}} N \sigma_{\mathrm{i}}(f) \tag{5.5}$$

where μ_i denoted molecule fraction, N represents the number of molecular density, and $\sigma_{\mathrm{i}}(f)$ denotes cross-section of ith molecular species absorption.

The spectral line shape and spectral line intensity together determine the absorption cross-section [31]. The absorption line strength is represented by spectral line intensity, and the width and shape of the spectral lines represent the effective area of absorption of a single molecule. The gas molecules act against the electromagnetic wave and make it travel more slowly by absorbing and scattering a part of its energy corresponding to the traveling distance. The molecular absorption loss can be neglected in a distance below 1 m, and it will become significant if distance exceeds 1 m. The temporal change of molecular fraction for water vapor and acetonitrile (CH_3CN) is shown in Figure 5.7. The path loss due to molecular gas absorption (A_{g}) can be calculated as follows. [32]:

$$A_{\mathrm{g}} = \gamma_{\mathrm{g}}.d \tag{5.6}$$

where γ_{g} represents gaseous attenuation and d denotes path distance.

Depending on the molecular absorption transmission windows, higher frequencies than 1 THz suffer from limitations of water vapor and oxygen molecules absorption in the atmosphere. The gaseous attenuation of THz waves depends on the interaction of molecules and radiation.

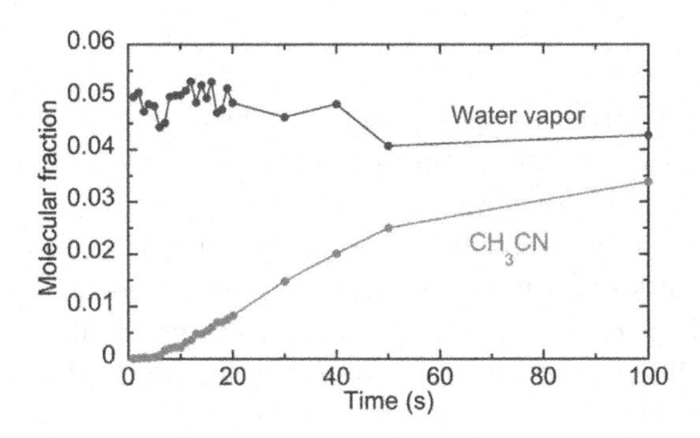

Figure 5.7 Molecular fraction temporal change.

5.5.2.2 THz Indoor Channel Characterization

In indoor propagation, tracing 60 GHz waves is suitable for modeling the propagation channel beyond 300 GHz. This channel scenario related to reflection and scattering processes for such indoor environment depends on reflection on smooth multilayers and scattering on rough surfaces. The other aspect of THz channel in an indoor environment is channel dynamics, which is related to the temporal variation in path loss, which occurs due to device or moving objects [6]. Meanwhile, any reflections from human body are considered a Doppler shifts, contributes to achieving channel dynamics. In outdoor propagation, channels may be affected by phenomena such as reflections from the ground or from the outdoor building walls. Moreover, other aspects like fog affect channel dynamics [6]. Considerations must be given to how these obstacles are encountered when modeling and designing the specifications of THz systems as done previously for wave channel propagation. The use of directional antennas can be one of the promising approaches to compensate for all mentioned kinds of loss. In addition, such antennas can dynamically choose the best radiation rays and avoid signal interference. The calculation of received power for THz communications can be expressed as in the following equation:

$$P_r = P_t + G_t + G_r + 20\log\left(\frac{\lambda c}{4\pi d}\right) - a(f,c)d - \alpha L_r \tag{5.7}$$

where P_t is the transmitting power, G_t and G_r are antenna gains of the transmitting and receiving antennas, d is the distance, and λ is the wavelength. $a(f,c)$ is the atmospheric attenuation determined by frequency f and molecular concentration c. L_r is the reflection loss. $\alpha=0$ when the propagation is in the line of sight (LOS) and $\alpha=1$ when it is in the non-line of sight (NLOS).

For the channels in indoor environment, a few clusters and many sub-multipaths in each cluster can be achieved. The clustering concept is related to indoor propagation modeling, by which the multipaths of wideband channels are grouped in a cluster. This clustering method can help in reducing the number of multipaths from thousands to less than 10. This means different multipath distances over impulse response would be defined to aggregating clusters of the multipath. For indoor characterization, the impulse response of THz channel can be derived by [7].

$$h(t) = \sum_{l}^{Nc} \sum_{k}^{Ns} \beta lk\, \delta(t - Tl - \tau kl) \tag{5.8}$$

The cluster sub-ray number is represented by k. The amplitude of each ray is denoted by βlk. Tl represents the lth cluster time delay. l denotes the cluster number, and τkl is the time delay of the kth ray in the lth cluster.

In general, there are limited numbers of clusters in indoor environments for the THz channel due to path loss arising from reflections. Multipaths in THz may be generated for paths reflected by ceiling, ground, walls, and furniture. It also depends on the type of reflection surface. Rough surface means several multipaths in the cluster. Smooth surface means only the specular reflection exists.

In THz channels, molecular absorption noise contributes primarily to the noise and it is caused by the vibration of molecules that can re-radiate a part of energy absorbed. This kind of noise is related to human body and composed of two types of noises: background denoted (N_b) and self-induced noise (N_s). Another type of THz channel noise is associated with the surrounding nano-scale devices and the energy radiating from the particles in the propagation medium. This kind of noise is known as the thermal noise, which results from a portion of the fracture energy absorbed by the molecules in the medium, as they increase the average temperature. Or they may be re-emitted in a random direction, which is actually adding noise to the received signals. The molecular absorption noise (N_m) can be found by summing the human body noise (N_b) and self-induced noise (N_s) which can be expressed as follows [8]:

$$N_b(r) = \int_B B(To, f) \left(\frac{C}{\sqrt{4\pi^* no^* fo}} \right)^2 df \tag{5.9}$$

$$N_s(r) = \int_B S(f)\left(1 - e^{-\alpha(f)r}\right) \left(\frac{C}{4\pi nfr} \right)^2 df \tag{5.10}$$

$$N_m(r) = \int_B [B(To, f) \left(\frac{C}{\sqrt{4\pi^* no^* fo}} \right)^2 + S(f)\left(1 - e^{-\alpha(f)r}\right) \left(\frac{C}{4\pi nfr} \right)^2]df \tag{5.11}$$

where B represents the bandwidth of the communication channel. n is the refractive index, and n_0 is THz wave medium's refractive index. f_0 denotes wave frequency, and $\alpha(f)$ represents the absorption coefficient. The Planck's function is denoted by $B(T_0,f)$. $S(f)$ represents the transmitter antenna's signal power density.

In particular, molecular absorption contributes more to the THz channel noise and its level tends to change according to the distances and frequencies. In indoor channel scenario, THz channel pulses generated by the transmitter are propagated through the air and received in the synchronized receiver depending on the detector time resolution, SNR, range of pulses, and THz frequency used. At the THz system terminals, a focus lens can be used in front of antennas to fully collect the signal and to produce parallel beams focusing on the entity to create a directional channel. Otherwise, if no lenses are used, the channel is not directional [19].

THz signals propagating through the indoor channel suffer from molecular resonance attenuation at specific frequencies. This resonance occurs because of rotational transitions corresponding to the THz frequency range, as a water vapor frequencies lie in THz domain and it also depends on air conditions such as dry or humid air. In case of long distances between THz system terminals, signals would be received at a delayed time due to longer signal travel time [19,20]. The estimation of propagation characteristics in the medium can be done by ray-tracing technique based on geometrical optics to determine the levels of electromagnetic fields and to find the signal behaviors through the paths [20].

In the LOS propagation mode, THz EM waves face a path loss due to absorption and spreading loss because of short-distance propagation. As mentioned before, absorption depends on the environment conditions. Three-line parameters would govern the computations. These are line intensity, shape, and position. The attenuation occurring for THz frequencies due to molecular absorption over r is calculated by the following equation [21]:

$$A(f,r) = e^{\alpha molec(f,T_k,P)r} \tag{5.12}$$

where α molec(f,T_k,p) is the molecular absorption coefficient. T_k represents the system temperature in Kelvin.

LOS propagation in the indoor scenario faces many kinds of barriers like people or any various objects. So NLOS propagation model provides the best option for stable transmission. The characteristics of THz channel through the NLOS is related to reflection properties of indoor materials and also with the behavior of EM waves scattering at surfaces in the THz band. The reflection characteristics are restricted by the scattering coefficient defined as follows [21]:

$$Cr = \frac{E_{se}}{E_{re}} \tag{5.13}$$

where E_{se} represents the scattered electric field. E_{re} denotes the reflected electric field. The reflection is considered to arise from a smooth conducting surface. The scattered power in a distance R_0 with respect to the incident power P_1 is given by

$$\mathrm{E}\{\mathrm{R}_{\mathrm{power}}\} = \left(\frac{f\,A\,\mathrm{Cos}\,\theta i}{CRo}\right)^2 \mathrm{E}\{\mathrm{Cr}\,Cr^*\}\,\mathrm{finite} \tag{5.14}$$

$$\mathrm{E}\{\mathrm{Cr}\,\mathrm{Cr}^*\} = \frac{E\{E_{se}\,E_{se}^*\}}{|E_{re}|^2} \tag{5.15}$$

where Θ_i is the incident angle. C is a light speed and f represents the THz frequency. The scattered power is described by the average power reflection coefficient of a surface area A. Non-specular scattering from rough surfaces will affect the behavior of the broadband channel.

In sub-THZ bands, the LOS path in the indoor propagation channel is not robust especially against shadowing. NLOS should be applied in such cases. In addition, dielectric mirrors for covering the walls and ceiling can be used to reduce reflection and scattering. In the case of low receiving sensitivity or low transmitted power, it would gain more degree of robustness [22].

5.5.2.3 THz Chip-to-Chip Channel Characterization

Chip-to-chip communication refers to communication between processors and memories within a computerized system. For such systems, THz channels can offer ultra-high data rate over short distances with an extreme narrow bandwidth. Studying THz waves in chip-to-chip scenario would open new opportunities to evaluate THz frequencies in computerized motherboard components and characterizing the path loss and multipath propagation in order to test the feasibility of realizing chip-to-chip communications in the THz band [23]. A few studies have been conducted in characterizing the THz channels in chip-to-chip communications, which are considered a first step in developing a new revolution in designing ultra-high data rate embedded systems. For example, Alenka and Prateek [24] consider THz chip-to-chip channel characterization with a metal case. They focused on the redundant cavity and took in account the other objects in the cover as conductive objects [24]. The authors proposed geometry-based statistical spread scenario that describes chip-to-chip spread in metal containers filled with conductive objects to characterize the multipath fading in cavity. The study concluded that multiple reflections in the resonance cavity will delay energy and affect the correlation function.

Other study conducted by Jinbang et al., characterized THz channel propagation in metal enclosures with a frequency up to 300 GHz [25]. The authors considered several scenarios, which included the study of LOS propagation inside a metal box and the study of the RNLOS model with and without an embedded memory unit as a reflective surface. In addition, they also studied LOS with a parallel plate structure as an obstacle, NLOS with a heat sink as an obstacle, and LOS on a programmable gate array (FPGA) plate. To clarify the concept of chip-to-chip THz channel characterization, here we consider only the general case of path loss and multipath characterization. According to the previous studies, the general case of path loss denoted by (PL) is related to transmitted power, gain of antennas used, and the received signal power. It also depends on the free space wavelength and distance of traveling signal. Inside the metal box, the metal box acts as a cavity resonator. New considerations are required for channel characterization related to resonant modes. For THz channel with a metal box cavity resonator, the electric field transverse components $E_z = 0$ for TE mode inside the cavity can be calculated as follows [25]:

$$E_x = \frac{jw_{\text{mnp}}\, \mu\, k_y\, H_o}{k_{\text{mnp}}^2 - k_z^2} \cos\frac{n\pi x}{a} \sin\frac{m\pi y}{b} \sin\frac{p\pi z}{z} \tag{5.16}$$

$$E_y = \frac{jw_{\text{mnp}}\, \mu k_x\, H_o}{k_{\text{mnp}}^2 - k_z^2} \sin\frac{m\pi x}{a} \cos\frac{n\pi y}{b} \sin\frac{p\pi z}{z} \tag{5.17}$$

where H_0 represents the arbitrary constant with units of A/m. The parameters m, n, and p are integers. K_{mnp} is an eigenvalue calculating from a set of scalars associated with a linear system of equations to represent the characteristic roots of the wave mode in the metal box cavity in three dimensions. The resonant frequency can be expressed by the following equation [25].

$$F_{\text{mnp}} = \frac{1}{2\sqrt{\mu \in}}\sqrt{\left(\frac{m}{a}\right)^2 + \left(\frac{n}{b}\right)^2 + \left(\frac{p}{c}\right)^2} \tag{5.18}$$

where parameters a, b, and c represent cavity height, length, and width, respectively. The time-invariant channel impulse response can be written as

$$h(t,d) = \sum_{K=1}^{L} a_k(d)\exp(j\theta_{\text{k}}(d))\delta(\tau - \tau_{\text{k}}) \tag{5.19}$$

where L represents the number of multipath components, d represents the signal traveling distance, a_k denoted for kth multipath component. θ_{k} and τ_{k} represent the associated phase and kth path delay respectively. δ is the Dirac delta function. The coherence bandwidth B_{c} of the channel can be expressed as follows [25]:

$$B_{\text{c}} = \frac{1}{2\pi\tau_{\text{rms}}} \tag{5.20}$$

where τ_{rms} represents the RMS delay spread, which depends on mean excess delay τ_{m} and the number of multipath components in addition to excess delay of the kth path. Through this general THz chip-to-chip channel characterization, it can be concluded that in LOS propagation inside the metal boxes for chips and motherboards components, the variation of received power respect to antenna height is related to the resonant TE modes. In this mode, the traveling waves between THz transceiver sides of the cavity would introduce a multipath, which would limit the coherence bandwidth [24]. In addition, the LOS THz propagation inside metal boxes would produces a variation in path loss due to different resonance modes that contribute to the received power. This can be considered as a combination of path loss of traveling waves and the variation of the received power due to the resonance modes. As an example of path loss variations, Figure 5.8 shows the comparison between measured and calculated theoretical path loss inside a metal box.

For NLOS propagation inside the metal box, the traveling wave dominated the channel. The structure of parallel plates introduces multipaths that affects the THz channel. This effect may decrease in case the distance is increased by a few centimeters. Since THz frequencies have a large electrical dimension in the casing, any objects inside the box will reflect the signals, which will create resonant cavity from the box sides [24,25]. In a characteristic comparison of signal behavior inside metal box and in free space, it found that the resonating modes inside the cavity will reduce the mean path loss. However, the frequency band would fluctuate strongly due to traveling wave reflections. Figure 5.9 shows the non-line-of-sight (NLOS) path loss variations in the THz frequency, where a very high

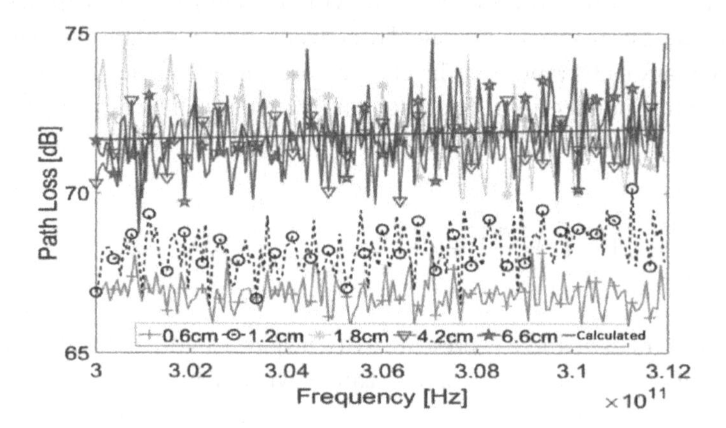

Figure 5.8 Line-of-sight (LOS) path loss [dB] for six measurements between 0.3 and 0.312 THz frequencies.

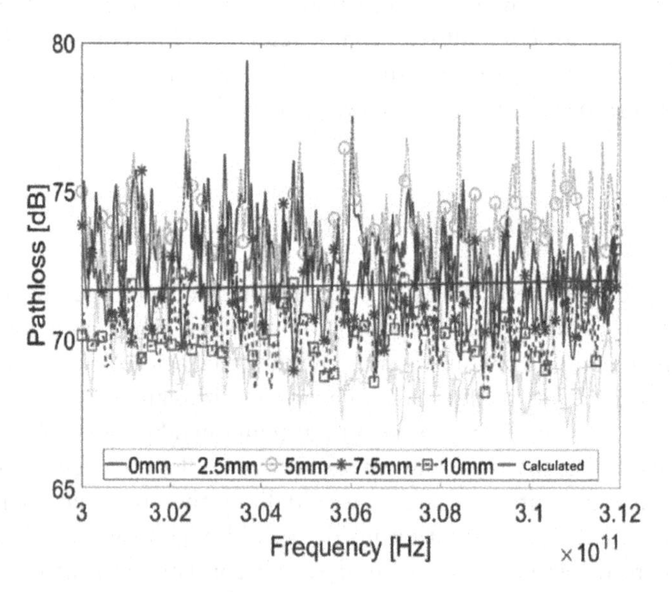

Figure 5.9 Non-line-of-sight (NLOS) path loss six measurements between 0.3 and 0.312 THz frequencies.

reflection/refraction/scattering loss is noticed, which depends on the material, roughness, and the shape of the affecting shallow influencing the THz wave propagation.

5.5.2.4 THz Channel Characteristics in Nano Communications

Nano communication networks have evolved considerably with the revolution of nanotechnology. This has led to the development of new device models on a nano-scale that are able to sense, compute, and store data, which would boost the nanotechnology applications in military, biomedicine, and industry. Nano-sensors in nano networks can interact with several environments such in biological and biomedical environment and or through what is known as molecular communication. Nanotechnology can be helpful to observe the biology of living organisms that can involve the reaction of specific molecules, which act to release other molecules in response to some processes [57]. Medical diagnostics and treatment through accessing small and delicate body sites is another application of nano communications. All these applications can be handled in THz bands, which can enable data exchange between nano-machines in some applications such in nano medicine. It provides benefits especially in medical applications such as through indirect diagnosis by propagating THz waves inside the real human tissue. Such applications encounter problems when using THz bands and the most important of them is path loss due to propagation through the skin. The path loss of THz channels in such an environment depends on water vapor. In nano-communication-based biomedical application, water vapor contributes to two path loss factors: the spreading path loss and absorption path loss. The THz wave's expansion inside the tissue represents the spreading path loss, while the waves absorbed by the tissue are described as absorption path loss. The water vapor loss can be expressed in the following equation [57]:

$$\mathrm{PL}_{\mathrm{total}} = \mathrm{PL}_{\mathrm{spr}}(f,d) + \mathrm{PL}_{\mathrm{abs}}(f,d);\ \text{all in dB} \tag{5.21}$$

$$\mathrm{PL}_{\mathrm{total}} = 20\log\left(\frac{4\pi d}{\lambda g}\right) + 10\,\alpha \times d \times \log e \tag{5.22}$$

where $\mathrm{PL}_{\mathrm{spr}}$ and $\mathrm{PL}_{\mathrm{abs}}$ represent the spreading path loss and absorption path loss respectively. f stands for the frequency and d the path length. K represents the extinction coefficient. And α is given by $4\pi K/\lambda$. The absorption loss calculation is based on the extinction coefficient K, which is measured with an assumption that electromagnetic power is spreading in a spherical mode in a specific distance.

Path loss in the THz channels on body or inside skin depends on the distance between Tx and Rx communication sides, where the distance is measured by millimeters [58]. The path loss due to propagation inside body can be expressed by Friis equation including random variations caused by different organs like lung, heart, or liver, which contribute to variations in dielectric properties along the propagation path. In such applications, the path loss would be increased dramatically with an increase in the propagation range in mm as shown in Figure 5.10. The figure shows how the variation of organs affects the propagation of THz waves between two scenarios of the deep tissue and body surface.

In these scenarios, where there is a concentration of nonhomogeneous molecules, the molecular channel depends on the adaptation of diffusion process. The molecular concentration at the transmission side can be modulated according to the Fick's diffusion laws. According to this law, the diffuse molecules follow the direction of the homogeneity of their concentration. Assuming this behavior, the signals propagate through the moving molecule by encoding the molecule concentration rate at the transmitter side. Channels would be able to delivering the signals at the receiver side according to the variations in molecular concentration [59].

Short-distance THz propagation problems are associated with molecular absorption loss and propagation path loss, in addition to the surrounding noise in the terahertz channel, which contributes to molecular noise. One side effect of concerns when utilizing the terahertz radiation for intra-body communication is the thermal effect on human

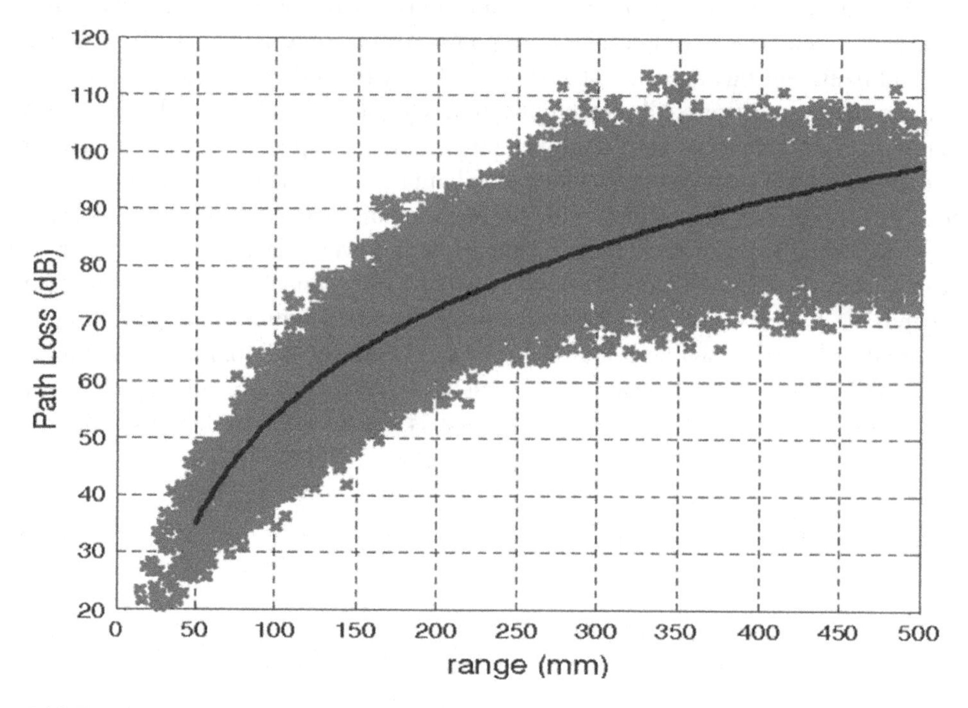

Figure 5.10 THz channel characterization for nano-communication-based human body tissues.

tissues caused by the absorption. By using on–off keying modulation scheme in a short-distance environment, the molecular absorption does not influence the transmission link, which means that the body would be exposed to such radiation intermittently [60]. THz path loss limitations in the human tissue related to the spreading, absorption, scattering, and reflection are reviewed below [61].

A. *Spreading*: The first limitation is due to spreading loss in THz bands, which would reduce the transmission range of the future nano-devices. But THz bands are suitable because of small transmission distances in millimeters, since increase in distance would increase the spreading loss.
B. *Absorption*: The vibration motion due to excitation of molecules in any medium in the THz bands would push the molecule to cause some of the propagating wave energy to be converted into a kinetic one, which would be considered as a loss due to molecular absorption of a fraction of the radiation.
C. *Scattering*: The attenuation scattering loss is measured according to the signal deflection from its directed trajectory to other routes. All variation in THz waves due to its propagation through the tissue like the size, shape, refractive index of each distinguished particle, the wavelength relevant to the incident beam, and their implications on the wave propagation should be taken in consideration.
D. *Internal Reflection*: The total internal reflection leads to reflection of the entire wave, preventing it from propagating to the desired medium. This limitation of THz band occurs when the wave is propagating from the wearable device into the implanted nano-sensors inside the human body.

For all the above limitations, the electrical properties of biological tissues in the operating terahertz band must be taken into consideration in order to evolve a multiple layer propagation model.

5.6 THz Channel Measurement Metrics

Channel measurement of THz radiation is an important issue, which helps to demonstrate the feasibility of using the THz band in telecommunications networks. In this section, we describe some of the required parameters for measurements in THz communication such as bandwidth, antenna directivity, distance, dynamic range, and Doppler frequency. These parameters must be accurately calculated when performing measurements to avoid misinterpretation of the results [33].

5.6.1 THz Antenna Measurement

Antenna measurement in THz bands is considered a sensitive metric especially for large and high-gain antennas. The far- and near-field measurements both play an important role in design requirements for THz band antennas. In addition, antenna tests using reflectors, lenses, holograms, and antenna arrays all determine the efficiency of the antenna and gain measurement. The propagation of THz waves from the antennas depends on the areas it propagates through, from reactive passing near the field zone to the far field, which is actually the target for measurements to drive the far-field distance. The measurement of THz antenna gains faces challenges due to high frequencies and power limitation. The measurements of the power density of the antenna radiation is the ratio of radiated power density $P(\theta,\varphi)$ to its average value over isotropic antenna. The directivity of antenna can be measured by the following equation [36]:

$$D(\theta,\varphi)=\frac{P(\theta,\varphi)}{p_r/(4\pi r^2)} \tag{5.23}$$

If no angular distribution is specified, then maximum directivity refers to maximum radiation power density, which can be calculated as follows:

$$D_{\max}=\frac{p_{\max}}{p_r/(4\pi r^2)} \tag{5.24}$$

The large distance can deteriorate THz signals even if the antenna gain and directivity remain at a high level, because of signal attenuation and distortion, which distort accurate antenna measurements at large distances. Antenna directivity measurements at THz bands can be done at different conditions. Different methodologies can be used to evaluate antenna directivity. In addition the different THz frequency bands would change in radiation pattern directivity as shown in Figure 5.11 when 500, 800 and 1000 THz frequency bands are considered.

The colorless rooms are used the study the effect of walls covered with absorbing materials so that they can be used to simulate free space conditions to calculate the absorption for normal incidence of a wave. The reflector and edge diffraction calculations are

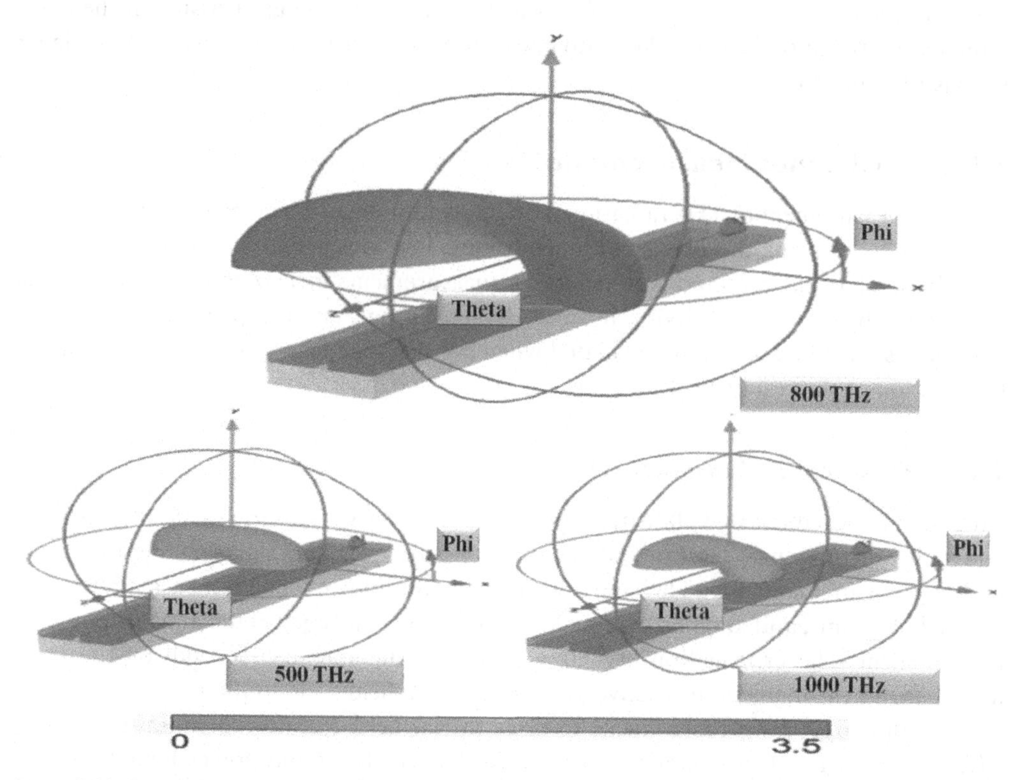

Figure 5.11 Directivity radiation patterns using two angles (phi, theta) at three different frequencies (500, 800, 1,000 THz).

both methodologies related to what is known as antenna test ranges and both would affect the antenna directivity. For reflector surface requirement, for example, the RMS deviation less than λ/100 and 4.6 μm at 650 GHz is required. The edge treatment of reflectors is required for reducing edge diffraction. In THz bands, the use of dielectric lens as a collimating element would contribute to an increase in antenna directivity [37].

5.6.2 THz Channel Distance Measurement

For THz applications, one of the most required metrics is measuring the pulsed THz radiation along distances. The use of THz radiation on rough surfaces through different objects enables to detect the effects of long propagation over long distances. The propagation of THz bands through the optically rough large-scale surfaces, that is, plastics or metals surfaces will impede the optical distance from providing accurate measurements. The use of corner reflector fixed on the target enables to avoid THz signal scattering and ensure specular reflection. However, this method will make difficult the measurement of distances in open fields and shaped environments [38]. Since electromagnetic wave for terahertz (THz) radiation from 1 to 10 THz frequency lies between optical and electric wave boundaries, it would possesses characteristics of both. The distance measurement for THz signals can be described by the THz impulse ranging possess, that is, by measurement of reflected THz pulse time. This measurement enables to count the precise distance because of very short pulse duration in picoseconds, which can used to measure the distance from objects [39]. In impulse ranging method, reflected pulse time delay relative to original signal time enables measurement of the target distance. For such calculations, the THz pulse train's spatial period denoted by L_{SP} can be determined by the following equation [39]:

$$L_{SP} = \frac{CT}{2Ng} \tag{5.25}$$

where T is the pulse period. N_g represents the refractive index number for the propagating THz medium. C denotes the velocity of light in vacuum y. Ng is assumed to be 1 in case of THz pulse propagation in air. In case the target distance becomes shorter than THz pulse train spatial period (L_{SP}), distance d can be determined as follows [39]:

$$d = \frac{C\phi}{2Ng} \tag{5.26}$$

where ϕ denotes time delay. In case the distance (d) exceeds the THz pulse train spatial period L_{SP}, the measured THz pulse order should be considered. Then ϕ changed by $(mT+\phi)$ to represents THz pulse time delay relative to sum frequency generation signal nearest in time.

5.6.3 THz Channel Spectrum Capacity

The bandwidth of THz band varies between several tens of gigahertz and several tens of terahertz. It provides a large spectrum density in wireless communication systems. The transmission speed of tens gigabits is much faster than any currently used technology [26]. The THz pulse bandwidth can be represented by quality factor (Q) or relative bandwidth (σf). The following equation is used to find the quality factor [34]:

$$Q = 1/\sigma f = f_c/\Delta f \tag{5.27}$$

where Δf denotes the bandwidth. f_c represents the frequency carrier. The relative energy bandwidth is given by $\sigma f/\sqrt{2}$.

Signal-to-noise ratio (SNR) is important to calculate bandwidth capacity by using the Shannon formula. To analyze capacity for the THz band, the capacity of the given bandwidth should be estimated. To calculate channel capacity by using signal-to-noise ratio (SNR), the following equation is used, which is based on the Shannon–Hartley theorem [35]:

$$C = B\log_2\left(1+S/N\right) \tag{5.28}$$

This formula shows the importance of SNR in calculating channel capacity. Channel capacity is defined with a given SNR value in a certain network. Shannon capacity is referred to speed of received messages in communication systems, and it depends on the bandwidth

(B) and SNR. Signal to interference plus noise ratio (SINR) is used to measure the percentage of signal power of interest received at receiver [35].

5.6.4 THz Channel Dynamic Range

Theoretically, channel dynamic range (DR) is known as the ratio between highest and lowest measurable signals. It is used to determine the bandwidth, while the SNR measures the sensitivity, which enables the maximum change in the signal to be measured. The channel dynamic range in practical measurement represents the ratio of maximum magnitude of amplitude to noise floor's root-mean-square (rms). The DR could be found by equation [40]

$$DR = \frac{|A_{\max}|}{NF_{\mathrm{rms}}} \tag{5.29}$$

where $|A_{\max}|$ represents the maximum magnitude of amplitude. $\mathrm{NF}_{\mathrm{rms}}$ denotes the noise floor rms.

The dynamic range for THz signals enables to determine the largest absorption coefficient in transmission measurements. The dynamic range system enhancement can be obtained by increasing the THz transmitted power or by decreasing noise floor of the system. When signal is attenuated to a level equal to the noise floor, the largest absorption coefficient can be measured with a given corresponding dynamic range [40]. In general, there is no direct relationship between the estimated parameters of DR and SNR in time-domain signals. Both the DR and SNR of the spectral data are strongly dependent on frequency, and they usually decline sharply with the frequency [40]. Estimations of THz dynamic range and SNR in time domain depend on the time-domain waveform, and noise signal in absence of THz. DR can be calculated as a ratio between the measured peak's mean value and noisy signal's standard deviation, while SNR is determined by the ratio between the measured peak's mean value and its standard deviation. The estimation of THz DR through amplitude spectrum calculation is based on the given DR relative to the ratio between mean amplitude and noise floor [41].

5.6.5 THz Doppler Frequency Characteristics

The Doppler effect appears when electromagnetic wave propagating with a frequency f_0 hits a movable object. The frequency of reflected wave can be found by $f_1 = f_0 + f_{\mathrm{D}}$ where f_{D} is Doppler shift. By Doppler effect, when using THz frequencies, wavelength (λ) is very small and any short distance traveled by a moving object would lead to a big shift in frequency. This effect would be observed as well when source of THz signal is moving. Doppler shift would have more impact on the receiver side than signals with lower frequencies [18]. Equation (5.30) is used for Doppler shift calculation when source wave is moving toward the receiver at rest [42]:

$$f_{\mathrm{D}} = \frac{c}{c-v} \times f \tag{5.30}$$

where f represents transmitted wave frequency, v is source velocity, and c represents the speed of light. The Doppler shift is proportional to transmitted wave frequency and this frequency is very high because it is in the THz band. So Doppler shift would be very high as well. The signal may fall in a peak due to attenuation and communication would be totally lost. To avoid this problem, the transmission window must be chosen carefully.

5.7 Methodologies of THz Channel Measurement

THz waves present many important properties that have an influence on many science applications. The characteristics of THz spectra are evaluated via different measurement methodologies. This section reviews various measurements related to THz channel and their performance in brief. Parameters such as dynamic range (DR) and signal-to-noise ratio (SNR) are considered as major factors affecting the accuracy of THz measurements. The THz channel measurement methodologies can be used to measure different factors and characteristics such as reflection and diffraction, special and broadband channel, MIMO and intra-device THz channels, which enable us to evaluate the use of THz bands in different applications.

5.7.1 Reflection and Diffraction Measurements

In THz bands, the prediction of THz system performance depends on ray tracing methods. In this method, the channel transfer function is measured to obtain an accurate information about reflection properties in addition to determining the proper modeling of scattering and diffraction effects. Reflection in transmission during dielectric objects is considered as an important propagation mechanism that depends on frequency, because frequency varies with dielectric properties of materials, which affect the reflection and transmission coefficients of the material object. The transmission during dielectric layer (T_{DL}) is determined by the following equation [18]:

$$T_{\mathrm{DL}} = \frac{T_1 T_2 e^{-j\alpha(f)}}{1+\rho_1\rho_2 e^{-2j\alpha(f)}} \tag{5.31}$$

where T and ρ represent the transmission and reflection coefficients respectively, subscripts 1 and 2 represent medium air and dielectric, respectively. $\alpha(f)$ represents the dielectric frequency-dependent electrical length with an angle θ_t with layer as shown in the equation [18].

$$\alpha = \frac{2\pi}{c_0} f \sqrt{\epsilon_r}\, d_{\mathrm{layer}} \cos\theta \tag{5.32}$$

where d_{layer} indicates the geometrical width of the layer. For the measurement of THz signal reflection, NLOS propagation can be demonstrated in the indoor surface, which can produce a directional reflected beam that relies on a specular reflection from rough surface. Figure 5.12 shows the implementation of the reflection channel characterization measurement system. In order to measure THz signal loss during reflection, transmitter

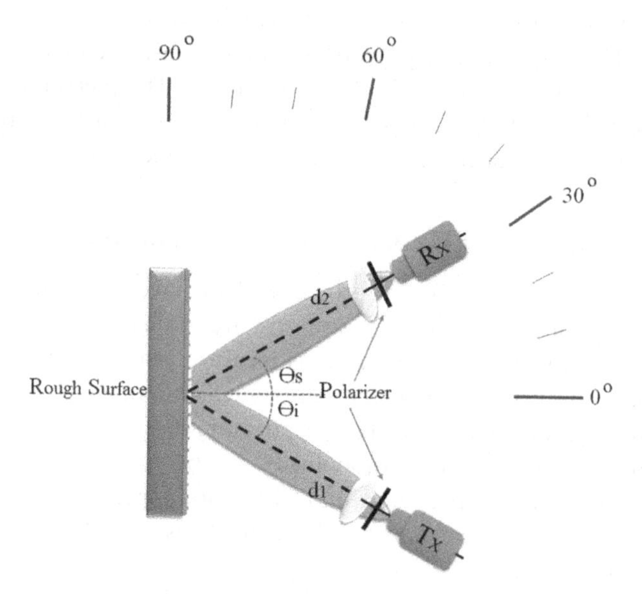

Figure 5.12 Reflection measurement setup configuration with rough surface and polarizers.

and receiver antennas can be mounted on moving bars with respect to the point of reflection on the surface to enable adjustment of the incident angle.

Fresnel analysis can be used to calculate reflection coefficient parameters according to surface roughness [20]. This measurement method can be used to characterize the THz beam pattern reflection from surface in LOS and specular NLOS paths.

Diffraction is another propagation mechanism, which would help in modeling the THz NLOS path. In diffraction, signals can travel in an NLOS environment and can be diffracted at edge of building, tree, window, or because of any other obstacle. When electromagnetic wave hits the obstacle between the transmitter and the receiver, then direction of wave would change into a shadow region. This phenomenon is known as the diffraction or knife-edge effect [18]. The diffraction by a half-plane, scattered field complex-value amplitude at distance r with a sinusoidal incident field $x_i = A_0 e^{(j\omega t)}$is given by equation [18].

$$x(j\omega) = A_0 e^{jkr\cos(\varnothing - \varnothing_0)} F\left(\sqrt{2kr}\cos(\varnothing - \varnothing_0)\right) \pm A_0 e^{jkr\cos(\varnothing + \varnothing_0)} F\left(\sqrt{2kr}\cos(\varnothing + \varnothing_0)\right) \tag{5.33}$$

where $F(x) = \left(e^{j}(\pi/4)/\left(\sqrt{\pi}\right)\right)\int_{x}^{\infty} e^{j\mu^2} d\mu$, $k = \omega / c_0$ is the wave number, and the minus signs indicate direction of polarization in the incident field. $\varnothing_0$ is an incident angle and $\varnothing$ is the observation angle. Figure 5.13 shows the geometry of the knife-edge measurements system, where in a knife-edge diffraction, obstacles would contribute to diffract THz signals and generate a secondary radiation source. The second radiation source puts THz signals in a shadow region of obstacles. The knife-edged path can be calculated depending on diffraction angle Θ and Fresnel parameters related to distances d_1 and d_2 between diffraction region and transmitter and receiver antennas, respectively, in addition to the height of knife-edge measured from LOS path.

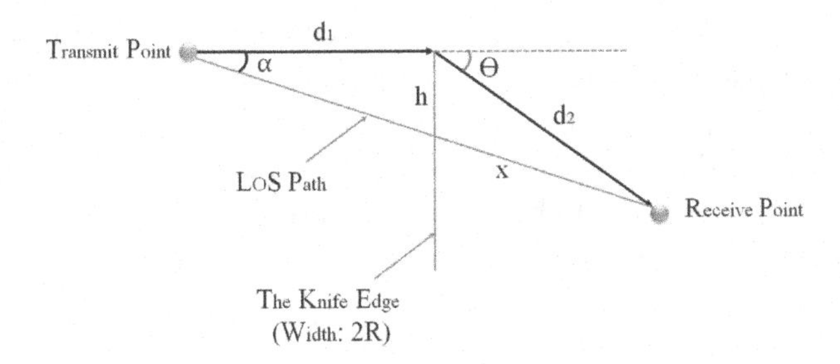

Figure 5.13 Knife-edge measurements system geometry with width $2R$ and theta angle.

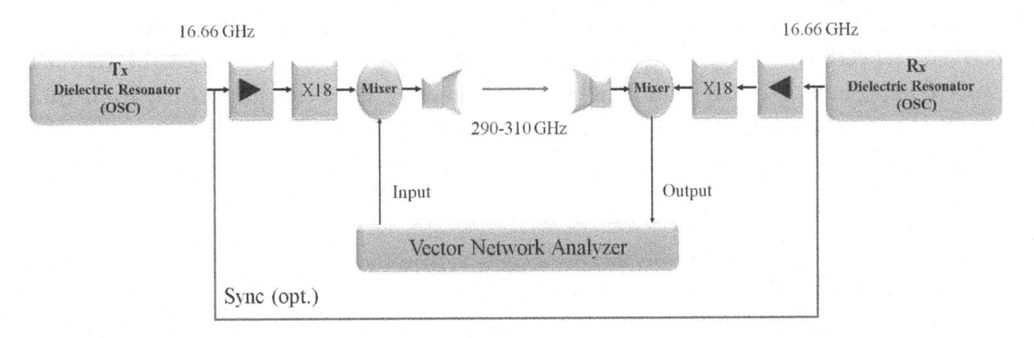

Figure 5.14 THz measurement for transceiver system block diagram with transmission frequency band 290–310 GHz.

5.7.2 *Spatial THz Channel Measurement*

For channel measurements and transmission experiments in the frequency of THz range band, the system contains a specific type of transmitter and receiver modules, which can be used to measure many spatial measurement tasks. Such system is designed to measure channel characteristics for indoor THz systems like multi-path propagation, scattering and diffraction analysis, and evaluation of appropriate modulation techniques. Figure 5.14 shows a THz measurement system's block diagram, which consists of two transmitters and detectors with a Schottky mixer operating on a second harmonic and local dielectric resonator oscillator, which is tripled in frequency twice [43].

Vector network analyzer (VNA) with the frequency extension module can be used to measure in THz frequency range between 27 and 325 GHz. VNA enable to measure the scattering parameter of transmission between TX output and RX input. The channel transfer function between transmission and reception sides can be measured at the frequency equal to 300 GHz + f_{test} according to the adjustment of stimulus signal frequency range (f_{test}) between 10 MHz and 10 GHz. To avoid multiple reflections in the transceiver modules, the modules can be equipped by an absorber panel in the front, and the horn antennas can be adjusted steering through it.

For spatial channel variation measurements in different sub-bands, it can be done at the field by observing different points within a specific area and evaluating the results observed from a group of measurement stations. In addition, the use of ray tracer will enable to obtain accurate spatial channel properties and details about path characteristics information. Ray tracer can enable very good prediction at low terahertz frequency [44]. In multipath propagation through rough surfaces, the THz waves propagating on the rough surface are composed of reflected or coherent rays in a specular direction in addition to other scattered rays with a diffraction phenomenon. According to ray-tracing methodology, measuring special channel can be done using the following equations for reflected ray, scattered ray, and diffracted ray respectively [45].

$$\mathrm{Href}(f)=\left(\frac{C}{r\pi f(r1+r2)}\right)e^{-j2\pi f\tau ref-0.5k(f)(r1+r2)}\cdot R(f) \tag{5.34}$$

where $H_{ref}(f)$ represents the scattered propagation frequency-dependent transfer function. r_1 and r_2 represent the distances between transmitter and reflector, between reflector and receiver, respectively. $\tau_{ref} = \tau_{LoS} + (r_1+r_2-r)/c$ denoted the time of arrival of the reflected ray. $R(f)$ represents the Rayleigh roughness factor [45]. The refection propagations that are described as a scattered ray propagation transfer function is given by the following equation:

$$H_{sca}(f)=\left(\frac{C}{r\pi f(s1+s2)}\right)e^{-j2\pi f\tau sca-0.5k(f)(s1+s2)}\cdot S(f) \tag{5.35}$$

where $\tau_{ref} = \tau_{LoS} + (s_1 + s_2 - r)/c$ denotes the scattered ray arrival time. $S(f)$ represents the rough surface scattering coefficient [45]. The diffracted propagation that describes the diffraction channel transfer function is given by the following equation:

$$H_{dif}(f)=\left(\frac{C}{r\pi(d1+d2)}\right)e^{-j2\pi f\tau dif-0.5k(f)(d1+d2)}\cdot \mathrm{L}(f) \tag{5.36}$$

where d_1 and d_2 represent A the distances between transmitter and diffracting point, between diffracting point and receiver, respectively. τ_{dif} represents the diffracted ray arrival time and equal to $\tau_{LoS} + \Delta d/c$. $L(f)$ represents the diffraction coefficient, which characterizes the LoS propagation attenuation.

According to above measurements, the direct rays in the LoS THz channel propagation will control the energy of the received signal, while in NLoS propagation, the reflected rays will take such a control. In the case of any increase in the frequencies, the propagation surface is considered more rough, which would result in THz signals spreading more energy out of the specular direction. Therefore, it is necessary to include scattered rays in ray tracing modules at both LOS and NLOS propagation conditions.

5.7.3 Measurement of Broadband Characteristics

Broadband radio channel measurements can be performed by lamellar interferometer, which has a variable depth binary grating. It operates in a diffraction pattern and uses a Fourier spectrometer at the beam splitter for the purpose of wave amplitude splitting [46].

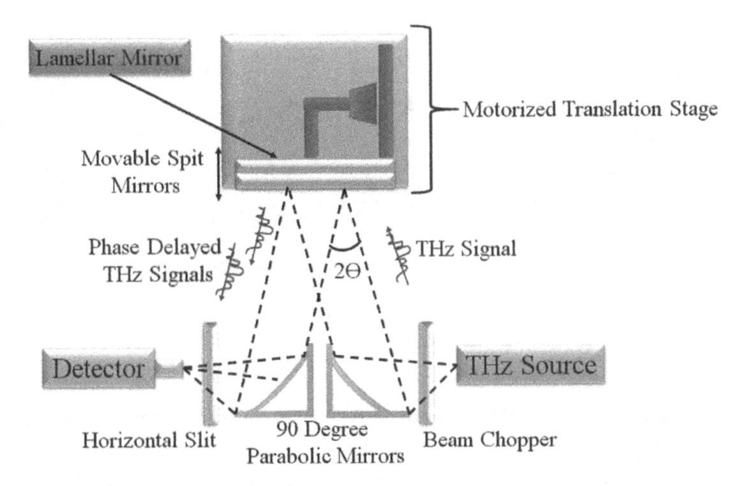

Figure 5.15 Moveable lamellar mirror acts as beam-splitter measurement with 90 degree parabolic mirrors.

As shown in Figure 5.15, a moveable lamellar mirror works to split the beams, where lamellar interferometer separates wavefront by letting one half of the beam to be reflected by front mirrors and the second half of beam to be reflected by back movable mirrors. The difference between the two separated waves can be determined by the distance between the two mirror sets.

This measurement system provides suitable detector for bandwidths up to 10 THz and is capable of detecting higher harmonics. It also measures the spectral emission profile for THz broadband as well as for free-space sources. In lamellar interferometer measurement systems, the design of mirrors must meet the considerations related to the range of wavelength bands that would be measured. These considerations depend on the estimations of minimum measurable wavelength according to diffraction orders. The sensitivity of estimation calculation depends on the ratio between the maximum and minimum wavelengths denoted by NP= $\lambda_{max}/\lambda_{min}$, and according to the number of points in each half of lamellar mirror. To avoid polarization effects in the long wavelength, the height of the mirrors should be bigger than double max wavelengths, $h > 2\lambda_{max}$ [47].

In general, lamellar interferometer enables to measure both the pulsed and continued THz broadband and is capable of characterizing the sources and detectors of the THz broadband. In order to characterize the broadband of wide-band THz beams, it must be ensured that the splitter has a wide bandwidth and low sensitivity to polarization. In addition, the characterization process should be free from any constant internal waves. The THz sources also should have reduced energy to avoid loss of throughput.

5.7.4 MIMO THz Channel Measurements

MIMO techniques can incorporate the THz band communication to enhance data rate, throughput, and system reliability. MIMO techniques provide high data rate without expanding the system bandwidth. The measurement of MIMO-based THz frequency band channel for testbed specification and MIMO measurements are important for system performance estimation [48]. The implementation of MIMO channel testbed can be performed

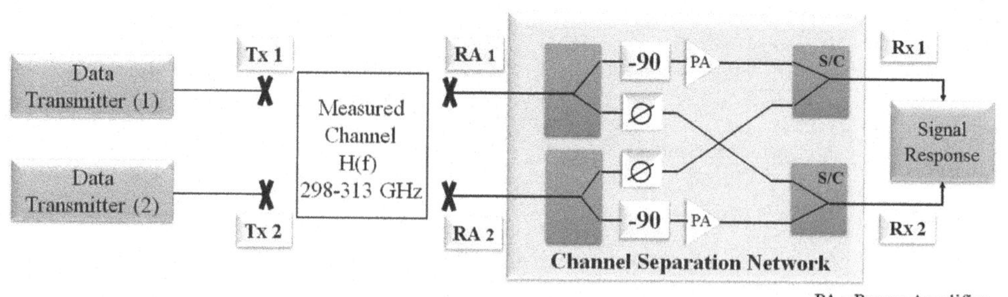

Figure 5.16 Measurement of 2×2 MIMO system with channel separation network and 298–313 GHz band.

by VNA. VNA acts like a broadband system extending the frequency up to 300 GHz by using extra modules based on sub-harmonic mixers. The mixers translate the frequency measurement of VNA in range from 298 to 313 GHz, interacting like an antenna array to characterize the MIMO channel. For example, Figure 5.16 shows the measurement of 2×2 THz band MIMO system by determining the transmitter channel responses in channel 1 and in channel 2 as cross-channel interference caused by transmitter 1. The transmitter 2 frequency response is recorded at both receivers' antenna locations, which act as a virtual array [49]. Figure 5.16 presents measured 2×2 THz band MIMO channel response, which can be observed by an eye probe at the end of channel separation network.

THz-band MIMO channels are more suitable for spatial LOS diversity [62] and conventional low-frequency MIMO is more convenient for NLOS multiple signals because in THz bands, multipath signals would attenuate the received signal. On the other hand, the small antenna size makes LoS MIMO more suitable for terahertz applications, where a large number of antenna elements can be placed next to a small area to create multiple parallel channels. Also, the high directivity of these antennas reduces the interference between channels from nearby and close transmitters to be very small. In THz band LOS MIMO, the separation between antennas ensures distinguishing multiple transmitters according to distances. Antenna separation can be measured by $D = \sqrt{\frac{R \times \lambda}{N}}$ where D is the antenna separation, R represents the distance between source and destination. λ denotes carrier frequency wavelength and N the number of antenna elements at the receiver [50].

5.7.5 THz Intra-Device Channel Measurement

For intra-device communications, the synchronization of effective memory chip components is challenging as the number of memory cores are growing in a multi-level memory hierarchy and multicore architectures. This challenge is solved by network on chip (NoC) technology. This technology enables to replace the network communication bus by a complex topology. The use of wireless network on chip (WNOC) enables the chip components to interact in efficient way. Moreover the existence of memory coherence protocols allows interaction as a broadcast [51]. THz-based short-range communications promise to conserve large resources and deal with bandwidth delay for critical communications within the chip. The short THz wavelengths enable to operate intra-device communications by integrated antennas for chip-to-chip applications. For such communication systems, there is a need to determine the diffusion properties with structures and materials typical of internal device connections, which help to develop an appropriate propagation model and software design tools [52]. The environment of intra-devices includes many features related to wavelength and antenna types used, which contribute significantly to setting limits for the use of high THz frequencies, especially in full-wave run time and memory requirements.

Intra-device THz channels consist of materials and coatings as well as electronic building parts throughout the surfaces of printed circuit boards (PCBs).The measurements of reflection processes at PCB surfaces can be obtained by a vector network analyzer, which is used to simulate propagation based on THz ray tracing in intra-device communications. In a THz intra-device communication, the most important measurement is to evaluate channel status in PCB scattering and reflection behaviors that depend on the superposition of multiple diffuse reflection processes on the surface of the PCB. Measurements can be handled by using ray tracing considering the inputs of the measurement system are the coordinates of transceiver positions as well as the specular reflection point of the considered propagation path. The output of measurements is reflected field components. These components are based on amplitude coefficient and frequency-dependent phase angle. The measurements depends on path delay in addition to azimuth and elevation angular coordinates relative to the propagation path according to ray-tracing methodology.

The PCB scattering can be calculated depending on the distance between transceiver projection and the scaling factor of reflection processes area, in addition to amplitude distribution in the scattering region and total reflectivity of the PCB denoted by (R_{PCB}). The total amplitude A_T of the generated scattering process can be calculated by the equation [53].

$$A_T = R_{PCB} A \frac{b}{a} \tag{5.37}$$

where A represents the relative amplitude over the radius of the semi-major axis length for the wave incidence plane. The factor *b/a* scales the amplitudes inversely proportional to the sine of the angle of the incident waves. A is calculated from two introduced amplitude modifiers depending on the distance between specular reflection and actual scattering point, in addition to shape parameters in certain direction according to the angle phase Φ. The angle phases of the scattering processes Φ can be found by $\exp(-2\,\pi f \frac{\Delta}{Co})$ where Δ denotes the path difference and Co represents the speed of light.

5.8 THz Channel Measurement Versus Channel Analysis

To evaluate the THz channel behavior based on the transmitted impulses, the received pulses should be investigated according to received pulse measurements and received pulse analysis. In measurements, the distance, temperature, and pressure all play an important role in the calculations. For molecular absorption, measurements should consider the composition of a medium control for setting the channel medium. Relative humidity (RH), together with the previously mentioned parameters, all show the relative degree of contribution to the selective fading of the broadband signal frequency. By taking several measurements to calculate the effects of different relative humidity values in the received pulse, analytical results can be obtained from the transmitted signal, or can be obtained from the impulse responses in minimum or linear phase [54]. In order to compare between THz frequency band with highest selectivity, measurements must be taken in higher frequencies, which may be difficult to measure because of limitations in the measurement equipment. By another way, it can be obtained by comparing the analysis of received pulses that are provided by impulse responses with minimum phase and linear phase responses.

Generally, experimentation and measurements in THz band is a bit tricky due to the very small wavelengths and strong signal attenuation that depends on propagations ranges. As illustrated before, in short distances difficulty in measurements arise due to the limitation of measurement devices. For outdoor THz scenario in the range of more than 100 m measurement of wideband, the channel characteristics are limited to path and atmospheric attenuation loss in LOS mode. The LOS requires more accurate testbed specifications with a measurement tools such like VNA [55]. For the analysis of channel characteristics, different aspects are important and should be taken into account. These aspects are power reception, which depends on path loss distance, multipath reflection delay and inter-symbol interference (ISI). In addition, K factor describes the ratio of fixed and fluctuating power components related to gain for radio environments [56].

5.9 Conclusions

Recently, THz bands have received increasing attention for many new applications that require ultra-wide-band communications and very high data rates in a short range. Implementation of THz communication requires various studies and evaluations in order to assess different THz fchannel and propagation issues. This chapter introduced THz band channel characterization and measurements, taking into consideration many parameters in detail such as path loss, attenuation, and fading, which are influenced by additional atmospheric and molecular absorption. The chapter also provided a brief review of the environmental aspects of THz wave propagation such as scattering, absorption, and refraction loss in indoor and outdoor environments. Channel capacity performance, spectral efficiency, and signal energy under the LOS and NLOS propagation conditions were also discussed. The THz band relatively has asymmetric parameters due to frequency-dependent behavior with resonant peaks at specific frequencies. In literature, there are many discussions related to THz channel impairments in nano communications. That is, some experiments were conducted for THz channel modeling in deterministic and probabilistic approaches.

References

[1] Ali Rıza Ekti et al. Statistical Modeling of Propagation Channels for Terahertz Band. IEEE Conference on Standards for Communications & Networking, 2017, pp. 1–7.

[2] Michael Taynnan Barros et al. Integrated Terahertz Communication with Reflectors for 5G Small-Cell Networks. *IEEE Transactions on Vehicular Technology*, vol. 66, no. 7, 2017, pp. 5647–5657.

[3] Sundeep Rangan et al. Millimeter-Wave Cellular Wireless Networks. Potentials and Challenges. Proceedings of the IEEE, vol. 102, no. 3, 2014, pp. 365–385.

[4] Hadeel Elayan. Terahertz Communication: The Opportunities of Wireless Technology Beyond 5G. International Conference on Advanced Communication Technologies and Networking, 2018, pp. 1–5.
[5] Hadeel Elayan et al. Terahertz Channel Model and Link Budget Analysis for Intrabody Nanoscale Communication. *IEEE Transactions on Nanobioscience*, vol. 6, 2017, pp. 491–503.
[6] Thomas Kürner & Sebastian Priebe. Towards THz Communications Status in Research, Standardization and Regulation, *Journal of Infrared, Millimeter, and Terahertz Waves*, vol. 35, 2014, pp. 53–62. https://doi.org/10.1007/s10762-013-0014-3
[7] Zheng Xu et al. A Statistical Model for the MIMO Channel with Rough Reflection Surfaces in THz band. *Nano Communication Networks*, vol. 8, 2016, pp. 25–34.
[8] Rui Zhang et al. Analytical Characterization of the Terahertz In-Vivo Nano-Network in the Presence of Interference Based on TS-OOK Communication Scheme. *IEEE Access*, vol. 5, 2017, pp. 10172–10181.
[9] Naoya Kukutsu and Yuichi Kado. Overview of Millimeter and Terahertz Wave Application Research. TT Microsystem Integration Laboratories, vol. 7, no. 3, 2009, pp. 1–6.
[10] Ian F. Akyildiz et al. Combating the Distance Problem in the Millimeter Wave and Terahertz Frequency Bands. *IEEE Communications Magazine*, vol. 56, no. 6, 2018, pp. 102–108.
[11] Nagatsuma, Ducournau, Renaud. Advances in Terahertz Communications Accelerated by Photonics. *Nature Photonics*, vol. 10, 2016, pp. 371–379.
[12] Elayan H, Shubair RM, Jornet JM, Johari P. Terahertz Channel Model and Link Budget Analysis for Intrabody Nanoscale Communication. *Trans Nano Bioscience*, vol. 16, no. 6, 2017, pp. 491–503.
[13] Yihong Yang et al. THz-TDS Characterization of the Digital Communication Channels of the Atmosphere and the Enabled Applications. *Journal of Infrared, Millimeter, and Terahertz Waves*, vol. 36, no. 2, 2015, pp. 97–129.
[14] Rudolf Mathar. Wireless Channel Modeling and Propagation Effects. RWTH Aachen University, 2009.
[15] Iury S. Batalha et al. Large-Scale Channel Modeling and Measurements for 10GHz in Indoor Environments. *International Journal of Antennas and Propagation*, vol. 2019, pp. 1–10. https://doi.org/10.1155/2019/9454271
[16] L. Pometcu, R. D'Errico. Large Scale and Clusters Characteristics in Indoor Sub-THz Channels. IEEE 29th Annual International Symposium on Personal, Indoor and Mobile Radio Communications, 2018, pp. 1405–1409.
[17] Mathew K. Samimi et al. 28 GHz Millimeter-Wave Ultra wideband Small-Scale Fading Models in Wireless Channels. IEEE 83rd Vehicular Technology Conference, 2016, pp. 1–6.
[18] Seunghwan Kim. THz Device to Device Communications: Channel Measurements, Modelling, Simulations, and Antenna Design. Georgia Institute of Technology, 2016.
[19] Seunghwan Kim et al. D-Band Channel Measurements and Characterization for Indoor Applications. *IEEE Transaction on Antennas and Propagation*, Vol. 63, no. 7, 2015, pp. 3198–3207.
[20] Ma, J., Shrestha, R., Moeller, L., & Mittleman, D. M. Channel Performance for Indoor and Outdoor Terahertz Wireless Links. *APL Photonics*, vol. 3, 2018, pp. 1–12.
[21] Anamaria Moldovan et al. LOS and NLOS Channel Modeling for Terahertz Wireless Communication with Scattered Rays. *2014 IEEE Globecom Workshops (GC Wkshps)*, Austin, TX, USA, 2014, pp. 388–392.
[22] Rashid A. Saeed, Raed A. Alsaqour, Ubaid Imtiaz, Wan Mohamad, Rania A. Mokhtar. Design of CMOS Power Amplifier for Millimeter Wave Systems at 70 GHz. *International Journal of Engineering and Technology (IJET)*, vol. 5 no. 1, 2013, pp. 498–503.
[23] Xavier Timoneda et al. Channel Characterization for Chip-scale Wireless Communications within Computing Packages. Twelfth IEEE/ACM International Symposium on Networks-on-Chip (NOCS), 2018, pp. 1–8.
[24] Alenka Zajic and Prateek Juyal. Modeling of THz Chip-to-Chip Wireless Channels in Metal Enclosures. 12th European Conference on Antennas and Propagation, EuCAP, 2018, pp. 1–5.

[25] Jinbang Fu et al. 300GHz Channel Characterization of Chip-to-Chip Communication in Metal Enclosure. 13th European Conference on Antennas and Propagation (EuCAP), 2019, pp. 1–11.

[26] Nagy Faroug Merghani, Rashid A. Saeed, Rania A. Mokhtar. Hard Handover Optimization for Co-channel WCDMA Heterogeneous Network. International Conference on Sustainable Research and Innovation, vol. 5, 2014, pp. 254–257.

[27] Jingye Sun et al. Predicting Atmospheric Attenuation Under Pristine Conditions between 0.1 and 100 THz. *IEEE Access*, vol. 4, 2016, pp. 9377–9399.

[28] Dan Li et al. Performance Analysis of Indoor THz Communications with One-Bit Precoding. IEEE Global Communications Conference (GLOBECOM), 2018, pp. 1–7.

[29] Gustavo A. Siles et al. Atmospheric Attenuation in Wireless Communication Systems at Millimeter and THz Frequencies. *IEEE Antennas and Propagation Magazine*, vol. 57, no. 1, 2015, pp. 48–61.

[30] Yi-Da Hsieh et al. Dynamic Terahertz Spectroscopy of Gas Molecules Mixed with Unwanted Aerosol Under Atmospheric Pressure Using Fiber-Based Asynchronous Optical-Sampling Terahertz Time Domain Spectroscopy. *Scientific Reports*, vol. 6, 2016, pp. 1–10.

[31] Qingfeng Jing, Danmei Liu. *Study of Atmospheric Attenuation Characteristics of Terahertz Wave Based on Line-By-Line Integration*. Wiley, vol. 31, no. 12, 2018.

[32] Milda Tamosiunaite et al. *Atmospheric Attenuation of the Terahertz Wireless Networks*. Intech Open, 2018.

[33] Leopoldo Angrisani, Giovanni Cavallo, Annalisa Liccardo, Gian Paolo Papari, and Antonello Anderone. *THz Measurement Systems*. InTech Open Science and Open Minds, 2018.

[34] Karl Baneb, Gennady Stupakov, Sergey Antipov and Dao Xiang. Measurements of Terahertz Radiation Generated using a Metallic, Corrugated Pipe. *Nuclear Instruments and Methods in Physics Research Section A: Accelerators, Spectrometers, Detectors and Associated Equipment*, vol. 844, 2017, pp. 121–128

[35] Xiao Nie. SINR and Channel Capacity in Terahertz Networks. Tampere University of Technology, 2015, pp. 1–35.

[36] Guillermo Carpintero et al. *Semiconductor Terahertz Technology Devices and Systems at Room Temperature Operation*. Wiley, 2015, pp. 1–408.

[37] John Federici and Lothar Moeller et al. Review of Terahertz and Sub Terahertz Wireless Communications. *Journal of Applied Physics*, 2010.

[38] S. S. Dhillon et al. The 2017 Terahertz Science and Technology Roadmap. *Journal of Physics D: Applied Physics*, vol. 50, 2017, pp. 1–49.

[39] Takeshi Yasui et al. Absolute Distance Measurement of Optically Rough Objects Using Asynchronous-Optical-Sampling Terahertz Impulse Ranging. *Applied Optics*, vol. 49, no. 28, 2010, pp. 5262–5270.

[40] Mira Naftaly and Richard Dudley. Methodologies for Determining the Dynamic Ranges and Signal-To-Noise Ratios of Terahertz Time-Domain Spectrometers. *Optical Society of America*, vol. 34, no. 8, 2009, pp. 1213–1215.

[41] Luigi Cocco. *New Trends and Developments in Metrology*. Intech Open 2016.

[42] Wafa Hedhly. Resource Allocation in Future Terahertz Networks. King Abdullah University of Science and Technology, 2019.

[43] T. Kleine-Ostmann et al. Measurement of Channel and Propagation Properties at 300 GHz. Conference on Precision electromagnetic Measurements, 2012, pp. 1–2.

[44] Antoine Jouadé. Millimeter-Wave Radar Imaging Systems: Focusing Antennas, Passive Compressive Device for MIMO Configurations and High-Resolution Signal Processing. University of Rennes 2017.

[45] Chong Han et al. Multi-Ray Channel Modeling and Wideband Characterization for Wireless Communications in the Terahertz Band. *IEEE Transactions on Wireless Communications*, vol. 14, no. 5, 2015, pp. 2402–2412.

[46] Omar Manzardo et al. Miniature Lamellar Grating Interferometer Based on Silicon Technology. *Optical Society of America*, vol. 29, no. 13, 2004, pp. 1437–1439.

[47] Mira Naftaly et al. A Simple Interferometer for the Analysis of Terahertz Sources and Detectors. *IEEE Journal of Selected Topics In Quantum Electronics*, vol. 14, no. 2, 2008, pp. 443–448.
[48] Alice Faisal et al. *Ultra-Massive MIMO Systems at Terahertz Bands: Prospects and Challenges.* Cornell University. 2019, pp. 1–8.
[49] Nabil Khalid et al. Experimental Throughput Analysis of Low-THz MIMO Communication Channel in 5G Wireless Networks. *IEEE Access*, vol. 5, no. 6, 2016, pp. 616–619.
[50] Hadi Sarieddeen et al. Terahertz-Band Ultra-Massive Spatial Modulation MIMO. arXiv 2019, pp. 1–13.
[51] Vitaly Petrov et al. Terahertz Band Intra-Chip Communications: Can Wireless Links Scale Modern x86 CPUs? *IEEE Access* 2017, pp. 6095–6109.
[52] A. Fricke et al. Characterization of Transmission Scenarios for Terahertz Intra-Device Communications. IEEE-APS Topical Conference on Antennas and Propagation in Wireless Communications, 2015, pp. 1137–1140.
[53] Alexander Fricke et al. A model for the reflection of terahertz signals from printed circuit board surfaces. *International Journal of Microwave and Wireless Technologies*, vol. 10, no. 2, 2018, pp. 179–186.
[54] Nelvia del Cisne Gonza Ajila. Channel Selection for Wearable Wireless Devices in THz Range. University in Erlangen, Germany 2017.
[55] Naveed A. Abbasi et al. Double Directional Channel Measurements for THz Communications in an Urban Environment. arXiv 2019, pp. 1–6.
[56] Zahed Hossain et al. Stochastic Multipath Channel Modeling and Power Delay Profile Analysis for Terahertz-band Communication. Proceedings of the 4th ACM International Conference on Nanoscale Computing and Communication, 2017, pp. 1–7.
[57] Qammer H. Abbasi et al. Terahertz Channel Characterization Inside the Human Skin for Nano-scale Body-Centric Networks. *IEEE Transactions on Terahertz Science and Technology*, vol. 6, no. 3, 2016, pp. 1–7.
[58] Muhammad Mahboob Ur Rahman et al. Physical Layer Authentication in Nano Networks at Terahertz Frequencies for Biomedical Applications. *IEEE Access*, vol. 5, 2017, pp. 7808–7815.
[59] Kamran Sayrafian-Pour et al. Channel Models for Medical Implant Communication. *International Journal of Wireless Information Networks*, vol. 17, 2010, pp. 105–112.
[60] Ian F. Akyildiz, et al. Propagation Models for Nano communication Networks. Proceedings of the Fourth European Conference on Antennas and Propagation, 2010, pp. 1–5.
[61] Ke Yang et al. Numerical Analysis and Characterization of THz Propagation Channel for Body-Centric Nano-Communications. *IEEE Transactions on Terahertz Science and Technology*, vol. 5, no. 3, 2015, pp. 419–426.
[62] Rashid A. Saeed, Esra B. Abbas. Performance Evaluation of MIMO FSO Communication with Gamma-Gamma Turbulence Channel using Diversity Techniques. International Conference on Computer, Control, Electrical, and Electronics Engineering (ICCCEEE), 2018, pp. 1–5.

Chapter 6

An Overview of the Terahertz Communication Networks and LOS and NLOS Propagation Techniques

Reinaldo Padilha França, Ana Carolina Borges Monteiro, Rangel Arthur, and Yuzo Iano

Contents

6.1 Introduction

In the age of telecommunications, numerous frequencies of the electromagnetic spectrum are used to transmit data and voice using telecommunication equipment. But that does not mean that we have already explored the whole spectrum. There is an untapped area in the wide range of frequencies – far-infrared, better known as terahertz radiation – allowing the construction of short-range data communication networks up to *n* times faster than today (Musey and Keener 2018; Reichel et al. 2019; Morohashi et al. 2016; Nakagawa et al. 2018; Kleine-Ostmann and Nagatsuma 2011; Song and Nagatsuma 2011).

The line that separates the two antennas is called line of sight (LOS), which is a type of propagation allowing transmission and reception of data. Here transmission and reception antennas do not have any type of obstacle between them. There are three main categories of sight lines: (i) the "full sight" where there are no obstacles between the two antennas; (ii) the "partial line of sight" or "partial sight" (nLOS) where there are partial obstacles such as a tree canopy between the two antennas; and (iii) the "covert line of sight" or "non-line of sight" (NLOS) where there is a total obstruction between the two antennas. Ascertaining line of sight environment determines the correct type of wireless system to use (Liu et al. 2016).

Radio transmission in a communication system may occur on an uneven ground. The terrain profile should be analyzed to estimate the received power attenuation between the two radios. The terrain profile may vary from a simple curved earth profile (due to the analysis of radio wave propagation in the atmosphere, the beam is considered to have no curvature, that is, the beam is represented in a straight line and the curvature of the earth) to a mountain profile. Several factors should also be considered, such as the presence of trees, buildings, and other obstacles in the line of sight (Popoola et al. 2018).

For wireless installations indoors, it is relevant to take into account obstacles such as ceilings, walls, and even furniture that will affect the line of sight and hence the reception of signals. In wireless transmissions (e.g., IEEE 802.11), reflection (wireless signals hit a physical object) and multipathing (digital signals travel in different paths, reaching the receiver at distinct times) are significant problems reducing the strength of the digital signal to be used and obstructing the successful installation of the system. A digital signal has peaks and valleys in its amplitude and change in its polarization (horizontal or vertical) when it propagates through ceilings, walls, and reflected in metallic physical objects. A "full line of sight" is a significant element for wireless installations (Kildal 2015; Popoola et al. 2018).

Line of sight can be understood as a characteristic of radio signal transmission and reception, typical of radio frequencies commonly used in aeronautical telecommunications and air traffic surveillance radars. It means that the transmission or reception takes place in a straight line and not hindered by any obstacles such as light or human vision. Therefore, buildings, hills, or the curvature of the earth itself may impose range limits or even prevent radar communication or detection. It is a line that joins the observer with the object or reference (Kildal 2015).

Path LOS is another field of interest when dealing with LOS. Although 2.4 GHz digital signals pass through the walls, they have trouble passing through leaves and trees due to water contained in them, and the trees and leaves by themselves consist of an obstruction. Still assuming that walls are dry and the trees contain high fluid levels and or even considering that radio waves in the 2.4 GHz digital band are easily absorbed by water, 900 MHz frequencies are best for "partial line of sight" and "covered line of sight" circumstances, with trees as obstacles, as waves at this frequency are poorly absorbed compared to 2.4 GHz (Yang 2019).

Point-to-multipoint systems have a central access point connected with various other points in distinct environments, generating an effective and wide data transmission network controlled by the principal station. This type of network enables direct communication between subordinate stations (subscriber modules), providing high-speed data links with ample coverage space. Using state-of-the-art technology, point-to-multipoint data transmission systems can operate over dedicated links, ensuring stability and speed of communication, supporting important business services such as real-time video transmission (videoconferencing) and voice over IP protocol (VoIP) services (Wu et al. 2018).

One of the main advantages of using a point-to-multipoint wireless data transmission system is its low cost of ownership. With a simple network design made up of small groups of equipment, these systems are quick and easy to install. System operation and maintenance also take place through simple processes, reducing costs for its users. Another advantage of this type of data-link connection is its great scalability. In this way, it is possible to adapt the system to possible evolutions, increasing its throughput capacity and its reach power (Taori et al. 2015).

In this type of scenario, they are expected to offer high performance in both LOS (direct line of sight), nLOS (near-line of sight), and NLOS (no line of sight) connections and are recommended for any kind of environment, from suburban areas and rural locations to urban environments that present a large amount of obstacles (Musey and Keener 2018).

WiMAX was developed as an IEEE 802.16 standard, providing wireless services, which is also known as unguided. Its frequency ranges from 10 to 66 GHz, allowing direct sighting, which in this case is very important to enable wireless communication in open environments, even in regions with obstacles such as vegetation. WiMAX has a baud rate of up to 70 Mbps and a range of up to 50 km, and its standard may use licensed or unlicensed spectrum, thus providing unguided Internet services to users using high data rates (Ahson and Ilyas 2018a).

Its structure is based on WMAN, enabling wireless broadband services within a building and can be shared with multiple Wi-Fi networks in different locations, regions, or even cities with the cell-like operation performed by a base station that serves to stabilize the wireless link to the subscriber, such as universal mobile telecommunication systems (UMTS) technology. Thus, WiMAX can be a point-to-point (LOS) connection, while between the subscriber and the base station, it is possible that this becomes a NLOS or even point to multipoint connection (Adediran et al. 2016).

In this scenario, efforts in pursuit of interoperability standards were made, where the Wireless National Electronics Systems Testbed (N-West) convened a meeting in 1998 on the requirements for an interoperability standard that was generated in the IEEE 802 standard. After much effort, the configuration of the IEEE 802.16 standard was defined. Primarily, the principal objective was to create the radio interface for using the 10–66 GHz band spectrum, which also supported the system's broadband LOS point-to-multipoint (PMP) wireless base (Kiokes et al. 2015; Kleine-Ostmann and Nagatsuma 2011; (Song and Nagatsuma 2011).

The standard has changed the IEEE 802.16 base using a 2–11 GHz frequency band that aggregates both free license and licensed frequency bands. Just the addition of low frequencies below 11 GHz, in this sense, makes NLOS digital communication possible. NLOS methods led to multipath propagation effects, which were overcome by adapting multipath modulation techniques in the physical layer (Ali and Hassanein 2018).

The terahertz band (0.3–10 THz) is the forthcoming mode in wireless digital communications since it allows the unlocking of significantly wider segments of unused bandwidth. Terahertz (THz) transmission is a complimentary wireless technology for communication networks, which enables the high-speed wireless extension of optical fibers to Beyond 5G (future generation of mobile telecommunication). This new technology promises faster data transfer. Wireless communication devices currently operate on microwave frequencies; however, as the demand for faster filters and larger data bands emerges, it is necessary to look for ways to alleviate or remove bottlenecks in the communication (Morohashi et al. 2016; Nakagawa et al. 2018; Kleine-Ostmann and Nagatsuma 2011; Song and Nagatsuma 2011; Sirenko and Velychko 2016; Cacciapuoti et al. 2018; Rappaport et al. 2019; Rangan et al. 2014; MacCartney and Rappaport 2019; Sun et al. 2017).

In the future, users in remote or rural regions, that is, in regions with difficult access such as mountains and/or islands, may be able to have a connection reaching high data rates of up to 10 Gbit/s for each user. THz communications are expected to allow continuous connection between ultra-high-speed wired networks, that is, fiber optic links and wireless devices (users), mobile devices such as tablets, smartphones, even laptops, achieving greater convergence and transparency between rates among wireless and wired links (Morohashi et al. 2016; Nakagawa et al. 2018; Kleine-Ostmann and Nagatsuma 2011; Song and Nagatsuma 2011; Sirenko and Velychko 2016; Cacciapuoti et al. 2018; Rappaport et al. 2019; Rangan et al. 2014; MacCartney and Rappaport 2019; Sun et al. 2017).

It is impracticable or very expensive to think that only fiber optic solutions will solve this emerging need for broadband, but complementary technology solutions are also available. Terahertz transmission as an extension of wireless optical fiber backhaul can and will be an important element to be employed in this type of scenario to ensure greater access to high-speed Internet anywhere beyond 5G. This will favor the employment of bandwidth-intensive applications between mobile and static users, especially in internal and local access scenarios (Morohashi et al. 2016; Nakagawa et al. 2018; Kleine-Ostmann and Nagatsuma 2011; Song and Nagatsuma 2011; Sirenko and Velychko 2016; Cacciapuoti et al. 2018; Rappaport et al. 2019; Rangan et al. 2014; MacCartney and Rappaport 2019; Sun et al. 2017).

This chapter provides a scientific discussion and overview of terahertz communication networks and propagation properties such as LOS, nLOS, and LOS. Addressing their key points assumes importance, as they are a complex and heterogeneous concept affecting the efficiency of wireless terahertz communication networks.

Therefore, this chapter provides an overview of terahertz communication networks, in particular LOS and NLOS propagation and related techniques. The relationship and integrations of LOS and NLOS are also discussed using the relevant bibliography that has explained the potential of both these technologies.

The present chapter is structured as follows: Section 6.2 discusses terahertz communication concepts. Section 6.3 presents a thematic discussion about fiber optics of terahertz communication. Section 6.4 discusses wireless terahertz communication networks. Section 6.5 presents an overview of LOS, nLOS, and NLOS. Section 6.6 presents the evolution of research focusing on terahertz communication over the past seven years, through a scientific review with research that highlighted the themes according to the view of the authors. Section 6.7 presents a discussion on the theme of Terahertz Communication Networks and LOS and NLOS Propagation Techniques. In Section 6.8 the conclusions are presented and finally in Section 6.9 future trends for terahertz communication are presented.

6.2 Concepts of Terahertz Communication

X-rays, gamma rays, ultraviolet rays, and microwaves are popular forms of electromagnetic radiation, but the infrared radiation also belongs to the electromagnetic spectrum and manifests itself in the form of heat. And between the microwave and infrared range, there is a zone that has been scientifically and technologically explored, called terahertz radiation, or T-rays (Sirenko and Velychko 2016; Kleine-Ostmann and Nagatsuma 2011; Song and Nagatsuma 2011).

The frequency range of this radiation, between 0.3 and 3 trillion hertz, is known to operate in terahertz part of the spectrum; therefore the T-rays have a frequency between 0.3 and 3 terahertz. Terahertz radiation is located between microwave and infrared radiation, at frequencies between 300 billion and 3 trillion cycles per second. In terms of wavelength, the T-rays range from 1,000 at 100 micrometers at the twin frequency units. Many of the analytical applications of T-radiation are similar to those obtained with infrared, but often in the gigahertz range. It is only capable of producing rotations in molecules, and the upper infrared, more often higher than 10 terahertz, is capable of producing vibrations resulting from intermolecular interactions (Sellers 2018; Reichel et al. 2019; Morohashi et al. 2016; Nakagawa et al. 2018; Kleine-Ostmann and Nagatsuma 2011; Song and Nagatsuma 2011).

In the past decade, advances in photonics and microelectronics have enabled many developments in this spectrum range. As a result, there has been a growing academic and industrial interest in exploring potential applications, since terahertz radiation has unique properties that enable multiple uses, among others. These properties include the broad spectral range and high bandwidth for communications. Terahertz radiation can accomplish both simultaneously, and the application of THz technology has become a reality supported by advances in the development of sources and detectors based on microelectronics and ultra-fast lasers (Png 2010; Reichel et al. 2019; Morohashi et al. 2016; Nakagawa et al. 2018; Nagatsuma et al. 2013, 2016).

For high-brightness, high-frequency T-beams to be produced, it is important to use a femtosecond laser (a femtosecond is one quadrillionth of a second). When a pulse emitted by the laser hits a photoconductive antenna, the material that emits electrical pulses when illuminated releases pulses with frequencies between 300 GHz and 10 THz. Just as interesting is the fact that small technical modifications to the circuit structure make a sender antenna a receiver. The sender and receiver are two important elements for the operation of T-ray equipment (Musey and Keener 2018; Sellers 2018; Reichel et al. 2019; Morohashi et al. 2016; Nakagawa et al. 2018; Nagatsuma et al. 2013, 2016).

Since the first demonstration of THz signal generation from ultrashort laser pulses, much advances have been made in the development of THz sources, where advanced quantum cascade laser-based THz sources with THz signal generation technique by optical grinding on special materials and in the design of photoconductive antennas allow THz signal generation ranging from about 0.1 to 5 THz with low cost and excellent portability. The boundary of research on THz sources currently resides in high-power signal generation. Since current techniques allow the generation of highly directional and ultra-wide intensity THz fields, approximately from 0.1 to 10 THz, as well as sources based on ultrashort laser pulses, lasting for sub-femtoseconds, they generate the possibility of bandwidths of approximately 0.1 to 100 THz (Wang et al. 2018) (Nagatsuma et al. 2013, 2016).

6.3 Terahertz's Communication Fiber Optics

In this section optical fibers, which are exceptional and allow a high-speed data transmission with minimal levels of data loss, are covered. Whereas in wireless terahertz, high data rate with latency tends to be minimal, and optical fibers consist of glass, the speed of light reduces in the fibers. This makes optical fibers unsuitable for certain applications that demand real-time responses since in certain situations it is necessary to balance between "minimum latency" (microwave links) and "high data rate" (optical fiber). However, with wireless terahertz communication, it is possible to have light-speed links with minimal latency, supporting fiber data rates. This shows the importance of fiber optics as a means of transmission within the terahertz communication networks.

Optical fiber is a cylindrical light guide made of two concentric crystalline materials. The two materials make up what is called the core and shell of the optical fiber, which differs in refractive indexes. It is the means by which the light power injected by the light emitter is transmitted to the photodetector. They are characterized by their operation in spectral regions where attenuation is minimal (Wang et al. 2018).

The high-transmission-capacity fibers have silica (SiO_2) as the raw material. In the first stage of manufacture using silica, the tube is made, forming the shell of the fiber. Inside this pure silica tube, gases ($SiCl_4$, $GeCl_4$) are injected to compose the core, where the concentration of these materials is controlled to obtain the desired refractive index (Yamamoto et al. 2016).

Today's optical fibers are widely used in short- and long-distance communication systems due to several advantages such as bandwidth, having a potential use in the range on the order of 1,012 Hz (1 THz). The capacity of optical fibers increases when used in transmission. Using wavelength division methods, the same fiber may carry different signals, each with a specific frame: (i) low attenuation, related to the characteristics of the fibers as the links of communication systems require few repeaters or regenerators, and the largest attenuations are due to couplers (connections) and splices; (ii) immunity to electromagnetic interference (EMI), where the optical fibers are not affected by nearby electromagnetic fields and may be applied to systems with the possibility of electric shock or near high-voltage installations; (iii) electrical insulation, in relation to the materials used for its manufacture, which do not require surge protection devices as they electrically isolate the communication terminals. Still considering the low weight and small size, since their cores are measured in microns (thousandths of a millimeter) compared to metal cables that are less than one-tenth of their volume, transmission safety is possible. It happens due to easy detection of possible "staples" because of the deviation of a considerable portion of the light output, which is not detected by electromagnetic or metal detectors. This makes optical fibers a potential material for applications in data, image, and voice (telephony) communication systems (Chesnoy 2015).

The propagation of light in the optical fiber is due to the difference in the refractive index between the shell and the core. With the calculation of the critical angle, it is possible to guarantee total or internal reflection, according to Snell's law. Therefore, the angle of incidence of the injected light must be greater than the critical angle (Venghaus and Grote 2017).

Transmission systems that use optical fibers require an emitter and an optical detector, which are transducers, because they convert both electrical and optical signals into electrical signals. Today, the most widely used optical emitter is a laser diode and the optical detector a photodiode. The transmission spectrum concerning the optical frequencies associated with these communication systems spans a wide range of wavelengths, within a spectral range, starting at the far-infrared region, approximately 100 mm, passing through the visible spectrum, around 390 to 770nm, and ending in the ultraviolet, which is close to 50 nm. Therefore, the optical frequencies offer fantastic possibilities for high transmission capacity. Considering only the range 100–1000 THz, theoretically transmission capacities can reach an order of 10 times higher than the current microwave systems (Venghaus and Grote 2017).

Thus, fiber optic transmission systems generally operate in the 0.6–1.6 mm spectral region, with a preference for 0.85, 1.3, and 1.55 mm transmission windows, which relates to this region's semiconductor materials (Si, Ge, InGaAsP, AlGaAs, among others). This range is suitable for the best performing photodetector light sources, as well as very low attenuating optical fibers, and it should be noted that these boundaries are not static but a result of continuous technological advances (Matias et al. 2016).

The transmission capacity of a fiber is limited to the carrier frequency, which with light signals, in theory, it is possible to operate in a baseband of one or two orders of magnitude below the light frequency, around 200 THz. At the same time, modern equipment has transmission rates of the order of 2.5 Gb/s at each wavelength, and up to 18 wavelengths can be allocated in a single fiber optic cable (Willner 2019).

6.4 Wireless Terahertz Communication Networks

Previously, wireless communication data rates were on the order of Mbps (megabits per second) or Gbps (gigabits per second), but nowadays it is approaching terabits per second. The fibers are exceptional, allowing high-speed data transmission with minimal data loss levels. Artificial satellites are also second to none, covering vast areas with no infrastructure, compensating for slower speed and a few moments of out of atmospheric optics (Willner 2019; Venghaus and Grote 2017; Rappaport et al. 2019; Rangan et al. 2014).

THz technology has been achieving data speeds *n* times faster than expected from fifth-generation (5G) cellular networks, which are only expected to come into operation in the coming years. Wireless data rates previously over Mbps or Gbps, are now reaching Tbps (terabits per second), utilizing a single simple communication digital channel, which makes possible applications including ultra-fast wireless connections between base stations. As well as fast download of content from servers to mobile devices wirelessly, terahertz is offering the high data rate with minimal latency (Cacciapuoti et al. 2018; Rappaport et al. 2019; Rangan et al. 2014).

Optical fibers are currently employed for "high data rate" (optical fiber) with "minimum latency" (links). But in contrast to terahertz wireless, it is possible to have links to light speed communications with minimum latency, sustaining the data rates of optical fibers. Demand for telecommunication bandwidth has grown dramatically in recent years, indicating the need to explore spectrum more effectively since wireless communication devices currently operate at microwave frequencies. However, as demand for faster speeds and higher bandwidths of data grows, terahertz radiation (THz), oscillating around 1 trillion times per second, is used (Binh 2015; Rappaport et al. 2019; Rangan et al. 2014; Reichel et al. 2019; Morohashi et al. 2016; Nakagawa et al. 2018).

Terahertz radiation is well suited for the development of a current-generation wireless telecommunications system that can operate at a faster speed of 100 Gb/s. The terahertz (THz) frequency ranging from 0.1 to 3 THz lies at the end of the entire spectrum of electromagnetic waves and is known in the scientific world as the THz gap. This terahertz gap has captured the imagination of the technology world at all new levels, opening up many new possibilities in optical communication technology. Terahertz technology is very promising for high-speed transmission of information between electronic devices, next-generation wireless personal networks (WPAN), and building WLAN (wireless local area networks) (Uddin 2017; Rappaport et al. 2019; Rangan et al. 2014; Reichel et al. 2019; Morohashi et al. 2016; Nakagawa et al. 2018).

Following this development, great efforts are being made in the development of transmission and reception techniques, taking advantage of the unique properties of THz waves, such as the high bandwidth, which in turn seeks to circumvent the issue of high THz signal attenuation. In a humid atmosphere, high-impact applications use THz wireless communication in data centers, taking advantage of the high bandwidth available at high frequencies, which could be accomplished through hybrid systems that can convert optical signals to the THz range. Just as new devices have been proposed for THz signal modulation based on new technologies such as graphene, which can be possibly used in short distance indoor communications, the THz system integrates with a mixed network of other communication technologies without thread. In the same way, THz waves are also studied for space applications focused on satellite and inter-satellite communication (Kazmierkowski 2019; Rappaport et al. 2019; Rangan et al. 2014; Reichel et al. 2019; Morohashi et al. 2016; Nakagawa et al. 2018).

The development of THz technology has led to the development of THz fibers manufactured from polymers or glass over time and they have been characterized based on previous experience with photonic crystal fiber optics with photonic crystal fiber (PCF) technology, or anti-resonant phenomenon and hollow-core fibers. Or even porous fibers that operate by the effect of photonic bandgap, further taking into consideration the development of components for the assembly of more complex systems such as power splitters and couplers, waveguides, lenses, and polarizers, gaining prominence in the development of low-loss dielectric waveguides for THz propagation (Cicerone et al. 2016; Rappaport et al. 2019; Rangan et al. 2014).

6.5 THz LOS and NLOS Propagation Techniques

This section assumes fundamental importance from the knowledge that the existence of certain properties cause harm to the signal transmitted over a wireless connection, even more at high speed as terahertz communication networks. If these hurdles are not addressed, there will be no high-frequency communication. This is important for the integrity of the link as a certain area around the line of sight would cause interference in the signal if it is blocked. This is of paramount importance for the perfect functioning of the system, so that both in transmission and reception there is no loss of the signal, as the poor quality of both transmission and reception of the signal, among other relatively important factors, impact the quality of service.

One of the main components present in a radio frequency circuit is the antennas, because it is through them that the signals are transmitted and received. For this reason, the antennas are positioned higher up, such as on top of buildings, which avoids obstacles in front of it. The principles of antenna alignment need to be followed to enable long-range transmission of high-frequency electromagnetic waves, where a signal emitted by one transmitter can accurately be emitted with full power to another transmitter at another localized point. For this to happen, it is necessary that both are perfectly aligned and "seeing" each other. Thus, when a frequency is generated that must be transmitted from one place to another through the free space, two antennas are required, one for each place. If this frequency is high, it will need to be aimed at the antennas, in a way one antenna can "see" the other, which is called sight. This is of fundamental importance, because without it, there will be no communication (link) between high-frequency antennas. However, it is not enough to "see" one antenna, or only the other antenna, but it is necessary to see a predetermined area, which should be greater than the distance between antennas (Morreale and Terplan 2018; Moldovan et al. 2014).

It is within this predetermined area that the Fresnel zone is found, where the propagation of the high frequencies happens around the line of sight (LOS), a field in elliptical form, through which most interacting data travels between antennas (Nandi et al. 2016; Moldovan et al. 2014).

The Fresnel zone is an electromagnetic phenomenon, in which radio signals or light waves are diffracted or deflected by solid physical objects next to the communication path. Radio waves reflected on physical objects arrive out of phase considering that the signal is traveling straight to the other antenna, thereby decreasing the strength of the received digital signal. So the area around the line of sight where signal interference is present becomes important. When this obstacle exists, it is necessary to avoid this degree of interference, which can be defined as a series of concentric ellipses around the line of sight. This type of communication obstacle determines the area around the line of sight where signal interference may be introduced if it is blocked (Xue et al. 2018).

When an obstruction is present, radio waves travel from the transmitter to the receiver, encountering the obstacles (impediment) near the communication path, because of which they are reflected by these objects and go out of phase with the traveling signals, causing a reduction of the received signal strength. In practice, 20% Fresnel zone blocking is acceptable by a suitable equipment, but above 40% signal loss is very significant for communication. If any point in the Fresnel zone is obstructed, whether by a tree or a building, the signal will no longer be the best. So the higher the antenna frequency, the smaller the Fresnel zone, thus decreasing the likelihood of any obstacle interfering with the link. Therefore, the straight line of sight should not be the only concern, but it is essential to analyze the entire Fresnel zone before installing the antennas. High frequency is relative to narrow Fresnel zone and low frequency is relative to wide Fresnel zone (Nefyodov and Smolskiy 2019).

6.5.1 LOS (Line of Sight)

IEEE 802.16 determines the frequency ranges of operations for different types of signal propagation. For example, a radio frequency beam is used to propagate digital signals between the nodes. This frequency beam is highly sensitive to radiofrequency obstacles. Thus, an unobstructed view between the nodes is required. This category of signal propagation is called the line of sight (LOS) (Figure 6.1), which can be divided into optical line of sight (Equation 6.1) and radio line of sight (Equation 6.2) (Stallings 2005), limited to fixed operations and employs a 10–66 GHz frequency range (Ali and Biradar 2016; Moldovan et al. 2014).

$$d = 3.57\sqrt{h} \tag{6.1}$$

$$d = 3.57\sqrt{Kh} \tag{6.2}$$

In these equations, d is the distance between the antenna and the horizon (km), h is the antenna height (m), and K is the adjust factor that considers refraction. The rule of thumb is $K = 4/3$.

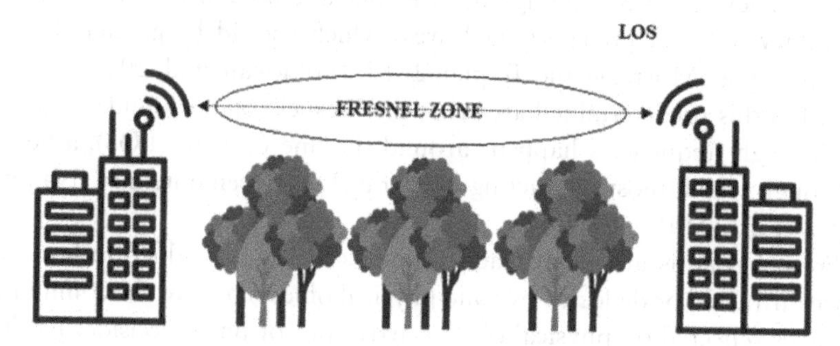

Figure 6.1 Illustrative example of LOS propagation.

When there is no obstacle, and one antenna can perfectly see the other, it is called direct sight, which means there will be no problem in the communication between the transmitter and receiver in the radio frequency circuit and the link is not affected. In designing a radio link, the goal is to ensure that the original digital signal carrying the information can be regenerated at the other end with an acceptable error rate. So that the carrier-to-noise ratio at reception must be greater than a specified minimum value, which is a function of the modulation and coding mechanisms used in the link. Transmitter power and antennas should, therefore, be sized to compensate for propagation and other losses due to cross-polarization and attenuation in connectors, coaxial cables, or waveguides. And a margin must be included to deal with interfering signals near the frequency band used by the link, which may increase the noise level at the receiver and therefore worsen the carrier-to-noise ratio (Morreale and Terplan 2018; Moldovan et al. 2014).

In a radio link, the signal is transmitted by the transmitting antenna and propagates in the form of radio waves, that is, electromagnetic waves, to the receiving antenna, where the signal is attenuated and subject to losses as loss in free space. Where the only piece of the energy transmitted over electromagnetic waves is captured by the receiving antenna, which is related to the higher the frequency and distance, and any loss is expressed in dB. Many wireless technologies face the problem of multipath propagation during transmission. In obstructed environments, the signal may be reflected in various obstacles and reach the receiver at a different time, with differences in module and phase. The propagation of radio waves are subject to reflections in the ground and the atmosphere causes changes in their amplitude and path traveled, resulting in variations in the received digital signal strength, called fading, which can be induced by obstacles in the direct line of sight, or by attenuation due to rainfall (Godara 2018).

6.5.2 nLOS (Near Line of Sight)

In near line of sight (nLOS), that is, "partial sight" (Figure 6.2), the transmitter and receiver are not connected by a straight line and may have obstacles in their path. This is a concept often employed when the transmitter and radio receiver are not in the direct line of sight, and this is handled by using multipath in signal propagation (Yuan and Ma 2016).

Radiofrequency technologies utilize the concept nLOS to describe a partially obstructed path between the signal transmitter location and the signal receiver location. This difficulty

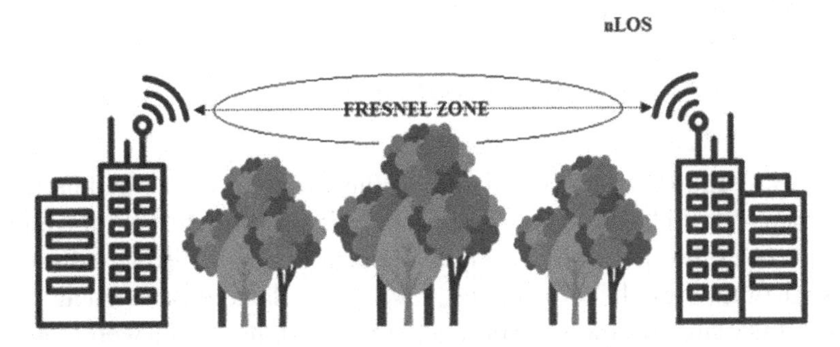

Figure 6.2 Illustrative example of nLOS propagation.

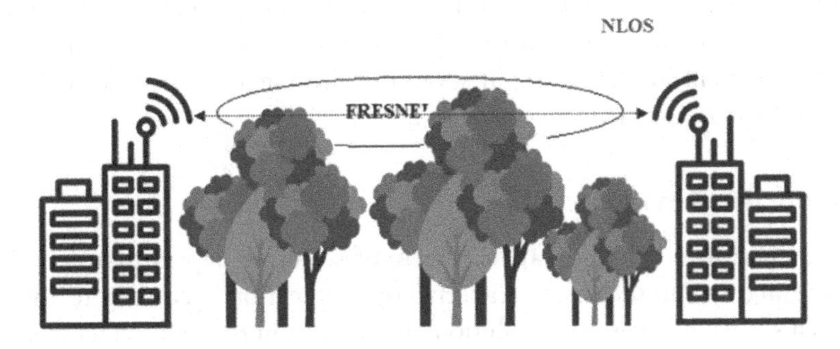

Figure 6.3 Illustrative Example of NLOS Propagation.

can be overcome by the use of antennas and other communication devices, which also play an important role in reducing the reception power of a signal, resulting in a bad transmission system. Obstacles that may cause a line of sight obstruction include buildings, trees, hills, mountains, and other natural formations or artificial structures, or even physical objects. The major concern of modern computer networks is to effectively reduce nLOS, and this is done over wireless networks using multipoint repeaters so that the signal is transmitted around the obstruction without data loss or transmission quality. The near line of sight is also known as the slackline of sight. Multipath signal propagation is also widely used (Yuan and Ma 2016).

6.5.3 NLOS (Non-Line of Sight)

The other category of signal propagation is known as non-line of sight (NLOS) (Figure 6.3). It uses advanced radiofrequency modulation techniques compensating for radio frequency signal changes generated by obstacles that avoid LOS communications. NLOS refers to the propagation path of a radio frequency obscured (completely) by obstacles that cause difficulty in the transmission of radio signal. The common obstacles between radio transmitters and receivers are trees, physical landscape, tall buildings, and conductors of high voltage. Some obstacles absorb and others reflect the radio signal. However, all of them cause limitation in signal transmission capacity (Wang et al. 2003; Moldovan et al. 2014).

$$Hl : \widehat{d_i} = d + w_{\text{los,i,}}\ i = 1,\ldots,N, LOScondition \tag{6.3}$$

$$Hn : \widehat{d_i} = d + b + w_{\text{nlos,i,}}\ i = 1,\ldots,N, LOS\,condition \tag{6.4}$$

Equations (6.3) and (6.4) are considered on the variation of the mobile terminal localization, where N is relative to measurements, d is the straight-line range related to the mobile and base station, $\widehat{d_i}$ is the i-th measurement of d, b is the extra distance regarding the blockage of the straight path, and $w_{\text{los,i,}}$ $w_{\text{los,i,}}$ are related measurement noise under the LOS and NLOS circumstances, respectively (Yu and Guo 2008; Woo et al. 2000; Cong and Zhuang 2005; Lee 1993; Alavi and Pahlavan 2003).

NLOS can be employed for both fixed WiMAX processes (below 11 GHz) and mobile telephony processes (below 6 GHz). NLOS signal propagation is usually employed than LOS due to obstacles interfering with LOS communications and strict rules for frequency antenna deployment and licensing in several environments that make LOS use difficult (Ahson and Ilyas 2018b) (Moldovan et al. 2014).

6.6 Scientific Review

This survey performs a bibliographic review of scientific papers that describe principal research related to the theme of terahertz communication according to the view of the authors. This is to showcase the evolution and interest in research on terahertz communication for the past seven years, affirming the importance of the study area in the science and development of innovative technologies that enable high-speed data transmission. Bibliographic material was taken from indexing and publications in renowned databases, such as Web of Science, IEEE, Scopus, and Elsevier.

In 2013, the growing interest in applying terahertz waves (THz) to broadband wireless communications was studied, with particular use at frequencies above 275 GHz since it is possible to employ extremely large bandwidths for wireless broadband networks in ultra-broadband communications as well as these frequency bands have not yet been allocated to specific active services. The photonic technologies were introduced for the generation, modulation, and detection of these signals. The effectiveness was not only seen to increase bandwidth and data rate but also to combine fiber optic (wired) and wireless networks. The study analyzed recent developments in THz wireless communications using electron-based photonic technologies toward 100 Gbit/s (Nagatsuma 2013).

In 2014, terahertz bandwidth communication in the range of 0.06 to 10 THz was seen as an essential technology that satisfies the growing demand for ultra-high-speed wireless links. In this sense, a unified multi-channel channel in the THz band was developed based on ray-tracing techniques, which incorporate the propagation models for the scattered, diffracted, and reflected line of sight paths. The developed theoretical model was validated with experimental measurements in the range of 0.06–1 THz from the literature, followed by the development of propagation models, and an in-depth analysis of the THz channel characteristics was performed. In particular, the selective nature of frequency and distance variable of the terahertz channel was analyzed, as well as the meaning of delay propagation and coherence bandwidth was studied. It also characterized the capacity of the broadband channel using flat energy allocation and water-filling strategies in conjunction with the study of the temporal magnification effects of the terahertz channel. And finally, adaptive, multi-carrier broadcasts were suggested that benefit from the unique relationship between distance and bandwidth. The analysis provided in the study set the foundation for efficient and reliable ultra-high-speed wireless communications in the 0.06–10 THz range (Han et al. 2014).

In 2015, a complete set of MMIC chips with front end transmission and reception capability and RF data transmission at a carrier frequency of 300 GHz and with data rates of up to 64 Gbit/s was presented. The radio was dedicated to future high-rate internal wireless communication, addressing application scenarios such as data centers, smart offices, and home theaters. The study analyzed the underlying high-speed transistor and MMIC process, quadrature transmitter and receiver performance, as well as local oscillator generation through frequency multiplication. Experiments with zero-IF transmission and reception scheme and initial transmission in a single input single output configuration achieved data rates of up to 64 Gbit/s with QPSK modulation. Also discussed, the current performance limitations of the RF front end describing the paths for improvement have reached a capacity of 100 Gbit/s (Kallfass 2015).

In 2016, photonic transmitters were presented showing communication link results using 0.2, 0.4, and 0.6 THz photonic-based emitters. Within the context of the rapid increase in mobile data transfers, wireless carrier frequencies had entered the millimeter-wave region and now are in the submillimeter or terahertz region (Ducournau 2016).

In 2017, a high spectral and efficient THz wireless transmission around 325 GHz was demonstrated experimentally using a 64-QAM-OFDM modulation format and a 10 GHz wide wireless channel. Achieving data rate of 59 Gbit/s, an efficient terahertz record spectral communication system was reported using a coherent radio over fiber (CRoF) approach (Hermelo 2017).

In 2018, the basic system architecture for THz wireless links with bandwidths of over 50 GHz in optical networks was discussed, where new technologies and design principles were needed to demonstrate near latency terabit data rates of zero. In a specific context, the concept of designing baseband signal processing for optical and wireless links was presented and using an E2E error correction approach for the combined link. Two possible architectures of electro-optical baseband interface – digital link and transparent optical link architectures – were provided. Research principles and guidelines for developing a new generation of transceiver front-ends that will be able to operate at ultra-high spectral efficiency along with THz wireless link requirements were presented, employing higher-order modulation schemes. Also discussed was the need to develop a new THz network information theory framework that takes into account the nature of THz band interference and channel characteristics. Also highlighted was the role of GMP, a resource needed to overcome propagation losses, as well as medium access control and physical layer challenges (Boulogeorgos 2018).

In 2019, the need for future wireless communication networks to handle data rates of tens or even hundreds of Gbit/s per link requiring carrier frequencies in the unallocated THz spectrum was studied. In this scenario, the integration of THz links into existing fiber optic infrastructures was considered of great importance to complement the advantages inherent to the flexibility and portability of wireless networks as well as the unlimited and reliable capacity of optical transmission systems. At the technological level, new device and signal processing concepts were required to perform the direct conversion of data streams between THz and optical domains. The study demonstrated that optical conversion to THz (O/T) on the wireless transmitter depends on the photo mixing in a single-traveler carrier photodiode by relating a THz link that was seamlessly integrated with a fiber optic network using direct THz conversion (T/O) on the wireless receiver. And an ultra-wide silicon-plasmonic modulator with a 3-dB bandwidth greater than 0.36 THz was explored for T/O conversion of a 50 Gbit/s data stream, which was transmitted on a 0.2885 THz carrier through a 16 m wireless link (Ummethala 2019).

6.7 Discussion

This section aims to present a discussion on the importance of high transmission rates obtained through terahertz communication concepts, explained in previous sections, seeking to address the relationship that optical fiber has with wireless terahertz communication, which has been the subject of research over the years, as previously presented. Even identifying the main technologies provided through it that will allow access to broadband of better quality and speed in a 5G network.

About 99% of the world's communications travel over submarine cables, which is an old structure. But this technology allows the sending of a cell phone message, a videoconference meeting, from a Latin country to an eastern country, such as Japan. The palpable infrastructure of submarine cables, the diameter of which can be held in the palm, lying on the seabed, carry optical fibers through which data, messages, and voices flow. For over a century, cables were usually underground or installed on high-voltage lines, performing telephony or data transmission, carrying the heavy communication infrastructure of the world (Mittleman 2017, 2018; Ma et al. 2018; Morohashi et al. 2019).

Besides, the fiber optics that currently fill submarine cables have much greater data traffic capacity than satellites, exceeding an average cable data transmission rate of approximately 4 Tb/s. Amplifiers that carry the function of maintaining the power of the light signal that travels inside the fibers along the path and, that too parametric amplifiers, can be the answer to the challenge of broadening bandwidth and hence network traffic.

Since most available systems guarantee a maximum bandwidth of 30 nm in the optical communication region, they can hold up to 80 lasers, limiting transmission-related factors to a few terabytes per second. However, the larger the bandwidth, the higher the number of lasers placed on a single fiber and consequently greater traffic capacity (Mittleman 2017, 2018; Ma et al. 2018; Morohashi et al. 2019).

To improve communication technology, the terahertz radiation spectrum range above microwaves is being used. With its systems, circuits, and components capable of communications at speeds above 100 Gbps per channel, the 300 GHz band, or 0.3 THz, is considered to be advantageous for low transmission latency (Mittleman 2017, 2018; Ma et al. 2018; Morohashi et al. 2019).

The development of standards that apply such high frequencies of communication always encounters physical problems, like the fact that terahertz transmission in today's cables is still difficult, because of the nature of the material used in the transmission wires. It eventually absorbs part of the digital information, attenuating the digital signal, and affecting the transfer (transmission). Another issue of transmitting data at these frequencies is related to the implemented structure of current networks. Nowadays, both wireless digital channels or optical cables carry hundreds of distinct information (Mittleman 2017, 2018; Ma et al. 2018; Morohashi et al. 2019).

In this sense, terahertz radiation is a solution since it may have higher speeds compared modern wireless digital communication standards. As the current technology is microwave-based in the gigahertz spectrum, this new technology promises to pave the way for high-frequency application in data transmission. Using multiplexers, this new technology allows hundreds of data to be transmitted over the same cable without any information being lost similar to how a single cable transmits uncountable phone talks at the same time, but at the end of the cable, each user only hears the conversation that is exclusive for him or her. The technology has an antenna consisting of two metal plates organized in parallel and free space, filled with air, which is where the emitted waves propagate, on a transmission pattern that creates waves at distinct frequencies. Each wave represents a digital information channel, meaning that one of these waves will be comprehended as an independent data chain (at the end of the cable), suppressing the possible issue of channel mixing (Mittleman 2017, 2018; Ma et al. 2018; Morohashi et al. 2019).

The THz band is also a vast new frequency feature not yet commercially exploited for wireless communications, where its frequencies are even higher than those used by the IEEE 802.11ay standard, ranging from 57 to 66 GHz. And the available bandwidths are much wider, and since the communication speed is proportional to the bandwidth in use, THz becomes ideal for ultra-high speeds. It is a huge new frequency feature that should be employed for future high-speed wireless digital communications. IEEE 802.15.3d, issued in October 2017, determines the utilization of the frequency range between 252 and 325 GHz, for example, "300 GHz," as high-speed wireless digital communication channels (Mittleman 2017, 2018; Ma et al. 2018; Morohashi et al. 2019; MacCartney and Rappaport 2019; Sun et al. 2017).

The 5G network is not yet standard anywhere in the world. However, developed countries such as the United States are gradually deploying the 5G technology. However, the country has allowed the opening of the spectrum of "terahertz waves" as wireless network technology. In the future, for possible 6G connections, the frequency band starting at 95 GHz, reaching up to 3 THz, is to be used and it will be released on an experimental basis to allow professionals in the field to explore (MacCartney and Rappaport 2019; Sun et al. 2017).

6.8 Challenges in Using THz Technology

The challenge regarding the use of THz technology is to develop a new type of network architecture employing terahertz frequencies for digital communications. It is necessary to manage many network and terminal equipment, terminal circuits, and the battery of communication protocol designs. At the same they have to be energy efficient as well (Han et al. 2019; Petrov et al. 2020; Yuan et al. 2020).

Challenges are also related to existing technologies at the same time, which have constrained communication distance and propagation loss. The THz gap should be closed by implementing a THz communication in this direction driven by global research activities (Han et al. 2019; Petrov et al. 2020; Yuan et al. 2020).

Another challenge concerns further decreasing the size of the equipment, since there is a growing interest in the THz band for its feasible application in nano- and micro-scale devices and its suitability for bandwidth-oriented applications is also being explored. High-sensitivity detectors working at room temperature and lack of compact high-power signal transmitters add to the challenge for THz communications in the future years (Zaman Chowdhury et al. 2019; Rodríguez-Vázquez et al. 2020; Yuan et al. 2020).

At the same time, device miniaturization is a trend when it comes to the practicability of miniaturized component manufacturing and even design, related to THz signal generators, transmitters (Tx), and receivers (Rx) with this technology together with antennas. Evaluating the feasibility of carrier-based communications, concerning directivity, that is, attenuation of high propagation losses and antenna arrays, and even massive architecture multiple-input and multiple-output (MIMO) is also essential (Zaman Chowdhury et al. 2019) (Rodríguez-Vázquez et al. 2020) (Yuan et al. 2020).

Another challenge of the THz technology concerns lower link for limitations of On/Off keying modulation and the suitability of full-duplex MAC as well as the practicality of IEEE 802.11ac-based standard signaling, that is, diminishing time-to-market (Zaman Chowdhury et al. 2019; Rappaport et al. 2019; Kato et al. 2020).

There are issues of higher layers and upper link, at a system level, where peer discovery, meaning directional antennas, or angle of arrival, among other aspects, in addition to security and privacy issues for a massive amount of devices that have to be addressed (Zaman Chowdhury et al. 2019; Rappaport et al. 2019; Kato et al. 2020).

Thus, it is possible to weigh the disadvantages of THz as part of the challenges that future 6G technology will have in wireless communication, in the same way as visible light frequencies posed disadvantages of visible light communication (VLC) (Zaman Chowdhury et al. 2019; Rappaport et al. 2019; Kato et al. 2020).

6.9 Conclusion

The main aim of this chapter was to provide an understanding of key questions about what is understood as terahertz communication networks, with a greater focus on wireless terahertz communication. To this end, this technology was characterized by pointing out its central aspects, merits, and certain weaknesses, such as expanding on the properties that hurt the integrity of the wireless transmission link around the line of sight causing both the loss of the signal and also poor quality of both transmission and reception of the signal. A discussion contextualizing some fundamental debates that arise around the theme of terahertz communication networks was also presented, since the flow of data and information traveling through the network today uses fixed and mobile technologies.

When designing a communication link for a wireless network, in a terahertz communication, the objective is to ensure that the original digital signal that carries the information can be regenerated at the other end with an acceptable error rate considering that this transmission in a communication system can occur on uneven terrain. The terrain profile must be analyzed, whether mountainous or not, and several factors must also be considered, such as the presence of trees, buildings, and obstacles in the line of sight, to estimate the attenuation of the power received between the transmitter and receiver.

Concerns regarding the propagation properties of this signal that involve LOS, nLOS, and NLOS need to be considered, since the line of sight is an indispensable feature when talking about communication without the Internet that applies to any type of communication. And when the waves propagate, they are subject to the presence of obstacles, giving rise to reflections, which cause changes in their amplitude and path followed, resulting in variations in the power of the received signal (fading).

THz technology is transformative and can lead to breakthrough technologies such as communication systems and networks with capacities many times higher than today's systems. This level of latency and capacity will be unparalleled and extending the performance of 5G solutions as well as extending the ambit of competence, supporting innovative tools in wireless detection, cognition, imaging, and even recognition.

It has to be kept in mind that terahertz radiation is safe because of its low-energy photons, as they consume on hundred times less energy that photons in the visible light range. With THz waves, it is possible to absorb moisture present in the air, which is useful for high-speed and even short-range wireless digital communications. Terahertz provides a better directivity and narrow beam, causing secure digital communication to be achieved and has a strong ability to block interference.

It is clear that there is a great effort of the scientific community focused on exploration and development to improve the potential of THz technology, generating innovations and practical applications. Though THz technology has s promising future involving major challenges, but its potential for knowledge generation, new applications, and products is equally great.

6.10 Future Trends

New advances are presented for the involvement of plasmons to increase the THz signal generation intensity. When a metal is subjected to an external electric field, the free electrons of the metallic material will move to the left side and positive ions will be formed on the right side, which is basically atomic nuclei that have lost their electronic coverage. Thus, these free-moving electrons, called electronic gas, make up a type of plasma. When the electric field is removed, the electrons will move to the right due to the mutual repulsion exerted between them and also due to the attraction exerted by the charges of the positive ions. That is, a left–right oscillation of these free electrons can be observed, which also generates their own oscillating electric field, synchronizing with the fluctuation of the external electric field (Akyildiz et al. 2014; Nagatsuma et al. 2016; Guzman et al. 2016; Dan et al. 2019; Ducournau 2018, 2019; Blin et al. 2017; MacCartney and Rappaport 2019; Sun et al. 2017).

Therefore, the quantization of these electronic oscillations can be characterized as plasmons. In other words, a plasmon would be a kind of energy cluster that sits on the surface of a metal. Thus, plasmons are being considered as a means of transmitting information between computer chips capable of accelerating internal information exchange processes and building interfaces between optical and electronic systems or even generating THz at the end of optical fiber (Ducournau 2018, 2019; Akyildiz et al. 2014; Nagatsuma et al. 2016; Guzman et al. 2016; Dan et al. 2019; Blin et al. 2017).

The research and design of a terahertz (THz) transceivers capable of transmitting or receiving digital data at 80 Gbit/s, considered an essential element for the next generation of telecommunications, involve the use of circuit technology that incorporate silicon-integrated CMOS, which offers a great advantage because of its industrial-scale production (Akyildiz et al. 2014; Nagatsuma et al. 2016; Guzman et al. 2016; Dan et al. 2019; Ducournau 2018, 2019; Blin et al. 2017).

References

Adediran, Y. A., Opadiji, J. F., Faruk, N., and Olayiwola, W. B. "On issues and challenges of rural telecommunications access in Nigeria". *African Journal of Education, Science and Technology*, April 2016 Vol. 3, No. 2.

Ahson, S. A., and Ilyas, M. (Eds.) *WiMAX: applications*. CRC Press, 2018a.

Ahson, S. A., and Ilyas, M. (Eds.). *WiMAX: Technologies, Performance Analysis, and QoS*. CRC Press, 2018b.

Akyildiz, I. F., Jornet, J. M., and Han, C. "Terahertz band: Next frontier for wireless communications." *Physical Communication*, 12, 16–32, 2014.

Alavi, B. and Pahlavan, K. "Modeling of the distance error for indoor geolocation," in Proc. IEEE Wireless Communications and Networking, pp. 668–672, Mar. 2003.

Ali, N. A., and Hassanein, H. S. *IEEE 802.16 Standards and Amendments. In WiMAX* (pp. 33–48). CRC Press, 2018.

Ali, T., and Biradar, R. C. "A compact multiband antenna using λ/4 rectangular stub loaded with metamaterial for IEEE 802.11 N and IEEE 802.16 E". *Microwave and Optical Technology Letters*, 59(5), 1000–1006, 2016.

Binh, L. N. *Advanced Digital Optical Communications*. Taylor & Francis, 2015.

Blin, S., Paquet, R., Myara, M., Chomet, B., Le Gratiet, L., Sellahi, M., ... & Lampin, J. F. "Coherent and tunable thz emission driven by an integrated iii–v semiconductor laser." *IEEE Journal of Selected Topics in Quantum Electronics*, 23(4), 1–11, 2017.

Boulogeorgos, A. A. "Terahertz technologies to deliver optical network quality of experience in wireless systems beyond 5G." *IEEE Communications Magazine* 56.6: 144–151, 2018.

Cacciapuoti, A. S., Sankhe, K., Caleffi, M., & Chowdhury, K. R. "Beyond 5G: THz-based medium access protocol for mobile heterogeneous networks". *IEEE Communications Magazine*, 56(6), 110–115, 2018.

Chesnoy, J. (Ed.). *Undersea Fiber Communication Systems*. Academic Press, 2015.

Cicerone, M. T., Lee, Y. J., Parekh, S. H., and Aamer, K. A. "Photonic crystal fiber-based broadband CARS microscopy". *In Coherent Raman Scattering Microscopy* (pp. 350–373). CRC Press, 2016.

Cong, L. and Zhuang, W. "Nonline-of-sight error mitigation in mobile location," *IEEE Trans. Wireless Communications*, 4, 560–573, Mar. 2005

Dan, I., Ducournau, G., Hisatake, S., Szriftgiser, P., Braun, R. P., & Kallfass, I. A superheterodyne 300 GHz wireless link for ultra-fast terahertz communication Systems. In 2019 49th European Microwave Conference (EuMC) (pp. 734–737). IEEE, 2019, October.

Ducournau, G. "Silicon photonics targets terahertz region." *Nature Photonics*, 12(10), 574–575, 2018.

Ducournau, G. "T-ray modulation of light for future THz radios." *Nature Photonics*, 13(8), 511–513, 2019.

Ducournau, G., "Terahertz wireless communications using photonic and electronic devices." 2016 18th International Conference on Transparent Optical Networks (ICTON). IEEE, 2016.

Godara, L. C. (Ed.). *Handbook of Antennas in Wireless Communications* (Vol. 4). CRC Press, 2018.

Guzman, R., Ducournau, G., Muñoz, L. E. G., Segovia, D., Cojocari, O., and Carpintero, G. Compact direct detection Schottky receiver modules for sub-terahertz wireless communications. In the 2016 41st International Conference on Infrared, Millimeter, and Terahertz waves (IRMMW-THz) (pp. 1–2). IEEE, 2016, September.

Han, C, Bicen, A. O., and Akyildiz, I. F. "Multi-ray channel modeling and wideband characterization for wireless communications in the terahertz band." *IEEE Transactions on Wireless Communications* 14(5): 2402–2412, 2014.

Han, C., Wu, Y., Chen, Z., & Wang, X. "Terahertz communications (TeraCom): Challenges and impact on 6G wireless systems." arXiv preprint arXiv:1912.06040, 2019.

Hermelo, M. F., "Spectral efficient 64-QAM-OFDM terahertz communication link." *Optics Express* 25(16): 19360–19370, 2017.

Kallfass, I. "Towards MMIC-based 300GHz indoor wireless communication systems." *IEICE Transactions on Electronics* 98(12): 1081–1090, 2015.

Kato, N., Mao, B., Tang, F., Kawamoto, Y., & Liu, J. "Ten challenges in advancing machine learning technologies toward 6G." *IEEE Wireless Communications* 27(3): 96–103, 2020.

Kazmierkowski, M. P. "Infrared and Terahertz Detectors". *IEEE Industrial Electronics Magazine*, 13(3), 53–54, 2019.

Kildal, P. S. *Foundations of Antenna Engineering: A Unified Approach for Line-of-Sight and Multipath. Artech House., A. Indoor Wireless Communications: From Theory to Implementation.* John Wiley & Sons, 2015.

Kiokes, G., Zountouridou, E., Papadimitriou, C., Dimeas, A., and Hatziargyriou, N. "Development of an integrated wireless communication system for connecting electric vehicles to the power grid. In 2015 International Symposium on Smart Electric Distribution Systems and Technologies (EDST) (pp. 296–301). IEEE, 2015.

Kleine-Ostmann, T., and Nagatsuma, T. A review on terahertz communications research. *Journal of Infrared, Millimeter, and Terahertz Waves*, 32(2), 143–171, 2011.

Lee, W. C. Y. *Mobile Communication Engineering*. McGraw-Hill, 1993.

Liu, P., Di Renzo, M., and Springer, A. "Line-of-sight spatial modulation for indoor mmWave communication at 60 GHz". *IEEE Transactions on Wireless Communications*, 15(11), 7373–7389, 2016.

Ma, J., Shrestha, R., Adelberg, J., Yeh, C. Y., Hossain, Z., Knightly, E., ... and Mittleman, D. M. "Security and eavesdropping in terahertz wireless links." *Nature*, 563(7729), 89–93, 2018.

MacCartney, G. R., and Rappaport, T. S. "Millimeter-wave base station diversity for 5G coordinated multipoint (CoMP) applications." *IEEE Transactions on Wireless Communications*, 18(7): 3395–3410, 2019.

Matias, I. R., Ikezawa, S., and Corres, J. (Eds.). *Fiber Optic Sensors: Current Status and Future Possibilities* (Vol. 21). Springer, 2016.

Mittleman, D. M. "Perspective: Terahertz science and technology." *Journal of Applied Physics*, 122(23), 230901, 2017.

Mittleman, D. M. "Twenty years of terahertz imaging." *Optics Express*, 26(8), 9417–9431, 2018.

Moldovan, A., Ruder, M. A., Akyildiz, I. F., and Gerstacker, W. H. LOS and NLOS channel modeling for terahertz wireless communication with scattered rays. In 2014 IEEE Globecom Workshops (GC Wkshps) (pp. 388–392). IEEE, 2014, December.

Morohashi, I., Irimajiri, Y., Kumagai, M., Kawakami, A., Sakamoto, T., Sekine, N., ... & Hosako, I. Terahertz source with broad frequency tunability up to 3.8 THz using MZ-FCG for frequency reference in phase-locking of THz source devices. In 2016 IEEE International Topical Meeting on Microwave Photonics (MWP) (pp. 126–128). IEEE, 2016 October.

Morohashi, I., Sekine, N., Kasamatsu, A., and Hosako, I. Full W-band frequency measurement of THz waves by electro-optic sampling using modulator-based optical comb source. In CLEO: Science and Innovations (pp. SW4F-4). Optical Society of America, 2019, May.

Morreale, P. A., AND Terplan, K. *CRC Handbook of Modern Telecommunications*. CRC Press, 2018.

Musey, J. A., and Keener, B. *The Spectrum Handbook 2018*. SSRN. Elsevier, 2018.

https://papers.ssrn.com/sol3/papers.cfm?abstract_id=3259782

Nagatsuma, T. "Terahertz wireless communications based on photonics technologies." *Optics Express* 21(20): 23736–23747, 2013.

Nagatsuma, T., Ducournau, G., and Renaud, C. C. "Advances in terahertz communications accelerated by photonics." *Nature Photonics*, 10(6): 371, 2016.

Nagatsuma, T., Horiguchi, S., Minamikata, Y., Yoshimizu, Y., Hisatake, S., Kuwano, S., ... and Takahashi, H. Terahertz wireless communications based on photonics technologies. *Optics Express*, 21(20): 23736–23747, 2013.

Nakagawa, D., Takizawa, K., Ikushima, K., Kim, S., Patrashin, M., Hosako, I., and Komiyama, S. "Terahertz response in the quantum-Hall-effect regime of a quantum-well-based charge-sensitive infrared phototransistor." *Japanese Journal of Applied Physics*, 57(4S): 04FK04, 2018.

Nandi, S., Thota, S., Nag, A., Divyasukhananda, S., Goswami, P., Aravindakshan, A., and Mukherjee, B. "Computing for rural empowerment: enabled by last-mile telecommunications". *IEEE Communications Magazine*, 54(6): 102–109, 2016.

Nefyodov, E. I., and Smolskiy, S. M. "Boundary conditions, integral and complex forms of electrodynamics equations, classification of electromagnetic phenomena". *In Electromagnetic Fields and Waves* (pp. 29–48). Springer, 2019.

Petrov, V., Moltchanov, D., Koucheryavy, Y., and Jornet, J. M. "Capacity and outage of terahertz communications with user micro-mobility and beam misalignment." *IEEE Transactions on Vehicular Technology*, 69(6): 6822–6827, 2020.

Png, G. M. W. N. *Terahertz Spectroscopy and Modelling of Biotissue* (Doctoral dissertation), 2010.

Popoola, S. I., Atayero, A. A., and Faruk, N. "Received signal strength and local terrain profile data for radio network planning and optimization at GSM frequency bands". *Data in Brief*, 16, 972–981, 2018.

Rangan, S., Rappaport, T. S., and Erkip, E. "Millimeter-wave cellular wireless networks: Potentials and challenges." *Proceedings of the IEEE*, 102(3): 366–385, 2014.

Rappaport, T. S., Xing, Y., Kanhere, O., Ju, S., Madanayake, A., Mandal, S., ... and Trichopoulos, G. C. "Wireless communications and applications above 100 GHz: Opportunities and challenges for 6G and beyond." *IEEE Access*, 7: 78729–78757, 2019.

Reichel, K. S., Lozada-Smith, N., Joshipura, I. D., Ma, J., Shrestha, R., Mendis, R., ... and Mittleman, D. M. "Terahertz waveguide signal processing: passive and active devices." In *Terahertz Emitters, Receivers, and Applications X* (Vol. 11124, p. 111240S). International Society for Optics and Photonics, 2019, September.

Rodríguez-Vázquez, P., Leinonen, M. E., Grzyb, J., Tervo, N., Parssinen, A., and Pfeiffer, U. R. Signal-processing Challenges in Leveraging 100 Gb/s Wireless THz. In 2020 2nd 6G Wireless Summit (6G SUMMIT) (pp. 1–5). IEEE, 2020, March.

Sellers, N. *Handbook of Terahertz Technology*. Scientific e-Resources, 2018.

Sirenko, Y., and Velychko, L. *Electromagnetic Waves in Complex Systems*. Springer, 2016.

Song, H. J., and Nagatsuma, T. "Present and future of terahertz communications." *IEEE Transactions on Terahertz Science and Technology*, 1(1): 256–263, 2011.

Stallings W., *Wireless Communications and Networks*, 2nd ed., Prentice-Hall, 2005.

Sun, S., MacCartney, G. R., and Rappaport, T. S. A novel millimeter-wave channel simulator and applications for 5G wireless communications. In 2017 IEEE International Conference on Communications (ICC) (pp. 1–7). IEEE, 2017.

Taori, R., and Sridharan, A. "Point-to-multipoint in-band mm-wave backhaul for 5G networks". *IEEE Communications Magazine*, 53(1): 195–201, 2015.

Uddin, J. *Terahertz Spectroscopy: A Cutting Edge Technology.* Books on Demand. IntechOpen, 2017.

Ummethala, S. "THz-to-optical conversion in wireless communications using an ultra-broadband plasmonic modulator." *Nature Photonics* 13(8): 519–524, 2019.

Venghaus, H., and Grote, N. (Eds.). *Fibre Optic Communication: Key Devices* (Vol. 161). Springer, 2017.

Wang, C., Yu, J., Li, X., Gou, P., and Zhou, W. "Fiber-THz-fiber link for THz signal transmission". *IEEE Photonics Journal*, 10(2): 1–6, 2018.

Wang, X., Wang, Z., and O'Dea, B. "A TOA-based location algorithm reducing the errors due to non-line-of-sight (NLOS) propagation". *IEEE Transactions on Vehicular Technology,* 52(1): 112–116, 2003.

Willner, A. *Optical Fiber Telecommunications* (Vol. 11). Academic Press, 2019.

Woo, S. S., You, H.-R. and Koh, J.-S. "The NLOS mitigation technique for position location using IS-95 CDMA networks," in Proc. IEEE Vehicular Technology Conf. (VTC), pp. 2556–2560, Sept. 2000.

Wu, H. C., Akamine, C., Rong, B., Velez, M., Wang, C., and Wang, J. "Point-to-multipoint communications and broadcasting in 5G". *IEEE Communications Magazine*, 56(3): 72–73, 2018.

Xue, S., Liu, Q., Liu, T., Yang, S., Su, P., Liu, K., and Wang, T. "Electromagnetic exploration of focusing properties of high-numerical-aperture micro-Fresnel zone plates." *Optics Communications*, 426: 41–45, 2018.

Yamamoto, Y., Kawaguchi, Y., and Hirano, M. "Low-loss and low-nonlinearity pure-silica-core fiber for C-and L-band broadband transmission". *Journal of Lightwave Technology*, 34(2): 321–326, 2016.

Yang, S. M. M. *Overview of Radio Communication Signals and Systems. In Modern Digital Radio Communication Signals and Systems* (pp. 1–28). Springer, 2019.

Yu, K., and Guo, Y. J. Non-line-of-sight detection based on TOA and signal strength. In 2008 IEEE 19th International Symposium on Personal, Indoor and Mobile Radio Communications (pp. 1–5). IEEE, 2008.

Yuan, R., and Ma, J. "Review of ultraviolet non-line-of-sight communication". *China Communications*, 13(6), 63–75, 2016.

Yuan, X., Zhang, Y. J., Shi, Y., Yan, W., and Liu, H. "Reconfigurable-intelligent-surface empowered 6G wireless communications: Challenges and opportunities." arXiv preprint arXiv:2001.00364, 2020.

Zaman Chowdhury, M., Shahjalal, M., Ahmed, S., and Jang, Y. M. 6G wireless communication systems: Applications, requirements, technologies, challenges, and research directions. arXiv preprint arXiv:1909.11315, 2019.

Part III

Terahertz Antenna Design

Chapter 7

Advancement in Terahertz Antenna Design and Their Performance

Sasmita Dash and Amalendu Patnaik

Contents

7.1 Introduction

The bandwidth requirements for wireless communications have increased rapidly over the last few decades. In order to tackle the situation, advanced modulation techniques have been applied to increase the spectral utilization efficiency [1]. This not only has increased the point-to-point data rates to a large extent but also enhanced the frequency reuse within a volume of space. However, the channel capacity upper limit is restricted by Shannon's formula, even with the use of multi-input multi-output (MIMO) strategy. Therefore, the only way to provide sufficient transmission capacity is by accessing transmission bands at higher carrier frequencies. This desire for higher carrier frequency or more bandwidth led the researchers to take advantage of the terahertz (THz) spectrum [2]. In addition to high bandwidth, THz wireless communication has other merits in comparison to microwave link or infrared (IR) based system, such as (i) more directional than microwave/millimeter links, (ii) secure, (iii) low attenuation compared to infrared (IR), and (iv) smaller scintillation effects compared to IR [3]. Because of these advantages, THz technology has grown dramatically over the last two decades and found its application in areas like communication, imaging, spectroscopy, biology/medicine, nondestructive evaluation, explosive detection, and radio astronomy. However, even today, these applications are not fully implemented due to the infancy stage of THz technology in terms of sources, detectors, antennas, and other essential components capable of working effectively in this frequency range. This chapter briefly discusses several THz source techniques and THz antenna fabrication techniques for the possible realization of the THz wireless communication system.

In the last few years, several metal antennas and array structures, including planar antennas, horn antennas, reflectarray antennas, and lens antennas, have been designed for THz applications [18–48]. The implementation of metal THz antennas suffers challenges from microfabrications to electromagnetic interaction at the nanoscale. The carbon-based nanomaterials (e.g., carbon nanotube and graphene) are emerging as promising materials for THz antenna designs [50–60, 65–78]. These antennas work on the principle of surface plasmons (SPs), which is a collective oscillation of electrons coupled to an electromagnetic (EM) field at a dielectric–metal interface [80]. The wave corresponding to the propagation of SPs is called the surface plasmon polariton (SPP) waves. The plasmonic antennas are such sub-wavelength structures on the surface of which SPs strongly confine. Properly designed plasmonic antennas can convert free-space EM radiation into SPs or can convert SPs into EM radiations. This chapter discusses several THz antennas made up of graphene, carbon nanotube, and copper material. For completeness from the THz communication point of view, different THz sources and some fabrication techniques have also discussed in this chapter.

7.2 THz Communication

The THz ($=10^{12}$ Hz) radiation refers to EM radiation in the 0.1–10 THz range, that is, the region between the microwave and IR frequencies. The THz band had not been properly investigated by the researchers for a long time due to the unavailability of suitable sources. Currently, the research interest increased toward the THz range due to the availability of THz sources and in the advancement in the laser, semiconductor, and optical photoconductive technology [81–107]. The different types of THz sources are discussed in the next section.

The THz EM spectrum has several benefits for wireless communications [3, 108]. The bandwidth requirements for wireless communications have been increasing rapidly since the last decade. There is the only way to provide sufficient transmission capacity, that is, by accessing transmission bands at higher carrier frequencies. THz spectrum opens the possibility for more bandwidth and high data rate transmission. Current wireless communications systems have bandwidths of a few GHz and data transmission rates up to 1 Gbps. The achievement of the data rates of 10 Gbps is comparatively difficult in the microwave band due to the narrow bandwidth. The data rate from 10 to 100 Gbps is realized by raising the carrier frequencies from 100 to 500 GHz [109]. The opportunity for large bandwidth in the THz band leads to the possibility of easy high data rate transmission [110]. THz communication is promising for wireless communications systems, particularly for the short-range indoor environment.

In spite of the advantages, THz communication has a few limitations. In the THz band, the free space path loss is more significant than at lower frequencies [111]. This is the main reason to have less received power than the transmitted power. The free space path loss obtained from the Friis transmission equation,

$$\text{Free space path loss} = 20\log\left(\frac{4\pi d}{\lambda}\right)[\text{dB}] \tag{7.1}$$

where λ is the wavelength and d is the distance. According to the Friis equation, the THz band leads to larger propagation loss. Furthermore, the THz signal suffers from both molecular absorption loss A_{ma} and spreading loss A_s. On account of the molecular absorption loss, several high attenuation levels are defined [31]. In the THz band, the spreading loss is 60 dB higher compared to the microwave band. Besides the path loss properties, the reflection properties in the THz band are different from the microwave band [112, 113]. The surface variations are on the order of THz wavelength (several hundred microns) for objects in indoor environments. Therefore, at the THz band, the indoor object surfaces are rough. We know from the scattering effect that the reflection angle can be different from the incidence angle for rough surface, and the receive antennas receive the signals from different spots, which leads to multi-path scattering [112–114]. At THz band, the multi-path scattering is significantly affected in both outdoor and indoor applications and makes the NLOS communication more challenging. Therefore, the THz band is merely appropriate for short-range communications where the range is in the order of a few tens of meters.

In the development of THz communication systems, the antenna is the most significant component. Antennas for THz communication have reported in the last few years. Presently, the focus is to design high-gain antennas because it enhances the performance of the overall system by compensating the substantial path loss at THz. The use of THz frequency allows for miniaturized antennas, which enables massive MIMO techniques for enhancement of spectral efficiency and directivity. In addition to high directive antennas, high power THz sources needed for the development of an effective THz wireless communication system.

7.3 THz Sources

In THz frequencies, the major problem is to get efficient sources. Different techniques for obtaining THz source found in the literature, namely quantum cascade laser(QCL), resonant tunneling diode (RTD), and optical photoconductive material (OPM) techniques, complementary metal-oxide–semiconductor (CMOS) and heterojunction bipolar transistor (HBT) [81–107].

QCL may be a potential THz source of communication at the upper THz band, even if with the high atmospheric attenuation. The output power in QCLs increases with the increase of signal frequency. The output power of 4.1 μW is achieved at 2.06 THz [81]. However, the QCL-based THz source is not appropriate in the lower THz band. Efficiencies of QCL-based sources are quite low at 1 THz region [82]. Several QCL-based source is developed to attain more output power [81–87]. Using the QCL technique, the achieved maximum output power is more than 1 W [84]. In QCL, the tunneling of the electron through the barrier is possible without any extra energy, whereas, in RTD, the energy levels are quantized. The tunneling process in the RTD technique is high-speed. Thus, this technique is one of the efficient THz sources [88–92].

Recently, OPM-based techniques have been explored. OPM-based THz sources are an efficient source in lower THz band and suitable substitutes to solid-state sources. This source also performs well in the upper THz band. The plasmonic electrode is now used for efficiency enhancement and achieves an efficiency of 7.5% [93]. Two different OPM-based THz sources techniques are commonly used, such as pulsed technique and photo mixing technique. In both cases, photoconductive antennas are used for the emission of THz radiation. In the pulsed technique, a femtosecond laser is used to obtain pulsed THz power, and in the photomixing technique, photo mixer is used to obtain continuous wave (CW) THz power.

CMOS-based sources are used in the lower THz band. In the CMOS device, an active multiplier chain or a voltage-controlled oscillator is inserted [94]. The multiplier chain technique is widely used to achieve higher frequency output powers. This technique is used to attain 2.58 THz source and output power as 0.018 mW in [95]. In lower THz frequency between 0.288 and 0.498 THz, the output power from 22 μW to 2.6 mW is achieved in Refs. [96–100]. These results reveal that low operational frequency leads to high output power and low efficiency. Solid-state diodes are used to obtain the output power of 2.82 mW and 0.11 THz source [101]. Higher output powers in higher frequency sources can be obtained using heterojunction bipolar transistor (HBT) technology. In Refs. [102–107], the output power from 12 μW to 10 mW, and the efficiency from 0.01% to 3.2% in frequency between 0.215 THz and 0.92 THz are measured. In solid-state devices, the output power reduces with the increase of THz frequency.

THz source techniques have several advantages and disadvantages. In view of the output power, CMOS is the best THz source in low THz frequencies. Above 0.3 THz, the output power in CMOS THz sources significantly reduces. In contrast, the QCL technique is not possible below 1 THz frequency. The OPM and RTD are the best THz source techniques in the 0.5–3 THz range. OPM THz source is the best suitable technique to achieve the best efficiency results. The CMOS and HBT THz sources also perform better at lower THz frequencies. Alternative to OPM, RTD is a suitable technique for THz bands. The insertion of plasmonic electrodes into the active photoconductive area improves the power and efficiency of THz radiation. The achievement of 1Tb/s wireless link using these efficient THz source techniques is not far away anymore.

7.4 THz Antennas

Copper is the well-known metal for the design of radio frequency (RF) and microwave antennas. The design of metal antennas in the RF and microwave frequency regime has thoroughly explored, and systematic antenna design methods exist [4]. However, the design of the copper metal antenna at THz band is entirely different from the classical microwave metal antenna. The classical microwave metal antennas are excited through various feed lines such as microstrip, coaxial cable, and CPW, whereas the laser excitation through fiber or the air is seen in THz antennas. Metal THz antenna and microwave antenna have another difference, that is, bias voltage. The design of the THz antenna needs bias voltage, but microwave antennas do not need any biasing. Furthermore, the fabrication of the metal THz antenna is more expensive and complicated than a metal microwave antenna. Over the past decades, several high directive metal THz antenna and array structures have been designed, including planar antenna and arrays [18–25], reflectarrays [26–36], lens antenna [37–43], horn antennas [44–48], carbon nanotube antennas [50–60], and graphene antennas [65–78].

7.4.1 *Planar Antennas and Arrays*

The unique features of the planar antenna, such as simple structure, low cost, and low profile, motivates the antenna community to design different kinds of planar antennas to meet the requirement of future wireless communication. However, the single-element antenna has a broad radiation pattern and small directivity values. In the wireless communication system, high directive antennas are needed for long-distance communication. An assembly of single element in an electrical and geometrical configuration has the capability to provide sharp beam and high directivity (gain). This multi-element antenna is known as an array. A plethora of antenna arrays at microwave frequencies is available in the literature for several applications [6–17].

Several planar antennas have been designed at THz frequency [18, 19]. One important issue has been addressed in most of these antennas, that is, low antenna directivity. By assembling of 1,000 single-element in the array, the antenna gain is enhanced to 31 dBi [20]. However, it is a sign of complex lossy networks. Planar antennas have designed at the low-THz frequency for beam scanning and high-gain applications [21, 22]. Nevertheless, the complexity in feeding and fabrication leads to more losses and shifts in the operational frequency band. The planar waveguide array antenna also has been developed at low THz frequency 0.12 THz [23]. The array provides 21.1 dBi gain of and 80% efficiency. A larger gain of 38 dBi and 43 dBi in 16×16 and 32×32 slot arrays are obtained in Ref. [24]. However, it is difficult to fabricate this multilayer waveguide slot array antenna, which has radiating waveguide and feeding network in the top layer and bottom layer, respectively. The fabrication process would be challenging for a large and complex structure. The planar Yagi–Uda antenna is also promising for communication. The planar Yagi–Uda configuration exhibits high gain and low cross-polarization, which is useful for wireless communication. Han et al. reported a design of the Yagi–Uda antenna at low THz frequency of 0.636 THz, which attained improved impedance matching using a photomixer [25].

7.4.2 *Reflectarrays*

In the RF and microwave frequency region, reflectarray antennas have their own unique place because of getting the best performance of reflector and array antenna [26, 27]. In recent years, the reflectarray antenna design at THz frequency has been found in the literature [28–34]. The most crucial factor from these THz reflectarray antennas is element loss. Moreover, low-cost designs with satisfactory performance cannot be achieved from these works. The main reason for this is the occurring of conductor loss in the resonant element at THz frequency, which leads to a significant loss of power and tuning range of phase.

In Ref. [35], a 220 GHz reflectarray antenna is reported. The reflectarray element consists of a dielectric block and a ground plane. The reflection phase is attained by controlling the block height. The 40×40 elements of reflectarray achieved 31.3 dBi gain, 27.6% aperture efficiency, and 20.9% 1-dB gain bandwidth. Another dielectric reflectarray antenna attained 22.5 dB of gain at 0.1 THz [36].

7.4.3 *Lens Antennas*

Lens antenna arrays consist of an EM lens with the potential of energy focusing and a matching array with antenna elements placed in the focal region of the lens. An EM lens is a transmissive device, which has the ability to alter the directions of EM ray propagation for the realization of beam collimation or energy focusing. EM lenses are employed by three different methods, such as (i) the dielectric lenses, (ii) the conventional planar lenses, and (iii) the modern compact planar lenses. Generally, the conductor losses, feeding losses, and fabrication precision avoid scaling the high-gain antennas from microwave to THz band. In this case, lens-based THz antennas are more suitable because of low losses. Several designs of lens antennas at THz frequency have been designed [37–43].

Llombart et al. reported a hemispherical silicon lens THz antenna at a low THz frequency 0.545 THz [37]. Mostly, the silicon lens is used at a higher frequency for high-speed communication [38]. The extended hemispherical lens with a double-slot antenna and all the gold-conducting layers deposit on a high-resistivity silicon wafer [39]. Extended hemispherical lens antennas are more suitable for beam scanning applications, especially with off-axis feeding [40, 41]. Bowtie THz antenna with a silicon lens improves the radiation characteristics in terms of directivity, gain, efficiency, and the front-to-back ratio [42]. The antenna at 1.05 THz attained directivity of 11.8 dB with the radiation efficiency of 96. Yi et al. reported 3-D printed dielectric lens based beam-scanning and high-gain THz antennas for THz communications and radar applications. Here, dielectric lenses are used to increase directivity [43]. However, the implementation of antennas at THz band demands a more complicated fabrication process, more cost, and more time.

7.4.4 Horn Antennas

Due to its unique performances, horn antenna has its existence since the late 1800s. The horn is a flaring metal waveguide and commonly employed as a feed element. The performance of the radiator element depends on the type, direction, and amount of taper. It has widespread application because of its simple structure, easy excitation, wide bandwidth, more power capacity, and high gain.

In Ref. [44], a pyramid horn antenna is studied at low THz frequency 0.3 THz. The low-temperature co-fired ceramic (LTCC) multilayer substrate is used to achieve wide bandwidth and high gain. The stepped profile horn is structured through drilling cavities by increasing the size step by step on each layer using substrate integrated waveguide (SIW) technology. The horn antenna is fabricated with the process of LTCC multilayer and provides 100 GHz bandwidth and 18 dBi gain. The optimization of the step height and corrugation slot depth allows wide bandwidth and high gain.

THz horn antennas are implemented using the metallic 3-D printing technologies in Ref. [45]. Antennas are printed using selective laser melting (SLM) technique on Cu–15Sn and sintering technique on 316L stainless steel. Horn antennas using the SLM technique on Cu–15Sn is developed at a low THz band from 110 to 320 GHz. The antenna exhibits around 22 dBi of gain. 3-D printed metallic antennas are more simple and durable than nonmetallic 3-D printed antennas. As compared to the traditional technique for metallic horn antenna implementation, the cost of a 3-D printed antenna is low.

Different kinds of horn antennas have been designed at THz [46–48]. The multiflare horn antenna is the best solution to achieve high directivity, and it provides directivity of 31.7 dBi at 1.9 THz [47]. However, at THz frequencies, the reduction of horn size leads to more difficulties in the fabrication, time, and cost.

7.4.5 CNT Antennas

Since the discovery of carbon nanotubes (CNTs) in 1991, CNT has gained more attention among the researchers. In the last decade, the use of CNT in numerous fields has been noted. Specifically, many antennas using CNT have designed at THz frequency due to its significant electrical properties. The current on a carbon nanotube antenna using a Fourier transform technique is investigated by Hanson [50]. The current on a CNT antenna is almost similar compared to current on copper antennas. For nanoscale radius, the CNT antenna has low losses than the copper antenna.

The theoretical study of the single-walled CNT (SWCNT) dipole antenna at THz frequency is found in Ref. [51]. Wavenumber-domain integral equation is formulated by combining the Boltzmann equation and Maxwell's equations for obtaining the distribution of current. The numerical result of SWCNT THz antennas provides broad bandwidth and higher efficiency. SWCNT THz antennas have merits in terms of miniaturization, directivity, biocompatibility, and output power compared to metal photoconductive THz antennas. CNT antenna provides slow wave propagation, high input impedance, and low radiation efficiency [52–55]. In the THz frequency, the CNT dipole antenna resonates at length $L \approx \lambda_p/2$, where λ_p is the plasmon wavelength [55]. Above THz frequency range, strongly damped current resonance is found, because of interband transitions in the optical frequency regime. The CNTs as both receiving and transmitting THz antennas has established in Ref. [56]. CNT antenna does not behave in the same manner as a nanowire antenna due to the variation in inductance [52]. The inductance of the CNT antenna is 10^4 times of the metal nanowire antenna. This brings a significant difference in the performance of CNT and metal nanowire antennas. The behavior of the CNT antenna and metal antenna is entirely different due to their kinetic inductance and quantum capacitance [53]. The high kinetic inductance of CNT reduces the size of the antenna as well as reduces the antenna radiation efficiency. Several CNT THz antenna designs have been designed to achieve the enhancement of the antenna efficiency using bundle structure [57–59]. In these antennas, the efficiency enhancement of the CNT antenna is achieved at a low THz band by using a bundle of SWCNT. Radiation resistance and radiation efficiency of bundled SWCNT are increased by controlling the permittivity of the metamaterial jacket. The radiation efficiency can also be enhanced using multi-walled CNT [60]. The radiation resistance and radiation efficiency increase with the number of layers and the frequency. However, recently graphene's unique properties at THz frequency open an exciting scenario in THz antenna application.

7.4.6 Graphene Antennas

The latest addition of carbon allotropes family, graphene, is widely considered as the mother of the carbon allotropes. Recently, graphene attracted significant attention in various research fields due to its unique properties. The most significant property is the propagation of SPP in graphene in the THz range. Graphene SPs exhibits strong confinement, low losses, and tunability. Graphene plasmons are more easily tunable by changing the doping level via chemical or electrostatic gating in both single layer and bilayer graphene structure [63, 64]. Recently, the research interest is increasing for the realization of graphene antenna design at THz frequencies. Due to SPP propagation at THz, graphene enables plasmonic antennas at THz, whereas metal antenna made of noble metals such as gold and silver shows plasmonic behavior at optical frequencies. Plasmonic antenna resonates at the sub-wavelength scale with a high near field and strong coupling between localized sources and far-field radiation.

Graphene was used as an antenna radiator at the THz frequency range for the first time in 2012 [65]. In this work, the propagation properties of transverse-magnetic (TM) SPP in graphene have been used to model graphene patch antennas in the THz band. In addition to the high miniaturization, the graphene-based THz antenna performs better than the metal antenna in terms of radiation efficiency. The radiation efficiency of the graphene antenna increases when the graphene chemical potential increases. Graphene plasmonic antenna enables high miniaturization and high directivity as compared to the metal antenna in the THz band [66]. Graphene-based antennas work at a much lower frequency than classical metallic antennas of the same size [67, 68]. Moreover, the performance of the graphene antenna at THz is enhanced by tuning the conductivity of graphene using an electric field effect. The performance merits of the graphene THz antenna are its high directivity, high miniaturization, stable impedance, and frequency reconfiguration [69]. Bilayer graphene provides dual-band reconfiguration with stable impedance, which avoids the need for a lossy and complex reconfigurable antenna [70]. Several graphene antenna designs at THz frequency have been studied further [71–78].

Although several reflectarray and transmitarray antennas have been designed using conventional metal microstrip patches, dipoles, and dielectric resonator antennas, the graphene reflectarray antenna provides a unique performance compared to metal reflectarray. Graphene reflective cells for THz reflectarray and graphene THz reflectarray antenna based on square graphene patches are designed at 1.3 THz [74, 75]. Graphene is used to control the phase of reflectarray at THz frequencies dynamically [75]. Graphene reflectarray allows drastically reduce the inter element spacing and wide bandwidth. The tunable conductivity behavior of graphene allows the reconfigurability in graphene reflectarrays.

MIMO technique in the wireless communications system is well known to increase spectral efficiency. However, the size of the antenna and the spacing between antennas are major obstacles for increasing the MIMO scale. Graphene antennas have the potential to reduce the size of antenna size and separation between antennas [77]. The spectral efficiency is increased by using the graphene-based MIMO antennas. The radiation patterns of graphene Yagi–Uda antenna can be easily reconfigured by controlling the properties of each graphene element [77]. Graphene material is suited for designing the reconfigurable directional antennas for THz communications.

7.5 Promising Material for THz Antenna

The materials so far used for antenna design in THz frequency are conventional metal copper, CNT, and graphene. However, the selection of promising material for THz antenna design is important for the THz community. This section provides the answer to this query by critically analyzing their properties at THz.

Graphene is a single layer of hexagonally arranged sp^2-bonded carbon atoms. Graphene has an electron mobility of 2×10^5 $cm^2V^{-1}s^{-1}$ [61] and a current density of 10^9 A/cm [62]. The surface conductivity of graphene based on Kubo formalism can be expressed as [79]

$$\sigma_g(\omega, E_f, \tau, T) = -j\frac{e^2\kappa_B T}{\pi\hbar^2(\omega - j\tau^{-1})}\left[\frac{E_f}{\kappa_B T} + 2\ln\left(e^{-E_f/\kappa_B T} + 1\right)\right] \tag{7.2}$$

where ω *is the* angular frequency, E_f is the Fermi energy, τ is the relaxation time, T is the temperature, j is the imaginary unit, e is the electron charge, $\hbar$ is the reduced Planck's constant, and k_B is the Boltzmann constant. At THz frequencies, graphene is highly inductive in nature [67]. The propagation of the plasmonic wave at THz is another important property of graphene. Due to the 2D nature of graphene, surface plasmons in graphene layer exhibit unique properties of low losses, strong confinement, and high tunability.

The wrapping of graphene sheet forms CNT, has an electron mobility of 8×10^4 $cm^2V^{-1}s^{-1}$, and a current density of 10^9 A/cm [49]. CNT supports plasmonic wave propagation at THz band. However, more plasmonic losses and less tunability occur in CNT in comparison to graphene. The surface conductivity and surface impedance of CNT of the small radius r expressed as [53]

$$\sigma_{CNT} = -j\frac{2e^2 v_f}{\pi^2\hbar r(\omega - j\tau^{-1})} \tag{7.3}$$

$$Z_{CNT} = \frac{1}{2\pi r\sigma_{CNT}} = \frac{j\pi\hbar(\omega - j\tau^{-1})}{4e^2 v_f} = \frac{\pi\hbar\tau^{-1}}{4e^2 v_f} + j\omega\frac{\pi\hbar}{4e^2 v_f} = R_{CNT} + j\omega L_{CNT} \tag{7.4}$$

Copper is a well-known and excellent electrical conductor in the microwave frequency regime. Copper has an electron mobility of 32 $cm^2V^{-1}s^{-1}$ and a current density of 10^6 A/cm. In the THz frequency regime, the surface impedance and conductivity of copper using Drude theory expressed as [5]

$$Z_{Cu} = \sqrt{\frac{j\omega\mu_0}{\sigma_D^{cu} + j\omega\varepsilon_0}} \text{ and } \sigma_D^{cu} = \frac{ne^2\tau}{m(1 + j\omega\tau)} \tag{7.5}$$

where m is the mass of the electron, and n is the conduction electron density. At low THz frequency range, the ohmic resistance of copper is dominant to the surface impedance. The conductivity of copper is less than the conductivity of graphene and CNT [119]. The conductivity of graphene is higher than that of CNT. Therefore, graphene is the best conductor than CNT and copper in THz frequency.

In order to compare the performance of THz antennas made up of these carbon-based materials with that of the traditional copper metal, a rigorous analysis has been done. In the first phase of this analysis, the physical size of the antenna is kept constant, whereas, in the second phase, the frequency is maintained at a constant value [66, 67]. The motive is to find the best material for the THz antenna design with superior performance.

In the first phase, THz dipole of length (L) 71 μm (arbitrary dimension for obtaining low THz operational frequency) is considered for comparison of the antenna performance for graphene, CNT, and copper material [67]. Antennas are placed over silicon dioxide (ε_r= 3.9) substrate in all the three cases. The graphene, CNT, and copper dipole antenna structures are shown in Figure 7.1. It found that the graphene antenna resonates at the

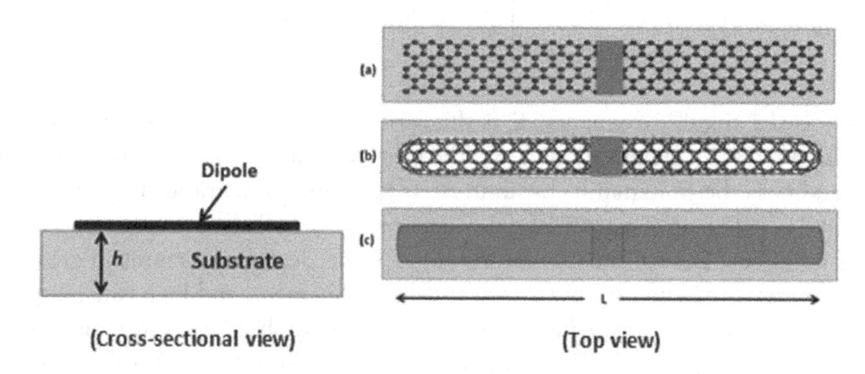

Figure 7.1 THz dipole antennas: (a) graphene, (b) CNT, and (c) copper. ©Springer Nature. [67]

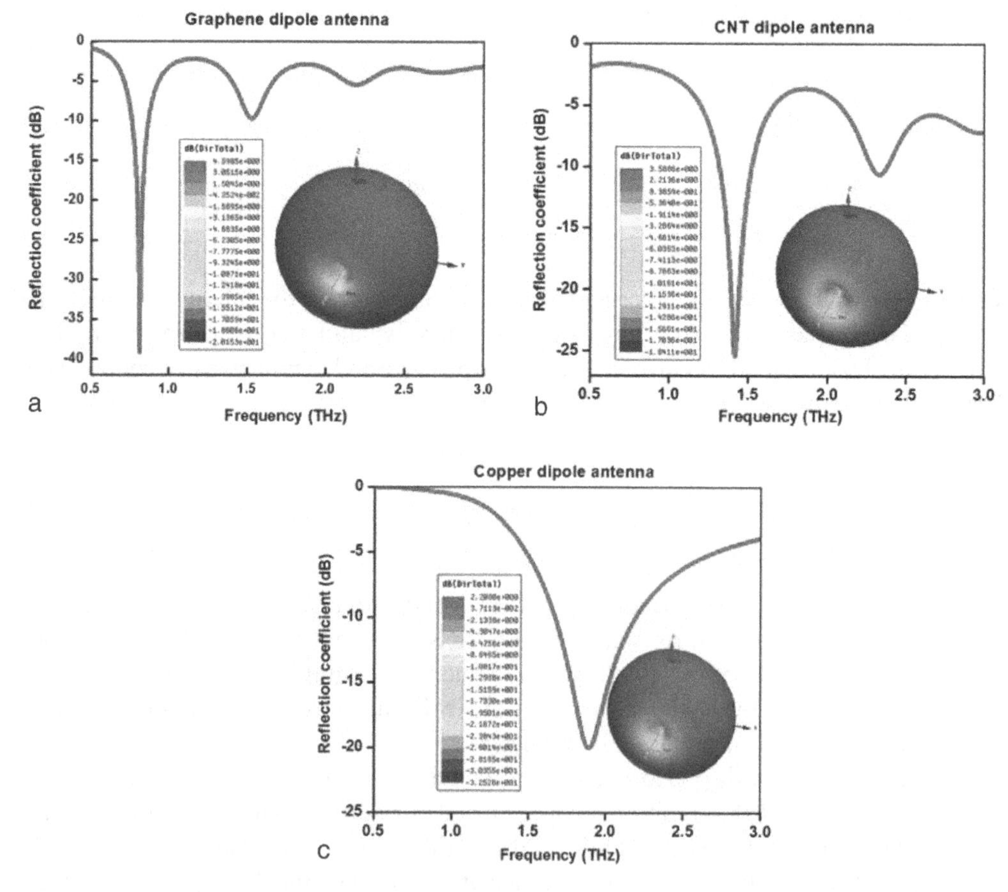

Figure 7.2 S11 parameter and the far-field pattern of THz antenna of the same length (a) graphene, (b) CNT, and (c) copper. © Springer Nature. [67]

lowest frequency compared to CNT and copper antenna at THz band, which is illustrated in Figure 7.2.

The radiation pattern of these THz dipole antennas at their resonant frequencies is shown in the inset of Figure 7.2. Although their radiation patterns are identical, they

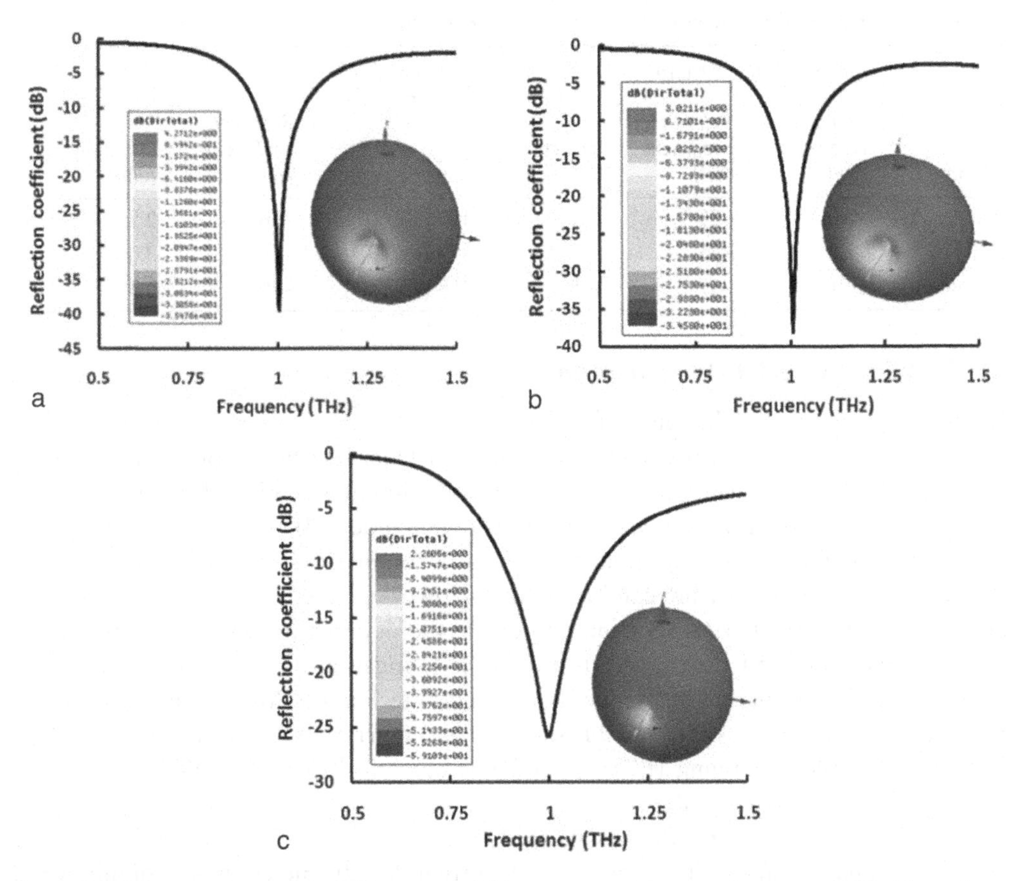

Figure 7.3 S11 parameter and far-field pattern of THz antennas at the same frequency 1 THz (a) graphene, (b) CNT, and (c) copper. ©Wiley. [66]

differ in their directivities. It has been observed that the graphene THz antenna has higher directivity than the CNT THz antenna, and the directivity of the CNT THz antenna is higher than that of the copper THz antenna.

In the second phase, the performances of graphene, CNT, and copper dipoles with the same resonant frequency of 1 THz are analyzed [66]. The S11 parameters and radiation patterns of graphene, CNT, and copper THz antennas are shown in Figure 7.3. Graphene THz antenna achieves high miniaturization compared to CNT and copper THz antenna. This analysis also shows similar radiation patterns for three antennas with graphene THz antenna having maximum directivity. The performance of graphene, CNT, and copper THz antenna is summarized in Table 7.1. Although these results are shown for dipole antennas, it has been found that the results are equally applicable to other varieties of antennas and hence general in nature.

The above analysis reveals that graphene THz antennas exhibit higher miniaturization and higher directivity than CNT and copper THz antennas. CNT THz antennas provide high miniaturization and larger directivity than copper THz antennas.

Table 7.1 Performance of graphene, CNT, and copper THz antenna

	Antenna (length 71 μm)			*Antenna (@1 THz)*		
Material	*Resonant Frequency (THz)*	*Reflection coefficient (dB)*	*Directivity (dBi)*	*Antenna length*	*Reflection coefficient (dB)*	*Directivity (dBi)*
Copper	1.90	–20	2.2	139 ($=\lambda_0/2$)	–26	2.2
CNT	1.42	–25	3.5	99 ($= \lambda_0/3$)	–38	3.0
Graphene	0.81	–39	4.5	68 ($= \lambda_0/4.4$)	–40	4.3

7.6 Fabrication of THz Antennas

Due to the persistent development of fabrication technologies, several new technologies are now ready to meet the processing requirements of THz. 3D printing or additive manufacturing [115–118] is a method used for the construction of 3D objects by printing them layer through the layer, primarily based on a 3D digital model. 3D printing is appropriate for structural components of large dimensions and complicated shapes. In 3D printing tools, a precision up to 0.01mm can be attained. However, there is a need for sintering of the powder metallurgy parts after being printed. In the sintering process, the deformation of high temperature and shrinkage rate exist. The fabrication of high-precision parts is needed using the machining technique. Now, several techniques for high-precision processing of electronic devices are available such as low-temperature co-fired ceramic(LTCC), electrical discharge machining (EDM), computer numerical control (CNC) machining, and printed circuit board (PCB).

The fabrications of THz antennas using new materials are also pivotal. The latest carbon-based nanomaterial graphene has high current density and electron mobility than conventional metal copper. The conductivity of graphene and CNT are more than copper [119]. Recently, several researchers demonstrated in other contexts that graphene over hexagonal boron nitride leads to significant improvement [120] [121]. Due to its properties, graphene has promising potential for the development of effective electronic devices. The tunability behavior of graphene conductivity also enables the design of reconfigurable antennas. The research on THz antennas is now more useful and meaningful using new fabrication technologies and new nanomaterials.

High conductive graphene nanomaterial easily integrates with different substrate materials [122, 123]. Graphene conductive ink is promising for printed electronics due to its unique mechanical and electrical properties. Recently, a few RF and microwave antennas are fabricated using graphene ink. An RFID dipole antenna is fabricated using graphene ink on the foam substrate by the rolling compression technique [122]. Another RFID antenna is fabricated using graphene ink by doctor-blading technique [124]. In this technique, graphene ink is spread on the substrate using a mechanical mask, and the doctor blade is used for flattening and controlling the thickness. Graphene ink is used as a conductor in antenna-electronics interconnection fabrication using the 3D direct-write dispensing method [125]. In this case, the graphene antenna is fabricated using graphene ink on a 100% cotton fabric using 3D-printing technology. In Ref. [126], the CNT antenna is printed on the substrate and followed by cutting by the milling machine. The transfer technique also used to laminate the CNT on the substrate [127]. The fabrication of graphene,

CNT, and metal antenna of the meander line dipole type is demonstrated using the direct ink-injecting technique at RF and microwave frequency [128]. The ink-injecting technology may have the potential for the fabrication of antenna for wireless communication applications. The fabrication of the graphene and CNT THz antenna is not yet fully explored. There is still a big room for improvement in the fabrication of graphene THz antennas, but the future prospect is certainly promising.

7.7 Conclusion

In the design of any wireless communication system, the antenna is the most significant component. In the THz wireless system especially, the focus is to design high directive antennas, because it enhances the performance of the overall system by compensating the large path loss that occurs at THz frequency and thus improves the signal-to-noise ratio. This chapter discussed different THz antennas, made up of traditional metals, and carbon-based materials.

Generally, conventional metal copper is the well-known and commonly used metal in the design of microwave antenna. The carbon-based nanomaterials such as graphene and CNT are recently explored in the design of THz antenna. In this chapter, we discussed about metal, graphene, and CNT THz antenna. Moreover, a critical comparison of the performance of THz antennas made up of copper, carbon nanotube, and graphene material have been made to find the best material for the design of THz antenna. It has been found that the graphene antenna provides superior performance in terms of miniaturization, directivity, and radiation efficiency.

THz band was least explored by the researchers for a long time, because of the unavailability of suitable THz antennas and sources. Currently, the research interest increased toward the THz frequency range due to the availability of the THz sources at THz band due to advancement in the laser and semiconductor technology. This chapter discussed the different THz sources and fabrication techniques for the THz antenna for the practical realization of THz wireless communication system.

References

[1] A. J. Paulraj, D. A. Gore, R. U. Nabar, and H. Bolcskei, "An overview of MIMO communications—A key to gigabit wireless," *Proceedings of the IEEE*, vol. 92, pp. 198–218, 2004.

[2] I. F. Akyildiz, J. M. Jornet, and C. Han, "Terahertz band: Next frontier for wireless communications," *Physical Communication*, vol. 12, pp. 16–32, 2014.

[3] J. Federici and L. Moeller, "Review of terahertz and sub terahertz wireless communications," *Journal of Applied Physics*, vol. 107, pp. 111101, 2010.

[4] C. A. Balanis, *Antenna Theory: Analysis and Design*, 3rd edn. New York: Wiley, 2005.

[5] M. P. Kirley and J. H. Booske, "Terahertz conductivity of copper surfaces," *IEEE Transactions on Terahertz Science and Technology*, vol. 5, no. 6, pp. 1012–1020, 2015.

[6] R. L. Haupt and Y. Rahmat-Samii, "Antenna array developments: A perspective on the past present and future," *IEEE Antennas and Propagation Magazine*, vol. 57, no. 1, pp. 86–96, 2015.

[7] R. J. Mailloux, *Phased Array Antenna Handbook*. Norwood, MA: Artech House, 2005.

[8] K. Y. Kapusuz, Y. Sen, M. Bulut, İ. Karadede and U. Oguz, "Low-profile scalable phased array antenna at Ku-band for mobile satellite communications," 2016 IEEE International Symposium on Phased Array Systems and Technology (PAST), Waltham, MA, 2016, pp. 1–4.

[9] H. Li, L. Kang, F. Wei, Y. Cai, and Y. Yin, "A low-profile dual-polarized microstrip antenna array for dual-mode OAM applications," *IEEE Antennas and Wireless Propagation Letters*, vol. 16, pp. 3022–3025, 2017,

[10] R. Lian, Z. Wang, Y. Yin, J. Wu, and X. Song, "Design of a low-profile dual-polarized stepped slot antenna array for base station," *IEEE Antennas and Wireless Propagation Letters*, vol. 15, pp. 362–365, 2016,

[11] Shih-Hsun Hsu, Yu-JiunRen, and Kai Chang, "A dual-polarized planar-array antenna for S-band and X-band airborne applications," *IEEE Antennas and Propagation Magazine*, vol. 51, no. 4, pp. 70–78, Aug. 2009.

[12] R. Di Bari, T. Brown, S. Gao, M. Notter, D. Hall, and C. Underwood, "Dual-polarized printed S-band radar array antenna for spacecraft applications," *IEEE Antennas and Wireless Propagation Letters*, vol. 10, pp. 987–990, 2011.

[13] H. A. Diawuo and Y. Jung, "Broadband proximity-coupled microstrip planar antenna array for 5G cellular applications," *IEEE Antennas and Wireless Propagation Letters*, vol. 17, no. 7, pp. 1286–1290, July 2018.

[14] Q. Liang, B. Sun and G. Zhou, "Multiple beam parasitic array radiator antenna for 2.4 GHz WLAN applications," *IEEE Antennas and Wireless Propagation Letters*, vol. 17, no. 12, pp. 2513–2516, Dec. 2018.

[16] L. Chioukh, H. Boutayeb, D. Deslandes and K. Wu, "Dual-band linear antenna array for harmonic sensing applications," *IEEE Antennas and Wireless Propagation Letters*, vol. 15, pp. 1577–1580, 2016.

[17] F. Tofigh, J. Nourinia, M. Azarmanesh, and K. M. Khazaei, "Near-field focused array microstrip planar antenna for medical applications," *IEEE Antennas and Wireless Propagation Letters*, vol. 13, pp. 951–954, 2014.

[18] H. Davoudabadifarahani and B. Ghalamkari, "High efficiency miniaturized microstrip patch antenna for wideband terahertz communications applications," *Optik*, vol. 194, p. 163118, 2019.

[19] A. Hocini, M. N. Temmar, D. Khedrouche, and M. Zamani, "Novel approach for the design and analysis of a terahertz microstrip patch antenna based on photonic crystals," *Photonics and Nanostructures –Fundamentals and Applications*, vol. 36, 100723, 2019.

[20] M. Koch, "Terahertz communications: A 2020 vision," in: R. E. Miles et al. (Eds.), *Terahertz Frequency Detection and Identification of Materials and Objects*. Berlin: Springer, 2007, pp. 325–338.

[21] K. Konstantinidis, A. P. Feresidis, Y. Tian, X. Shang, and M. J. Lancaster, "Micromachined terahertz Fabry–Perot cavity highly directive antennas," *IET Microwaves, Antennas and Propagation*, vol. 9, no. 13, pp. 1436–1443, 2015.

[22] A. Jam and K. Sarabandi, "A horizontally polarized beam-steerable antenna for sub-millimeter-wave polarimetric imaging and collision avoidance radars," *IEEE International Symposium on Antennas and Propagation (APSURSI)*, pp. 789–790, 2016.

[23] D. Kim, "4_4-element corporate-feed waveguide slot array antenna with cavities for the 120 GHz-band," *IEEE Transactions on Antennas and Propagation.*, vol. 61, no. 12, pp. 5968_5975, Dec. 2013.

[24] D. Kim, J. Hirokawa, M. Ando, J. Takeuchi, and A. Hirata, "64_64-element and 32_32-element slot array antennas using doublelayer hollow-waveguide corporate-feed in the 120 GHz band," *IEEE Transactions on Antennas and Propagation*, vol. 62, no. 3, pp. 1507_1512, Mar. 2014.

[25] K. Han, T. K. Nguyen, I. Park, and H. Han, "Terahertz Yagi-Uda antenna for high input resistance," *Journal of Infrared, Millimeter and Terahertz Waves*, vol. 18, pp. 441–454, 2010.

[26] J. Huang, and J. A. Encinar, *Reflectarray Antennas*, New York: Wiley, 2008.

[27] P. Nayeri, F. Yang, and A. Z. Elsherbeni, "Beam-scanning reflectarray antennas: A technical overview and state of the art," *IEEE Antennas and Propagation Magazine*, vol. 57, no. 4, pp. 32–47, 2015.

[28] P. Nayeri et al., "3D printed dielectric reflectarrays: Low-cost high gain antennas at sub-millimeter waves," *IEEE Transactions on Antennas and Propagation*, vol. 62, no. 4, pp. 2000–2008, 2014.

[29] J. C. Ginn, B. A. Lail, and G. D. Boreman, "Phase characterization of reflectarray elements at infrared," *IEEE Transactions on Antennas and Propagation*, vol. 55, no. 11, pp. 2989–2993, 2007.
[30] F. Yang et al., "Reflectarray design at infrared frequencies: Effects and models of material loss," *IEEE Transactions on Antennas and Propagation,* vol. 60, no. 9, pp. 4202–4209, 2012.
[31] A. Tamminen, S. Makela, J. Ala-Laurinaho, J. Hakli, P. Koivisto, P. Rantakari, J. Saily, A. Luukanen, A. V. Raisanen, "Reflectarray design for 120-GHz radar application: Measurement results," *IEEE Transactions on Antennas and Propagation*, vol. 61, no. 10, pp. 5036–5047, 2013.
[32] J. C. Ginn, B. A. Lail, G. D. Boreman, "Phase characterization of reflectarray elements at infrared," *IEEE Transactions on Antennas and Propagation*, vol. 55, no. 11, pp. 2989–2993, 2007.
[33] F. Yang, P. Nayeri, A. Z. Elsherbeni, J. C. Ginn, D. Shelton, G. Boreman, Y. Rahmat-Samii, "Reflectarray design at infrared frequency: Effects and models of material loss," *IEEE Transactions on Antennas and Propagation*, vol. 60, no. 9, pp. 4202–4209, 2012.
[34] T. Niu, W. Withayachumnankul, B. SY. Ung, H. Menekse, M. Bhaskaran, S. Sriram, C. Fumeaux, "Experimental demonstration of reflectarray antennas at terahertz frequencies," *Optics Express*, vol. 21, no. 3, pp. 2875–2889, 2013.
[35] M. D. Wu, B. Li, Y. Zhou, D. L. Guo, Y. Liu, F. Wei, and X. Lv, "Design and measurement of a 220 GHz wideband 3-D printed dielectric reflectarray," *IEEE Antennas and Wireless Propagation Letters*, vol. 17, no. 11, pp. 2094_2098, Nov. 2018.
[36] P. Nayeri, M. Liang, R. A. Sabory-Garcia, M. Tuo, F. Yang, M. Gehm, H. Xin, and A. Z. Elsherbeni, "3D printed dielectric reflectarrays: Lowcost high-gain antennas at sub-millimeter waves," *IEEE Transactions on Antennas and Propagation*, vol. 62, no. 4, pp. 2000–2008, Apr. 2014.
[37] N. Llombart, G. Chattopadhyay, A. Skalare, and I. Mehdi, "Novel terahertz antenna based on a silicon lens fed by a leaky wave enhanced waveguide," *IEEE Transactions on Antennas and Propagation*, vol. 59, no. 6, pp. 2160–2168, 2011.
[38] B. Jalali, D. Solli, and S. Gupta, "Silicon Photonics: Silicon's time lens," *Nature Photonics*, vol. 3, no. 1, pp. 8–10, 2009.
[39] D. F. Filipovic, S. S. Gearhart, and G. M. Rebeiz, "Double-slot antennas on extended hemispherical and elliptical silicon dielectric lenses," *IEEE Transactions on Microwave Theory and Techniques*, vol. 41, no. 10, pp. 1738–1749, 1993.
[40] D. F. Filipovic, G. P. Gauthier, S. Raman, and G. M. Rebeiz, "Off-axis properties of silicon and quartz dielectric lens antennas," *IEEE Transactions on Antennas and Propagation*, vol. 45, no. 5, pp. 760–766, 1997.
[41] A. P. Pavacic, D. L. del Rio, J. R. Mosig, and G. V. Eleftheriades, "Three dimensional ray-tracing to model internal reflections in off axis lens antennas," *IEEE Transactions on Antennas and Propagation*, vol. 54, no. 2, pp. 604–612, 2006.
[42] N. Zhu and R. W. Ziolkowski, "Photoconductive THz antenna designs with high radiation efficiency, high directivity, and high aperture efficiency," *IEEE Transactions on Terahertz Science and Technology*, vol. 3, no. 6, pp. 721–730, 2013.
[43] H. Yi, S. W. Qu, K. B. Ng, C. H. Chan, and X. Bai, "3-D printed millimeter-wave and terahertz lenses with fixed and frequency scanned beam," *IEEE Transactions on Antennas and Propagation*, vol. 64, no. 2, pp. 442–449, 2016.
[44] T. Tajima, H. J. Song, K. Ajito, M. Yaita, and N. Kukutsu, "300-GHz step-profiled corrugated horn antennas integrated in LTCC," *IEEE Transactions on Antennas and Propagation*, vol. 62, no. 11, pp. 5437–5444, Nov. 2014.
[45] B. Zhang, Z. Zhan, Y. Cao, H. Gulan, P. Linner, J. Sun, T. Zwick, and H. Zirath, "Metallic 3-D printed antennas for millimeter- and submillimeter wave applications," *IEEE Transactions on Terahertz Science and Technology*, vol. 6, no. 4, pp. 592–600, Jul. 2016.
[46] P. Kittara, A. Jiralucksanawong, G. Yassin, S. Wangsuya, and J. Leech, "The design of potter horns for THz applications using a genetic algorithm," *International Journal of Infrared and Millimeter Waves volume*, vol. 28, no. 12, pp. 1103–1114, 2007.

[47] N. Chahat, T. J. Reck, Cecile Jung-Kubiak, T. Nguyen, R. Sauleau, and Goutam Chattopadhyay, "1.9 THz multiflare angle horn optimization for space instruments," *IEEE Transactions on Terahertz Science and Technology* vol. 5, no. 6, pp. 914–921, 2015.

[48] E. Peytavit, J. F. Lampin, T. Akalin, and L. Desplanque, "Integrated terahertz TEM horn antenna," *Electronics Letters*, vol. 43, no. 2, pp. 73–75, 2007.

[49] P. L. McEuen, M. S. Fuhrer, and H. K. Park, "Single-walled carbon nanotube electronics," *IEEE Transactions on Nanotechnology*, vol. 1, pp. 78–85, 2002.

[50] G. W. Hanson, "Current on an infinitely-long carbon nanotube antenna excited by a gap generator," *IEEE Transactions on Antennas and Propagation*, pp. 76–81, 2006.

[51] M. Zhao, M. Yu, and H. Robert Blick, "Wavenumber-domain theory of terahertz single-walled carbon nanotube antenna," *IEEE Journal of Selected Topics in Quantum Electronics*, vol. 18, no. 1, pp. 166–175, 2012.

[52] P. J. Burke, S. Li, and Z. Yu, "Quantitative theory of nanowire and nanotube antenna performance," *IEEE Transactions on Nanotechnology*, vol. 5, pp. 314–334, 2006.

[53] G. W. Hanson, "Fundamental transmitting properties of carbon nanotube antennas," *IEEE Transactions on Antennas and Propagation*, vol. 53, pp. 3426–3435, 2005.

[54] M. V. Shuba, S. A. Maksimenko, A. Lakhtakia, "EM wave propagation in an almost circular bundle of closely packed metallic carbon nanotubes," *Physical Review B*, vol. 76, no. 15, pp. 155407, 2007.

[55] J. Hao, G. W. Hanson, "Infrared and optical properties of carbon nanotube dipole antennas," *IEEE Transactions on Nanotechnology,* vol. 5, no. 6, pp. 766–774, 2006.

[56] G. Y. Slepyan, S. A. Maksimenko, A. Lakhtakia, O. Yevtushenko, and A. V. Gusakov, "Electrodynamics of carbon nanotubes: Dynamic conductivity, impedance boundary conditions, and surface wave propagation," *Physical Review B*, vol. 60, pp. 17136–17149, 1999.

[57] Y. Huang, W.-Y. Yin, and Q. H. Liu, "Performance prediction of carbon nanotube bundle dipole antennas," *IEEE Transactions on Nanotechnology*, vol. 7, no. 3, pp. 331–337, 2008.

[58] M. V. Shuba, S. A. Maksimenko, and A. Lakhtakia, "EM wave propagation in an almost circular bundle of closely packed metallic carbon nanotubes," *Physical Review B*, vol. 76, pp. 155407–155415, 2007.

[59] S. F. Mahmoud and A. R. Alajmi, "Characteristics of a new carbon nanotube antenna structure with enhanced radiation in the sub-terahertz range," *IEEE Transactions on Nanotechnology*, vol. 11, no. 3, pp. 640–646, 2012.

[60] A. R. Alajmi and S. F. Mahmoud, "Investigation of multiwall carbon nanotubes as antennas in the sub terahertz range," *IEEE Transactions on Nanotechnology*, vol. 13, no. 2, pp. 268–273, 2014.

[61] K. I. Bolotin, K. J. Sikes, Z. Jiang, M. Klima, G. Fudenberg, J. Hone, P. Kim, and H. L. Stormer, "Ultrahigh electron mobility in suspended graphene," *Solid State Communications*, vol. 146, pp. 351–355, 2008.

[62] J. Yu, G. Liu, A. V. Sumant, V. Goyal, A. A. Balandin, "Graphene-on-diamond devices with increased current-carrying capacity: carbon sp2-on-sp3 technology," *Nano Letters*, vol. 12, pp. 1603–1608, 2012.

[63] S. M. Farzaneh and S. Rakheja, "Voltage tunable plasmon propagation in dual gated bilayer graphene," *Journal of Applied Physics*, vol. 122 (15), pp. 153101, 2017.

[64] S. Rakheja, and P. Sengupta, "Gate–voltage tunability of plasmons in single-layer graphene structures– Analytical description, impact of interface states, and concepts for terahertz devices," *IEEE Transactions on Nanotechnology*, vol. 15(1), pp. 113–121, 2016.

[65] M. Tamagnone, J. S. Gómez-Díaz, J. R. Mosig, and J. Perruisseau-Carrier, Analysis and design of terahertz antennas based on plasmonic resonant graphene sheets," *Journal of Applied Physics*, vol. 112, pp. 114915, 2012.

[66] S. Dash, and A. Patnaik, "Material selection for THz antennas," *Microwave and Optical Technology Letters*, vol. 60, pp. 1183–1187, 2018.

[67] S. Dash, and A. Patnaik, "Performance of graphene plasmonic antenna in comparison with their counterparts for low-terahertz applications," *Plasmonics*, vol. 13, no. 6, pp. 2353–2360, 2018.

[68] J. M. Jornet and I. F. Akyildiz, "Graphene-based plasmonic nano-antenna for terahertz band communication in nanonetworks," *IEEE Journal on Selected Areas in Communications Part 2*, vol. 31, No. 12, pp. 685–694, 2013.
[69] S. Dash, A. Patnaik, and B. K. Kaushik, "Performance enhancement of graphene plasmonic nanoantenna for THz communication," *IET Microwaves, Antennas and Propagation*, vol. 13, no. 1, pp. 71–75, Jan 2019.
[70] S. Dash, and A. Patnaik, "Dual-band reconfigurable plasmonic antenna using bilayer graphene," Proc. of IEEE International Symposium on Antennas and Propagation AP-S 2017, San Diego, California, USA, pp. 921–922, July 9–14, 2017.
[71] M. Tamagnone, J. S. Gómez-Díaz, J. R. Mosig, and J. Perruisseau-Carrier, "Reconfigurable THz plasmonic antenna concept using a graphene stack," *Applied Physics Letters*, vol. 101, pp. 214102, 2012.
[72] S. Dash and A. Patnaik, "Sub-wavelength Graphene Planar nanoantenna for THz Application," *Materials Today: Proceedings*, vol. 18, Part 3, pp. 1336–1341, 2019.
[73] S. Prakash, S. Dash, and A. Patnaik, "Reconfigurable circular patch THz antenna using graphene stack based SIW technique," 2018 IEEE Indian Conference on Antennas and Propagation, Hyderabad, India, pp. 1–3, 2018.
[74] E. Carrasco and J. Perruisseau-Carrier, "Tunable graphene reflective cells for THz reflectarray and generalized law of reflection," *Applied Physics Letters*, vol. 102, pp. 104–103, 2013.
[75] E. Carrasco and J. Perruisseau-Carrier, "Reflectarray antenna at terahertz using graphene," *IEEE Antennas and Wireless Propagation Letters*, vol. 12, pp. 253–256, 2013.
[76] H. A. Malhat, S. H. Zainud-Deen, and S. M. Gaber, "Graphene based transmitarray for terahertz applications," *Progress In Electromagnetics Research M*, vol. 36, pp. 185–191, 2014.
[77] Z. Xu, X. Dong, and J. Bornemann, "Design of a reconfigurable MIMO system for THz communications based on graphene antennas," *IEEE Transactions on Terahertz Science and Technology*, vol. 4, pp. 609–617, 2014.
[78] S. Dash, and A. Patnaik, "Grapheneplasmonic bowtie antenna for UWB THz application," IEEE 24th National Conference on Communications, Hyderabad, India, pp. 1–4, 2018. DOI: 10.1109/NCC.2018.8599940
[79] V. Gusynin, S. Sharapov, and J. Carbotte, "Magneto-optical conductivity in graphene," *Journal of Physics: Condensed Matter*, vol. 19, pp. 026222(1–28), 2006.
[80] W. L. Barnes, A. Dereux, and T. W. Ebbesen, "Surface plasmon sub wavelength optics," *Nature*, vol. 424, pp. 824–830, 2003.
[81] Q. Lu, D. Wu, S. Sengupta, S. Slivken, and M. Razeghi, "Room temperature continuous wave, monolithic tunable THz sources based on highly efficient mid-infrared quantum cascade lasers," *Nature Scientific Reports*, vol. 6, p. 23595, 2016.
[82] B. S. Williams, "Terahertz quantum-cascade lasers," *Nature Photonics*, vol. 1, pp. 517–525, 2007.
[83] Y. Jiang, K. Vijayraghavan, S. Jung, A. Jiang J. H. Kim, F. Demmerle, G. Boehm, M. C. Amann, and M. A. Belkina, "Spectroscopic study of terahertz generation in mid-infrared quantum cascade lasers," *Nature Scientific Reports*, vol. 6, p. 21169, 2016.
[84] L. Li, L. Chen, J. Zhu, J. Freeman, P. Dean, A. Valavanis, A. G. Davies, and E. H. Linfield, "Terahertz quantum cascade lasers with >1 W output powers," *Electronics Letters*, vol. 50, no. 4, pp. 309–311, 2014.
[85] Y. Irimajiri, M. Kumagai, I. Morohashi, A. Kawakami, S. Nagano, N. Sekine, S. Ochiai, S. Tanaka, Y. Hanado, Y. Uzawa, and I. Hosako, "Precise evaluation of a phase-locked THz quantum cascade laser," *IEEE Transactions on Terahertz Science and Technology*, vol. 6, no. 1, pp. 115–120, 2016.
[86] S. Khanal, L. Gao, L. Zhao, J. L. Reno, and S. Kumar, "High-temperature operation of broadband bidirectional terahertz quantum-cascade lasers," *Nature Scientific Reports*, vol. 6, 32978, 2016.
[87] A. Albo and Y. V. Flores, "Temperature-driven enhancement of the stimulated emission rate in terahertz quantum cascade lasers," *IEEE Journal of Quantum Electronics*, vol. 53, no. 1, pp. 1–5, 2017.

[88] M. Kim, J. Lee, J. Lee, and K. Yang, "A 675 GHz differential oscillator based on a resonant tunneling diode," *IEEE Transactions on Terahertz Science and Technology*, vol. 6, no. 3, pp. 512–516, 2016.

[89] T. Maekawa, H. Kanaya, S. Suzuki, and M. Asada, "Oscillation up to 1.92 THz in resonant tunneling diode by reduced conduction loss," *Applied Physics Express*, vol. 9, no. 2, 024101, 2016.

[90] M. Asada, and S. Suzuki, "Room-temperature oscillation of resonant tunneling diodes close to 2 THz and their functions for various applications," *Journal of Infrared, Millimeter, and Terahertz Waves*, vol. 137, no. 12, pp. 1185–1198, 2016.

[91] S. Kitagawa, S. Suzuki, and M. Asada, "650-GHz resonant-tunneling- diode VCO with wide tuning range using varactor diode," *IEEE Electron Device Letters*, vol. 35, no. 12, pp. 1215–1217, 2014.

[92] T. Maekawa, H. Kanaya, S. Suzuki, and M. Asada, "Frequency increase in terahertz oscillation of resonant tunnelling diode up to 1.55 THz by reduced slot-antenna length," *Electronics Letters*, vol. 50, no. 17, pp. 1214–1216, 2014.

[93] Shang-Hua Yang, M. R. Hashemi, C. W. Berry, and M. Jarrahi, "7.5%Optical-to-terahertz conversion efficiency offered by photoconductive emitters with three-dimensional plasmonic contact electrodes," *IEEE Transactions on Terahertz Science and Technology*, vol. 4, no. 5, p. 57581, 2014.

[94] S. Jameson and E. Socher, "A 0.3 THz radiating active x27 frequency multiplier chain with 1 mW radiated power in CMOS 65-nm," *IEEE Transactions on Terahertz Science and Technology*, vol. 5, no. 4, pp. 645–648, 2015.

[95] A. Maestrini, I. Mehdi, J. V. Siles, J. S. Ward, R. Lin, B. Thomas, C. Lee, J. Gill, G. Chattopadhyay, J. Pearson, and P. Siegel, "Design and characterization of a room temperature all-solid-state electronic source tunable from 2.48 to 2.75 THz," *IEEE Transactions on Terahertz Science and Technology*, vol. 2, no. 2, pp. 177–185, 2012.

[96] S. Jameson and E. Socher, "A 0.3 THz radiating active x27 frequency multiplier chain with 1 mW radiated power in CMOS 65-nm," *IEEE Transactions on Terahertz Science and Technology*, vol. 5, no. 4, pp. 645–648, 2015.

[97] Y. Yang, O. D. Gurbuz, G. M. Rebeiz, "An eight-element 370–410-GHz phased-array transmitter in 45-nm CMOS SOI with peak EIRP of 8–8.5 dBm," *IEEE Transactions on Microwave Theory and Techniques*, vol. 64, no. 12, pp. 4241–4249, 2016.

[98] J. Grzyb, Y. Zhao, and U. R. Pfeiffer, "A 288-GHz lens-integrated balanced triple-push source in a 65-nm CMOS technology," *IEEE Journal of Solid-State Circuits*, vol. 48, no. 7, pp. 1751–1761, 2013.

[99] M. Adnan, and E. Afshari, "14.8 A 247-to-263.5GHz VCO with 2.6mW peak output power and 1.14% DC-to-RF efficiency in 65nm Bulk CMOS," IEEE International Solid-State Circuits Conference (ISSCC), pp. 262–263, 2014.

[100] T. Chi, J. Luo, S. Hu, and H. Wang, "A multi-phase sub-harmonic injection locking technique for bandwidth extension in silicon-based THz signal generation," *IEEE Journal of Solid-State Circuits*, vol. 50, no. 8, pp. 1861–1873, 2015.

[101] J. Zhao, Z. Zhu, W. Cui, K. Xu, B. Zhang, D. Ye, C. Li, and L. Ran, "Power synthesis at 110-GHz frequency based on discrete sources," *IEEE Transactions on Microwave Theory and Techniques*, vol. 63, no. 55, pp. 1633–1644, 2015.

[102] J. Yun, J. Kim, and J.-Sung Rieh, "A 280-GHz 10-dBm signal source based on InP HBT technology," *IEEE Microwave and Wireless Components Letters*, vol. 27, no. 2, pp. 159–161, 2017.

[103] M. Seo, M. Urteaga, J. Hacker, A. Young, Z. Griffith, V. Jain, R. Pierson, P. Rowell, A. Skalare, A. Peralta, R. Li, D. Lin, and M. Rodwell, "InP HBT IC technology for terahertz frequencies: fundamental oscillators up to 0.57 THz," *IEEE Journal of Solid-State Circuits*, vol. 46, no. 10, pp. 2203–2214, 2011.

[104] H.-C. Lin, and G. M. Rebeiz, "A SiGe multiplier array with output power of 5–8 dBm at 200–230 GHz," *IEEE Transactions on Microwave Theory and Techniques*, vol. 64, no. 7, pp. 2050–2058, 2016.

[105] H. Aghasi, A. Cathelin and E. Afshari, "A 0.92-THz SiGe power radiator based on a nonlinear theory for harmonic generation," *IEEE Journal of Solid-State Circuits*, vol. 52, no. 2, pp. 406–422, 2017.

[106] J. Yun, D. Yoon, S. Jung, M. Kaynak, Bernd Tillack, and J. S. Rieh, "Two 320 GHz signal sources based on SiGe HBT technology," *IEEE Microwave and Wireless Components Letters*, vol. 25, no. 3, pp. 178–180, 2015.

[107] J. Yun, D. Yoon, H. Kim, and J. S. Rieh, "300-GHz InP HBT oscillators based on common-base cross-coupled topology," *IEEE Transactions on Microwave Theory and Techniques*, vol. 62, no. 12, pp. 3053–3064, 2014.

[108] R. Piesiewicz, T. Kleine-Ostmann, N. Krumbholz, D. Mittleman, M. Koch, J. Schoebel, and T. Kurner, "Short-range ultra-broadband terahertz communications: Concepts and perspectives," *IEEE Antennas and Propagation Magazine*, vol. 49, no. 6, pp. 24–39, 2007.

[109] K. C. Huang, and Z. Wang, "Terahertz terabit wireless communication," *IEEE Microwave Magazine*, vol. 12(4), pp. 108–116, 2011.

[110] I. Akyildiz, J. Jornet, and C. Han, "TeraNets: Ultra-broadband communication networks in the terahertz band," *IEEE Wireless Communications*, vol. 21, no. 4, pp. 130–135, Aug. 2014.

[111] J. M. Jornet and I. F. Akyildiz, "Channel modeling and capacity analysis for EM wireless nanonetworks in the terahertz band," *IEEE Transactions on Wireless Commun*ications, vol. 10, no. 10, pp. 3211–3221, Oct. 2011.

[112] C. Jansen, S. Priebe, C. Moller, M. Jacob, H. Dierke, M. Koch, and T. Kurner, "Diffuse scattering from rough surfaces in THz communication channels," *IEEE Transactions on Terahertz Science and Technology*, vol. 1, no. 2, pp. 462–472, Nov. 2011.

[113] R. Piesiewicz, C. Jansen, D. Mittleman, T. K. Ostmann, M. Koch, and T. Kurner, "Scattering analysis for the modeling of THz communication systems," *IEEE Transactions on Antennas and Propagation*, vol. 55, no. 11, pp. 3002–3009, Nov. 2007.

[114] P. Beckmann and A. Spizzichino, *The Scattering of EM Waves from Rough Surfaces*. New York: Macmillan, 1963.

[115] S. Y. Jun, B. Sanz-Izquierdo, E. A. Parker, D. Bird, and A. McClelland, "Manufacturing considerations in the 3-D printing of fractal antennas," *IEEE Transactions on Components, Packaging and Manufacturing Technology*, vol. 7, no. 11, pp. 1891–1898, Nov. 2017.

[116] R. Sorrentino and O. A. Peverini, "Additive manufacturing: A key enabling technology for next-generation microwave and millimeter-wave systems [point of view]," *Proceedings of the IEEE*, vol. 104, no. 7, pp. 1362–1366, Jul. 2016.

[117] E. A. Rojas-Nastrucci, J. T. Nussbaum, N. B. Crane, and T. M. Weller, "Ka-band characterization of binder jetting for 3-D printing of metallic rectangular waveguide circuits and antennas," *IEEE Transactions on Microwave Theory and Techniques*, vol. 65, no. 9, pp. 3099–3108, Sep. 2017.

[118] S. Verploegh, M. Coffey, E. Grossman, and Z. Popovic, "Properties of 50–110 GHz waveguide components fabricated by metal additive manufacturing," *IEEE Transactions on Microwave Theory and Techniques*, vol. 65, no. 12, pp. 5144–5153, Dec. 2017.

[119] M. C. Lemme, T. J. Echtermeyer, M. Baus, and H. Kurz, "A graphene field effect device," *IEEE Electron Device Letters*, vol. 28, no. 4, pp. 282–284, Apr. 2007.

[120] W. Gannett, W. Regan, K. Watanabe, T. Taniguchi, M. F. Crommie, and A. Zettl, "Boron nitride substrates for high mobility chemical vapor deposited graphene," *Applied Physics Letters*, vol. 98, no. 24, pp. 99–102, 2011.

[121] A. Woessner et al., "Highly confined low-loss plasmons in graphene–boron nitride heterostructures," *Nature Materials.*, vol. 14, no. 4, pp. 421–425, 2014.

[122] X. Huang, T. Leng, X. Zhang et al., "Binder-free highly conductive graphene laminate for low cost printed RF applications," *Applied Physics Letters*, vol. 106, no. 20, Article ID 203105, 2015.

[123] M. Akbari, J. Virkki, L. Sydanheimo, and L. Ukkonen, "Toward graphene-based passive UHF RFID textile tags: A reliability study," *IEEE Transactions on Device and Materials Reliability*, vol. 16, no. 3, pp. 429–431, 2016.

[124] M. Akbari, M. W. A. Khan, M. Hasani, T. Björninen, L. Sydänheimo, and L. Ukkonen, "Fabrication and characterization of graphene antenna for low-cost and environmentally friendly RFID tags," in *IEEE Antennas and Wireless Propagation Letters*, vol. 15, pp. 1569–1572, 2016.

[125] H. He, M. Akbari, L. Sydänheimo, L. Ukkonen, and J. Virkki, "3D-printed graphene antennas and interconnections for textile RFID tags: Fabrication and reliability towards humidity," *International Journal of Antennas and Propagation*, vol. 1386017, pp. 1–5, 2017.

[126] S. D. Keller, A. I. Zaghloul, V. Shanov, M. J. Schulz, D. B. Mast, and N. T. Alvarez, "Radiation performance of polarization selective carbon nanotube sheet patch antennas," *IEEE Transactions on Antennas and Propagation*, vol. 62, no. 1, pp. 48–55, Jan. 2014.

[127] M. Kubo, X. Li, C. Kim, M. Hashimoto, B. J. Wiley, D. Ham, and G. M. Whitesides, "Stretchable microfluidic radiofrequency antennas." *Advanced Materials*, 22, pp. 2749–2752, 2010.

[128] H. Qiu, H. Liu, X. Jia, X. Liu, Y. Li, T. Jiang, B. Xiong, Y. Yangab and T. Ren, Ink-injected dual-band antennas based on graphene flakes, carbon nanotubes, and silver nanowires, *RSC Advances*, vol. 8, pp. 37534, 2018.

Chapter 8

Antenna Misalignment and Blockage in THz Communications

Alexandros-Apostolos A. Boulogeorgos and Angeliki Alexiou

Contents

8.1 Introduction

By taking into account that the used frequency spectrum for the fifth generation (5G) applications has limited capacity (Boulogeorgos, Chatzidiamantis, and Karagiannidis, 2016; Boulogeorgos and Karagiannidis, 2017; Cherry, 2004; Staple and Werbach, 2004), terahertz (THz) wireless systems were considered a promising complementing technology to the expensive optical-fiber systems (Petrov, Pyattaev, Moltchanov, and Koucheryavy, 2016) and to shorter range setups (Huq, Jornet, Gerstacker, and et al., 2018; Piesiewicz, Kleine-Ostmann, Krumbholz, and et al., 2007). The THz wireless systems aim to deliver the fiber quality of experience (QoE) in the wireless world (Akyildiz, Jornet, and Han, 2014; Boulogeorgos, Alexiou, Merkle, and et al., 2018; Federici and Moeller, 2010; Seeds, Shams, Fice, and Renaud, 2015; Song and Nagatsuma, 2011). In other words, as illustrated in Figure 8.1, THz wireless system development is expected to address both the spectrum scarcity and capacity limitations of the current cellular systems as well as to create the opportunity for commercializing a plethora of life-changing technologies, including device-to-device (D2D) communications, Internet of vehicles (IoV), backhaul, small/femto-cell access, nomadic connectivity, as well as applications, such as augmented and virtual reality, and virtual presence (Machado, Boulogeorgos, Escher, and et al., 2019; Salgado, Silva, Elschner, Boulogeorgos, Katsiotis, Kritharidis, Alexiou, Kokkoniemi, and Lehtomaki, 2017; Zhang, Ota, Jia, and Dong, 2018). As a consequence, they will influence the main technology trends in the wireless world in the decades to follow (Boulogeorgos, Alexiou, Kritharidis, and et al., 2018; Koch, 2007).

The propagation environment of THz systems suffers from sparse scattering (Boulogeorgos et al., 2018; Kokkoniemi, Boulogeorgos, Lehtomaki, Ntouni, Juntti, and Alexiou, 2018; Papasotiriou et al., 2018). This causes to the main signal directions of arrivals (DoAs) to be below the noise floor. Thus and in order to achieve a sufficient coverage, THz wireless links should be established in a directional manner (Boulogeorgos, Papasotiriou, Kokkoniemi, and et al., 2018). In this direction, high-gain antennas, with low beamwidth, are employed in both the transceiver (TX) and receiver (RX). Directional THz wireless links demands *perfectly aligned TX and RX antennas*. However, in practice, this may not be possible (Boulogeorgos and Alexiou, 2019a; Boulogeorgos, Papasotiriou, and Alexiou, 2019; Wildman, Nardelli, Latva-aho, and Weber, 2014). Moreover, THz and sub-THz wireless links experience high penetration losses and reduced diffraction (Bai and Heath, 2015; Boulogeorgos, Papasotiriou, and Alexiou, 2018; Jain, Kumar, and Panwar, 2018). In particular, even human body can decrease the signal power by approximately 20–40 dB (Boulogeorgos, Goudos, and Alexiou, 2018). In other words, THz wireless systems are sensitive to blockages. To sum up, although THz wireless systems can open the possibility of unprecedented performances, we first have to deal with two limiting factors, namely, antenna misalignment and blockage.

Usually, in the technical literature, the transceivers' antennas are considered to be static and the technical challenge is in acquiring the appropriate direction that maximizes the received signal strength (see, for example, Ekti, Boyaci, Alparslan, and et al., 2017; Petrov et al., 2016, and references therein). However, in realistic scenarios, the antennas may not be static even in backhaul or fronthaul applications because of (1) environmental impacts, such as small earthquakes and winds (Boulogeorgos and Alexiou, 2019a; Boulogeorgos et al., 2019) or (2) stochastic tracking estimation error, which results in estimation errors in the DoA or angle of departure (Lee, Liang, Kim, and Park, 2017; Zhang, Ge, Li, Guizani, and Zhang, 2017); or (3) antenna array imperfections that include mutual coupling and array perturbation (Pradhan, Li, Zhuo, and et al., 2019). This is not a major problem at lower frequencies due to the use of large beamwidth antennas, that is, low antenna gains. Likewise, THz bands experience very large path loss and therefore maintaining sufficient link conditions require large antenna gains or equivalently antennas with extremely low beamwidth at both the TX and RX. As a consequence, antenna misalignment is expected to influence the design of THz systems. Modeling antenna misalignment, quantifying its effect on directional communications, and proposing mitigation solution

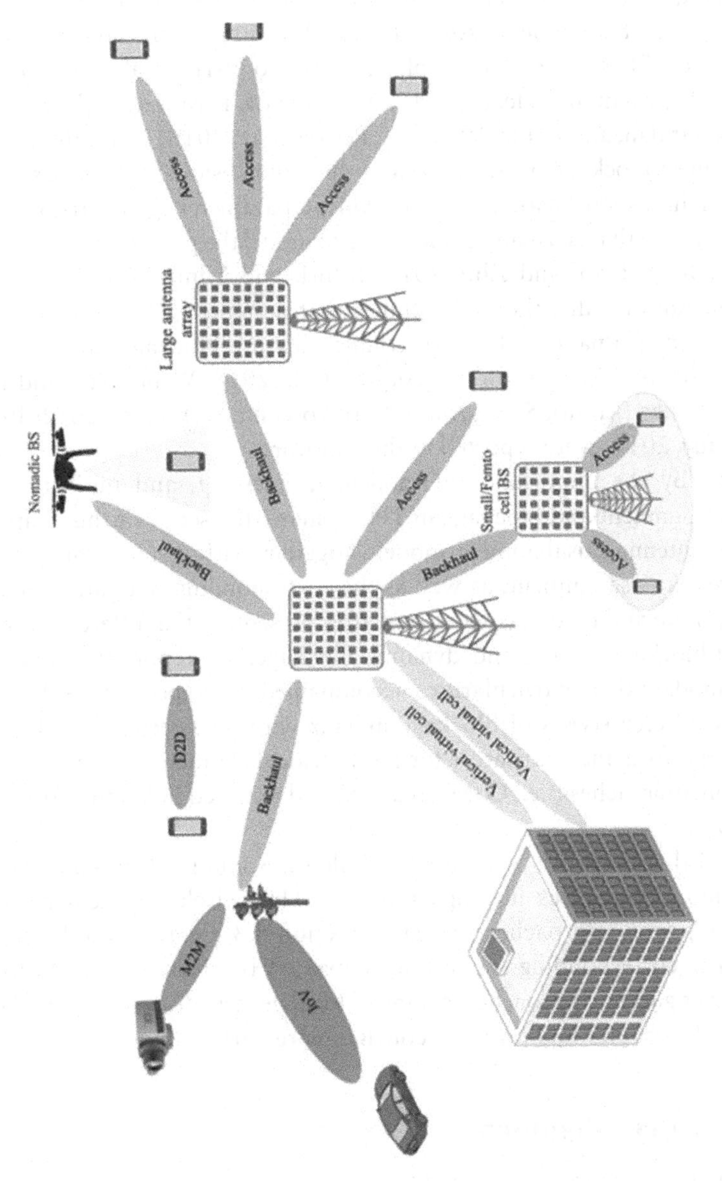

Figure 8.1 THz massive MIMO applications in the beyond 5G era.

become hot topics in the technical literature (see, for example, Boulogeorgos et al., 2019; Boulogeorgos, Papasotiriou, and Alexiou, 2019; Boulogeorgos et al., 2018; Kokkoniemi et al., 2019, and references therein).

Regarding blockage, there have been several published works in the technical literature, modeling its effect THz wireless systems, and proposing mitigation schemes. The use of stochastic geometry is the commonly used approach to model and assess the impact of blockage in a THz wireless system. In this direction, most authors employed Poisson point processes (PPPs) (see for example, Gapeyenko 2016, et al., 2017; Han, Wang, and Schotten, 2017; Khan and Heath, 2017; Moltchanov, Kustarev, and Koucheryavy, 2018; Niknam, Natarajan, and Barazideh, 2018; Petrov et al., 2016) to accommodate the impact of human/object blockage in two (2D) and three dimensional (3D) cases (considering the blocker's height, as stochastic or a deterministic parameter) geometrical representations. Another approach that is usually employed is to model the position of the blockers through a uniform (Dong, Liao, and Zhu, 2012; Erturk and Yilmaz, 2018), or an exponential distribution. Additionally, there are studies that assume a deterministic blockage probability (Lin and Weitnauer, 2014). To countermeasure blockage, the use of coordinated multipoint (CoMP), relaying (Boulogeorgos et al., 2018; Wang, Niu, and H. Wu, 2019), and reflected links (Monti, Soric, Alu, Bilotti, Toscano, and Vegni, 2012; Renzo, Debbah, and Phan-Huy, 2019) were reported in the literature.

Motivated by the importance of modeling, assessing, and mitigating the effect of antenna misalignment and blockage in THz systems, the scope of this chapter is to report appropriate antenna misalignment models, together with results that quantify the influence of antenna misalignment, as well as the corresponding mitigation approaches, such as beam-tracking and relaying. Moreover, after presenting the different types of blockage, namely self-blockage, static, and dynamic blockage, we report theoretical models that can accommodate their particularities accompanied by respective results that reveal the influence of different types of blockage in THz wireless systems. Finally, we deliver the appropriate performance indicators for the statistical characterization of blockage as well as mitigation approaches (i.e., CoMP, relaying and reflected links) that are currently under investigation.

The rest of this chapter is structured as follows: Section 8.2 presents the antenna misalignment models, assesses its impact on the THz wireless system performance, and discusses mitigation approaches. Similarly, Section 8.3 identifies the blockage types and provides their corresponding models accompanied by their statistical characterization. Moreover, mitigation approaches against blockage are discussed. Finally, Section 8.4 summarizes the chapter and provides concluding remarks.

8.2 Antenna Misalignment

As illustrated in Figure 8.2, antenna misalignment may occur in both backhaul (Boulogeorgos et al., 2019) and fronthaul (Bai and Heath, 2015) wireless THz links. In the former case, the main cause of antenna misalignment is wind, small earthquakes, and other environmental phenomenon, while, in the latter case, it is the result of tracking estimation errors and antenna array imperfections. Next, we present a number of different antenna misalignment models, identify the advantages and disadvantages as well as the suitability of each model for each use case, and assess the influence of antenna misalignment on the

Figure 8.2 Antenna misalignment in backhaul (a) and fronthaul (b) application scenarios.

THz wireless system performance. Finally, we report and explain the most commonly used antenna misalignment mitigation approaches.

8.2.1 Antenna Misalignment Modeling and Impact Assessment in THz Wireless Systems

Section 8.2.1.1 presents the Gaussian distributed beamsteering errors model, which is widely used to model the DoA and angle of departure (AoD) estimation error, accompanied by novel closed-form expressions that quantify the performance degradation, due to this type of antenna misalignment, in terms of the expected value of the total directional gain. Moreover, Section 8.2.1.2 discusses the 2D Gaussian shaking of a single node, which is usually used to accommodate the impact of antenna shaking due to wind, small earthquakes or other environmental phenomena. Finally, in Section 8.2.1.3, the wind vibration antenna misalignment model is presented, which accommodates the effect of antenna shaking because of wind.

8.2.1.1 Gaussian Distributed Beamsteering Errors

This model was introduced in Bai and Heath (2015) and accommodates the stochastic antenna beamsteering error. In more detail, by respectively denoting the beamsteering errors of the BS and the user equipment (UE) as ϵ_z with $z \in \{B, U\}$ (see Figure 8.3), and assuming that ϵ_B and ϵ_U are independent and identical zero-mean Gaussian distributed random processes with variance σ_B and σ_U[1], respectively, their cumulative density functions (CDFs) can be expressed as (Papoulis and Pillai, 2002)

$$F_{\epsilon z}(x) = \frac{1}{2}\left(1 + \operatorname{erf}\left(\frac{x}{\sigma_z \sqrt{2}}\right)\right) \tag{8.1}$$

where $z \in \{B, U\}$ and erf $(\cdot)$ stands for the error function. Notice that this is a 2D model.

To quantify the influence of antenna misalignment in the link budget, we approximate the actual beamforming pattern based on the sectored model. Note that this model has

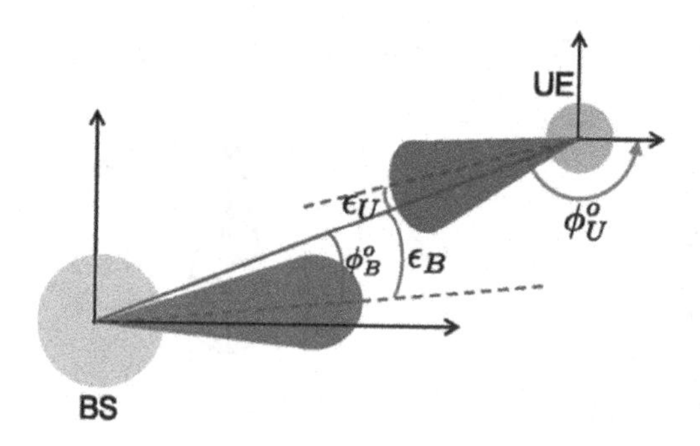

Figure 8.3 Beamsteering errors.

been also employed in several previously published contributions, including Alkhateeb et al. (2014a) due to the fact that it is able to capture the key characteristics of the actual beamforming pattern, such as the half-power beamwidth as well as the front-to-back ratio. According to this model, the antenna gain can be obtained as

$$G_z(\theta_z, \phi) = \frac{2\pi}{\theta_z}\frac{\gamma_z}{\gamma_z + 1}U\left(\frac{1}{2}\theta_z|\phi|\right) + \frac{2\pi}{2\pi - \theta_z}\frac{1}{\gamma_z + 1}U\left(\frac{1}{2}\theta_z - |\phi|\right) \tag{8.2}$$

where $U(\cdot)$ represents the unit step function, $z \in \{B, U\}$. Additionally, θ_z represents the beamwidth of the node $z \in \{B, U\}$ main lobe, φ stands for the angle of the boresight direction, and γ_z is the forward-to-backward power ratio that can be expressed as (Bai and Heath, 2015)

$$\gamma_z = \frac{2\pi}{\alpha_z(2\pi - \theta_z)} \tag{8.3}$$

with α_z being an antenna-specific constant.

From (8.2), we can straightforwardly obtain the total directional gain between the BS and the UE as

$$\mathcal{D} = G_B(\theta_B, \phi_B^o + \epsilon_B)G_U(\theta_U, \phi_U^o + \epsilon_U), \tag{8.4}$$

where $\varphi^o{}_B$ and $\varphi^o{}_U$ are the error-free boresight directions of the BS and UE, respectively.

Moreover, since ϵ_B and ϵ_U are independent, the expected value of D can be expressed as

$$\mathbb{E}[\mathcal{D}] = \mathbb{E}\left[G_B(\theta_B, \phi_B^o + \epsilon_B)\right]\mathbb{E}\left[G_U(\theta_U, \phi_U^o + \epsilon_U)\right]. \tag{8.5}$$

Note that

$$\mathbb{E}\left[G_Z\left(\theta_z, \phi_z^0 + \epsilon_z\right)\right] = \int_{-\pi}^{\pi} G_z\left(\theta_z, \phi_z^0 + \epsilon_z\right) f_{T_z}(\epsilon_z)\, d\phi, \tag{8.6}$$

with $z \in \{B,U\}$ and $f_{\epsilon_z}(\epsilon_z)$ being the PDF of ε_z, which, with the aid of (8.1), can be evaluated as

$$f_{\epsilon_z}(x) = \frac{\mathrm{d}F_{\epsilon_z}(x)}{dx}. \tag{8.7}$$

or equivalently

$$f_{\epsilon_z}(x) = \frac{1}{\sqrt{2\pi\sigma_z}} e^{-\frac{x^2}{2\sigma_z^2}} \tag{8.8}$$

By substituting (8.2) and (8.8) into (8.6), we can derive

$$\begin{aligned}\mathbb{E}\left[G_z\left(\theta_z, \phi_z^o + \epsilon_z\right)\right] = &\frac{2\pi}{\theta_z}\frac{\gamma_z}{\gamma_z+1}\int_{-\frac{1}{2}\theta_z}^{\frac{1}{2}\theta_z}\frac{e^{-\frac{x^2}{2\sigma_z^2}}}{\sqrt{2\pi\sigma_z}}\mathrm{dx} + \frac{2\pi}{2\pi-\theta_z}\frac{1}{\gamma_z+1}\int_{-\pi}^{-\frac{1}{2}\theta_z}\frac{e^{-\frac{x^2}{2\sigma_z^2}}}{\sqrt{2\pi\sigma_z}}\mathrm{dx} \\ &+ \frac{2\pi}{2\pi-\theta_z}\frac{1}{\gamma_z+1}\int_{\frac{1}{2}\theta_z}^{\pi}\frac{e^{-\frac{x^2}{2\sigma_z^2}}}{\sqrt{2\pi\sigma_z}}\mathrm{dx}.\end{aligned} \tag{8.9}$$

Additionally, by performing the integration, (8.9) can be rewritten as

$$\mathbb{E}\left[G_z\left(\theta_z, \phi_z^o + \epsilon_z\right)\right] = \frac{2\pi \mathrm{erf}\left(\frac{\theta_z}{2\sqrt{2}\sigma_z}\right)}{\frac{\theta_z}{\gamma_z}+\theta_z} + \frac{2\pi\gamma_z\left(\mathrm{erf}\left(\frac{\pi}{\sqrt{2}\sigma_z}\right) - \mathrm{erf}\left(\frac{\theta_z}{2\sqrt{2}\sigma_z}\right)\right)}{(1+\gamma_z)(2\pi-\theta_z)}. \tag{8.10}$$

Finally, by applying (8.10) into (8.5), we extract the expected value of the total directional gain as

$$\begin{aligned}\mathbb{E}[\mathcal{D}] = &\left(\frac{2\pi \mathrm{erf}\left(\frac{\theta_B}{2\sqrt{2}\sigma_B}\right)}{\frac{\theta_B}{\gamma_B}+\theta_B} + \frac{2\pi\gamma_B\left(\mathrm{erf}\left(\frac{\pi}{\sqrt{2}\sigma_B}\right) - \mathrm{erf}\left(\frac{\theta_B}{2\sqrt{2}\sigma_B}\right)\right)}{(1+\gamma_B)(2\pi-\theta_B)}\right) \\ &\times\left(\frac{2\pi \mathrm{erf}\left(\frac{\theta_U}{2\sqrt{2}\sigma_U}\right)}{\frac{\theta_U}{\gamma_U}+\theta_U} + \frac{2\pi\gamma_U\left(\mathrm{erf}\left(\frac{\pi}{\sqrt{2}\sigma_U}\right) - \mathrm{erf}\left(\frac{\theta_U}{2\sqrt{2}\sigma_U}\right)\right)}{(1+\gamma_U)(2\pi-\theta_U)}\right).\end{aligned} \tag{8.11}$$

From (8.11), it is evident that the expected value of the total directional gain depends on the type of the BS and UE antennas, their beamwidths, as well as σ_B and σ_U.

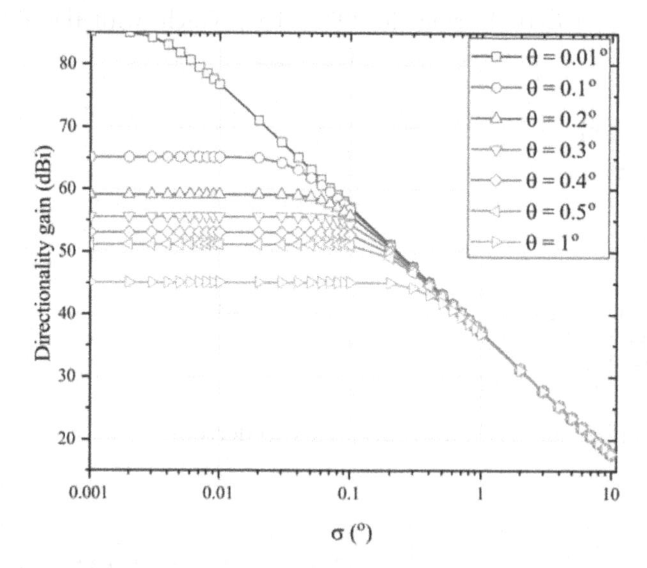

Figure 8.4 Directional gain vs angular misalignment standard deviation for different values of antenna beamwidth.

In Figure 8.4, the expected value of the total directional antenna gain is plotted against the angular misalignment standard deviation, σ, for different values of antenna beamwidth, θ. Of note, in this figure, it is considered that $\sigma_B = \sigma_U = \sigma$ and $\theta_B = \theta_U = \theta$. Moreover, $a = a_B = a_U = 1$. As expected, for a given $\theta > 0.1\sigma$, where the beamsteering error falls outside the half-power beamwidth, as σ increases, the expected value of the total directional antenna gain decreases, whereas, for $\theta < 0.1\sigma$, the beamsteering error falls within the half-power beamwidth of both the transceivers antennas; hence, there is no effect on the system performance. Likewise, we see that as θ increases, the expected value of the total directional gain decreases; however, its tolerance to antenna misalignment increases. The reason behind this is that as the half-power-beamwidth increases, the antenna gain decreases; thus, the directionality gain also decreases. On the other hand, as the half-power-beamwidth increases, the probability for the two beams to be partially or fully aligned increases. Finally, Figure 8.4 reveals the importance of modeling the angular misalignment when computing the system performance. For example, for $\theta = 0.01^{\circ}$ and σ=0.1°, an approximately 10 dB total directional gain evaluation error occurs, if we neglect the impact of antenna misalignment.

The Gaussian distributed beamsteering errors model is tractable and suitable for modeling tracking estimation errors. Its main disadvantage is that it accommodates only the horizontal angular error and it totally neglects the vertical one. In other words, it is a one dimensional (1D) model.

8.2.1.2 Two-Dimensional Gaussian Shaking of a Single Node

As demonstrated in Figure 8.5, the 2D Gaussian shaking of a single-node model accommodates scenarios in which either the BS or the UE experience antenna misalignment (Kokkoniemi, Boulogeorgos, Aminu, Lehtomaki, Alexiou, and Juntti, 2020). Without loss

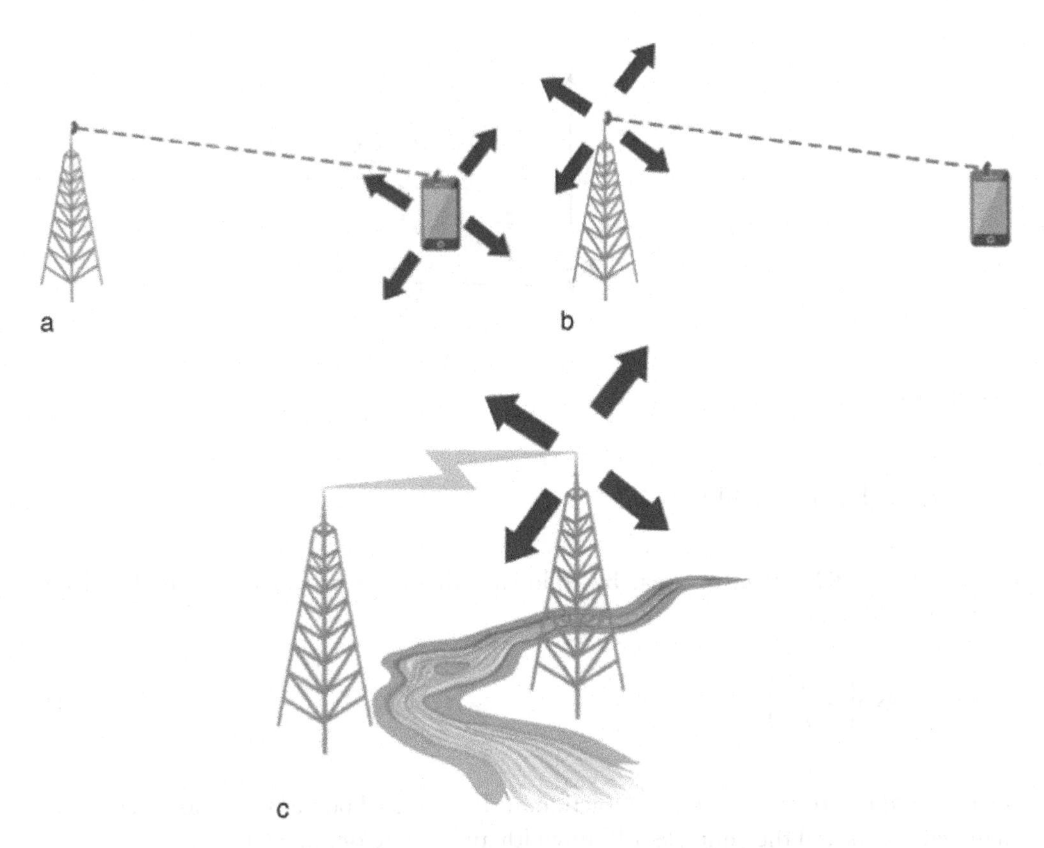

Figure 8.5 Two-dimensional Gaussian shaking of (a) the UE, (b) the BS in fronthaul scenarios, and (c) a single BS in backhaul scenarios.

of generality, in what follows, we examine a downlink scenario in which the UE shakes. Likewise, it is assumed that the RX antenna pattern provides a circular effective area. The radius of this area is a. Moreover, the TX beam is circular and at a transmission distance, d, has a radius ρ, with $\rho \in [0,w_{\mathrm{d}}]$. Of note, w_{d} is the maximum radius of the TX beam footprint at d. Moreover, as illustrated in Figure 8.6, both the RX antenna effective area and the TX beam footprint are in the x–y plane. Finally r represents the pointing error.

The normalized spatial distribution of the TX signal at distance d can be obtained as (Farid and Hranilovic, 2007)

$$P(\rho,d) = \frac{2}{\pi w_{\mathrm{d}}}\exp\left(-\frac{2|\rho|}{w_{\mathrm{d}}^{2}}\right) \tag{8.12}$$

where

$$w_{\mathrm{d}} = d\tan\left(\frac{\theta_{\mathrm{t}}}{2}\right) \tag{8.13}$$

Additionally, the power collected by the effective area of the RX antenna can be obtained as

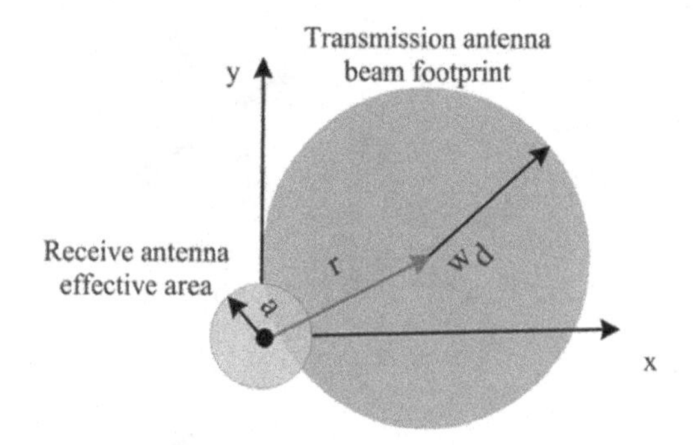

Figure 8.6 RX plane.

$$h_p(\mathrm{r};d)=\int_{\varepsilon}P(\rho-\mathrm{r},d)\mathrm{d}\rho, \tag{8.14}$$

where ε is the RX effective area. By following the same procedure as in (Farid and Hranilovic, 2007),

$$h_p \approx S_0 \exp\left(-\frac{2r^2}{w_e^2}\right) \tag{8.15}$$

where S_0 and w_e are respectively the fraction of the collected power when no antenna misalignment exists and the equivalent beamwidth and can be obtained as

$$S_0=\left(\mathrm{erf}(u)\right)^2 \tag{8.16}$$

and

$$w_e^2=w_d^2\frac{\sqrt{\pi}\mathrm{erf}(u)}{2u\exp(-u^2)} \tag{8.17}$$

with

$$u=\frac{\sqrt{\pi}a}{\sqrt{2}w_d}. \tag{8.18}$$

By further assuming that

$$r_x=\mathbf{r}\mathbf{x}_o \tag{8.19}$$

and

$$r_y=\mathrm{r}\mathrm{y}_{o,} \tag{8.20}$$

where $\mathbf{x}_o$ and $\mathbf{y}_o$ are respectively the unit vectors of the x and y axis, are independent and identical Gaussian distributed random processes, then r is a Rayleigh distributed random process with PDF given by

$$f_r(x)=\frac{x}{\sigma_r^2}\exp\left(-\frac{x^2}{2\sigma_r^2}\right), x>0 \tag{8.21}$$

where σ_r stands for the standard deviation of r (spatial jitter). Additionally, we can obtain the PDF of h_p, with the aid of (15), as

$$f h_p(x)=\frac{\xi^2}{S_0^{\xi^2}}x^{\xi^2-1}, \quad \text{for } x \in [0, S_0], \tag{8.22}$$

where

$$\xi=\frac{w_e}{2\sigma_e} \tag{8.23}$$

In this case, the total directional gain can be expressed as

$$\mathcal{D}=G_B G_U h_p. \tag{8.24}$$

Thus, the expected value of D can be obtained as

$$\mathbb{E}[\mathcal{D}]=G_B G_U \mathbb{E}[h_p] \tag{8.25}$$

or equivalently

$$\mathbb{E}[\mathcal{D}]=G_B G_U \int_0^{S_0} x f h_p(x)\,\mathrm{d}x \tag{8.26}$$

Finally, by performing the integration, the expected value of D can be rewritten as

$$\mathbb{E}[\mathcal{D}]=G_B G_U \frac{S_0\xi^2}{1+\xi^2} \tag{8.27}$$

From (8.27), it becomes apparent that the expected value of the total directional gain depends both on the transceivers antenna gains and on the TX antenna beamwaist at the RX plane as well as the spatial jitter variance.

Next, we consider the backhaul scenario, in which both BSs are equipped with Cassengrain antennas. In this case, w_d can be evaluated as $w_d = d\tan\left(\frac{\theta_t}{2}\right)$ and $\theta_t = 70\frac{\lambda}{d_t}$, where λ and d_t are respectively the transmission wavelength and the diameter of the TX antenna. In what follows, we set d = 30 m, $G_t = G_r$ = 45 dBi. Moreover, the transmission frequency is f = 275 GHz.

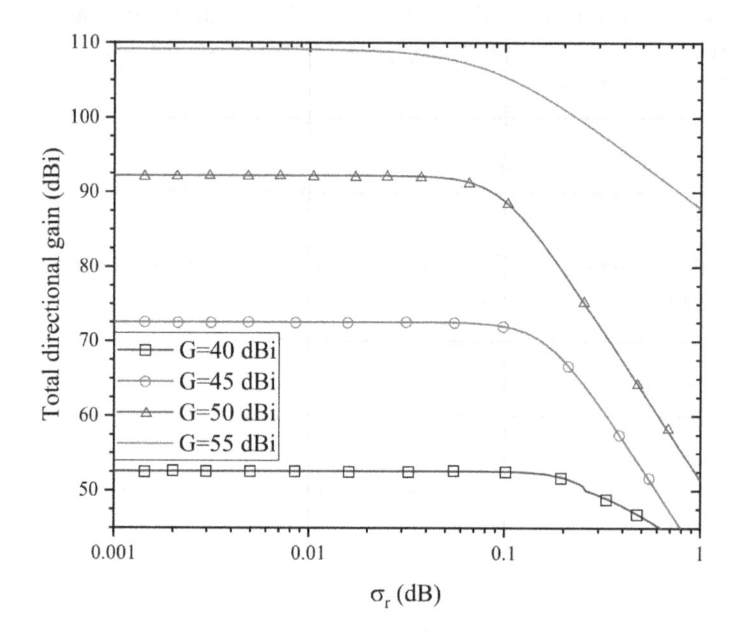

Figure 8.7 Directional gain vs spatial jitter standard deviation for different values of antenna gains.

Figure 8.7 plots the expected value of the total directional gain vs σ_r, for different $G_t = G_r$=G. For a given G, as σ_r increases, the expected value of the total directional gain decreases. Interestingly, for low values of σ_s, where the antenna misalignment is not quite important, we observe a total directional gain loss, due to the beam waist at the RX plain. Finally, from Figure 8.7, we observe that, for a given σ_s, as G increases, the expected value of the total directional gain also increases.

This model can accommodate the impact of antenna misalignment that is due to natural phenomena, such as wind, and small earthquakes. It is tractable; hence, it can be used for theoretical analysis purposes. Another important characteristic of this model is that it takes into account the TX beamwaist at the RX plane, thus, the transmission characteristics, namely transmission distance and frequency, as well as the type of the TX and RX antennas.

8.2.1.3 Wind Vibration Antenna Misalignment Model

Following the approach proposed in Piersol and Paez (2009) and Hur, Kim, Love, Krogmeier, Thomas, and Ghosh (2013), a wind vibration antenna misalignment model that takes into account the wind mean characteristics, such as the mean velocity, was developed. Specifically, after assuming that (i) the mean wind vector velocity, **v**, is unchangeable for small variations in the surface variables; (ii) the antenna height variable, z, is constant; and (iii) the z component of **v** is 0, the wind velocity can be expressed as

$$v(t) = \overline{\mathrm{v}} + \mathbf{c}_1 \upsilon(t) + \mathbf{c}_2 \upsilon_2(t), \tag{8.28}$$

where $\mathbf{c}_1$ and $\mathbf{c}_2$ are respectively unit vectors in the direction and orthogonal to mean wind direction. Likewise, $v_1(t)$ and $v_2(t)$ respectively stand for along-wind and across-wind directions turbulence, which can be modeled as independent zero-mean Gaussian random variables with power spectral densities (PSDs) that can be respectively obtained as (Hur et al., 2013; Piersol and Paez, 2009)

$$S_{\upsilon_1}(fw) = \frac{500\tilde{\upsilon}^{-2}}{\pi\bar{\upsilon}}\left(1+\frac{500f_{\mathrm{w}}}{2\pi\bar{\upsilon}}\right)^{-5/3} \tag{8.29}$$

and

$$S_{\upsilon_2}(f_w) = \frac{75\tilde{\upsilon}^2}{2\pi\bar{\upsilon}}\left(1+\frac{95f_{\mathrm{w}}}{2\pi\bar{\upsilon}}\right)^{-5/3} \tag{8.30}$$

with f_{w} being the wind frequency in Hz, $v = |\mathbf{v}|$, and

$$\tilde{\upsilon} = \frac{\bar{\upsilon}}{2.5}\ln\left(\frac{z}{z_0}\right) \tag{8.31}$$

is the shear velocity at height z with a terrain roughness length parameter z_0. Of note, z_0 has been characterized in (Wieringa, 1986).

Two mechanisms are responsible for the wind force that the antenna mounting pole experiences:

- *Orthogonal wind speed turbulence*: The orthogonal wind speed turbulence is due to the wind drag resulting from the mean wind speed. The mean wind speed can be analyzed into two orthogonal wind speed components. In consequence, the orthogonal wind speed turbulence can be also written as the sum of two orthogonal components. Its impact is mainly on misalignment in the x–y plane, and it can be modeled by two aero-admittance functions $H_{a,1}$, $H_{a,2}$ that can be respectively obtained as

$$\left|H_{\mathrm{a},1}\right| = 2K_{\mathrm{w}}\bar{\upsilon} \tag{8.32}$$

and

$$\left|H_{\mathrm{a},2}\right| = K_{\mathrm{w}}\bar{\upsilon}. \tag{8.33}$$

where

$$K_{\mathrm{w}} = \frac{1}{2}\rho_{\mathrm{a}}C_{\mathrm{D}}A_{\mathrm{e}}. \tag{8.34}$$

is the air density in kg/m^3. Of note, in (8.34), ρ_{a}, C_{D}, and A_{e} respectively stand for the air density, the drag coefficient, and the effective area in m^2.

Thus, the time-varying drag forces caused by the orthogonal wind speed turbulence components can be modeled as zero-mean random variables F_1 and F_2 with PSDs that can be expressed as (Hur et al., 2013)

$$S_{F_1}(f_w) = |H_{a,1}|^2 S_{u_1}(f_w) \tag{8.35}$$

and

$$S_{F_2}(f_w) = |H_{a,2}|^2 S_{u_2}(f_w). \tag{8.36}$$

- *Vortex shedding*: This phenomenon is created when the wind passes by the transceiver antenna supporting pole and results to a lifting force in the acrosswind direction, which can be approximately periodic. It is characterized by means of cortex-shedding-frequency that can be calculated as

$$f_\upsilon = S\frac{\overline{\upsilon}}{d_p} \tag{8.37}$$

where S stands for the Strouhal number,[2] and d_p represents the pole diameter. Additionally, the PSD due to vortex shedding can be evaluated as

$$S_{F_A}(f_w) = K_w^2 \frac{1.125}{\sqrt{\pi f_w f_\upsilon}} \exp\left(-\left(\frac{1-\frac{f_w}{f_\upsilon}}{0.18}\right)^2\right). \tag{8.38}$$

The aforementioned forces compose a mechanical model that accommodates the antenna-mounting pole with a transfer function that can be obtained as

$$H_m(f_w) = \left(4m\pi^2 f_n^2\left(1-\frac{f_w^2}{f_n^2}\right) + 4\delta^2\frac{f_w^2}{f_n^2}\right)^{-1/2} \tag{8.39}$$

where m denotes the mass of the antenna on the pole's top, while δ and f_n respectively represent the damping coefficient of the natural frequency. From (8.39), we can derive the pole-displacement PSDs as

$$S_1(f_w) = |H_m(f_w)|^2 S_{F_1}(f_w) \tag{8.40}$$

and

$$S_2(f_w) = |H_m(f_w)|^2 S_{F_2}(f_w) \tag{8.41}$$

As illustrated in Figure 8.8, we define $\Delta L_1(t)$ and $\Delta L_2(t)$ as the relative displacements of the tops of the poles at the two links' ends. Moreover, let us assume that the link transmission distance equals d. Next, we examine the worst case scenario, in which the expected

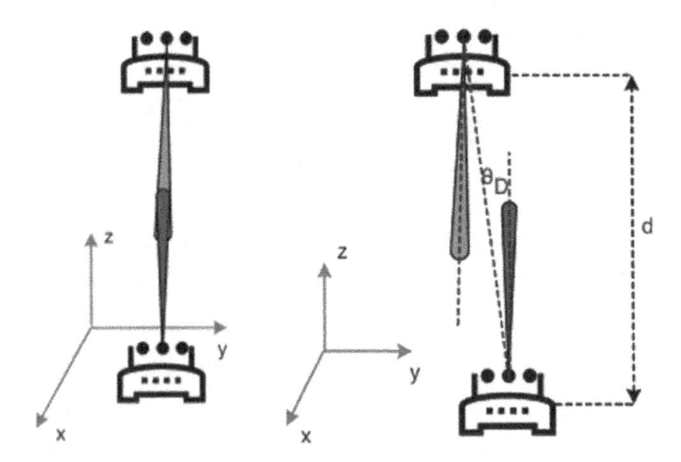

Figure 8.8 An indicative example of (a) aligned beams and (b) pole movement due to wind.

value of the wind-direction is perpendicular to the beam direction. As a consequence, the sway angle can be obtained as

$$\theta_{\mathrm{D}}(t) = \tan^{-1}\left(\frac{\Delta L_1(t)}{d + \Delta L_2(t)}\right). \tag{8.42}$$

Finally, the beam outage probability can be mathematically defined as

$$P_{\mathrm{b,o}}(\theta_{3\mathrm{dB}}) = \Pr\left(\left|\theta_{\mathrm{D}}(t)\right| \geq \theta_{\max}\right), \tag{8.43}$$

where $\theta_{\max}$ is the maximum allowable deflection angle and is connected with the half power beamwidth of the RX antenna, $\theta_{3\mathrm{dB}}$, through

$$\theta_{\max} = b\theta_{3\mathrm{dB}}. \tag{8.44}$$

with b being the fraction ratio, which is beamforming gain loss specific. Notice that when the absolute value of the sway angle surpasses the maximum allowable deflection angle, the communication is interrupted and an outage phenomenon due to antenna misalignment occurs. Note that (8.43) can be evaluated through simulation, according to the process described in (Shinozuka and Deodatis, 1991).

Next, we provide simulation results that quantify the beam outage probability for the following insightful scenario. We assume that $f_{\mathrm{n}} = 1$ Hz, $f_{\mathrm{w}} = 20$ Hz, $\delta = 0.002$, $m = 5$ kg, $A_{\mathrm{e}} = 0.09$ m^2, $b = 0.3578$, $\theta_{3\mathrm{dB}} = 2^{\mathrm{o}}$, $\rho_{\mathrm{a}} = 1.22$ kg/m^3 and $C_{\mathrm{D}} = 0.5$. Moreover, $z_0 = 2$ m, and $d = 30$ m. Notice that the simulation parameters that are used are realistic and correspond to a THz wireless fiber extender application.

Figure 8.9 demonstrates the beam outage probability as a function of the wind velocity for different $\theta_{3\mathrm{dB}}$. From this figure, it becomes apparent that, for a given $\theta_{3\mathrm{dB}}$, as v

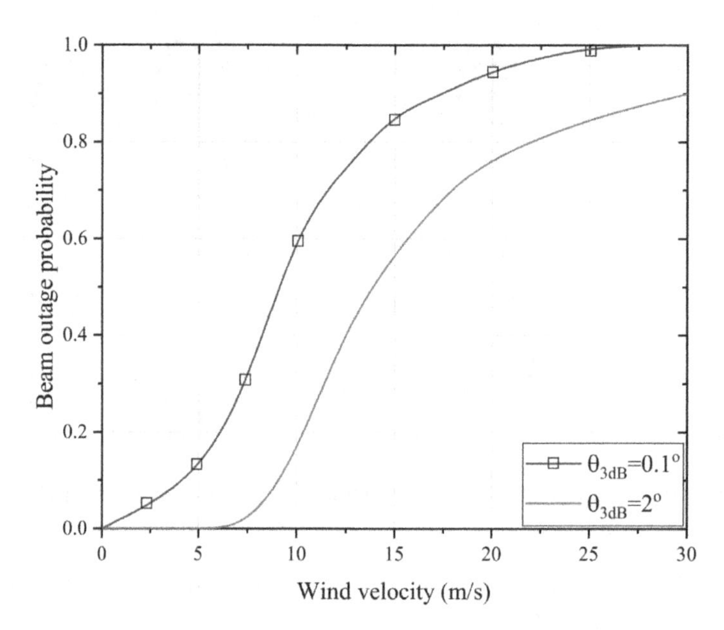

Figure 8.9 Beam outage probability vs wind velocity for different half-power beamwidths.

increases, the beam outage probability also increases. Additionally, for a given v, as θ_{3dB} increases, the impact of pointing errors becomes less severe; thus, the beam outage probability decreases.

The model presented in this section accommodates the impact of antenna misalignment due to the wind. It takes into account the winds characteristics, namely wind velocity and frequency, as well as its density. However, it does not account for the beam waist and it is not tractable for theoretical analysis.

8.2.2 Beam Misalignment Mitigation Approaches

This section discusses the key antenna misalignment mitigation approaches that are reported in the current technical literature. In particular, Section 8.2.2.1 presents the different beam-tracking approaches and identifies the advantages and disadvantages of each one of them, while Section 8.2.2.2 explains how relaying strategies can be used as countermeasures to antenna misalignment.

8.2.2.1 Beam-Tracking

One of the critical tasks of wireless systems is channel estimation. As a result, a significant amount of effort was put on developing channel estimation approaches in systems operating in low-RFs (see, for example, Kim, 2015 and references therein). However, in general, these approaches cannot be adopted in millimeter wave (mmW) and THz systems due to the following issues:

(1) High-frequency transceivers are equipped with a limited number of RF chains in order to constrain the implementation cost and the power consumption (Alkhateeb, Mo,

Gonzalez-Prelcic, and Heath, 2014b). This leads to several limitations concerning the beamforming and combining types that can be employed (Zhang, Guo, and Fan, 2016).

(2) Because of the limited scattering of the THz channel, it becomes possible to provide accurate estimation of channel parameters, like AoD and AoA instead of each element in a large channel matrix (Zhou and Ohashi, 2015).

Motivated by the above, a number of beam-tracking approaches have been reported. In more detail, the following type of beam-tracking approaches are investigated:

- *Exhaustive search-based methods*: As proposed in IEEE 802.11ad standard and adopted by IEEE 802.11ay (Ghasempour, da Silva, Cordeiro, and Knightly, 2017), a simple approach to estimate the channel is to exhaustively search the AoD and AoA, and choose the direction with the largest gain. In particular, the following procedure is performed. The time is divided into a number of beam-training cycles. Each cycle has a duration T, which is used by the BS to initiate a cell search (CS) procedure. During the CS, the BS uses different codebooks to sweep through N_b ideally non-overlapping identical directions in order to virtually broadcast synchronization signals. At the same time, the UE sweeps through N_u RX beamforming identical directions and performance energy detection. This procedure is followed by a random access (RA) phase. In the RA phase, the UE sends a connection request to the BS, and the BS station replies with a random access response (RAR). Finally, the data transmission (DT) phase follows. This procedure was analyzed and its effectiveness was quantified in Boulogeorgos and Alexiou (2019b) and Machado et al. (2019). In these works, the authors have theoretically shown the high accuracy of this approach, explained that it can significantly reduce the tracking estimation error, and highlighted that it comes with the cost of unacceptably high tracking overhead that is translated into a beam-training latency in the order of ms. Another disadvantage of this approach is that it obtains only coarse knowledge of the channel and as a consequence spatial multiplexing cannot be exploited.
- *Hierarchical beam-searching methods*: To break the barriers of exhaustive search-based methods, beamwidth-adaptive algorithms that essentially scan the beamdirections by employing beamforming codebooks with different beamwidths in a hierarchical manner have been presented (see, for example, Alkhateeb, El Ayach, Leus, and Heath, 2014a and references therein). Specifically, each codebook consists of a predetermined number of levels, which have different number of codewords and different resolution. In other words, the beam coverage varies among different levels. As the index of the codebook level increases, the beamwidth of the corresponding codewords decreases. This indicates that higher levels of the codebook return a higher number of codewords with narrower beamwidth. The BS and UE first perform an exhaustive search over wide beams, that is, using the lowest level of the codebook, and then refine to search narrow beams, where higher levels of the codebook are used. In comparison with exhaustive search, hierarchical search generally has smaller beam-training latency, however, it is unable to provide the coverage performance to the cell-edge users of the exhaustive search approach (Giordani, Mezzavilla, Barati, Rangan, and Zorzi, 2016).
- *Perturbation methods*: These approaches are based on the basic assumption that the UE position does not significantly vary from one transmission to the next;

hence, it can be predicted. In this sense, these methods design the training beams by perturbing the present beam's weights and selecting the training-beam that provides the highest gain for weight update (He, Kim, Ghauch, Liu, and Wang, 2014). A similar approach was discussed in Gao, Dai, Zhang, Xie, Dai, and Wang (2017), where a priori aided channel tracking scheme was presented. In particular, by considering a linear motion model, the physical direction's temporal variation law the between the BS and the UE was discovered. Then, building upon this law and the sparse nature of THz channels, the beamspace channels in the previous time slots was employed in order to predict the information of the beamspace THz channel in the next time slot. Evidently, the aforementioned methods can significantly reduce the beam-tracking latency with a cost on the estimation accuracy. Moreover, they require a priori knowledge of the UE motion, which, in several realistic scenarios, may not be possible.

- *Lower-frequency-based tracking methods*: The use of lower-frequencies for tracking purposes was discussed in several works, including Boulogeorgos et al. (2018), Tong and Han (2017), Yao and Jornet (2016). These methods enable the use of well-investigated UE localization techniques in THz wireless systems; however, it requires the BS and UE to be equipped with lower-frequency transceivers, which increases the implementation cost and the transceiver's complexity; hence, it results to a significant reduction of the mobile UE energy autonomy. Moreover, due to the blockage and antenna misalignment, the utilization of a microwave channel does not necessarily result in the establishment of the corresponding THz link.

To sum up, we observe that there exists a trade-off between tracking accuracy and latency. Low beam-tracking accuracy may cause significant angular misalignment errors, which may be detrimental for the system reliability. On the other hand, approaches with high beam-tracking accuracy, in most cases, are accompanied by high beam-tracking latency, which can significantly reduce the effective throughput.

8.2.2.2 Relaying

A possible approach to mitigate the effect of antenna misalignment is the use of relaying selection (RS) strategies. In order to present the effectiveness of this approach we consider a dual-hop THz decode-and-forward (DF) RS system depicted in Figure 8.10. The system consists of one source (S), N relays and one destination (D). S, R, and D nodes are equipped with high-gain antennas and operate in half-duplex mode, that is, in each timeslot a node can either transmit or receive a message. Moreover, due to the use of high-gain antennas, in each timeslot, S can only transmit in one R node and only one R node can communicate with D. Each transmission period is composed of two phases. In the first phase, the S transmits to a predetermined relay, R∗, which is selected in accordance to the following rule:

$$R^{*} = \arg\max\left(\gamma_{\mathrm{DF}}^{i}\right). \tag{8.45}$$

where $\gamma_{\mathrm{DF}}{}^{i}$ is the equivalent end-to-end SNR for the link between the S, the i-th relay and the D, which can be obtained as

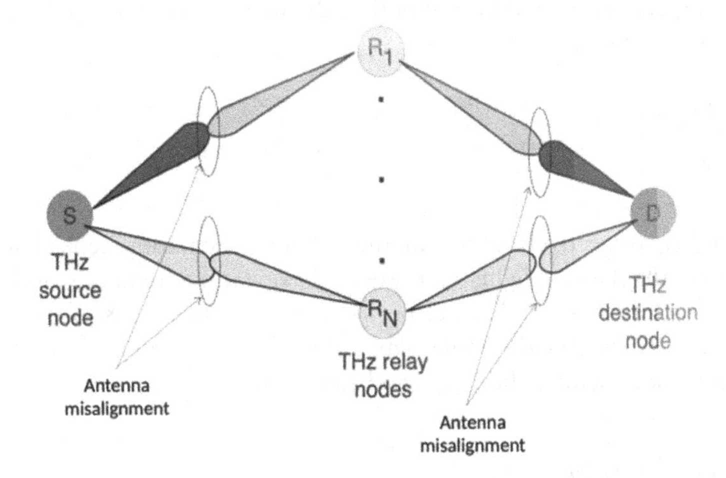

Figure 8.10 System model.

$$\gamma_{\mathrm{DF}}^{i} = \min\left(\gamma_1^i, \gamma_2^i\right), \tag{8.46}$$

with γ_1^i and γ_2^i respectively being the SNRs at R$_i$ and D, which can be obtained as

$$\gamma_1^i = \frac{\left|h_1^i\right|^2 P_S}{N_o} \tag{8.47}$$

and

$$\gamma_2^i = \frac{\left|h_2^i\right|^2 P_i}{N_o}. \tag{8.48}$$

In (8.47) and (8.48), P_s, P_i, and N_o are respectively the S, R$_i$ transmission powers, and the noise power. Moreover, h_1^i and h_2^i are the S-R$_i$ and R$_i$-D channel coefficients and can be respectively obtained as

$$h_1^i = h_{l1}^i h_{m1}^i \tag{8.49}$$

and

$$h_2^i = h_{l2}^i h_{m2}^i, \tag{8.50}$$

where $\left|h_{l1}^i\right|^2$ and h_{m1}^i respectively represent the deterministic path-gain and the antenna misalignment coefficients of the S-R$_i$ link, while $\left|h_{l2}^i\right|^2$ and h_{m2}^i the corresponding ones of the R$_i$-D link.

The path-gain coefficients can be expressed as

$$h_{l,j}^i = h_{fl}^j h_{al}^j, \tag{8.51}$$

where h_{fl}^{j} models the propagation gain of the j-th link and, according to Friis equation, can be modeled as (Balanis, 2008)

$$h_{fl}^{j} = \frac{c\sqrt{G_t^j G_r^j}}{4\pi f_j d_j} \tag{8.52}$$

where G_t^j and G_r^j respectively stand for the antenna orientation-dependent TX and RX gains, c denotes the speed of light, f_j represents the operating frequency of the j-th link and d_j stands for the transmission distance between the TX and the RX of the j-th link.

Additionally, h_{al}^{j} denotes the molecular absorption gain in the j-th link and can be evaluated as (Kokkoniemi, Lehtomaki, and Juntti, 2018)

$$h_{al}^{j} = \exp\left(-\frac{1}{2}k_{\alpha}\left(f_j\right)d_j\right), \tag{8.53}$$

where $\kappa_{\alpha}(f_j)$ is the absorption coefficient describing the relative area per unit of volume, in which the molecules of the medium are capable of absorbing the electromagnetic wave energy (Boronin et al., 2014; Jornet and Akyildiz, 2011). To evaluate the molecular absorption coefficient, we use the model presented in Kokkoniemi et al. (2018). Based on this this model the absorption coefficient can be evaluated as

$$k_{\alpha}\left(f_j\right) = y_1\left(f_j, v\right) + y_2\left(f_j, v\right) + g\left(f_j\right) \tag{8.54}$$

where the parameters $y_1(f_j,v)$, $y_2(f_j,v)$ and $g(f_j)$ are defined as

$$y_1\left(f_j, v\right) = \frac{A(v)}{B(v) + \left(\frac{f_j}{100c} - C_1\right)^2}, \tag{8.55}$$

$$y_2\left(f_j, v\right) = \frac{C(v)}{D(v) + \left(\frac{f_j}{100c} - C_2\right)^2} \tag{8.56}$$

and

$$g\left(f_j\right) = p_1 f_j^3 + p_2 f_j^2 + p_3 f_j + p_4 \tag{8.57}$$

where c represents the speed of light, while $c_1 = 10.835$ cm^{-1}, $c_2 = 12.664$ cm^{-1}, $p_1 = 5.54 \times 10^{-37}$ Hz^{-3}, $p_2 = -3.94 \times 10^{-25}$ Hz^{-2}, $p_3 = 9.06 \times 10^{-14}$ Hz^{-1}, $p_4 = -6.36 \times 10^{-3}$ and

$$A(v) = g_1 v\left(g_2 v + g_3\right), \tag{8.58}$$

$$B(v) = (g_4 v + g_5)^2, \tag{8.59}$$

$$C(v) = g_5 v (g_6 v + g_7), \tag{8.60}$$

$$D(v) = (g_8 v + g_9)^2, \tag{8.61}$$

with $g_1 = 0.2205, g_2 = 0.1303, g_3 = 0.0294, g_4 = 0.4093, g_5 = 0.0925, g_5 = 2.014, g_6 = 0.1702, g_7 = 0.0303, g_8 = 0.537$ and $g_9 = 0.0956$. Likewise, v denotes the volume mixing ratio of the water vapor and can be obtained as (Kokkoniemi et al., 2018)

$$v = \frac{\phi}{100} \frac{p_w(T,p)}{p}, \tag{8.62}$$

where φ and p are the relative humidity and atmospheric pressure, respectively. Moreover, $p_w(T,p)$ denotes the saturated water-vapor partial-pressure in temperature T, which according to Alduchov and Eskridge (1996), can be calculated as

$$p_w(T,p) = q_1 (q_2 + q_3 \phi_h) \tag{8.63}$$

with $q_1 = 6.1121, q_2 = 1.0007, q_3 = 3.46 \times 10^{-6}\,\text{hPa}^{-1}, q_4 = 17.502, q_5 = 273.15\,^{o}\text{K}, q_6 = 32.18\,^{o}\text{K}$ and φ_h standing for the pressure in hPa. Of note the accuracy of the presented model was validated for up to 1 km links in standard atmospheric conditions.

By assuming that all the links are independent, the outage probability can be obtained as

$$P_O = \prod_{i=1}^{N_r} P_r(\gamma_{DF}^i \le \gamma_{th}), \tag{8.64}$$

where γ_{th} stands for the SNR threshold. With the aid of (46), $P_r(\gamma_{DF}^i \le \gamma_{th})$ can be evaluated as

$$P_r(\gamma_{DF}^i \le \gamma_{th}) = P_r(\min(\gamma_1^i, \gamma_2^i) \le \gamma_{th}), \tag{8.65}$$

or equivalently

$$P_r(\gamma_{DF}^i \le \gamma_{th}) = 1 - P_r(\min(\gamma_1^i, \gamma_2^i) > \gamma_{th}), \tag{8.66}$$

which can be rewritten as

$$P_r(\gamma_{DF}^i \le \gamma_{th}) = 1 - P_r(\{\gamma_1^i > \gamma_{th}\} \cap \{\gamma_2^i > \gamma_{th}\}) \tag{8.67}$$

Since γ_1^i and γ_2^i are independent random variable, (67) can be rewritten as

$$P_{\rm r}\left(\gamma_{\rm DF}^{i} \leq \gamma_{\rm th}\right) = 1 - P_{\rm r}\left(\gamma_{1}^{i} > \gamma_{\rm th}\right)\Pr\left(\gamma_{2}^{i} > \gamma_{\rm th}\right), \tag{8.68}$$

which can be equivalently written as

$$P_{\rm r}\left(\gamma_{\rm DF}^{i} \leq \gamma_{\rm th}\right) = 1 - \left(1 - \Pr\left(\gamma_{1}^{i} \leq \gamma_{\rm th}\right)\right)\left(1 - \Pr\left(\gamma_{2}^{i} \leq \gamma_{\rm th}\right)\right) \tag{8.69}$$

Finally, by substituting (8.69) into (8.64), we can write the outage probability as

$$P_{\rm o} = \prod_{i=1}^{N_{\rm r}}\left(1 - \left(1 - \Pr\left(\gamma_{1}^{i} \leq \gamma_{\rm th}\right)\right)\left(\left(1 - \Pr\left(\gamma_{2}^{i} \leq \gamma_{\rm th}\right)\right)\right)\right) \tag{8.70}$$

Note that a similar expression to (8.70) has been obtained in several previously published works, including Boulogeorgos (2016) and Mokhtar, Boulogeorgos, Karagiannidis, and Al-Dhahir (2014). Notice that (8.70) is independent of the path-loss and antenna misalignment model that is employed.

Next, we provide illustrative results that quantify the impact of antenna misalignment in THz wireless relaying systems and reveal the efficiency of the RS strategy. For convenience, the path-loss is provided by the model presented in Boulogeorgos et al. (2019), while for the antenna misalignment, the model discussed in Section 8.2.1.2 is used. The system parameters that are considered are summarized in Table 8.1. Of note, for the simulated scenario, we assumed that the TX and RX antenna have the same gain. This is because, in practice, relays usually use the same antenna for transmission and reception. As a consequence, the TX and RX antenna gains are the same. However, notice that we also take into consideration case in which the transceivers have different TX and RX antennas.

Figure 8.11 depicts the outage probability as a function of the source transmission SNR to SNR threshold ratio, for different values of relay transmission SNR to SNR threshold ratio and spatial jitter standard deviation, assuming that $N_{\rm r} = 1$. We observe that, for a given relay transmission SNR to SNR threshold ratio and jitter standard deviation, as the source transmission SNR to SNR threshold ratio increases, the outage performance improves. Similarly, for a fixed source transmission SNR to SNR threshold ratio and jitter standard deviation, as the relay transmission SNR to SNR threshold ratio increases, the outage probability decreases. Finally, it is evident that the effect of misalignment fading

Table 8.1 Relaying system parameters.

Parameter	*Value*
Relative humidity	50%
Atmospheric pressure	101,325 Pa
Air temperature	296 K
Transmission frequency	275 GHz
TX antenna gains (in both S and R_i nodes with $i = 1,\cdots,N_r$)	55 dBi
RX antenna gains (in both S and R_i nodes with $i = 1,\cdots,N_r$)	55 dBi
S-R_i distance with $i = 1,\cdots,N_r$	3 m
R_i-D distance with $i = 1,\cdots,N_r$	3 m

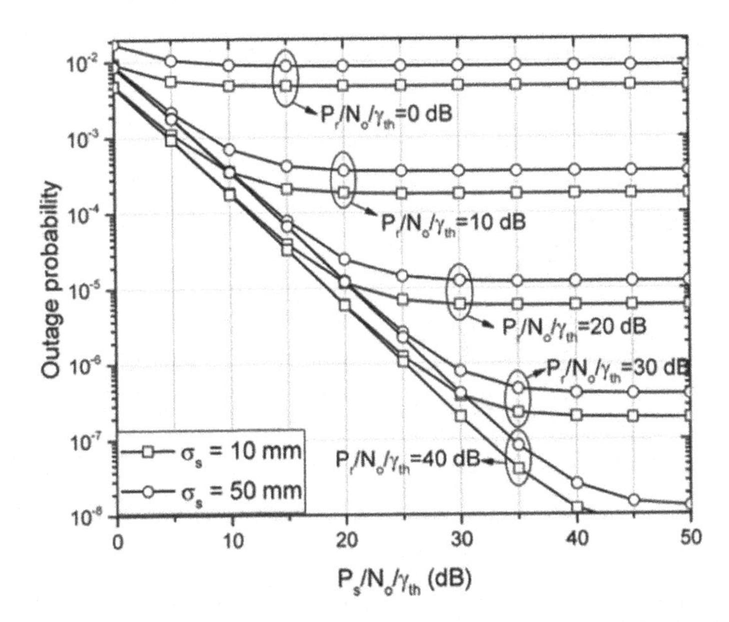

Figure 8.11 Outage probability as a function of the source transmission SNR to SNR threshold ratio, for different values of relay transmission SNR to SNR threshold ratio and spatial jitter standard deviation, assuming one relay node.

on the outage performance of the THz relaying system is independent from the level of transmission power of both the source and relay.

Figure 8.12 demonstrates the outage performance and the robustness of the THz DF SR system to antenna misalignment. In more detail, it illustrates the outage probability as a function of the transmission SNR to SNR threshold ratio, for different values of number of relays and spatial jitter standard deviation, assuming a fixed relay transmission SNR to SNR threshold ratio that equals 10 dB. As expected for a given number of relays and spatial jitter standard deviation, as the source transmission SNR to SNR threshold ratio increases, the outage probability decreases. Interestingly, there is an outage probability floor, in the high-source transmission SNR to SNR threshold ratio regime. In more detail, when source transmission SNR to SNR threshold ratio is greater than the corresponding relay transmission SNR to SNR threshold ratio, the outage probability is constrained by the minimum performance of the R*-D link. Likewise, we observe that, for a given source transmission SNR to SNR threshold ratio and number of relays, as the spatial jitter standard deviation increases, the outage performance significantly degrades. Finally, for a given source transmission SNR to SNR threshold ratio and spatial jitter standard deviation, as the number of relays increases, the outage performance improves. This indicates that the use of THz DF SR can counterbalance the impact of antenna misalignment.

To sum up, our results revealed that the effect of antenna misalignment in THz wireless SR system is detrimental. However, by increasing the number of relays, we can significantly limit the impact of antenna misalignment. Other alternatives are to increase the transmission power at both the source and relay nodes, or equivalently to decrease the spectral efficiency of the transmission scheme.

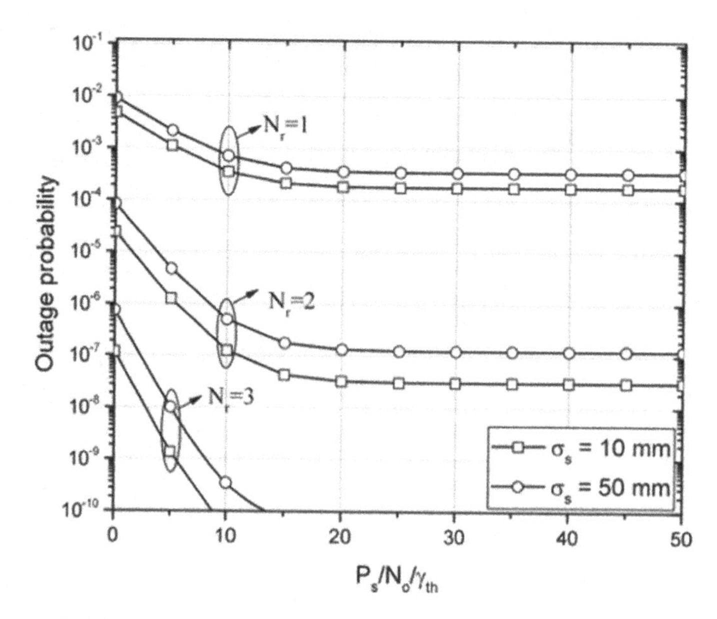

Figure 8.12 Outage probability as a function of the source transmission SNR to SNR threshold ratio, for different values of number of relays and spatial jitter standard deviation, assuming that the relay transmission SNR to SNR threshold ratio equals 10 dB.

8.3 Blockage

This section is organized as follows: Section 8.3.1 reports the type of blockage and conducts a brief literature review on the models that are used to accommodate their particularities, while Section 8.3.2 is focused on presenting the statistical characterization of some of the most-commonly used blockage models. Finally, Section 8.3.3 reports blockage mitigation approaches.

8.3.1 Blockage Types and Models

A THz wireless link may experience three different types of blockage, namely static, dynamic, and self-blockage (Jain et al., 2018):

- *Static blockage* is caused by buildings and permanent structures (Bai, Vaze, and Heath, 2012) and it can be modeled by means of random shape theory and stochastic geometry principles. In more detail, let us assume the 2D problem, that is, we neglect the buildings height, and a set of bounded 2D objects S. By randomly placing objects from S on the 2D plane at points, which are created by a Poisson point process (PPP) and with an orientation that follows a uniformly distribution with range $[0,2\pi)$, we can model the urban propagation environment. This type of blockage can cause a permanent interruption of the line-of-sight (LoS) link. However, due to the short range of the THz link, it is not expected to play an important role on the system performance.
- *Dynamic blockage* is caused by mobile humans and vehicles, which may result in interruptions to the LoS THz link. This type of blockage was considered of high importance by the third generation partnership project (3GPP) and was analyzed in Release

14 of the technical report (TR) 38.901 (3GPP, 2017). Dynamic blockage was modeled in Gapeyenko et al. (2016), where a static user was assumed to be served by a single BS, which is located in a fixed position, while the location and number of blockers were generated through a PPP. Similarly, in Wang, Venugopal, Molisch, and Heath (2017), the authors used the Manhattan Poisson point (MPP) process model in order to accommodate the blockage effect in road intersection scenarios. In MacCartney, Rappaport, and Rangan (2017), the authors fitted measurements of a BS-UE link to a Markov model. A similar approach was considered in Raghavan, Akhoondzadeh-Asl, Podshivalov, Hulten, Tassoudji, Koymen, Sampath, and Li (2019), where the authors fitted the blockage measurements in a variety of different statistical models. Finally, 3D blockage models were presented in Han et al. (2017) and Jain et al. (2018).

- *Self-blockage* occurs when the UE transmit or receive signal is obstructed by the user's own body. In Raghavan et al. (2019), its detrimental effect was revealed through experiments and a statistical self-blockage model was presented, while in (Abouelseoud and Charlton (2013), simulation models were employed to quantify this effect. Finally, in Bai and Heath (2014), stochastic geometry tools were used to compute the received signal strength due to self-blockage.

To quantify the effect of blockage on the THz wireless system performance, in what follows, we use the blocking probability that is defined as the LoS link is interrupted due to the existence of an obstacle (3GPP, 2017; Zafer and Modiano, 2006).

8.3.2 Statistical Characterization

8.3.2.1 Urban Outdoor Micro-Cellular Model

The urban outdoor micro-cellular model was introduced by 3GPP for mmW wireless system (3GPP, 2017). According to this model, the probability of a link of length d to be LoS can be obtained as

$$P_{\mathrm{L}}(d) = \min\left(\frac{d_{\mathrm{o}}}{2d}, 1\right)\left(1 - \exp\left(-\frac{d}{d_{\mathrm{o}}}\right)\right) + \exp\left(-\frac{d}{d_{\mathrm{o}}}\right), \tag{8.71}$$

where $d_{\mathrm{o}} = 36$ m. Hence, the blocking probability can be expressed as

$$P_{\mathrm{B}}(d) = 1 - P_{\mathrm{L}}(d), \tag{8.72}$$

or equivalently

$$P_{\mathrm{L}}(d) = \min\left(\frac{d_{\mathrm{o}}}{2d}, 1\right)\left(1 - \exp\left(-\frac{d}{d_{\mathrm{o}}}\right)\right) + \exp\left(-\frac{d}{d_{\mathrm{o}}}\right). \tag{8.73}$$

From (8.73), it is evident that, according to this model, the blocking probability only depends on the length of the BS-UE link. This is based on the logical assumption that as the transmission distance increases, the probability that at least one static one moving obstacle to exist between the TX and RX increases.

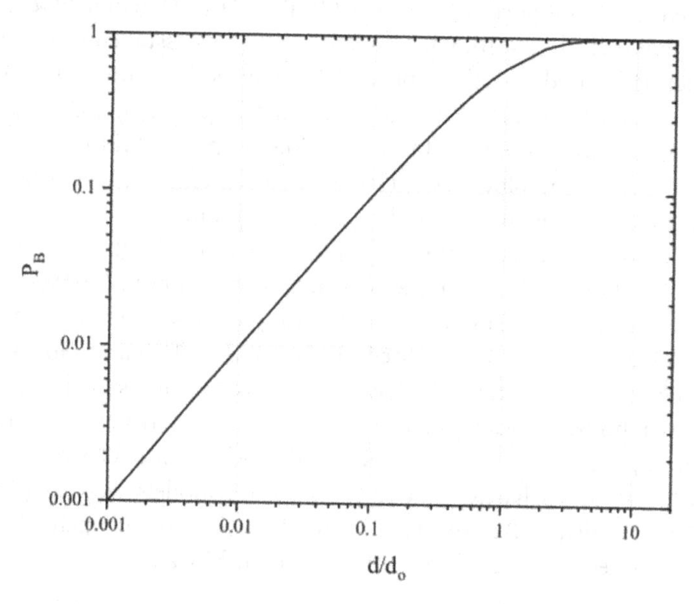

Figure 8.13 Blocking probability as a function of d/d_0 for urban outdoor micro-cellular scenarios.

Figure 8.13 illustrates the probability that the link is blocked as a function of d/d_o assuming an urban outdoor micro-cellular scenario. As expected, as d/d_o increases, the probability that a blocker will interrupt the LoS link increases; hence, the blocking probability also increases.

8.3.2.2 Random Shape Theory-based Model

By employing random shape theory described in Section 8.3.1, in Bai, Vaze, and Heath (2014), it was proven that the probability of LoS link under dynamic blockage can be obtained as

$$P_L(d) = \exp(-gd), \tag{8.74}$$

where

$$g = -\frac{r_B \ln(1 - k_A)}{\pi A}, \tag{8.75}$$

where k_A is the fraction of the area covered by buildings, r_B is the average building perimeter, and A is the average building area. From (8.74), we can straightforwardly obtain the blocking probability as

$$P_B(d) = 1 - \exp(-gd). \tag{8.76}$$

Notice that this model takes not only into account the link length, but also the mean characteristics of the blockers. Finally, it is worth noting that this model is tractable; hence, it can be used in the theoretical analysis of the THz system performance.

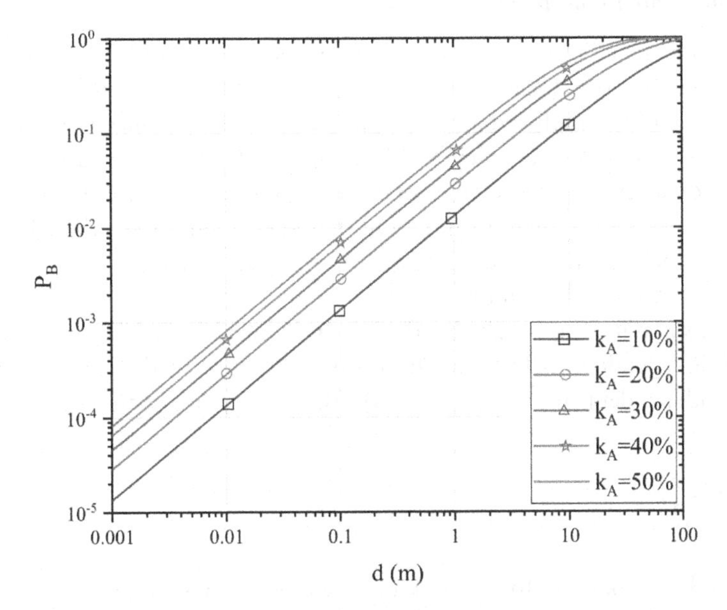

Figure 8.14 Blocking probability as a function of the transmission distance of the link for different values of the portion of the area covered by buildings.

Figure 8.14 depicts the blocking probability as a function of k_A, assuming $r_B = 50$ m and $A = 100$ m^2. As expected, for a given k_A, as d increases, P_B also increases. Finally, for a given d, as k_A increases, the number of blockers also increases; hence, P_B increases.

8.3.2.3 LoS Ball Model 1

According to the LoS ball model, which was introduced in Bai and Heath (2015), the LoS probability can be written as

$$P_L(d) = \begin{cases} 1, \text{for } x < D_0, \\ 0, \text{otherwise} \end{cases} \tag{8.77}$$

where D_0 can be evaluated by fitting the LOS association probability with $1 - \exp(-\pi \mu D_o^2)$. Of note, μ represents the density of the BS, which are generated by means of a PPP.

8.3.2.4 LoS Ball Model 2

The probability of LoS link based on the LoS ball model 2 is also given by (8.77) (Bai and Heath, 2015). The difference with the LoS ball model 1 is that the parameter D_0 is selected such that the mean number of LoS BSs that are in LoS with the UE under investigation, is matched.

8.3.3 Blockage Mitigation Approaches

Two types of blockage mitigation approaches have been identified, namely CoMP and (active/relaying and passive/reconfigurable intelligent surface (RIS)-assisted) reflected links.

8.3.3.1 Coordinated Multipoint

To mitigate the impact of blockage in THz wireless systems, it is expected that a single UE will be simultaneously associated with several BSs and joint transmission (JT) or selection diversity (SD) will be employed as a downlink coordinated multipoint (CoMP) technique (Boulogeorgos, 2016; MacCartney and Rappaport, 2019). JT requires the UE to be simultaneously connected to more than one BS. In other words, the UE should be able to employ a type of hybrid beamforming. On the other hand, in SD, the UE is connected to the BS from which the maximum throughput can be achieved. For both techniques, the BS needs to share channel state information.

Next, let us assume the insightful scenario, in which a single UE is connected to N BS through N independent THz link. The probability that all the link are simultaneously blocked is independent of the type of CoMP that is used, and can be obtained as

$$P_{\mathrm{B}}^{\mathrm{CoMP}} = \prod_{i=1}^{N} P_{\mathrm{B}_i}, \tag{8.78}$$

where P_{Bi} is the probability for the link between the i-th BS and the UE to be blocked.

In order to evaluate the efficiency of CoMP, let us now assume that blockage can be modeled according to the random shape theory based model presented in Section 8.3.2.2, and that the UE is at the center of a circle of radius d and that N BSs are located on the circle perimeter. Moreover, we set $r_{\mathrm{B}} = 50$ m, $k_{\mathrm{a}} = 50\%$, $A = 100$ m^2. For this scenario, Figure 8.15 plots the blocking probability as a function of d, for different values of N. As expected, for a given N, as d increases, P_{B} also increases. This indicates the significance of taking into consideration the impact of blockage in CoMP systems.

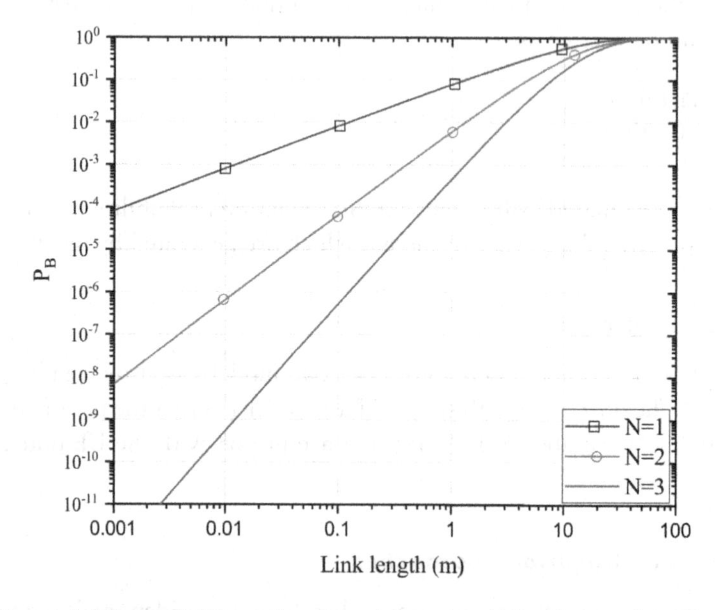

Figure 8.15 Blocking probability in CoMP as a function of the transmission distance for different values of available BSs.

Additionally, we observe that, for a given d, as N increases, the number of alternative links also increases; hence, P_B decreases. This reveals the effectiveness of the CoMP approach.

8.3.3.2 *Reflected Links*

Another approach to mitigate the impact of blockage is to create alternative links through relays or RIS. Let us consider the same system model as in Section 8.2.2.2, that is, a single S equipped with a high-gain antenna communicates with the D through 1 of the N available relays or by using 1 out of the N available metasurfaces of the RIS. In this scenario, the S-R$_i$-D path will be blocked if and only if one at least one of the S-R$_i$ or/and R$_i$-D is blocked. In other words, the S-R$_i$-D path will be unblocked if and only if both the S-R$_i$ and R$_i$-D links are unblocked. This can be mathematically expressed as

$$\bar{P}_{\mathrm{B,i}} = \left(1 - P_{\mathrm{B}}^{s_i}\right)\left(1 - P_{\mathrm{B}}^{r_i}\right), \tag{8.79}$$

where $\bar{P}_{\mathrm{B,i}}$ represents the probability that both the involved links are unblocked. Hence, the probability that at least one link is blocked, can be obtained as

$$P_{\mathrm{B},i} = 1 - \bar{P}_{\mathrm{B,i}}, \tag{8.80}$$

or equivalently

$$\bar{P}_{\mathrm{B,i}} = 1\left(1 - P_{\mathrm{B}}^{s_i}\right)\left(1 - P_{\mathrm{B}}^{r_i}\right), \tag{8.81}$$

where $P_{\mathrm{B}}^{s_i}$ and $P_{\mathrm{B}}^{r_i}$ are respectively the probabilities that the S-R$_i$ and R$_i$-D link is blocked. Finally, by assuming that all the paths are independent, the blocking probability of the relaying or RIS-assisted THz system can be expressed as

$$P_{\mathrm{B}} = \prod_{i=1}^{N} P_{\mathrm{B,i}}. \tag{8.82}$$

In order to quantify the efficiency of this approach, let us use the blockage model presented in Section 8.3.2.2 and let us set $r_B = 50$ m, $k_A = 50\%$, $A = 100$ m^2. In this direction, Figure 8.16 depicts the blocking probability as a function of the distance between S and R_i/RIS, d, which is assumed to be equal to the R_i/RIS-D distance, for different values of N. From this figure, we observe that, for a given N, as d increases, the blocking probability increases. Moreover, for a fixed d, as N increases, the blocking probability decreases. This indicates that relaying and RIS-assisted systems can be used as countermeasures to blockage. Finally, by comparing the results of Figures 8.15 and 16, it becomes evident that CoMP outperforms relaying/RIS-assisted systems in terms of blockage mitigation efficiency.

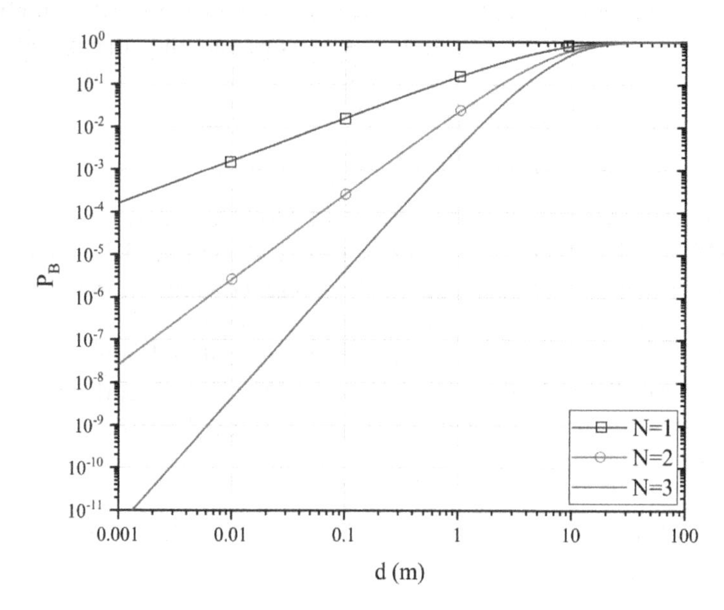

Figure 8.16 Blocking probability as a function of the transmission distance (*d*) for different values of *N* in relaying and RIS-assisted systems.

8.4 Conclusions

This chapter focused on presenting antenna misalignment and blockage models suitable for THz wireless systems. In more detail, the following antenna misalignment models were reported: (i) Gaussian distributed beamsteering errors, (ii) 2D Gaussian shaking of a single node, and (iii) wind vibration antenna misalignment model. For each of these models, the suitability of accommodating the impact of different sources of antenna misalignment was reported and their mathematical tractability was discussed. Additionally, the theoretical framework for the evaluation and quantification of the performance of this system under antenna misalignment, in terms of the expected value of the directional gain, was presented. Next, the chapter focused on presenting beam misalignment mitigation approaches. Two fundamental types of such approaches were identified, namely (i) beam-tracking and (ii) relaying. Beam-tracking approaches were further classified in exhaustive search, hierarchical beam-searching, perturbation, and lower-frequency-based tracking methods. The advantages and disadvantages of each method were highlighted and the trade-off between tracking accuracy, complexity, and latency was discussed. In this sense, the need of presenting novel low-complexity beam-tracking approaches that can achieve specific latency requirements was recognized. To increase the THz wireless network robustness to beam-tracking errors, a relay-based antenna misalignment approach was presented. The performance of this approach was quantified in terms of outage probability. Our results revealed that, although the effect of antenna misalignment is detrimental in SR THz systems, by employing a relative small number of relays, it can be significantly suppressed.

In the second part of this chapter, the impact of blockage on the performance of THz wireless systems was discussed accompanied by a number of different mitigation approaches. In particular, we identified three blockage types, namely: (i) static, (ii) dynamic, and self-blockage. Moreover, different blockage models were reported and their suitability to accommodating the particularities of each blockage type was identified. Moreover, the use of CoMP and reflected link methods in order to mitigate the impact of blockage was discussed. To quantify the efficiency of these methods, we used the blocking probability. Our results revealed that in general CoMP outperforms relaying/RIS-assisted systems in terms of blockage mitigation efficiency. However, both CoMP and relaying/RIS-assisted systems require the investigation and development of channel estimation and predictions techniques.

Other possible future directions may include, the investigation of the robustness of different types of beamforming approaches in beam misalignment and tracking, as well as the quantification of the joint impact of hardware imperfections, beam misalignment, and blockage. In more detail, it would be interesting to study the joint impact of beam misalignment and blocking in wireless communications systems in which the one of the following types of beamforming are employed: (i) analog baseband beamforming with amplitude and phase shifters, (ii) analog RF beamforming, (iii) analog intermediate frequency beamforming, and (iv) digital baseband beamforming, in the presence of different types of hardware imperfections. Finally, adaptive beamforming schemes need to be designed in order to improve the system robustness to beam misalignment and blockage.

Notes

1 Note that ϵ_B and ϵ_U are channel estimation errors and their distribution depends on the type of beam steering estimator that is used. A commonly acceptable distribution that is used for this type of errors is the Gaussian one.
2 The Strouhal number is a body-shape-specific constant. According to (Yam, Leung, Li, and Xue, 1997), for a pole of circular cross-section S = 2.

References

3GPP (2017, July). Study on channel model for frequencies from 0.5 to 100 GHz (Release 14). Tech Report 38.901, 3GPP. V14.1.1.

Abouelseoud, M., and G. Charlton (2013, June). The effect of human blockage on the performance of millimeter-wave access link for outdoor coverage. In *IEEE* 77th Vehicular Technology Conference (VTC Spring), pp. 1–5.

Akyildiz, I. F., J. M. Jornet, and C. Han (2014, September). Terahertz band: Next frontier for wireless communications. *Phys. Commun. 12*, 16–32.

Alduchov, O. A. and R. E. Eskridge (1996). Improved magnus form approximation of saturation vapor pressure. *J. Appl. Meteorol. 35*(4), 601–609.

Alkhateeb, A., O. El Ayach, G. Leus, and R. W. Heath (2014a, October). Channel estimation and hybrid precoding for millimeter wave cellular systems. *IEEE J. Sel. Topics Signal Process. 8*(5), 831–846.

Alkhateeb, A., J. Mo, N. Gonzalez-Prelcic, and R. W. Heath (2014b, December). MIMO precoding and combining solutions for millimeter-wave systems. *IEEE Commun. Mag. 52*(12), 122–131.

Bai, T. and R. W. Heath (2014, November). Analysis of self-body blocking effects in millimeter wave cellular networks. In 48th Asilomar Conference on Signals, Systems and Computers, pp. 1921–1925.

Bai, T. and R. W. Heath (2015, February). Coverage and rate analysis for millimeter-wave cellular networks. *IEEE Trans. Wireless Commun. 14*(2), 1100–1114.

Bai, T., R. Vaze, and R. W. Heath (2012, July). Using random shape theory to model blockage in random cellular networks. In International Conference on Signal Processing and Communications (SPCOM), pp. 1–5.

Bai, T., R. Vaze, and R. W. Heath (2014, September). Analysis of blockage effects on urban cellular networks. *IEEE Trans. Wireless Commun. 13*(9), 5070–5083.

Balanis, C. A. (2008). *Modern Antenna Handbook*. New York: Wiley-Interscience.

Boronin, P., V. Petrov, D. Moltchanov, Y. Koucheryavy, and J. M. Jornet (2014, September). Capacity and throughput analysis of nanoscale machine communication through transparency windows in the terahertz band. *Nano Commun. Networks 5*(3), 72–82.

Boulogeorgos, A.-A. A. (2016, September). Interference mitigation techniques in modern wireless communication systems. Ph. D. thesis, Aristotle University of Thessaloniki, Thessaloniki, Greece.

Boulogeorgos, A.-A. A., and A. Alexiou (2019a). Error analysis of mixed THz-RF wireless systems. *IEEE Commun. Lett.*, 1–1.

Boulogeorgos, A.-A. A. and A. Alexiou (2019b, August). Performance evaluation of the initial access procedure in wireless THz systems. In 16th International Symposium on Wireless Communication Systems (ISWCS), pp. 422–426.

Boulogeorgos, A.-A. A., A. Alexiou, D. Kritharidis, and et al. (2018, July). Wireless terahertz system architectures for networks beyond 5G. White Paper 1.0, Terranova Consortium.

Boulogeorgos, A.-A. A., A. Alexiou, T. Merkle, et al. (2018, June). Terahertz technologies to deliver optical network quality of experience in wireless systems beyond 5G. *IEEE Commun. Mag. 56*(6), 144–151.

Boulogeorgos, A.-A. A., N. D. Chatzidiamantis, and G. K. Karagiannidis (2016, July). Energy detection spectrum sensing under RF imperfections. *IEEE Trans. Commun. 64*(7), 2754–2766.

Boulogeorgos, A.-A. A., S. Goudos, and A. Alexiou (2018, June). Users association in ultra dense THz networks. In IEEE International Workshop on Signal Processing Advances in Wireless Communications (SPAWC), Kalamata, Greece.

Boulogeorgos, A.-A. A. and G. K. Karagiannidis (2017, November). Low-cost cognitive radios against spectrum scarcity. *IEEE Technical Committee on Cognitive Networks Newsletter 3*(2), 30–34.

Boulogeorgos, A.-A. A., E. N. Papasotiriou, and A. Alexiou (2018, July). A distance and bandwidth dependent adaptive modulation scheme for THz communications. In 19th IEEE International Workshop on Signal Processing Advances in Wireless Communications (SPAWC), Kalamata, Greece.

Boulogeorgos, A.-A. A., E. N. Papasotiriou, and A. Alexiou (2019, January). Analytical performance assessment of THz wireless systems. *IEEE Access 7*(1), 1–18.

Boulogeorgos, A.-A. A., E. N. Papasotiriou, and A. Alexiou (2019, September). Analytical performance evaluation of THz wireless fiber extenders. In IEEE 30th Annual International Symposium on Personal, Indoor and Mobile Radio Communications (PIMRC), pp. 1–6.

Boulogeorgos, A.-A. A., E. N. Papasotiriou, J. Kokkoniemi, and et al. (2018, June). Performance evaluation of THz wireless systems operating in 275–400 GHz band. In IEEE 87th Vehicular Technology Conference (VTC Spring), pp. 1–5.

Cherry, S. (2004, July). Edholm's law of bandwidth. *IEEE Spectrum 41*(7), 58–60.

Dong, K., X. Liao, and S. Zhu (2012, November). Link blockage analysis for indoor 60 GHz radio systems. *Electron. Lett. 48*(23), 1506–1508.

Ekti, A. R., A. Boyaci, A. Alparslan, and et al. (2017, September). Statistical modeling of propagation channels for terahertz band. In IEEE Conference on Standards for Communications and Networking (CSCN), Helsinki, Finland, pp. 275–280.

Erturk, O. and T. Yilmaz (2018, July). A hexagonal grid based human blockage model for the 5G low terahertz band communications. In IEEE 5G World Forum (5GWF), pp. 395–398.

Farid, A. A., and S. Hranilovic (2007, July). Outage capacity optimization for free-space optical links with pointing errors. *J. Lightwave Technol. 25*(7), 1702–1710.

Federici, J. and L. Moeller (2010, June). Review of terahertz and subterahertz wireless communications. *J. Appl. Phys. 107*(11), 111101.

Gao, X., L. Dai, Y. Zhang, T. Xie, X. Dai, and Z. Wang (2017, July). Fast channel tracking for terahertz beamspace massive MIMO systems. *IEEE Trans. Veh. Technol. 66*(7), 5689–5696.

Gapeyenko, M., A. Samuylov, M. Gerasimenko, et al. (2017, November). On the temporal effects of mobile blockers in urban millimeter-wave cellular scenarios. *IEEE Trans. Veh. Technol. 66*(11), 10124–10138.

Gapeyenko, M., A. Samuylov, M. Gerasimenko, D. Moltchanov, S. Singh, E. Aryafar, S. Yeh, N. Himayat, S. Andreev, and Y. Koucheryavy (2016, May). Analysis of human-body blockage in urban millimeter-wave cellular communications. In *IEEE International Conference on Communications (ICC)*, pp. 1–7.

Ghasempour, Y., C. R. C. M. da Silva, C. Cordeiro, and E. W. Knightly (2017, December). IEEE 802.11ay: Next-generation 60 GHz communication for 100 Gb/s Wi-Fi. *IEEE Commun. Mag. 55*(12), 186–192.

Giordani, M., M. Mezzavilla, C. N. Barati, S. Rangan, and M. Zorzi (2016, March). Comparative analysis of initial access techniques in 5G mmWave cellular networks. In *Annual Conference on Information Science and Systems (CISS)*, pp. 268–273.

Han, B., L. Wang, and H. D. Schotten (2017, December). A 3D human body blockage model for outdoor millimeter-wave cellular communication. *Phys. Commun. 25*(P2), 502–510.

He, J., T. Kim, H. Ghauch, K. Liu, and G. Wang (2014, December). Millimeter wave MIMO channel tracking systems. In IEEE Globecom Workshops (GC Wkshps), pp. 416–421.

Huq, K. M. S., J. M. Jornet, W. H. Gerstacker, and et al. (2018, June). THz communications for mobile heterogeneous networks. *IEEE Commun. Mag. 56*(6), 94–95.

Hur, S., T. Kim, D. J. Love, J. V. Krogmeier, T. A. Thomas, and A. Ghosh (2013, October). Millimeter wave beamforming for wireless backhaul and access in small cell networks. *IEEE Trans. Commun. 61*(10), 4391–4403.

Jain, I. K., R. Kumar, and S. Panwar (2018, September). Driven by capacity or blockage? A millimeter wave blockage analysis. In 30th International Teletraffic Congress (ITC 30), Volume 1, pp. 153–159.

Jornet, J. M. and I. F. Akyildiz (2011, October). Channel modeling and capacity analysis for electromagnetic wireless nanonetworks in the terahertz band. *IEEE Trans. Wireless Commun. 10*(10), 3211–3221.

Khan, T. A. and R. W. Heath (2017, October). Analyzing wireless power transfer in millimeter wave networks with human blockages. In IEEE Military Communications Conference (MILCOM), pp. 115–120.

Kim, H. (2015). Channel estimation and equalization. In Wireless Communications Systems Design, pp. 279–299. New York: Wiley.

Koch, M. (2007). Terahertz communications: A 2020 vision. In R. E. Miles, X.-C. Zhang, H. Eisele, and A. Krotkus (Eds.), *Terahertz Frequency Detection and Identification of Materials and Objects*, pp. 325–338. Dordrecht: Springer.

Kokkoniemi, J., A.-A. A. Boulogeorgos, M. Aminu, J. Lehtomaki, A. Alexiou, and M. Juntti (2020, April). Impact of beam misalignment on THz wireless systems. *Nano Communications Networks*. 24, 100302.

Kokkoniemi, J., A.-A. A. Boulogeorgos, M. U. Aminu, et al. (2019, June). Stochastic antenna misalignment impact on the antenna gains in the THz band. In Third International Balkan Conference on Communications and Networking, Skopje, North Macedonia.

Kokkoniemi, J., A.-A. A. Boulogeorgos, J. Lehtomaki, G. Ntouni, M. Juntti, and A. Alexiou (2018, August). D3.2 channel and propagation modelling and characterization. Research Report, Terranova Consortium.

Kokkoniemi, J., J. Lehtom¨aki, and M. Juntti (2018, April). Simplified molecular absorption loss model for 275–400 gigahertz frequency band. In 12th European Conference on Antennas and Propagation (EuCAP), London, UK.

Lee, J., J. Liang, M. Kim, and J. Park (2017, September). Millimeter-wave beam misalignment analysis based on 28 and 38 GHz urban measurement. In IEEE 86th Vehcile Technology Conference (VTC-Fall), pp. 1–5.

Lin, J. and M. A. Weitnauer (2014, December). Pulse-level beam-switching MAC with energy control in picocell terahertz networks. In IEEE Global Communications Conference, pp. 4460–4465.

MacCartney, G. R. and T. S. Rappaport (2019, July). Millimeter-wave base station diversity for 5G coordinated multipoint (CoMP) applications. *IEEE Trans. Wireless Commun. 18*(7), 3395–3410.

MacCartney, G. R., T. S. Rappaport, and S. Rangan (2017, Dec). Rapid fading due to human blockage in pedestrian crowds at 5G millimeter-wave frequencies. In IEEE Global Communications Conference (GLOBECOM), pp. 1–7.

Machado, J., A.-A. A. Boulogeorgos, E. Escher, and et al. (2019, March). Wireless Terahertz System Applications for Networks beyond 5G. White Paper 1.0, Terranova Consortium.

Mokhtar, M., A.-A. A. Boulogeorgos, G. K. Karagiannidis, and N. Al-Dhahir (2014, May). OFDM opportunistic relaying under joint transmit/receive I/Q imbalance. *IEEE Transactions on Communications 62*(5), 1458–1468.

Moltchanov, D., P. Kustarev, and Y. Koucheryavy (2018). Analytical approximations for interference and sir densities in terahertz systems with atmospheric absorption, directional antennas and blocking. *Phys. Commun. 26*, 21–30.

Monti, A., J. Soric, A. Alu, F. Bilotti, A. Toscano, and L. Vegni (2012). Overcoming mutual blockage between neighboring dipole antennas using a low-profile patterned metasurface. *IEEE Antennas Wireless Propag. Lett. 11*, 1414–1417.

Niknam, S., B. Natarajan, and R. Barazideh (2018, April). Interference analysis for finite-area 5G mmWave networks considering blockage effect. *IEEE Access 6*, 23470–23479.

Papasotiriou, E. N., J. Kokkoniemi, A.-A. A. Boulogeorgos, and et al. (2018, September). A new look to 275 to 400 GHz band: Channel model and performance evaluation. In IEEE International Symposium on Personal, Indoor and Mobile Radio Communications (PIMRC), Bolonia, Italy.

Papoulis, A. and S. Pillai (2002). *Probability, Random Variables, and Stochastic Processes*. McGraw-Hill Series in Electrical Engineering: Communications and Signal Processing. New Delhi: Tata McGraw-Hill.

Petrov, V., M. Komarov, D. Moltchanov, J. M. Jornet, and Y. Koucheryavy (2016, Dec). Interference analysis of EHF/THF communications systems with blocking and directional antennas. In IEEE Global Communications Conference (GLOBECOM), pp. 1–7.

Petrov, V., A. Pyattaev, D. Moltchanov, and Y. Koucheryavy (2016, October). Terahertz band communications: Applications, research challenges, and standardization activities. In International Congress on Ultra Modern Telecommunications and Control Systems and Workshops (ICUMT), Lisbon, Portugal, pp. 183–190.

Piersol, A. and T. Paez (2009). *Shock and Vibration Handbook*, 6th ed. New York: McGraw Hill.

Piesiewicz, R., T. Kleine-Ostmann, N. Krumbholz, and et al. (2007, December). Short-range ultra-broadband terahertz communications: Concepts and perspectives. *IEEE Antennas Propag. Mag. 49*(6), 24–39.

Pradhan, C., A. Li, L. Zhuo, and et al. (2019, October). Beam misalignment aware hybrid transceiver design in mmWave MIMO systems. *IEEE Trans. Veh. Technol. 68*(10), 10306–10310.

Raghavan, V., L. Akhoondzadeh-Asl, V. Podshivalov, J. Hulten, M. A. Tassoudji, O. H. Koymen, A. Sampath, and J. Li (2019, July). Statistical blockage modeling and robustness of beamforming in millimeter-wave systems. *IEEE Trans. Microw. Theory Techn. 67*(7), 3010–3024.

Renzo, M. D., M. Debbah, and e. a. Phan-Huy, Dinh-Thuy (2019, May). Smart radio environments empowered by reconfigurable ai meta-surfaces: An idea whose time has come. *EURASIP Journal on Wireless Communications and Networking 2019*(1), 129.

Salgado, J., N. Silva, R. Elschner, A.-A. A. Boulogeorgos, A. Katsiotis, D. Kritharidis, A. Alexiou, J. Kokkoniemi, and J. Lehtomaki (2017, December). D2.1 Terranova system requirements. Tech Report, Terranova Consortium.

Seeds, A. J., H. Shams, M. J. Fice, and C. C. Renaud (2015, February). Terahertz photonics for wireless communications. *J. Lightwave Technol. 33*(3), 579–587.

Shinozuka, M. and G. Deodatis (1991, April). Simulation of stochastic processes by spectral representation. *Appl. Mech. Rev. 44*(4), 191–204.

Song, H. and T. Nagatsuma (2011, September). Present and future of terahertz communications. *IEEE Trans. THz Sci. Technol. 1*(1), 256–263.

Staple, G. and K. Werbach (2004, March). The end of spectrum scarcity [spectrum allocation and utilization]. *IEEE Spectrum 41*(3), 48–52.

Tong, W. and C. Han (2017, December). MRA-MAC: a multi-radio assisted medium access control in terahertz communication networks. In IEEE Global Communications Conference (GLOBECOM), pp. 1–6.

Wang, Y., Y. Niu, and e. a. H. Wu (2019, July). Relay assisted concurrent scheduling to overcome blockage in full-duplex millimeter wave small cells. *IEEE Access 7*, 105755–105767.

Wang, Y., K. Venugopal, A. F. Molisch, and R. W. Heath (2017, May). Blockage and coverage analysis with mmwave cross street bss near urban intersections. In IEEE International Conference on Communications (ICC), pp. 1–6.

Wieringa, J. (1986, July). Roughness-dependent geographical interpolation of surface wind speed averages. *Quart. J. Roy. Meteor. Soc. 112*(473), 867–889.

Wildman, J., P. H. J. Nardelli, M. Latva-aho, and S. Weber (2014, Dec). On the joint impact of beamwidth and orientation error on throughput in directional wireless poisson networks. *IEEE Trans. Wireless Commun. 13*(12), 7072–7085.

Yam, L., T. Leung, D. Li, and K. Xue (1997). Use of ambient response measurements to determine dynamic characteristics of Slender structures. *Eng. Struct. 19*(2), 145–150.

Yao, X.-W. and J. M. Jornet (2016). TAB-MAC: Assisted beamforming MAC protocol for terahertz communication networks. *Nano Commun. Networks 9*, 36–42.

Zafer, M. and E. Modiano (2006, April). Blocking probability and channel assignment in wireless networks. *IEEE Trans. Wireless Commun. 5*(4), 869–879.

Zhang, C., D. Guo, and P. Fan (2016, May). Tracking angles of departure and arrival in a mobile millimeter wave channel. In IEEE International Conference on Communications (ICC), pp. 1–6.

Zhang, C., K. Ota, J. Jia, and M. Dong (2018, June). Breaking the blockage for big data transmission: Gigabit road communication in autonomous vehicles. *IEEE Commun. Mag. 56*(6), 152–157.

Zhang, J., X. Ge, Q. Li, M. Guizani, and Y. Zhang (2017, April). 5G millimeter-wave antenna array: Design and challenges. *IEEE Wireless Commun. 24*(2), 106–112.

Zhou, L. and Y. Ohashi (2015, September). Fast codebook-based beamforming training for mmwave MIMO systems with subarray structures. In IEEE 82nd Vehicular Technology Conference (VTC2015-Fall), pp. 1–5.

Chapter 9

Hybrid Beamforming in Wireless Terahertz Communications

Hang Yuan, Nan Yang, Kai Yang, and Jianping An

Contents

9.1 Introduction

Terahertz (THz) communication has been widely acknowledged as an avant-garde technology in the next generation wireless networks, due to its ultra-large usable bandwidth and huge potential to support ultra-high-data-rate transmissions (Akyildiz et al. 2014). However, severe path loss of THz waves affects the propagation intensely, which restricts the transmission distance within tens of meters. Despite such path loss, the extremely short wavelength at the THz band has a superiority, that is, large-scale or even massive antennas can be made into small size at THz devices. Thus, regular and massive MIMO techniques can be utilized to enable the high beamsteering gain and combat the severe path loss. Notably, one of the most important challenges to realize massive MIMO in THz wireless communication systems is to deal with hardware limitations on THz devices. If not carefully designed, the transmission strategies and signal processing algorithms for THz systems may incur large power consumption and high complexity. Undoubtedly, this will hinder the effectiveness of using THz communications in the next generation wireless networks.

Against this background, novel low-complexity architectures are mandated to facilitate the deployment of regular and massive MIMO in THz communications. In this chapter, we introduce hybrid beamforming architectures in THz wireless systems, where the number of radiofrequency (RF) chains at the transmitter can be much less than that of transmit antennas. Notably, compared to the conventional fully digital beamforming, hybrid beamforming reduces hardware complexity significantly by decreasing the number of RF chains, thus being of great practicality for implementation.

Hybrid beamforming was first developed for achieving multiplexing and diversity tradeoff in conventional MIMO systems (Zhang et al. 2005). Due to its low hardware complexity and high spectrum efficiency, hybrid beamforming has recently been investigated in millimeter wave (mmWave) (Heaths et al. 2016) and THz communication systems. For example, considering the sparsity of mmWave channels, the hybrid beamforming problem was formalized as a sparse approximation problem, where an orthogonal matching pursuit (OMP)-based solution was proposed (Ayach et al. 2014). The findings showed that mmWave hybrid beamforming systems are promising to approach their unconstrained performance limits, even with strict hardware constraints at transceivers (Ayach et al. 2014). In indoor THz systems, two typical hybrid beamforming architectures, that is, the array-of-subarray architecture and the fully connected architecture, were investigated (Lin and Li 2016a). Specifically, it was shown that the energy efficiency of the array-of-subarray architecture is higher than that of the fully connected architecture (Lin and Li 2016a). Furthermore, to explore the unique frequency–distance-dependent characteristics of THz channels, an adaptive distance-aware scheme with antenna selection and power allocation strategy were proposed, which achieved significant gains over conventional schemes (Lin and Li 2015a). Despite the importance, these studies barely considered the THz frequency selective fading. THz communication is very promising to be operated over tens-of-GHz bandwidth. Hence, developing novel frequency selective hybrid beamforming transmission schemes is an urgent research work for practical THz wireless communications.

This chapter provides an overview of hybrid beamforming techniques in wireless THz communication systems. The fundamental knowledge of hybrid beamforming in wireless THz systems and its ergodic capacity analysis are presented in Section 9.2. Next, Section 9.3 describes the concept of frequency selective hybrid beamforming over the ultra-wide THz band. Specifically, a statistical eigen-based hybrid beamforming design with digital compensation beamformer is presented to approach the fully digital beamforming design. Furthermore, the multiuser wideband hybrid beamforming and distance-aware multi-carrier (DAMC) modulation are discussed. The summary of this chapter is finally presented in Section 9.4.

9.2 Basics and Ergodic Capacity Analysis of Hybrid Beamforming in THz Wireless Systems

Complex THz devices impose strict hardware constraints on THz wireless systems such that conventional transceiver architectures, for example, the fully digital processing at the baseband, are no longer practical. Hence, it is imperative to develop a novel low-complexity

THz transmission architecture which exploits the capabilities of THz transceivers and antennas as well as the peculiarities of THz channels.

In this section, we introduce the basic hybrid beamforming architecture in THz wireless systems and evaluate its performance. Particularly, a close-form upper bound on the capacity of using hybrid beamforming in the system is provided.

9.2.1 System Model

We consider a THz wireless system, where a multi-antenna base station (BS) communicates to a multi-antenna user equipment (UE) in the THz band. The BS adopts the hybrid beamforming architecture which drives massive antennas via only a few numbers of RF chains. This subsection presents some important descriptions of our considered system model, including the block diagram of the transmitter architecture and the THz channel model.

9.2.1.1 Hybrid Beamforming Architecture

A block diagram of the narrow-band hybrid beamforming architecture is depicted in Figure 9.1. The BS equipped with N_{BS} antennas and N_{RF} RF chains transmits to the UE via N_s data streams, such that $N_s \leq N_{RF} \leq N_{BS}$. The transmit signal vector $\boldsymbol{s}$ is first precoded by the $N_{RF} \times N_s$ digital baseband beamforming $\boldsymbol{F}_{BB}$. Then, the precoded digital vector is transformed into the analog vector before applying the $N_{BS} \times N_{RF}$ analog RF beamforming $\boldsymbol{F}_{RF}$. Here, high-resolution digital-to-analog converters (DACs) are assumed and the quantization error is ignored. Hence, the discrete-time transmit signal at the BS is expressed as

$$\boldsymbol{x} = \boldsymbol{F}_{RF}\boldsymbol{F}_{BB}\boldsymbol{s}, \tag{9.1}$$

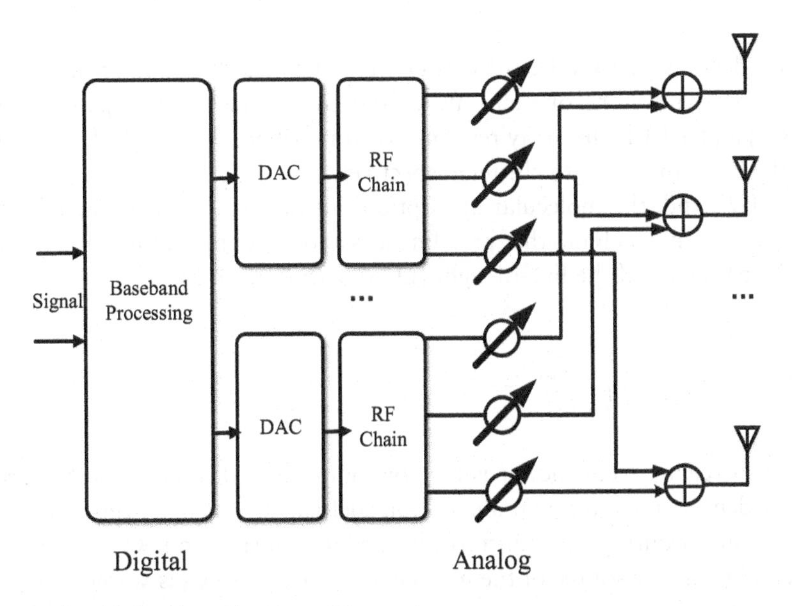

Figure 9.1 Narrow-band hybrid beamforming architecture.

where s is an $N_s \times 1$ signal vector transmitted to the UE. Here, s is normalized such that the transmit power constraint which is given by $\mathbb{E}[ss^H] = I_{N_s}$, where $(\cdot)^H$ denotes the Hermitian transpose and $\mathbb{E}[\cdot]$ denotes the mathematical expectation.

In hybrid beamforming architecture, the analog RF beamforming F_{RF} is implemented by employing some analog components such as THz phase shifters, switching networks, and lens antennas. In this case, a constant-modulus constraint is imposed on all entries of the RF beamforming matrix, such that $[F_{RF}]_{i,j} = e^{j\phi_{i,j}}$, where $\phi_{i,j}$ is a beam-steering angle implemented by phase shifters and $[\cdot]_{i,j}$ denotes the entry in the ith row and the jth column.

At the UE, the hybrid architecture with N_{UE} antennas and a single RF chain is considered. The received signal is given by

$$y_u = w^H H_u x + n_u, \tag{9.2}$$

where w is the combiner at the UE, H_u is an $N_{UE} \times N_{BS}$ matrix that denotes the THz channel, and n_u represents the additive Gaussian white noise (AWGN).

9.2.1.2 THz Channel Model

The geometric channel model associates clustering and stochastic angles of departure (AoD) and angles of arrival (AoA) to each path. Based on THz-band measurement results, a modified geometric model to reflect the stochastic characteristics of THz channels was established (Priebe and Kurner 2013), which is described as follows:

In the THz channel model, arrival paths consist of several clusters, each of which is composed of several paths caused by reflection. Therefore, the channel response matrix for one antenna array is given by

$$H_u = \sum_{i=1}^{N_{clu}} \sum_{l=1}^{L_{ray}^i} \alpha_{i,l} G_t G_r a_r(\theta_{i,l}) a_t^H(\phi_{i,l}), \tag{9.3}$$

where N_{clu} denotes the number of clusters, L_{ray}^i denotes the number of paths in the ith cluster, $\alpha_{i,l}$ are path gains of each ray, G_t and G_r are the antenna gains. Moreover, we denote $a_t(\cdot)$ and $a_r(\cdot)$ as the array response vectors, where $\theta_{i,l}$ and $\phi_{i,l}$ denote the AoD and AoA of the lth path in the ith cluster, respectively.

In the THz band, the molecular absorption drastically affects the signal transmission. The path loss mainly includes the spreading loss and the unique molecular absorption loss. Hence, the path gain of the line-of-sight (LOS) path is given by

$$|\alpha^L(f,d)|^2 = \left(\frac{c}{4\pi f d}\right)^2 e^{-k_{abs}(f)d}, \tag{9.4}$$

where f is the center frequency, d is the communication distance, c is the speed of light, and $k_{abs}(f)$ denotes the molecular absorption coefficient, which depends on the molecular-level composition during the transmission medium (Jornet and Akyildiz 2011). At THz frequencies, the major source of the molecular absorption loss is water vapor, which also indicates that the THz signal transmission is highly dependent on the humidity.

Considering the severe reflection loss, only the second order reflection is modelled in the THz channel. The path loss of non-line-of-sight (NLOS) paths can be obtained with two Fresnel reflection coefficients (Han et al. 2014). Indeed, THz channel only has several NLOS paths and exhibits unique channel sparsity. We note that the path loss gap between the THz LOS path and NLOS paths is generally up to 15 dB. This gap is much larger than the LOS–NLOS gap in mmWave channels. Thus, THz transmission is absolutely LOS-dominant. In particular, THz channels are much easier to be blocked than mmWave channels.

9.2.2 Analysis of Ergodic Capacity

With the transmission signal model and the THz channel model described before, we next investigate the hybrid beamforming design and analyze its associated ergodic capacity.

As aforementioned, the hybrid beamforming consists of digital baseband beamforming and analog RF beamforming. For the analog beamforming, the beam-steering vector, $\boldsymbol{a}_r(\phi_0)$, is the optimal analog beamforming vector with a target AoA, ϕ_0. Similarly, at the transmitter side, the beam-steering vector, $\boldsymbol{a}_t(\theta_0)$, is the optimal analog beamforming vector with a target AoD, θ_0. In this case, the effective channel response at the baseband is presented as

$$\hat{h}_{\mathrm{n}} = \boldsymbol{a}_{\mathrm{r}}^{\mathrm{H}}(\phi_0)\boldsymbol{H}_{\mathrm{u}}\boldsymbol{a}_{\mathrm{t}}(\theta_0) = \sum_{i=1}^{N_{\mathrm{clu}}}\sum_{l=1}^{L_{\mathrm{ray}}^{i}} \alpha_{\mathrm{i,l}} G_{\mathrm{t}} G_{\mathrm{r}} \mathcal{A}_{\mathrm{r}}^{\mathrm{eq}}(\phi_{\mathrm{i,l}}) \mathcal{A}_{\mathrm{t}}^{\mathrm{eq}}(\theta_{\mathrm{i,l}}), \tag{9.5}$$

where

$$\begin{aligned}\mathcal{A}_r^{eq}(\phi) &= \frac{1}{\sqrt{N_{\mathrm{UE}}}}\sum_{n_r=0}^{N_{\mathrm{UE}}-1} e^{j\frac{2\pi a n_r}{\lambda_c}(\sin(\phi)-\sin(\phi_0))} \\ &= \frac{1-e^{j\pi N_{\mathrm{UE}}(\sin\phi-\sin\phi_0)}}{\sqrt{N_{\mathrm{UE}}}\left(1-e^{j\pi(\sin\phi-\sin\phi_0)}\right)} \approx \frac{\sin\left(\pi N_{\mathrm{UE}}(\sin\phi-\sin\phi_0)\right)}{\sqrt{N_{\mathrm{UE}}}\sin\left(\pi(\sin\phi-\sin\phi_0)\right)}\end{aligned} \tag{9.6}$$

and

$$\begin{aligned}\mathcal{A}_t^{eq}(\theta) &= \frac{1}{\sqrt{N_{\mathrm{BS}}}}\sum_{n_t=0}^{N_{\mathrm{BS}}-1} e^{j\frac{2\pi a n_t}{\lambda_c}(\sin(\theta)-\sin(\theta_0))} \\ &= \frac{1-e^{j\pi N_{\mathrm{BS}}(\sin\theta-\sin\theta_0)}}{\sqrt{N_{\mathrm{BS}}}\left(1-e^{j\pi(\sin\theta-\sin\theta_0)}\right)} \approx \frac{\sin\left(\pi N_{\mathrm{BS}}(\sin\theta-\sin\theta_0)\right)}{\sqrt{N_{\mathrm{BS}}}\sin\left(\pi(\sin\theta-\sin\theta_0)\right)}.\end{aligned} \tag{9.7}$$

Given the analog beamforming vector, the effective channel response, $\hat{\boldsymbol{h}}$, has a MISO structure. This channel vector is expressed as

$$\hat{\boldsymbol{h}} = w^H H_u F_{\mathrm{RF}} = \left[\hat{h}_1, \hat{h}_{\mathrm{n}}, \ldots, \hat{h}_{N_{\mathrm{RF}}}\right]. \tag{9.8}$$

Then, the optimal digital beamforming vector for the effective MISO channel is expressed as (Vu 2011)

$$f_{BB} = \frac{\hat{\boldsymbol{h}}^H}{\|\hat{\boldsymbol{h}}\|}, \tag{9.9}$$

where $\|\cdot\|$ denotes the norm of a vector. Hence, we assume the channel is flat in the bandwidth, B, the ergodic capacity, denoted by $C(f,d)$, with the hybrid beamforming is given by (Lin and Li 2015b)

$$\begin{aligned}
C(f,d) &= \mathbb{E}\left[B\log_2\left(1+\frac{P}{N_0}\left\|\hat{\mathbf{h}}\right\|^2\right)\right] \\
&= \mathbb{E}\left[B\log_2\left(1+\frac{P}{N_0}\sum_{n=1}^{N_{RF}}|\hat{h}_n|^2\right)\right] \\
&\le B\log_2\left(1+\frac{P}{N_0}\sum_{n=1}^{N_{RF}}\mathbb{E}\left[|\hat{h}_n|^2\right]\right) \\
&= B\log_2\left(1+\frac{PN_{RF}}{N_0}\mathbb{E}\left[\left|\sum_{i=1}^{N_{clu}}\sum_{l=1}^{L_{ray}^i}\alpha_{i,l}G_tG_r\mathcal{A}_r^{eq}(\phi_{i,l})\mathcal{A}_t^{eq}(\theta_{i,l})\right|^2\right]\right) \\
&\le B\log_2\left(1+\frac{PN_{RF}}{N_0}\mathbb{E}\left[\sum_{i=1}^{N_{clu}}\sum_{l=1}^{L_{ray}^i}|\alpha_{i,l}|^2\right]\mathbb{E}\left[\sum_{i=1}^{N_{clu}}\sum_{l=1}^{L_{ray}^i}\left|G_tG_r\mathcal{A}_r^{eq}(\phi_{i,l})\mathcal{A}_t^{eq}(\theta_{i,l})\right|^2\right]\right),
\end{aligned} \tag{9.10}$$

where P denotes the transmit power and N_0 denotes the AWGN power.

The ergodic capacity of the indoor wireless THz system using narrow-band hybrid beamforming is upper bounded by (9.10). Based on the statistical characteristics of the THz channel, a close-form solution for $C(f,d)$ is given by (Lin and Li 2015b):

$$C(f,d) \le B\log_2\left(1+\frac{PN_{RF}N_{BS}N_{UE}}{N_0}\beta(f,d)\right), \tag{9.11}$$

where β is a frequency and distance dependent variable. Hence, the upper bound on the ergodic capacity is jointly determined by the communication frequency and distance, the number of antennas, and the number of RF chains.

9.2.3 Numerical Results

We next investigated a practically important question: "*How large is the capacity gap between the novel hybrid beamforming architecture and the conventional fully digital beamforming architecture?*" To this end, we provide a numerical example in Figure 9.2 where a 64-antenna BS transmits with a 16-antenna UE. In this figure, we plot the achievable spectral efficiencies of schemes against the transmit SNR for comparison: (1) fully digital beamforming scheme with 64 RF chains, (2) hybrid beamforming with 3 RF chains, and (3) hybrid beamforming with only 1 RF chain. We observe from this figure that the spectral efficiency of the hybrid beamforming approaches that of the fully digital beamforming when the number of RF chains is very large. Moreover, the hybrid beamforming with only

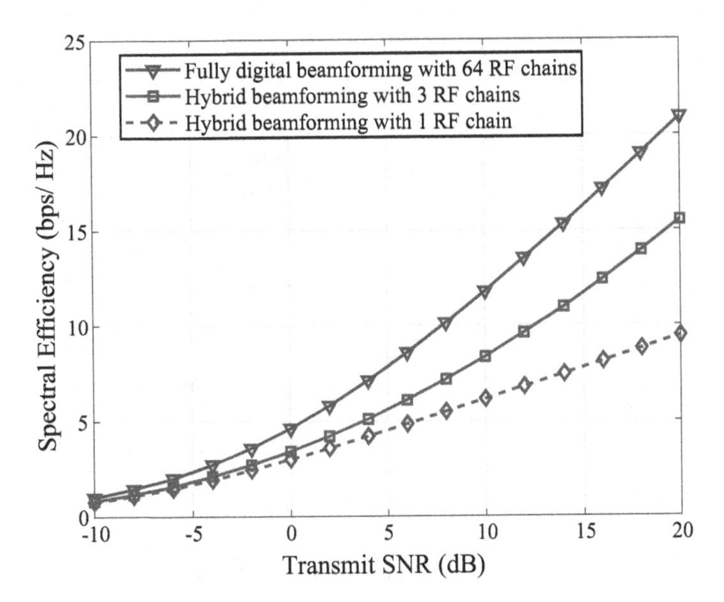

Figure 9.2 Spectral efficiencies versus transmit SNR with $N_{BS} = 64$ and $N_{UE} = 16$. The transmission distance is 1 m. The transmission frequency is 0.3 THz.

1 RF chain achieves almost half spectral efficiency of the fully digital beamforming. This numerical result demonstrates the crucial superiority of the hybrid beamforming architecture for practical implementation, since the hybrid beamforming brings about much lower hardware and computational complexity.

In Figure 9.3, we show the spectral efficiencies of the fully digital and hybrid beamforming schemes with 3 RF chains versus the transmission distance. Here, we consider two cases: (1) the BS is equipped with 144 antennas and (2) the BS is equipped with 64 antennas. We first observe that, as the transmission distance increases, the achievable spectral efficiencies rapidly decrease, which is caused by the severe spread loss in the THz band. We also observe that when the number of antennas at the BS increases, the performance of the hybrid beamforming improves. Specifically, the hybrid beamforming with three RF chains and 144 antennas achieves almost the same spectral efficiencies as the fully digital scheme with 64 RF chains and 64 antennas. This observation indicates that the THz system performance can be boosted by increasing the number of antennas rather than RF chains. The finding is of particular importance for green communications as the utilization of RF chains occurs high power consumption.

9.3 Frequency Selective Hybrid Beamforming in Wideband THz Wireless Systems

The ultra-wide-band THz channel experiences unique frequency selective fading. Actually, the width of THz sub-band, approximately 200 GHz (Lin and Li 2016b), is much larger than the channel coherence bandwidth (Han et al. 2015). Hence, hybrid beamforming scheme considering frequency selective fading is of great significance for wideband THz wireless systems. In this section, the concept of frequency selective hybrid beamforming and corresponding design guidelines are presented.

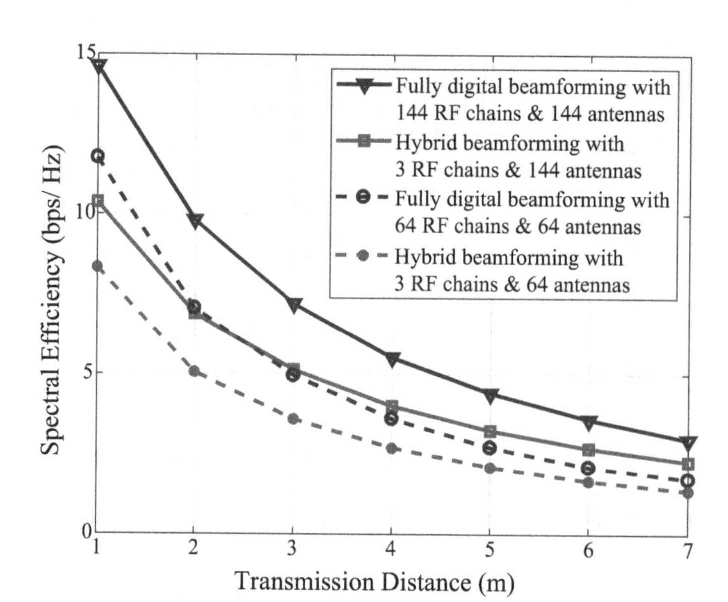

Figure 9.3 Spectral efficiencies versus the transmission distance with $N_{\mathrm{UE}} = 16$. The transmit SNR is 10dB. The transmission frequency is 0.3 THz.

9.3.1 Statistical Eigen Scheme with Digital Compensation Beamforming

We consider a frequency selective hybrid beamforming architecture for a THz system. In this system, the analog beamforming operated in RF chains is frequency-flat, while the digital beamforming is variable for different carriers. The block diagram of the frequency selective hybrid beamforming architecture is depicted in Figure 9.4.

In the frequency selective hybrid beamforming, the BS transmits signal on multiple carriers. The transmit data is assumed to be modulated across K carriers. At the transmitter, an $N_s \times 1$ vector $\boldsymbol{s}[k]$ is the data transmitted at the kth carrier, where $k \in \{1,2,\ldots,K\}$. We note that $\boldsymbol{s}[k]$ is limited by the normalized transmit power, which is given by $\mathbb{E}\left[\boldsymbol{s}[k]\boldsymbol{s}^H[k]\right] = \boldsymbol{I}_{N_s}$.

At the baseband of the BS, $\boldsymbol{s}[k]$ is first precoded by an $N_{\mathrm{RF}} \times N_s$ digital baseband beamforming matrix $\boldsymbol{F}_{\mathrm{BB}}[k]$ in each carrier. The precoded signal is then transformed to the time domain through K-point inverse fast Fourier transform (IFFT). After this, the signal is transformed by high-resolution DACs, before applying the $N_{\mathrm{BS}} \times N_{\mathrm{RF}}$ analog RF beamforming $\boldsymbol{F}_{\mathrm{RF}}$. Here, the common analog beamforming needs to serve all carriers. Hence, the transmit data vector on the kth carrier is presented as

$$\boldsymbol{x}[k] = \boldsymbol{F}_{\mathrm{RF}}\boldsymbol{F}_{\mathrm{BB}}[k]\boldsymbol{s}[k]. \tag{9.12}$$

We also impose a power constraint on $\boldsymbol{F}_{\mathrm{BB}}[k]$ such that $\left\|\boldsymbol{F}_{\mathrm{RF}}\boldsymbol{F}_{\mathrm{BB}}[k]\right\|_F^2 = P$, where $\|\cdot\|_F$ denotes the Frobenius norm.

We clarify that the design of an appropriate frequency selective hybrid beamforming scheme is challenging, due to the following factors:

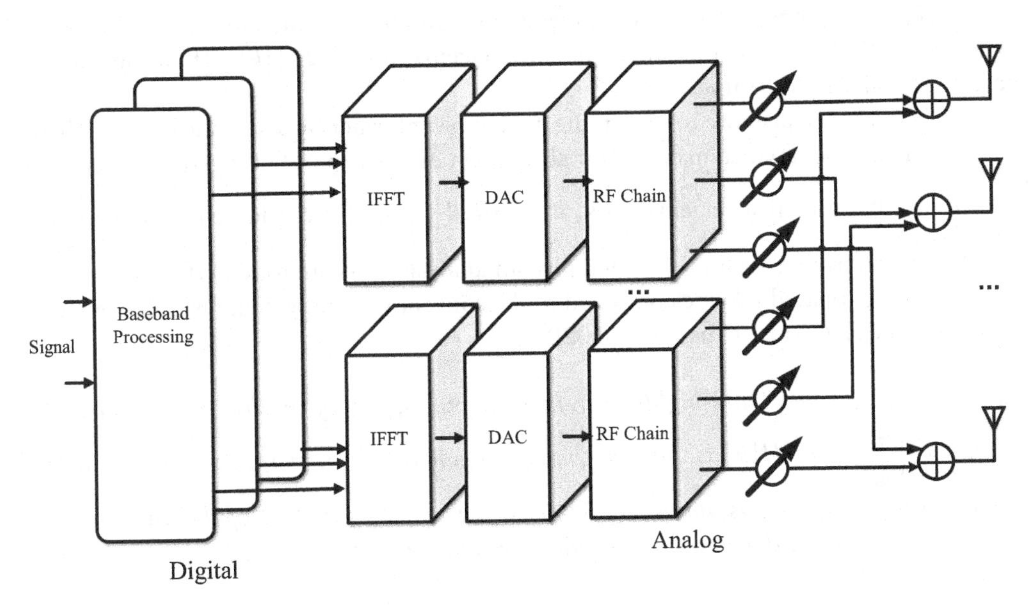

Figure 9.4 Frequency selective hybrid beamforming architecture.

1. The coupling effect between the analog beamforming matrix and digital beamforming matrices, that is, $\left\|\boldsymbol{F}_{\mathrm{RF}}\boldsymbol{F}_{\mathrm{BB}}[k]\right\|_F^2 = P$.
2. The non-convex constraints on the entries of analog beamforming matrix (Gao et al. 2018), that is, $[\boldsymbol{F}_{\mathrm{RF}}]_{i,j} = e^{j\phi_{i,j}}$.
3. The consistency of analog beamforming for all carriers, which is a unique characteristic of the multi-carrier hybrid beamforming (Alkhateeb and Heath 2016).

Against this background, we address all the three factors and provide a statistical eigen scheme with digital compensation beamforming, aiming to approach the fully digital beamforming design. We propose a two-step design as follows (Yuan et al. 2018a):

1. We design the analog RF beamforming, $\boldsymbol{F}_{\mathrm{RF}}$, to maximize the received power under the hardware constraints, where the statistical eigenvalue decomposition (EVD) of channel covariance matrices across all subcarriers is used.
2. We develop the digital baseband beamforming, $\boldsymbol{F}_{\mathrm{BB}}[k]$, to compensate for the performance loss associated with the RF beamforming, which is resulting from the hardware constraint and the multi-carrier channel difference.

9.3.1.1 Analog Beamforming Design

In conventional hybrid beamforming, the optimal analog beamforming matrix consists of eigenvectors of $\boldsymbol{H}^H[k]\boldsymbol{H}[k]$. In the presence of frequency selective fading, different carriers have different channel coefficients, which leads to different optimal analog beamforming and makes the design very challenging. However, we note that the channels across carriers are highly correlated in THz hybrid beamforming. Specifically, antenna array response vectors, $\boldsymbol{a}_r(\phi)$ and $\boldsymbol{a}_t(\theta)$, are the same across all carriers. Also, the eigenvectors of

$H^H[k]H[k]$ are approximately the same as $a_t(\theta)$ in massive antenna scenarios (Sohrabi and Yu 2017). Hence the dominant statistical eigenvectors of $H^H[k]H[k]$ for all carriers are approximately the same.

In this work, we propose to design the analog beamformer by a statistical EVD scheme where channel covariance matrices on all carriers are incorporated together. Specifically, the dominant statistical eigenvectors of $\frac{1}{K}\sum_{k=1}^{K} H^H[k]H[k]$ are adopted as the optimal analog beamforming vectors. We then give an analysis in asymptotic performance of the statistical eigen-based scheme compared to the conventional fully digital scheme, which is presented in Theorem 1 (Yuan et al. 2020).

Theorem 1. For wideband highly correlated channels, if they are low-rank, that is, the rank of $\frac{1}{K}\sum_{k=1}^{K} H^H[k]H[k]$ *is smaller than or equal to* N_{RF}*, the statistical eigen hybrid beamforming scheme has almost the same ergodic capacity as that of the unconstrained digital scheme when the constant-modulus constraint is not considered.*

Proof: Let the rank of $\frac{1}{K}\sum_{k=1}^{K} H^H[k]H[k]$, be N_H, such that $N_H \le N_{RF}$. The eigenvectors of $\frac{1}{K}\sum_{k=1}^{K} H^H[k]H[k]$ corresponding to non-zero eigenvalues are expressed as $\tilde{U}$. Hence, $\frac{1}{K}\sum_{k=1}^{K} H^H[k]H[k]$ is expressed as $\tilde{U}\tilde{\Lambda}\tilde{U}^H$. Here, the rank of $\frac{1}{K}\sum_{k=1}^{K} H^H[k]H[k]$ is not full.

In the frequency selective hybrid beamforming scheme, the analog beamforming matrix ignoring the constant-modulus constraint and other hardware constraints is given by $F_{RF} = [\tilde{U}0]$, where only N_H vectors are effective. Thus, the channel covariance matrix is presented by

$$\begin{aligned}\hat{R}^{HB} &= \frac{1}{K}\sum_{k=1}^{K}(H[k]F_{RF})^H H[k]F_{RF}\\ &= F_{RF}^H\left(\frac{1}{K}\sum_{k=1}^{K} H^H[k]H[k]\right)F_{RF}\\ &= \begin{bmatrix}\tilde{U}^H\\ 0\end{bmatrix}\tilde{U}\tilde{\Lambda}\tilde{U}^H[\tilde{U}0]\\ &= \begin{bmatrix}\tilde{\Lambda} & 0\\ 0 & 0\end{bmatrix} \in C^{N_{RF}\times N_{RF}}.\end{aligned} \tag{9.13}$$

The ergodic capacity for the frequency selective hybrid beamforming scheme is presented by

$$\begin{aligned}C^{HB} &= \mathbb{E}\left[Blog_2\left(1+\frac{P}{N_0}\mathrm{tr}\{\hat{R}^{HB}\}\right)\right]\\ &= Blog_2\left(1+\frac{P}{N_0}\mathbb{E}\left[\mathrm{tr}\{\hat{\Lambda}\}\right]\right),\end{aligned} \tag{9.14}$$

where $\mathrm{tr}\{Z\}$ denotes the trace of the matrix.

In the fully digital scheme, the eigenvectors of $H^H[k]H[k]$ on each carrier corresponding to non-zero eigenvalues are denoted by $\tilde{U}[k]$. We assume that $F_{RF}[k]=\left[\tilde{U}[k]V[k]\right]$ is a unitary matrix, where $V[k]$ is the null space of $\tilde{U}[k]$ such that $\tilde{U}^H[k]V[k]=0$ and $V^H[k]V[k]=I_{N_{BS}-N_H}$. The channel covariance matrix is presented by

$$\begin{aligned}\hat{R}^{FD} &= \frac{1}{K}\sum_{k=1}^{K}\left(H[k]F_{RF}[k]\right)^H H[k]F_{RF}[k] \\ &= \frac{1}{K}\sum_{k=1}^{K}F_{RF}^H[k]\left(H^H[k]H[k]\right)F_{RF}[k] \\ &= \frac{1}{K}\sum_{k=1}^{K}\begin{bmatrix}\tilde{U}^H[k]\\ V\ [k]\end{bmatrix}\tilde{U}[k]\tilde{\Lambda}[k]\tilde{U}^H[k]\left[\tilde{U}[k]V[k]\right] \\ &= \frac{1}{K}\sum_{k=1}^{K}\begin{bmatrix}\tilde{\Lambda}[k] & 0\\ 0 & 0\end{bmatrix}\in \mathcal{C}^{N_{BS}\times N_{BS}}.\end{aligned} \tag{9.15}$$

For wideband THz channels, we note that $a_t(\theta)$ is approximately the same for all carriers. The channel matrices are highly correlated. Correspondingly, the eigenvalues of $H^H[k]H[k]$, demoted by $\tilde{\Lambda}[k]$, are also approximately the same for all carriers.

Thus, we obtain

$$\left\|\frac{1}{K}\sum_{k=1}^{K}\tilde{\Lambda}[k]-\tilde{\Lambda}\right\|_F^2=0. \tag{9.16}$$

The ergodic capacity of the fully digital beamforming scheme is given by

$$\begin{aligned}C^{FD} &= \mathbb{E}\left[B\log_2\left(1+\frac{P}{N_0}\text{tr}\left\{\hat{R}^{FD}\right\}\right)\right] \\ &= B\log_2\left(1+\frac{P}{N_0}\mathbb{E}\left[\text{tr}\left\{\hat{\Lambda}[k]\right\}\right]\right) \\ &\approx B\log_2\left(1+\frac{P}{N_0}\mathbb{E}\left[\text{tr}\left\{\hat{\Lambda}\right\}\right]\right).\end{aligned} \tag{9.17}$$

Since (9.14) is identical to (9.17), the proof is completed.

The condition of the rank of $\frac{1}{K}\sum_{k=1}^{K}H^H[k]H[k]$ being smaller than or equal to the number of RF chains is justifiable in THz systems. Indeed, THz channels contain only a few numbers of NLOS rays and are particular low-rank channels. We emphasize that the RF beamforming cannot be the optimal one, as its amplitude is limited to be constant. Thus, we develop a greedy method that the RF beamforming has the same direction with the optimal beamforming vector, which is given by

$$\left(f_{RF,n}\right)_i = e^{j\measuredangle\left(\text{eig}_n\left(\frac{1}{K}\sum_{k=1}^{K}H^H[k]H[k]\right)_i\right)}, \tag{9.18}$$

where $i \in \{1,2,\ldots,N_{BS}\}$ and $\text{eig}_n(\boldsymbol{Z})$ denotes the eigenvector of the matrix associated with the nth largest eigenvalue.

9.3.1.2 Digital Beamforming Design

The performance of only analog RF beamforming cannot approach the fully digital beamforming because of the strict hardware constraints. This motivates us to propose a digital beamforming at the baseband to compensate the performance loss. We design the compensation digital beamforming matrix utilizing the Corollary 2 in (Alkhateeb and Heath 2016). This digital beamforming at the baseband is designed jointly with the RF beamforming to approach fully digital beamforming, which is presented as

$$F_{BB}[k] = (F_{RF}^H F_{RF})^{-\frac{1}{2}} [\bar{V}[k]]_{:,1:N_{RF}}, \tag{9.19}$$

where $\bar{V}[k]$ comes from the singular value decomposition (SVD) of the matrix, $H[k]F_{RF}(F_{RF}^H F_{RF})^{-\frac{1}{2}} = \bar{U}[k]\bar{\Sigma}[k]\bar{V}[k]$.

To evaluate the performance of the statistical eigen hybrid beamforming scheme with digital compensation beamforming, we provide a numerical example in Figure 9.5. In this figure, we plot the spectral efficiencies versus the transmit power at the BS, P. We consider three benchmark schemes: (1) unconstrained fully digital beamforming, (2) unconstrained hybrid beamforming where the constant modulus constraint is not considered, and (3) constrained analog beamforming without digital compensation beamforming. In this figure, we consider the operating frequency from 350 to 450 GHz and the bandwidth $B = 5$ GHz. Thus, the number of carriers is 20. The THz channels are composed of a LOS

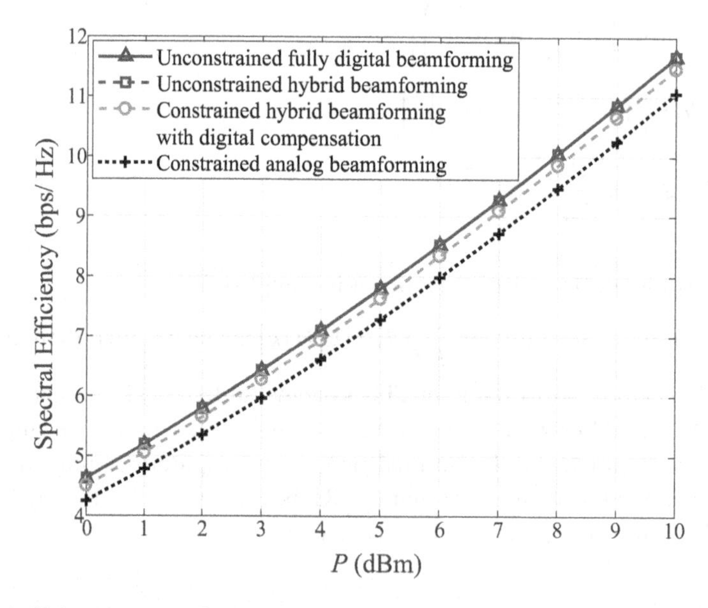

Figure 9.5 **Spectral efficiencies against the transmit power with $N_{BS} = 128$, $N_{RF} = 4$, and $N_{UE} = 16$. The transmission distance is 5 m.**

path and three NLOS paths. The AWGN power is $\sigma_n^2 = -75$dBm. From this figure, we first observe that the statistical eigen-based hybrid beamforming scheme, especially the hybrid beamforming scheme ignoring the constant-modulus constraint, has almost the same performance as that of the fully digital scheme, which shows the accuracy of Theorem 1. Also, the constrained statistical eigen-based hybrid beamforming scheme outperforms the constrained analog only beamforming scheme. The performance gap is resulting from the additional digital beamforming and demonstrates the effectiveness of $F_{BB}[k]$ in (9.19). We note that $F_{BB}[k]$ is low-dimensional, but $F_{RF}F_{BB}[k]$ can sufficiently approach the optimal unconstrained digital beamforming, thanks to the unique sparse property of THz channels.

9.3.2 *Multiuser Hybrid Beamforming*

In this subsection, we consider the hybrid beamforming for a multi-carrier multiuser MISO system where the BS with N_{BS} antennas and N_{RF} RF chains communicates with U noncooperative single-antenna users. The hybrid beamforming design problem in multiuser systems differs from that in single-user systems in two respects:

1. The UEs in multiuser systems are not cooperative, which causes a multiuser interference term in the signal-to-interference-plus-noise ratio (SINR) formulation.
2. It is very challenging to design a common RF beamforming, which needs to serve all users even channels among users are different.

These two differences make the hybrid beamforming design in multiuser systems much more complicated compared to that in single-user systems. To tackle this problem, we present the following transmission scheme:

1. The same group of UEs are scheduled to be served at the entire band for each time slot (Kong et al. 2015).
2. The analog beamforming and the digital compensation beamforming are jointly designed first, assuming that all UEs are cooperative. This indicates that the common analog beamforming and the digital compensation beamforming are designed to increase the receive power of the all UEs when the multiuser interference is ignored.
3. An additional digital beamforming matrix at the baseband is designed to eliminate the multiuser interference.

We emphasize that it has been shown that the performance of multiuser hybrid beamforming can be close to that of the fully digital beamforming only for the case where the same group of users are scheduled at the entire band (Kong et al. 2015). This is because in this case, the channels over all carriers can be guaranteed to be highly correlated (Sohrabi and Yu 2017). Otherwise, it is impossible to design a common analog beamforming suitable for all carriers and all UEs.

Ignoring the multiuser interference, we incorporate the optimal beamforming for all carriers together to derive the optimal common analog beamforming. Following the similar procedure in Ayach et al. (2014), the hybrid beamforming design problem is approximately rewritten as

$$\{F_{RF}, F_{bb}\} = \text{argmin} \, \| F_{opt} - F_{RF}F_{bb} \|_F . \tag{9.20}$$

where $F_{opt} = \left[F_{opt}[1],\cdots,F_{opt}[K]\right] \in \mathbb{C}^{N_{BS}\times KU}$ includes all the optimal beamforming matrices across subcarriers. In addition to F_{opt}, $F_{bb} = \left[F_{BB1}[1],\cdots,F_{BB1}[K]\right] \in \mathbb{C}^{N_{RF}\times KU}$ includes all the digital compensation beamforming matrices.

We present an alternating minimization (AM)-based hybrid beamforming algorithm in Algorithm 1 (Yuan et al. 2018b). The main idea is to solve two individual optimization problems iteratively. First, we fix F_{RF} and optimize F_{bb} to minimize the gap between two matrices, F_{opt} and $F_{RF}F_{bb}$. The solution to this problem is given by

$$F_{bb} = V_1\bar{U}_1^H, \tag{9.21}$$

where $\bar{U}_1 = [U_1]_{:,1:N_s}$ is from the SVD of $F_{opt}^H F_{RF} = U_1\Sigma_1 V_1^H$ and $[U_1]_{:,\alpha}$ is the αth column of U_1. Second, we fix F_{bb} and optimize the analog beamforming by

$$F_{RF} = F_{opt}F_{bb}^H. \tag{9.22}$$

Considering the constant modulus constraint, each element of F_{RF} should be normalized by $F_{RF} = F_{RF} \oslash |F_{RF}|$, where $F_{RF} \oslash |F_{RF}|$ denotes that the elements of F_{RF} are normalized individually.

Algorithm 1. AM-based hybrid beamforming

1. **Input: The optimal digital beamforming matrices** $F_{opt}[k], k = 1,\ldots,K$
2. **Initialization:** $F_{RF} = F_{opt}[1] \oslash |F_{opt}[1]|$.
3. **For each iteration do**

   Compute the SVD of $F_{opt}^H F_{RF} = U_1\Sigma_1 V_1^H$, where $F_{opt} = \left[F_{opt}[1],\ldots,F_{opt}[K]\right]$.
   Update $F_{bb} = V_1\bar{U}_1^H$, where $\bar{U}_1 = [U_1]_{:,1:N_s}$.
   Update $F_{RF} = F_{opt}F_{bb}^H \oslash |F_{opt}F_{bb}^H|$.

4. **End for**
5. **While** $k = 1,\ldots,K$ **do**

   Calculate the SVD of $H[k]F_{RF}\left(F_{RF}^H F_{RF}\right)^{-\frac{1}{2}} = \bar{U}[k]\bar{\Sigma}[k]\bar{V}[k]$.
   Let $F_{BB1}[k] = \left(F_{RF}^H F_{RF}\right)^{-\frac{1}{2}}[\bar{V}[k]]_{:,1:N_{RF}}$.
   Normalize $F_{BB1}[k]$.
   Compute $He[k] = H[k]F_{RF}F_{BB1}[k]$.
   Design $F_{BB2}[k] = \left(He[k]^H He[k]\right)^{-1} He[k]^H$.
   Normalize $F_{BB2}[k]$.
   Let $F_{BB}[k] = F_{BB1}[k]F_{BB2}[k]$.

6. **End while**
7. **Output:** F_{RF} and $F_{BB}[k], k = 1,\cdots,K$.

Combined with the aforementioned digital compensation beamformer, the obtained hybrid beamformer, $F_{RF}F_{BB1}[k]$, can be considered sufficiently close to the optimal beamforming $F_{opt}[k]$ for the kth subcarrier.

Moreover, the multiuser interference will degrade the system performance. Hence, an additional baseband beamforming $F_{BB2}[k]$, apart from $F_{BB1}[k]$, is designed to eliminate

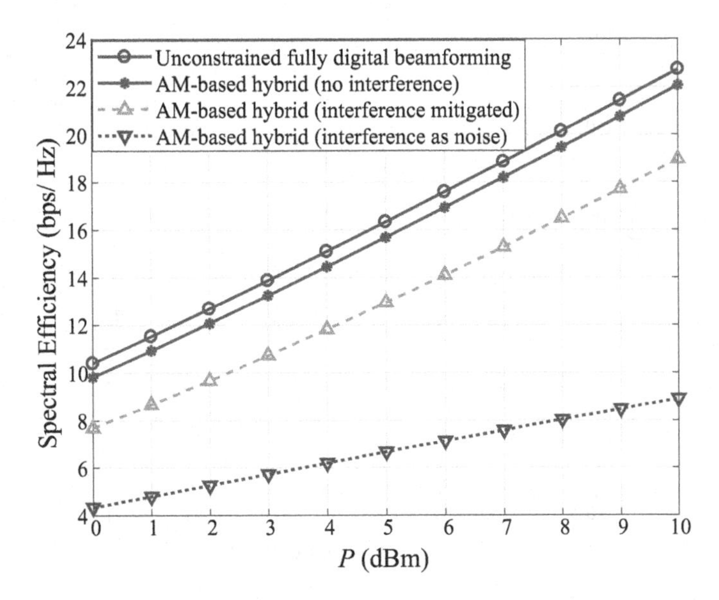

Figure 9.6 Spectral efficiencies versus the transmit power at the BS with $N_{BS} = 64$, $N_{RF} = 4$, and $N_{UE} = 16$. The transmission distance is 5 m. The BS serves 4 single-RF-chain users.

the multiuser interference. We use adopt the conventional zero-forcing method to design $F_{BB2}[k]$ for the sake of low complexity. Specifically, the BS calculates the effective channel $He[k] = H[k]F_{RF}F_{BB1}[k]$ and designs its zero-forcing digital beamforming as

$$F_{BB2}[k] = \left(He[k]^H He[k]\right)^{-1} He[k]^H. \tag{9.23}$$

The complete baseband beamforming matrix is then given by $F_{BB}[k] = F_{BB1}[k]F_{BB2}[k]$.

In Figure 9.6, we plot the spectral efficiencies of the AM-based hybrid beamforming against the transmit power at the BS, P, in the multi-carrier multiuser scenario, where the operating frequency, bandwidth of each carrier, the THz channels, and the AWGN power are the same as those considered in Figure 9.5. In this figure, we consider three cases. In the first case, the multiuser interference is regarded as noise. In the second case, the multiuser interference is neglected. In the third case, the proposed method is used to mitigate the multiuser interference. We can observe from this figure, the performance of the AM-based hybrid beamforming approaches that of the unconstrained fully digital beamforming. We also observe that the spectral efficiency achieved in the third case is significantly higher than that achieved in the first case, which clearly demonstrates the effectiveness of the proposed interference mitigation method.

9.3.3 Distance-Aware Multi-Carrier Modulation

In the THz band, the molecular absorption effect highly depends on the frequency, which results in many high attenuation peaks and defines multiple transmission windows (Jornet and Akyildiz 2011). Figure 9.7 plots the path loss of the LOS ray with different transmission distances. We observe from Figure 9.7 that each transmission window has

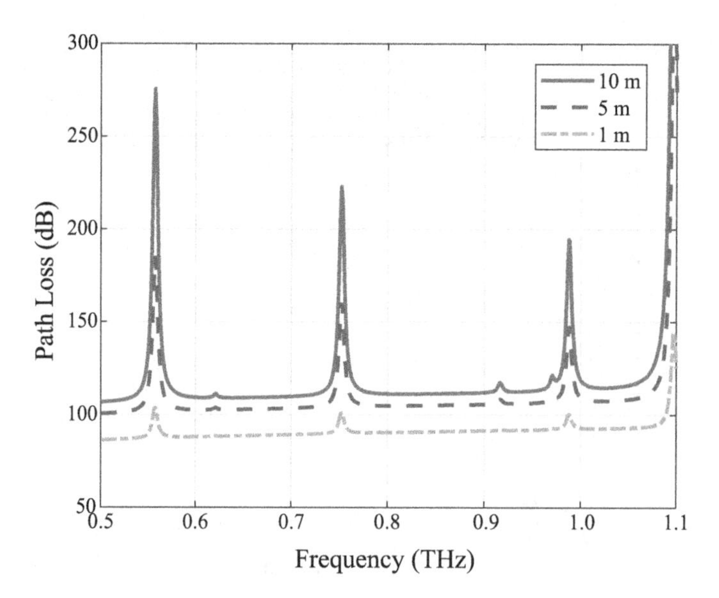

Figure 9.7 Path loss of the LOS ray against frequency with different transmission distances.

a changeable bandwidth, which is determined by the communication distance and frequency. The design of novel distance-aware transmission scheme is motivated by the unique distance-and-frequency-dependent characteristic in the THz band.

The ultra-wide bandwidth in the THz band can be utilized adaptively in the DAMC modulation (Han and Akyildiz 2014). As the communication distance increases, the path gain will decrease significantly, which will reduce the signal power at the receiver and usable bandwidth. In the THz band, the side of transmission window has much more severe path loss than the central part and can only be used for short-distance transmissions. For the sake of fairness, the central frequency is first used for spectrum allocation to long-distance UEs in DAMC modulation. Then, short-distance UEs are served using the sides of the transmission window, which has no interference on long-distance communications. Hence, different parts of the THz transmission window are occupied by UEs with different distances.

Before the transmission, a subset of antennas is activated and performs a fast pre-scanning at an angular interval. These UEs located at the same sector are allocated to the same group. Thus, those UEs in different groups are very likely to be divided in space, as they have completely different AoDs for the LOS rays. In this case, the beamsteering vectors are nearly orthogonal for massive MIMO systems (Lin and Li 2015a). Hence, under the beam division, the same THz frequency resources are shared among different user groups, and UEs with similar AoD LOS rays in the same group are served by the same beam and differentiated by the DAMC modulation.

We consider the transmission band ranging from 550 GHz to 1 THz and the carrier bandwidth being 5 GHz. The adaptive spectrum allocation scheme for different-distance UEs is listed in Table 9.1 (Lin and Li 2016b). According to the principle of the DAMC modulation, long-distance (5–10 m) UEs are allocated to the central band, while short-distance (within 1m) UEs are allocated to the sides.

Table 9.1 Adaptive spectrum allocation for different-distance UEs

Distance (m)	*Allocated sub-bands (THz)*
0–1	0.55–0.575, 0.725–0.775, 0.95–1
1–5	0.575–0.6, 0.7–0.725, 0.775–0.8, 0.9–0.95
5–10	0.6–0.7, 0.8–0.9

Based on the above-mentioned distance-aware spectrum allocation scheme, THz frequency resource is multiplexed among multiple user groups, which can greatly increase the spectrum efficiency. In addition, each RF chain is shared by multiple UEs who are located at the same group, thereby reducing complexity and realizing large-scale connections. After the distance-aware spectrum allocation, the BS can perform accurate analog beamforming searches to enhance the system performance.

9.4 Summary

A key technology of wireless THz communications, namely, hybrid beamforming, has been investigated in this chapter. Performance analysis has shown that the ergodic capacity of using hybrid beamforming in wireless THz systems exhibits an upper bound, which is jointly determined by the communication frequency and distance, the number of antennas, and the number of RF chains. Moreover, the frequency selective hybrid beamforming has been studied, as THz communications is very promising to operate over ultra-wideband channels. This study has shown that the statistical eigen-based hybrid beamforming scheme with digital compensation beamformer has great potential to approach the performance of high-complexity fully digital beamforming scheme. In multiuser scenarios, apart from the consideration of multiuser interference, the transmission strategy that the same group of UEs are scheduled at the entire band has been discussed. Furthermore, it has shown that the unique distance-frequency-dependent transmission bandwidth of the THz band provides an opportunity for enabling DAMC-based spatial multiplexing. Overall, this chapter has investigated the superiority of hybrid beamforming in wireless THz systems, especially considering its low complexity and high spectrum efficiency, which mainly benefits from the unique sparsity of THz channels.

References

Akyildiz, I. F., Jornet, J. M., and Han, C. 2014. Terahertz band: Next frontier for wireless communications. *Physical Communication* 12: 16–32.

Zhang, X., Molisch, A. F., and Kung, S. Y. 2005. Variable-phase-shift-based RF-baseband codesign for MIMO antenna selection. *IEEE Transactions on Signal Processing* 53(11): 4091–4103.

Heath, R. W., Gonzalez-Prelcic, N., Rangan, S., Roh, W., and Sayeed, A. M. 2016. An overview of signal processing techniques for millimeter wave MIMO systems. *IEEE Journal of Selected Topics in Signal Processing* 10(3): 436–453.

El Ayach, O., Rajagopal, S., Abu-Surra, S., Pi, Z., and Heath, R. W. 2014. Spatially sparse precoding in millimeter wave MIMO systems. *IEEE Transactions on Wireless Communications* 13(3): 1499–1513.

Lin, C. and Li, G. Y. 2016a. Energy-efficient design of indoor mmWave and sub-THz systems with antenna arrays. *IEEE Transactions on Wireless Communications* 15(7): 4660–4672.

Lin, C. and Li, G. Y. 2015a. Adaptive beamforming with resource allocation for distance-aware multi-user indoor terahertz communications. *IEEE Transactions on Communications* 63(8): 2985–2995.

Priebe, S. and Kurner, T. 2013. Stochastic modeling of THz indoor radio channels. *IEEE Transactions on Wireless Communications* 12(9): 4445–4455.

Jornet, J. M., and Akyildiz, I. F. 2011. Channel modeling and capacity analysis for electromagnetic wireless nano networks in the terahertz band. *IEEE Transactions on Wireless Communications* 10(10): 3211–3221.

Han, C., Bicen, A. O., and Akyildiz, I. F. 2014. Multi-ray channel modeling and wideband characterization for wireless communications in the terahertz band. *IEEE Transactions on Wireless Communications* 14(5): 2402–2412.

Vu, M. 2011, December. MIMO capacity with per-antenna power constraint. in Proceedings of 2011 IEEE Global Telecommunications Conference, pp. 1–5. IEEE.

Lin, C. and Li, G. Y. 2015b. Indoor terahertz communications: How many antenna arrays are needed? *IEEE Transactions on Wireless Communications* 14(6): 3097–3107.

Lin, C. and Li, G. Y. L. 2016b. Terahertz communications: An array-of-subarrays solution. *IEEE Communications Magazine* 54(12): 124–131.

Han, C., Bicen, A. O., and Akyildiz, I. F. 2015. Multi-wideband waveform design for distance-adaptive wireless communications in the terahertz band. *IEEE Transactions on Signal Processing* 64(4): 910–922.

Gao, X., Dai, L., and Sayeed, A. M. 2018. Low RF-complexity technologies to enable millimeter-wave MIMO with large antenna array for 5G wireless communications. *IEEE Communications Magazine* 56(4): 211–217.

Alkhateeb, A. and Heath, R. W. 2016. Frequency selective hybrid precoding for limited feedback millimeter wave systems. *IEEE Transactions on Communications* 64(5): 1801–1818.

Yuan, H., Yang, N., Yang, K., Han, C., and An, J. 2018a, December. Hybrid beamforming for MIMO-OFDM terahertz wireless systems over frequency selective channels. In Proceedings of 2018 IEEE Global Communications Conference pp. 1–6. IEEE.

Sohrabi, F. and Yu, W. 2017. Hybrid analog and digital beamforming for mmWave OFDM large-scale antenna arrays. *IEEE Journal on Selected Areas in Communications* 35(7): 1432–1443.

Yuan, H., Yang, N., Yang, K., Han, C., and An, J. 2020. Hybrid beamforming for terahertz multi-carrier systems over frequency selective fading. *IEEE Transactions on Communications* 68(10): 6186–6199.

Kong, L., Han, S., and Yang, C. 2015, December. Wideband hybrid precoder for massive MIMO systems. In Proceedings of 2015 IEEE Global Conference on Signal and Information Processing, pp. 305–309. IEEE.

Yuan, H., An, J., Yang, N., Yang, K., and Duong, T. Q. 2018b. Low complexity hybrid precoding for multiuser millimeter wave systems over frequency selective channels. *IEEE Transactions on Vehicular Technology* 68(1): 983–987.

Han, C. and Akyildiz, I. F. 2014, June. Distance-aware multi-carrier (DAMC) modulation in terahertz band communication. In Proceedings of 2014 IEEE International Conference on Communications, pp. 5461–5467. IEEE.

Chapter 10

Ultra-Massive MIMO in THz Communications

Concepts, Challenges and Applications

Mona Bakri Hassan, Elmustafa Sayed Ali, and Rashid A. Saeed

Contents

10.1 Introduction

Wireless THz technology is expected to be a key spectrum in the near future that can meet the high bandwidth demand and data rates of 5th Generation (5G). THz wireless band communication suffers from severe path loss and transmission attenuation in transceivers, which lead to distance constraints [1]. Ultra-massive multiple-input multiple-output (UM-MIMO) technology promises to achieve distance enhancement for THz band channel [2]. UM-MIMO integrates a large number of graphene-based plasma nano-scale arrays into very small space where these arrays create an effective beamforming (BF). BF enhances THz channel capacity. However, the BF is sometimes inappropriate for spatial multiplexing MIMO [2,3]. The performance of UM-MIMO in THz bands depends on the design of nano-antenna arrays and capabilities of plasmonic elements. The individual plasmonic nano-antennas suffer from mutual coupling effects between arrays, other impacts such as spreading, molecular absorption, reflection and scattering, which need to be taken into UM-MIMO design considerations [3].

Given the incredible demand of THz communications applications, i.e., 5G communications, biomedical, ultra-dense indoor networks, and nano communication, it is important to understand THz signal propagation challenges in the considerations of system design. In this chapter, we present the emerging UM-MIMO technology in THz band with a particular focus on the THz communications. The fundamentals of UM-MIMO nano-antenna array and its fabrication issues using plasmonic antenna based on graphene material are discussed. THz UM-MIMO band has been considered for dense and nano communications as it is characterized with ultra-high data rates and ultra-bandwidth, which may lead to extreme nano device communications. UM-MIMO in THz band provides different opportunities for 5G applications and beyond. The chapter provides an extensive literature about UM-MIMO communication systems with configurations related to THz signal processing in addition to the challenges in plasmonic antenna fabrication, channel modelling, and network design. The chapter also gives an example of the use of UM-MIMO in THz communication applications.

The chapter is arranged as follows. In Section 10.2, a brief review of MIMO in THz band and related works is presented. A background description of ultra-massive MIMO communications for dynamic and multiband environments is discussed in Section 10.3. The concept of using UM-MIMO in THz band is reviewed in Section 10.4. In Section 10.5, the plasmonic nano-antenna array structure is presented with different scenarios related to MIMO antenna characteristics. In Section 10.6, UM-MIMO signal processing of beamforming, modulation and multicarrier configurations are presented. Section 10.7 provides different challenges and problematic issues for the UM-MIMO system implementation. An extensive review about THz Ultra-Massive MIMO array of sub-Arrays design is discussed in Section 10.8. A general review for THz UM-MIMO applications is presented in Section 10.9. Finally, a conclusion and summary for the chapter is given in Section 10.10.

10.2 MIMO Gigahertz to Terahertz Era

Sub-microwaves and millimeter waves were considered as operating in a gigahertz frequency for a radio system that can support a point-to-point fixed digital links and operate in a duplex mode where each radio channel consists of a pair of frequencies for both transmit and receive directions. The bandwidth of GHz is quite symmetric and wide, which can be determined by link capacity and modulation scheme [4]. The widespread use of various proximity electronic devices ranging from laptops, mobile phones, and tablets, has dramatically increased connectivity across the world, and results in an extensive media sharing, that is, videos, music, and social networks. This requires a massive increase in data rate between devices over a large bandwidth capacity. Gigahertz has been an exciting development in modern technology and one of the most important cofactors of the rapidly expanding information age [5]. In gigahertz communications, the use of massive MIMO technology enables the deployment of base stations (BSs) and access points with a large number of antennas to increase the 5G network capacity.

Recently, most applications need ultra-high data rates due to big data exchanges, which requires tens of tera bit/seconds communication rate as well as lower energy consumption. The terahertz waves would provide higher capacity performance and ultra-high data rates because of higher spectrum availability in THz band ranging between 30 GHz and 3 THz. The communications devices operating in the low frequency of millimeter wave band would give low performance and be degraded due to their cut-off frequencies, even with the use of complementary metal–oxide–semiconductor (CMOS) technologies in mm-wave circuits; it gives typical cut-off frequencies of 200 GHz [6]. In moving from gigahertz to terahertz systems, investigations have been focused in the photonics domain up to 3 THz and more toward infrared bands. It is found that a significant proportion of the electromagnetic spectrum is unexplored starting from 50 GHz to 3 THz and is popularly known as the terahertz gap. The use of this range is highly suitable for communications data rate in indoor and outdoor environments, and even for providing high-definition (HD) video applications and gigabit wireless (Gi-Fi) networks [7]. In THz band, a large bandwidth is obtained but with a number of limitations related to propagation losses and power losses. The frequencies in this band allow for short-range communications. UM-MIMO depends on plasmonic nano-antenna arrays fabricated by graphene. The UM-MIMO technology allows grouping hundreds of antenna elements in very small distance measured by millimeters. These elements will provide an efficient beamforming to enhance the communications in THz band.

10.2.1 Related Works

Since THz band UM-MIMO is fairly new technology, relatively few studies exist that cover experimental aspects of the coverage and propagation properties for channel modelling, characterization and beamforming system design. The following are the most related researches conducted in THz UM-MIMO systems. Rodrigo et al. (2013) presented time division duplexing mode massive MIMO system where resource allocation and calibration for data models are examined [42]. Moreover, the authors reviewed various massive MIMO challenges and future directions related to massive MIMO system applications Marco et al. (2014) introduced a comprehensive tutorial and survey for spatial modulation MIMO (SM-MIMO) using various aspects related to potential advantages and research challenges related to the SM–MIMO deployment [55]. Lu et al. (2014) presented theoretical study of massive MIMO in the THz range and the effects that appear when non-orthogonal experimental sequences appear in neighboring cell systems in terms of energy and single-carrier spectral efficiency [58].

A research study focusing on evaluating the channel state information (CSI) at the receiver MIMO systems was presented by Dushyantha and Harald [85]. The study showed that the CSI improves the link capacity and energy efficiency of spatial modulation MIMO at the expense of computational cost. A capacity of open-loop MIMO is compared to sub-array antenna for every channel used. Various scenarios sub-array antennae were considered by activating and deactivating some elements of the array.

Akyildiz and Josep (2016) presented an extensive literature for THz UM-MIMO communications [19]. They discussed the concept of using hundreds elements of plasmonic nano-antennas in small square centimeters in 0.06–1 THz frequencies. The study also presented the operation method for UM-MIMO communication while illustrating potential challenges and possible solutions to them. Lin et al. (2016) presented the hardware constraints of an indoor multiuser array of sub-array antenna architecture and channel characteristics for both THz and mm-waves bands [65]. The study used array of sub-array structure for multi-carrier wideband THz communications with distance-aware approach. The proposed approach explained the possibility of reserving a part of allocated bandwidth for short-distance users in order to be used by long-distance users while maintaining as little path loss as possible. The idea is that the central part of the bandwidth can be used for long-distance communications, while the central part and sides are used for short-distance communications.

Massive MIMO characterization with many parameters related to non-wide-sense stationary channel is presented by Olabode et al. (2017). The study is focused on 3D plane channel characterization and measurement methodologies. The paper discussed the issues related to capacity enhancement in 5G networks using massive MIMO scale characterization (SSC) and beamforming for 2D and 3D channel models [12].

In the study presented by Shahar Stein et al. (2017), various data phases connected to massive MIMO communication is examined, taking the consideration of hybrid beam formers to minimize data estimation error [86]. In addition, an optimal fully digital precoder MIMO matrix to minimize the approximation gap is presented an evaluated. Several direction of arrival (DOA) and beamforming algorithms were extensively studied by Stepanets et al. (2017) [48]. The authors discussed a few scenarios of DOA-based adaptive beamforming with a massive MIMO antenna with 128 elements in 5G communications. The study achieved beamforming accuracy up to 4° resolution, which suppressed the interference up to 340 dB.

A study by Chong et al. (2018) is focused on intergradation of large number (1024 × 1024) of nano-THz UM-MIMO antennas in nano meter footprints [23]. The study gave a detailed description of graphene-based plasmonic nano-antenna array properties and its application for THz propagation in three dimensions. The performance of 1024 × 1024 UM-MIMO channel with different measurements, that is, path gain, array factor, and capacity multi-terabit/second links in distance up to 20 meters are presented. Irfan et al. (2018) proposed hybrid beamforming digital and analog of massive MIMO transceivers for resource management of heterogeneous wireless networks (Het-NET). The authors provided a broad scope of discussion on the inclusion issues related to hybrid beamforming resource management. The study also reviewed the massive MIMO system context for hybrid transceiver architectures, digital and analogue beamforming arrays, and potential heterogeneous network scenarios that rely on configuring hybrid beamforming and massive MIMO antenna system [87].

Alice Faisal et al. (2019) reviewed the research and development in recent of THz UM-MIMO channel modelling, channelization, and signal processing, where the system performance is examined for effective array dimensions [52]. Resource allocation enhancement at THz bands with respect to spatial modulation, multicarrier configuration, and waveform design is also discussed. Sherif et al. (2019) evaluated the performance of 5G network based on THz-enabled massive MIMO access point to serve pedestrian users equipment [11]. In the proposed study, the network is evaluated with three pre-coding schemes known as analogue-only beamsteering. The authors present a methodology to measure network performance depending on the energy efficiency and spectral carrier frequency, in addition to considerations of antenna gain and bandwidth.

Hadi et al. (2019) proposed theoretical concepts for spatial modulation (SM) techniques, which depend on the arrays of sub-array nano-antenna configuration, in addition to presenting the evaluation of capacity and spectral efficiency performance [13]. The authors verified and validated that the spatial modulation in massive MIMO systems is feasible at THz with promising performance. Also, in Arkady et al. (2019), the authors evaluated massive MIMO energy and spectral efficiency for different channel models at 6 GHz and mm-waves [84]. The paper simulated hybrid beamforming with respect to power consumption and transceiver cost reduction for both multi- and single-carrier UM-MIMO system. Shuai et al. (2019) proposed mm-waves and THz band for ultra-broadband communications [31]. An introduction to intelligent plasmonic antenna array ultra-massive MIMO physical model-based mm-wave and THz band environments is presented. Moreover, the developed model was evaluated in different scenarios and performance improvements were achieved according to distance and data rate.

10.3 Ultra-Massive MIMO Communications

The demands of 5G and beyond applications require large bandwidth with ultra-broadband frequency, which can be achieved by using THz band gap. The use of ultra-broadband frequencies can be affected by much impairment due to high propagation loss, in addition to transceiver distance-based power limitations in both millimeter waves (mm-wave) and THz-band. To overcome these limitations, ultra-massive MIMO platforms are preferred, which add value to increase the THz band data rates and communication distance [31]. The use of plasmonic antenna arrays both at the transmitting and receiving nodes enable operation in different THz communication modes. THz band UM-MIMO communication systems can be aided by creating massive controllable nano-antenna arrays. The main aim of utilizing UM-MIMO technique is to increase data rate and enhance the communication distance, in addition to obtaining maximum achievable capacity. These aims can be achieved if the THz system has an ability to overcome the problems of molecular absorption loss and spreading loss [32].

10.3.1 Dynamic UM-MIMO

Dynamic UM-MIMO helps to adjust the phase and amplitude of plasmonic signals at each nano-antenna to provide many types of operation mode in UM-MIMO for both

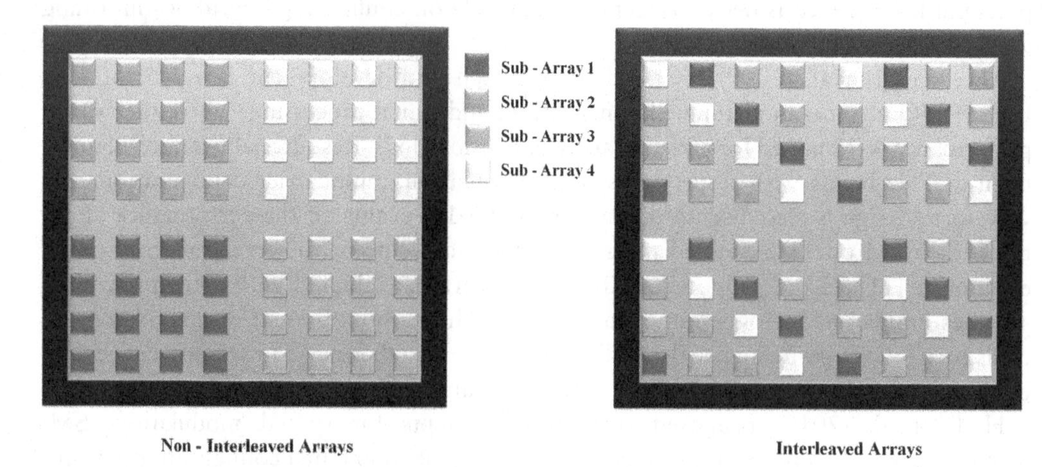

Figure 10.1 Plasmonic nano-antenna sub-arrays interleaving.

beamforming and spatial multiplexing [33]. The main feature of dynamic UM-MIMO is the ability to combine a number of nano antennas into one group, in addition to the ability to integrate a plasmonic signal source inside each nano antenna, as this method helps to increase the output power [34]. However, this would be different from conventional architectures, which allow distributing one signal among all AoSA architecture elements [35]. Theoretically, the plasmonic nano-antenna array would give ultra-high gain. In real-life applications, the spaces of the nano-antennas are much closer to each other, which reduce the beamforming performance.

In a dynamic UM-MIMO, to support multiple wider beams with lower gain in different directions, it can be achieved by splitting the dense antenna arrays [36]. Those beams are used for spatial diversity and increased system capacity. Furthermore, the independent signal control of each nanoscale antenna in plasmonic nano transmitters provides a special way to aggregate array elements to increase the simultaneous beams [37]. Sub-arrays interleaving could be used instead of dividing the array in separate sub-arrays (see Figure 10.1). For interleaved sub-arrays, virtual sub-arrays consist of separate sub-elements, which can be scaled up without affecting the physical components of the system. In a non-interleaved sub-array, it is possible to adjust the beam gain because each subset of sub-array contains fewer active elements, where the potential for beam formation appears. The sub-array elements interleaving enables to obtain beamforming gain due to the capability of increasing distance between elements to $\lambda/2$ [19].

The dynamic UM-MIMO beam gain can be defined as a beam number function for both separate and interleaved sub-arrays. The use of interleaved sub-arrays would contribute by way of more gain when compared to separate sub-arrays beam gain (see Figure 10.2). For example, in case of using plasmonic nano-antenna array with 1024 nano-antennas at 1 THz to create a single beam as an ultra-massive beamforming, each beam in nano-antenna can be used to transmit separately [19]. In separated sub-arrays, each beam gain may take the value of 15 dB, while in an interleaving sub-array, the gain may become 25 dB per beam.

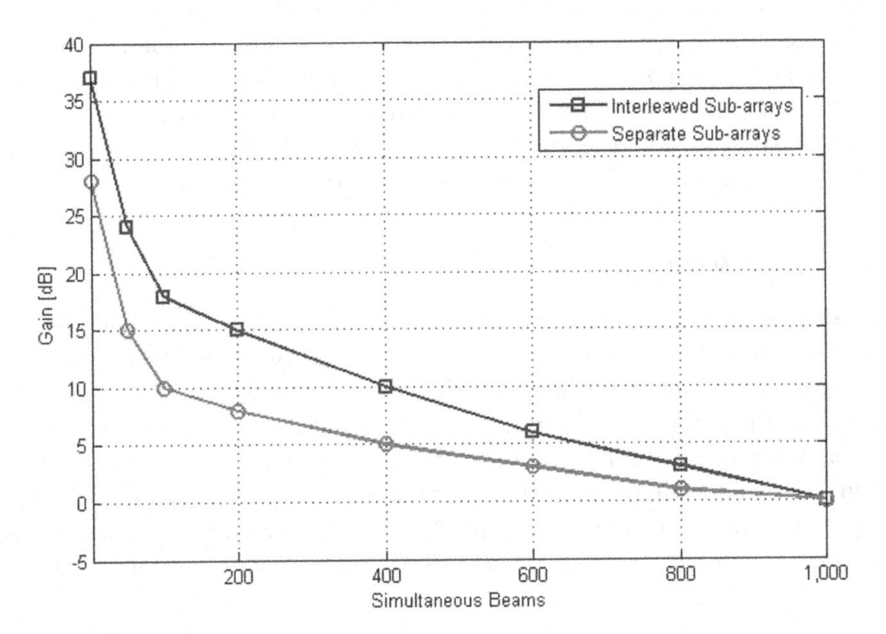

Figure 10.2 Gain per beam for separated and interleaved sub-array.

10.3.2 Multi-Band UM-MIMO

In THz communication, the utilization of THz channel enables to obtain up to tera bit/s speed by using multiband UM-MIMO system. The multiband UM-MIMO has the capability to operate simultaneously in various transmission windows to facilitate the use of plasmonic nano antenna arrays [38]. In multiband UM-MIMO, single nano-antenna would be divided virtually to work at different center frequencies, which can provide an effective narrow bandwidth at transmission windows. To simplify the design of nano-antenna and facilitate the dynamic control to the nano-antenna arrays, different frequencies can be used to interleave the sub-arrays. The frequency response of a single plasmonic nano antenna can be adjusted electronically [39]. The response of individual elements in the array can be modified dynamically and independently. Furthermore, to utilize multiband UM-MIMO, massive number of nano-array elements must have a space much shorter than wavelength of free space, which is required to create the convenient space at the desired frequency [40].

10.4 UM-MIMO in THz Band

In 5G networks, a large bandwidth is essential to support high-speed communications [8]. UM-MIMO provides the possibility of not only increasing the data rate but also improving the reliability of communication systems. It is able to achieve communication links with higher data rate. In a MIMO system, multiple antennas are placed in the transceivers to create parallel communication channels that enhance the productivity and reliability of the system [9]. The THz band MIMO structure can be described as "N×N" LOS MIMO system and can be considered as a conventional low-frequency communication system from 2 to 5 GHz. The LOS MIMO system in THz band communication enables to design small compact antennas to obtain higher directional gain. Moreover, the use of large MIMO arrays mitigates link breakage caused by path obstacle [10]. Massive MIMO in THz system provides large array gains to reduce much high path loss and enable highly directional beams to overcome the impairments of inter-cell interference. Moreover, massive MIMO can offer a special multiplexing gain to boost system capacity and reduce the multiuser interference for outdoor 5G communications [11].

10.4.1 UM-MIMO Channel Condition

The system performance of THz UM-MIMO depends on the exact channel conditions and channel state information accuracy. Efficient THz band utilization depends on the accuracy of UM-MIMO channel modelling and should take into account many factors such as propagation loss, molecular absorption loss, LOS and NLOS reflection, and scattering in addition to treating the static and time variant environments [12]. The signal propagation in THz band is almost optical. Because of large reflection loss, THz channel would possibly lose energy. The frequency-dependent LOS path gain is based on two losses: the spreading and molecular absorption loss, which can be obtained as follows [13]:

$$\alpha_{\mathrm{SAs}}^{\mathrm{LoS}} = \frac{\mathrm{c}}{4\pi \mathrm{fd}} \times \mathrm{e}^{\left(-\frac{1}{2}\kappa(\mathrm{f})\times d\right)} \times \mathrm{e}^{(-\mathrm{j}2\pi\mathrm{f}\times T_{\mathrm{LOS}})} \tag{10.1}$$

where f represents the operation frequency, c is for the speed of light in free space, d denotes the distance between the source and destination based on sub-array (SAs) structure, and $\kappa(\mathrm{f})$ stands for the absorption coefficient. $\mathrm{T_{LOS}}$ denotes the LOS path gain time-of-arrival (ToA) given by d/c. The high reflection losses in NLOS scenarios make the THz channel dominate a few NLOS paths. The frequency-dependent NLOS path gain for reflected rays can be obtained as follows [13]:

$$\alpha_{\mathrm{SAs}}^{\mathrm{NLoS}} = \frac{c}{4\pi f(r_1 + r_2)} \times e^{\left(-\frac{1}{2}\kappa(f)(r_1 + r_1)\right)} \times R(f) \times e^{(-j2\pi f \times T_{\mathrm{Ref}})} \tag{10.2}$$

where $R(f)$ represents the reflection coefficient. r_1 denotes the distance between the transmitter and the reflector, and r_2 is for the distance between the reflector and the receiver. T_{Ref} is equal to the T_{Los}+ $(r_1+r_2 - d)/c$, which stands for the reflected ray arrival time. Equations (10.1) and (10.2) are based on the subarray (SAs) structure according to the dimension matrix of UM-MIMO in which LOS and NLOS path gains are calculated between the (m_t, n_t) transmit sub-array and (m_r, n_r) receive sub-array.

10.4.2 Graphene-Enabled Terahertz-Band

The use of plasmonic nano-antenna in nano-antenna arrays design will support the surface plasmon polariton (SPP) wave's propagation and minimize the traditional metallic antenna arrays. The surface plasmon polariton (SPP) waves will appear between the metal and dielectric that interact like electrical charge oscillations interface [20]. Many SPP waves can be generated at different frequencies depending on the plasmonic materials.

The graphene material can provide unique SPP propagations waves, which are very favorable in developing plasmonic nano-antennas. SPP waves in free space have lower propagation speed than the EM waves [21]. The wavelength of SPP (λ_{spp}) is much smaller than the wavelength of vacuum (λ). The ratio between these respective wavelengths is known as a confinement factor denoted by $\gamma = \lambda/\lambda_{spp}$, which depends on the system frequency and plasmonic material. The confinement factor could be determined by solving the equation of SPP wave dispersion [22]. The calculation of dispersion the equation of SPP waves for graphene structures with almost static pattern can be given by the following equation [23]:

$$\frac{\varepsilon_1 + \varepsilon_2 \coth\left(k_{spp} d\right)}{k_{spp}} + i\frac{\sigma_g}{\omega\epsilon_o} = 0 \tag{10.3}$$

where σ_g represents graphene's conductivity. The equation condition is $k_{spp} \gg \omega / c$, while c denotes speed of light. ε_1 represents the dielectric relative permittivity that is placed in above the graphene layer, and ε_2 denotes the dielectric relative permittivity that is placed between the two plates separated by a graphene layer distance d and the metallic ground plane. The complex wave vector k_{spp} can be determined by solving Equation (10.2). λ_{spp} can be found by the real part of the wave vector $\mathcal{R}e\{\mathrm{k}_{spp}\}$as in following equation [23]:

$$\lambda_{spp} = \frac{2\pi}{\mathcal{R}e\left\{k_{spp}\right\}} \tag{10.4}$$

While the imaginary part ($\mathrm{Im}\{\mathrm{k}_{spp}\}$), defines the SPP decay or it is inversely proportional with propagation length L as shown in the equation,

$$\mathcal{L} = \frac{1}{2Im\left\{k_{spp}\right\}} \tag{10.5}$$

The resonance length of a plasmonic antenna for different metallic of nano-antennas can be given by $l_p \approx \frac{\lambda_{spp}}{2}$. According to the resonance of plasmonic antenna and due to the resonance length for metallic of nano-antennas, UM-MIMO system can be designed as a group of plasmonic nano-antenna arrays and controlled by system beamforming mechanism, which would be discussed in Section 10.7.1 [23,24]. Figure 10.3 describes a system design for an AoSA architecture based on the UM-MIMO system. In AoSA architecture, the analogue pre-coding array is a diagonal block array, which means that the analogue hybrid pre-coding is independent of each sub-array. In addition, the digital pre-coding array denoted by H is the THz channel array in which the row and column numbers correspond to the numbers of RF chains.

In UM-MIMO antenna subarrays, each subarray is derived by a single RF-chain baseband as well as each subarray consists of very small-packet antenna elements. The antenna element is connected to a broadband phase shifter to implement digitally controlled graphene integrated gates. The beamforming gain in a single subarray helps to solve the very high path loss problem at THz frequencies.

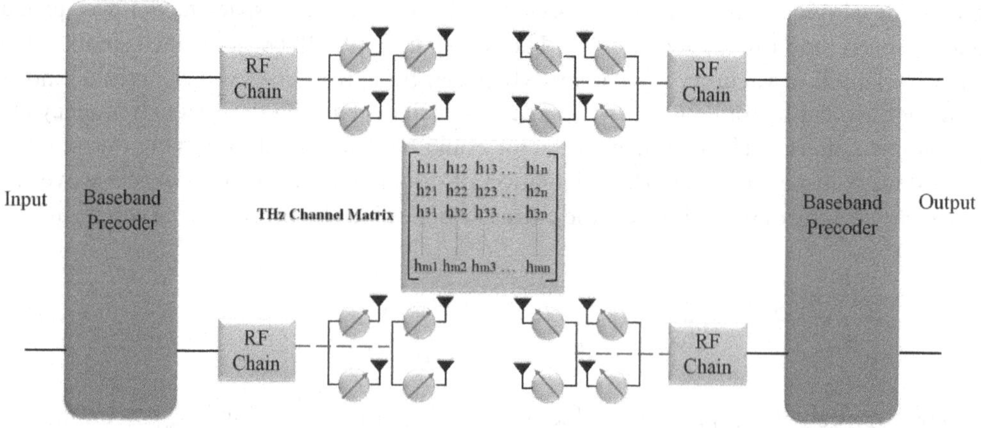

Figure 10.3 AoSA architecture-based UM-MIMO.

10.5 Plasmonic Nano-Antenna Array

THz waves are sensitive to detection and their propagation depends on the efficiency of the antenna design. The use of plasmonic photoconductive electrodes would provide a good sensitivity for THz signals by enhancing the concentration of carriers near the antenna electrodes. Plasmonic detectors can efficiently enhance THz signal absorption in nano-antennas and can provide a good alignment of THz waves between two transmission sides. The plasmonic electrode increases THz signal intensity, which would reduce the signal carrier transport path length [14]. Despite the aforementioned benefits of plasmonic photoconductive technology for THz waves, the detection bandwidth is limited due the terahertz detector constraints. The use of nano-antenna arrays enables to increase the interaction between optical pump and incident THz beams at the near end for the THz device operation. In addition, the polarization of optical pump beam must be normal to nano-antenna and THz beam polarization must be parallel to nano-antennas to obtain an optimal operation [15]. The operation of THz beam detection through plasmonic photoconductive nano-antenna arrays is shown in Figure 10.4.

The structure of plasmonic nano-antenna arrays consists of nano-antennas connected to bias lines. The gaps between the cathode and anode bias are shaded by a metal layer on top of Si_3N_4. AU array on top of Si_3N_4 is used for shadowing the horizontal gaps between the nano-antennas [15]. By reducing the gaps between the nano-antenna sides, the induced THz electric field can be enhanced. On the other hand, the gaps must be greater than the effective wavelength of the photovoltaic pump to maintain a high absorption. In addition, the large length of nano-antennas would provide an increase in the induced THz fields between nano antenna sides, but it should be smaller than the THz wavelength to maintain the operation of antenna broadband [14,15].The performance of generating and detecting pulses in THz systems depends on array antenna design and characterization of photoconductive sources and detectors, which they must be deployed with a large-area plasmonic nano-antenna arrays [16]. The use of plasmonic nano-antenna arrays can serve as broadband THz radiating elements to generate high power and broadband-pulsed THz

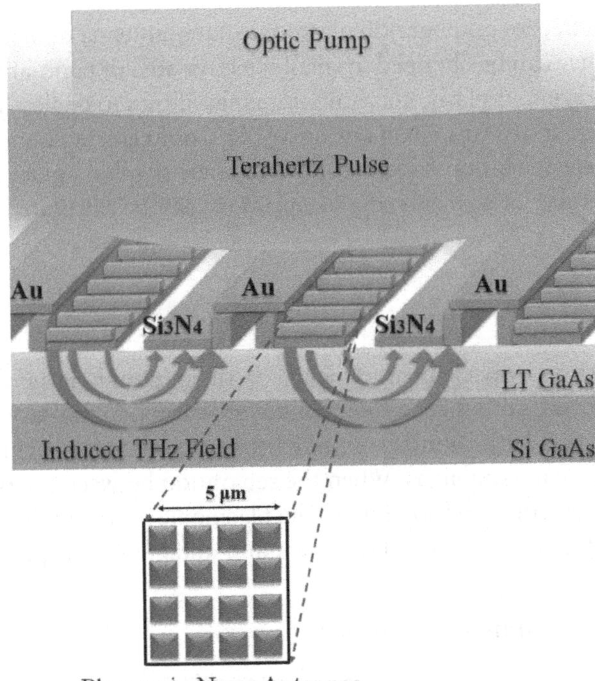

Figure 10.4 Photonic photoconductive nano-antenna array structure.

radiation and this based on the capabilities of plasmonic nano cavities formed near the plasmonic Nano antenna arrays.

10.5.1 Antenna Miniaturization

Nanotechnology provides a new tool to design and manufacture miniature devices, capable of performing various tasks on the micro and nano scales. Graphene plasmons can be miniaturized and reconfigured in the THz platform. This is because THz band in micro or nano scale applications, such as in Internet of Nano Things, needs antenna miniaturization considering designing requirements. In the THz band, the use of surface plasmon polaritons can provide a graphene antenna that can be effectively miniaturized [17]. The use of hybrid plasmonic structures provides perfect equilibrium between confinement mode and propagation loss in waveguides that operate in the THz band with grapheme. This will aid in designing THz antenna elements by combining plasmonic miniaturization with dielectric resonator antenna characteristics [18]. In antenna miniaturization, resonant antenna length can be approximated to equal half wavelength at the resonance frequency to obtain wavelength range in THz band from 5 mm at 60 GHz to 30 µm at 10 THz. For example, the resonance frequency for metallic antenna at 1 THz needs to be approximately the resonant antenna length (l_m) equal to 150 µm. This result already demonstrates the potential for developing very large THz antenna arrays and more substantial gains could be achieved through the use of plasmonic materials to develop nano-antennas and nano-scale transmitters and receivers. [19].

10.5.2 Antenna Integration

By studying the effect of graphene on plasmonic nano-antenna design, it is found that high radiation efficiency can be obtained in small effective area of nano-antennas. However, the highly effective region of plasmonic nano-antennas allows to deploy a very dense group of nano-antenna arrays in a very small size area [25]. Confinement factor γ will enable to integrate the array elements into fixed footprint (δ). For regular square nanoplanar antenna array with N number of elements, the footprint (δ) can be calculated as follows [23]:

$$\delta = \left(\frac{N\lambda}{\gamma}\right)^2 \tag{10.6}$$

The total number of array elements depends on the minimum requirement for separation of antennas, their size, and the maximum allowable array. In the design of nano-antennas, there should not be a large mutual coupling between antennas to achieve minimum separation between the nano-antennas. When the separation between the two nano-elements is close to the wavelength corresponding to the plasmonic wavelength λ_{spp}, the mutual coupling that takes place between the plasmonic nano-antennas is dropped quickly.

10.5.3 Antenna Feeding and Control

For array design requirements, capacitance generation must be controlled to be likely in the time or phase delay of the surface plasmon polariton (SPP) wave in each nano-antenna. At present, several alternative studies have focused on how to generate the THz plasmonic signals [26]. At frequencies below 1 THz, it could use different material technologies such as silicon germanium (SiGe), standard silicon (Si), and semiconductor technologies like gallium arsenide (GaAs) and gallium nitride (GaN). These material technologies enable to generate high-frequency signals [27]. For more than 1 THz frequencies, plasmonic grating structure creates SPP wave to generate the metamaterial-based antennas. For higher frequencies, optical pumping techniques are able to excite the SPP waves [28].

The nano-antenna array feeding depends on the mechanism of excitation of the plasmonic signal to the array elements. The optical pumping enables to excite all the nano-antennas simultaneously by SPP waves generated from pumped single laser. This optical pumping mechanism simplifies the process of nano-antenna feeding, but it requires to feed the elements at the same phase and time delay, which would limit the array antenna applications [29]. In electrical pumping, the control of signal generation depends on the plasmonic controllers to adjust the plasmonic waveguide, phase, and delay. These controllers are enabling to distribute the signals to the different nano-antennas in a specific phase. The low power resulting from a single nano transceiver with limited SPP propagation length can tolerate the performance of the nanoscale antenna array [30]. Generally, the very small size of individual plasmonic sources can enable integration with each nanoscale antenna, therefore allowing digitizing the equivalent geometry.

10.6 UM-MIMO Signal Processing

Since the challenges of high propagation loss and power limitations in THz band has been overcome by ultra-massive MIMO (UM-MIMO) technology, signal processing problem is considered one of the limitations in UM-MIMO THz communications. This problem is related to beamforming, pre-coding and modulation, low-cost MIMO channel estimation, and configuration [41]. Massive MIMO signal processors provide unique solutions in dealing with weak physical medium as well as bestow efficient information processing capabilities. In the MIMO antenna in which the mathematical cost of transmitting and receiving processing grows with the cubic function of the number of antennas, it found that signal processing is permanently affected, especially by adding a large number of antenna elements. The computations considerations related to signal processing in the massive MIMO network is coupled to a large extent with complexity, scalability, and delay, in addition to other aspects, that is, radio frequency limitations and coupling effects. [42]. UM-MIMO antenna arrays operate in an adaptive manner by dynamic modulation and operating frequency tuning, which requires highly flexible and efficient signal processing techniques at receivers. The configuration of UM-MIMO cognitive radio for efficient THz spectrum would be obtained through altering the operating frequencies and modulation modes for each antenna element [41,43]. The processing technique depends on the estimation of the received spectral parameters related to data detected at receivers in addition to enabling spatial modulation (SM) and index modulation (IM). Spatial modulation in THz can contribute effectively to gain benefits from the properties of dense arrays of sub-array (AoSAs) configuration to ensure high spectral efficiency and efficient beamforming [43].

In signal processing, interference cancellation technique with the power loading algorithm and user scheduling process aid in obtaining an optimum multiuser massive MIMO downlink channel transmission [41]. For transmission processing in massive MIMO systems, channel state information (CSI) in time-varying with TDD mode is obtained in order to make uplink channel signal processing estimates on the transmission side [42]. The signal processing mechanisms used for massive MIMO transmission are algorithms based on cost-effective scheduling, that is, greedy algorithms and discrete optimization techniques. Filters and amplifiers work to operate TDD, which needs a mechanism to calibrate links efficiently according to their transmission characteristics. For receiving processing parameters that is, channels gain, filter coefficients, and transmitted symbol detection should be estimated carefully by training sequences and signal processing algorithms. To provide more accurate estimates of terahertz signals and to track differences in node mobility in application scenarios for UM-MIMO networks, non-orthogonal training sequences can be used [42]. Another important recovery processing is related to decoding strategies with reduced delay, especially in high streaming rates and sensitivity against delay such as in video applications that require message passing algorithms with smarter information exchange strategies.

10.6.1 Hybrid Beamforming

The THz UM-MIMO system requires beamforming to be optimized multi-carrier waveform. In the MIMO systems, the distance- and frequency-dependent THz channel characteristics must be taken into account carefully to obtain an improved transmission capacity, and the sub-window assignment calculations must be verified, as these requirements play a critical role in forming an effective beam capable of overcoming the high path loss [44].The beamforming in UM-MIMO can be formed by feeding every nano-antenna on the integrated array by a plasmonic signal, which requires high active power

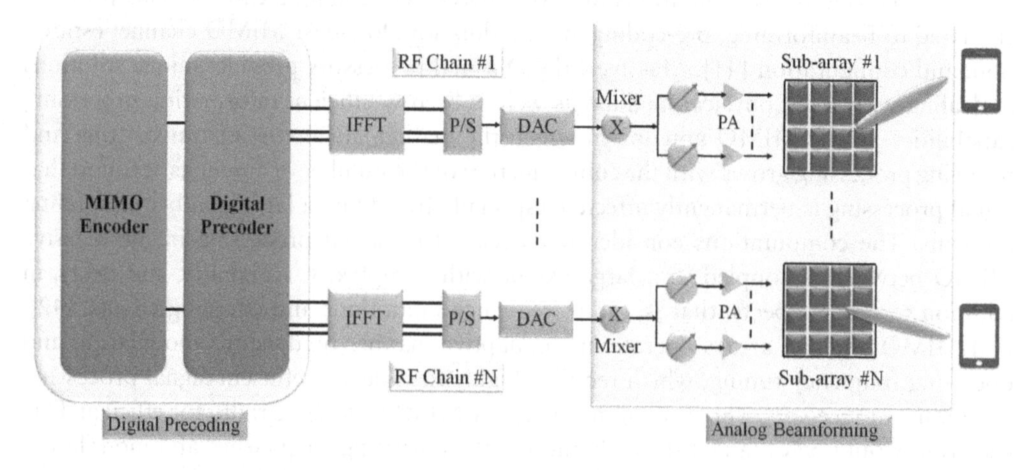

Figure 10.5 block diagram of hybrid beam former.

but with higher gain. The estimation of the direction of arrival of beamforming is based on the measurements of time delays of incoming wave front to the different antenna elements [45]. Many methods can be used to accomplish beamforming, but the two most popular are switched-beam and adaptive array. Switched-beam system realization utilizes a predefined number of lobes in a beampattern and switches between them during a connection [45,46]. In adaptive array systems, there are no predefined beams, but the antenna diagram changes its shape and direction toward each dedicated receiver adaptively, providing more degrees of freedom.

UM-MIMO beamforming can be designated by analog or digital platforms. The advantages of analog beamforming are related to power and computations, which entail lower power consumption and lower computation complexity when compared with digital beamforming [47]. One RF signal can be applied to all antenna elements in an analog scheme, which is possibly forming a beam only in a single direction. Digital scheme can allow to form a beam in many directions with more flexibility in beamforming design, which requires that each antenna element should by applied a dedicated RF signal, but it would consume much power [48]. Hybrid beamforming can be used to improve power consumption in addition to benefit of multiple directions beamforming scheme, which combines the advantages and features of both digital and analog beamforming. Figure 10.5 shows a hybrid beam system for MIMO systems.

Beamforming design in THz communications is based on graphene material as it has an efficient capability to take control of the insertion loss with THz radiation and also able to overcome low absorption. The use of tunable plasmonic material and the cavity, which are used to concentrate electromagnetic energy at the active graphene region, both would provide a suitable beamforming for THz applications [49]. Plasmonic antennas in MIMO have a very small structure in nanometers, especially those based on grapheme. Due to the small antenna structure, MIMO system consists of thousands of elements in denser antennas arrays in millimetric area [47,49]. The use of ultra-massive antennas arrays with very small size for THz plasmonic nano-antennas would offer a great opportunity to develop fully digital beamforming system. Beamforming uses very broadband

MIMO systems to generate THz signals that are able to focus their energy in the direction of transmission and achieve large channel gain. For MIMO system consisting of a number of transmitting antennas denoted by N_t and receiving antennas by N_r, the received signal vector R_s at receiving antennas can be given by [50]

$$R_s = H\psi + N \tag{10.7}$$

where ψ represents the vector of the transmitted signal. N denotes $N_r \times 1$ vector in zero-mean independent noises and variance (σ^2). H stands for normalized channel matrix. The MIMO capacity can be calculated by the following equation [50]:

$$C = \log_2 \det\left(\text{In}_r + \frac{Pt}{N_t \sigma^2} \right) HH^H \ldots \tag{10.8}$$

where P_t represents the total transmitting power. The identity matrix determinant of $\left(In_r + \frac{P}{N_t \sigma^2} \right)$ is computed by the product of the eigenvalues of the matrix HH^H. The MIMO capacity can therefore be written as follows.

$$C = \sum_{i=1}^{k} \log_2 (1 + \frac{P\lambda_i^2}{k\sigma^2}) \tag{10.9}$$

where λ_i denotes singular values of the matrix. λ_i^2 denotes the eigenvalues of the matrix HH^H. Each of the λ_i^2 characterizes an equivalent information channel, where $\frac{P\lambda_i^2}{k\sigma^2}$ is the corresponding channel signal-to-noise ratio (SNR) at the receiver. k represents the number of non-zero λ_i^2. The beamforming technique improves the channel's SNR by maximizing the wavelength (λ). For UM-MIMO THz communications, the design of arrays of subarrays (AoSAs) depends on the dimensions of antenna subarrays $M_t \times N_t$ and $M_r \times N_r$ at the transmitter and receiver side, respectively (see Figure 10.6). The subarrays are driven by separated basebands and composed of Q×Q tightly packed antenna elements and each is controlled by graphene gateways due to the broadband phase shift.

In AOSA structure, the main THz system component is sub-array that can generate beamforming with an acceptable gain contributing to reduce the high path loss at THz frequencies [51]. For such structure with perfect beamforming angle alignment, the SNR per stream can be expressed as in the following equation [13].

$$\gamma = \frac{Gt.Gr.\ Q^2 |\alpha|^2}{\sigma^2} \tag{10.10}$$

where σ^2 represents noise power, Q^2 is an antenna element (AE) index in sub-arrays (SA) represents (Q × Q AE's in a SA), and α is a path gain. A large number of antenna arrays can be divided into sub-arrays as it helps to provide low gain beams in different directions while increasing the capacity. The beamforming can be performed if the array elements

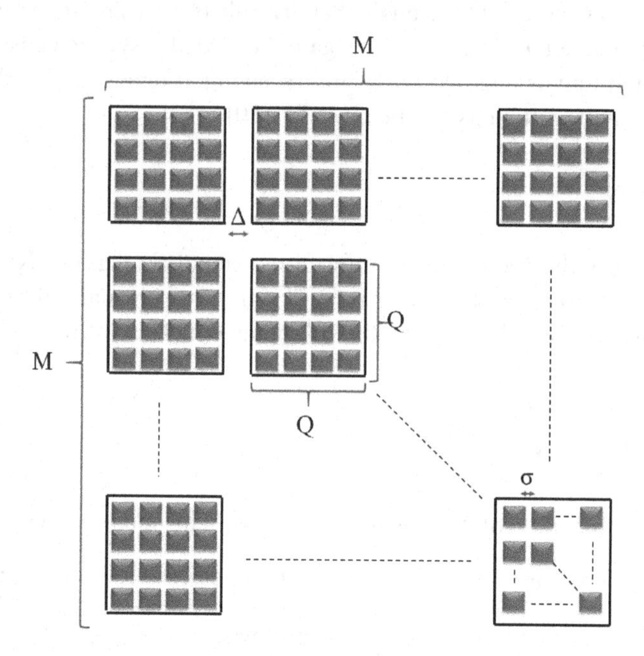

Figure 10.6 Arrays of sub-arrays structure for UM-MIMO THz communications.

extend in an area within the half wavelength and not exceed the full wavelength to avoid grating the wavelenths (λs). The increase in gain beamforming can be achieved by interleaving sub-array elements with more than λ/2 separation length between elements. The benefits of sub-array interleaving will stimulate the development of new methodologies for fitting array patterns [51]. The methodology of dividing one nano-antenna array into multiple sub-arrays operated at different center frequencies facilitate the extraction of the capabilities of plasmonic nano-antenna arrays in generating beamforming for multi-band UM-MIMO THz communications.

10.6.2 Spatial Modulation

In UM-MIMO THz communications, very short pulses can be generated at higher THz frequencies with the corresponding power on the order of a few milliwatts, which is insufficient for long-range communications and limits the ability to generate continuous bus-based modifications. Pulse-based asymmetric time-spread on–off keying modulation (TS-OOK) can support a large number of nano-devices and enables to communicate in range between Gbps and Tbps [52]. The use of MIMO system with such modulation requires a means of modification methodology to account for pulse-based modulations. Cognitive systems can be used to adapt the modulation modes according to the condition of THz channel. OOK modulation has been proposed for wide bands over 100 GHz and recently demonstrated for 300 GHz in less than one-meter transmission. For THz band, it gives a simple and cost-effective configuration wireless communication, but with low sensitivity. It can be improved by vector modulation, which provides more improvement [52].

Spatial modulation (SM) can be considered as a spectral and energy-saving model of THz UM-MIMO. SM is a novel way to transmit information by means of MIMO transmitting antenna indicators in addition to traditional M-array signal combinations. For multicarrier, the antenna design would have large array dimensions that would allow for a large number of bit information to be set [53]. Spatial modulation in UM-MIMO depends on the distribution of the sub-array antennas controlled by inter-antenna separations and communication range according to the channels orthogonally, which is governed by optimal spacing value Δ. The optimal spacing value Δ can be calculated as follows [13]:

$$\Delta_{\text{optimal}} = \sqrt{Z\frac{DC}{Mf}}\ldots \tag{10.11}$$

where D represents the communication distance between the receiving antenna arrays and the centers of transmitting. M is matrix index for array-of-subarrays antenna. The integer value of Z denotes the condition guaranteeing the optimal AoSA design [53]. Optimal Δ depends on λ and D. The small separation optimal antennas can be obtained by a shorter λ and smaller d. In short communication ranges, optimal Δ is large compared to D. For large communication distances, channel paths become highly correlated where $D > \Delta$ [53,54].

UM-MIMO systems would obtain high data rates and improve error performance through transmitting many data streams, which would increase signal processing complexity at the receiver. SM in UM-MIMO would contribute to mapping more information bits on the SM installation consisting of a single element or a subset of the antenna elements. This feature would offer high rate with reducing the processing complexity. The use of SM is very suitable for unbalanced MIMO configurations and has the ability to be implemented for the UM-MIMO uplink and downlink transmission [55].

10.6.3 Multi-Carrier Configuration and Control

Ultra-massive multi-carrier MIMO communication relies on ability of tuning the ultra-dense frequency plasmonic nano-antenna arrays. UMMC MIMO would mitigate the effects of water vapor absorption loss that divides the THz band into multiple transmission windows. The tuning of different nano-antennas at different THz frequency windows would enable the use of UMMC MIMO, which maximizes the THz band utilization [56]. The configuration and control of multicarrier in UM-MIMO depends on nano-antenna spacing. Placing the array elements close to each other would limit the beamforming gain and reduce the capability of spatial sampling. Multiplexing gain for multicarrier can be obtained with much larger same THz frequency in the UM-MIMO sub-arrays with an adequate separation and it requires dynamic tuning for each array elements in sub-arrays to a specific frequency. For lower THz frequencies, software-defined plasmonic metamaterials should be considered. To obtain such operation, the separation between two array elements operated by same frequency is equal to half wavelength and that for operating in different frequencies is equal to wavelength of the surface plasmon polariton [57].

10.7 UM-MIMO System Challenges

One of the UM-MIMO THz communication challenges is how to obtain maximum THz channel utilization. Maximum utilization of THz UM-MIMO channels can be achieved by using multiple spectral windows simultaneously and letting UM-MIMO simultaneously utilize various bands of frequency and using graphene-based plasmonic nano-antenna (PNA) response frequency that can be tuned electrically. A new and innovative frequency–time–space and modulation coding is required for UM-MIMO systems to take advantage of spatial frequency diversities [57]. In addition, UM-MIMO communications require developing an accurate channel model to analyze the statistics of antenna arrays between the transmitter and the receiver in various scenarios for plasmonic nano-antennae. In addition the PDF profile for the THz signals coming from the nano-antenna arrays, particularly those propagated in high-frequency range needs to be analyzed. It is well known that the existing low frequency is difficult to be re-used for channel model of UM MIMO due absorption and reflection loss [58].

10.7.1 Fabrication of Plasmonic Nano-antenna Arrays

The need to move from RF antennas to THz antennas with similar functionality is becoming a challenge in designing THz nano-antennas. Many methods and tools have been used for THz nano-antennae fabrication to achieve acceptable results that looks similar to RF. Many improved radio frequency antennas have been investigated in the THz band, and for directivity, nano loop antennas and travelling-wave (TW) or wave-train antennas find obvious applications [58,59]. The antenna directivity can be controlled by phased array, but it faces challenge of matching source/antenna impedance in between. The fabrication of THz antenna arrays pose main challenges in designing UM-MIMO system feeding and control. The construction of arrays of sub-arrays also would be a challenge for balancing operations between analog and digital schemes and of course with the difficulties of building PNAs array from nanomaterials or metamaterials [59].

10.7.2 UM-MIMO Channel Modeling

In UM-MIMO system, many studies characterized channel modeling for two and three dimensions of antenna arrays. The characterization of channel modelling in the other forms, that is, spherical or cylindrical is one of UM-MIMO channel modeling challenges [60]. Even antennas were deployed arbitrarily; array response vector would be absent for receivers located in the far field. Accordingly, for modeling UM-MIMO channel and collecting statistics for THz channel are quite difficult for many deployment scenarios [38]. There is a need to do more investigations on the physical characteristics and improvements in ray-tracing of channel modelling. The statistical modeling of THz bands model is a challenging issue especially in near-electromagnetic field.

In THz UM-MIMO channel modeling, neighboring PNAs can be mutually coupled between each other, and a control circuit is needed for array network to control phase or time delay for THz channels, taking into account the influence of reflection loss, molecular absorption loss, and the spreading loss of tera hertz various bands. Other consideration is related to PNA arrays with the emission of many parallel signals in the channel which operate in complex with multipath, scattered, and highly attenuated environment that requires an efficient estimation mechanism. This challenge can be reduced by using a spatial correlation between neighboring plasmonic nano-antennas (PNAs) separated by a less than free space wavelength [60].

10.7.3 Network Layer Design

For THz UM-MIMO systems, the physical layer must be designed with an optimal control algorithm to control operation frequencies, as it is a challenge to exploit the terahertz channel abilities of many NPA arrays. The resource allocation with different optimizations is also considered as challenge in order to control multiband communication and dynamic beamforming or spatial multiplexing [61]. So, such limitations require an efficient algorithm to have an optimum resolution in RT for practical THz propagation scenarios. THz bandwidth depends on distance. Developing a modulation technique that is aware of distance becomes a challenge especially for operations over multiple separated bands in UM-MIMO, which requires new kinds of coding mechanisms for signals across various bands periods in order to improve the strength of THz signal to travel longer distances. Dynamic modulation and coding in THz band are two important parameters that can lead to capitalizing on the employment this technology.

At the layer two (data link layer), the timing and clock management is a problematic issue related to UM-MIMO transmission at narrow bands. THz with very high data rates exist for different kinds of THz noises [61]. To tackle this problem, new network protocols are needed that are capable to overcoming synchronization in UM-MIMO communication systems. Optimal timing and high accurate clock for oscillator matching to reduce delay of acquisitions process is required to enhance the capacity of the channel. Beamsteering also can pose a delay challenge to the link layer, which is related to technology of building the UM-MIMO arrays and considered a problem affecting the system throughput due to low beamsteering process. The usage of bandwidth at surface plasmon polariton (SPP) wave and plasmonic nano-antennas (PNAs) arrays may modulate the phase in THz band by thousands of GHz, which may allow fast speed of steering the beam at the arrays [62]. Due to fast beamsteering, in with THz narrow transmission beams would causes high values of instantaneous interference, which needs to be analyzed. There is also a need to develop a mechanism to overcome it according to design considerations.

In network layer, obtaining high gain directional antenna for UM-MIMO systems in Tx and Rx transmission ends would increase the complexity of network design for broadcasting and relaying communications. Broadcasting dynamic beamsteering at very high speeds would require an optimal relaying strategy to consider the behavior of THz bands channels in very large arrays in addition to synchronization requirements. For previous layer design challenges, there is a need to solve all of them in a cross-layer fashion to ensure consistency of THz networks and communication for device-to-device transmissions [63].

10.8 Ultra-Massive MIMO Array of Sub-Array Design

In the THz UM-MIMO system, the massive plasmonic antenna would lower the gain of multiplexing and beamforming, due to inaccurate sampling of temporal and spatial

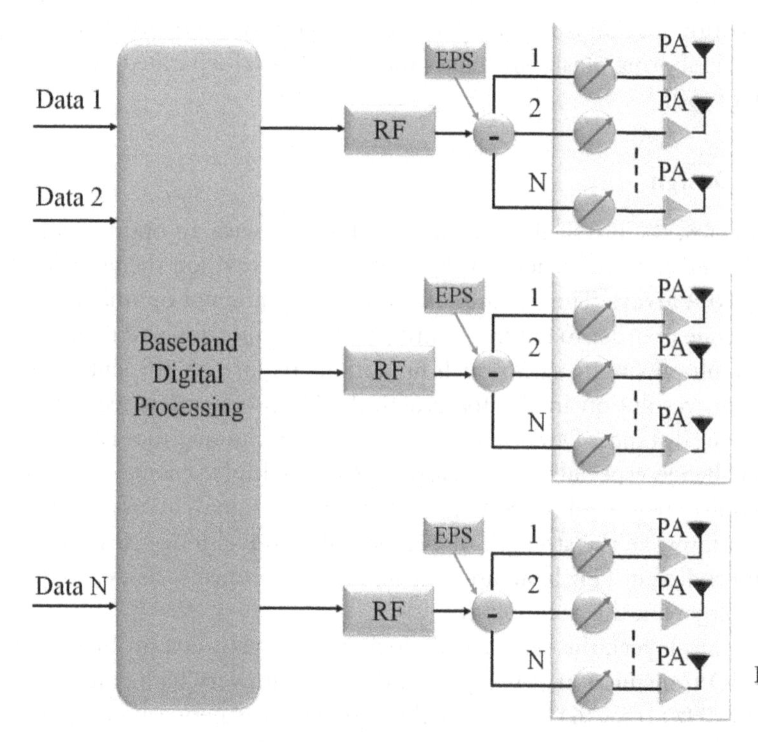

Figure 10.7 Arrays of subarrays system for RF beam optimization.

aspects in addition to complexity in antenna array control. The dividing of large antenna arrays to multiple sub-arrays would contribute to improving beamforming gain and transmission power consumption [64]. A structure of sub-array-based coordinated beam training with time delay phase shift would be promising for the optimization of RF beam directions across multiple sub-arrays. The complexity of building UM-MIMO systems is a trade-off between analog or digital schemes. The feature performance of the two schemes can be solved by building a hybrid digital analog architecture. The architecture of arrays of sub-arrays structure can consist of many RF chains (see Figure 10.7). The RF chains are used to drives the disjoint subset of antennas in sub-arrays and each subset is attached to an exclusive phase shifter. The RF chains can be fully connected by sharing it among antennas, but for arrays of sub-arrays structure, only accessibility is available to one exact radio frequency array-of-sub-arrays chain. Digital processing is implemented for controlling the data streams and for many users' side interfaces baseband [64,65].

The structure of array of sub-arrays is operated by signal processing performed at sub-array level due to an adequate number of antennas. In addition to using a few phase shifters with disjoint structure, the complexity of architecture design and power consumption would be reduced. Multiplexing capability and signal energy distribution among different sub-arrays can be provided due to use of digital baseband cooperation, which would in turn lead to higher beamforming and spatial diversity gains. Power amplifier in THz bands would dynamically compensate the overall power consumption due to processing, which is generally calculated as a function of the output power as $\eta = P_{out}/P_{AC}$, where P_{out} is output power and P_{AC} is the power amplifier consumption [64]. The THz total energy consumption of power amplifier of can be achieved using a model with non-linear characteristics along RF N antenna arrays chains.

Loss in an array of sub-arrays of UM-MIMO would occur according to insertion loss that comes from the power splitting among each subarray [65]. The use of time delay phase shifters in multiuser wideband communications can contribute to designing a hybrid beamforming. The shifted phase Φ_0 for hybrid beamforming depends on the carrier frequency and antennas element spacing in addition to angle of steering direction. Other considerations are important while designing the UM-MIMO array of sub-arrays such as losses due to feeding and mutual coupling. Large feeding loss would appear due to larger sub-arrays used and the neighbor array antenna elements coupling would reduce efficiency. To improve this degradation, a careful installation, configuration, and distribution of frequencies are needed.

10.9 UM-MIMO Applications

UM-MIMO systems can be used in many applications with different communication bands and technologies. THz UM-MIMO opens new opportunity to develop more interesting applications in the areas of localization, imaging, and sensing. Gas sensing can be developed by using UM-MIMO systems, using which gases can be sensed over distances and gas behavior due to molecular absorptions can be extracted for different kinds of measurements such as monitoring air pollution. The benefits of using UM-MIMO in THz arise from higher data rate for connecting data centers of big data exchangers together wirelessly by supporting multiple inter- or intra-rack communication links. Transceivers using UM-MIMO can contribute to optimizing the operation of such communication environment and supporting ultra-high-speed broadband for wireless body area network (WBAN) or IEEE802.15.7 and wireless personal area network (WPAN) or IEEE802.15. Adding to that, UM-MIMO would provide high data rates for vehicular and mobile networks and minimize the effect of Doppler phenomena in such networks by moving to higher frequencies like THz bands. Higher throughput utilization, reliability, and latency would make the use of THz UM-MIMO a best solution for unmanned aerial vehicles or drones. UM-MIMO is promising to develop a new communication generation in 5G networks, medical imaging, ultra wideband (UWB), and ultra-dense wireless networks.

10.9.1 5G Backhaul

The 5G backhaul for future communication demands wireless broadband due to its contribution in providing huge data services. The use of UM-MIMO is considered as a prime support to 5G, which would offer higher data rates, and coordinating the communication among MIMO transceivers in should be in such a way to mitigate the interference [66]. UM-MIMO with mm-wave technologies would offer important solutions for several technical issues of 5G and future Het-Net. The UM-MIMO would promise to offer massive capacities to the 5G systems due to antenna structure, which reduces the transmission power, hardware complexity, and the power consumed by signal processing [66,67]. The use of massive MIMO would provide an efficient way to secure spectrum competence for broadband networks. In addition, the massive MIMO would enable to equip the eNodeBs

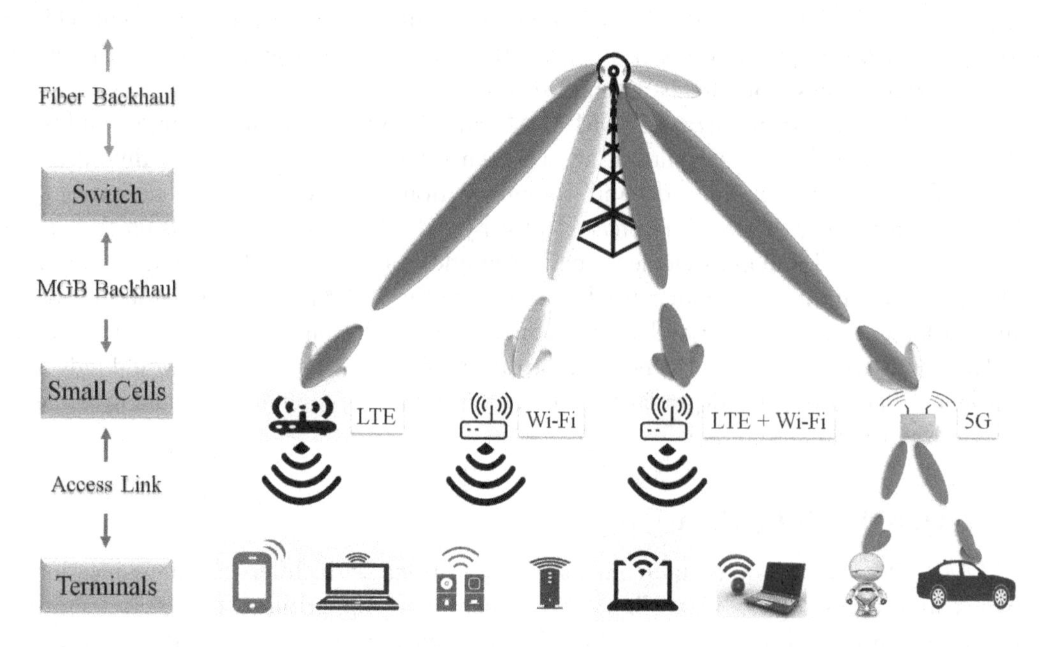

Figure 10.8 MGB system for 5G backhaul.

with large antennas arrays and attain to improved spectral and energy efficiency in wireless network base stations (BSs).

Moving to much higher frequency bands closed to mm-wave band or low THz bands, UM-MIMO would be able to solve the problem of capacity that exists in the lower frequency backhaul systems by providing large available bandwidths and ultra-high data rates. In addition, the use of high steerable antennas would reduce interference and penetration losses in propagating through walls and obstacles. mm-waves with UM-MIMO would offer a suitable coverage for small cell backhaul in an outdoor environment for a few hundred meters [67]. Other contributions of massive MIMO is in the use of millimeter-wave Gbps broadband (MGB) that is presented as a solution to provide Gbps for small cells and fixed broadband accessing. The use of MGB would allow small cells to operate within a coverage area in Gbps without a limitation of access to wired backbone and would offer coverage with high data rates for mobile services and vehicle applications as shown in Figure 10.8.

The MGB backhaul would provide a communication between the fiber backhaul and small cells by using mm-waves and sub-THz waves through hybrid analogy beamforming and digital MIMO processing, which would help to balance the power consumption and digital processing for channel status. This means MIMO processing schemes have to be adapted according to the channel conditions. Antennas in fiber backhaul side and small cells can achieve a sufficient range by a large massive array structure [68]. By using antenna sub-arrays, analogy beamforming, and digital MIMO, MGB system would provide a flexible, scalable, and cost-effective solution for small cell backhaul mobile broadband. In THz band, UM-MIMO can provide a hybrid radio and optical system backhaul, which would allow two communication links interacting according to weather conditions. The backhaul system is based on RF in the THz band and free space optics. Optical link is preferred for clear weather while in case of fog or wind conditions, the link could be taken through RF THz frequencies.

10.9.2 Medical UWB Imaging

In medical diagnostics, the use of ultra-wideband (UWB) sensing would provide a novel technology specially in fields of cancer detection. With the help of using UM-MIMO systems, it would be possible to develop an appropriate and fast measurement system with a sufficient recordable channel [69].The imaging technology of UWB UM-MIMO would provide a means of processing scattered signals collected from sub-array antennas to identify the tumors. Using massive MIMO technology with UWB applications, estimation of volumetric pixel measurements can be achieved for tumor detection processes [70]. Another application related to use of UWB with massive MIMO is to detect humans through multiple sight vision and two-dimension imaging as a human radar can used for medical purposes and military or surveillance applications to improve rescue efficiency. Massive MIMO UWB would provide also high-range resolution and distance accuracy for human detection, with the capability to track small movements [71]. For such applications, MIMO array system can greatly reduce the cost of radar system by combining high-range resolution by UWB signaling with high directional property of multiple array antennas. The use of UWB imaging system with digital beamforming would be able to deliver higher resolution three-dimensional imaging for short-range applications, which can contribute significantly to developing a new system for detecting medications that are packaged or stored inside boxes without disturbing the medical-saving cans [72]. UM-MIMO array can also enable new concepts, such as digital wave coding, providing opportunities to drive such system performance even more to a higher level.

The technology of UWB with massive MIMO can open new opportunities to develop even amazing applications such as imaging behind walls to image the building interiors. Massive MIMO antennas in UWB can used as a sensor according to the system design requirements for various applications of sensing such as moisture, temperature, tumor, crack, or imaging, which depends on suitable spectrum [72,73]. Microwave radar imaging based on UWB sensors is another application system that can be used for detecting cancer in the human body. In such systems, MIMO with electrical steering would produce 3D images by an array-based sensor consisting of single transmitter and multiple receivers (see Figure 10.9).

Since small cancerous tumors cannot be detected by X-ray mammography, radar-based microwaves UWB with higher THz bands could detect them. Deeper accuracy in detecting cancers that can more precisely differentiate between different tumor tissues and determining their types would also be possible [73]. The benefit of using UM-MIMO THz band in medical imaging is due to it being non-ionizing radiation for medical imaging and being safe, being sensitive to water components, and being non-destructive to the tissues [74]. In addition, THz waves using UM-MIMO can also penetrate many materials.

10.9.3 Ultra-Dense Wireless Networks

IoT Industry 4.0, vehicle-to-vehicle (V2V), and M2M communications all are applications requiring significant multimedia services and wide broadband capacity, therefore requiring

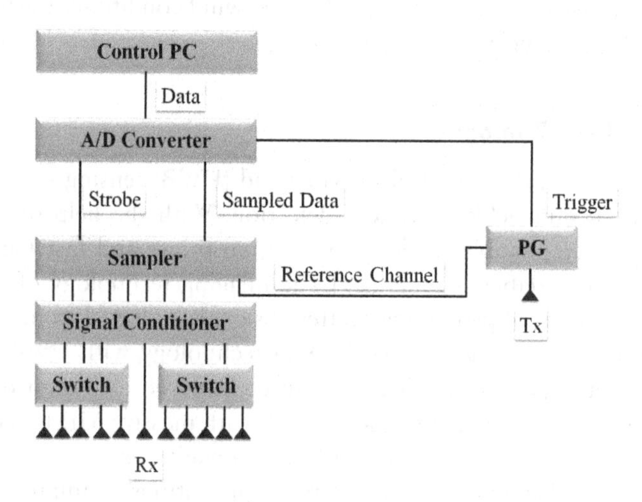

Figure 10.9 Radar imaging based on UWB sensors system.

new 5G wireless standard. The use of continuous ultra-dense networks (CUDNs) can contribute to adding an interactive method for attaining abovementioned goals of 5G [75]. The utilization of UM-MIMO system with micro cells would provide more efficient communication systems that can achieve a degree of freedom system and large array gain with low transmission power consumption.UM-MIMO is considered as a strategy for boosting capacity in wireless networks. Developing massive MIMO antenna with millimeter waves of near to THz waves for communications can contain a huge numbers of picocells and femtocells for 5th generation in dense cities and urban areas. For ultradense wireless networks, a means of distributed network architecture is used due to relaying gateways by multiple hop link to support wireless backhaul traffic [75]. Due to dense backhaul gateway configurations, there are two types of networking architecture for controlling hand off between the cellular stations. In dense cities and urban area 5G networks, single or multiple gateways can be used.

In ultradense network with a single gateway, gateways are configured only at microcell base stations (BSs) that have space for UM-MIMO array of antennae for sending and receiving traffic from/to small cells inside the microcell. Communications between adjacent small cells can be facilitated by multi-hop millimeter wave links [75,76]. For 5G networks, multiple gateways can be used for forwarding traffic from dense cities and urban areas to backhaul and to core network. Each small cell base station can be deployed by a gateway based on geographical scenarios and each adjacent cell can communicate with each other by millimeter wave link, that is, traffic of small cells can be distributed into multiple gateways in microcell [76]. The use of massive MIMO with millimeter waves would provide enough resource for small-cell base stations and help optimize resource allocation for 5G ultra-dense cellular networks. The use of massive MIMO as in cell-free scenario (see Figure 10.10) would improve both the network coverage and energy efficiency. The power control parameters exchanged due to backhaul network would lead to utilizing the performance of defined wireless networks, in addition to obtaining macro-diversity gain for massive MIMO distributed antennas among deployed small cells over a large geographical area.

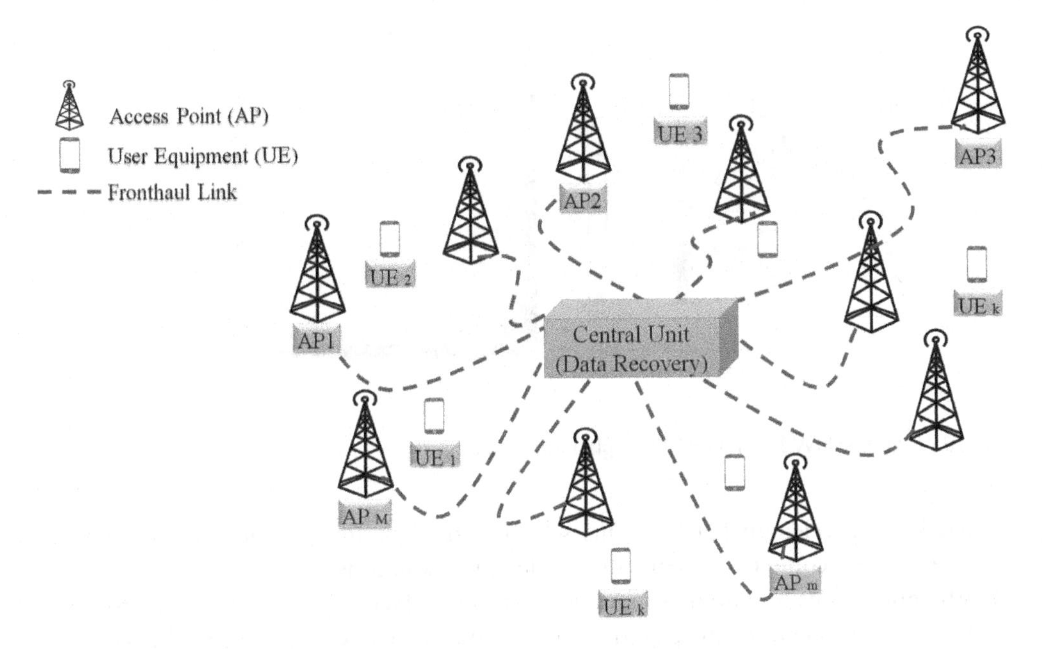

Figure 10.10 Cell-free massive MIMO network.

In a free UM-MIMO cell, the services can be distributed to the users via several APs at the same time. The AP can be attached to central server that can process and cooperate for detecting signals from the user. The use of distributed detection and pre-coding cell-free massive MIMO would reduce the complexity of the system and improve CSI channel scalability and acquisition at processing site [77]. The deployment of small cells with and AP for users would represent important key requirement for ultra-dense networks. The usage of free UM-MIMO cell as a technology for ultra-massive network would help develop a smart new-generation networks with UM-MIMO holding promising for future dense networks [78].

10.9.4 Indoor UWB

The combination of using UWB with MIMO would provide higher capacity and data rate for long transmission range. The use of massive MIMO in UWB would open new opportunities for UWB communications in different applications such indoor positioning and wireless sensor networks [79]. The combination of UWB and MIMO would be a good choice for indoor transmission. In an indoor scenario, UWB signals would encounter several reflection and scattering due to dense multipath environment, which can be resolved for all multipaths by using temporal and spatial resonance at a prespecified location at a spatial moment, which is considered as a virtual antenna to realize massive MIMO for better multiplexing [80]. More multipaths mean more bandwidth availability for indoor UWB communications. The UWB with UM-MIMO system platforms can contain reconfigurable PNA arrays at the transceiver node, working as plasmonic transceiver arrays as shown in Figure 10.11. The PNA arrays control the physics of SPP waves to capably radiate at the frequency of the target resonant at the same time being much lower than

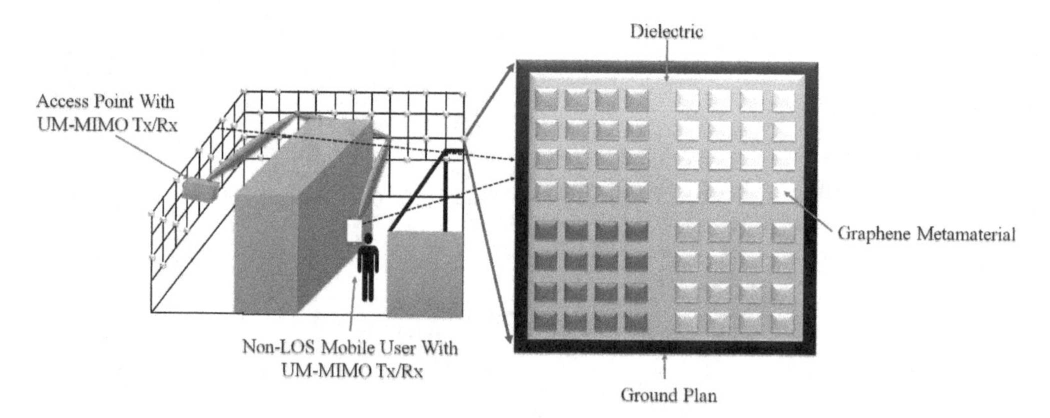

Figure 10.11 Indoor UWB with UM-MIMO plasmonic antenna array.

the matching wavelength [31,81]. This scheme would permit the antenna arrays to be combined in very compact area at sub-wavelength resolutions.

UWB applications in indoor communications might face huge fading and inter-symbol interference (ISI) due to its propagation in LOS, but MIMO can be used in the scattering environment to overcome these drawbacks. The use of massive MIMO with UWB in indoor environment would provide high throughput, reduce multipath fading and ISI, and enhance capacity and an ability to extend the communication range [82]. Indoor localization applications are used for positioning of mobile devices in the indoor scenario, such as in office halls, industrial rooms, or those can be communicated by short UWB pulses. The use of positioning methods would describe the synchronization between access points (APs) and mobile users [83]. According to AoA architecture, the MIMO of antenna arrays can be used in each access points. The use of massive MIMO would provide higher accuracy and precision than outdoor localization services to work with fairly small spaces and amidst blocks and obstacles. And the use of UWB would allow to transmit large data with low energy consumption, which is most favorable for indoor applications [82,83].

10.10 Conclusion

The demands of higher bandwidth and transmission speeds for future wireless communications motivates the researchers to move toward higher frequencies than regular RF. THz bands can meet the demand for higher bandwidth but has the drawback of higher attenuation and loss. The design of THz transceivers faces some barriers related to power constraints, leading to a very short communication of only a few meters. Ultra-massive MIMO (UM-MIMO) is a new technology that could improve the traveling distance of the signal and add much higher capacity for THz communication networks. The UM-MIMO system depends on antenna arrays designed with graphene and plasmonics, which contain of large number of nano arrays in very compact and small space. This enables deployment of massive elements in arrays for transmission and reception, that is, 1024 × 1024 UM-MIMO at 1 THz that can provide different transmission modes. The signal processing problem is considered one of limitations in UM-MIMO THz communications.

The chapter presented an extensive range of concepts about THz UM-MIMO systems and antenna array fabrication, in addition to many problems encountered in feeding and control. Moreover, the chapter provided a detailed construction for arrays of sub-arrays using metamaterials and nanomaterials to build plasmonic nano-antenna (PNA) arrays with various challenging issues and related applications.

Notes

[1] Vidal et al. Photonic Technologies for Millimetre- and Submillimetre-Wave Signals. *Advances in Optical Technologies*, vol. 2012, pp. 1–18.

[2] Siva Viswanathan Thiagarajan. Millimetre-Wave/Terahertz Circuits and Systems for Wireless Communication. Electrical Engineering and Computer Sciences University of California at Berkeley 2016, pp. 1–190.

[3] Robert T. Schwarz et al. MIMO Applications for Multibeam Satellites. *IEEE Transaction on Broadcasting*, vol. 65, no. 4, 2019, pp. 664–681.

[4] Renato Lombardi. Microwave and Millimetre-wave for 5G Transport. ETSI 06921 Sophia Antipolis CEDEX, France, White Paper No. 25, 2018, pp. 1–16.

[5] Min Liang et al. Microwave to Terahertz: Characterization of Carbon-Based Nanomaterials. *IEEE Microwave Magazine*, vol. 15, no. 1, 2014, pp. 40–51.

[6] Roger Appleby, Rupert N. Anderton. Millimetre-Wave and Submillimetre-Wave Imaging for Security and Surveillance. *Proceedings of the IEEE*, vol. 95, no. 8, August 2007, pp. 1683–1690.

[7] Zhi Ning Chen et al. Microwave, Millimetre Wave, and Terahertz Technologies in Singapore. Proceedings of the 41st European Microwave Conference; Manchester, UK, October 2011, pp. 1348–1351.

[8] Hadeel Elayan et al. Terahertz Communication: The Opportunities of Wireless Technology Beyond 5G. International Conference on Advanced Communication Technologies and Networking, CommNet 2018, pp. 1–5.

[9] Kürşat Tekbiyik et al. Terahertz Band Communication Systems: Challenges, Novelties and Standardization Efforts. *Physical Communication*, vol. 35, May 2019, pp. 1–18.

[10] Nabil Khalid and Ozgur B. Akan. Experimental Throughput Analysis of Low-THz MIMO Communication Channel in 5G Wireless Networks. *IEEE Wireless Communication Letters*, vol. 5, no. 6, 2016, pp. 616–619.

[11] Sherif Adeshina Busari et al. Terahertz Massive MIMO for Beyond-5G Wireless Communication. IEEE International Conference on Communications (ICC), 2019, pp. 1–6.

[12] Olabode Idowu-Bismark et al. Massive MIMO Channel Characterization and Modelling: The Present and the Future. *International Journal of Applied Engineering Research*, vol. 12, 2017, pp. 13742–13754.

[13] Hadi Sarieddeen et al. Terahertz-Band Ultra-Massive Spatial Modulation MIMO. *IEEE Journal on Selected Areas in Communications*, vol. 37, no. 9, 2019, pp. 2040–2052.

[14] Nezih Tolga Yardimci and Mona Jarrahi. High Sensitivity Terahertz Detection through Large-Area Plasmonic Nano-Antenna Arrays. Electrical Engineering Department, University of California, vol. 7, 42667, 2017, pp. 1–8.

[15] Nezih Tolga Yardimci and Mona Jarrahi. Plasmonic Nano-Antenna Arrays for High-Sensitivity and Broadband Terahertz Detection. IEEE International Symposium on Antennas and Propagation & USNC/URSI National Radio Science Meeting 2017, pp. 1747–1748.

[16] S. I. Lepeshov et al. Optimization of Nanoantenna-Enhanced Terahertz Emission from Photoconductive Antennas. *IOP Conference Series: Journal of Physics: Conference Series*, vol. 917, no. 6, 2017, pp. 1–5.

[17] Seyed Ehsan Hosseininejad1 et al. Terahertz Dielectric Resonator Antenna Coupled to Graphene Plasmonic Dipole. 12th European Conference on Antennas and Propagation, EuCAP 2018, pp. 1–6.

[18] Taha A. Elwi. A Miniaturized Folded Antenna Array for MIMO Applications. *Wireless Pers Communcation*, Springer, vol. 98, 2017, pp. 1871–1883.

[19] Ian F. Akyildiz, Josep Miquel Jornet. Realizing Ultra-Massive MIMO (1024× 1024) communication in the (0.06–10) Terahertz band. *Nano Communication Networks*, vol. 8, 2016, pp. 46–54.
[20] Josep Miquel Jornet et al. Graphene-based Plasmonic Nano-Antenna for Terahertz Band Communication in Nanonetworks. *IEEE Journal on Selected Areas in Communications*, vol. 31, no. 12, 2013, pp. 685–694.
[21] Xiaoguang Luo et al. Plasmons in graphene: Recent progress and applications. *Materials Science and Engineering R Reports*, vol. 1, 2013, pp. 1–84.
[22] Farzaneh Fadakar Masouleh et al. Nano-Structured Metal–Semiconductor–Metal Photodetector for Sensor Network Systems. *Biomaterials and Tissue Engineering Bulletin*, vol. 4, 2017, pp. 50–65.
[23] Chong Han, Josep Miquel Jornet, and Ian F. Akyildiz. Ultra-Massive MIMO Channel Modeling for Graphene-Enabled Terahertz-Band Communications. IEEE 87th Vehicular Technology Conference, VTC Spring, 2018, pp. 1–5.
[24] Shobhit K. Patel, Christos Angelopoulos. Plasmonic nanoantenna: enhancing light-matter interactions at the nanoscale. EPJ Applied Metamaterials, Advanced Metamaterials in Microwaves, Optics and Mechanics, vol. 2, 2015, pp. 1–15.
[25] Ignacio Llatser et al. Characterization of Graphene-based Nano-antennas in the Terahertz Band. Conference: 6th European Conference on Antennas and Propagation (EuCAP) 2012, pp. 1–5.
[26] T. Harter et al. Silicon–Plasmonic Integrated Circuits For Terahertz Signal Generation and Coherent Detection. *Nature Photonics*, vol. 12, 2018, pp. 625–633.
[27] Mladen Božanić, Saurabh Sinha. Emerging Transistor Technologies Capable of Terahertz Amplification: A Way to Re-Engineer Terahertz Radar Sensors. Sensors (Basel), vol. 19, no. 11, 2019, pp. 1–32.
[28] Tianjing Guo et al. Hybrid Graphene-Plasmonic Gratings to Achieve Enhanced Nonlinear Effects at Terahertz Frequencies. *Physical Review Applied*, vol. 11 (2), 2018, pp. 1–14.
[29] Sebastian Kosmeier et al. Coherent Control of Plasmonic Nanoantenna Using Optical Eigenmodes. *Scientific Reports, Nature Journal*, vol. 3, 2013, pp. 1–7.
[30] Jihua Zhang. Enhancement of Nonlinear Effects Using Silicon Plasmonic Structures. Université Paris-Saclay 2015, 99, 1–169.
[31] Shuai Nie et al. Intelligent Environments Based on Ultra-Massive MIMO Platforms for Wireless Communication in Millimetre Waves and Terahertz Bands. IEEE International Conference on Acoustics, Speech and Signal Processing (ICASSP), 2019, pp. 1–5.
[32] Ian F. Akyildiz et al. Combating the Distance Problem in the Millimetre Wave and Terahertz Frequency Bands. *IEEE Communication Magazine*, 2018, pp. 1–8.
[33] Seyed Ehsan Hosseininejad et al. MAC-Oriented Programmable Terahertz PHY via Graphene-based Yagi-Uda Antennas. IEEE Wireless Communications and Networking Conference (WCNC), 2018, pp. 1–6.
[34] Anil Kumar. Optical Nano-Antennas: Fabrication, Characterization and Applications. University of Illinois at Urbana-Champaign, 2011, pp. 1–93.
[35] Hadi Sarieddeen et al. Next Generation Terahertz Communications: A Rendezvous of Sensing, Imaging and Localization. arXiv:1909.10462v2, 2019, pp. 1–7.
[36] Xiaohang Song et al. Two-Level Spatial Multiplexing using Hybrid Beamforming Antenna Arrays for mmWave Communications. IEEE Wireless Communications and Networking Conference (WCNC), 2016, pp. 1–29.
[37] Manar Mohaisen et al. Multiple Antenna Technologies. arXiv:0909.3324, Cornell University, 2009, pp. 1–20.
[38] Emil Björnson et al. Massive MIMO is a Reality—What is Next? *Digital Signal Processing*, arXiv:1902.07678v2 94, 2019, pp. 1–20.
[39] Bin Zhang et al. Graphene-Based Frequency Selective Surface Decoupling Structure for Ultra-Dense Multi-Band Plasmonic Nano-Antenna Arrays. Proceedings of the 5th ACM International Conference on Nanoscale Computing and Communication, 2018, pp. 1–6.

[40] Cheng-Ming Chen et al. Finite Large Antenna Arrays for Massive MIMO: Characterization and System Impact. *IEEE Transactions on Antennas and Propagation*, vol. 65, no. 12, 2017, pp. 6712–6720.
[41] Santiago et al. Multi-User Ultra-Massive MIMO for very high frequency bands (mm Wave and THz): a resource allocation problem. North-Eastern University 2018, pp. 1–58.
[42] Rodrigo C. de Lamare. Massive MIMO Systems: Signal Processing Challenges and Future Trends. *URSI Radio Science Bulletin*, vol. 2013, pp. 8–20.
[43] Mohamed Habib Loukil et al. Terahertz-Band MIMO Systems: Adaptive Transmission and Blind Parameter Estimation. King Abdullah University of Science and Technology 2019, pp. 1–20.
[44] Chong Han et al. Multi-Wideband Waveform Design for Distance-Adaptive Wireless Communications in the Terahertz Band. *IEEE Transaction on Signal Processing*, vol. 64, no. 4, 2016, pp. 910–922.
[45] Aqeel Hussain Naqvi and Sungjoon Lim. Review of Recent Phased Arrays for Millimetre-Wave Wireless Communication. Sensors (Basel), vol. 18, no. 10, 2018, pp. 1–31.
[46] Monthippa Uthansakul et al. Performance Evaluation of Automatic Switched-Beam Antennas for Indoor WLAN Systems. *WSEAS Transaction on Communications*, vol. 9, no. 12, 2010, pp. 782–792.
[47] Kilian Roth et al. A Comparison of Hybrid Beamforming and Digital Beamforming with Low-Resolution ADCs for Multiple Users and Imperfect CSI. *IEEE Journal of Selected Topics in Signal Processing*, vol. 12, no. 3, June 2018, pp. 1–11.
[48] Irina Stepanets et al. Beamforming Techniques Performance Evaluation for 5G massive MIMO Systems. *Frontiers of Information Technology & Electronic Engineering*, vol. 18, 2017, pp. 753–772.
[49] Berardi et al. Graphene for Reconfigurable Terahertz Optoelectronics. *Proceedings of the IEEE* , vol. 101, no. 7, 2013, pp. 1705–1716.
[50] Sayed Amir Hoseini et al. Massive MIMO Performance Comparison of Beamforming and Multiplexing in the Terahertz Band. arXiv:1710.09031v2, 2017, pp. 1–6.
[51] Chong Han et al. Ultra-Massive MIMO Channel Modelling for Graphene-Enabled Terahertz-Band Communications. IEEE 87th Vehicular Technology Conference (VTC Spring), 2018, pp. 1–5.
[52] Alice Faisal et al. Ultra-Massive MIMO Systems at Terahertz Bands: Prospects and Challenges. arXiv:1902.11090v4, 2019, pp. 1–8.
[53] Alwyn J. Seeds et al. Tera Hertz Photonics for Wireless Communications. *Journal of Light Wave Technology*, vol. 33, no. 3, 2015, pp. 579–587.
[54] Ertugrul Basar et al. Index Modulation Techniques for Next-Generation Wireless Networks. *IEEE Access,* vol. 5, 2017, pp. 16693–16746.
[55] Marco Di Renzo et al. Spatial Modulation for Generalized MIMO: Challenges, Opportunities, and Implementation. *Proceedings of the IEEE*, vol. 102, no. 1, 2014, pp. 56–103.
[56] Zakrajsek, Luke M et al. Design and Performance Analysis of Ultra-Massive Multi-Carrier Multiple Input Multiple Output Communications in the Terahertz Band. *Proceedings of the SPIE*, vol. 10209, 2017, pp. 1–12.
[57] Xiaohang Song et al. Two-Level Spatial Multiplexing Using Hybrid Beamforming for Millimeter-Wave Backhaul. *IEEE Transactions on Wireless Communications*, vol. 17, no. 7, 2018, pp. 1–29.
[58] Lu et al. An Overview of Massive MIMO: Benefits and Challenges. *IEEE Journal of Selected Topics in Signal Processing*, vol. 8, no. 5, 2014, pp. 722–758.
[59] Abu Sulaiman Mohammad Zahid Kausar et al. Optical Nano Antennas: State of the Art, Scope and Challenges as a Biosensor Along with Human Exposure to Nano-Toxicology. *Sensors*, vol. 15, no. 4, 2015, pp. 8788–8831.
[60] Kan Zheng et al. Survey of Large-Scale MIMO Systems. *IEEE Communications Surveys & Tutorials*, vol. 17, no. 3, 2015, pp. 1738–1760.
[61] Hakim Mabed et al. A Flexible Medium Access Control Protocol for Dense Terahertz Nanonetworks. Proceedings of the 5th ACM International Conference on Nanoscale Computing and Communication, 2018, pp. 1–7.

[62] Khondokar FidaHasan et al. Time Synchronization in Vehicular Ad-Hoc Networks: A Survey on Theory and Practice. arXiv:1811.04580 2018, pp. 1–13.
[63] Yong Niu et al. A Survey of Millimetre Wave (mmWave) Communications for 5G: Opportunities and Challenges. *Wireless Networks*, vol. 21, 2015, pp. 2657–2676.
[64] Rashid A. Saeed, Esra B. Abbas. Performance Evaluation of MIMO FSO Communication with Gamma-Gamma Turbulence Channel using Diversity Techniques. International Conference on Computer, Control, Electrical, and Electronics Engineering (ICCCEEE), 2018, pp. 1–5.
[65] Cen Lin and Geoffrey Ye Li et al. Terahertz Communications: An Array-of-Subarrays Solution. *IEEE Communications Magazine*, vol. 54, December 2016, pp. 124–131.
[66] MONA JABER et al. 5G Backhaul Challenges and Emerging Research Directions: A Survey. *IEEE Access*, vol. 4, 2016, pp. 1743–1766.
[67] Raveena K R et al. A Survey on Massive MIMO Antennas for 5G Applications. National Conference on Recent Trends in VLSI, Communication and Networks RTVCN2K18, 2018.
[68] Rashid A. Saeed, Raed A. Alsaqour, Ubaid Imtiaz, Wan Mohamad, Rania A. Mokhtar. Design of CMOS Power Amplifier for Millimetre Wave Systems at 70 GHz. *International Journal of Engineering and Technology (IJET)*, vol. 5 (1), 2013, pp. 498–503.
[69] M. Helbig et al. Development and Test of a Massive MIMO System for Fast Medical UWB Imaging. International Conference on Electromagnetics in Advanced Applications (ICEAA), 2017, pp. 1331–1334.
[70] Wang and Bin Huang. Design of Ultra-Wideband MIMO Antenna for Breast Tumor Detection. *International Journal of Antennas and Propagation*, Hindawi, vol. 2012, pp. 1–7.
[71] Fulai Liang et al. Detection of Multiple Stationary Humans Using UWB MIMO Radar. *Sensors (Basel)*, vol. 16, no. 11, 2016, pp. 1–17.
[72] Xiaodong Zhuge et al. A Sparse Aperture MIMO-SAR-Based UWB Imaging System for Concealed Weapon Detection. *IEEE Transactions on Geoscience and Remote Sensing*, vol. 49, no. 1, 2010, pp. 509–518.
[73] Md. Zulfiker Mahmud et al. Ultra-Wideband (UWB) Antenna Sensor Based Microwave Breast Imaging: A Review, *Sensors*, vol. 18, no. 9, 2018, pp. 1–15.
[74] A. A. Yassin, Rashid A. Saeed, R. A. Alsaqour and R. A. Mokhtar. Design of Fixed Microstrip Patch Antenna for WLAN Applications. *International Journal of Applied Engineering Research*, vol. 9, no. 19, 2014, pp. 6231–6238.
[75] Petteri Kela et al. Flexible Backhauling with Massive MIMO for Ultra-Dense Networks. *IEEE Access*, vol. 4, 2016, pp. 1–10.
[76] Rashid A. Saeed. Femtocell Communications: Business Opportunities and Deployment Challenges. IGI Global, USA, January, 2012, pp. 295.
[77] Dick Maryopi and Alister Burr. Few-Bit CSI Acquisition for Centralized Cell-Free Massive MIMO with Spatial Correlation. IEEE Wireless Communications and Networking Conference (WCNC), 2019, pp. 1–6.
[78] Aldo Petosa. Engineering the 5G Environment. IEEE 5G World Forum (5GWF), Silicon Valley, CA, USA, 2018, pp. 478–481.
[79] Yasser Zahedi et al. Characterization of Massive MIMO UWB Channels for Indoor Environments. IEEE 13th Malaysia International Conference on Communications (MICC), 2017, pp. 57–62.
[80] Yi Han, Yan Chen, Beibei Wang, and K. J. Ray Liu. Time-Reversal Massive Multipath Effect: A Single-Antenna "Massive MIMO" Solution. *IEEE Transactions on Communications*, vol. 64, no. 8, 2016, pp. 3382–3394.
[81] Rashid A. Saeed, S. Khatun. Design of Microstrip Antenna for Wireless Local Area Network (WLAN). Asian Network for Scientific Information (ANSI), *Journal of Applied Science (JAS)*, vol. 5, no. 1, Pakistan, 2005, pp. 47–51.
[82] sonika et al. A Compact Dual Port UWB-MIMO/Diversity Antenna for Indoor Application. *International Journal of Microwave and Wireless Technologies*, vol. 10, no. 3, 2017, pp. 360–367.
[83] Lukasz Zwirello et al. UWB Localization System for Indoor Applications: Concept, Realization and Analysis. *Journal of Electrical and Computer Engineering*, hindawi, vol. 2012, pp. 1–11.

[84] Arcady Molev Shteiman. Ultra-Massive MIMO and an Alternative to Ultra Dense Network: Benefits and Challenges. *Journal of Radio Electronics*, vol. 5, 2019, pp. 1–25.
[85] Dushyantha A. Basnayaka and Harald Haas. Spatial Modulation for Massive MIMO. IEEE International Conference on Communications (ICC), London, UK, 2015, pp. 1945–1950.
[86] Shahar Stein et al. A Family of Hybrid Analog-Digital Beamforming for Massive MIMO Systems. *IEEE Transactions on Signal Processing*, vol. 67, no. 12, 2019, pp. 3243–3257.
[87] Irfan Ahmed et al. A Survey on Hybrid Beamforming Techniques in 5G: Architecture and System Model Perspectives. *IEEE Communications Surveys & Tutorials* , vol. 20, no. 4, 2018 , pp. 3060–3097.

Chapter 11

Design of Passive Components for Microwave Photonics-based Millimetre Wave Systems

Bilal Hussain, Luís M. Pessoa, and Henrique M. Salgado

Contents

11.1 Introduction

The ever-increasing demand for high-speed wireless communications has pushed the communication technologies toward sub-terahertz operating frequencies (0.1–0.5 THz). The recent introduction of millimetre wave (mm-wave) band in 5G standard has ushered a new era of design and development in sub-THz communication. Millimetre-wave systems can provide a data rate of 10 Gb/s with the possibility of dedicated antenna beams for each user in a micro-cell structure. Although the path loss experienced by sub-THz frequencies is very high (82 dB for 1 m @ 300 GHz), yet the small size and large bandwidth make them suitable for in-door high-speed communication. In the mm-wave transceiver chain, the antenna plays a crucial role in determining the overall performance of the link. Due to very high path loss, sub-THz systems require high gain directive antennas. Conventional bulky high-gain antennas such as horn, parabolic dish and slot can be designed for mm-wave frequencies. Yet, the large footprint (in comparison with device size) and lack of interconnects make them less favourable for commercial indoor communications. As in the case of mobile communications, patch antennas are the preferred choice for realizing a hand-held device. Planar patch antennas can be used at mm-wave frequencies, although they cannot alone provide very high directivities due to their broad radiation pattern. The directivity of a planar structure can be increased by utilizing a planar array. An array consisting of planar patch antennas not only improves the directivity and gain of the structure but also opens up the possibility of steering the radiation pattern by adjusting the phase of each element. Furthermore, an electronic beam steering mechanism allows for automatic reconnection between transmitter (Tx)/receiver (Rx) nodes, ensuring reliable operation if the direct link becomes entirely blocked by an obstacle. This can be guaranteed by non-line-of-sight path scenarios using direct reflections from the walls, floor or ceiling in indoor environments. Moreover, the use of sub-THz and THz frequencies provides the opportunity for on-wafer integration with active devices, enabling compact mm-wave transceivers.

Conventional complementary metal-oxide-semiconductor (CMOS) technologies fail to provide complete mm-wave systems beyond 100 GHz. The reason lies in the fact that the crystalline structure of Silicon is noisy and thereby hinders the development of an efficient device operating at the sub-THz range. Alternative technologies such as Silicon–Germanium (SiGe), Indium-Phosphide (InP) and Gallium-Arsenide (GaAs) have been exploited to fabricate sub-THz devices operating beyond 100 GHz with an output power of around 100 μW [1–3]. Millimetre wave frequencies can be generated using an *N*-step frequency multiplication process with the phase noise of the generated signal being strongly dependent on the number of multiplication stages. A single step signal generation is often achieved by biasing a transistor in the unstable region and by using a filter at the input/output, it is possible to create a mm-wave oscillator. Electronics-based mm-wave signal generation is power hungry due to loss of energy either in multiplication stages or filtering. On the other hand, such techniques are strongly dependent on operating temperature and crystalline structure of the material.

Another limitation of electronic-based mm-wave generation systems is the lack of tuneable phase shifters. Conventionally, electronically implemented phase shifters are based either on switched delay lines using micro-electromechanical systems (MEMS) or on monolithic microwave integrated circuits (MMICs). The main limitations of these solutions lie in their limited operating bandwidth (BW), or low power handling, often with the additional constraint of relatively narrow scanning angles [4–7]. In particular, they do not guarantee the same performance, for example, in terms of minimum step of scanning angle and maximum transmission range over their operating bandwidth. Even the appealing and promising technology of Rotman lenses allowing for (discrete only) true-time-delay (TTD)-based beam steering, may pose issues related to size, scalability and modelling complexity [8]. The electronic analogue or digital approaches to phase shift, on the other hand, can operate on large BWs. Still, they usually do not offer the same performance at every frequency, introducing non-negligible losses and signal distortions [4]. For operation beyond 100 GHz, switched delay lines are mostly used to realize phase shifters. Such TTD-based approach is limited in minimum steering step and the operating bandwidth.

An interesting solution to both problems – (i) signal generation and (ii) tuneable phase shifter – comes from the field of optics. In early 2000, photonics systems performing microwave operations were reported. Optical frequencies range from 180 to 800 THz (1600–400 nm) while sub-THz frequencies range from 0.1 to 0.3 THz. The complete mm-wave range is just 16% of optical C-band (1530–1565 nm). In turn, this means that an optical device having a bandwidth of 2% (around 1550 nm) can provide a bandwidth of hundreds of GHz without adding any complexity to the design. Before microwave photonics (MWP) systems, optical fibres were used widely to transport RF signals. This technology is commonly known as radio-over-fibre (RoF) [9]. A simple description of a RoF system is presented in Figure 11.1. It consists of an optical modulator which modulates the incoming RF signal onto an optical carrier from a laser source. At the receiving end, a photodiode is used to convert the optical signal back to the electrical domain. The bandwidth of a RoF system is mainly limited by the speed of the modulator and the photodiode cut-off frequency. Instead of just transporting RF signals over fibre, it is possible to perform operations in the optical domain over the RF signals [9]. These operations include filtering, phase shifting, modulation, up/down conversion and analogue to digital conversion. The main advantage of optical processing is that thermal noise does not affect optical signals. Moreover, the tunability of optical devices is relatively easy to achieve. Another attractive property of photonic processing is that stable mm-wave oscillators can be constructed using optical devices. Instead of heterodyning a local optical oscillator (LO) with an optical data signal, two inter-locked laser signals spaced in frequency are mixed in a photodiode. The difference of the laser frequencies generates a highly stable sub-THz LO signal. By varying the frequency of one of the lasers, it is possible to tune

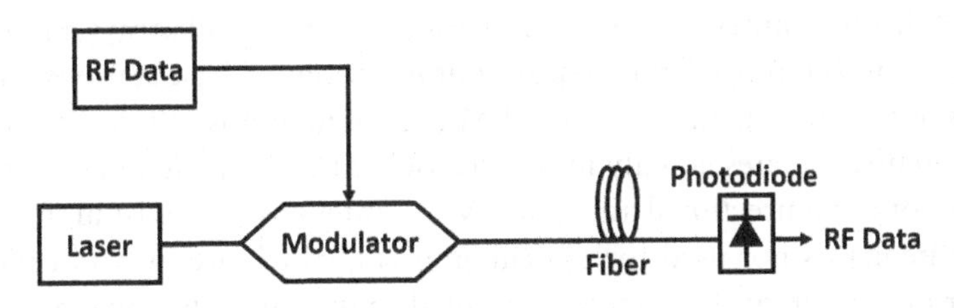

Figure 11.1 Block diagram of radio-over-fibre (RoF).

the LO frequency. Earlier MWP systems were developed using fibre-based components. Although these systems have provided a proof of concept, their dimensions were not suitable for large-scale manufacturing. Recent developments in photonic integrated circuit (PIC) technology has made it possible to realize mm-wave signal generation using microwave photonics techniques.

Using MWP techniques, it is possible to generate sub-THz signals with precision. Also, a tuneable phase shifter can be implemented in the optical domain such that the mm-wave signal is phase delayed prior to conversion to the electrical domain. For such operations, the limitation of the optical domain comes from the limited cut-off frequency of photodiodes and limited modulation speed of optical modulators. Recently introduced Silicon photonics technology cannot provide the necessary modulation speed and cut-off frequency for sub-THz signal processing. III–V compounds are the only choice for realizing mm-wave systems. There exist optical as well as electrical components in III–V materials independently, yet the interconnects needed for realization of complete opto-electronic systems such as waveguides and transmission lines are not standardized. The reason lies in the fact that III–V technologies constitute a very small portion of the integrated device market. A lot of focus has been given to Silicon-based technologies; therefore III–V technologies lack necessary fabrication techniques for realizing a complete opto-electronic system operating in the mm-wave range. The difficulties in realizing vias and cavities in III–V materials are a design bottleneck for realizing passive components such as antennas, filters and circulators. In the following sections, we present the development of an antenna array in InP (commonly used III–V material in optics and electronics) using a sacrificial layer of BCB. An alternative to this technique is also presented where a waveguiding structure is developed on InP using a via-less bed of nails architecture. These two techniques can provide the necessary interconnects for the realization of a complete opto-electronic system operating in the sub-THz region.

11.2 Unit Radiating Cell or Antenna

Electromagnetic radiation is created whenever a charge accelerates or deaccelerates. This effect can be enhanced by introducing asymmetric magnetic fields. Such asymmetries are introduced by adding bends and gaps to a conductor surface. The simplest form of antenna is a wire antenna where a leaking magnetic field generates a travelling electromagnetic (EM) wave. Radiation is an inherent phenomenon in all the matter above 0 K. Here we focus only on conductor surfaces that are optimized to enhance the radiation towards a particular direction in space. Antennas can be classified into two categories: (i) directive antennas and (ii) omni-directional antennas. Directive antennas focus electromagnetic (EM) energy in a particular direction across a three-dimensional space. Common examples of directive antennas are horns, parabolic dishes and arrays, among others. Such antennas are often employed in high power applications or base stations. Non-directive or omnidirectional antennas radiate equal EM energy in almost all directions. A theoretical isotropic antenna radiates equally in all directions. Dipoles, patches and helical wires are all examples of omnidirectional antennas. While antennas also exist in the optical domain, the term 'antenna' is not used for optical emitters, which are instead called mirrors and lenses. Our discussion in this chapter is limited only to non-directive patch antennas and directive arrays.

11.2.1 Patch Antenna

Since the 1970s, microstrip antennas are utilized for many applications due to their design flexibility and cost effectiveness. They allow for compact solutions since they can be integrated into a single printed circuit broad (PCB) along with active components allowing compact wireless transceivers.

A simple microstrip antenna consists of a sandwiched dielectric material between two metal layers where the top metal layer houses a feeding line and radiating patch while the bottom metal layer serves as a ground plane. The shape of the radiating patch is only limited by the imagination of the designer. The most commonly used shapes are shown in Figure 11.2.

As an example a square patch fed by a microstrip line is shown in Figure 11.3. Its resonance frequency is mainly determined by the physical length of the radiating element, L,

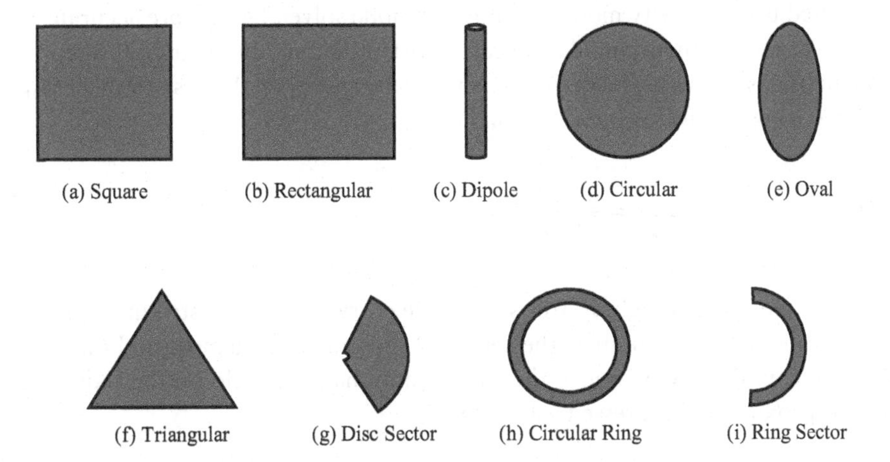

Figure 11.2 Different shapes possible for microstrip antennas (adapted from [10]).

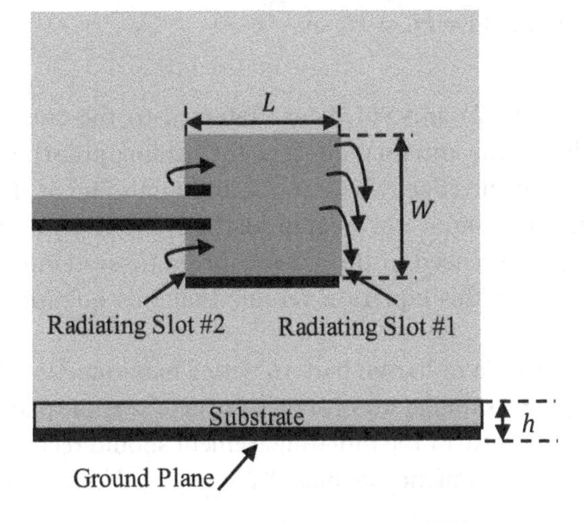

Figure 11.3 Illustration of a microstrip patch antenna (adapted from [10]).

which is calculated as a function of the substrate properties, ε_r, h, the relative dielectric constant and thickness, respectively, to satisfy the equation $L = \frac{\lambda}{2} + \Delta l$, where λ is the wavelength given by $\frac{1}{f\sqrt{\mu_0\mu_r\,\varepsilon_o\varepsilon_r}}$ in which ε_o and μ_0 are free space dielectric constant and magnetic permeability, respectively. The Δl factor is used to compensate the fringing fields effect and, although theoretically quantified, in terms of design it is often obtained using parametric simulations or optimizations algorithms. The width of the patch, W, does not have a significant impact on the resonance frequency, but it allows for improvements on the bandwidth and efficiency since it affects the losses along the radiating element. The variation in the width of radiating patch can improve the performance to a certain limit beyond which undesired radiation effects take place. The microstrip antenna can exhibit different radiation patterns depending on the excitation caused by current distribution. Despite the complexity involved in the theoretical study, the optimal current distribution is calculated using a cavity model or an Eigen mode solver. For a more accurate analysis, a full-wave electromagnetic simulation can be performed to obtain an optical current mode for the desired structure. The desired resonance frequency of the first mode (TM_{010}) can be calculated using the following equation [10]:

$$(f_r)_{010} = \frac{1}{2L\sqrt{\varepsilon_r}\sqrt{\mu_0\varepsilon_0}} = \frac{v_0}{2L\sqrt{\varepsilon_r}} \tag{11.1}$$

where v_0 is the speed of light in free space. The current density distribution, $\overline{J_s} = \hat{n} \times \overline{H}$, of the TM_{010} mode is maximum at the centre of the radiating element and minimum at the ends, and from the cavity model, the electric and magnetic field on the radiating element can be approximately expressed as[10, 11]

$$\text{vertical electrical field} = E_x = E_0 \cos\frac{\pi y}{L}, \tag{11.2}$$

$$\text{transverse magnetic field} = H_z = H_0 \sin\frac{\pi y}{L}. \tag{11.3}$$

Note that Equations (11.2) and (11.3) are referred to the coordinate system shifted towards the left edge of the antenna (close to the feeding point). For example, the electric field is maximum at the edge, where $y = 0$, and minimum at the centre of the patch, where $y = L/2$, and maximum once again at the right edge, corresponding to $y = L$. The magnetic field is referenced in the same coordinate system being 90° out of phase of the electric field. Also, this equation reveals that this quantity remains constant in the z-axis.

The TM_{010} mode creates a radiation pattern with a maximum in the normal plane of the antenna (xy-plane). For example, an observer in the elevation plane (xy-plane), positive y-axis aligned with the centre of the radiating element should receive a maximum whereas an observer in the azimuth plane (zy-plane) aligned with the (y-axis) should receive a minimum.

The patch antenna can also be modelled as an open-ended transmission lime, in which the voltage and current can be approximated as

$$\text{voltage} = V(y) = V_0 \cos\frac{\pi y}{L}, \tag{11.4}$$

$$\text{current} = I(y) = \frac{V_0}{Z_0}\sin\frac{\pi y}{L}, \tag{11.5}$$

and, since $Z_{in} = \frac{V(y)}{I(y)}$, this shows that the impedance varies along the length of the patch, remaining constant across its width. This equation is also referred to the coordinate system shifted to the left edge of the antenna (close to the feeding point). For example, the impedance is maximum at the edges and is minimum at the centre. The optimal feed point, considering a 50-ohm source matched to a 50-ohm transmission line, is often found around $y = \frac{L}{4}$ from the edge (along y-axis). This is the reason why in Figure 11.3, the microstrip line goes into the radiation patch for better impedance matching.

11.2.2 Coupling of Microstrip Antenna

There exist several feeding mechanisms for microstrip antenna, namely, microstrip feed, aperture feed, magnetic coupling and capacitive coupling. Illustrations of the different feed mechanisms are presented in Figure 11.4 [10, 11]. The simplest microstrip feed consists of a microstrip transmission line that feeds the antenna as illustrated in Figure 11.4 (a).

The input impedance of the TM_{010} mode can be optimized by varying the position of the feeding point of radiation patch, which is commonly known as inset feed. As mentioned before, this technique benefits from the fact that the impedance is the function of the length of the patch, so by moving the position of feeding point into the radiating patch, it is possible to obtain a desired input impedance over a certain bandwidth. However, as the substrate thickness increases, the microstrip line cannot support the wave propagation thereby making the feeding mechanism not suitable for very thick substrates. Furthermore, this feeding technique is not suitable for antennas whose feeding point is not on the same side of PCB as radiating patch.

Another commonly used feeding technique is the coaxial feed, as shown in Figure 11.4(b). Coaxial feed is constructed by connecting the out sheath of coaxial connector to the ground plane and the connector pin to the radiation patch. The parasitic inductance resulting from the feed line length limits the application of this technique for very high frequency applications, and therefore it is not recommended for antenna arrays. Similarly, to the microstrip feed, the input impedance can be controlled adjusting the feed position.

The limitations presented by microstrip and coaxial feeds can be mitigated by using aperture-coupled feed and proximity coupled feed. The aperture feed shown in Figure 11.4 (c) is constructed using two different substrates separated by a metal layer. The feeding line is then designed on the bottom metal layer of the first substrate layer and the radiation patch is housed on the top metal layer of the second substrate. The feeding is reached by means of an aperture in the sandwiched metal layer between two substrates,

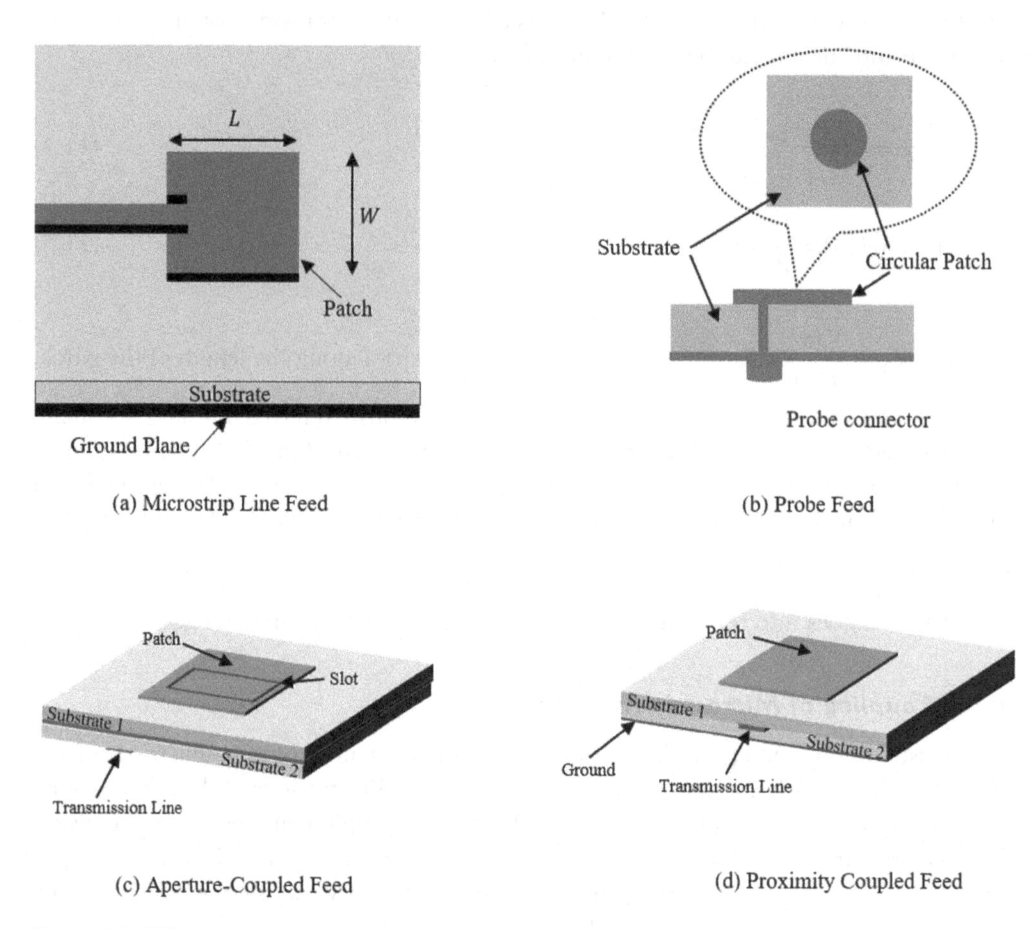

Figure 11.4 Different microstrip antenna feed typologies (adapted from [10]).

which serves as the shared ground plane between the radiation patch and the microstrip feed line. The aperture can be designed for any geometrical shape; however, a rectangular aperture is preferred due to its wide bandwidth. The input impedance can be controlled varying the length of the feed line, the position and size of the slot. The proximity feed, also known as magnetic coupling feed, often consists of two substrate layers as shown in Figure 11.4 (d). The feeding line consists of a microstrip line between the two layers, which induces a current distribution on the patch. The input impedance is optimized by varying the length of the transmission line and the width ratio between radiation patch and microstrip feed.

The equivalent circuit models for the different feeding methods are presented in Figure 11.5. It is evident that the microstrip and coaxial feed approaches present similar equivalent models, exhibiting a dominant inductive behaviour. The proximity feed presents a dominant capacitive behaviour while the aperture feed presents both capacitive and inductive coupling mechanisms. It is important to note that all the equivalent models correspond only to the fundamental mode TM_{010}. In case of more than one mode, for example, if a parasitic element is added for bandwidth enhancement, these equivalent models are no longer valid, and more complex models must be derived.

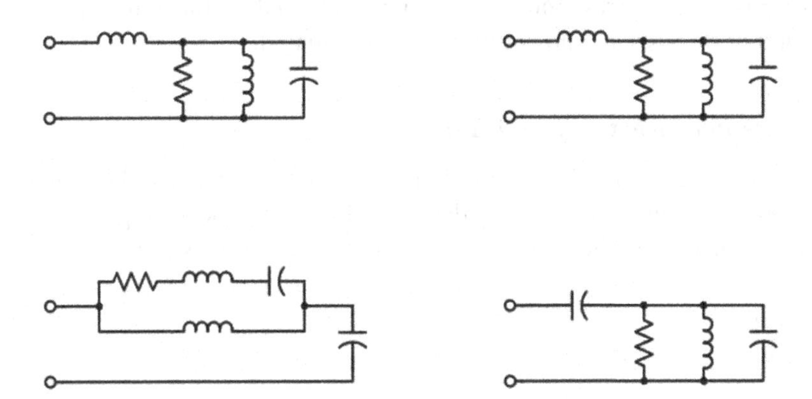

Figure 11.5 First-order equivalent model for the different feed techniques (adapted from [10]).

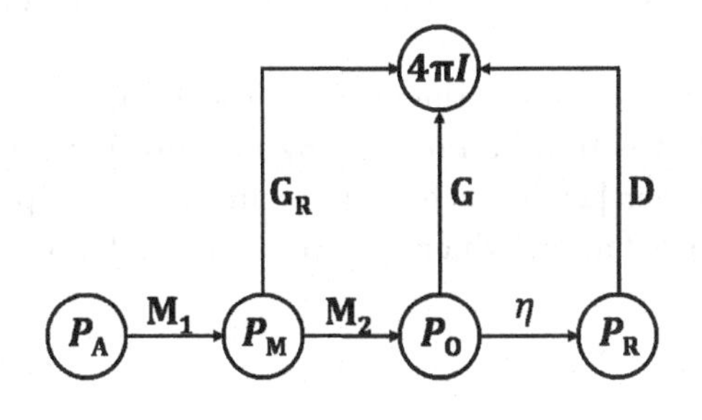

P_A = power available from the generator
P_M = power to matched transmission line
P_O = power accepted by the antenna
P_R = power radiated by the antenna
M_1 = impedance match factor 1
M_2 = impedance match factor 2

η = radiation efficiency
G_R = realized gain
G = gain
D = directivity
I = radiation Intensity

Figure 11.6 Gain and directivity flow chart (adapted from [12]).

11.2.3 Antenna Efficiency

Figure 11.6 [11] presents an intuitive view of some of the most important antenna parameters extracted from the IEEE standard definition of terms for antennas [12], in which efficiency is included, and the same original source also provides definitions for all the illustrated parameters. One of the most important parameters is antenna efficiency especially when dealing with very small output powers (< 1mW) and very high operating frequency (> 100GHz), which is the case of the milli-meter wave sources considered here. The antenna efficiency depends on several factors including substrate loss, conduction loss and surface waves as these do not contribute to the useful radiated power, P_R. The efficiency can be described mathematically as

$$\eta = \frac{P_R}{P_R + P_{Loss}}, \tag{11.6}$$

where $P_O = P_R + P_{Loss}$ corresponds to the power accepted by the antenna, in which P_{Loss} is the fraction of the accepted power which is not radiated.

11.2.4 Antenna Directivity and Gain

As described in IEEE standard definition of terms for antennas [12], directivity is the radiated power intensity in a certain direction divided by the average radiated power by the same antenna. In mathematical form, it can be expressed as [10]

$$D = \frac{4\pi U}{P_R}, \tag{11.7}$$

where U denotes power intensity in a certain direction. The typical directivity of a single element microstrip antenna is 5 – 7 dBi. In most of the cases, the directivity bandwidth is greater than the impedance bandwidth thus the antenna bandwidth is mostly defined by the impedance bandwidth.

The gain of an antenna defined as the radiated power intensity in a certain direction over the total power that would be radiated by an isotropic (ideal) antenna receiving the same amount of power [2]. Therefore, the gain (G) corresponds to the directivity affected by the efficiency factor, which can be expressed, for a certain direction, as follows:

$$G = \eta \frac{4\pi U}{P_R}, \tag{11.8}$$

11.2.5 Antenna Bandwidth

Antenna bandwidth is defined as 'the range of frequencies within which the performance of the antenna, with respect to some characteristic, conforms to a specified standard'. The concept of antenna bandwidth extends beyond the impedance bandwidth since the directivity and efficiency must be considered. For example, an antenna is wideband if it has wideband impedance and both directivity and efficiency fulfil the requirements in the entire impedance bandwidth. The figure of merit that considers all these parameters is known as realized gain (RG). The realized gain in a certain direction takes into account the power reflected by the antenna towards the generator, conduction loss, substrate loss, surface wave radiation and radiated power.

In a microstrip antenna, the realized gain is dominated by the efficiency and the impedance bandwidth, since the directivity remains almost constant in a broad frequency range compared to the impedance bandwidth or efficiency. Typically, a microstrip antenna in its simple form has a fractional bandwidth (FB) of 5 %. Higher bandwidth can be achieved considering thicker substrates, employing parasitic elements or slot configurations. The bandwidth resultant from thick substrates is derived from the natural dependence of the Q factor with the substrate height, while parasitic elements and slot configurations allow the excitation of more than one current mode, which results in multiple resonance frequencies, increasing the overall bandwidth.

11.3 Phased Array Theory

The antenna array is the most appropriate solution for many applications where high directivity is necessary. Besides the directivity, antenna arrays are also used to combine radiated fields when each array element is fed by an independent source. This is especially interesting to increase communications range and data rates when the available source power is low. By properly arranging of the elements of the array and controlling the phase excitation of the different elements, the beam can be oriented in a well-established observation angle. The mechanism of beam orientation by controlling the relative phase or time delay of each element is known as beam scanning and the array is known as phased array. In a phased array system, the inter-element distance takes an important role, since it can give rise to grating lobes, which in its turn limits the scanning range. Grating lobes are beams of equal strong radiation as the main beam but in an unwanted direction, which leads to power spread out of the desired observation angle. To avoid grating lobes, the inter-element spacing must be less than or equal to $\lambda / 2$. However, reducing the inter-element spacing increases the mutual coupling, which can affect the field distribution of the elements in the array and consequently the reflection coefficient and resonance frequency. Furthermore, the inter-element distance corresponding to maximum directivity is not the same as the distance that optimizes the scanning range. Therefore, a trade-off must be established between bandwidth, scanning range, mutual coupling and gain. Nevertheless, phased arrays are likely one of the most attractive solutions for THz communications.

The total radiated field of an array arrangement is obtained by multiplying the radiation field of the single element by the array factor (AF) – pattern multiplication rule – as stated by (11.9), which is valid under the consideration of equal antenna elements,

$$\begin{gathered} E_{\text{total}} = E_{\text{element}} \times A \\ E_{\text{total}}(\text{dB}) = 10\log_{10}\left(E_{\text{totallinear}}\right) + 10\log_{10}\left(AF_{\text{linear}}\right) \end{gathered} \tag{11.9}$$

The array factor of both linear and planar arrays is described in the succeeding sections.

11.3.1 Array Factor of Linear Array

The array factor of N equally spaced elements, with non-uniform amplitude and progressive phase shift is given by

$$AF = \sum_{n=1}^{N} a_n e^{j(n-1)(kd\cos\theta+\beta)} = \sum_{n=1}^{N} a_n e^{j(n-1)\psi}, \tag{11.10}$$

where

$$\Psi = kd\cos\theta + \beta,$$

$$\beta = -kd\cos\theta_0,$$

N corresponds to the number of elements, a_n is the amplitude of the nth antenna element, k is the wave number, d is the inter-element distance and θ_0 the observation angle.

In the case of uniform amplitude excitation, it is possible to show by means of the Euler formula that (11.10) can be written in an alternative form as

$$AF = \left[\frac{\sin\left(\frac{N}{2}\psi\right)}{\sin\left(\frac{1}{2}\psi\right)} \right], \tag{11.11}$$

which in logarithmic corresponds to

$$AF(\text{dB}) = 10\log_{10}\left(AF_{\text{linear}}\right).$$

The maximum of the array factor obtained from (11.10) is given by:

$$AF\,|_{\theta_0,\max} = \sum_{n=1}^{N} a_n = a_1 + a_2 + \ldots + a_n, \tag{11.12}$$

which in the case of uniform amplitude, can be obtained applying the L'Hospital's formula in Equation (11.11) as follows:

$$AF\,|_{\theta_0,\max} = \lim \frac{d}{d\theta} AF = \lim N \frac{\cos(0)}{\cos(0)} = N \Rightarrow AF\,|_{\theta_0,\max}(\text{dB}) = 10\log_{10}(N) \tag{11.13}$$

Also, it is possible to demonstrate that the sidelobe levels can be controlled through a proper arrangement of the different components of the source excitation magnitude (referring to Equation (11.10)). Moreover, the uniform source excitation is the one that offers higher gain; however, it presents the worse level of sidelobes. This typically leads to a trade-off when low sidelobe levels (> 13.2 dB) are required.

11.3.2 Grating Lobes

In the case of uniform arrays, the maxima of the array factor occur when the denominator of Equation (11.11) goes to zero and it can be calculated as follows:

$$\sin\left(\frac{1}{2}\Psi\right) = 0 \Rightarrow \frac{1}{2}\Psi = \pm m\pi$$

$$\frac{1}{2}\left(kd\cos\left(\theta_{\max}\right) + \beta\right) = \pm m\pi$$

$$\theta_{\max} = \operatorname{acos}\left[\frac{k}{2\pi d}(-\beta \pm 2m\pi)\right], m = 0,1,2,\ldots \tag{11.14}$$

Therefore, the maximum inter-element distance to avoid grating lobes in the domain $]0 : 2\pi[$ is $d = \lambda / 2$. Also, this equation shows that the grating lobes do not depend on the

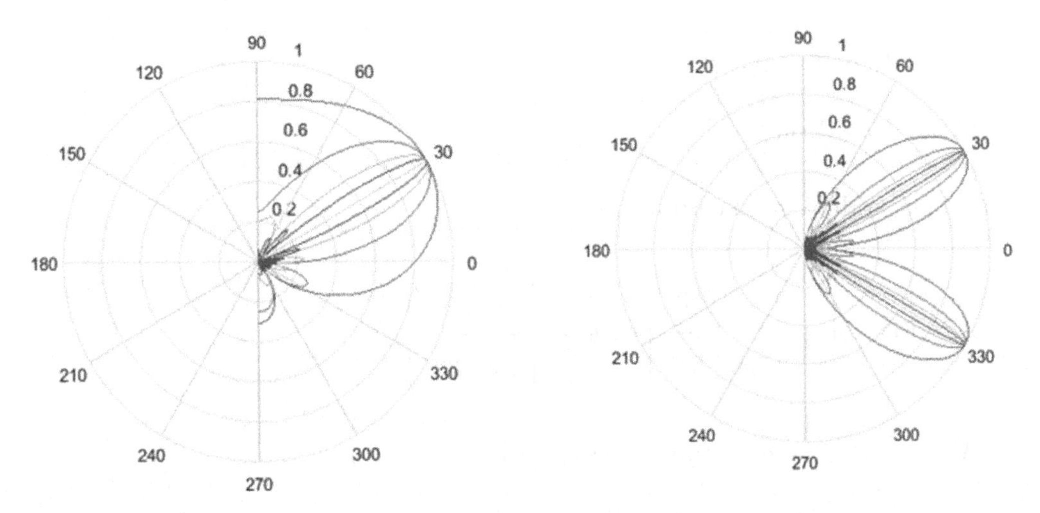

Figure 11.7 Normalized polar plot of linear array factor for $N = \{2, 4, 8 \text{ and } 16\}$ an observation angle $\theta_0 = 30^0$.

number of elements, which means that, independently of the number of elements, if the inter-element distance is greater than $\lambda/2$, grating lobes will exist. Figure 11.7, illustrates the normalized AF for different number of elements $N = \{2, 4, 8 \text{and} 16\}$ and an observation angle $\theta_0 = 30^0$, for case (a), considering an inter-element distance of $\lambda/2$, which shows no grating lobes, and for case (b), considering an inter-element distance of λ, which shows two different beams with equal intensity. Considering this last case, the radiated power will be equally divided between the desired angle and the direction pointed by the grating lobe.

11.3.3 Scan Blindness

Besides grating lobes, the scan blindness phenomenon is another important limitation in phased array antennas with major effect in the refection coefficient. This fact is motivated by the presence of surface waves – characterized by poor radiation performance – that occur as a function of the scanning angle. These surface waves affect the input impedance of the array, and consequently the reflection coefficient. Theoretically, total reflection can occur at certain scanning angles which are known as blindness angle. Both planar and linear arrays are affected by scan blindness.

11.3.4 Planar Array

In addition to arranging elements along a line (to form a linear array), antenna elements can be positioned along a rectangular grid to form a rectangular or planar array. The array factor of a rectangular array with uniform excitation is given by:

$$AF = \left\{ \frac{\sin \frac{M}{2} \Psi_x}{\sin \frac{1}{2} \Psi_x} \right\} \left\{ \frac{\sin \frac{N}{2} \Psi_y}{\sin \frac{1}{2} \Psi_y} \right\}, \tag{11.15}$$

where:

$$\Psi_x = kd_x \cos\theta_0 \sin\phi_0 + \beta_x,$$

$$\Psi_y = kd_y \cos\theta_0 \sin\phi_0 + \beta_y,$$

which in logarithmic form corresponds to:

$$AF(\mathrm{dB}) = 10\log_{10}\left(\frac{\sin\frac{M}{2}\Psi_x}{\sin\frac{1}{2}\Psi_x}\right) + 10\log_{10}\left(\frac{\sin\frac{N}{2}\Psi_y}{\sin\frac{1}{2}\Psi_y}\right)$$

$$AF_{max} = M \times N \Rightarrow AF_{max}(\mathrm{dB}) = 10\log_{10}(M) + 10\log_{10}(N).$$

All the limitations presented in the case of linear arrays, namely grating lobes and scan blindness, are also extended to planar arrays and therefore we don't need to repeat that analysis this section.

11.4 Millimetre Wave Generation and Phase Shifting Techniques

As mentioned earlier, the conventional electrical approaches for generation of mm-wave signals is not feasible due to limited precision, temperature dependence and reduced frequency agility. Similarly, designing tunable phase shifters using electrical resonant components for a wider bandwidth is also challenging due to unavailability of variable capacitor and inductors at sub-THz frequencies. Using MEM switches or switched transmission line does not provide needed speed and bandwidth. The following section presents microwave photonic techniques suitable for generation and processing of mm-wave signals in the optical domain. By utilizing the high-frequency optical signal, it is possible to achieve a wider bandwidth in the sub-THz region.

11.4.1 Sub-THz Signal Generation Using Microwave Photonic (MWP) Techniques

Microwave photonics systems were introduced almost two decades ago. Earlier MWP systems were designed with the purpose of performing certain functions such as filtering, beam forming and analogue to digital conversion. These earlier systems have provided the possibility to devise more complex systems. Furthermore, these operations are necessary not only for sub-THz but also for RF systems. This section describes earlier photonic-based systems along with the current state of art for MWP based communication systems. The technical challenges faced by MWP systems are highlighted particularly for the implementation of mm-wave systems.

11.4.1.1 Microwave Photonic Systems

Microwave photonics is an interdisciplinary area that studies the interaction between microwave and optical signals. Since the first demonstration of lasers in 1960 by Bell

Laboratories, the era of optical communication started. Due to very large instantaneous bandwidth and low loss transmission, optical systems attracted huge interest from the research community. The development of semiconductor laser has provided the ability to directly modulate an RF signal onto the optical carrier. The introduction of optical modulator has provided the ability to extend the modulation speeds above 100 GHz. The breakthroughs in the field of optical detectors has made it possible to demodulate a high-frequency RF signal modulated over an optical carrier. Such optical communication systems inspired the use of photonics for microwave signal generation, processing and distribution. All optical signal processing is implemented using optical delay lines and filters. Phased array antennas may also be implemented using the TTD technique [13, 14]. The following section presents a detailed state of art of various MWP systems.

A low phase noise microwave/mm-wave signal is desirable for many applications such as radars and wireless communication. In the sub-THz domain, signals are generated using frequency multiplication circuits. For the generation of stable mm-wave signals, such techniques are costly and not feasible using integrated circuit technology. Photonics-based microwave signal generation relies on heterodyning two laser signals $E_1(t)$ and $E_2(t)$ at optical angular frequencies ω_1 and ω_2, respectively [15], as shown in Figure 11.8. The two optical signals beat in a photodiode (PD). The generated RF signal is at the difference frequency of the optical sources:

$$I_{\mathrm{RF}} = A\cos\left[(\omega_1 - \omega_2) + (\varnothing_1 - \varnothing_2)\right] \tag{11.16}$$

where A is the amplitude of the RF signal, ω_1 and ω_2 are the optical frequencies and $\phi_{1,2}$ denotes the optical phase of the signals.

A simple heterodyning operation, shown in Figure 11.8, cannot generate a stable sub-THz signal due to fluctuation of the phase of optical signals. Different techniques are employed in order to decrease the jitter of generated mm-wave signal.

11.4.1.1.1 Optical Phase Lock Loop

The phase of two optical sources can be locked using an optical phase lock loop (OPLL). The idea is similar to an electrical phase locked loop. In the optical domain, two laser sources with narrow linewidth are used to create a beat note in the PD. Narrow linewidth ensures that the phase fluctuations are only at low frequencies. The generated RF/mm-wave signal at the PD is mixed with a reference RF/mm-wave signal. A low-pass loop filter

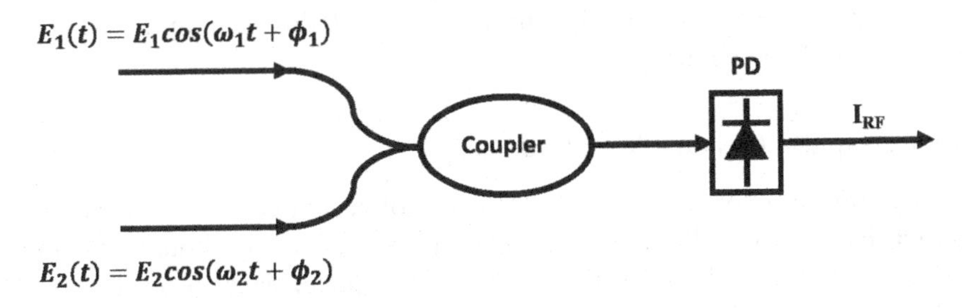

Figure 11.8 Heterodyning of two optical signals.

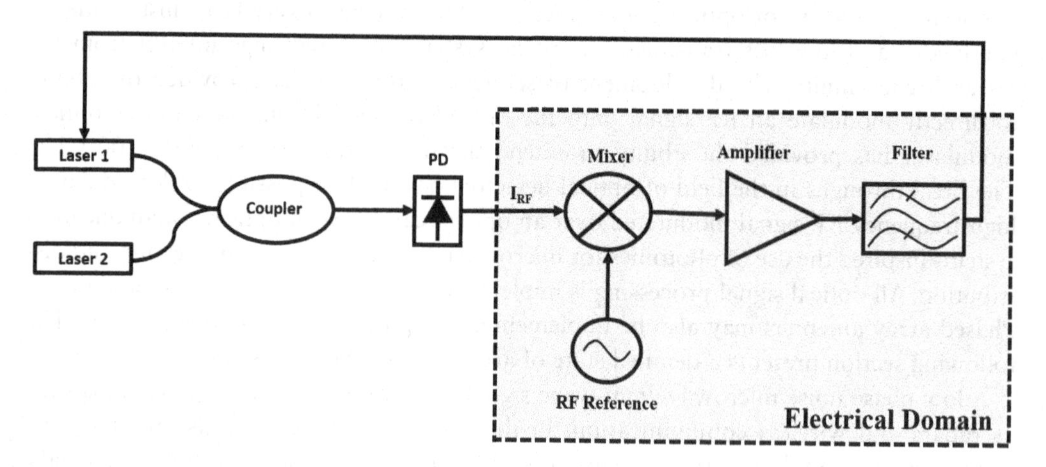

Figure 11.9 Schematic of optical phase lock loop (OPLL) (adapted from [9]).

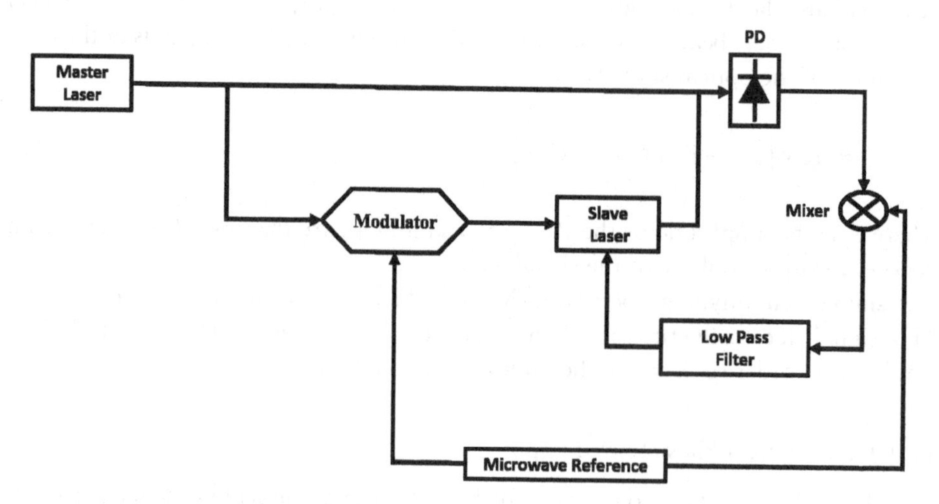

Figure 11.10 Schematic of optical injection phase locking (adapted from [9]).

is used to select the difference signal generated due to phase difference between reference and mm-wave signal. An amplifier is often used to amplify this error signal. The error signal is fed back to one of the laser sources as shown in Figure 11.9. The cycle repeats until the error signal becomes zero. Various implementations of OPLL have been reported during the past few years [16–20]. The drawback of this technique is that it requires narrow linewidth laser sources.

11.4.1.1.2 Optical Injection Phase Locking

In order to further improve the phase coherence of generated RF/mm-wave signal, it was proposed that the optical injection locking and phase locking should be combined in a single scheme [18]. As shown in Figure 11.10, the optical signal from the master laser is equally divided into two signal paths. On one of the optical paths, a modulator is used

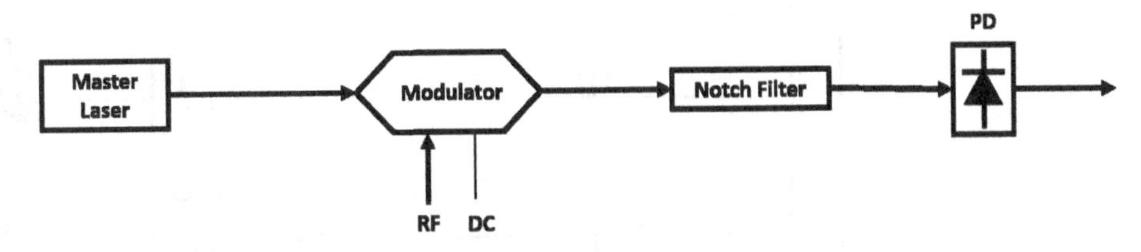

Figure 11.11 Schematic of RF signal generation using external modulation (adapted from [9]).

prior to a slave laser. The modulator uses a microwave reference signal to modulate the optical carrier. The slave laser is locked to the one of the sidebands of the modulated optical carrier. The output of the modulator is injected into the slave laser. The optical signals from both optical paths generate a beating signal in the PD. The generated RF signal is mixed with the reference microwave signal. A low pass filter is used to filter the output of the mixer. The filtered electrical signal is used to achieve phase matching at the slave laser. Such a scheme can provide a stable microwave carrier signal with low phase noise.

11.4.1.1.3 Microwave Signal Generation Based on External Modulation

Instead of using multiple laser sources, a single optical source can be used to generate microwave signals. Using a Mach-Zehnder modulator (MZM), an optical carrier is modulated with an applied RF/mm-wave signal [21] (Figure 11.11). The DC bias of the MZM can be used to suppress odd or even modes by driving the modulator at maximum/minimum transmission point. The optical carrier is filtered out using a notch filter. The filter is often implemented using a fibre Bragg grating (FBG). The optical sidebands generate a phase coherent RF carrier. This technique does not require a tuneable filter. The MZM is sensitive to bias drift and thus can affect the phase noise of the generated RF signal. Instead an optical phase modulator (PM) can be used to generate optical sidebands. Use of phase modulator limits the tunability of this technique as the PM cannot suppress even nor odd modes [22].

11.4.1.2 Photonics-Based Phase Shifters for Millimetre-Wave Applications

Optical waveguides/fibres can provide a phase delay based on the applied optical frequency. This technique is used to provide phase delay to antenna elements and is commonly known as true time delay (TTD). True-time delay phase shifter based on photonic technologies has been extensively researched in the past few years and a large number of papers have been published [23–49]. The techniques reported in the literature can be classified into two categories: true-time delay phase shifter based on free-space optics and true-time delay phase shifter based on fibre or guided-wave optics. In [38], a true-time delay phase shifter based on free space optics was proposed and experimentally demonstrated. Since the system was based on bulky optics, it is massive and bulky. Most of the reported systems were based on fibre optics. The realization of tuneable true-time delays based on a fibre-optic prism consisting of an array of dispersive delay lines was demonstrated in 1993 [25]. To reduce the size of the fibre-optic prism, the dispersive delay

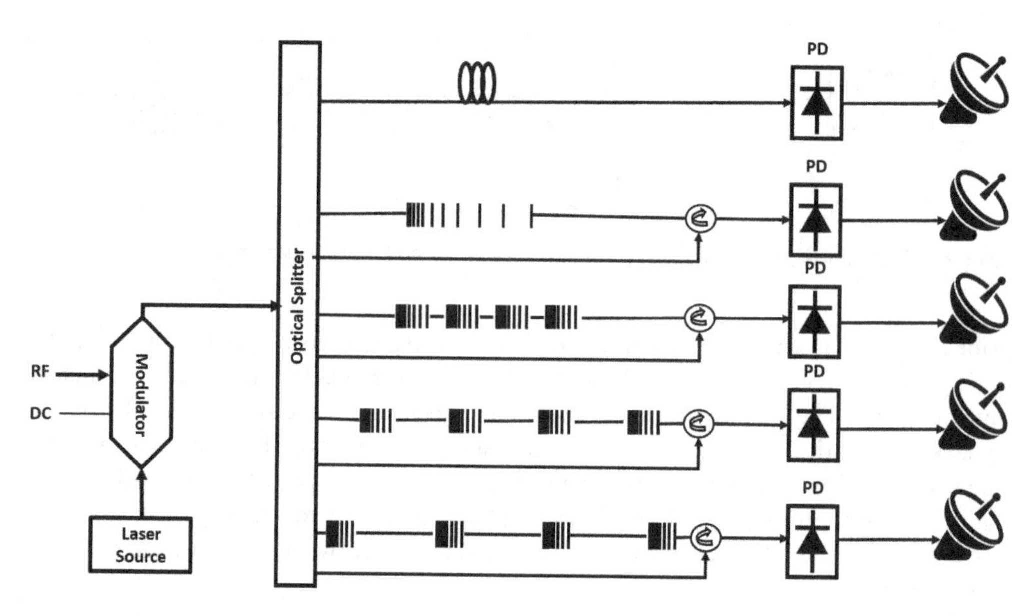

Figure 11.12 Schematic TTD based beam forming system (adapted from [9], [54]).

lines could be replaced by fibre Bragg gratings (FBGs) based delay lines [50]. A FBG prism consisting of five channels of FBG delay lines is shown in Figure 11.12 [42]. Here the beam pointing direction can be steered by simply tuning the wavelength of the laser source. Since the grating spacing in the second delay line is very small, to simplify the fabrication, the discrete FBGs can be replaced by a single chirped Bragg grating. In fact, if all the discrete grating delay lines were replaced by chirped grating delay lines, a true time delay beamforming system with continuous beam steering would be feasible [51]. The architecture shown in Figure 11.12 has the advantage of using a single tuneable laser source, which is easy to implement with fast beam steering capability by tuning the wavelength of the tuneable laser source. However, the prism consists of many discrete FBGs, which may make the system bulky, complicated and unstable. A solution is to use a single chirped Bragg grating [33]. Different time delays are achieved by reflecting the wavelengths from a tuneable multiwavelength laser source at different locations of the chirped Bragg grating. To achieve tuneable time delays, the wavelength spacing should be tuneable. Therefore, a multiwavelength laser source with tuneable wavelength spacing is required [52]. To use a light source with fixed multi-wavelengths, a chirped Bragg grating is proposed [42]. A technique to tune the chirp rate of a chirped FBG without central wavelength shift was demonstrated in Ref. [53].

In the above section, various techniques for generation and phase shifting of mm-wave signals using photonic components were presented. These days optical components are implemented in integrated circuit technology. Although MWP-based techniques can provide efficient solution towards generation and signal processing of sub-THz signals, but due to unavailability of suitable interconnection, the extraction and transportation of mm-wave signal is not possible. In Sections 11.5 and 11.6 two different techniques for realizing interconnects for mm-wave frequencies are presented. The first technique relies on deposition a superficial layer of BCB on InP such that the guiding mechanisms can be realized while avoiding the fabrication challenges of InP. The second technique implements waveguiding structure by making use of an inverted bed of nail architecture. Both techniques are used to realize passive components in order to provide a proof of concept. Using a combination of such techniques, it will be possible to realize a MWP based mm-wave system.

11.5 A Millimetre-Wave Antenna Design Based on BCB Deposition

Sub-THz technology differs from conventional RF design techniques due to the fact that interconnections such as coaxial cables, SMA connectors and waveguide components are either not possible or too costly to implement. Furthermore, the conventional PCB technology cannot be used to implement substrate integrated waveguide (SIW) structures due to standard lithography limit of 100 μm. As discussed earlier, CMOS technologies are not suitable for mm-wave applications. For realizing active electronic as well as photonic components, III–V compounds such as InP, GaAs and InGaAsP are utilized. InP-based technologies are usually preferred due to the availability of laser sources. These fabrication platforms provide the flexibility of developing active and passive optical components on a single wafer. Yet for InP-based technologies, there exists a severe bottleneck in terms of interconnects. InP inherently is a high dielectric constant material with $\varepsilon_r = 12$. A high dielectric constant introduces dispersion for mm-wave signals with large bandwidth. Moreover, the basic EM structures such as transmission line and strip line cannot be fabricated without using expensive wafer thinning processes due to the large thickness of InP wafer. Another fabrication challenge for InP devices comes from the fact that InP has an isotropic etching profile, which in turn means that depth and width of a via cannot controlled efficiently. While it is possible to overcome this issue using laser etching, the cost of fabrication increases. The lack of a suitable via-hole implementation makes it impossible to realize a rectangular waveguide structure like SIW. The fabrication challenges of InP hinders a suitable implementation of on-wafer antenna array.

Recently, new approaches have been proposed, which consist of stacked layers of different substrates. In this approach, the active circuit element (photodiode, amplifier) is fabricated using the high dielectric constant semiconductor (InP, GaAs), which is separated from the antenna using a sacrificial substrate. The antenna layer typically consists of benzocyclobutene (BCB) or polyimide with typical dielectric constants of 2.6 and 3.5, respectively. Using this technique, a radiation (in the plane normal to the radiating element) can be achieved using a simplified fabrication process. An antenna on a 7 μm-thick BCB substrate operating at 510 GHz was presented in [55, 56], integrated with a resonate-tunnelling-diodes (RTD). Additionally, simulation results of a 50 μm BCB thickness bow-tie antenna operating at 300 GHz was presented in Ref. [57]. It was concluded that the antenna performance can be improved by increasing the power coupling efficiency between the sources and the antennas. The coupling efficiency can be improved by matching the input impedance of antenna to the output impedance of active element (which is different from typical 50Ω).

Recently, microstrip antennas have been applied in the THz frequency range combined with state-of-the-art THz sources such as RTDs and (uni-travelling-carrier) UTC-PDs. In these emerging technologies, microstrip antennas offer the interesting features of being compact and at the same time allowing the antenna to be designed in low dielectric constant ($\varepsilon_r \leq 3$) substrates, which is isolated from high dielectric constant ($\varepsilon_r \geq 10$) substrates by means of a ground plane. This approach allows the cancellation of substrate wave modes and supports a radiation (in the normal direction to the antenna plane). Many configurations have been proposed to enhance microstrip antenna bandwidth, such as stacked parasitic layers, employment of coplanar parasitic elements and slot configurations, achieving fractional bandwidths (FW) of about 30% [58–63], when normal values reside between 5% and 7%. Although a single resonant element could be used with an additional matching network to extend the bandwidth, the multi-resonance techniques are predominantly preferred. The most direct method for efficiency improvement consists of increasing the substrate thickness, leading to 60–80% efficiency.

11.5.1 Design of a Unit Cell

The antenna element presented in this section corresponds to an extension of the work presented in Ref. [64], originally introduced to eliminate the mismatch caused by the parasitic coaxial probes instead of resorting to an external matching network. The application of the embedded substrate capacitive feed presented in Ref. [64] to cancel source inductance is extended to mm-wave range and at the same time to tune the real part of the antenna input impedance in order to achieve complex conjugate matching. This antenna finds applications in terahertz technologies where the sources usually have complex output impedance like uni-travelling carrier photodiodes (UTC-PD) and resonant tunnelling diode (RTD). We are especially interested in the frequency range centred around 300 GHz, but the analysis could be extended beyond this frequency. Stacked microstrip antennas have been used extensively in the literature to improve antenna characteristics such as bandwidth or to achieve multiband operation by combining radiating modes at different wavelengths. However, here the capacitive feed patch (Figure 11.13) is considerably small compared with the radiating patch (top patch) and therefore it does not support any radiating modes being exclusively used as a capacitor to cancel the source impedance.

A single patch antenna (unit-cell) is created using a capacitive feed as shown in Figure 11.13(a). The UTC-PD (uni-travelling carrier photodiode) operating at 300 GHz generates the signal using a co-planar waveguide (CPW) structure. Using a GSG (ground-signal-ground) pad formation, the signal is coupled to the feed patch using a via-hole through the dielectric layer of BCB (benzocyclobutene) Layer 1 (Figure 11.14(a)). Due to a practical limitation on the BCB etching process, the depth of the via-hole cannot exceed 4 μm, in order to limit the height to width ratio of the via-hole to approximately 1. The feed patch is capacitively coupled to the top radiating patch through a second BCB layer. The thickness of BCB Layer 2 strongly influences the radiation efficiency of antenna. For a radiation efficiency of 50% or higher, the thickness of BCB layer 2 should be greater than 9μm. The layer stack-up of antenna structure is presented in Figure 14(b).

The antenna structure in Figure 11.14 (a) is simulated using a 3D electromagnetic solver. The input impedance of the antenna is set equal to the measured output impedance of UTC-PD, that is, $Z_{in} = 15 - j8\ \Omega$. Using the parametric analysis, the height of via-hole was varied between 3 and 4 μm, and it was observed that although the unit-cell can be designed using a via-hole height of 3 μm, it will require a very large ground plane, with a

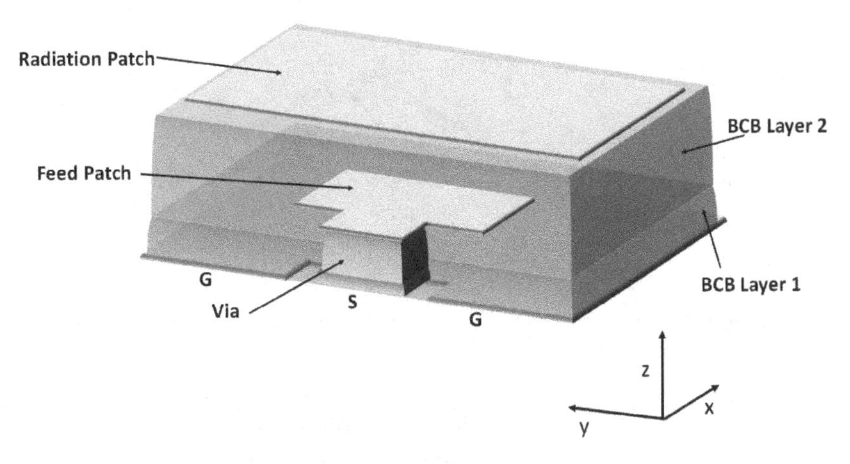

Figure 11.13 Proposed capacitively coupled antenna element over InP substrate.

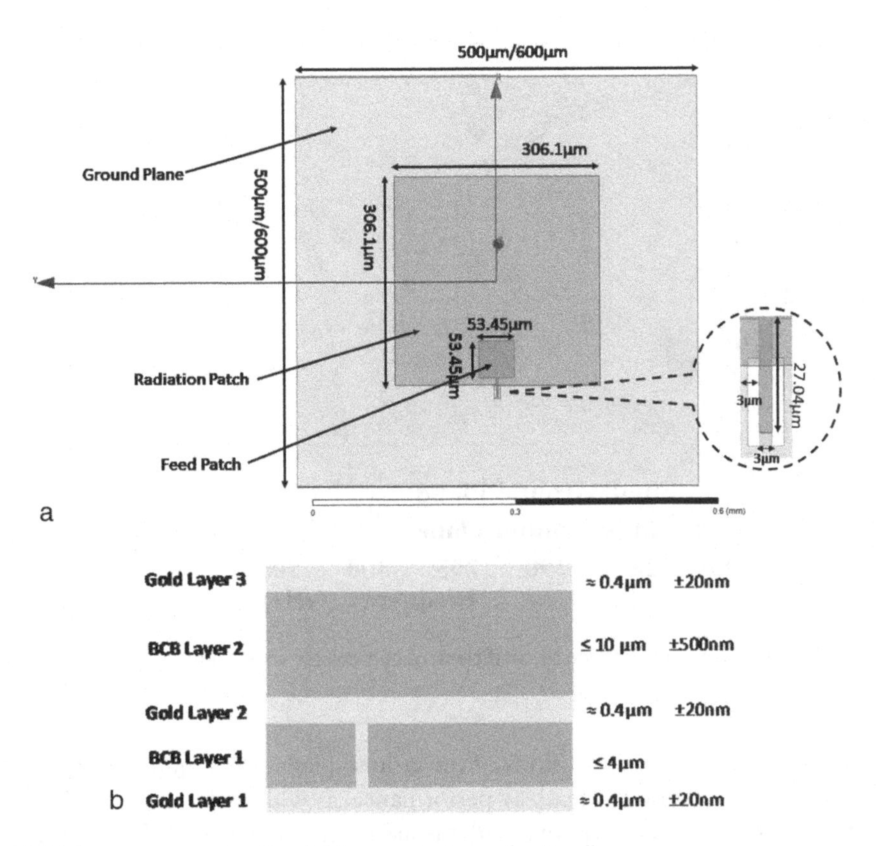

Figure 11.14 (a) Structure of unit-cell and (b) Layer stack-up of unit-cell.

side dimension of approximately 800 µm. For an efficient array implementation, the size of the ground plane must be close to half of the free space wavelength at 300 GHz, that is, 500 µm. Therefore, a 4 µm height via-hole is chosen due to its suitability for the realization of antenna array. It is worth noting that the performance of the unit-cell can be enhanced

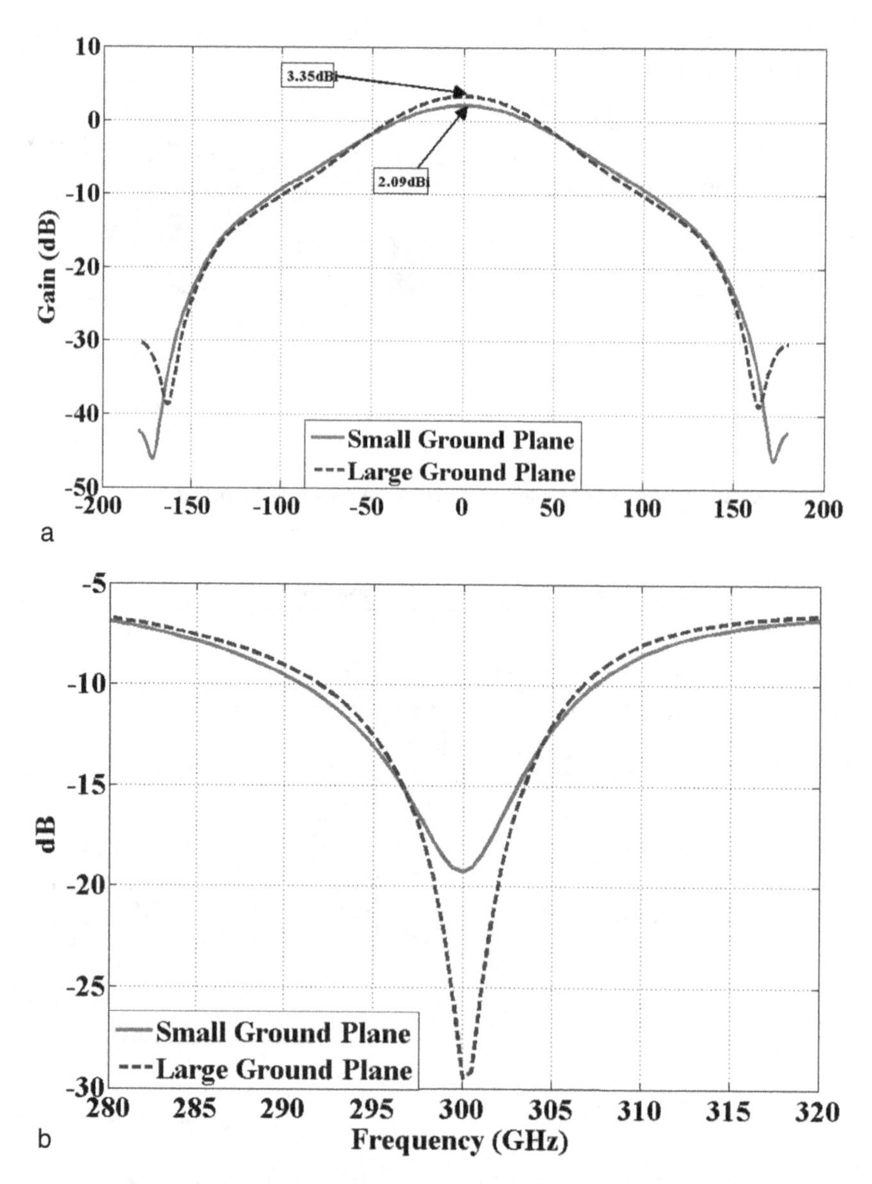

Figure 11.15 (a) Gain and (b) Reflection coefficient, of unit-cell for small and large ground plane sizes.

by slightly increasing the dimensions of the ground plane to 600 μm × 600 μm, although with a slight degradation in the array performance, as will be shown next.

For the current design, we analysed the unit-cell using two different ground plane sizes: (1) small ground plane of size 500 μm × 500 μm and (2) large ground plane of size 600 μm × 600 μm. The simulated results of two antenna structures are presented in Figure 11.15. It can be observed that the large ground plane provides better matching and higher gain as compared to the small ground plane. The radiation efficiency for both structures is slightly greater than 50%.

11.5.2 Design of the One-Dimensional Antenna Array

A one dimensional 1 × 4 array is formed using the aforementioned unit-cell. The radiating elements are placed at half the free space wavelength apart from each other, i.e., $d = \frac{\lambda_0}{2} = 500$ µm. The layout of the array structure is presented in Figure 11.16. The antenna array is simulated for two different sizes of the ground plane, that is, large ground plane (600 µm × 600 µm) and small ground plane (500 µm × 500 µm) as described in the preceding section. The main beam of the array is steered from –30° to +30° in the $+z$ -direction. The two dimensional and three dimensional gain patterns of the simulated antenna array are presented in Figures 11.17 and 11.18, respectively.

From the presented results of Figures 11.17 and 11.18, it can be observed that the one-dimensional 1 × 4 array with a large ground plane provides a gain of 9.52 dBi but the side lobe level (SLL), analysed at the extreme steering angles (worst case), that is, –30° and +30°, reaches approximately –10 dB (at 30° offset), –12 dB (at 60° offset) and –11.5 dB at 100° offset from the main beam. On the other hand, using a small ground plane reduces the gain by ~1 dB (to approximately 8.13 dBi), yet the SLL level for extreme steering angles, that is, –30° and +30°, is approximately –10.5 dB (at 40° offset), –14 dB (at 70° offset) and reaching –20 dB at 110° offset from the main beam. Therefore, the antenna with small ground plane provides a better SLL performance throughout all the radiation pattern angles.

11.5.3 Design of Two-Dimensional Array

A two-dimensional array can be constructed using a similar layout of that in Figure 11.16. In the most simplified manner, a two-dimensional arrays consist of multiple images of a one-dimensional array, yet due to a MWP implementation the antenna structures are not symmetric along both planar dimensions. The added biasing structure needed for a heterodyne generation of sub-THz signal creates asymmetry along one of the planar dimensions. Nevertheless, a two-dimensional array implementation is possible provided that unequal side lobe levels across any planar dimension is acceptable for the desired application.

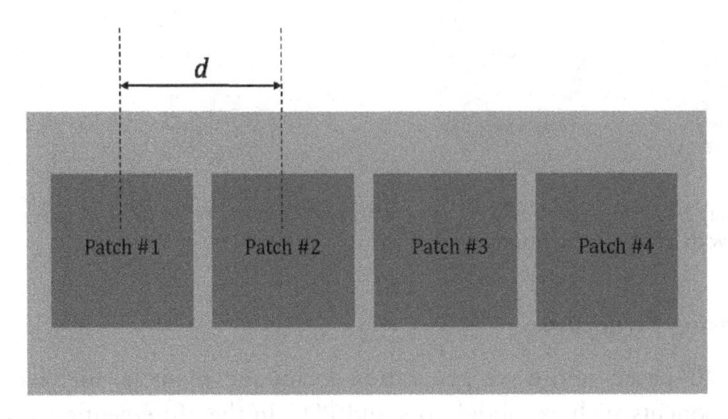

Figure 11.16 1×4 antenna array structure.

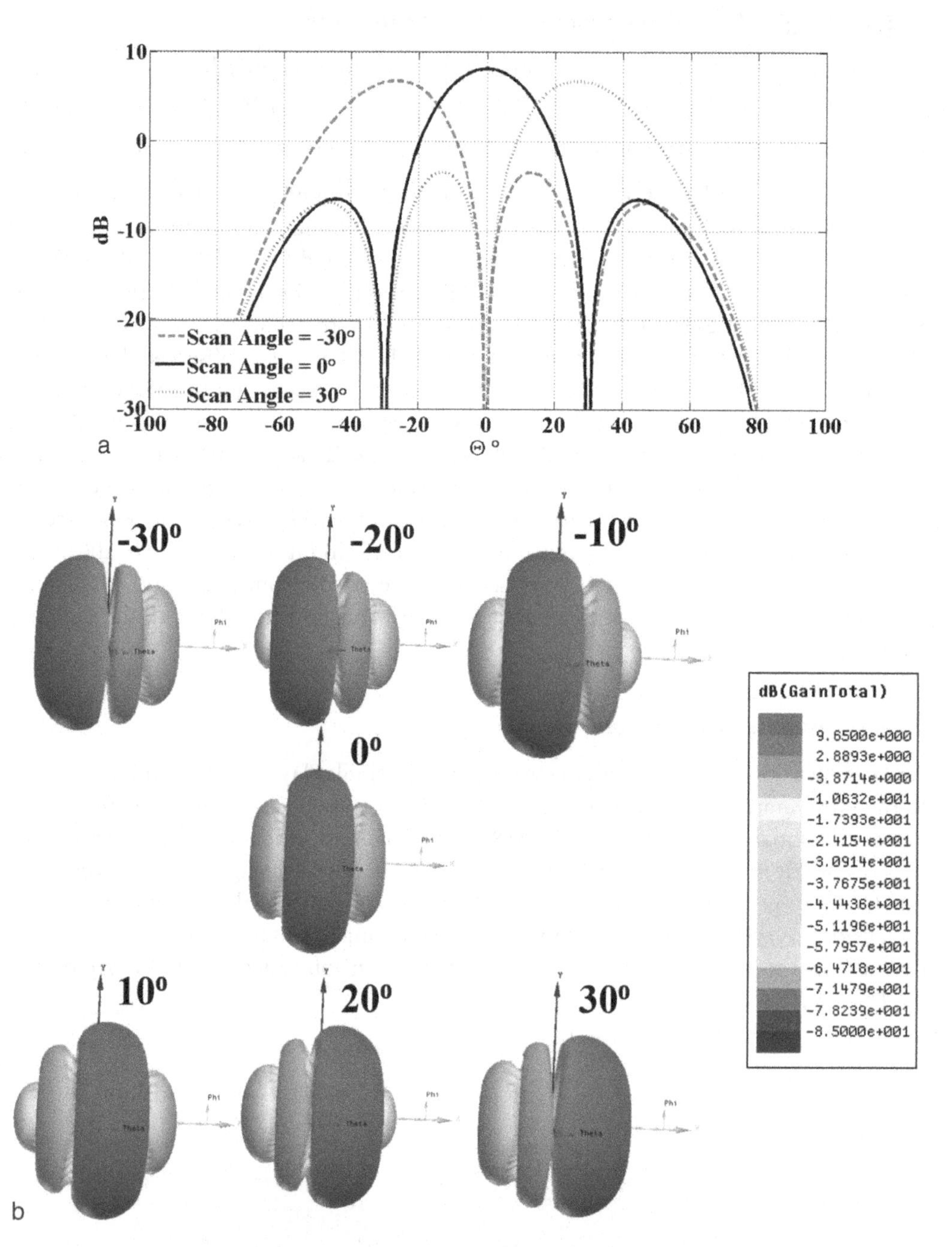

Figure 11.17 (a) 2-D gain pattern of 1×4 array with small ground plane (b) 3-D gain pattern of 1×4 array with small ground plane.

11.5.4 Biasing Structure

Using a MWP-based mm-wave generation technique requires effective DC biasing of optical components such as modulators and PDs. In the aforementioned design example, the UTC-PD needs to be DC biased while effectively not disturbing the conjugate matching

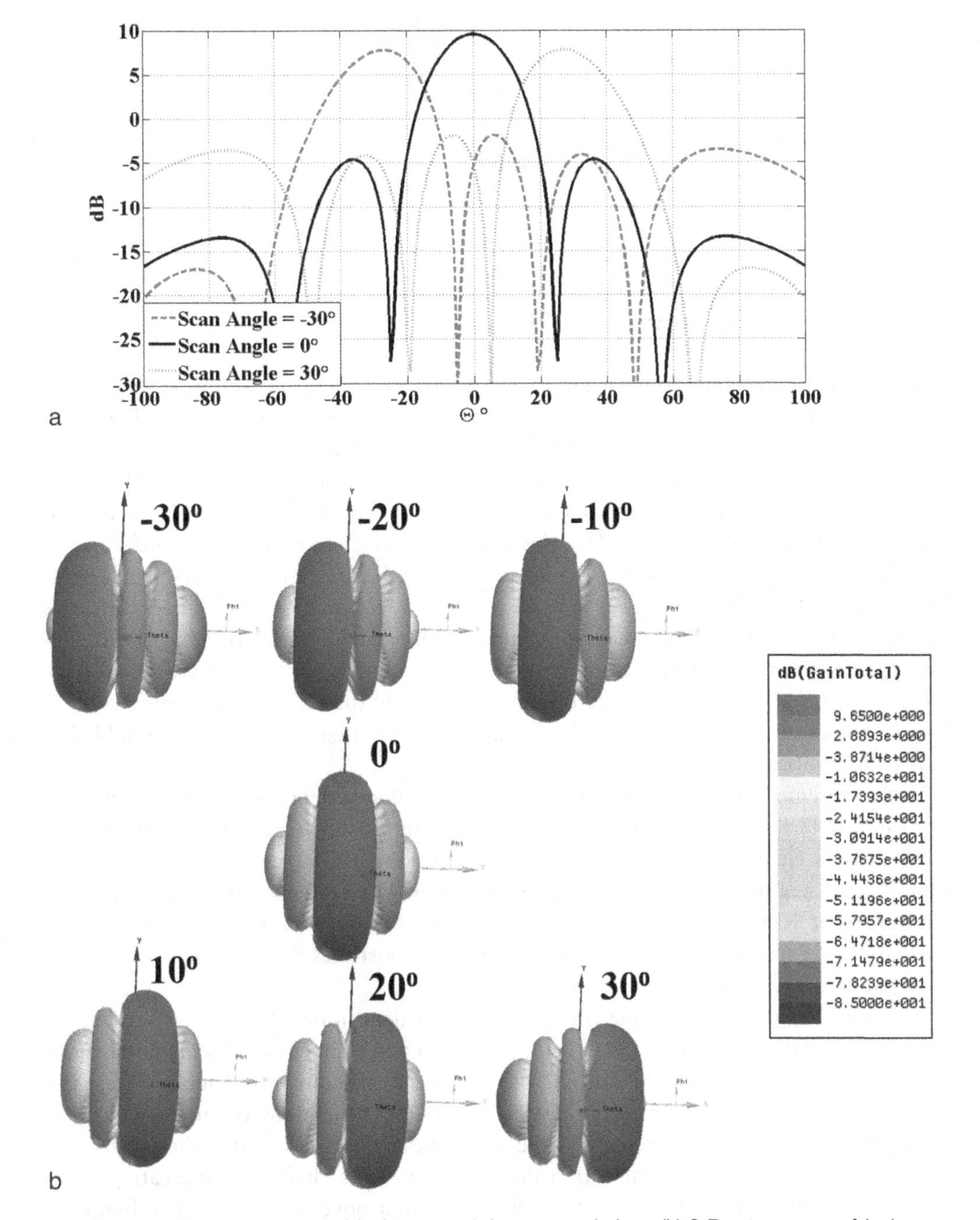

Figure 11.18 (a) 2-D gain pattern of 1×4 array with large ground plane; (b) 3-D gain pattern of 1×4 array with large ground plane.

of antenna. From conventional RF design, we know that such functionality can be achieved by using a bias-tee. The schematic representation of an RF bias-tee is shown in Figure 11.19.

For sub-THz implementation, a wirebond can be used to connect the PD to a DC power supply. In the simplest small signal model, the wirebond presents a series inductance

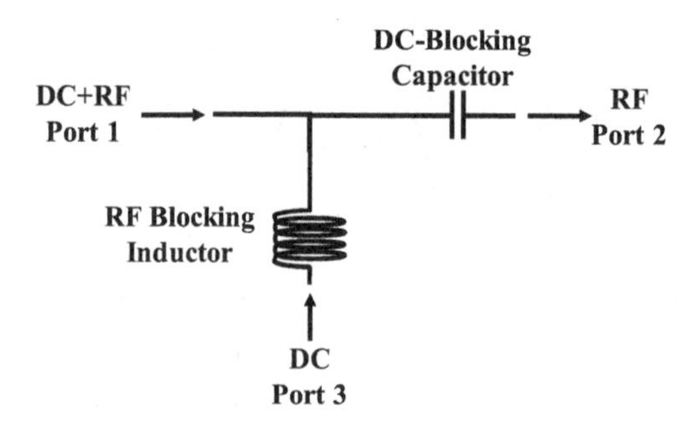

Figure 11.19 Schematic representation of an RF bias-tee.

whose value is dependent on the length. For a gold wire of diameter 25 μm, the minimum achievable length is 0.5 mm, which presents approximately an inductance of 0.5 nH. This wirebond inductance is sufficient to provide the necessary RF blocking (@300 GHz). The implementation of a series capacitor is a design challenge and can be implemented using an interdigitated planar capacitor design. For this particular antenna structure, as the antenna is an open circuit at DC, the series capacitor is not needed. In order to isolate the DC bias from the ground plane, an epitaxial layer of Silicon-Oxy-Nitride must be deposited. This bias structure is shown in Figure 11.20(a). Instead of three ports, this bias-tee has two ports only, since the antenna is connected directly to the UTC-PD. The radial stubs are used to provide a perfect ground for any leaking mm-wave signal from the UTC-PD. The choice of radial stubs is motivated by the fact that they are shorter in length thus can be compactified. Such a structure can be used to realize a reasonable DC bias with suitable isolation.

Using aforementioned, design techniques, a 1×4 antenna array is realized. Each array element is fed by a UTC-PD, which as an optical input (as shown in Figure 11.20(b)). The delay at each element is introduced in the optical domain. The heterodyning of optical signals generates the phase-delayed mm-wave signal. The presented array can support a bandwidth of 10 GHz with a maximum gain of 8 dB (after considering all losses). The realization of capacitively coupled antenna array demonstrates the possibility of realizing passive microwave components on InP.

Although the design technique using a sacrificial deposition of BCB layer is feasible using current fabrication techniques, yet it requires a three-dimensional transition instead of a planar transition. The via size is large and for a structure with multiple vias, this technique may not feasible. Instead, there is a need for waveguiding technique, which can be implemented on InP wafer without requiring the deposition of other dielectrics. The deposition of dielectric material with a different *k* presents challenges in creating adhesion among different layers. In the following section, a novel waveguiding technique is presented for InP substrates.

11.6 Via-Less Planar Interconnect for Integrated Circuits on InP

The choice of a semiconductor material for mm-wave devices is limited to GaAs and InP. Due to diverse properties of InP in electronics as well as in the optical domain, it is commonly used for the fabrication of active devices such as amplifiers, photodiodes and lasers,

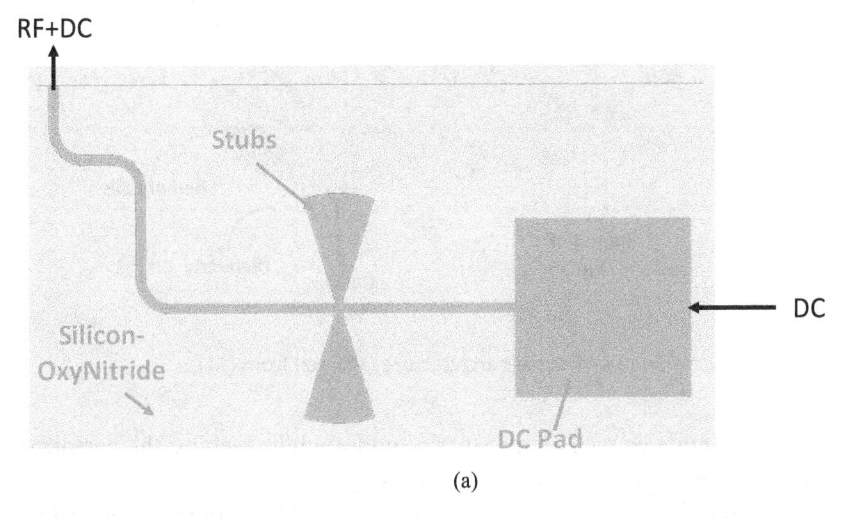

(a)

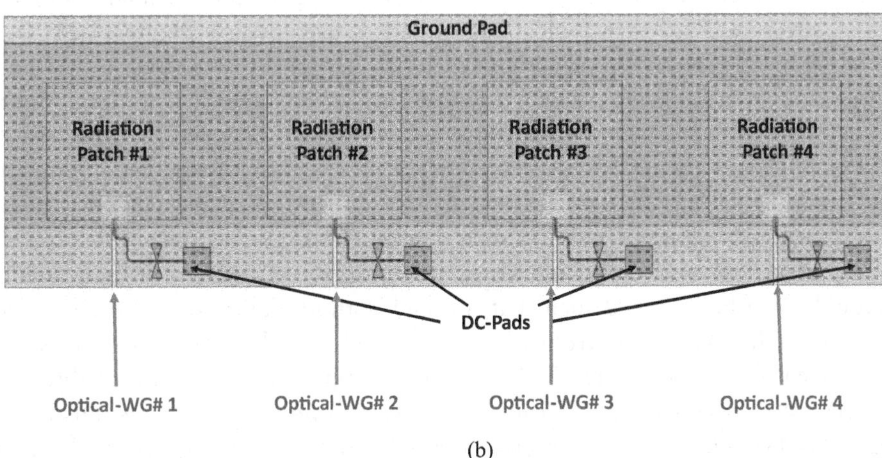

(b)

Figure 11.20 (a) Realized biasing structure for UTC-PD; (b) Complete layout of 1×4 antenna array.

among others. On the other hand, the fabrication of guiding structures such as microstrips, co-planar waveguides (CPWs) and rectangular waveguides is quite challenging on InP due to its isotropic etching profile [65–68]. The isotropic response of InP to most etchants results in vias having a diameter equal to the depth, which is not desirable for guiding structures. Moreover, the typical thickness of InP substrates is in the order of 125 μm, which is not suitable for microstrip structures for G-band frequencies (220–320 GHz). A CPW structure without a ground plane can be realized on InP, albeit the air–dielectric boundary not only introduces losses but also dispersion (due to the higher dielectric constant of InP) limiting high-data rate communications. Grounded-CPW (GCPW) is often implemented using either inefficient thinning and etching of through-via holes (wet or plasma etching [69, 70]) or by using a lossy hybrid platform. Also rectangular waveguides are designed using substrate integrated waveguide structure, but since these technologies are not tailored to integrated circuits, provision of through via-holes remains a bottleneck in InP substrates. Thus, there is a need for a via-less guiding structure which can support

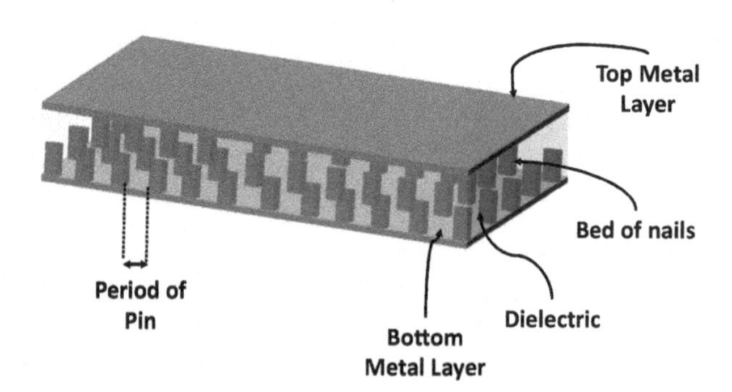

Figure 11.21 Representation of bed of nails architecture (adapted from [71]).

rectangular waveguide as well as microstrip modes while keeping the performance comparable to or better than through-via structures.

Via-less waveguides were first introduced by Prof. Per Simon Kildal using a bed of nails architecture [72–75]. A typical bed of nails is presented in Figure 11.21. Two non-contacting metal layers with or without dielectric are used to construct a two-conductor system. On one metal layer, periodic rows of pins are introduced. The size and period of pins determine the frequency of operation. The distance between the pins and the top metal plate should always be less than $\lambda/4$. The wavelength λ is the maximum operating wavelength. Kildal has shown that by carefully dimensioning the pins, it is possible to create a perfect boundary condition such that no electromagnetic wave can travel in the gap (between pins and top metal). Using this property of the bed of nails architecture, waveguides can be designed such that top and bottom plates are similar to rectangular waveguides while side walls are presented by a bed of nails. A single bed of nails row is not sufficient for blocking the EM fields, thus multiple rows (2–3) are used to construct a perfect metal boundary [76]. Various microwave components such as filters, antennas, couplers and cavities are designed using this technology. So far, the designed components are operating below 100 GHz. In the following section via-less waveguides are presented suitable for III–V technologies especially InP for frequency ranges approaching 320 GHz.

11.6.1 Via-Less Planar Technology for InP

Inspired in the bed of nail architecture, via-less waveguides on InP can be created by simply inverting the structure of Figure 11.21 [77]. As the fabrication process starts using a bottom-up approach, the bed of nails can be created using a common etchant, provided that the aspect ratio of pins (area/depth) is on the order of 2–3. This ensures that the size of pins can be controlled while maintaining the isotropic etching profile. After etching, the metal can be deposited using any deposition process such as vapour deposition or plasma-assisted deposition. The thinning of substrate is usually the last step and back metal can be deposited at the end of the process. This can overcome the conventional challenges of InP fabrication and can provide a planar guiding structure. Via-less waveguides are of two kinds: (i) gap waveguide and (ii) ridge waveguide. Gap waveguide is similar to conventional rectangular waveguide. Slots/gaps of size similar to rectangular waveguides are provided among rows of pins to support waveguide modes. The cutoff frequency of such waveguides is similar to conventional metallic waveguides. Ridge waveguides are similar to microstrip. A metal ridge is used to guide the wave along the structure.

11.6.2 Microstrip to Rectangular Waveguide Transition

In order to prove the feasibility of the proposed via-less technology, a transition from microstrip to reduced height rectangular waveguide is designed. From microwave theory, it is known that a typical thickness (125 μm) of InP substrate cannot support a microstrip structure as the thickness is much larger than 10% of the wavelength (λ = 0.25–0.3 mm in G-band). For this reason, the thickness h (Figure 11.22(a)) of InP substrate should be reduced to 50 μm using thinning as the last fabrication step. Using the rectangular waveguide alike structure of Figure 11.22(a), the bed of nails is optimized such that it will support only single mode (TE_{10}) propagation. The bed of nails is implemented using square shaped pins with g = 7 μm (Figure 11.22(a)). The depth d_p of each pin is optimized such that the aspect ratio ($AR = d_p/g$) is less than or equal than 3. It is worth noting that a smaller aspect ratio simplifies the fabrication process by minimizing the etching and metal deposition time. For rectangular waveguide-like propagation, three rows of pins are used to construct the side walls of a rectangular waveguide. The wider dimension k of the rectangular waveguide is equal to a conventional dielectric filled WR-3 waveguide. The period (p_1 = 40 μm) is constant along the transversal and longitudinal directions.

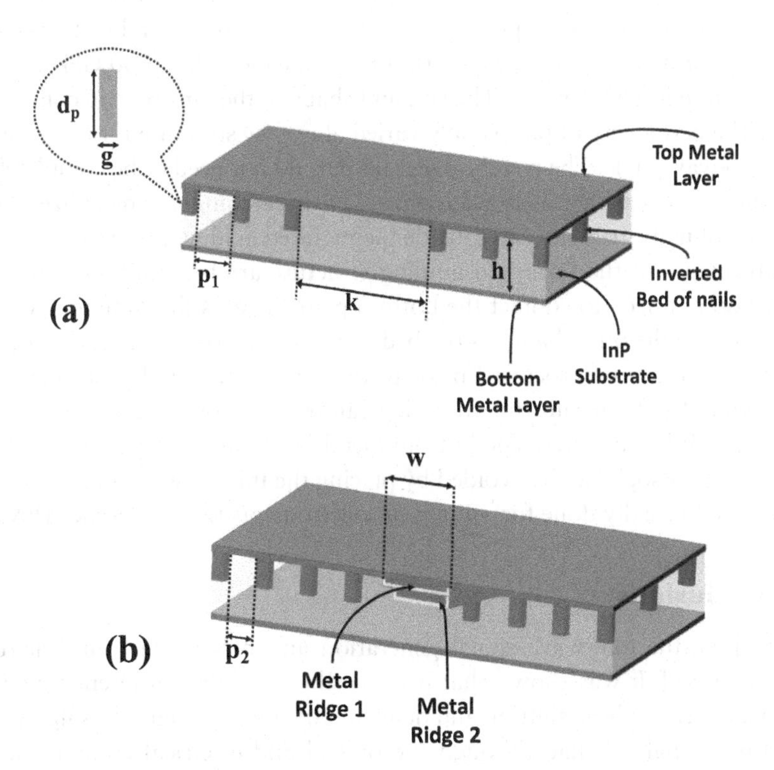

Figure 11.22 (a) Modified bed of nails architecture for gap-waveguide with reduced height; (b) Ridge-based transition in bed of nails.

Table 11.1 Dimensions of designed microstrip transition

Dimension	*Value (μm)*
Substrate thickness (*h*)	50
Microstrip width (*b*)	50
Size of pin	7×7×21
Width of waveguide (*k*)	210
Period of pins (p_1,p_2)	15,40
Width of ridge (*w*)	210

After optimizing the bed of nails structure, a transition from microstrip to the gap-waveguides is constructed. From a microstrip architecture (with b = 45 μm), a transition is created using two metal ridges with depth d_{p1} = 15 μm and d_{p2} = 35 μm, respectively, as shown in Figure 11.22(b). The width w of each ridge is restricted to 15 μm. The pins are of the same dimension as used for the aforementioned gap-waveguide structure. By optimizing the periodicity of pins (p_2 = 15 μm), a transition from microstrip to rectangular waveguide is designed. The length of the transition section is 100 μm where each metal ridge is 50 μm long. The dimensions of rectangular wave-guide transition can be found in Table 11.1.

After optimizing the ridge and gap-waveguide structures, the complete structure of Figure 11.23(a), where the input is microstrip and the output is a gap-waveguide, is simulated. The obtained S-parameters are presented in Figure 11.23(b). It can be observed that such a transition can provide a return loss of –18 dB over a bandwidth of 30 GHz (240–270 GHz) with an insertion loss of 0.15 dB. The size and shape of the pins remain constant across the structure. The periodicity of pins is only varied along the structure for designing the transition and gap-waveguide. The metal ridges and pins do not need to be completely filled with metal, rather as long as the cavity walls (created using etching) are metallized, the proposed inverted bed of nails structure can provide the required guiding properties.

The advantages of the proposed guiding structure are two-fold. On one hand, as the metal pins do not need to contact the bottom plate, it eases the etching and metal deposition process. On the other hand, as the bed of nails is deposited using the top metal layer, the InP substrate can be processed prior to thinning. Once the BEOL (back-end of line) process is completed, the substrate thinning can be performed. This provides further ease in handling of InP substrates. The bottom metal layer can be deposited as the last metal deposition step. Also, it can be avoided by placing the InP substrate in a metallic container which is conventionally done for mitigating electromagnetic interference (EMI).

11.7 Conclusions

In this chapter, the mm-wave signal generation and processing using microwave photonics is discussed. It was shown that it is possible to realize different functions such as generation, filtering, phase shifting and detection of sub-THz signals using optical devices. It was also highlighted that although the optical and electrical components have been developed using III–V compounds, yet the lack of suitable interconnections hinders the realization of a complete mm-wave system. The lack of suitable waveguiding technique at

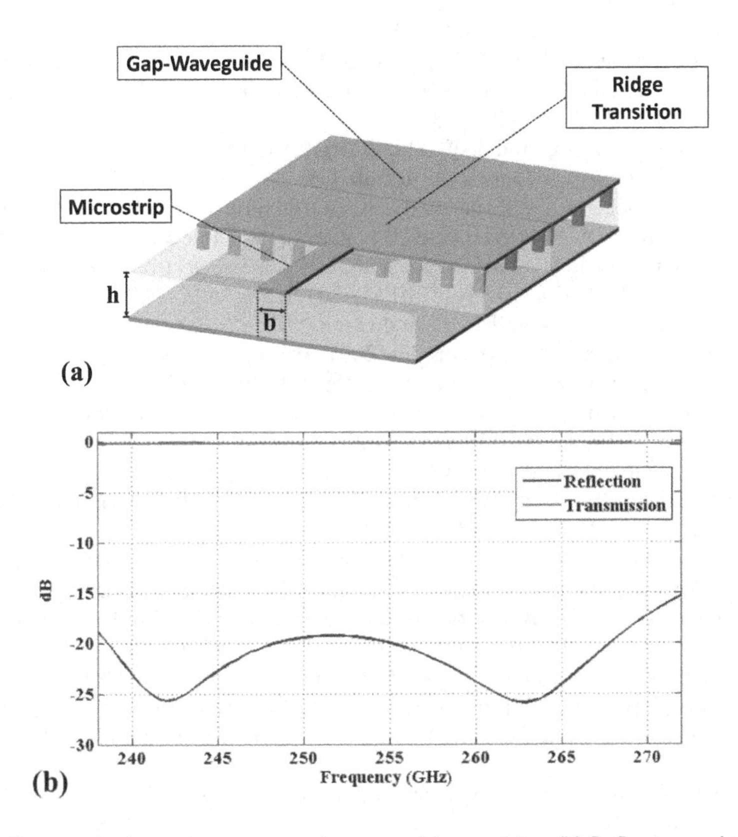

Figure 11.23 (a) Proposed microstrip to rectangular waveguide transition; (b) Reflection and transmission coefficients of transition [77].

sub-THz frequencies results in the unavailability of necessary passive components such as antennas, waveguides and connectors. The problem of interconnections in III–V, especially InP, can be solved using two different approaches. The first technique relies on depositing an additional dielectric layer (BCB in this case). Instead of guiding the THz signal on InP, the newly deposited material is used for guiding the mm-wave signal. The newly formed dielectric layer can be used to fabricate passive components such as antennas. This technique is limited due to added fabrication steps and the need for three-dimensional transitions for realizing any passive component. The second technique relies on realizing an inverted bed of nails architecture. This technique is planar and does not require through via-holes. The drawback of this technique is that the InP substrate needs an additional thinning step such that it can support conventional transmission line modes. In conclusion, the challenges faced by opto-electronic systems operating in the sub-THz region can be solved using innovative fabrication and design techniques, thereby realizing a complete mm-wave system with superior performance than their sole electronic counterparts.

Notes

[1] J .-M. Wun, R.-L. Chao, Y.-W. Wang, Y.-H. Chen, and J.-W. Shi, "Type-II GaAs0.5Sb0.5/InP Uni-Traveling Carrier Photodiodes With Sub-Terahertz Bandwidth and High-Power

Performance Under Zero-Bias Operation," *J. Lightwave Technol.*, vol. 35, no. 4, pp. 711–716, Feb. 2017, doi: 10.1109/JLT.2016.2606343.

[2] J. Xu, X. Zhang, and A. Kishk, "InGaAs/InP evanescently coupled one-sided junction waveguide photodiode design," *Opt. Quant Electron.*, vol. 52, no. 5, p. 266, May 2020, doi: 10.1007/s11082-020-02392-8.

[3] J.-M. Wun, Y.-L. Zeng, and J.-W. Shi, "GaAs0.5Sb0.5/InP UTC-PD with Graded-Bandgap Collector for Zero-Bias Operation at Sub-THz Regime," in Optical Fiber Communication Conference, Anaheim, California, 2016, p. Tu2D.4, doi: 10.1364/OFC.2016.Tu2D.4.

[4] S. Dey, S. K. Koul, A. K. Poddar, and U. L. Rohde, "Reliable and Compact 3- and 4-Bit Phase Shifters Using MEMS SP4T and SP8T Switches," *J. Microelectromech. Syst.*, vol. 27, no. 1, pp. 113–124, Feb. 2018, doi: 10.1109/JMEMS.2017.2782780.

[5] Y. Liang, C. W. Domier, and N. C. Luhmann, "MEMS Based True Time Delay Technology for Phased Antenna Array Systems," in 2007 Asia-Pacific Microwave Conference, Bangkok, Thailand, Dec. 2007, pp. 1–4, doi: 10.1109/APMC.2007.4554725.

[6] F. Ellinger *et al.*, "Integrated Adjustable Phase Shifters," *IEEE Microwave*, vol. 11, no. 6, pp. 97–108, Oct. 2010, doi: 10.1109/MMM.2010.937730.

[7] K. Van Caekenberghe and T. Vaha-Heikkila, "An Analog RF MEMS Slotline True-Time-Delay Phase Shifter," *IEEE Trans. Microwave Theory Techn.*, vol. 56, no. 9, pp. 2151–2159, Sep. 2008, doi: 10.1109/TMTT.2008.2002236.

[8] M. A. Hassanien, M. Jenning, and D. Plettemeier, "Beam Steering System Using Rotman Lens for 5G Applications at 28 GHz," in 2017 IEEE International Symposium on Antennas and Propagation & USNC/URSI National Radio Science Meeting, San Diego, CA, USA, Jul. 2017, pp. 2091–2092, doi: 10.1109/APUSNCURSINRSM.2017.8073088.

[9] J. Capmany and D. Novak, "Microwave Photonics Combines Two Worlds," *Nature Photonics*, vol. 1, no. 6, pp. 319–330, Jun. 2007, doi: 10.1038/nphoton.2007.89.

[10] C. A. Balanis, *Antenna Theory: Analysis and Design,* 4th ed. ISBN: 978-1-118-64206-1, Wiley, 2016.

[11] C. A. Balanis, "Antenna theory: a review," *Proc. IEEE*, vol. 80, no. 1, pp. 7–23, Jan. 1992, doi: 10.1109/5.119564.

[12] *IEEE Std 145–2013 (Revision of IEEE Std 145–1993) – Redline.* IEEE, 2014.

[13] D. T. K. Tong and M. C. Wu, "Multiwavelength Optically Controlled Phased-Array Antennas," *IEEE Transactions on Microwave Theory and Techniques*, vol. 46, no. 1, Art. no. 1, Jan. 1998, doi: 10.1109/22.654929.

[14] Yunqi Liu, Jianliang Yang, and Jianping Yao, "Continuous True-Time-Delay Beamforming For Phased Array Antenna Using a Tunable Chirped Fiber Grating Delay Line," *IEEE Photonics Technology Letters*, vol. 14, no. 8, Art. no. 8, Aug. 2002, doi: 10.1109/LPT.2002.1022008.

[15] U. Gliese, T. N. Nielsen, S. Norskov, and K. E. Stubkjaer, "Multifunctional Fiber-Optic Microwave Links Based On Remote Heterodyne Detection," *IEEE Transactions on Microwave Theory and Techniques*, vol. 46, no. 5, pp. 458–468, May 1998, doi: 10.1109/22.668642.

[16] J. Harrison and A. Mooradian, "Linewidth and Offset Frequency Locking of External Cavity GaAlAs Lasers," *IEEE Journal of Quantum Electronics*, vol. 25, no. 6, pp. 1152–1155, Jun. 1989, doi: 10.1109/3.29240.

[17] R. T. Ramos and A. J. Seeds, "Fast Heterodyne Optical Phase-Lock Loop Using Double Quantum Well Laser Diodes," *Electronics Letters*, vol. 28, no. 1, pp. 82–83, Jan. 1992, doi: 10.1049/el:19920050.

[18] A. C. Bordonalli, C. Walton, and A. J. Seeds, "High-Performance Phase Locking of Wide Linewidth Semiconductor Lasers by Combined Use of Optical Injection Locking and Optical Phase-Lock Loop," *Journal of Lightwave Technology*, vol. 17, no. 2, pp. 328–342, Feb. 1999, doi: 10.1109/50.744252.

[19] Z. F. Fan and M. Dagenais, "Optical Generation of a mHz-Linewidth Microwave Signal Using Semiconductor Lasers and a Discriminator-Aided Phase-Locked Loop," *IEEE Transactions on Microwave Theory and Techniques*, vol. 45, no. 8, p. 5, 1997.

[20] H. R. Rideout, J. S. Seregelyi, S. Paquet, and J. Yao, "Discriminator-Aided Optical Phase-Lock Loop Incorporating a Frequency Down-Conversion Module," *IEEE Photonics Technology Letters*, vol. 18, no. 22, pp. 2344–2346, Nov. 2006, doi: 10.1109/LPT.2006.885212.

[21] J. O'Reilly and P. Lane, "Remote Delivery of Video Services Using mm-Waves and Optics," *Journal of Lightwave Technology*, vol. 12, no. 2, pp. 369–375, Feb. 1994, doi: 10.1109/50.350584.

[22] Guohua Qi, Jianping Yao, J. Seregelyi, S. Paquet, and C. Belisle, "Optical Generation and Distribution of Continuously Tunable Millimeter-Wave Signals Using an Optical Phase Modulator," *Journal of Lightwave Technology*, vol. 23, no. 9, pp. 2687–2695, Sep. 2005, doi: 10.1109/JLT.2005.854067.

[23] W. Ng, A. A. Walston, G. L. Tangonan, J. J. Lee, I. L. Newberg, and N. Bernstein, "The First Demonstration of an Optically Steered Microwave Phased Array Antenna Using True-Time-Delay," *Journal of Lightwave Technology*, vol. 9, no. 9, pp. 1124–1131, Sep. 1991, doi: 10.1109/50.85809.

[24] P. M. Freitag and S. R. Forrest, "A Coherent Optically Controlled Phased Array Antenna System," *IEEE Microwave and Guided Wave Letters*, vol. 3, no. 9, pp. 293–295, Sep. 1993, doi: 10.1109/75.244857.

[25] R. D. Esman et al., "Fiber-Optic Prism True Time-Delay Antenna Feed," *IEEE Photonics Technology Letters*, vol. 5, no. 11, pp. 1347–1349, Nov. 1993, doi: 10.1109/68.250065.

[26] G. A. Ball, W. H. Glenn, and W. W. Morey, "Programmable Fiber Optic Delay Line," *IEEE Photonics Technology Letters*, vol. 6, no. 6, pp. 741–743, Jun. 1994, doi: 10.1109/68.300180.

[27] E. H. Monsay, K. C. Baldwin, and M. J. Caccuitto, "Photonic True Time Delay for High-Frequency Phased Array Systems," *IEEE Photonics Technology Letters*, vol. 6, no. 1, pp. 118–120, Jan. 1994, doi: 10.1109/68.265909.

[28] A. Molony, Lin Zhang, J. A. R. Williams, I. Bennion, C. Edge, and J. Fells, "Fiber Bragg-Grating True Time-Delay Systems: Discrete-Grating Array 3-b Delay Lines and Chirped-Grating 6-b Delay Lines," *IEEE Transactions on Microwave Theory and Techniques*, vol. 45, no. 8, pp. 1527–1530, Aug. 1997, doi: 10.1109/22.618470.

[29] I. Frigyes and A. J. Seeds, "Optically Generated True-Time Delay in Phased-Array Antennas," *IEEE Transactions on Microwave Theory and Techniques*, vol. 43, no. 9, pp. 2378–2386, Sep. 1995, doi: 10.1109/22.414592.

[30] Ligeng Xu, R. Taylor, and S. R. Forrest, "The Use of Optically Coherent Detection Techniques for True-Time Delay Phased Array And Systems," *Journal of Lightwave Technology*, vol. 13, no. 8, pp. 1663–1678, Aug. 1995, doi: 10.1109/50.405308.

[31] D. Dolfi, D. Mongardien, S. Tonda, M. Schaller, and J. Chazelas, "Photonics for Airborne Phased Array Radars," in Proceedings 2000 IEEE International Conference on Phased Array Systems and Technology (Cat. No.00TH8510), Dana Point, CA, USA, 2000, pp. 379–382, doi: 10.1109/PAST.2000.858979.

[32] Ligeng Xu, R. Taylor, and S. R. Forrest, "True Time-Delay Phased-Array Antenna Feed System Based On Optical Heterodyne Techniques," *IEEE Photonics Technology Letters*, vol. 8, no. 1, pp. 160–162, Jan. 1996, doi: 10.1109/68.475812.

[33] J. L. Corral, J. Marti, J. M. Fuster, and R. I. Laming, "True Time-Delay Scheme for Feeding Optically Controlled Phased-Array Antennas Using Chirped-Fiber Gratings," *IEEE Photonics Technology Letters*, vol. 9, no. 11, pp. 1529–1531, Nov. 1997, doi: 10.1109/68.634731.

[34] H. Zmuda, R. A. Soref, P. Payson, S. Johns, and E. N. Toughlian, "Photonic beamformer for phased array antennas using a fiber grating prism," *IEEE Photonics Technol. Lett.*, vol. 9, no. 2, Art. no. 2, Feb. 1997, doi: 10.1109/68.553105.

[35] R. A. Minasian and K. E. Alameh, "Optical-Fiber Grating-Based Beamforming Network for Microwave Phased Arrays," *IEEE Transactions on Microwave Theory and Techniques*, vol. 45, no. 8, pp. 1513–1518, Aug. 1997, doi: 10.1109/22.618466.

[36] P. J. Matthews, M. Y. Frankel, and R. D. Esman, "A Wide-Band Fiber-Optic True-Time-Steered Array Receiver Capable of Multiple Independent Simultaneous Beams," *IEEE Photonics Technology Letters*, vol. 10, no. 5, pp. 722–724, May 1998, doi: 10.1109/68.669401.

[37] D. T. K. Tong and M. C. Wu, "Common Transmit/Receive Module for Multiwavelength Optically Controlled Phased Array Antennas," in OFC '98. Optical Fiber Communication Conference and Exhibit. Technical Digest. Conference Edition. 1998 OSA Technical Digest Series Vol.2 (IEEE Cat. No.98CH36177), San Jose, CA, USA, 1998, pp. 354–355, doi: 10.1109/OFC.1998.657467.

[38] S. Tonda-Goldstein, D. Dolfi, A. Monsterleet, S. Formont, J. Chazelas, and Jean-Pierre Huignard, "Optical Signal Processing in Radar Systems," *IEEE Transactions on Microwave Theory and Techniques*, vol. 54, no. 2, pp. 847–853, Feb. 2006, doi: 10.1109/TMTT.2005.863059.

[39] B. Ortega, J. L. Cruz, J. Capmany, M. V. Andres, and D. Pastor, "Variable Delay Line for Phased-Array Antenna Based on a Chirped Fiber Grating," *IEEE Transactions on Microwave Theory and Techniques*, vol. 48, no. 8, pp. 1352–1360, Aug. 2000, doi: 10.1109/22.859480.

[40] Y. Wang, S. C. Tjin, J. Yao, J. P. Yao, L. He, and K. A. Ngoi, "Wavelength-Switching Fiber Laser For Optically Controlled Phased-Array Antenna," *Optics Communications*, vol. 211, no. 1–6, pp. 147–151, Oct. 2002, doi: 10.1016/S0030-4018(02)01888-6.

[41] Yunqi Liu, Jianliang Yang, and Jianping Yao, "Continuous True-Time-Delay Beamforming for Phased Array Antenna Using a Tunable Chirped Fiber Grating Delay Line," *IEEE Photonics Technology Letters*, vol. 14, no. 8, pp. 1172–1174, Aug. 2002, doi: 10.1109/LPT.2002.1022008.

[42] Y. Liu, J. Yao, and J. Yang, "Wideband True-Time-Delay Unit for Phased Array Beamforming Using Discrete-Chirped Fiber Grating Prism," *Optics Communications*, vol. 207, no. 1–6, pp. 177–187, Jun. 2002, doi: 10.1016/S0030-4018(02)01529-8.

[43] B. Vidal et al., "Photonic True-Time Delay Beamformer for Broadband Wireless Access Networks at 40 GHz Band," in 2002 IEEE MTT-S International Microwave Symposium Digest (Cat. No.02CH37278), Seattle, WA, USA, 2002, vol. 3, pp. 1949–1952, doi: 10.1109/MWSYM.2002.1012246.

[44] S. T. Winnall and D. B. Hunter, "A Fibre Bragg Grating Based Scanning Receiver for Electronic Warfare Applications," in 2001 International Topical Meeting on Microwave Photonics. Technical Digest. MWP'01 (Cat. No.01EX476), Long Beach, CA, USA, 2001, pp. 211–214, doi: 10.1109/MWP.2002.981833.

[45] Yihong Chen and R. T. Chen, "A Fully Packaged True Time Delay Module for a K-Band Phased Array Antenna System Demonstration," *IEEE Photonics Technology Letters*, vol. 14, no. 8, pp. 1175–1177, Aug. 2002, doi: 10.1109/LPT.2002.1022009.

[46] R. Rotman, O. Raz, and M. Tur, "Requirements for True Time Delay Imaging Systems with Photonic Components," in IEEE International Symposium on Phased Array Systems and Technology, 2003, Boston, MA, USA, 2003, pp. 193–198, doi: 10.1109/PAST.2003.1256980.

[47] O. Raz, R. Rotman, Y. Danziger, and M. Tur, "Implementation of Photonic True Time Delay Using High-Order-Mode Dispersion Compensating Fibers," *IEEE Photonics Technology Letters*, vol. 16, no. 5, pp. 1367–1369, May 2004, doi: 10.1109/LPT.2004.826263.

[48] S.-S. Lee, Y.-H. Oh, and S.-Y. Shin, "Photonic Microwave True-Time Delay Based on a Tapered Fiber Bragg Grating With Resistive Coating," *IEEE Photonics Technology Letters*, vol. 16, no. 10, pp. 2335–2337, Oct. 2004, doi: 10.1109/LPT.2004.833874.

[49] B.-M. Jung, J.-D. Shin, and B.-G. Kim, "Optical True Time-Delay for Two-Dimensional X-Band Phased Array Antennas," *IEEE Photonics Technology Letters*, vol. 19, no. 12, pp. 877–879, Jun. 2007, doi: 10.1109/LPT.2007.897530.

[50] H. Zmuda, R. A. Soref, P. Payson, S. Johns, and E. N. Toughlian, "Photonic Beamformer for Phased Array Antennas Using a Fiber Grating Prism," *IEEE Photonics Technology Letters*, vol. 9, no. 2, pp. 241–243, Feb. 1997, doi: 10.1109/68.553105.

[51] Y. Liu, J. Yao, and J. Yang, "Wideband True-Time-Delay Beam Former That Employs a Tunable Chirped Fiber Grating Prism," *Applied Optics*, vol. 42, no. 13, p. 2273, May 2003, doi: 10.1364/AO.42.002273.

[52] Jianping Yao, Jianliang Yang, and Yunqi Liu, "Continuous True-Time-Delay Beamforming Employing a Multiwavelength Tunable Fiber Laser Source," *IEEE Photonics Technology Letters*, vol. 14, no. 5, pp. 687–689, May 2002, doi: 10.1109/68.998726.

[53] Y. Liu, "Tunable Chirping of a Fiber Bragg Grating Without Center Wavelength Shift Using a Simply Supported Beam," *Optical Engineering*, vol. 41, no. 4, p. 740, Apr. 2002, doi: 10.1117/1.1461834.

[54] I. Gasulla, J. Lloret, J. Sancho, S. Sales, and J. Capmany, "Recent Breakthroughs in Microwave Photonics," *IEEE Photonics Journal*, vol. 3, no. 2, Art. no. 2, Apr. 2011, doi: 10.1109/JPHOT.2011.2130517.

[55] K. Okada, S. Suzuki, and M. Asada, "Resonant-Tunneling-Diodeterahertz Oscillator Integrated With Slot-Coupled Patch Antenna," *Conference Proceedings – International Conference on Indium Phosphide and Related Materials*, pp. 14–15, 2014, doi: 10.1109/ICIPRM.2014.6880525.

[56] K. Okada, K. Kasagi, N. Oshima, S. Suzuki, and M. Asada, "Resonant-Tunneling-Diode Terahertz Oscillator Using Patch Antenna Integrated on Slot Resonator for Power Radiation," *IEEE Transactions on Terahertz Science and Technology*, vol. 5, no. 4, pp. 613–618, 2015, doi: 10.1109/TTHZ.2015.2441740.

[57] K. H. Alharbi, A. Ofiare, M. Kgwadi, A. Khalid, and E. Wasige, "Bow-tie Antenna For Terahertz Resonant Tunnelling Diode Based Oscillators On High Dielectric Constant Substrate," 2015 11th Conference on Ph.D. Research in Microelectronics and Electronics, PRIME 2015, pp. 168–171, 2015, doi: 10.1109/PRIME.2015.7251361.

[58] H. Xu, J. Zhou, K. Zhou, and Q. Wu, "Planar Wideband Circularly Polarized Cavity-backed Stacked Patch Antenna Array for Millimeter-Wave Applications," vol. 66, no. 10, pp. 5170–5179, 2018, doi: 10.1109/TAP.2018.2862345.

[59] X. Ding, Z. Zhao, Y. Yang, Z. Nie, and Q. H. Liu, "A Compact Unidirectional Ultra-wideband Circularly Polarized Antenna Based on Crossed Tapered Slot Radiation Elements," *IEEE Transactions on Antennas and Propagation*, vol. 66, no. 12, pp. 7353–7358, 2018, doi: 10.1109/TAP.2018.2867059.

[60] N. W. Liu, L. Zhu, W. W. Choi, and J. D. Zhang, "A Low-Profile Differentially Fed Microstrip Patch Antenna with Broad Impedance Bandwidth under Triple-Mode Resonance," *IEEE Antennas and Wireless Propagation Letters*, vol. 17, no. 8, pp. 1478–1482, 2018, doi: 10.1109/LAWP.2018.2850045.

[61] Z. Tang, X. Wu, J. Zhan, Z. Xi, and S. Hu, "A Novel Miniaturized Antenna with Multiple Band-Notched Characteristics for UWB Communication Applications," *Journal of Electromagnetic Waves and Applications*, vol. 32, no. 15, pp. 1961–1972, 2018, doi: 10.1080/09205071.2018.1486235.

[62] A. K. Bhattacharyya, Y. M. M. Antar, and A. Ittipiboon, "Spectral domain analysis of aperture-coupled microstrip patch antennas," *IEE Proc. H Microw. Antennas Propag.*, vol. 139, no. 5, p. 459, 1992, doi: 10.1049/ip-h-2.1992.0081.

[63] C. Tsao, Y. Hwang, F. Kilburg, and F. Dietrich, "Aperture-Coupled Patch Antennas with Wide-Bandwidth and Dual-Polarization Capabilities," Antennas and Propagation Society International Symposium, 1988. AP-S. Digest, pp. 936–939, 2002, doi: 10.1109/APS.1988.94241.

[64] G. A. E. Vandenbosch, "Capacitive Matching of Microstrip Antennas," *Electronics Letters*, vol. 31, no. 18, pp. 1535–1536, 2002, doi: 10.1049/el:19951095.

[65] T. R. Hayes, "Reactive Ion Etching of InP using CH4/H2 Mixtures: Mechanisms of Etching and Anisotropy," *Journal of Vacuum Science & Technology B: Microelectronics and Nanometer Structures*, vol. 7, no. 5, p. 1130, Sep. 1989, doi: 10.1116/1.584564.

[66] R. Khare, "CH4/H2/Ar/Cl2 Electron Cyclotron Resonance Plasma Etching of via Holes for InP-based Microwave Devices," *Journal of Vacuum Science & Technology B: Microelectronics and Nanometer Structures*, vol. 12, no. 5, p. 2947, Sep. 1994, doi: 10.1116/1.587541.

[67] S. J. Pearton, F. Ren, and C. R. Abernathy, "Optical Emission End Point Detection For Via Hole Etching in InP and GaAs Power Device Structures," p. 5.

[68] S. Trassaert, "Bromine/Methanol Wet Chemical Etching of Via Holes for InP Microwave Devices," *Journal of Vacuum Science & Technology B: Microelectronics and Nanometer Structures*, vol. 16, no. 2, p. 561, Mar. 1998, doi: 10.1116/1.589863.

[69] L. G. Hipwood, "Dry Etching of Through Substrate Via Holes for GaAs MMIC's," *Journal of Vacuum Science & Technology B: Microelectronics and Nanometer Structures*, vol. 3, no. 1, p. 395, Jan. 1985, doi: 10.1116/1.583271.

[70] E. W. Sabin, "Estimation of the Activation Energy for Ar/Cl 2 Plasma Etching of InP Via Holes Using Electron Cyclotron Resonance," *J. Vac. Sci. Technol. B*, vol. 16, no. 4, p. 6, 1998.

[71] B. Hussain, "Short-Slot Hybrid Coupler in Gap Waveguides at 38 GHz," Chalmers University of Technology, Goteborg, Sweden, 2011.

[72] P.-S. Kildal, "Artificially Soft and Hard Surfaces in Electromagnetics and Their Application to Antenna Design," in 23rd European Microwave Conference, 1993, Madrid, Spain, Oct. 1993, pp. 30–33, doi: 10.1109/EUMA.1993.336763.

[73] A. A. Brazalez, E. Rajo-Iglesias, J. L. Vazquez-Roy, A. Vosoogh, and P.-S. Kildal, "Design and Validation of Microstrip Gap Waveguides and Their Transitions to Rectangular Waveguide, for Millimeter-Wave Applications," *IEEE Transactions on Microwave Theory and Techniques*, vol. 63, no. 12, pp. 4035–4050, Dec. 2015, doi: 10.1109/TMTT.2015.2495141.

[74] A. A. Brazalez, A. U. Zaman, and P.-S. Kildal, "Design of a Coplanar Waveguide-To-Ridge Gap Waveguide Transition Via Capacitive Coupling," in 2012 6th European Conference on Antennas and Propagation (EUCAP), Prague, Czech Republic, Mar. 2012, pp. 3524–3528, doi: 10.1109/EuCAP.2012.6206372.

[75] A. Uz Zaman, T. Vukusic, M. Alexanderson, and P.-S. Kildal, "Design of a Simple Transition from Microstrip to Ridge Gap Waveguide Suited for MMIC and Antenna Integration," *IEEE Antennas and Wireless Propagation Letters*, vol. 12, pp. 1558–1561, 2013, doi: 10.1109/LAWP.2013.2293151.

[76] B. Hussain, "Short-Slot Hybrid Coupler in Gap Waveguides at 38 GHz," Chalmers University of Technology, Goteborg, Sweden, 2011.

[77] B. Hussain, G. Serafino, P. Ghelfi, A. Bogoni, and A. Stohr, "Via-Less Microstrip to Rectangular Waveguide Transition on InP," in 2019 44th International Conference on Infrared, Millimeter, and Terahertz Waves (IRMMW-THz), Paris, France, Sep. 2019, pp. 1–2, doi: 10.1109/IRMMW-THz.2019.8874316.

Part IV

Terahertz Links, Application, and Deployment

Chapter 12

Terahertz Band Intersatellite Communication Links

Meltem Civas, Turker Yilmaz, and Ozgur B. Akan

Contents

12.1 Introduction

More than 14 years passed between the launch of the first ever satellite, Sputnik 1, and demonstration of the first intersatellite communications link (ISL) in January 1975 by the radio amateurs within AMSAT. AMSAT/OSCAR-7 sent a signal to AMSAT/OSCAR-6 at a frequency of 432.15 MHz and these were repeated back to Earth at 145.95 and 29.50 MHz, respectively. The National Aeronautics and Space Administration established an ISL in the April of the same year between the low-Earth orbit (LEO) satellite GEOS-3 and ATS-6. Although the possible use of optical beams for ISL was first mentioned in 1945 by Arthur C. Clarke [1], due to the unavailability of necessary components, the first optical ISL could only be established in November 2001 between ARTEMIS of the European Space Agency and French SPOT-4 [2].

There are many criteria that need to be considered when designing an ISL system. For a microwave system, the main performance criterion, data rate, is directly proportional to the available bandwidth; thus, higher frequency bands are preferred over the lower ones. Higher frequency also translates to lower antenna diameters since effective aperture area of an antenna is directly proportional to the square of the wavelength of the transmitted wave. Smaller antennas are advantageous in terms of weight and cost too. Atmospheric attenuation is also considered during the ISL design: Frequency bands with high atmospheric attenuation are preferable since it helps ISL to be isolated from any possible interference from the Earth. Iridium and teledesic systems are two LEO constellations that are employed in microwave ISLs.

Optical ISL systems necessitate different selection criteria compared to legacy microwave technologies. The selection of laser source depends on many parameters including output power, operating temperature, and lifetime. On the receiver (RX) side, direct or heterodyne detection can be chosen. Acquisition and tracking are part of the design issue considering the very narrow beamwidths over very long distances. Due to the advantages in size, power requirement, and data rate compared to the microwave ISL systems, optical technologies are currently the choice and many laser communication terminals are currently used in satellites including NFIRE and TerraSAR-X.

Due to the fact that components required for possible ISLs in the THz band (0.3–1 THz) were not available at the time when the performance of microwave technologies became insufficient and optical components started to provide necessary performance to realize ISLs, research was focused optical links completely by-passing the THz band. On account of the advancements in THz technologies since the beginning of 1980s, it is now possible to manufacture transceivers (TRXs) working at the frequency range of 2–3 THz. This means that it is now possible to overcome many shortcomings of the microwave systems compared to the optical systems, and competitive ISL systems that use electromagnetic (EM) radiation can be assembled.

In line with these, in this chapter first a review on the development of ISL technologies, namely microwave and optical solutions, is presented. State-of-the-art technologies are then compared in terms of their suitability for ISL adoption using many criteria including size, power consumption, and data rate. Then, keeping in mind that THz band is left out of discussion for ISL without even really being considered, a THz band ISL system is proposed in light of the developments that occurred in the area during the last decade, by also considering the “NR above 52.6 GHz” and “non-terrestrial networks” study items of the future Release 17 of the 3rd Generation Partnership Project (3GPP). Subsequently, link budgets for a typical usage scenario are calculated, helping to explain the propagation mechanisms. The chapter concludes by setting the directions for future research and challenges they may face.

12.2 Intersatellite Communications Links

Satellites can be classified into three categories based on the orbit altitude: geostationary Earth orbit (GEO) satellites operating at 35,786 km, medium Earth orbit (MEO) satellites operating at the orbits between 2,000 and 35,786 km, and LEO satellites with the orbit altitude varying between 400 and 2000 km. GEO satellites have been widely used for satellite communications since they can provide wide coverage and consequently minimal handoff. On the other hand, GEO satellites have the longest propagation latency compared to LEO and MEO; thus, they are not suitable for applications requiring low delay. Recent trends show that there is an increasing interest in LEO constellations, which can provide high throughput, lower latency, and launching cost. Multiple companies such as SpaceX, OneWeb, and Amazon are currently working on or planning construction of LEO satellite constellations for broadband Internet connectivity, or scientific or military purposes. SpaceX aims to deploy thousands of small LEO satellites to build the Starlink. Due to lower altitude, much more LEO satellites are required for global coverage compared to GEO. Several disadvantages arise from deploying high number of satellites including larger Doppler spread and transmission loss due to frequent handoff [3]. MEO satellite constellation, which is exemplified by O3b, also is appealing owing to its unique characteristics in terms of constellation size and latency. Satellite miniaturization is another aspect of the "new space era" [4]. In contrast to costly traditional GEO, MEO and LEO satellites, low-cost cube-, micro-, and nano-satellites with short deployment cycles have also emerged [5]. These satellites can pave the way for new paradigms including Internet of Space Things (IoST).

Existing missions use X- (8–12 GHz), K_u- (12–18 GHz) and K_a-bands (27–40 GHz). Currently V- (40–75 GHz) and W-bands (75–110 GHz) are being considered for high throughput satellite links; however, it is apparent that increasing demand for the spectrum usage requires the employment of even higher bands of the spectrum. Thus, a paradigm shift is needed for the next-generation space services.

12.3 Terahertz Band Communications

Different application areas of THz communication have been examined in various studies. The application areas covered include Internet of Nano-Things (IoNT) [6], Internet of Things (IoT) [7,8], vehicular networks [9], nanocommunication and nanonetworking [10], THz communication networks [11–28] and space-based THz systems [29]. In [6], a reference architecture for IoNT paradigm and communication challenges for THz nanonetworks are discussed. Early examinations on and the opportunities and challenges of the utilization of the THz band for IoT usage models are illustrated in [7,8]. Research challenges in THz communication and the opportunities that THz band can offer for vehicular networks operating in beyond the fifth generation (5G) of mobile telecommunication systems are discussed in [9]. THz band nanocommunication and nanonetworking are detailed in [10] by identifying applications of THz nanonetworks, challenges, and reviewing different aspects including channel models and experimentation tools. The novel applications of THz communications and challenges are surveyed in [11,12]. In [13,14], overviews for device technologies and channel models for THz communication are given together with technical issues that need to be addressed. In [15], an overview

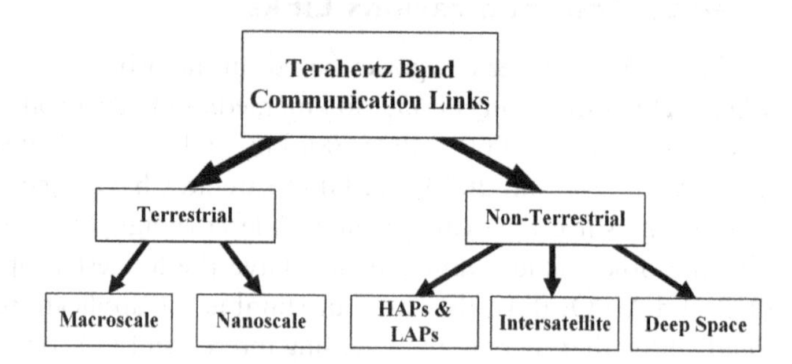

Figure 12.1 Classification of THz links.

of device and antenna technologies is given, and standardization activities are discussed. Standardization activities are also discussed in [16] along with novel applications for THz communication and research challenges. In addition to opportunities for vehicular networks, a number of opportunities through THz wireless communication including hybrid-optical communication links, THz-based data centers with superior performance, and deployment of femtocells for mobile heterogeneous networks are detailed in [17–19]. An in-depth survey of THz medium access protocols is presented in [20]. Summaries of THz-enabled wireless systems, scenarios, applications, and challenges for the deployment of beyond 5G mobile systems are presented in [21,22]. The state-of-the-art in THz signal generation and channel modeling is presented in [24,25], as well as a comparison with the other wireless technologies, THz standardization activities, and possible applications of THz communications. Various modulation schemes are analyzed for THz band communication over systems simulating real-world devices, and optimum modulation and physical layer (PHY) parameters in terms of energy consumption are published in [26,27]. Key enabling technologies for THz communication and application scenarios are discussed in [28]. The current state in THz communications concerning THz channel, devices, and space-based systems from the perspective of device technology is summarized in [29]. However, very few studies covering the THz band ISLs are available in the literature. In [5], the authors propose a new CubeSat design with multi-band radios and briefly study link budget for LEO constellation of CubeSats. In [30], the feasibility of THz band for intersatellite communications is studied. In [31], the authors examine the viability of using 0.75–10 THz among ISLs.

THz band communication links, as illustrated in Figure 12.1, can be classified into two categories: terrestrial and non-terrestrial links. Terrestrial links comprise macroscale and nanoscale links. Non-terrestrial links are further classified into ISLs, deep space links, and satellite/airborne platform links, that is, high-altitude and low-altitude platforms (HAPs and LAPs) such as aircraft, unmanned aerial vehicles (UAVs), and high-altitude balloons (HABs). In the following sections, we compare THz band communications with the other state-of-the-art technologies and identify further benefits of using THz band for ISLs.

12.3.1 Comparison with Other Technologies

THz communication has several key advantages over millimeter wave (mmWave) and free-space optical communication (FSO), which are summarized in Table 12.1.

Table 12.1 Comparison of Different Technologies [20,25]

Parameter	*mmWave*	*Terahertz*	*FSO*
Frequency range	30–300 GHz	0.3–10 THz	0.01–30 PHz
Data rate (Gbps)	$<10^2$	$\in[10^2, 10^3)$	$\in[10^2, 10^3)$
Transmission range	Short	Short	Short
Power consumption	Medium	Low	High
Weather conditions	Robust	Robust	Sensitive
Security	Medium	High	High
System size	Large	Small	Bulky

12.3.1.1 Millimeter Wave Band Communication

Millimeter-wave band covers the frequencies from 30 to 300 GHz. The mmWave wireless communication can offer multi gigabit-per-second (Gbps) data rates and enable many applications including vehicular and short-range wireless communications and high-quality data streaming. Due to strong molecular absorption at frequencies around 60, 120, and 180 GHz, several special bands are chosen for utilization, namely 35, 94, 140 and 220 GHz, where the attenuation is relatively small [32,33]. However, mmWave band communication cannot provide data rates as high as THz band communication, which can offer data rates on the orders of terabit-per-second (Tbps) owing to the extremely large bandwidth available.

THz communication has other advantages over mmWave communication apart from its higher bandwidth. Another advantage is THz links are inherently more directional compared to mmWave links because THz waves experience less free-space diffraction due to their shorter wavelengths [34,35]. For space applications, especially for autonomous missions, security is essential. High directionality also provides more secure communication compared to mmWave communication since it is harder to detect THz beams. This significantly reduces the possibility of eavesdropping. Finally, THz frequency band is still largely not regulated whereas 275–300 GHz is allocated for mobile communications [36,37].

12.3.1.2 Free-Space Optical Communication

Free-space optical (FSO) communication systems operate at infrared (10–430 THz), visible light (430–790 THz), and ultraviolet (0.79–30 PHz) spectral ranges. Considering the well-known fact that information capacity increases with the bandwidth, FSO can enable high-capacity links via high bandwidths. To illustrate, allowable bandwidth is on the order of hundreds of THz in a typical optical communication link [38]. Moreover, highly directional and narrow beams reduce the probability of eavesdropping; thus, they provide highly secure communication. Although FSO can provide high bandwidth and secure links, it has many disadvantages too. FSO links heavily depend on the atmospheric conditions such as fog, pollution, dust, rain, and snow. Variations of temperature and pressure of the atmosphere, that is, atmospheric turbulence, also impacts the propagation of optical waves. In optical communication, very narrow beams are used. As the communication distance increases, beam alignment issue becomes critical to maintain reliable communication between the transmitter (TX) and RX. Since THz waves are longer than optical wavelengths, THz communication is advantageous over FSO in terms of tolerance for beam alignment. Moreover, THz communication is also favorable in terms of power and weight requirements, which are inherently large for laser-based systems [39]. Furthermore, relative position of the Sun to the laser TX and RX can degrade the system performance by increasing the solar background radiation. In case of superior solar conjunction, which occurs when the Sun, Earth, and planet considered are aligned, optical communication is not possible. Lastly, due to the safety limits for eye, at certain wavelengths laser power level is limited to relatively low values, which limits the transmission range and data rates.

12.3.2 Terahertz Band Intersatellite Links

THz band communication is beneficial in space compared to existing radio frequency technologies in terms of data rate, size, interference, and security. Higher data rates on the order of hundreds of Gbps is achievable owing to the available ultra-wide bandwidth. The space applications are expected to have low-complexity designs because of the size restrictions. At the THz frequencies, antenna size is significantly small. Moreover, using simple modulation and coding schemes in THz communication, complexity can be further reduced, and high data rates can be supported for satellites and spacecrafts. THz band antennas inherently have high directivity; hence, the interference between the satellites is relatively small. Very narrow THz beams reduce the probability of eavesdropping too.

THz band communications can pave the way for new opportunities in space by enabling huge bandwidths. In terrestrial networks, electromagnetic (EM) waves at the frequencies above 300 GHz experience high gaseous attenuations due to the molecular absorption loss caused mainly by the water vapor, oxygen-, and pressure-induced nitrogen found in the atmosphere. Conversely, the effect of molecular absorption in the space and higher layers of the Earth's atmosphere is absent due to the lack of these absorbing molecules. So, THz band communication can potentially be employed in the links among LEO, MEO, GEO, and high Earth orbit (HEO) satellites.

12.4 Link Budget Analysis

Investigating the link performance is critical to understand the feasibility of THz links. In this respect, research efforts focused on the link budget analysis of ground-to-satellite and intersatellite links. In [40], a ground-to-geostationary satellite link has been modeled. The authors report that based on their analyses, THz systems are superior to the optical communication when fog, cloud cover or, clear-air turbulence exists. In [41], link budgets of both intersatellite (GEO-to-LEO) and GEO satellite-to-Earth station systems have been considered. The results suggest that due to the absence of water vapor, high carrier frequencies (f_c) above 300 GHz can be employed in ISLs. In [42], the authors propose a THz satellite communication system with massive antenna array to be located at Tanggula, Tibet, where the precipitable water density is low. The link budget analysis in this study shows that 1 Tbps can be achieved in the low-THz band (0.275–0.37 THz). In [5], another link budget analysis for CubeSats, which are 10 cm^3 miniaturized satellites with a mass of up to 1.33 kg, that operate ISLs at mmWave and THz bands was also simulated.

In this section, we first consider the propagation of THz waves in ISLs and then conduct link budget analyses for GEO-to-GEO and LEO CubeSat ISLs to examine the feasibility of THz ISLs.

12.4.1 Terahertz Propagation in Space

The propagation of THz EM waves in ISLs suffers from several types of attenuation and noise sources, which are described in this section.

12.4.1.1 Loss in Terahertz Frequencies

The propagation of THz waves can be described by the familiar Friis' transmission equation as follows [43]:

$$P_{rx}[dB] = P_{tx}[dB] + G_{tx}[dB] + G_{rx}[dB] - A_{tot}[dB] \tag{12.1}$$

where P_{rx}, P_{tx}, G_{tx}, G_{rx} and A_{tot} are the received signal power, transmitted signal power, TX antenna gain, RX antenna gain, and total path loss, respectively.

Spreading loss $A_{spread}(f)$, molecular absorption loss $A_{abs}(f)$, and scattering loss $A_{sca}(f)$ are the three main contributors to the total path loss $A_{tot}(f)$ in THz frequencies [44,45]. Therefore, the total path loss in dB is given by

$$A_{tot}(f)[dB] = A_{spread}(f)[dB] + A_{abs}(f)[dB] + A_{sca}(f)[dB] \tag{12.2}$$

Spreading loss describes the attenuation caused by the expansion of the wave through the propagation medium. It can be expressed, in dB, as

$$A_{spread}(f)[dB] = 20\log(4\pi fd / c) \tag{12.3}$$

where c is the speed of light and d is the distance in m. Considering that the GEO-to-GEO and GEO-to-LEO link distances are on the orders of 10^4 km, the spreading loss is substantial [46].

Molecular absorption loss occurs when a part of the wave energy is transformed into the internal energy of molecules in the propagation environment by molecular vibration. Water vapor molecules are the main sources of molecular absorption in the Earth's atmosphere. Above 300 GHz molecular absorption is highly strong, such that the band above 300 GHz is not feasible for use in terrestrial systems. On the other hand, most of the satellites occupy the two outermost layers of Earth's atmosphere, that is, the thermosphere and exosphere. LEO satellites inhabit the thermosphere, MEO and HEO satellites operate at the exosphere, and GEO satellites revolve at an altitude of 35,786 km, where the atmosphere does not exist. Water vapor density is very low in the thermosphere and exosphere too. Therefore, the molecular absorption loss is negligible regarding the propagation of EM waves within ISLs. This means that higher frequencies that are not feasible for use in terrestrial networks can be employed in ISLs, which paves the way for the opportunity of performing communication in ISLs at extremely high data rates thru the available huge bandwidths.

Scattering loss occurs when the propagating EM wave is redirected into one or several paths away from the intended path, hence decreasing the received signal energy. Depending on the diameter of the atmospheric scattering particles, examples for which are gases, aerosols, hydrosols, and hydrometeors such as haze, cloud, fog, rain, snow and hail, THz waves can be affected by three types of scattering: Rayleigh, Mie, and, wavelength independent. Rayleigh scattering occurs when the wavelength of the EM wave is much longer than the diameter of the scattering particles and due to polar molecules of atmospheric gases. Mie scattering is generated if the sizes of the scattering particles are comparable to the wavelength of the EM wave. Otherwise, wavelength independent scattering, which is modeled by diffraction theory, is produced [38]. The amount of signal strength loss on account of Rayleigh scattering is insignificant in the THz band [47]. Regarding ISLs, we consider the layers of the Earth's atmosphere where the air molecule density is very low, namely the thermosphere and exosphere. Therefore, the effect of scattering due to atmospheric particles can be neglected for ISLs.

12.4.1.2 System Noise

The system noise power at the RX terminals is contributed by the antenna and RX noises [48]. Antenna noise depends on the frequency and direction where the antenna is pointed. For instance, if the antenna is aimed to the sky, antenna temperature is mainly contributed by the sky noise, also termed as the background noise, and extraterrestrial noise called the cosmic microwave background radiation, which is bounded by 2.725 K [49]. The sky noise is contributed by the temperature of the atmosphere, which acts as an absorbing medium [50]. Furthermore, RX noise temperature, which is caused by thermal noise in the RX components, contributes to the overall system temperature. Thus, system noise power at the RX terminals is given by

$$P_{\mathrm{n}} = kT_{\mathrm{sys}}B \tag{12.4}$$

where k is Boltzmann's constant (1.38×10^{-23} J/K), T_{sys} is the effective system noise temperature, and B is the bandwidth. The effective system noise temperature at the RX terminals can be expressed as

$$T_{\mathrm{sys}} = T_{\mathrm{ant}} + T_{\mathrm{rec}} \tag{12.5}$$

where T_{ant} is the antenna temperature and T_{rec} is the temperature at the RX.

12.4.2 Geostationary Earth Orbit-to-Geostationary Earth Orbit Links

In this section, the link budget analysis of a typical GEO-to-GEO ISL at the THz band is examined. The distance between GEO satellites is taken to be 83,043 km [46]. Moreover, the effect of the Sun is assumed to be absent. As described in [40], the spectrum is separated into six bands from which we consider only the center frequencies (f_{cen}) of the bands B (186–327 GHz), C (327–555 GHz) and D (567–747 GHz). We set the link budget to ensure a signal-to-noise ratio (SNR) of 10 dB at the RX. SNR can be expressed in dB as follows:

$$(\mathrm{SNR})[dB] = (P_{\mathrm{tx}} + G_{\mathrm{tx}} + G_{\mathrm{rx}} - A_{\mathrm{tot}}) - P_{\mathrm{n}}. \tag{12.6}$$

Table 12.2 GEO-to-GEO Communications Link Case for an SNR of 10 dB

	Band B	*Band C*	*Band D*
f_{cen} (GHz)	225	350	675
P_{rx} (dBm)	–59.1	–58.3	–58.3
T_{rec} (°K)	3000	3600	3600
T_{ant} (°K)	0.6	0.6	0.7
B (GHz)	3	3	3
P_n (dBm)	–69.1	–68.3	–68.3

Table 12.3 GEO-to-GEO Communications Link Budget for an SNR of 10 dB

	Band B	*Band C*	*Band D*
f_{cen} (GHz)	225	350	675
P_{tx} (dBm)	10	10	0
G_{tx} (dBi)	88	92	95
G_{rx} (dBi)	80.8	81.5	94.2
A_{tot} (dB)	237.9	241.7	247.4

Table 12.2 shows the necessary received power for an SNR of 10 dB at different frequency bands, whereas Table 12.3 presents the corresponding link budget to obtain such a received power. The results suggest that high gain transmit and receive antennas are required for GEO-to-GEO links. Moreover, the gains of the transmit and receive antennas need to increase with the rising f_c. Finally, due to the absence of molecular absorption, path loss is the main impairment regarding the EM wave propagation.

12.4.3 Low-Earth Orbit CubeSat Links

In this section, we conduct a link budget analysis for ISLs between LEO CubeSats operating at THz frequencies. In [5], the authors propose a CubeSat design with reconfigurable multi-band radios covering mmWave, THz, and optical frequencies. In line with this study, we consider the analysis of THz ISLs at the frequencies 0.35, 0.675, and 1 THz. The bandwidth, temperature, and transmit power are taken to be 0.5 GHz, 1500 K and 10 W, respectively [5]. The simulation results shown in Figure 12.2 illustrate the minimum gains required at the TX and RX, and beamwidth required at the RX to obtain 10 dB SNR. Figure 12.2a demonstrates that minimum antenna gain is less than 50 dB up to an f_c of 1 THz, which is potentially achievable by ultra-massive multiple input multiple output (UM MIMO) antennas. As expected, increasing transmission distance results in higher required antenna gain. The required beamwidth to obtain 10 dB SNR at the RX ranges from 0.6° to 8.3°, as presented in Figure 12.2b. Narrower beamwidth means highly directional antenna, which can lead to outages due to relative motion. Mitigation techniques are needed for this challenge, including fast antenna beam alignment and faster handover management techniques [51]. Moreover, gain required at the RX and TX increases by 5 dB as the bandwidth is doubled, as displayed in Figure 12.2c.

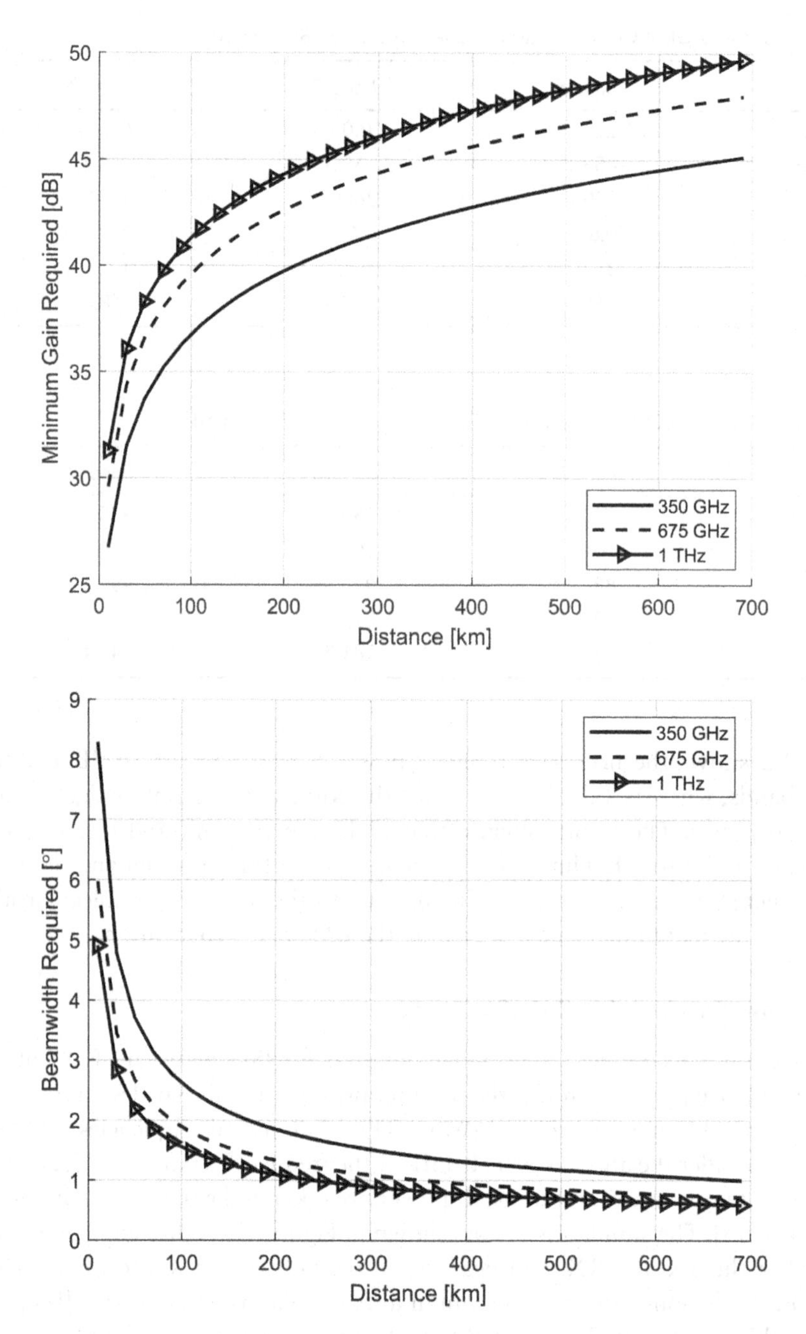

Figure 12.2 Required gain and beamwidth for an ISL between LEO CubeSats. (a) Minimum gain required. (b) Beamwidth required. (c) Minimum gain required for several bandwidths, f_c = 1 THz.

12.5 Future Research Directions

In this section, we discuss the future research directions, namely THz UM MIMO, 5G satellite networks, IoST, airborne military and defense applications, and the major challenges ISL encounters.

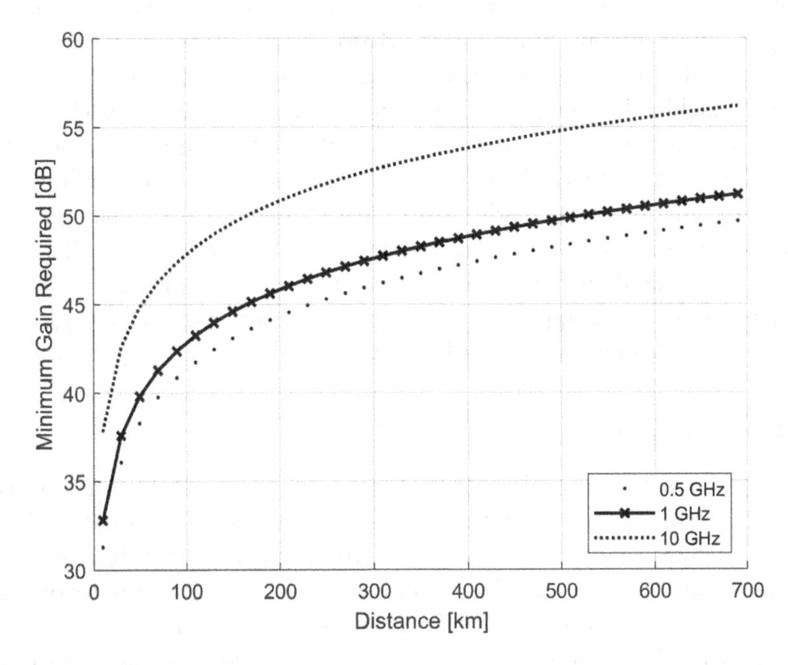

Figure 12.2 Cont.

12.5.1 Terahertz Ultra-Massive Multiple Input Multiple Output

EM waves experience high spreading loss and the transmission power of TRXs is limited in THz frequencies. These impairments limit the transmission distance into a few tens of meters. Thus, highly directional antennas with high gains are required to increase the transmission distance. To combat the distance problem, the concept of UM MIMO communication in the THz band, which leverages many nanoantennas, has been introduced. As an enabling technology, plasmonic nanoantennas have been proposed. It has been estimated that using graphene-based plasmonic antenna arrays, thousands of antennas can be embedded in a few mm^2. On the other hand, the fabrication of nanoantenna arrays and PHY design still poses challenges for this concept [52].

12.5.2 5G and Beyond 5G Satellite Networks

IoT is a vital part of the 5G systems. The need for Internet provisioning to rural and remote areas, and ubiquitous coverage for mission critical applications focus the attention on satellites, which are also recognized as a part of 5G connectivity. The future Release 17 of the 3GPP also considers satellite networks as a study item under "Non-Terrestrial Networks," which proposes satellite systems as either a standalone solution or a part of 5G connectivity. This creates the opportunity to integrate the terrestrial and satellite networks. Challenges faced by traditional satellite networks such as latency and cost per data rate are expected to be overcome by 5G systems. Thus, satellites are envisioned to provide backhaul connectivity for the IoT [53]. Yet, there are still challenges such as high path losses, delays, and Doppler shifts [54] that need to be addressed to realize satellite-based new radio platforms. 5G communication systems have service requirements including enhanced mobile broadband (eMBB), massive machine-type communications (mMTC), and ultra-reliable low-latency communications (URLLC). eMBB service's use cases include backhauling, which is an emergent need in disaster areas where large amount of bandwidth is abruptly demanded. mMTC and URLLC can pave the way for plethora of IoT applications. However, terrestrial infrastructure has certain limitations including coverage to ensure full network utilization. Hence, satellites, being able to provide ubiquitous coverage to rural and remote areas such as oceans and deserts, will be integrated to both 5G and beyond 5G networks. Satellite networks can enhance network reliability for especially mission critical applications and moving platforms such as trains and planes, and enable service scalability by their broadcasting and multicasting capabilities [4].

12.5.3 Internet of Space Things

An LEO satellite constellation-based IoT system, which can enable the IoST concept, for the ubiquitous coverage of the Earth's remote areas has been proposed in [55]. However, traditional satellites suffer from several limitations such as long development time and high cost. Considering these drawbacks, another IoST concept that relies on CubeSats has been introduced in [56]. The CubeSat design proposed in [5] also supports wireless communication at the THz frequency band. IoST can be used as an alternative backhaul to terrestrial infrastructure or employed in monitoring activities and aerial reconnaissance. Yet, several challenges in the placement of the hubs and resource allocation need to be overcome.

12.5.4 Hybrid Space–Air Network

Space networks containing GEO, MEO, or LEO satellites can also be integrated into air networks comprising UAVs and HAPs such as HABs. Hybrid network, which is illustrated in Figure 12.3, can pave the way for many applications including Earth observation and secure military missions. In [57], a space-based network architecture containing hybrid links between spacecraft and satellites for Earth observation purposes has been discussed. In these links, THz communication can also be employed. For military functions such as reconnaissance, surveillance, and patrol, UAV and HABs are being used. Many single UAV networks, which are already in service for almost 30 years, use satellite communications [58]. High atmospheric attenuation can create an advantage regarding secure communication links for these military and defense applications. THz waves of orbital links cannot penetrate the troposphere so that interception or disruption of the airborne links, such as satellites-to-HAB or UAVs, from the ground is not possible [39]. Moreover, terrestrial spectral noise will be minimal since THz waves from the Earth cannot penetrate space too. These eliminate the chances of interference, jamming, and eavesdropping.

12.6 Challenges

In addition to the characteristic challenges of THz communication, the challenges in THz ISLs include the following subjects.

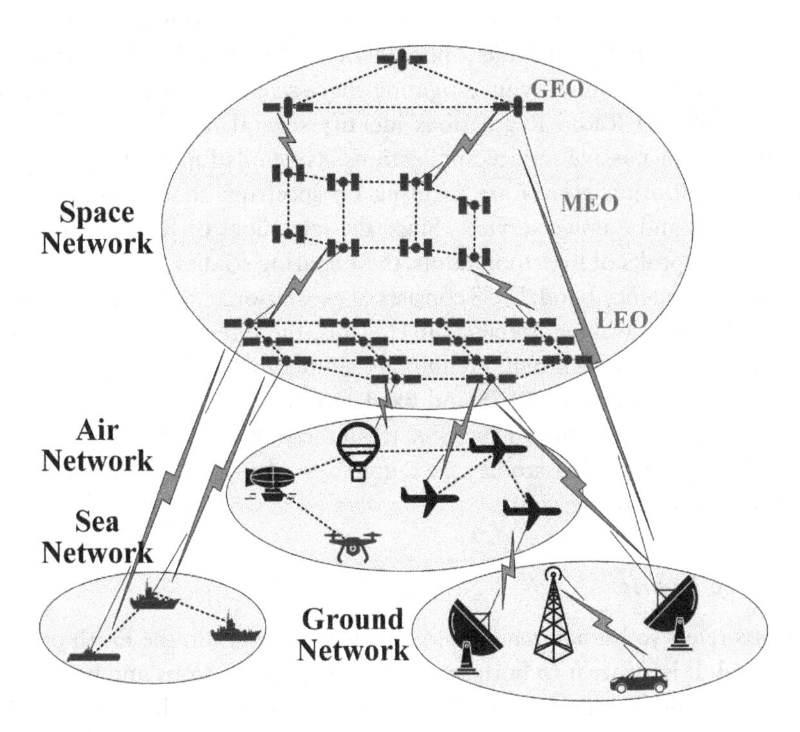

Figure 12.3 Hybrid network.

12.6.1 High Path Loss

High propagation loss is a major limitation for the THz communications. Due to higher f_c, amplified spreading loss restricts the THz communication into few tens of meters. To overcome this limitation for terrestrial networks, the authors in [59] introduce several solutions including distance-aware PHY design by considering the strong relationship between the distance and transmission windows, UM MIMO communications to augment the communication distance and reflect arrays to enhance signal coverage when line-of-sight is not available. Moreover, software-defined metamaterials controlling the propagation environment for terrestrial networks have been proposed to overcome the distance limitation in the THz band [60]. Yet, further research on the methods for combating high spreading loss is still necessary regarding ISLs. Another possible solution is benefiting from the radio astronomy implementations where large-aperture THz optics are employed, two exemplary systems of which are the 12.5 m Atacama Large Millimeter/Submillimeter Array (ALMA) in San Pedro de Atacama, Chile, and 3.5 m Herschel Space Observatory. The authors in [40] have shown that data rates of more than 1 Tbps is possible for a ground-to-GEO satellite link, where the ground station is in the dry regions of Earth. They have further shown that aircraft and balloons with smaller apertures can even exceed 1 Tbps.

12.6.2 Interference

Earth exploration-satellite service (EESS) and radio astronomy service (RAS) use several bands in the 275–1000 GHz range. The International Telecommunication Union (ITU) identifies 275–1000 GHz to be potentially used by active communication services, even though this spectrum is still largely unregulated. On other hand, to protect the activities of the scientific community investigating space science from the interference of active services, current ITU Radio Regulations identify several frequency bands between 275 and 1,000 GHz for passive service applications, as detailed in the footnote 5.565. In this respect, standardization efforts are focusing on spectrum sharing studies between THz communications and passive services. Since the telescopes of RAS are located in remote areas such as the peaks of high mountains, these sharing studies are targeted to EESS in the 275–450 GHz frequency band. EESS consists of geostationary and non-geostationary orbit satellites with down-looking antennas [61]. Thus, the interference from fixed communication services can be substantial. Preliminary simulation results reveal that the following bands can be shared between EESS and fixed services: 275–296, 306–313, 319–333 and 354–450 GHz [62]. Regarding THz ISLs, the interference from both EESS and RAS can be critical. Therefore, further studies are required to identify the bands that can be shared by THz ISLs and passive services.

12.6.3 Space Debris

Space debris refers to human-made objects that are present in the Earth orbit and mostly nonfunctional. It is a threat to both satellites and space missions and pose a challenge for THz communication. Reflection and scatterings result in non-line-of-sight (NLoS) communication paths. Due to space debris, multipath components can be observed at the RX. As of February 2020, 34,000 objects larger than 10 cm, 900,000 objects between 1 and 10 cm, and 128 million objects ranging from greater than 0.1 to 1 cm are estimated to be in the Earth's orbit [63]. Thus, geometric scattering occurs due to these objects. In [30], the author simulates the effect of space debris on the EM wave propagation in THz ISLs for a simple network topology, and shows that space debris can result in intersymbol interference. On the other hand, strong scattering effect of space debris can also be used to compensate the losses at RX, since any of the NLoS paths may behave as the dominant path in THz frequencies [64].

12.6.4 Doppler Spread

Relative movement between a TX and RX results in Doppler shift in the f_c of the received signal. If the RX is moving toward the TX with a speed of v, the maximum Doppler shift is calculated by $f_D = v/\lambda$, where $\lambda = c/f_c$ is the signal wavelength. If the relative speeds of satellites are high, the impact of the Doppler shift on the THz ISLs can be severe. At 300 GHz f_c, the Doppler spread reaches 10 MHz for a v of 10 km/s, whereas the coherence time, in which the channel impulse response remains unvaried, is 0.018 μs [65]. The symbol period can be chosen to be less than the coherence time to prevent fast fading or the bandwidth can be adjusted accordingly. In any case, the Doppler shift should be estimated and compensated at the RX. However, Doppler shift estimation approaches for traditional wireless systems do not produce satisfactory performance for the high-speed systems with high f_c [66]. Therefore, new estimation methods and compensation techniques are needed for satellite systems operating at THz frequencies.

12.6.5 Device Linearity

System linearity is an important criterion to support a wide range of power levels in satellite communications. Due to only newly emergent THz device technologies, the state-of-the-art THz TRXs based on complementary metal oxide semiconductor (CMOS) have several drawbacks including non-linearity of devices. CMOS technology is preferred because it can offer low-cost solutions. However, high-frequency operation is still a challenge for CMOS devices. Linear amplification is possible up to only about 300 GHz [67]. Since the cut-off frequency for transistor or maximum oscillation frequency is higher than the target frequency, the non-linearity of devices has to be utilized at THz frequencies, yet optimizing these devices increases the complexity of the designs [68]. Moreover, it degrades the power efficiency. Higher order modulations such as quadrature amplitude modulation also necessitate linearity.

12.7 Conclusions

THz band can pave the way for new opportunities by enabling huge bandwidth regarding THz ISLs. Satellite networks, which are one of the study items of 3GPP, can be employed as a part of 5G connectivity to provide ubiquitous coverage. Moreover, the constellation of small satellites with THz communication capability can enable the IoST concept. Large-aperture THz optics such as the ALMA in San Pedro de Atacama, Chile are already being used by radio astronomy, which has the potential to drive THz ISLs. On the other hand, there are major challenges that need to be addressed such as high-power TXs to overcome the high spreading loss occurring between satellites.

Acknowledgment

This work was supported in part by the Huawei Graduate Research Scholarship.

Notes

[1] A. C. Clarke. Extraterrestrial Relays. *Wireless World*, 51(10):305–308, 1945.

[2] Z. Sodnik, B. Furch, and H. Lutz. Optical Intersatellite Communication. *IEEE Journal of Selected Topics in Quantum Electronics*, 16(5):1051–1057, 2010.

[3] P. Wang, J. Zhang, X. Zhang, Z. Yan, B. G. Evans, and W. Wang. Convergence of Satellite and Terrestrial Networks: A Comprehensive Survey. *IEEE Access*, 8:5550–5588, 2020.

[4] O. Kodheli, E. Lagunas, N. Maturo et al. Satellite Communications in the New Space Era: A Survey and Future Challenges. *IEEE Communications Surveys & Tutorials*, 2020. doi: 10.1109/COMST.2020.3028247.

[5] I. F. Akyildiz, J. M. Jornet, and S. Nie. A new CubeSat design with Reconfigurable Multi-Band Radios for Dynamic Spectrum Satellite Communication Networks. *Ad Hoc Networks*, 86:166–178, 2019.

[6] I. F. Akyildiz, and J. M. Jornet. The Internet of Nano-Things. *IEEE Wireless Communications*, 17(6):58–63, 2010.

[7] T. Yilmaz, G. Gokkoca, and O. B. Akan. *Millimetre Wave Communication for 5G IoT Applications*, pp. 37–53. Springer International Publishing, Cham, Switzerland, 2016. doi:10.1007/978-3-319-30913-2_3.

[8] T. Yilmaz, and O. B. Akan. On the Use of the Millimeter Wave and Low Terahertz Bands for Internet of Things. In *Proc. IEEE 2nd World Forum on Internet of Things (WF-IoT)*, pp. 177–180, 2015.

[9] S. Mumtaz, J. M. Jornet, J. Aulin, W. H. Gerstacker, X. Dong, and B. Ai. Terahertz Communication for Vehicular Networks. *IEEE Transactions on Vehicular Technology*, 66(7):5617–5625, 2017.

[10] F. Lemic, S. Abadal, W. Tavernier et al. Survey on Terahertz Nanocommunication and Networking: A Top-Down Perspective. *arXiv preprint arXiv:1909.05703v2*, 2020.

[11] I. F. Akyildiz, J. M. Jornet, and C. Han. Terahertz Band: Next Frontier for Wireless Communications. *Physical Communication*, 12:16–32, 2014.

[12] T. Yilmaz, and O. B. Akan. State-of-the-Art and Research Challenges for Consumer Wireless Communications at 60 GHz. *IEEE Transactions on Consumer Electronics*, 62(3):216–225, 2016.

[13] H.-J. Song, and T. Nagatsuma. Present and Future of Terahertz Communications. *IEEE Transactions on Terahertz Science and Technology*, 1(1):256–263, 2011.

[14] T. Yilmaz, and O. B. Akan. On the 5G Wireless Communications at the Low Terahertz Band. *arXiv preprint arXiv:1605.02606*, 2016.

[15] A. Hirata, and M. Yaita. Ultrafast Terahertz Wireless Communications Technologies. *IEEE Transactions on Terahertz Science and Technology*, 5(6):1128–1132, 2015.

[16] V. Petrov, A. Pyattaev, D. Moltchanov, and Y. Koucheryavy. Terahertz Band Communications: Applications, Research Challenges, and Standardization Activities. In Proc. 8th International Congress on Ultra Modern Telecommunications and Control Systems and Workshops (ICUMT), pp. 183–190, 2016.

[17] H. Elayan, O. Amin, R. M. Shubair, and M. Alouini. Terahertz Communication: The Opportunities of Wireless Technology Beyond 5G. In Proc. International Conference on Advanced Communication Technologies and Networking (CommNet), pp. 1–5, 2018.

[18] S. Ahearne, N. O'Mahony, N. Boujnah et al. Integrating THz Wireless Communication Links in a Data Centre Network. In Proc. IEEE 2nd 5G World Forum (5GWF), pp. 393–398, 2019.

[19] N. Boujnah, S. Ghafoor, and A. Davy. Modeling and Link Quality Assessment of THz Network Within Data Center. In Proc. European Conference on Networks and Communications (EuCNC), pp. 57–62, 2019.

[20] S. Ghafoor, N. Boujnah, M. H. Rehmani, and A. Davy. MAC Protocols for Terahertz Communication: A Comprehensive Survey. *IEEE Communications Surveys & Tutorials*, 2020. doi:10.1109/COMST.2020.3017393.

[21] K. M. S. Huq, S. A. Busari, J. Rodriguez, V. Frascolla, W. Bazzi, and D. C. Sicker. Terahertz-Enabled Wireless System for Beyond-5G Ultra-Fast Networks: A Brief Survey. *IEEE Network*, 33(4):89–95, 2019.

[22] T. Yilmaz, and O. B. Akan. On the Use of Low Terahertz Band for 5G Indoor Mobile Networks. *Computers & Electrical Engineering*, 48:164–173, 2015.

[23] T. Yilmaz. *Advanced Image Coding Algorithms: Beyond JPEG2000*. MSc thesis, Department of Electronics and Electrical Engineering, University College London, London, 2009.

[24] T. Yilmaz. *On the Use of Low Terahertz Band for Wireless Communications*. Ph.D. dissertation, Department of Electrical and Electronic Engineering, Koç University, Istanbul, Turkey, 2018.

[25] H. Elayan, O. Amin, B. Shihada, R. M. Shubair, and M. Alouini. Terahertz Band: The Last Piece of RF Spectrum Puzzle for Communication Systems. *IEEE Open Journal of the Communications Society*, 1:1–32, 2020.

[26] N. Khalid, T. Yilmaz, and O. B. Akan. Energy-efficient modulation and physical layer design for low terahertz band communication channel in 5G femtocell Internet of Things. *Ad Hoc Networks*, 79:63–71, 2018.

[27] N. Khalid, T. Yilmaz, and O. B. Akan. Energy-efficient Modulation Scheme for THz-band 5G Femtocell Internet of Things. In Proc. International Balkan Conference on Communications and Networking (BalkanCom), 2017.

[28] Z. Chen, X. Ma, B. Zhang et al. A Survey on Terahertz Communications. *China Communications*, 16(2):1–35, 2019.

[29] J. F. O'Hara, S. Ekin, W. Choi, and I. Song. A Perspective on Terahertz Next-Generation Wireless Communications. *Technologies*, 7(2):43, 2019.

[30] P. Hanswal. *Terahertz Communication for Satellite Networks*. Thesis, 2018.
[31] A. Saeed, O. Gurbuz, and M. A. Akkas. Terahertz Communications at Various Atmospheric Altitudes. *Physical Communication*, 41:101113, 2020.
[32] T. Yilmaz, and O. B. Akan. *Millimeter-Wave Communications for 5G Wireless Networks*, pp. 425–440. CRC Press, Boca Raton, FL, 2016. doi:10.1201/b19698-20.
[33] X. Wang, L. Kong, F. Kong et al. Millimeter Wave Communication: A Comprehensive Survey. *IEEE Communications Surveys & Tutorials*, 20(3):1616–1653, 2018.
[34] T. Yilmaz, N. A. Abbasi, and O. B. Akan. *Millimeter-Wave 5G-Enabled Internet of Things*, pp. 163–181. CRC Press, Boca Raton, FL, 2019. doi:10.1201/9780429199820-8.
[35] T. Yilmaz, and O. B. Akan. Utilizing Terahertz Band for Local and Personal Area Wireless Communication Systems. In Proc. IEEE 19th International Workshop on Computer Aided Modeling and Design of Communication Links and Networks (CAMAD), pp. 330–334, 2014.
[36] O. Erturk, and T. Yilmaz. A Hexagonal Grid Based Human Blockage Model for the 5G Low Terahertz Band Communications. In Proc. IEEE 5G World Forum (5GWF), pp. 395––398, 2018.
[37] T. Yilmaz, E. Fadel, and O. B. Akan. Employing 60 GHz ISM band for 5G wireless communications. In Proc. IEEE International Black Sea Conference on Communications and Networking (BlackSeaCom), pp. 77–82, 2014.
[38] H. Kaushal, and G. Kaddoum. Optical Communication in Space: Challenges and Mitigation Techniques. *IEEE Communications Surveys & Tutorials*, 19(1):57–96, 2017.
[39] I. Mehdi, J. Siles, C. P. Chen, and J. M. Jornet. THz Technology for Space Communications. In Proc. Asia-Pacific Microwave Conference (APMC), pp. 76–78, 2018.
[40] J. Y. Suen, M. T. Fang, S. P. Denny, and P. M. Lubin. Modeling of Terabit Geostationary Terahertz Satellite Links From Globally Dry Locations. *IEEE Transactions on Terahertz Science and Technology*, 5(2):299–313, 2015.
[41] M. Saqlain, N. M. Idrees, X. Cao, X. Gao, and X. Yu. Feasibility Analysis of Opto-Electronic THz Earth-Satellite Links in the Low- and Mid-Latitude Regions. *Applied Optics*, 58(25):6762–6769, 2019.
[42] R. Zhen, and C. Han. Link Budget Analysis for Massive-Antenna-Array-Enabled Terahertz Satellite Communications. *Journal of Shanghai Jiaotong University (Science)*, 23(1):20–27, 2018.
[43] H. T. Friis. A note on a simple transmission formula. *Proc. IRE*, 34(5):254–256, 1946.
[44] H. Elayan, R. M. Shubair, J. M. Jornet, and P. Johari. Terahertz Channel Model and Link Budget Analysis for Intrabody Nanoscale Communication. *IEEE Transactions on NanoBioscience*, 16(6):491–503, 2017.
[45] T. Yilmaz, and O. B. Akan. Attenuation Constant Measurements of Clear Glass Samples at the Low Terahertz Band. *Electronics Letters*, 56(25):1423–1425, 2020. DOI: 10.1049/el.2020.1593, IET Digital Library, https://digital-library.theiet.org/content/journals/10.1049/el.2020.1593
[46] *NASA 60 GHz Intersatellite Communication Link Definition Study*. Report, Ford Aerospace & Communications Corp., Palo Alto, CA, 1986.
[47] *Propagation data required for the design of Earth-space systems operating between 20 THz and 375 THz*. Rec. ITU-R P.1621-2, International Telecommunication Union, Geneva, CHE, 2015.
[48] C. A. Balanis. *Antenna Theory: Analysis and Design*. 4th ed. Hoboken, NJ: Wiley. 2016.
[49] A. Straiton. The Absorption and Reradiation of Radio Waves by Oxygen and Water Vapor in the Atmosphere. *IEEE Transactions on Antennas and Propagation*, 23(4):595–597, 1975.
[50] J. Kokkoniemi, J. Lehtomaki, and M. Juntti. A Discussion on Molecular Absorption Noise in the Terahertz Band. *Nano Communication Networks*, 8:35–45, 2016.
[51] R. Singh, and D. Sicker. Beyond 5G: THz Spectrum Futures and Implications for Wireless Communication. In Proc. 30th European Regional International Telecommunication Society Conference, 2019.
[52] I. F. Akyildiz, and J. M. Jornet. Realizing Ultra-Massive MIMO (1024x1024) Communication in the (0.06–10) Terahertz Band. *Nano Communication Networks*, 8:46–54, 2016.

[53] E. Yaacoub, and M.-S. Alouini. A Key 6G Challenge and Opportunity – Connecting the Remaining 4 Billions: A Survey on Rural Connectivity. *arXiv preprint arXiv:1906.11541v1*, 2019.
[54] A. Guidotti, A. Vanelli-Coralli, M. Conti et al. Architectures and Key Technical Challenges for 5G Systems Incorporating Satellites. *IEEE Transactions on Vehicular Technology*, 68(3):2624–2639, 2019.
[55] Z. Qu, G. Zhang, H. Cao, and J. Xie. LEO Satellite Constellation for Internet of Things. *IEEE Access*, 5:18391–18401, 2017.
[56] I. F. Akyildiz, and A. Kak. The Internet of Space Things/CubeSats. *IEEE Network*, 33(5):212–218, 2019.
[57] J. Du, C. Jiang, Q. Guo, M. Guizani, and Y. Ren. Cooperative Earth Observation through Complex Space Information Networks. *IEEE Wireless Communications*, 23(2):136–144, 2016.
[58] X. Cao, P. Yang, M. Alzenad, X. Xi, D. Wu, and H. Yanikomeroglu. Airborne Communication Networks: A Survey. *IEEE Journal on Selected Areas in Communications*, 36(9):1907–1926, 2018.
[59] I. F. Akyildiz, C. Han, and S. Nie. Combating the Distance Problem in the Millimeter Wave and Terahertz Frequency Bands. *IEEE Communications Magazine*, 56(6):102–108, 2018.
[60] C. Liaskos, A. Tsioliaridou, A. Pitsillides et al. Design and Development of Software Defined Metamaterials for Nanonetworks. *IEEE Circuits and Systems Magazine*, 15(4):12–25, 2015.
[61] M. J. Marcus. WRC-19 Issues: Agenda Item 1.15 and the Use of 275–450 GHz. *IEEE Wireless Communications*, 23(6):2–3, 2016.
[62] T. Kurner. Regulatory Aspects of THz Communications and Related Activities towards WRC 2019. In European Conference on Networks and Communications (EuCNC), Special Session 2: Advanced THz Technologies Towards Terabit/s Wireless Communications, 2019.
[63] The European Space Agency. Space debris by the numbers. www.esa.int/Safety_Security/Space_Debris/Space_debris_by_the_numbers (accessed June 1, 2020).
[64] S. Ju, S. H. A. Shah, M. A. Javed et al. Scattering Mechanisms and Modeling for Terahertz Wireless Communications. In Proc. IEEE International Conference on Communications (ICC), pp. 1–7, 2019.
[65] T. S. Rappaport. *Wireless Communications: Principles and Practice*. 2nd ed. Upper Saddle River, NJ: Prentice Hall, 2001.
[66] C. Zhang, G. Wang, M. Jia, R. He, L. Zhou, and B. Ai. Doppler Shift Estimation for Millimeter-Wave Communication Systems on High-Speed Railways. *IEEE Access*, 7:40454–40462, 2018.
[67] K. K. O, W. Choi, Q. Zhong et al. Opening Terahertz for Everyday Applications. *IEEE Communications Magazine*, 57(8):70–76, 2019.
[68] S. M. H. Naghavi, M. T. Taba, R. Han, M. A. Aseeri, A. Cathelin, and E. Afshari. Filling the Gap With Sand: When CMOS reaches THz. *IEEE Solid-State Circuits Magazine*, 11(3):33–42, 2019.

Chapter 13

Terahertz Front End Technology and Deployment for Ultra-High Capacity Links

Quang Trung Le, Roberto Llorente, François Magne, Claudio Paoloni, and Antonio Ramírez

Contents

13.1 Introduction

The 5G today and the 6G tomorrow can be enabled only by availability of wireless networks with multigigabit per second data rate with high density. This needs substantial technological progresses. The terahertz portion of the spectrum (100–1000 GHz) offers wide frequency bands for potentially achieving data rates in the order of hundreds of Gigabits per second [1–15]. However, the major challenge for enabling links at THz frequency is the opposing trend at the increase of the frequency of attenuation and available transmission power. The attenuation increases due to path loss, high atmosphere attenuation, and rain attenuation. The transmission power available from amplifiers decreases at the increase of the frequency at rate of 10 or 20 dB per decade. A second challenge is the availability of affordable wide band microwave monolithic integrated circuits (MMICs) and subsystems for multigigabit per second transmission. In addition, as the frequency, transitions, assembly, and losses increase, non-linearities pose substantial technological issues, due to the short wavelength that defines the dimensions of the parts, resulting in production difficulty and increased costs. An intense research activity has been devoted worldwide to design and fabricate THz wireless links. Numerous wireless systems covering the frequency range from 90 to 400 GHz were reported [16–20]. In most of the cases those systems are limited to short transmission distance, not suitable for outdoor test. However, promising results and achieved data rate are opening new perspectives to the exploitation of the THz spectrum. Presently THz front ends or wireless links are not commercially available yet. The availability of enabling components is a fundamental step to produce affordable THz wireless systems for the development of new network concepts.

The electronic technology is only one aspect of the wireless communications. New network architectures have to be implemented to respond to the needs of different features for the different applications or use cases, such as low latency, high reliability, and long links.

This chapter explores the challenges for electronic components and systems above 100 GHz, describing the state-of-the-art technology for design, fabrication, testing, and deployment of THz links for multigigabit data rate. New network concepts will be discussed to exploit the spectrum above 100 GHz and how these are integrated in a network architecture capable of supporting the 5G radio access networks (RANs) that are designed targeting support for future Beyond 5G (B5G) networks.

The chapter introduces the electronic technology at THz for wireless links in the first three sections. Section 13.2 provides the overview on the status of microwave monolithic integrated circuits at THz frequency. A brief discussion on antennas will highlight their importance for the transmission section of the radio. Section 13.3 is on packaging, transition, and assembly that represent one of the main technology challenges at those frequencies due to the small size of the parts. Section 13.4 describes the challenge for long-range propagation due to the high atmosphere attenuation that has so far prevented a wider use of THz frequencies. The last two sections are focused on novel THz wireless networks and their implementation and future perspectives. Section 13.5 discusses architecture, performance and technology of a new ultra-high capacity layer at D-band (141–148.5 GHz) with links at G-band (275–305 GHz) for supporting data distribution in urban areas. The chapter concludes with Section 13.6 where an analysis of the future concepts beyond the 5G is proposed, highlighting the relevance of the THz spectrum for the implementation of new services and applications.

13.2 MMIC and Antenna Technology for THz Links

The radio is the enabling system for any wireless link. Its structure in principle simply consists of a transmitter and receiver (transceiver). Typically, modems operate at sub-6 GHz frequency, so the radio has to upconvert and downconvert the signal at the link frequency. In Figure 13.1 is shown a typical topology of a radio. The transmitter section includes an upconverter that raises the frequency from the modem to the operation link frequency, a filter to reduce spurious content, a power amplifier to produce a signal with proper level to be transmitted by the antenna. The receiver section includes an antenna connected to a low-noise amplifier to raise the signal to noise ratio (SNR) and then the signal is down converted to sub-6GHz frequency and transferred to the modems. A high-stability local oscillator is needed to drive the mixers, usually a frequency synthesiser.

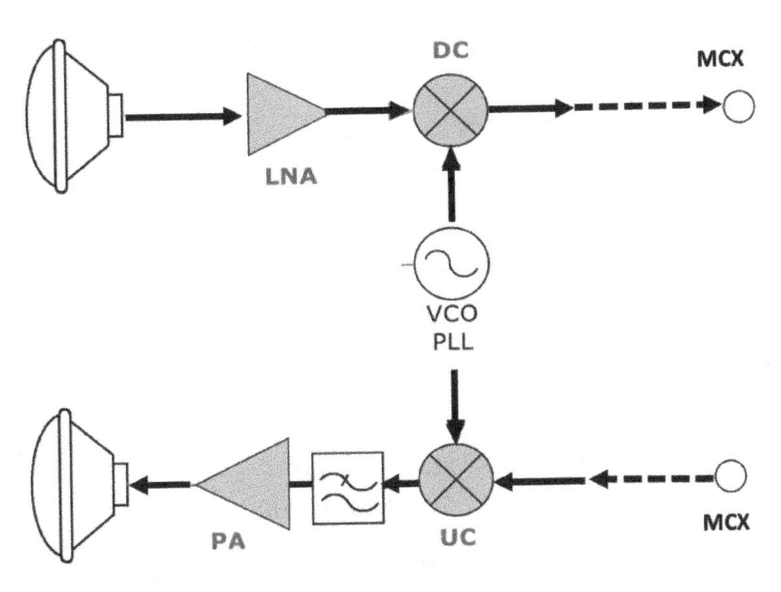

Figure 13.1 Schematic of a radio.

Depending on the difference between the input frequency at sub-6GHz and the link frequency, one or two levels of frequency conversion are needed. The topology of a radio is substantially the same for any operational frequency. The radio topology can be split in a low power section (mixers, LNA) and a high-power section (power amplifier).

Above 100 GHz, the fabrication of a radio is challenging due to the low availability of components and difficult assembly.

Presently a wide range of active devices (high electron mobility transistor [HEMT], heterojunction bipolar transistors [HBT], double HBT, metamorphic HEMT) with different processes (InP, GaAs, GaN, SiGe, complementary metal-oxide-semiconductor CMOS, BiCMOS) have been reported with maximum frequency of oscillation (f_{MAX}) up to 1.5 THz [21]. These devices are used to design and build monolithic integrated circuit (MMICs) with different functions and performance in a wide frequency range, higher than 850 GHz [22]. A wide range of components of the low power section of the radio, such as low-noise amplifiers (LNA), mixers, filters, diplexers, and multipliers has been reported with promising results over multi-gigahertz frequency bands below the 400 GHz, portion of the spectrum suitable for outdoor wireless links [23]. Several initiatives aimed to develop the integrated circuit technology required for THz radio transceivers [24–26]. Nevertheless, this is an ongoing effort that will surely succeed in the years to come.

13.2.1 THz Power Amplifiers

The power amplifier is the most critical component because it determines the range of the link and the SNR to support the desired modulation scheme. The survey in [27] includes a plot of the output power for state-of-the-art power amplifier up to 0.5 THz (Figure 13.2). On average, the power is below 25 dBm. It is notable that the power above 200 GHz reduces below 10 dBm. If this value is compared to tens of watt (more than 40 dBm) at

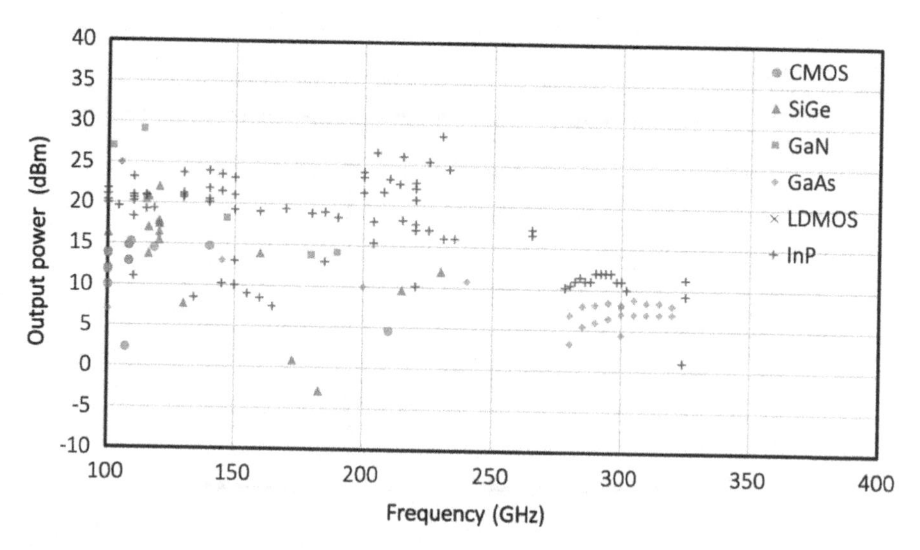

Figure 13.2 Output power vs. frequency for different fabrication processes [27].

microwaves, it gives the sense of the technology gap for establishing long links above 100 GHz.

For wireless applications, in addition to power, requirements for linearity, efficiency and power consumption increase the design challenges. The main problem is that power amplification reduces approximately with the square of the frequency ($1/f^2$). On the contrary, attenuation increases with an increase in frequency. The low transmission power available can be partially compensated by high-gain antennas, but for very high-gain antennas (> 40 dBi) a narrow beam could make difficult the alignment with the receiving antennas.

Both silicon and III–V technologies are used for millimetre wave power amplifiers [27–30]. CMOS process has the advantage of low cost, but the output power is limited [28]. A D-band 16 nm FinFET CMOS power amplifier with a peak gain of 25.6 dB and Psat of 15dBm is reported in [31]. Higher output power can be achieved with III–V MMIC processes. In Figure 13.3 the layout of a 800 nm InP DHBT power amplifier in the band 140–148.5 GHz with 19 dBm output power is shown in [32]. A 35 nm GaAs mHEMT 250 GHz PA is reported in [33], with about 11 dBm output power over 55 GHz bandwidth.

More research is needed to increase the output power of solid state amplifiers. Investigation on the use of gallium nitride (GaN) is in progress. Its high breakdown voltage potentially can provide high power, but above 100 GHz the technology is still in development.

13.2.2 THz Low-Noise Amplifiers

Low-noise amplifiers (LNAs) are important components to improve the SNR of the receiver. Their main parameter is the noise figure, which is also a function of the bandwidth, with impact on data rate for wide band links. The design of LNAs has to balance gain, noise figure and input matching. The best performance has been achieved by using indium phosphide (InP) HEMT and short gate length [34]. The input transistor has to have a very low noise. HEMTs are typically used. A GaAs 140–150 GHz LNA with

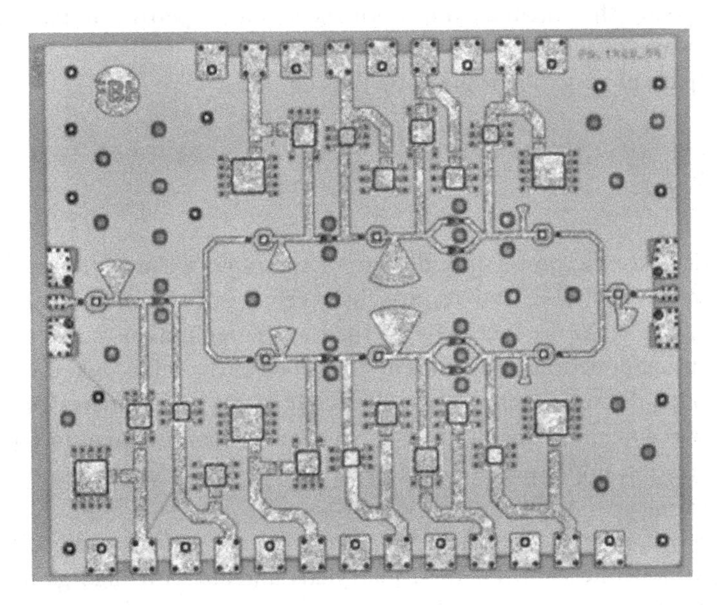

Figure 13.3 D-band DHBT power amplifier [32].

5 dB NF is reported in [28]. A 215–225 GHz LNA in InP HEMT with about 4 dB NF is reported in [35].

13.2.3 Mixers

Downconverter and upconverter are available in a mixer fed by a high-stability local oscillator, usually a frequency synthesiser, to produce by an intermediate frequency (IF) (Figure 13.1). Mixers can be active or passive [36]. Passive mixers can use both diodes either FETs. FETs are preferred for linearity. In general, passive mixer at millimetre waves are preferable since they are less sensitive to terminations. In [32] a downconverting mixer with RF at 141–148.5 GHz and IF at 92–95 GHz is reported.

The availability of complete MMIC chipsets above 100 GHz is still limited, but the growing interest to exploit that portion of the spectrum is fostering a rapid technology development.

13.2.4 THz Antennas

Antennas are key components in THz radio for wireless systems. THz bands operate at extremely short wavelengths, which make the antenna fabrication challenging, but gives the advantage of producing high-gain antennas or antenna arrays with small physical dimensions.

It is important to note that, for antenna systems operating at higher frequencies, considering the same physical size, the antenna will exhibit higher directionality and gain, as given by the equation,

$$G = \frac{A_e 4\pi}{\lambda^2} \tag{13.1}$$

where G stands for the antenna gain, A_e for the antenna aperture and λ is the wavelength of the radio wave.

The free-space propagation is given by Friis equation,

$$P_r = P_t \times \frac{A_{et} A_{er}}{d^2 \lambda^2} \tag{13.2}$$

where P_r and P_t are the power flux density of the radio wave at the receiver and at the transmitter respectively, A_{er} and A_{et} are the receiver and transmitter antenna apertures respectively, and d is the distance between the transmitter and the receiver. Equation (13.2) implies that using directional antennas at both ends of the radio-link, the path loss in free space at a given distance decreases quadratically as frequency increases when the physical size of the antenna is maintained. In this way, radio attenuation and other atmospheric propagation impairments could be effectively overcome when operating at such high frequencies.

Many different antenna configurations are used or are under investigation, such as horn antennas for moderate gain (Figure 13.4a), lens antennas for high gain (Figure 13.4b), planar antennas for compactness. Each of them has specific features and performance. Additive manufacturing and planar processes are the new fabrication approaches for affordable high performance antennas [37].

13.3 Sub-Assembly, Transition and Packaging

Once the chipset is available, it has to be connected to power supply and external circuits. Packaging of THz devices is facing several technological challenges mainly related to signal interconnection [38]. For operating frequencies up to 40 GHz, components are usually available in surface-mount technology (SMT) packages. The core devices are mounted in the package frame and signal pads are wire-bonded to metal leads of pins. Two fundamental issues arise with this signal interconnection technique when the frequency increases: the inductance of the bond wire can no longer be ignored and the size of SMT

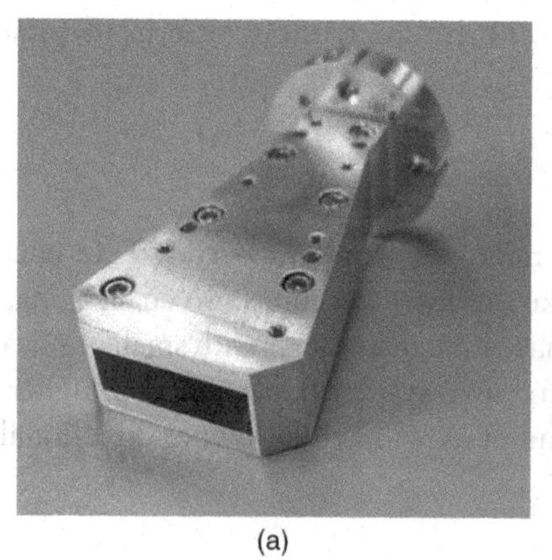

(a)

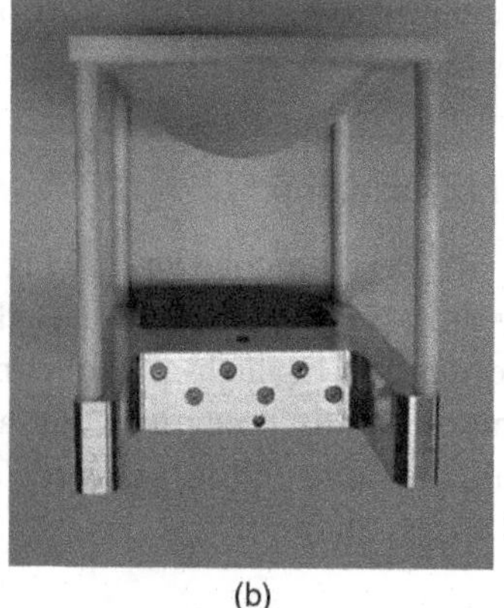

(b)

Figure 13.4 D-band antennas: (a) horn and (b) lens antenna developed in the frame of the ULTRAWAVE H2020 project [58].

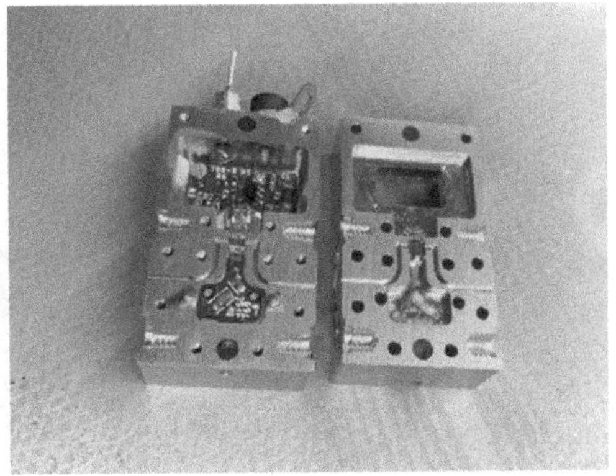

Figure 13.5 A D-band low-noise amplifier module with WR-6 waveguide interfaces developed in the frame of H2020 ULTRAWAVE (courtesy of HF Systems Engineering GmbH).

pads or pins is too large, which leads to the excitation of multimode, radiation and reflection. Development effort is underway to push the practical operating frequency of SMT to the millimetre waves, notably the efforts driven by 77-GHz automotive radar [39,40] and E-band (71–76 GHz and 81–86 GHz) wireless communication [41] industries, with focus on wafer-level chip-scale packaging.

However, with operating frequency beyond 100 GHz, SMT can no longer provide efficient interconnection without the excitation of multimode, radiation and reflection. So far, rectangular hollow metallic waveguide with associated transition has proven to be the most suitable for THz devices, because of its low loss, excellent shielding and durability characteristics. The production of rectangular-hollow-metallic-waveguide-based packages is usually done with split-block technology. The split blocks are made one by one with computer numerical control (CNC) machining, which could offer tolerance as low as a couple of micrometres and surface roughness (Ra) lower that 100 nm. Figure 13.5 shows a D-band (110–170 GHz) low-noise amplifier module with WR-6 waveguide interfaces.

The main drawback of waveguide split block housing is the high production cost incompatible with mass production. Some efforts were invested in inexpensive production technique such as silicon waveguide, 3D-printing, LTCC and plastic moulding technology.

In addition to the fabrication challenge, rectangular-waveguide modules must incorporate a transition from/to the core chip with signal path using microstripline or coplanar waveguide. Transition circuits such as E-plane probes on a separate substrate with a bonding wire in the RF path have been used for millimetre wave frequencies (Figure 13.6). The interconnection from chip to transition is done here again with wire bonding. Above 100 GHz, the wires start to behave as a distributed component. Shortening them can effectively reduce wire inductances. The use of wire bonding has been demonstrated up to 300 GHz [42,43]. The choice of dielectric material for transition is also narrowed down when the frequency increases. Materials such as quartz and GaAs offer low loss and excellent surface roughness, however in order to avoid radiation and excitation of multimode, the substrates must be thinned down to 50 μm for quartz and to 25 μm for GaAs for operation at 300 GHz.

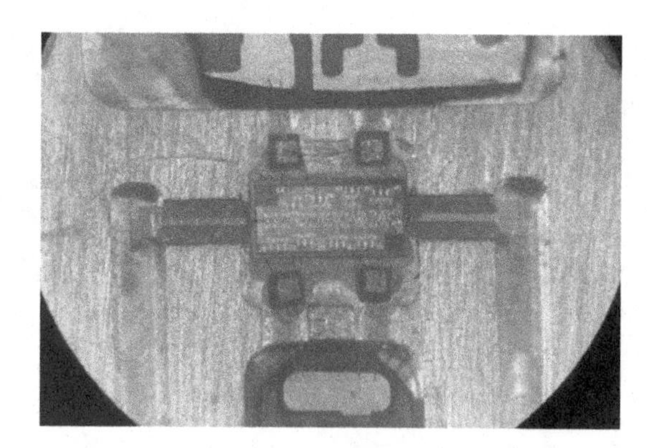

Figure 13.6 Integrated D-band LNA MMIC with 50-μm quartz E-plane probe developed in the frame of H2020 ULTRAWAVE (courtesy of HF Systems Engineering GmbH and OMMIC).

Alternative to wire bonding, flip-chip mounting technique is an attractive technique for interconnecting millimetre wave and THz signals, which uses solder bumps deposited on the chip pads. To be mounted, the chip is flipped top face-down, so that its pads are in contact with the external circuit, and then the solder is reflowed to complete the interconnect. The bump transitions offer smaller intrinsic parasitics compared to wire bonding. There have been demonstrations of the flip-chip bonding technique at frequencies close to 300 GHz [44,45]. The potential use of this approach for 500 GHz [46] interconnection has been proven. The main disadvantage of flip-chip technique is the risk of fracture due to the difference of thermal expansion between the chip and the supporting circuit.

Another approach to avoid the use of wire bonding is to integrate the transitions with the core chip in a single MMIC. The MMICs are mounted in the split blocks so that the transitions is inserted in the rectangular waveguide and aligned to the fields on the E-plane [47]. However, when the chip width is large, the required waveguide opening could cause leakage and result in oscillations. To overcome the issue, it is possible to tailor the chip geometry so that the on-chip transitions are electrically narrow. Such nonrectangular MMIC geometry can be accomplished by chemical etching [48] or laser dicing processes [49]. Nevertheless, the integration of transitions with the core chips inhibits on-wafer testing. The MMICs can only be individually tested after packaging, and the dedicated dicing processes of nonrectangular MMICs are both cost-intensive.

In conclusion, the recent progress in THz devices has enabled the use of THz waves in various applications. However, because of the millimetre and sub-millimetre wavelength we are facing an important problem associated with efficient interconnections and packaging. Metallic split-block rectangular waveguide provides excellent performance. However, the production process using CNC machining is cost-intensive and incompatible with large-scale applications. Silicon machining, low-temperature co-fired ceramic, 3D-printing and plastic moulding are technologies which will potentially reduce the production cost.

13.4 Challenges in the THz Propagation

Propagation of electromagnetic waves are ruled by the Maxwell equations [50] valid for the full spectrum, so in principle no difference at the change of frequency. The free space

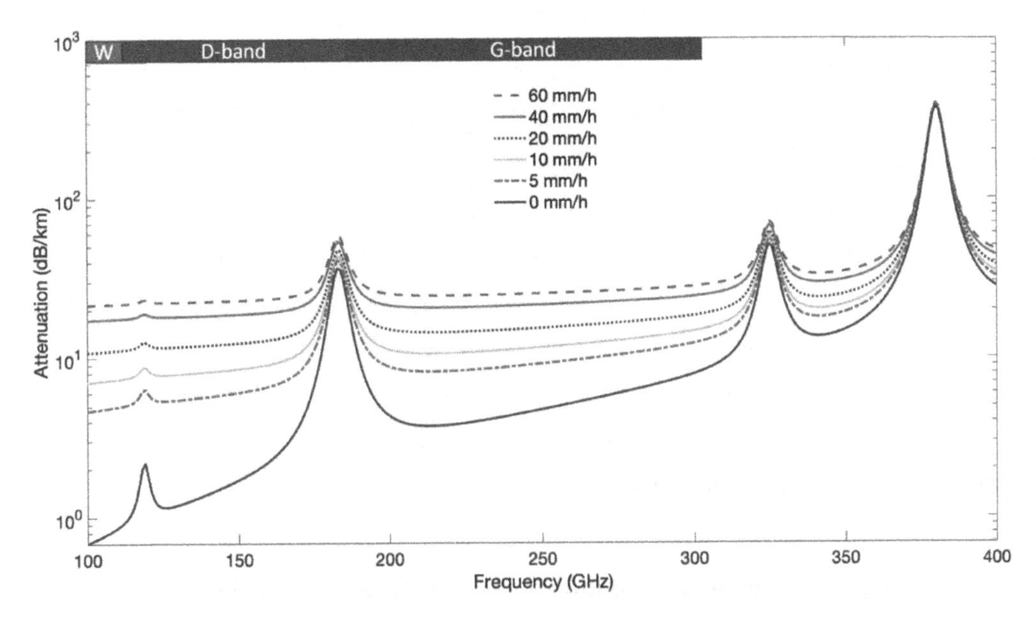

Figure 13.7 **Frequency bands and atmospheric attenuation beyond 100 GHz as a function of rain intensity assuming 10 g/m³ water vapour density.**

path loss (FSPL) depends only on frequency and distance. If the distance remains the same, the FSPL increases at the increase of the frequency. Rain and humidity are random disturbances that are gaining increasing importance with the increase of the frequency (Figure 13.7). The International Telecommunications Union (ITU) provided a map of world divided in a rain zone from A to Q, each of them is characterised by a percentage of time the rainfall intensity exceeds a certain value, corresponding to a given level of attenuation per kilometre. Most of the Western countries lay in the zone H and K with level of attenuation per kilometre at millimetre wave up to 23 dB per kilometre. The gaseous constitution of the atmosphere adds attenuation that increases with the frequency [51,52]. A number of attenuation peaks provoked by the absorbance of specific molecules alternate with region of lower attenuations. Peaks of attenuation are present at about 103, 180, 330, and 380 GHz. However, it is true that those peaks of higher attenuation represent regions to avoid for long-range links, but in reality, no matter the frequency, above 100 GHz rain is the major factor to deal with for ensuring the link availability to satisfy network specifications.

The typical specification of availability of service has to be better than 99.99% of the time. It means that any link has to be dimensioned to work against the maximum attenuation level of the ITU zone.

The main link parameters are the data rate associated to the bandwidth and the SNR. If an increase of attenuation, for example, due to very high rain rate, reduces the SNR and consequently in the supported modulation scheme the link quality is degraded. This does not interrupt the link, but a mechanism to reduce the data rate has to be introduced (automatic control modulation [ACM]) by reducing the modulation scheme order suitable for lower SNR.

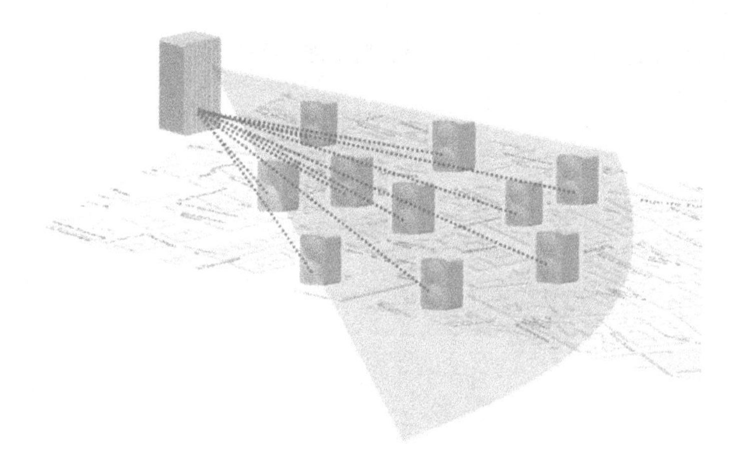

Figure 13.8 Point to multipoint distribution.

In summary, the design of outdoor links above 100 GHz substantially depends on the different components of attenuation in addition to the FSPL.

At microwave, intended below 10–15 GHz, the availability of high-power amplifiers and the intrinsic low attenuation and FSPL make the link quite robust in long range (several kilometres). Differently, at THz frequencies, the low power level from solid state amplifiers, below 1 W around 100 GHz and in the range of tens of milliwatt or less above 100 GHz, limits the useful range of links, even by the use of high-gain antennas [27].

Point-to-point links are available in the market at E-band [53], with Gb/s capability, but they partially cover the needs due to high gain antenna with large footprint, suitable for transport of internet, but not for the capillary backhaul that the 5G needs to support cell densifications. Above E-band, no front end is presently available in the market.

Point-to-multipoint (PmP) distribution is the solution when a high number of links is needed to cover a certain area (Figure 13.8). However, PmP at THz is even more arduous due to the need of low gain antennas to produce a wide beam.

In both the case of THz PtP and PmP, the availability of adequate transmission power, in the watt range, is needed to allow high-capacity long links with multi Gb/s data rate exploiting the available wide frequency bands [54,55].

An emerging solution is the use of travelling wave tubes (TWTs), a high power amplifier, usually adopted in the niche sector of satellite communications [56].

TWTs produce power level up to two orders of magnitude higher than solid state amplifiers at the same frequency. TWTs have a structure fully scalable in frequency, as demonstrated by prototypes realised up to 1 THz (1000 GHz). TWTs at 100 GHz are able to produce tens of watts. About 1 W was demonstrated at G-band (in the range 200–305 GHz) [57]. So far, the production process is not yet mature and still at laboratory level. On that basis, two European Commission Horizon 2020 projects (TWEETHER [58] and ULTRAWAVE [59]) propose for the first time the use of TWTs as power amplifiers for transmission hub in PmP at W-band (92–95 GHz) [60] and D-band (141–148.5 GHz) and for PtP G-band (275–305) high transport capacity links.

At W-band a 92–95 GHz TWT was designed and realised, with 40 W saturated power [61]. At D-band, a 141–148.5 GHz TWT is in fabrication phase to produce about 10–12 W saturated power [62].

To note that the reported best D-band solid state power amplifier delivers around 100 mW (20 dBm) [27].

The exploitation of the spectrum above 100 GHz is progressing, and the electronic technology is in development. The next step is a production strategy to bring on the market affordable and high specifications THz front end.

13.5 Network Design and Performances for Ultracapacity Distribution

The previous sections described the technology development at THz frequency for enabling wireless links. Ultra-high capacity networks are expected to play an important role in our society and in the industrial framework supporting it. The use of new spectrum bands permits the transmission of larger bandwidth signals, enabling higher bitrate services, and also enabling latency reduction by the use of hybrid analogue–digital schemes with reduced digital signal processing associated with data compression, error correction and wireless modulation, which are placed close to the final user in a small-cell scenario.

In the following, the European Commission Horizon 2020 ULTRAWAVE 'Ultra capacity wireless layer beyond 100 GHz based on millimetre wave Travelling Wave Tubes' project will be described as example of novel network solutions for ultracapacity at THz frequencies [57,58]. ULTRAWAVE develops a network architecture based on two THz systems operating above 100 GHz: a point to multipoint (PmP) wireless distribution network at D-band (141–148.5 GHz), and a point-to-point (PtP) link at G-band (275–305 GHz) [57]. The PmP D-band systems will implement the backhaul of small cell sites, while the G-band links will transport the aggregated capacity of the D-band sectors toward the operator's core network. The combination of the two THz systems allows flexible pay-as-you-grow deployments of end-to-end IP transport networks to backhaul 4G and 5G small cells as well as providing fixed wireless access in dense urban and suburban areas. Figure 13.9 illustrates a scenario where a 5G operator provides wireless backhaul to a

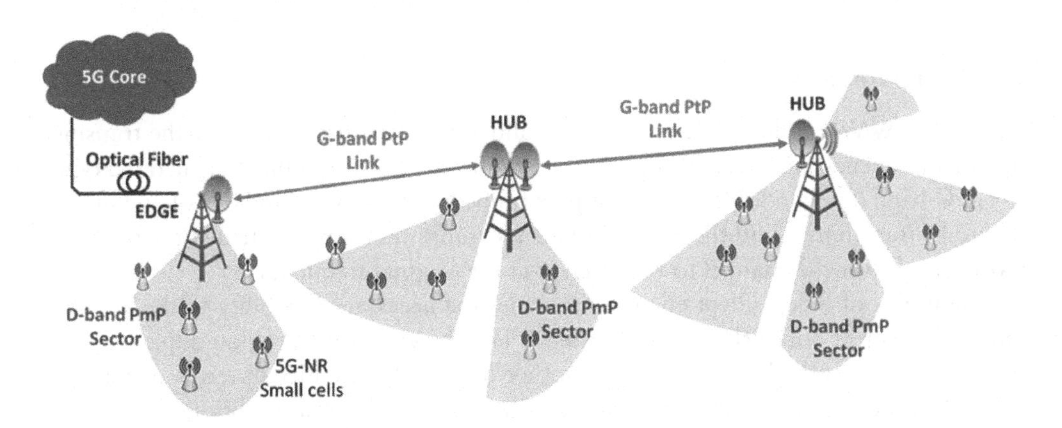

Figure 13.9 ULTRAWAVE architecture example with two concatenated G-band PtP links and seven D-band sectors providing wireless backhaul to different 5G-NR small cells.

wide number of small cells employing seven different D-band PmP sectors, and a concatenation of two G-band PtP links carries the aggregated traffic to a fibre point of presence connected to the 5G core.

In the following sections the specifications, system architecture and performances of ULTRAWAVE system are described.

13.5.1 Specifications

ULTRAWAVE system aims at exploiting frequency bands above 100 GHz to distribute and transport traffic to and from 4G and 5G small cells in scenarios with a high density of eNB (evolved NodeB, typical of 4G) and gNB (gNodeB, typical of 5G) units and unsuitable or unaffordable optical fibre deployments [63]. Around 30 GHz of spectrum are available both in D-band and in G-band. The high attenuation and line of sight (LOS) requirement to operate at those bands makes possible a high reutilisation of the extensive amount of available spectrum. In Europe, the regulation states that the available spectrum for fixed service applications in D-band is fragmented in four different blocks between 130 GHz and 174.8 GHz (130–134 GHz, 141–148.5 GHz, 151.5–164 GHz and 167–174.8 GHz) [64]. In G-band, a continuous block comprised between 275 and 305 GHz is targeted for fixed service.

The enabling technologies to achieve significant operating distances with the required availability at those frequency bands have been described in previous sections as well as in [65].

The targeted functionalities and performances expected are the following:

- Distribution and transport of relevant capacity: 30 Gbps
- Adaptation to the various topologies that operators would deploy considering the site acquisition constraints and their legacy
- Efficiency to backhaul the main applications classes: extreme mobile broadband (eMBB), ultra-reliable low latency communications (URLLC) and massive machine-type communication (mMTC).
- Minimum latency impact
- Load balancing: between application classes, DL/UL ratio and small cell load.

13.5.2 Architecture

The ULTRAWAVE architecture is based an innovative data distribution by the transmission of a frequency multiplex of time division duplexing (TDD) channels, generated typically at C-band (about 5 GHz). The multiplex is upconverted to D-band or G-band in each case to be transmitted, and the inverse frequency translation is performed in the reception process to deliver the channel to the receiver at the original frequency.

A multiplex of independent channels provides the necessary flexibility to the architecture to implement both PtP and PmP systems. The choice of time division duplexing over frequency division duplexing provides another degree of flexibility in terms of the variation of the downlink/uplink traffic ratio depending on each cell's conditions.

In the D-band PmP system, a D-band terminal receives and demodulates each of the channels of the multiplex serving a small cell site. In the G-band PtP link, each of

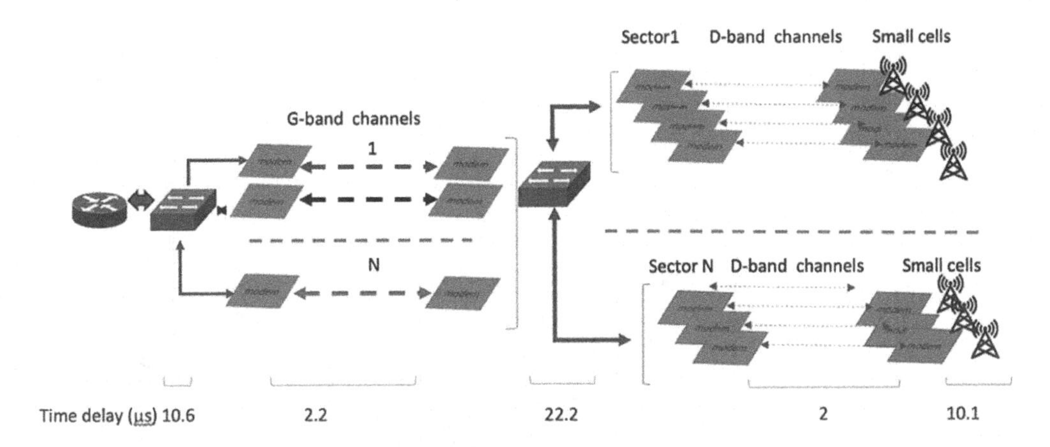

Figure 13.10 ULTRAWAVE architecture at modems stacks and switching level. On the left, the PtP G-band link is composed of *N* high speed channels. Each of the *N* channels is associated with a D-band sector (on the right), dividing its capacity among *n* D-band terminals per sector.

multiplexed channels will have enough capacity to aggregate the capacity of a D-band sector, therefore requiring higher speed modems.

Figure 13.10 illustrates the described architecture. *N* high-capacity channels in the G-band PtP link, with *N* from 1 to 12, feed data to *N* different D-band transmission hubs that serve *n* D-band Terminals each, with *n* comprises between 1 to 10, depending on the capacity and range each D-band sector has to provide. At the interconnection of both systems, the required capacity for the PmP hubs is switched to the D-band systems. A cascaded PtP G-band link can transport any of the *N* channels to a different location where additional D-band sectors could be deployed. The difference between the modems employed in the G-band PtP and the ones employed in the D-band PmP is that the capacity of the G-band channels needs to be significantly larger.

As for modem choices for this type of architecture, the D-band PmP sectors could be implemented with IEEE 802.11ac (Wi-Fi 5) modems arranged between 5 and 6 GHz (C-band). In the case of the G-band PtP links, the choice would lie on IEEE 802.11ad modems (60 GHz). However, this approach could be revised to meet the channelisation imposed by spectrum regulations and achieve higher bandwidth efficiency through custom high-speed modems. Figure 13.11 provides an example of block diagram of the implementation of the D-band PmP system with three IEEE 802.11ac modems [65].

13.5.3 Capacity

The global capacity of both the systems is given by the achievable link budget enabling the use of modulation schemes high enough to support the required throughput. The capacity has been simulated for several link ranges and ITU areas (rain absorption); for instance, in H zone, the capacity of the G-band PtP system varies from 28 Gbps over 700 m to 38 Gbps over 450 m with 99.99% availability. In D-band, at a range of 600 m, ULTRAWAVE system provides around 30 Gbps/km^2 available area capacity [66]. In a public deliverable available on ULTRAWAVE project's website [67], the system performances are described in detail. Tables 13.1 and 13.2 summarise the most relevant aspects of both systems.

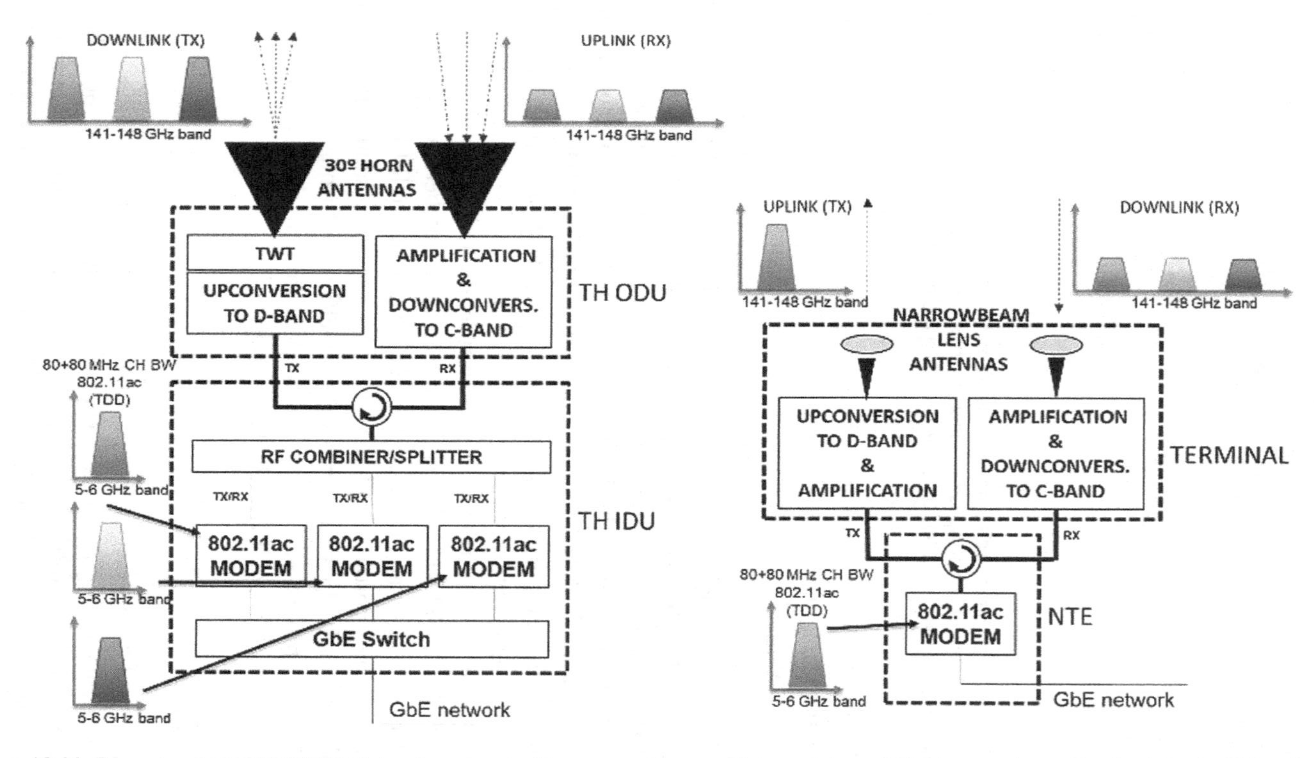

Figure 13.11 Example of ULTRAWAVE D-band system architecture with a multiplex of three 802.11ac modems. On the left, the D-band transmission hub (TH) containing the three modems and the up/down-conversion and amplification stages. On the right, a D-band terminal equipped with one 802.11ac modem [68].

Table 13.1 G-band PtP link performances [57]

G-band PtP Link	
Services, Frequency, Architecture and Performances	
Function	Connects D-band hubs to core network
Frequency band	275–305 GHz
Frequency radio block allocation	Per 10 GHz up to 30GHz
Capacity	Around 30 Gbps (depends on range and number of channels)
Range in K rain area 99.99% availability	600 m
Range in H rain area 99.99% availability	700 m
ACM Margin for 99.999%	+5 dB on modulation, +6 dB on bandwidth
Max throughput per channel	4 Gbps
Modems	
Channels multiplex	Up to 12 channels spaced 2.5 GHz
Channel sizes	440, 880, 1760 MHz
Transmission mode	Dynamic TDD per channel
Waveform standards	802.11ad
Modulations	16QAM-3/4; QPSK-(3/4, 1/2) & BPSK-1/2
Controls	ACM on MCS, channels BW

Table 13.2 D-band PmP backhaul performances [57]

D-band PmP BACKHAUL	
Services, Frequency, Architecture and Performances	
Function	Backhaul of 5G NR small cells
Frequency Band	141–148.5 GHz
Frequency radio block allocation	Per 1GHz up to 8GHz
Architecture PmP	HUB with several sectors providing coverage in different directions
TH sector aperture	30° typical (horn antenna kit 45°, 30°, 15°)
Sector capacity (*n* channels)	$n \times 528$ Mbps effective (FEC and MAC)
Number of terminals connected/sector	*n* up to 12
Range in K rain area 99.99% availability	600 m
Range in H rain area 99.99% availability	650 m
ACM Margin for 99.997%	+5 dB on modulation
Effective throughputs/sector	Up to 5.3Gbps
Modems	
Channels multiplex	Up to 12 channels spaced by 20MHz
Channel sizes	160–80–40 MHz
Transmission mode	Dynamic TDD per channel
Waveform standards	802.11ac wave 2
Modulations	OFDM with variable number of subcarriers; 64QAM-2/3; 16QAM-3/4; QPSK -(4/5, 3/4, 1/2) & BPSK-1/2
ACM control	Increased TX power in lower modulation schemes

13.5.4 Latency

Latency is a key feature for massive machine-type communications (mMTC) and ultra-reliable low latency communication (URLLC) applications. Examples of URLLC applications include remote robot-based medical assistance [69], the connected autonomous car [70] and also highly valued economic segments as high-frequency trading communications [71]. The latency contribution of the elements composing ULTRAWAVE system can be calculated accordingly to the architecture presented on Figure 13.13, where the result of adding the various delays over the chain, is 50 microseconds. Whereas 5G specification for URLLC is about one millisecond, ULTRAWAVE transport and distribution latency contributes for less than 10% of the overall value.

13.6 Future Concepts beyond 5G

Cellular radio networks are a key element of the connected society. Their role is expected to become more and more important in the coming years due to the intensive development of data centre networking as the basis of a cloud-based modern goods and services economy. It is extremely important then to increase the capacity of the cellular network and the associated backbone to match the optical backhaul. Recent studies indicate that cellular networks comprise more than half of the power consumption of the telecommunications operator network [65].

State-of-the-art 3GPP 5G new radio (5G NR) supports the connected society most relevant use cases in three defined profiles. These are: enhanced mobile broadband (eMBB), massive machine-type communications (mMTC) and ultra-reliable low latency communications (URLLC) [72]. In particular, eMBB targets to fulfil the capacity demands of the connected society, URLLC is of special relevance for the transportation sector, where low latency, ultra-reliable connectivity enables autonomous driving, and the mMTC profile is intended to support high connection density scenarios, like autonomous vehicle interaction with the city infrastructure, for example, intelligent lights, signals, pedestrian walks, etc.

Looking to future concepts beyond 5G NR, work has been already done by 3GPP as in the June 2018 Release 16 study [73,74], where the target applications and technologies to be considered in the evolution of 5G (sometimes named 5G-Phase 2 or Beyond-5G, B5G) have been identified. These can be summarised as:

- Vehicular to anything (V2X) communications in support of full autonomous driving
- Enhanced MIMO technologies for massive MIMO antenna systems
- Support for seamless operation in unlicensed bands
- Support for factory automation and other vertical use cases
- Support of frequency bands beyond 52.6 GHz
- Convergence of fixed, wireless and satellite radio services.

To address these targets, system- level specification of Beyond-5G (B5G) wireless, also called 6G cellular, is under development, but some key performance indicators have been already established. These are summarised in Figure 13.12 [75], where an organic evolution to smart remote antenna units (RAUs) capable of providing fibre-like capacity

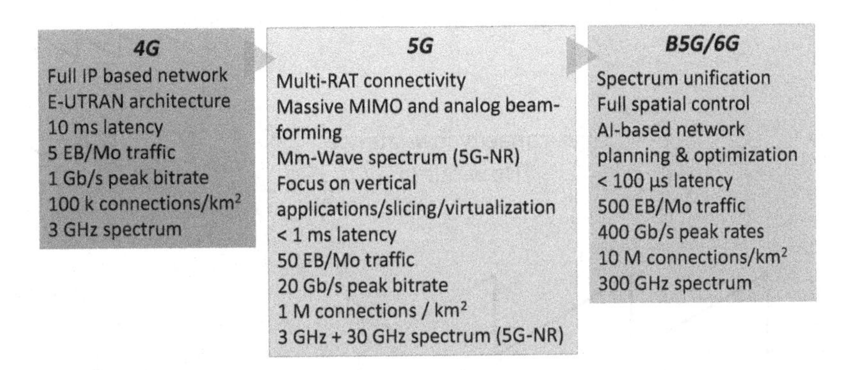

4G
Full IP based network
E-UTRAN architecture
10 ms latency
5 EB/Mo traffic
1 Gb/s peak bitrate
100 k connections/km²
3 GHz spectrum
5G
Multi-RAT connectivity
Massive MIMO and analog beam-forming
Mm-Wave spectrum (5G-NR)
Focus on vertical applications/slicing/virtualization
< 1 ms latency
50 EB/Mo traffic
20 Gb/s peak bitrate
1 M connections / km²
3 GHz + 30 GHz spectrum (5G-NR)
B5G/6G
Spectrum unification
Full spatial control
AI-based network planning & optimization
< 100 µs latency
500 EB/Mo traffic
400 Gb/s peak rates
10 M connections/km²
300 GHz spectrum
E-UTRAN: Evolved UMTS Terrestrial Radio Access Network
RAT: Radio-access technology
MIMO: Multiple-input multiple-output
5G-NR: 5G New-Radio specification
AI: Artificial Intelligence

Figure 13.12 Evolution of generations from 4G to 6G.

operating in the mm-wave and sub-THz range, being scalable enough to seamless operate at medium and higher frequencies is clearly depicted, paving the way for next-generation Beyond-5G (B5G) wireless connectivity. These RAUs should support the key novel concepts introduced in B5G: spectrum unification, full spatial control of radio signals and Artificial Intelligence (machine learning)-based network operation.

Spectrum unification implies that the same baseband digital processing can be applied to wireless signals radiated in different frequency bands from the same cellular base station, thus enabling seamless operation at different frequency bands. Optical up-conversion schemes play a key role here, enabling a flexible generation of the wireless signals [75]. Full spatial control of radio signals also benefits from photonic beamforming architectures which permit multi-band operation [76] [77]. Finally, artificial intelligence based on machine learning algorithms are expected to play an important role controlling data flow in the networks, and also at device-level enhancing the performance of optical and wireless transceivers [78].

13.6.1 The Future of THz Networks

In dense areas 5G already faces big challenges with regard to spectrum management under 3.5 GHz (5G-WiFi6 cohabitation) as interferences, hand-over and mitigation. These issues on access have now extended to X-HAUL: multiplication of wireless threads over the RAN from small cells to the EDGE with huge capacity makes impractical the use of standard microwave links in SHF. Moreover, frequency plans, regulation, footprint, roof-site acquisition, leasing and maintenance become a nightmare for operators. Then mm-wave PtP links offer has flourished in V and E band. Nevertheless, above issues remain with the numerous applications and clients in a city.

Thinking of prospectives beyond 5G, one can see that most applications have been envisioned by 5G, but it is clear that the future RANs will expand and will require more VR in eMBB, more private networks in professional mMTC and strong secure uRLLC for V2X.

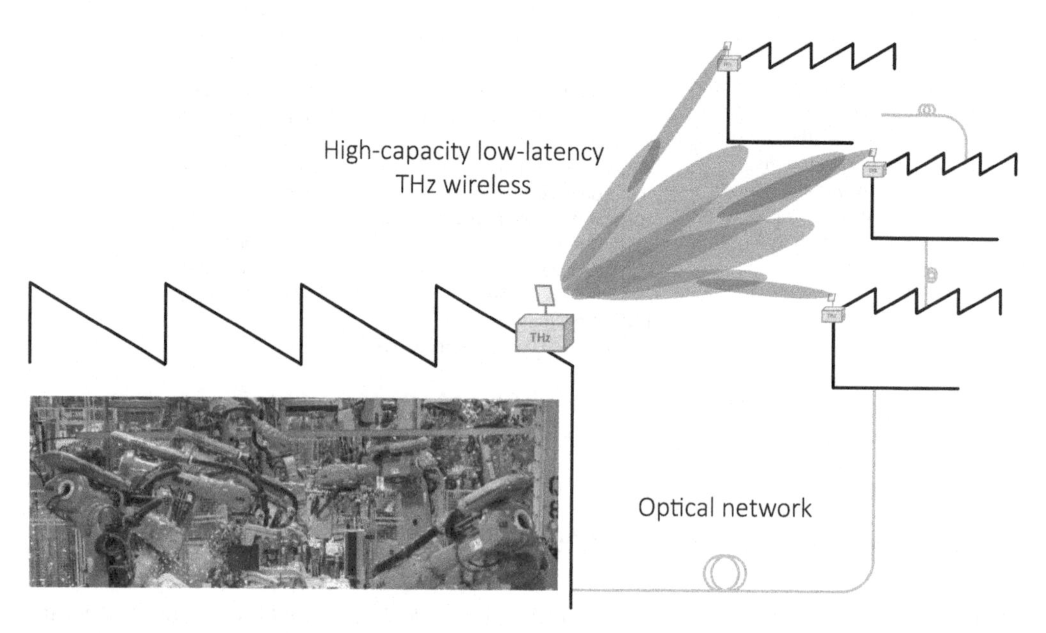

Figure 13.13 Industry 4.0 scenario of multiple links at high capacity.

Then new challenges would appear with robot dissemination: the machine will become a new user requiring secured communication. Another evolution would come from low-altitude UAV (flying robots). UAV applications are presently generating some momentum in transport and inspection.

UAV operation is a fundamental part of the Industry 4.0 paradigm, that is, a fully automatised connected Internet-of-Things factory [79]. Large-scale verticals, such as the automotive industry, comprise several companies operating in proximity in order to quickly react to manufacturing chain requirements or to correct issues detected in almost real time. These closely collaborating industries are usually interconnected by optical fibre-based networks with wireless point-to-multipoint backup radio links, which must match the optical transmission capacity as depicted in Figure 13.13, where THz wireless links in support of the optical infrastructure is depicted. It should be noted that industry verticals as the 5G Automotive Association (5GAA) [80] and the 5G Alliance for Connected Industries and Automation (5GACIA) [81] have become contributors to 5G standardisation organisations, primarily through the Third Generation Partnership Project (3GPP) [82].

Considering these applications, unified 5G networks will certainly develop themselves, in hand of Tier 1 operators but, many ad hoc private networks will locally be deployed by small specialised operators or fibre wholesalers but for the UAV where ATC operators will take the lead for efficiency and security.

Among the advantages of using THz frequencies in addition to capacity, suppleness (Load balancing) and latency, there is an efficient deployment competitive to PtP micro-wave links:

- Huge spectrum, reusable thanks to limited propagation and very low side lobe that lens antenna can provide. Therefore, and particularly in PmP, frequency plan is easy for dozens of mm-Wave threads over some circular areas with some kilometre radius.

- Small footprint with a reduction of 50% of antenna radius comparatively with E-band.
- Low weight of the terminals (most equipment): modem on SoC, photonic direct multiplexing, GaAs and HBT MMICs. All these small components integrated in SIP will provide light and compact equipment easily carried and installed by one person with no visual incidence on the roof.
- Compact TWTs, which have been developed for the hubs and PtP terminals to counter the terahertz drawback due to rain.

13.6.2 Configurations and Optimisation

The density of X-HAUL threads to connect huge number of small cells or machines implies that ULTRAWAVE system can distribute dozens of Gbps on any kind of topology for any type of application class with best optimisation of coverage-capacity and spectrum use. Therefore, deployment design and engineering become essential for efficiency.

Given a set of terminals in a sector the hub at the distribution unit (DU) for fronthaul should be configured to optimise terminal throughputs and global capacity. The configuration of each channel pertains mainly to the frequency and the power of the channel. These two parameters should be settled considering internal properties of the hub (as frequency gain over the bandwidth used) and external contingencies as the distance, the bearing offset, rain attenuation for each terminal and potential jamming of terminals in neighbours' sectors. As an example, to better understand the issue, the power delivered to a terminal would depend on its range, its off set in the sectors' beam and of its local rain attenuation; moreover its frequency should not jeopardise neighbour's links. Optimisation of the links ensemble is possible in ULTRAWAVE thanks to its architecture in which each modem will be configurated through the management software. Such software can be designed upon the complete modelling of the system.

X-HAUL optimisation widely depends on antenna technology.

So far X-HAUL PtP links are using more and more high-gain antennas (Cassegrain type) thanks to alignment tools, but this simple strategy leads to important footprints.

ULTRAWAVE uses a kit of antennas for hubs in order to adapt antenna aperture to the topology of the terminals. Highly directive small-size lenses are well fitted for terminals. Lenses (digitally designed) can be shaped with desired directivity (and particularly for low side lobes).

Evolution of antenna technology in beamforming will greatly foster PmP systems like D-band ULTRAWAVE segment.

The technologies developed in ULTRAWAVE, for example, micro mechanic guides for the TWT and lenses could be combined to obtain multi-beams and phase shifting.

Because of its huge spectrum, terahertz technology, by using ULTRAWAVE architecture, can distribute dozens of Gb/s to high densities of terminals ensuring the X-HAUL of future RANs beyond 5G. The innovation and progress in terahertz technology as in TWT, antennas and management software will fully alleviate the rain drawback.

13.7 Conclusions

The terahertz technology for wireless networks is progressing fast. The THz spectrum provides the greatest opportunity to satisfy the huge increase of wireless traffic with wide bandwidth to support multigigabit per second data rate. However, substantial nodes have to be resolved in terms of technology availability, regulations, transmission power and economy.

The new network concept described in the chapter is one of the most novel approach to respond to the quest of high-capacity wireless. The deployment of this wireless system will solve the X-haul for very dense networks core of 5G and beyond 5G. The future challenges highlight the need for multiple features that for which further research efforts and overcoming of technology limitations are required. Affordable high-performance MMIC chipset, easy assembly and packaging, high linearity and power for power amplifiers and small antennas are the elements for the future of terahertz wireless networks.

Notes

[1] T. S. Rappaport et al., 'Wireless communications and applications above 100 GHz: Opportunities and challenges for 6G and beyond,' in *IEEE Access*, vol. 7, pp. 78729–78757, 2019. doi: 10.1109/ACCESS.2019.2921522

[2] I. F. Akyildiz, J. M. Jornet, and C. Han, 'Terahertz band: Next frontier for wireless communications,' *Physical Communication*, vol. 12, pp. 16–32, Sept. 2014.

[3] S. S. Dhillon, et. al., 'The 2017 terahertz science and technology roadmap', *Journal of Physics D: Applied Physics* 2017, *50*, 043001.

[4] J. Takeuchi et. al. '10-Gbit/s Bi-directional wireless data transmission system using 120-GHz-band ortho-mode transducers,' 2012 IEEE Radio and Wireless Symposium, 2012, pp. 63–66.

[5] R. Taoriand and A. Sridharan. 'Point-to-multipoint in-band mmwave backhaul for 5G networks,', *IEEE Communications Magazine*, pp. 195–201, Jan. 2015.

[6] I. F. Akyildiz, JM. Jornet, and C. Han, 'TeraNets: Ultra-broadband communication networks in the terahertz band,' *IEEE Commun. Mag.*, vol. 21, no. 4, pp. 130–135, Aug. 2014.

[7] H. Elayan, O. Amin, B. Shihada, R. M. Shubair and M. Alouini, 'Terahertz Band: The Last Piece of RF Spectrum Puzzle for Communication Systems,' *IEEE Open Journal of the Communications Society*, vol. 1, pp. 1–32, 2020, doi: 10.1109/OJCOMS.2019.2953633

[8] Z. Zhang et al., '6G wireless networks: Vision, requirements, architecture, and key technologies,' *IEEE Vehicular Technology Mag.*, vol. 14, no. 3, Sept. 2019, pp. 28–41.

[9] E. Calvanese Strinati et al., '6G: The next frontier: From holographic messaging to artificial intelligence using sub- terahertz and visible light communication,' *IEEE Vehicular Technology Mag.*, vol. 14, no. 3, Sept. 2019, pp. 42–50.

[10] Khaled B. Letaief, Wei Chen, Yuanming Shi, Jun Zhang, Ying-Jun Angela Zhang, 'The roadmap to 6G: AI empowered wireless networks', *Communications Magazine IEEE*, vol. 57, no. 8, pp. 84–90, 2019.

[11] N. Rajatheva et al., 'White paper on broadband connectivity in 6G,' June 2020. [Online]. Available: http://urn.fi/urn:isbn:9789526226798

[12] P. Yang, Y. Xiao, M. Xiao, and S. Li, '6G wireless communications: Vision and potential techniques,' *IEEE Network*, vol. 33, no. 4, pp. 70–75, July 2019.

[13] Z. Pi and F. Khan, 'An introduction to millimeter-wave mobile broadband systems,' *IEEE Comm. Magazine*, pp. 101–107, June 2011.

[14] T.S. Rappaport et al., 'Millimeter wave mobile communications for 5G cellular: It will work!', *IEEE Access*, pp. 335–349, May 2013.

[15] T. Ku ̈rner, 'THz communications – An overview and options for IEEE 802 standardization,' IEEE 802.15-18-0516-02-0thz, Nov. 2018.

[16] T. Nagatsuma et al. 'Advances in terahertz communications accelerated by photonics,' *Nature Photonics* 10, 371–379 (2016).

[17] X. Li, et al., 'Fiber-wireless-fiber link for 100-Gb/s PDM-QPSK signal transmission at W-band,' *IEEE Photon. Technol. Lett.*, July 2014.

[18] W. Steyaert and P. Reynaert, 'A 0.54 THz signal generator in 40 nm bulk CMOS with 22 GHz tuning range and integrated planar antenna,' *IEEE J. Solid-State Circuits*, vol. 49, no. 7, pp. 1617–1626, July 2014.

[19] I. Kallfass, F. Boes, T. Messinger, J. Antes, A. Inam, U. Lewark, A. Tessmann, and R. Henneberger, '64 Gbit/s transmission over 850 m fixed wireless link at 240 GHz carrier frequency,' *J. Infrared Millim. Terahertz Waves*, vol. 36, no. 2, pp. 221–233, 2015.

[20] I. Ando, M. Tanio, M. Ito, T. Kuwabara, T. Marumoto, and K. Kunihiro, 'Wireless D-band communication up to 60 Gbit/s with 64QAM using GaAs HEMT technology,' Radio and Wireless Symposium (RWS), 2016 IEEE. IEEE, 2016, pp. 193–195.

[21] X. Mei, W. Yoshida, M. Lange, J. Lee, J. Zhou, P.-H. Liu, K. Leong, A. Zamora, J. Padilla, S. Sarkozy et al., 'First demonstration of amplification at 1 THz using 25-nm InP high electron mobility transistor process,' *IEEE Electron Device Lett.*, vol. 36, no. 4, pp. 327–329, 2015.

[22] K. M. K. H. Leong et al., '850 GHz receiver and transmitter front-ends using InP HEMT,' *IEEE Transactions on Terahertz Science and Technology*, vol. 7, no. 4, pp. 466–475, July 2017, doi: 10.1109/TTHZ.2017.2710632.

[23] A. B. Amado-Rey et al., 'Analysis and development of submillimeter-wave stacked-FET power amplifier MMICs in 35-nm mHEMT technology,' *IEEE Transactions on Terahertz Science and Technology*, vol. 8, no. 3, pp. 357–364, May 2018, doi: 10.1109/TTHZ.2018.2801562.

[24] I. Dan, S. Hisatake, P. Szriftgiser, R. Braun, I. Kallfass and G. Ducournau, 'Towards superheterodyne THz links pumped by photonic local oscillators,' 2019 44th International Conference on Infrared, Millimeter, and Terahertz Waves (IRMMW-THz), Paris, France, 2019, pp. 1–2, doi: 10.1109/IRMMW-THz.2019.8873762.

[25] P. Roux, A. Singh, F. Jorge, M. Moretto and Y. Baeyens, 'Wideband variable gain amplifier for D-band backhaul transceiver in 55nm BiCMOS technology,' 2019 IEEE Asia-Pacific Microwave Conference (APMC), Singapore, Singapore, 2019, pp. 1313–1315, doi: 10.1109/APMC46564.2019.9038683.

[26] I. Dan et al., 'A 300-GHz wireless link employing a photonic transmitter and an active electronic receiver with a transmission bandwidth of 54 GHz,' *IEEE Transactions on Terahertz Science and Technology*, vol. 10, no. 3, pp. 271–281, May 2020, doi: 10.1109/TTHZ.2020.2977331.

[27] H. Wang (2019). Power Amplifiers Performance Survey 2000-Present. [Online]. Available: https://gems.ece.gatech.edu/PA_survey.html

[28] V. Camarchia, R. Quaglia, A. Piacibello, D. P. Nguyen, H. Wang, and A. Pham, 'A review of technologies and design techniques of millimeter-wave power amplifiers,' *IEEE Transactions on Microwave Theory and Techniques*, doi: 10.1109/TMTT.2020.2989792.

[29] L. Zhang, J. Wen, L. Sun, and T. Wu, 'A three stage, fully differential D-band power amplifier,' 2014 12th IEEE International Conference on Solid-State and Integrated Circuit Technology (ICSICT), Guilin, 2014, pp. 1–3, doi: 10.1109/ICSICT.2014.7021512.

[30] R. Cleriti et al., 'D-band LNA using a 40-nm GaAs mHEMT technology,' 2017 12th European Microwave Integrated Circuits Conference (EuMIC), Nuremberg, 2017, pp. 105–108, doi: 10.23919/EuMIC.2017.8230671.

[31] B. Philippe and P. Reynaert, '24.7 A 15 dBm 12.8%-PAE compact D-band power amplifier with two-way power combining in 16 nm FinFET CMOS,' 2020 IEEE International Solid- State Circuits Conference – (ISSCC), San Francisco, CA, USA, 2020, pp. 374–376, doi: 10.1109/ISSCC19947.2020.9062920.

[32] C. Paoloni et al., 'Technology for D-band/G-band ultra capacity layer,' 2019 European Conference on Networks and Communications (EuCNC), Valencia, Spain, 2019, pp. 209–213, doi: 10.1109/EuCNC.2019.8801983.

[33] L. John, A. Tessmann, A. Leuther, P. Neininger, T. Merkle and T. Zwick, 'Broadband 300-GHz Power Amplifier MMICs in InGaAs mHEMT Technology,' *IEEE Transactions on Terahertz Science and Technology*, vol. 10, no. 3, pp. 309–320, May 2020, doi: 10.1109/TTHZ.2020.2965808.

[34] D. Yang, J. Wen, M. He and R. He, 'A D-band monolithic low noise amplifier on InP HEMT technology,' 2018 12th International Symposium on Antennas, Propagation and EM Theory (ISAPE), Hangzhou, China, 2018, pp. 1–4, doi: 10.1109/ISAPE.2018.8634087.

[35] C. M. Cooke et al., 'A 220 GHz dual channel LNA front-end for a direct detection polarimetric receiver,' 2019 IEEE MTT-S International Microwave Symposium (IMS), Boston, MA, 2019, pp. 508–511, doi: 10.1109/MWSYM.2019.8701101.
[36] F. Thome et al., 'Frequency multiplier and mixer MMICs based on a metamorphic HEMT technology including Schottky diodes,' *IEEE Access*, vol. 8, pp. 12697–12712, 2020, doi: 10.1109/ACCESS.2020.2965823.
[37] Rui Xu, Steven Gao, Benito Sanz Izquierdo, Chao Gu, Patrick Reynaert, Alexander Standaert, Gregory J. Gibbons, Wolfgang Bösch, Michael Ernst Gadringer and Dong Li, 'A review of broadband low-cost and high-gain low-terahertz antennas for wireless communications applications,' *IEEE Access*, vol. 8, pp. 57615–57629, 2020. doi: 10.1109/ACCESS.2020.2981393.
[38] L. Devlin, 'The future of mm-wave packaging,' *Microw. J.*, vol. 57, p. 24, Feb. 2014.
[39] H. Knapp et al 'Three-channel 77 GHz automotive radar transmitter in plastic package', Proceedings of the 2012 RFIC Symposium, pp. 119–122.
[40] W. Hartner, M. Fink, G. Haubner, C. Geissler, J. Lodermeyer, M. Niessner, F. Arcioni and M. Wojnowski, 'Reliability and performance of wafer level fan out package for automotive radar,' 2019 International Wafer Level Packaging Conference (IWLPC), San Jose, CA, USA, 2019, pp. 1–11.
[41] K. Tsukashima, M. Kubota, O. Baba, T. Kawasaki, A. Yonamine, T. Tokumitsu and Y. Hasegawa, 'E-band receiver and transmitter modules with simply reflow-soldered 3-D WLCSP MMIC's,' 2013 European Microwave Integrated Circuit Conference, Nuremberg, 2013, pp. 588–591.
[42] A. Tessmann et al., 'A 300 GHz mHEMT amplifier module,' 2009 IEEE International Conference on Indium Phosphide & Related Materials, Newport Beach, CA, 2009, pp. 196–199.
[43] L. Samoska et al., 'Miniature packaging concept for LNAs in the 200–300 GHz range,' 2016 IEEE MTT-S International Microwave Symposium (IMS), San Francisco, CA, 2016, pp. 1–4.
[44] Y. Kawano et al., 'Flip chip assembly for sub-millimeter wave amplifier MMIC on polyimide substrate,' 2014 IEEE MTT-S International Microwave Symposium (IMS2014), Tampa, FL, 2014, pp. 1–4.
[45] S. Monayakul et al., 'Flip-Chip Interconnects for 250 GHz Modules,' *IEEE Microwave and Wireless Components Letters*, vol. 25, no. 6, pp. 358–360, June 2015.
[46] S. Sinha et al., 'Flip-chip approach for 500 GHz broadband interconnects,' *IEEE Transactions on Microwave Theory and Techniques*, vol. 65, no. 4, pp. 1215–1225, April 2017.
[47] W. R. Deal et al., 'Demonstration of a 0.48 THz amplifier module using InP HEMT transistors,' *IEEE Microwave and Wireless Components Letters*, vol. 20, no. 5, pp. 289–291, May 2010.
[48] M. Urteaga et al., 'InP HBT integrated circuit technology for terahertz frequencies,' 2010 IEEE Compound Semiconductor Integrated Circuit Symposium (CSICS), Monterey, CA, 2010, pp. 1–4.
[49] A. Tessmann et al., 'Metamorphic HEMT MMICs and modules operating between 300 and 500 GHz,' *IEEE Journal of Solid-State Circuits*, vol. 46, no. 10, pp. 2193–2202, Oct. 2011.
[50] Robert E. Collin, *Foundations for Microwave Engineering*, 2nd ed., Wiley-IEEE Press, 2001.
[51] Recommendation ITU-R P.838-2, Specification attenuation model for rain for use in prediction methods, 2003.
[52] Recommendation ITU-R P.676: Attenuation by atmospheric gases, 2013.
[53] H. Linpu, Z. Zhigang and Z. Jian, 'A low-cost 10-Gbit/s millimeter-wave wireless link working at E-band,' *Journal of Communications and Information Networks*, vol. 1, no. 2, pp. 109–114, Aug. 2016, doi: 10.11959/j.issn.2096-1081.2016.035.
[54] F. Magne, A. Ramirez, C. Paoloni, 'Millimeter wave point to multipoint for affordable high capacity backhaul of dense cell networks', IEEE Wireless Communications and Networking Conference 2018, WCNC 2018, Barcelona, Spain, April 2018.
[55] Shi, I., Ni, Q., Pervaiz, H., Paoloni, C., 'Modeling and analysis of point-to-multipoint millimeter-wave backhaul networks', *IEEE Transactions on Wireless Communications*, vol. 18, no. 1, pp. 268–285, January 2019.

[56] R. J. Barker et al., *Modern Microwave and Millimeter-Wave Power Electronics*, Wiley-IEEE Press, 2005.
[57] www.ultrawave2020.eu – website [on line].
[58] C. Paoloni et al. "D-band point to multi-point deployment with G-band transport,' European Conference on Networks and Communications 2020, EUCNC 2020 – Online.
[59] R. Basu, L. R. Billa, R. Letizia and C. Paoloni, 'Design of sub-THz traveling wave tubes for high data rate long range wireless links', *Semicond. Sci. Technol.* vol. 33, p. 124009, 2018.
[60] C. Paoloni, F. Magne, F. Andre, J. Willebois, T.L. Quang, X. Begaud, G. Ulisse, V. Krozer, R. Letizia, M. Marilier, A. Ramirez and R. Zimmerman, 'Transmission hub and terminals for point to multipoint W-band TWEETHER system', EUCNC 2018, Ljubljana, June 2018.
[61] C. Paoloni, F. Magne, V. Krozer, R. Letizia, E. Limiti, M. Marilier, S. Boppel, A. Ramírez, B. Vidal, T. Le and R. Zimmerman, 'Technology for D-band/G-band ultra capacity layer', European Conference on Networks and Communications (EuCNC), Valencia (Spain), 18–21 June 2019.
[62] F. André et al., 'Technology, assembly, and test of a w-band traveling wave tube for New 5G high-capacity networks,' *IEEE Transactions on Electron Devices*, vol. 67, no. 7, pp. 2919–2924, July 2020, doi: 10.1109/TED.2020.2993243.
[63] R. Basu, L. Billa, J. Mahadev Rao, R. Letizia, and C. Paoloni, 'Design and fabrication of a D-band traveling wave tube for millimeter wave communications', IRMMW-THZ2019, Paris, Sept. 2019.
[64] ECC Recommendation (18)01 on 'Radio frequency channel/block arrangements for Fixed Service systems operating in the bands 130–134 GHz, 141–148.5 GHz, 151.5–164 GHz and 167–174.8 GHz'.
[65] E. Khorov, A. Kiryanov, A. Lyakhov and G. Bianchi, 'A tutorial on IEEE 802.11ax high efficiency WLANs,' *IEEE Commun. Surv. Tut.*, vol. 21, no. 1, pp. 197–216, 1st Quarter 2019.
[66] C. Paoloni et al., 'Toward 100 Gbps wireless networks enabled by millimeter wave Traveling Wave Tubes,' 2018 IEEE International Vacuum Electronics Conference (IVEC), Monterey, CA, 2018, pp. 417–418, doi: 10.1109/IVEC.2018.8391528.
[67] ULTRAWAVE Deliverable 2.2 'Final System and Components Specifications based on the evolution of the technological processes' http://ultrawave2020.eu/wp-content/uploads/2018/12/D2-2-ULTRAWAVE-Final-system-and-components-specifications-_public.pdf
[68] A. Ramirez et al., 'D-band Point to MultiPoint Wireless Testbed, European Conference on Networks and Communications 2020, EUCNC 2020. Online
[69] C.S. Pattichis, E. Kyriacou, S. Voskarides, M.S. Pattichis, R. Istepanian and C.N. Schizas (2002). 'Wireless telemedicine systems: an overview.' *IEEE Antennas and Propagation Magazine*, vol. 44, no. 2, pp. 143–153.
[70] F. Giust, V. Sciancalepore, D. Sabella, M.C. Filippou, S. Mangiante, W. Featherstone and D. Munaretto (2018). 'Multi-access edge computing: The driver behind the wheel of 5G-connected cars.' *IEEE Communications Standards Magazine*, vol. 2, no. 3, pp. 66–73.
[71] E. Budish, P. Cramton and J. Shim. 'The high-frequency trading arms race: Frequent batch auctions as a market design response.' *The Quarterly Journal of Economics*, vol. 130, no. 4, pp. 1547–1621, 2015.
[72] Siva Muruganathan, et al. 'On the system-level performance of coordinated multi-point transmission schemes in 5G NR deployment scenarios.' 2019 IEEE 90th Vehicular Technology Conference (VTC2019-Fall). IEEE, 2019.
[73] www.3gpp.org/release-16
[74] M. H. Alsharif, et al. 'Green and sustainable cellular base stations: An overview and future research directions', *Energies* 10(5), 587, 2017.
[75] S. Rommel, T. R. Raddo, U. Johannsen, C. Okonkwo and I. Tafur Monroy, 'Beyond 5G - wireless data center connectivity', Photonics West 2019, San Francisco, CA.
[76] M. Beltran, M. Morant, J. Perez, R. Llorente and J. Marti. 'Photonic generation and frequency up-conversion of impulse-radio UWB signals.' In LEOS 2008–21st Annual Meeting of the IEEE Lasers and Electro-Optics Society, pp. 498–499). IEEE, 2008, November.

[77] M. Morant, A. Trinidad, E. Tangdiongga, T. Koonen and R. Llorente, 'Dual-wavelength integrated K-band multi-beamformer operating over 1-km 7-core multicore fiber,' Optical Fiber Communication Conference (OFC) 2020, OSA Technical Digest (Optical Society of America, 2020), paper M4I.4.
[78] J Zhao, Y. Liu and T. Xu. 'Advanced DSP for coherent optical fiber communication.' *Appl. Sci.* vol. 9, p. 4192, 2019.
[79] G. Peralta, M. Iglesias-Urkia, M. Barcelo, R. Gomez, A. Moran and J. Bilbao, 'Fog computing based efficient IoT scheme for the Industry 4.0,' 2017 IEEE International Workshop of Electronics, Control, Measurement, Signals and their Application to Mechatronics (ECMSM), Donostia-San Sebastian, 2017, pp. 1–6, doi: 10.1109/ECMSM.2017.7945879.
[80] 5G Automotive Association (5GAA). Available online: www.5gaa.org
[81] 5G Alliance for Connected Industries and Automation (5GACIA). Available online: www.5gaa.org
[82] Third Generation Partnership Project (3GPP). Available online: www.3gpp.org

Chapter 14

Terahertz Waveguides for Next Generation Communication Network

Needs, Challenges and Perspectives

Kathirvel Nallappan,[*] Hichem Guerboukha,[*] Yang Cao, Guofu Xu, Chahé Nerguizian, Daniel M. Mittleman, and Maksim Skorobogatiy

[*] Equal contribution

Contents

14.1 Introduction

The terahertz (THz) frequency band (0.1–10 THz) is considered by many as the next frontier of ultra-high-speed communication systems [1–4]. Until now, most of the works have been focusing on demonstrating free-space wireless links that take advantage of several modestly low-loss atmospheric transmission windows [5,6]. Wireless THz communications hold many advantages, including convenience in mobility for the end-user, ease in scaling up the network, and flexibility of device interconnectivity. They can be deployed for static (fixed transmitter and receiver) and dynamic (moving transmitter and/or receiver) operations, for both long-term (e.g., workspace networks) and short-term (e.g., live sports event telecasting) applications. Many challenges remain, however. For example, the THz beam's directional nature requires careful alignment of the transmitter and receiver antennas to maintain the reliability of the communication link. This requirement is even more crucial when the line-of-sight link is easily compromised, for example, in complex building architectures, around corners, and between moving objects. Moreover, atmospheric weather conditions such as rain and snow negatively impact the performance and reliability of the wireless link, requiring a constant reevaluation of the transmitted link power. At the same time, this power must be maintained low to minimize the effect of interference to the neighboring wireless THz devices, especially in ultra-dense networks [7]. Compared to radio/microwave communication, the highly directional THz beam is, by itself, a great defense against security threats. However, recent studies have shown that eavesdropping in free-space THz links is still possible [8] and therefore it should be taken into account when establishing secured THz networks.

Given these limitations, short-range THz fiber/waveguide links can provide additional opportunities when designing future THz networks. In this context, it is important to note that THz fibers are not meant to replace optical fiber communication at infrared frequencies. They can, however, strengthen in various ways the performance, reliability, and stability of THz networks by giving them the choice between wired and wireless communication links. Indeed, THz waveguides hold many advantages to some of the challenges mentioned earlier. They can cover complex geometrical paths while offering robust coupling to the receiver and transmitter in both static and dynamic operations. They enclose the THz radiation in a highly controlled propagation environment, thus immunizing the data stream from environmental factors and eavesdropping threats. For very short distances, the fibers outperform wireless-based communications links, which is also efficient in terms of power consumption. Finally, in addition to the simple transport of information, the waveguides can also be remarkable tools for signal processing, beam steering, routing, multiplexing and so on.

In this chapter, we cover the most recent developments of THz waveguides applied to communications. In more than two decades of THz science and technology, THz waveguides have been thoroughly studied in numerous research works, including several excellent comprehensive reviews [9–13]. However, only recently, THz waveguides for communications has been the subject of increased interest [14–18]. This chapter intends to provide an up-to-date state-of-the-art on the topic. We begin by introducing in Section 14.2 a few practical examples where THz fibers can bring additional benefits to wireless links. Then, in Section 14.3, we highlight key parameters to consider when designing THz waveguides for communications, including losses, dispersion, excitation efficiency, and fabrication techniques. In Section 14.4, we then study several types of waveguides developed in the past years. These include metallic waveguides (wire waveguides and parallel plate waveguides) as well as dielectric waveguides (hollow-core, porous, and solid core waveguides). Finally, in Section 14.5, we discuss additional applications of THz waveguides as integrated devices designed to support both wired and wireless communications networks.

14.2 Applications of THz Waveguides/Fibers for Communication

In this section, we present potential applications of THz waveguides within the context of high-speed communication networks. As the deployment of these networks is still in its infancy, it is difficult to give a complete picture of future applications. As such, the following list of examples is by no means exhaustive but can provide general ideas on how THz wireless networks can benefit from the addition of THz waveguides.

14.2.1 Telecommunication Applications

Current wireless technologies require few access points to cover large areas. For example, in workspaces and shopping malls, few wireless routers are necessary to provide internet access to a large and scattered number of moving users. This is mainly due to the use of carrier frequencies below 6 GHz that can cover relatively large areas. In contrast, when increasing the operating frequency to the THz range, the effect of propagation losses becomes higher, and the beam can span shorter distances. Therefore, more access points are necessary to cover the entire space. These access points are generally physically fixed, preferably in the ceilings [19,20]. In this context, THz waveguides can be used to bridge the different access points from one another while ensuring a direct communication link to a single central server. Similar architectures could be used in home and office networks for high-speed wireless internet access.

14.2.2 Data Centers

Data centers contain a collection of high-performance servers, computers, and associated devices that are today predominantly linked together using wired networks. In some scenarios, it can be beneficial to use high-speed wireless data transmission, for example, to reduce the cabling cost and/or to connect distant server rooms [21,22]. THz waves are particularly well suited for the task due to their ability to transfer large amounts of data. However, the use of THz fibers may be unavoidable in certain situations. For example, they can be necessary to route the signals between devices that lack direct line-of-sight communication links.

14.2.3 Vehicular Communication and Distributed Antennas/Sensors

Intelligent vehicles, including self-driving cars, will play a significant role in future transportation systems for both moving passengers and cargo. These autonomous vehicles include a variety of sensors meant to perceive their surroundings and identify safe and optimal paths. In some approaches, the computing brain of this future transportation system is located in a cloud network that is physically separated from the vehicle. The increased number of sensors requires a large volume of data to be transmitted to achieve centimeter-level precision for efficient and safe transportation. In such scenarios, the information transmission between the cloud network and a signal processing unit embedded in the vehicle could be handled by a THz wireless link, but a THz waveguide could be used to establish the links between various on-board sensors and this processing unit [23–25]. Similarly, remote communication in airplanes and naval ships requires several antennas that must be placed at different parts to receive the highly directional THz wireless signals from ground stations or geostationary satellites. THz fibers can thus be used to interface between the antennas and the signal processing unit located deep inside the vehicles.

14.2.4 Transmission of Uncompressed Ultra-High-Definition Videos

Transmission of uncompressed ultra-high-definition videos is one of the most notable applications of THz communications [26,27]. Such uncompressed videos can capture an object's in-depth information at 4K/8K resolutions (and beyond) that can be analyzed in detail using ultra-high zoom features. Traditional wireless systems cannot handle the required amount of data transfer, especially for live streaming. THz wireless communications are thus highly relevant for these large-scale video systems for industrial, entertainment, and security applications. For example, a central room can be remotely connected to high-quality video cameras placed in various fixed places, such as around a stadium for a live sports event, in harsh industrial environments for quality control, or in unmanned aerial vehicles for surveillance purposes. While the traditional wireless THz link can be used for point-to-point communication, the THz fibers could support non-line-of-sight links.

14.2.5 Intra/Inter-Chip Communications

With the growing demand for speed and performance in the electronics industry, the input/output pins cannot keep up with the increased bandwidth requirement. At the junction between traditional microwaves and optical frequencies, THz interconnects hold high promise for better performance by taking advantage of both high-speed electronic devices and low-loss optical channels [27–29]. These THz data channels can be considered as short waveguides, and they can be fabricated using silicon, for example, which is a low-loss material in the THz range while being readily available in the semiconductor industry.

14.3 Challenges and Important Parameters of THz Waveguides

In this section, we detail important challenges when designing THz waveguides for communication applications. In particular, we highlight the need to control the losses, excitation efficiency, dispersion, flexibility, bending losses, and fabrication techniques. These parameters can be summarized in the following equation that relates the electric field at the output $E_{out}(\omega)$ to the input electric field $E_{in}(\omega)$ for a waveguide of length L [12]:

$$E_{out}(\omega) = C_1(\omega)C_2(\omega)E_{in}(\omega)\exp\left(-\alpha_{eff}(\omega)L/2\right)\exp\left(-j\beta_{eff}(\omega)L\right) \tag{14.1}$$

where $\omega = 2\pi\nu$ is the angular frequency and ν the frequency. $C_1(\omega)$ and $C_2(\omega)$ are the input and output coupling coefficients; $\alpha_{eff}(\omega)$ is the effective power absorption losses; $\beta_{eff}(\omega) = \frac{\omega}{c}n_{eff}(\omega)$ is the propagation constant where $n_{eff}(\omega)$ is the effective refractive index and c the speed of light.

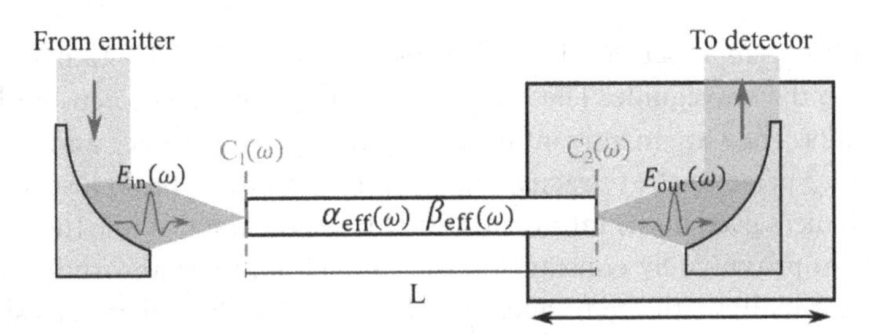

Figure 14.1 The typical experimental setup used to measure the power losses and effective refractive index of a THz waveguide. The input end of the fiber is placed at a focal distance from a parabolic mirror. The output radiation is collected by a movable parabolic mirror that can accommodate different lengths of fibers.

In the THz range, both amplitude and phase of the THz radiation can be measured using particular experimental systems: THz time-domain spectroscopy (THz-TDS) and continuous-wave spectroscopy (CW-THz) [30,31]. The cut-back method is one of the useful techniques used to extract the absorption losses (α_{eff}) and the propagation constant (β_{eff}) from measurements involving these systems [9] (Figure 14.1). Within this method, transmission measurements are performed on two waveguides of lengths L_1 and L_2. Ideally, the same waveguide is used and cut to obtain a measurement for a second length. We are interested in computing the complex transmission ratio of both transmission lengths $E^1_{out}(\omega) / E^2_{out}(\omega)$, where the superscripts indicate the fiber. Because we assume that the same waveguide is used within the same experimental configuration, we can assume that the electric field inputs ($E^1_{in}(\omega) = E^2_{in}(\omega)$) are equal as well as the coupling coefficients ($C^1_1 = C^2_1, C^1_2 = C^2_2$). Therefore, we can write:

$$\frac{E^1_{out}(\omega)}{E^2_{out}(\omega)} = \exp\left[-\frac{\alpha_{eff}}{2}(L_1 - L_2)\right]\exp\left[-j\beta_{eff}(L_1 - L_2)\right] \tag{14.2}$$

from which we can directly compute the power losses

$$\alpha_{eff}(\omega) = -\frac{2}{L_1 - L_2}\ln\left|\frac{E^1_{out}(\omega)}{E^2_{out}(\omega)}\right| \tag{14.3}$$

and the propagation constant

$$\beta_{eff}(\omega) = -\frac{1}{L_1 - L_2}\text{phase}\left[\frac{E^1_{out}(\omega)}{E^2_{out}(\omega)}\right] \tag{14.4}$$

In the following subsections, we study in detail the different parameters presented here and how they specifically impact the data transmission when applied to communications.

14.3.1 Losses

The parameter $\alpha_{\text{eff}}(\omega)$ describes the power losses incurred by THz radiation when propagating through the waveguide. These losses must be minimized to ensure highly efficient waveguide links. They are mainly influenced by the material choice and waveguide geometry. Figure 14.2 presents the refractive index and power absorption losses of conventional dielectric polymers used in the fabrication of THz waveguides. While the refractive indices are generally approximately constant on the THz band, their absorption losses typically increase polynomially with the frequency [32]. Although these materials show the lowest losses in the THz range, they still can exhibit quite significant losses (~ cm^{-1}), thus limiting the maximal propagation length before complete attenuation. Assuming a propagation length L and power absorption α, the output-to-input power ratio is given $\gamma = e^{-\alpha L}$. For example, considering a material with power losses of 1 cm^{-1}, the power drops to half ($\gamma = 0.5$, 3 dB losses) after propagating only ~6.9 mm.

In comparison, dry gases (such as air) show negligible absorption in THz atmospheric transmission windows far from absorption lines [6,33]. Therefore, an important strategy

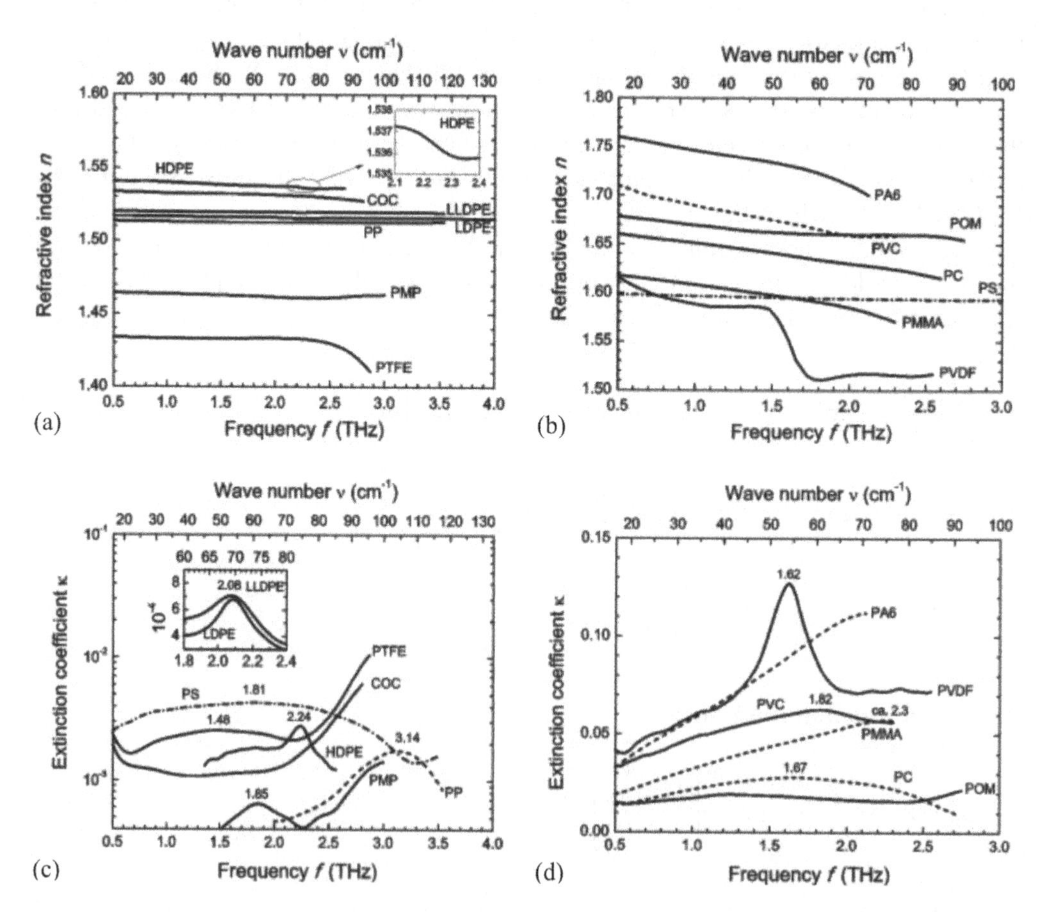

Figure 14.2 (a, b) Refractive index and (c, d) extinction coefficient of common dielectric polymers. The extinction coefficient κ is related to power absorption losses α by $\alpha = 2\kappa / \omega c$. Reprinted from Wietzke S. et al., Terahertz spectroscopy on polymers: A review of morphological studies, *Journal of Molecular Structure*, 1006, 41–51, Copyright (2011), with permission from Elsevier.

used in the design of THz waveguides is to maximize the power guided in air. As we show in Section 14.4, this can be realized using metallic waveguides, subwavelength single-core fibers, hollow-core waveguides, and porous structures.

14.3.2 Excitation Efficiency

The excitation (or coupling) efficiency to (and from) the waveguide is a crucial parameter that dictates the waveguide's ability to transfer energy and therefore the data. In Equation (14.1), $C_1(\omega)$ and $C_2(\omega)$ are the frequency-dependent coupling coefficients, whose values are bounded between 0 and 1, where 0 indicates a null energy transfer, while 1 indicates a complete energy transfer. $C_1(\omega)$ is the input coefficient (from air to the waveguide), while $C_2(\omega)$ is the output coefficient (from waveguide to air). Typically, one wants to increase both coupling coefficients to improve the transmission link.

In general, when light reaches an interface of dissimilar media (different refractive index), a portion of the energy is reflected by the interface while the remainder is transmitted. Fresnel transmission coefficients characterize the amount of energy transferred through the interface. Assuming an interface between medium 1 ($\tilde{n}_1$) and medium 2 ($\tilde{n}_2$) and a normal incidence from medium 1 to 2, the Fresnel reflection ($r_{1\to2}$) and transmission ($t_{1\to2}$) coefficients for the amplitude of the fields are [34]:

$$\begin{aligned} r_{1\to2} &= \frac{\tilde{n}_1 - \tilde{n}_2}{\tilde{n}_1 + \tilde{n}_2} \quad \text{Reflection coefficient (amplitude)} \\ t_{1\to2} &= \frac{2\tilde{n}_1}{\tilde{n}_1 + \tilde{n}_2} \quad \text{Transmission coefficient (amplitude)} \end{aligned} \tag{14.5}$$

where ~ indicates the complex material refractive index: $\tilde{n}_{1,2} = n_{1,2} - jk_{1,2}$. The imaginary part $k_{1,2}$ of the refractive index is related to the power losses $\alpha_{1,2}$ through $k_{1,2} = \alpha_{1,2}c/4\pi\nu$. For example, for coupling from air ($\tilde{n}_1 = \tilde{n}_{air} = 1$) to the waveguide ($\tilde{n}_2 = \tilde{n}_{eff} = n_{eff} - jk_{eff}$), Equation (14.5) reduces to:

$$\begin{aligned} r_{\text{air}\to\text{waveguide}} &= \frac{1 - \tilde{n}_{eff}}{\tilde{n}_{eff} + 1} \quad \text{Reflection coefficient (amplitude, air to waveguide)} \\ t_{\text{air}\to\text{waveguide}} &= \frac{2}{\tilde{n}_{eff} + 1} \quad \text{Transmission coefficient (amplitude, air to waveguide)} \end{aligned} \tag{14.6}$$

In general, the material losses for waveguide applications are intended to be small, such that $n_{eff} \gg k_{eff}$. Thus, one can avoid using complex refractive indices and simply replace $\tilde{n}_{1,2}$ by $n_{1,2}$. Due to energy conservation, the sum of the transferred and reflected energy follows $|r_{1\to2}|^2 + |t_{1\to2}|^2 = 1$. From Equation (14.6), the reflection is minimized (maximized transmission) when $n_{eff} \to 1$, which can be achieved by maximizing the fraction of power propagating in air. On the other hand, when $n_{eff} \to \infty$, the waveguide essentially acts as a mirror and reflects the incident power entirely. For a general formulation considering a non-perpendicular incident beam, see [35]. There, the Fresnel coefficients depend on the polarization of the incident beam. In particular, one can show that the reflection is zero for the parallel polarization at the Brewster's angle $\theta_B = \text{atan}(n_2 / n_1)$. The ends of optical fibers are often cut at this angle to minimize back-reflections.

Under the waveguide mode theory, the coupling coefficients can be computed using a mode-matching method. Knowing the input and waveguide modes, one can obtain numerical solutions and, in some cases, analytical expressions of the coupling coefficients [36]. Consider $\vec{E}_{\text{in,m}}$ and $\vec{H}_{\text{in,m}}$ the electric and magnetic vector fields of the incident beam (index *in*) and the waveguide mode (index *m*). The electromagnetic fields of the two modes must be continuous across the coupling interface (waveguide's input end), thus leading to the following mode-matching integral [36]:

$$C_{\text{m}} = \frac{\iint dxdy\left(\vec{E}^*_{\text{m}} \times \vec{H}_{\text{in}} + \vec{E}_{\text{in}} \times \vec{H}^*_{\text{m}}\right)}{2\sqrt{\iint dxdyRe\left(\vec{E}^*_{\text{m}} \times \vec{H}_{\text{m}}\right)}\sqrt{\iint dxdyRe\left(\vec{E}^*_{\text{in}} \times \vec{H}_{\text{in}}\right)}} \tag{14.7}$$

where C_{m} is the excitation efficiency for the mode m and * denotes the complex conjugate. From Equation (14.7), we deduce that the fields must be geometrically superimposed to ensure high coupling. This is true for the amplitude, but also the polarization. This means that, in general, one needs to make sure that both the input and waveguide modes have vector fields of similar orientations. THz radiation is commonly generated from emitters with linear polarization; therefore, a linearly polarized waveguide mode is preferable.

In a typical numerical experiment, a mode-solver algorithm computes the excited modes of the waveguide, namely the complex effective refractive indices (eigenvalues) and the corresponding electromagnetic fields modal distributions (eigenvectors). Examples of such calculations can be found in [37,38]. There, Equation (14.7) is used to calculate the coupling coefficients, assuming $\left(\vec{E}_{\text{in}}, \vec{H}_{\text{in}}\right)$ to be known. For example, typically, the input mode is assumed to be a linearly polarized Gaussian beam propagating in air and focused using a spherical lens. In this case, since the input is treated as a free-space beam, this approach is known as "quasi-optic" coupling in analogy with the typical treatment of optical waveguides. Assuming that the propagation is in the z direction and that the electric field is polarized in the x direction, the Gaussian input beam can be written as:

$$\vec{E}_{\text{in}}(x, y) = \exp\left[-\frac{x^2 + y^2}{2\sigma^2}\right]\vec{x} \tag{14.8}$$

The corresponding magnetic field can be computed from the electric field using one of Maxwell's equations $\vec{H}_{\text{in}} = \frac{i\mu_0}{\omega}\vec{\nabla} \times \vec{E}_{\text{in}}$, where μ_0 is the vacuum permeability. In Equation (14.8), the parameter σ is the standard deviation of the Gaussian beam width and can be related to its full width at half maximum (FWHM) through $\text{FWHM} = 2\sqrt{2\ln 2}\sigma$. According to Gaussian beam theory [39], σ is related to the wavelength and the focusing optics through

$$\sigma = \frac{\sqrt{2}}{\pi}\frac{F}{D}\lambda \approx 0.4502\frac{F}{D}\lambda \tag{14.9}$$

where F and D are the focal length and diameter of the focusing optics. Therefore, to ensure a high quasi-optic coupling, the waveguide mode must be compatible in size with the input beam by selecting the appropriate focal length, diameter, and wavelength.

14.3.3 Dispersion

Dispersion is another crucial parameter to control while designing fibers for long-distance and high bitrate communications. The origin of dispersion lies in the frequency dependence of the effective refractive index $n_{\text{eff}}(\omega)$. In simple terms, each wavelength contained in a pulse propagates at different velocities inside the waveguide, causing the pulse to spread in time-domain which reducing the THz spectral bandwidth. In turn, this lowers the effective data rate by scrambling the pulse train when adjacent "bits" overlap with each other (Figure 14.3).

The dispersion is typically quantified using a Taylor expansion of the propagation constant $\beta_{\text{eff}} = \frac{\omega}{c} n_{\text{eff}}(\omega)$. The second-order term β_2 is the group velocity dispersion (GVD):

$$\beta_2 = \frac{d^2\beta_{\text{eff}}}{d\omega^2} = \frac{2}{c}\frac{dn_{\text{eff}}}{d\omega} + \frac{\omega}{c}\frac{d^2 n_{\text{eff}}}{d\omega^2} \tag{14.10}$$

When used in the THz range, the GVD is generally expressed in $\text{ps}/(\text{THz}\cdot\text{cm})$, that is ps of time broadening at a given THz frequency and for a given fiber length in cm [40]. A positive β_2 characterizes a waveguide with a normal dispersion (GVD increases with the frequency or red shift), while a negative β_2 indicates an anomalous dispersion (GVD decreases with the frequency or blue shift).

To give a simple idea on how β_2 affects the pulse width, consider a Gaussian pulse in time-domain $E_{\text{in}}(t) \propto \exp(-t^2/2T_0^2)$, where T_0 is the characteristic time width related to the FWHM of the pulse by $T_{\text{FWHM}} = 2\sqrt{\ln 2}T_0$. After propagating through a waveguide of length L and dispersion β_2, the initial Gaussian pulse is transformed into another Gaussian of larger width:

$$T_1 = T_0\sqrt{1 + \frac{L^2}{L_D^2}} \tag{14.11}$$

where $L_D = T_0^2/|\beta_2|$ is defined as the dispersion length. The broadening of the pulse caused by the fiber dispersion can be used to estimate the maximal bit rate. For on–off keying

Figure 14.3 Impact of the dispersion on a pulse train. Pulses are broadened in such a way that adjacent bits overlap with each other. In this example, the second "0" bit is mistakenly interpreted as "1" at the output of the waveguide.

(OOK) and amplitude shift keying (ASK) modulation format with a source of small spectral width [41], the maximal bitrate is:

$$B_{\max} = \frac{1}{4\sqrt{|\beta_2|L}} \tag{14.12}$$

where the factor 4 is selected to ensure that 95% of the pulse energy remains inside the unit bit time slot. For example, considering a THz fiber dispersion of $\beta_2 = 1\,\text{ps}/(\text{THz}\cdot\text{cm})$, a bit rate of 100 Gbit/s can be transmitted up to ~6 m. On the other hand, if $\beta_2 = 0.01\,\text{ps}/\text{THz}$, the same bit rate could be transmitted up to 600 m. Therefore, it is crucial to reduce the fiber dispersion as much as possible. In Section 14.5.1, we present a dispersion compensation waveguide used to compress the pulse train.

Between the regions of normal and anomalous dispersion, there is a zero-dispersion wavelength for which $\beta_2 = 0$. There, the third-order dispersion β_3 in the fiber still occurs and needs to be considered:

$$\beta_3 = \frac{d^3\beta_{\text{eff}}}{d\omega^3} = \frac{3}{c}\frac{d^2n_{\text{eff}}}{d\omega^2} + \frac{\omega}{c}\frac{d^3n_{\text{eff}}}{d\omega^3} \tag{14.13}$$

This means that even when $\beta_2 = 0$, the third-order dispersion may still limit the maximal bitrate to [41]:

$$B_{\max}^{\text{ZD}} = \frac{0.324}{\sqrt[3]{|\beta_3|L}} \tag{14.14}$$

14.3.4 Flexibility, Bending Losses, and Fabrication Considerations

One of the main requirements of THz waveguides is the possibility to reach distant areas despite complex geometries. This requires flexible waveguides with low bending losses. The material and geometry of the waveguide dictate its flexibility while bending losses can be mitigated when the THz radiation is well confined in the core. Moreover, the handling of the waveguide must be simple enough not to compromise the propagating wave's stability. When necessary, one can use low-index low-loss foam as a cladding material to isolate the mode from the environment. Several such examples are shown in Section 14.4.

Furthermore, the waveguide must be relatively easy to manufacture. The conventional fabrication method is fiber drawing, which is borrowed from standard glass fiber technology. A preform (~cm-diameter cylinder) is vertically introduced in a furnace which is heated to a glass transition temperature that is determined by the material. When the preform becomes soft enough, it is pulled down at a constant speed. Several meter-long μm-diameter fibers can be drawn this way. The fiber drawing technique has been extensively used in the THz range. Complex waveguide geometries can also be obtained using that method. For example, drilling holes in the preform and pressurizing them during the drawing process can fabricate fibers with porous cross-section [42]. Two materials in the preform can also be drawn at the same time to realize waveguide with alternating refractive index layers [43] or to realize porous waveguides by removing the second material with a solvent [44]. In another approach, the preform is directly fed into a microstructured extruder to get the desired geometry [45].

Recently, advanced prototyping technologies have been getting more attention to design THz components such as lenses, beam splitters, and waveguides [46]. These techniques can outperform fiber drawing for complex geometries (3D models). The fabricated waveguides can be metalized using wet chemistry or simple metallic spray painting [37]. Stereolithography and fused deposition modeling (FDM) are conventional 3D printing technologies used in the fabrication of THz waveguides. In stereolithography, patterned UV light hardens a liquid photopolymer layer by layer to construct a 3D model [30,37,47,48]. Stereolithography can achieve high resolution (~50 μm) but is limited by the available photopolymers, which have significant losses in the THz range [49]. In fused deposition modeling, a mm-sized filament is heated to a temperature close to its melting point and passes through an extruder nozzle. Mounted on a three-axis stage, this extruder moves to deposit the filament layer by layer to construct the 3D object. The resolution is mainly determined by the extruder nozzle size (~0.5 mm). The most used filaments are acrylonitrile butadiene styrene (ABS) and polylactic acid (PLA). Low-loss THz plastics such as cyclic olefin copolymer (TOPAS) and polypropylene (PP) have also been used to print high-performance THz waveguides. However, in both stereolithography and conventional fused deposition modeling, the length of fiber is limited to a few tens of centimeters due to the limited printing volume. Recently, Xu et al. showed that polypropylene waveguides of infinite length could be 3D-printed using a specially designed 3D printer (BlackBelt, Figure 14.4) [50].

14.4 Types of Waveguides

As we saw in the previous section, parameters such as transmission losses, dispersion, excitation efficiency, and ease of handling play significant roles in the design of efficient THz fiber communication links. While the fiber loss and coupling efficiency limit the communication link distance, the maximum achievable bit rate can be reduced by the fiber dispersion. Therefore, low transmission loss and low dispersion are the primary concerns when

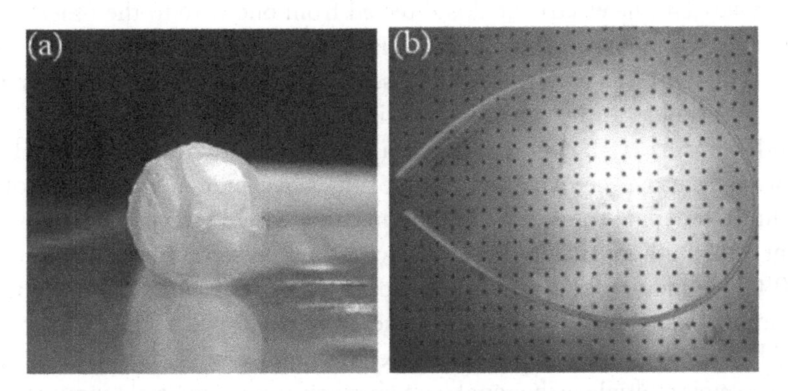

Figure 14.4 (a) Cross-section and (b) top view of the "infinite" 3D printed wagon wheel fiber (2 m is shown here) using BlackBelt printer.

designing THz fibers. In this section, we review several types of existing THz waveguides. We cover both metallic and dielectric waveguides.

14.4.1 Metallic Waveguides

In metallic THz waveguides, THz radiation propagates at the boundary between a metal (support) and a dielectric (propagation medium). Generally, the air is selected as the dielectric for its low loss and low dispersion properties. In the following, we focus on the single-wire and two-wire waveguides as well as the parallel-plate metallic waveguide.

14.4.1.1 Single-Wire and Two-Wire Waveguides

The single-wire waveguide is the simplest form of transmission line that guides electromagnetic radiation as surface waves. As the guided mode propagates in air, there is virtually no dispersion, and the attenuation is mainly caused by metallic ohmic losses [51]. In practice, however, efficient excitation is difficult, since the fundamental mode is radially polarized with an electric field perpendicular to the transverse section of the bare metal wire (Figure 14.5a). This polarization is incompatible with commonly used THz sources that tend to produce linearly polarized THz light. To match the guided mode symmetry, Jeon et al. and Deibel et al. proposed to use a radially symmetric photoconductive antenna as the THz source, achieving excitation efficiencies higher than 50% [52,53] (Figure 14.5b). Another proposed approach uses a tapered coplanar waveguide at the input end [54]. Additionally, due to the delocalization of the field around the wire, any disturbance caused by nearby objects and micro/macro bends can lead to high radiation losses as well as difficulties in mechanical handling. Nonetheless, numerical studies have shown that transmission of over terabits per second data stream at a distance of a few hundred meters is theoretically achievable (Figure 14.5c) [55].

To overcome the challenges posed by the single-wire waveguide, Mbonye et al. proposed to position a second metal wire parallel to the first one (Figure 14.5d) [56]. This two-wire waveguide geometry ensured that the propagating plasmonic mode is very well confined in the interwire gap, granting a better immunization against environmental disturbances while maintaining a low-loss and low-dispersion propagation over a broad frequency range (Figure 14.5e) [57]. Furthermore, the two-wire waveguide's fundamental mode is linearly polarized (the electric field is directed from one wire to the other), which makes it compatible with most commonly used THz sources. In [9], a porous dielectric cladding (polystyrene foam) was used to further ease the handling and mechanical stability of the two-wire waveguide, at the expense of additional losses and dispersion.

In [58], Shrestha et al. explored the idea of using the two-wire waveguide to transmit high data rates using the modal diversity of the waveguide in the context of digital subscriber line (DSL) data transmission. To mimic the twisted wire pairs used in conventional DSL lines, the authors enclosed the two wires in a metallic sheath designed to support the 200 GHz radiation and eliminate bending losses. The authors noted the presence of significant mode mixing as the radiating energy was reflected at the metallic interface and coupled back in the waveguide. In their case, this mode mixing was an advantage as it was used to transmit multiple orthogonal data channels using a multiple-input multiple-output (MIMO) strategy. However, additional ohmic losses caused by the metallic cladding lead

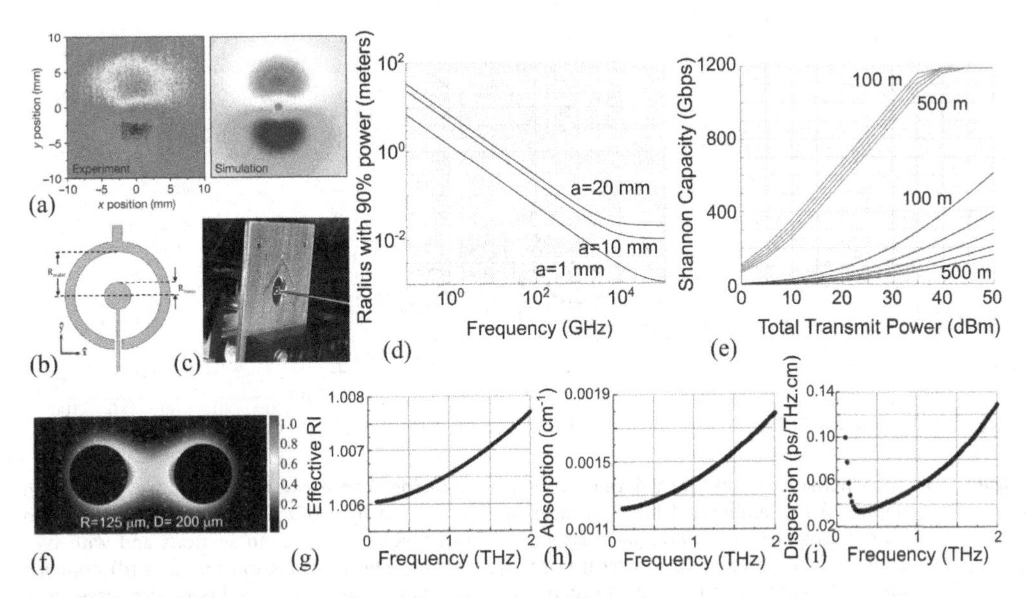

Figure 14.5 (a–e) Single-wire waveguide. (f–i) Two-wire waveguide. (a) Measurement and simulation of the amplitude of the electric field $|E_y|$ around a single-wire waveguide. The detector could only measure the electric field in the y direction. Wang. K et al., Metal wires for terahertz wave guiding, *Nature*, 2004 [51] (b) Design and (c) picture of the special radial antenna designed to excite the radial mode of the single-wire waveguide. [52] (d) Radius including 90% of the power of a single-wire waveguide for three different wire radiuses a. (e) Calculated Shannon capacity of the single-wire waveguide as a function of the transmitted power for different link distances in the 1–100 GHz band. © [2018] IEEE. Reprinted, with permission, from [55] (f) Fundamental mode of the two-wire waveguide, where the electric field is confined and linearly polarized in the gap. (g) Effective refractive index, (h) absorption losses, and (i) dispersion of the fundamental mode of the two-wire waveguide [57].

the authors to conclude that such an approach is limited to short-range (~10 m) communication scenarios.

In general, long two-wire waveguide (>1 m) are inconvenient in practice because the inter-wire gap needs to be precisely maintained (at the subwavelength scale), which is challenging to achieve for long fiber lengths. For shorter links (<1 m), 3D printing of two-wire plasmonic waveguide is a promising fabrication method with high accuracy and reproducibility (Figure 14.6). The penetration of THz waves into the metal is only hundreds of nanometers. Therefore, a μm-thick metal layer can be coated on the surface of the 3D printed polymer to fabricate THz metallic waveguides. Furthermore, complex structures (e.g., curved two-wire waveguides) can be obtained in a single-step process, in contrast with metallic components fabricated with CNC machining. In [59], Cao et al. used 3D printing to fabricate a two-wire waveguide (Figure 14.6a,b). Long waveguides could be assembled by connecting multiple waveguide sections via the excavations and extrusions on both ends of the cages that support the wires. The transmission loss (by field) and coupling loss (coupling between two waveguide sections) are shown in Figure 14.6c and d respectively, revealing remarkable low-loss behavior in the 120–155 GHz band. Curved waveguides were also fabricated and connected to other waveguide sections. Due to the anisotropic field distribution of the supported THz wave, two kinds of bends were possible

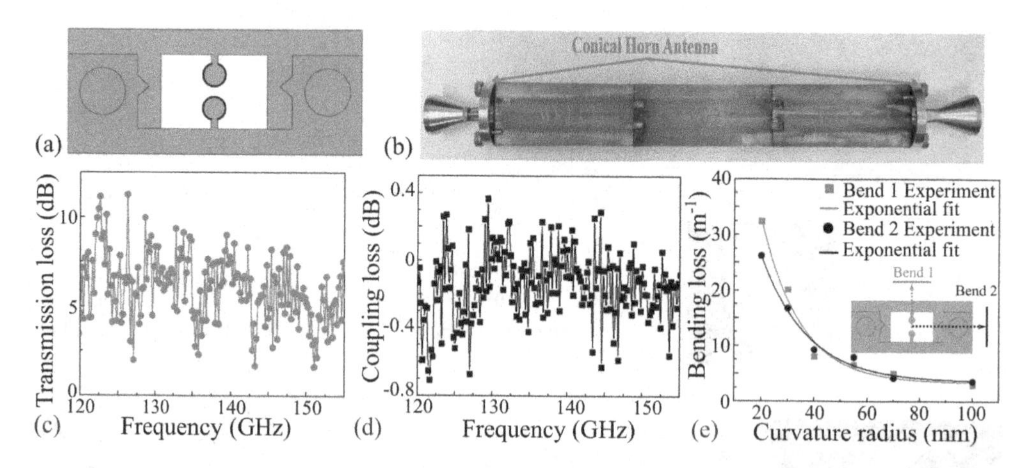

Figure 14.6 (a) Schematic of the 3D printed two-wire waveguide where the two wires are covered in metal (black), embedded in air (white), and supported by a 3D printed cage (gray). (b) An 18-cm long two-wire waveguide assembled from three distinct 6-cm sections and with two ends connected to WR6.5 conical horn antennas. (c) The transmission loss and (d) coupling loss of two connected waveguide sections. (e) Bending losses of a curved two-wire waveguide. Inset shows the geometry for the two bends. Bend 1: the axis of rotation is perpendicular to the line connecting the wire centers. Bend 2: The axis of rotation is parallel to the line connecting the wire centers [59].

(see inset in Figure 14.6e in which the line connecting the wire centers are parallel or perpendicular to the axis of rotation). The measured bending losses (by field) were shown to be less than 10 m^{-1} for a bending radius of 10 cm (Figure 14.6e).

14.4.1.2 Metallic Parallel Plate Waveguides

Metallic plates placed parallel to each other are simple waveguides that have attracted much attention in the THz range [60]. The parallel plate waveguide (PPWG) supports two types of modes. Its fundamental mode is the TEM mode, which is the lowest order TM mode (TM_0) and has no cut-off frequency and hence no dispersion. This TEM mode can be excited using a linearly polarized incident light with an electric field normal to the parallel plates (direction y in Figure 14.7a). On the other side, the TE_1 mode can be excited when the electric field is polarized parallel to the PPWG (direction x in Figure 14.7a). Compared to the TEM mode with no cut-off frequency, the TE_1 mode has a low-frequency cut-off at $f_c = c/(2nb)$, where b is the plate separation and n the refractive index of the medium between the plates [61]. Consequently, this cut-off frequency introduces spectral filtering at low frequencies and high dispersion for the spectral components near the cut-off. However, a careful comparison of the two types of modes shows that the TE_1 mode has low ohmic losses when increasing the frequencies, in opposition to the TEM mode [61] (Figure 14.7c). The remaining losses of the TE_1 mode are attributed to diffraction losses and can be mitigated using a concave structure in the plates along the propagation direction [62] (Figure 14.7d). To benefit from the low-loss nature of the TE_1 mode while also maintaining low dispersion, one can decrease the value of the cut-off frequency f_c by increasing the plate separation b. However, this comes with the disadvantage of

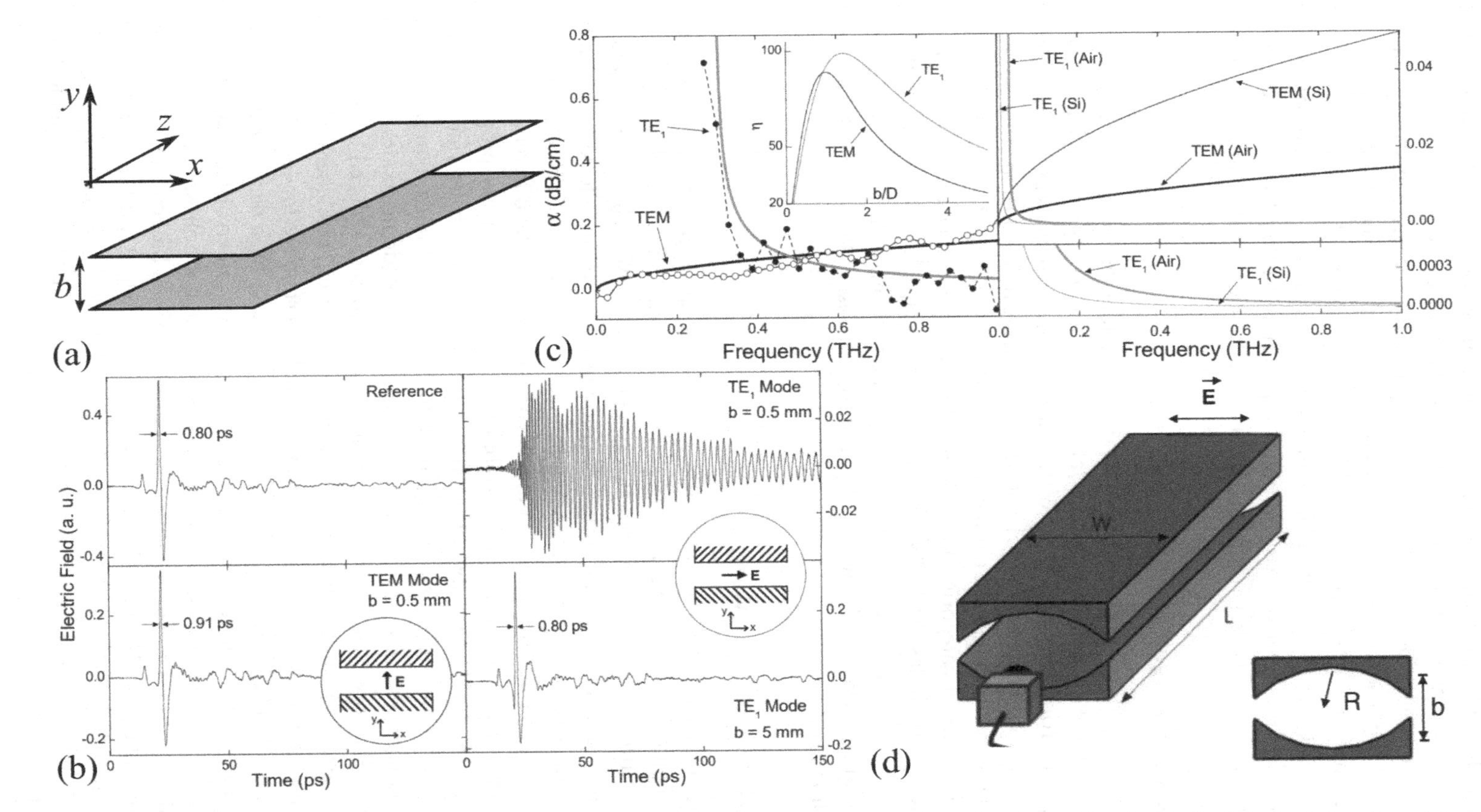

Figure 14.7 (a) Schematic of the metallic parallel plates separated by a distance *b*. The TEM modes can be excited by electric fields in the *y* direction, while the TE modes can be excited with electric fields in the *x* direction. (b) Comparison of the pulse propagation through empty space (top left), through the PPWG with a TEM mode (bottom left), the PPWG with a TE$_1$ mode and $b = 0.5$ mm (top right) and the PPWG with a TE$_1$ mode and $b = 5$ mm (bottom right). (c) Comparison of the losses of the TE$_1$ mode and TEM mode for $b = 0.5$mm (left) and $b = 5$ mm (right). Inset in the left figure is the excitation efficiency. Inset in the right figure is a zoom on the lower part of the right figure [61]. (d) Concave structure to mitigate the diffraction losses of the TE$_1$ mode. Reprinted by permission from Springer Nature [72] Copyright (2013).

multimode propagation and requires careful mode-matching to selectively excite the TE_1 mode [38]. The PPWG modes can be excited using simple focusing optics (quasi-optic coupling) and the coupling efficiency can be improved up to 80% with adiabatically tapered coupler [63]. Flexible thin copper strips can be used for short-link interconnects [64]. The PPWG can also be used for signal processing applications by introducing resonant Bragg cavities [65–68]. To modulate the THz pulses externally, one can insert high resistivity silicon between the parallel plates and modulate the pulse using photoexcitation [69–71]. In Section 14.5.2, we discuss additional applications of the PPWG in the context of THz communications, namely frequency division multiplexing and couplers/splitters.

14.4.2 Dielectric Waveguides

Dielectric THz waveguides are the second class of waveguides developed in the literature. They are made of low-loss dielectric materials such as polytetrafluoroethylene (PTFE), polyethylene, polypropylene, and cyclic olefin copolymer. Their waveguide structure is engineered in such a way that the mode is mostly propagating in low-loss air, thus allowing to establish long THz communication links (>10 m). Typical dielectric THz waveguides fall under one of three main categories: (1) hollow-core waveguides, including antiresonant reflecting optical waveguides (ARROW) and photonic bandgap (PBG) waveguides [48,73–78]; (2) porous core waveguides that use total internal reflection or PBG guidance [42,79–83]; and (3) solid core waveguides with TIR guiding mechanism [84–86]. We discuss each type in the following.

14.4.2.1 Hollow-Core Dielectric Waveguides

In hollow-core dielectric fibers, a large fraction of the modal energy propagates in the air core, thus significantly reducing the losses and dispersion. A simple hollow tube is an example of an antiresonant reflecting optical waveguide (ARROW). The ARROW guiding mechanism can be understood as follows. A fraction of the core mode leaks as radiation at the multiple core–cladding interfaces. At every interface, a portion is "reflected" back in the core. By carefully designing the cladding and core geometries, the different reflections from the interfaces can constructively interfere and increase the energy propagating in the core (Figure 14.8a). The bandwidth of the low-loss window is determined by the finite thickness of the thin tubular cladding. Introducing a lossy tubing material such as poly(methyl methacrylate) (PMMA) can help to suppress the cladding modes in favor of the core mode [74]. Moreover, the propagation loss can be further reduced by increasing the hollow core size [75]. However, this comes with the excitation of higher-order modes, which may cause intermodal dispersion and intermodal interference for long fibers [74]. Several communication links based on the tube waveguides have been demonstrated in the past years [14,16,87]. In [16], Van Thienen et al. demonstrated an error-free transmission of 7.6 and 1.5 Gbps over a link distance of 8 and 15 m using a hollow-core waveguide made of PMMA at 120 GHz. The measured eye pattern and the bit error rate are shown in Figure 14.8b and c [16]. The modal loss of the waveguide was 2.5 dB/m. The maximal bit rate here is mainly limited by the propagation loss. Losses due to bending were also shown to be negligible at higher carrier frequencies due to confinement of the mode in the core (Figure 14.8d) [87].

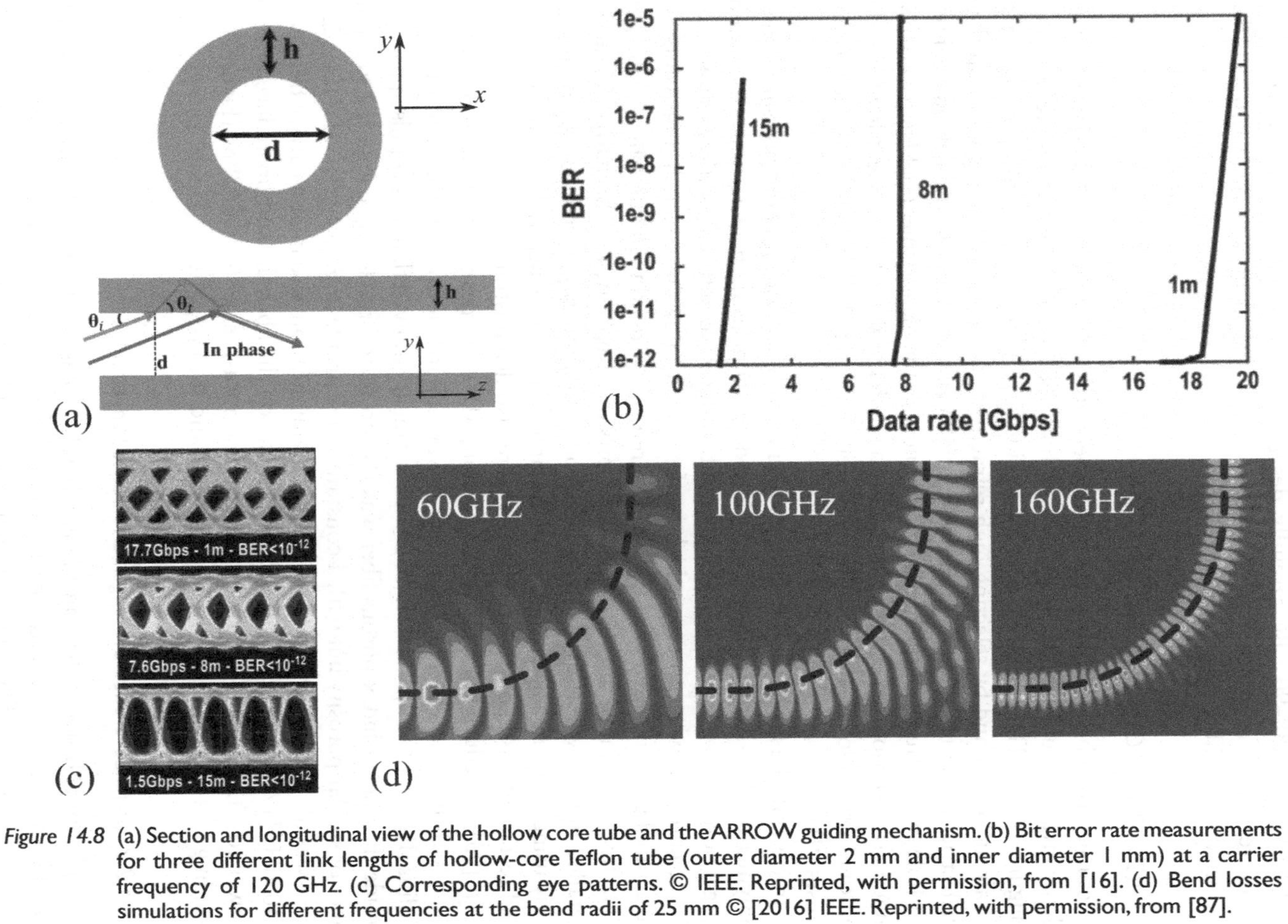

Figure 14.8 (a) Section and longitudinal view of the hollow core tube and the ARROW guiding mechanism. (b) Bit error rate measurements for three different link lengths of hollow-core Teflon tube (outer diameter 2 mm and inner diameter 1 mm) at a carrier frequency of 120 GHz. (c) Corresponding eye patterns. © IEEE. Reprinted, with permission, from [16]. (d) Bend losses simulations for different frequencies at the bend radii of 25 mm © [2016] IEEE. Reprinted, with permission, from [87].

On the other hand, by arranging alternating layers of high and low refractive index cladding material (Bragg fibers) or by introducing judiciously designed arrays of air inclusions in the cladding (photonic band gap fibers) around the hollow core, the losses and the transmission bandwidth can be improved compared to tube-based ARROW fibers [30,40,48,77,78,88–90]. By carefully designing the alternating layers, effective single-mode regimes can be achieved in long sections of such fibers, which can significantly reduce effective fiber dispersion.

14.4.2.2 Porous Core Dielectric Waveguides

In porous core fibers, low-loss and low dispersion can be achieved using spatially variable dense arrays of subwavelength air holes in both the core and the cladding regions [80,82] (Figure 14.9a). The losses of such waveguides can be reduced by using a high fraction of air inclusions and by selecting high refractive index contrast between air and the host dielectric. The latter allows a more substantial presence of the electric field in the low-loss air hole by continuity of the normal component of the displacement field across the waveguide cross-section: $E_a \propto E_m \varepsilon_m / \varepsilon_a$ [81]. This enhanced field presence allows the design of low-loss and low-dispersion fibers, as well as various THz components such as porous lenses [91].

In general, the fabrication of porous fibers involves drawing under pressure a thermo-polymer or glass-based preform with drilled or 3D printed air inclusions. Maintaining a target porosity over an extended length of fiber requires careful calibration and monitoring of the entire drawing process, which is often challenging due to small dimensions of the structured preforms used in the design. Therefore, the maximum link length using porous THz fibers is limited by the fabrication process. Recently, an alternative method was detailed in [89], where over meter-long monocrystalline sapphire fibers were grown directly using structured dies. Apart from the circular porous structure [80], honey-comb [92] and rectangular [45] porous geometries were also demonstrated.

In [93], to reduce the effect of dispersion, Ma et al. designed a graded-index porous fiber with holes of different diameters to vary the refractive index as a function of the radius (Figure 14.9ab). Time-domain measurements showed that a THz pulse propagating inside the graded-index porous fiber was less prone to broadening compared to a porous fiber of similar porosity but with a constant index profile (Figure 14.9c). The same 22 cm-fiber was also characterized using a THz communication system at a carrier frequency of 140 GHz [17]. To observe the effect of bending on the fiber, the authors reduced the input power to ~12 μW and recorded bit error rate measurements at 6 Gbps (Figure 14.9d). Bending had little effect on the bit error rate and the eye patterns (Figure 14.9e).

14.4.2.3 Solid Core Subwavelength Dielectric Waveguides

In solid core THz fibers, the transmission bandwidth is much larger than that in the hollow core fibers as the propagation mechanism is total internal reflection. However, the transmission loss in such fibers is much higher and generally comparable to the fiber material absorption loss. To minimize the transmission loss, one usually resorts to a subwavelength dielectric that is simply maintained in an air cladding [94,95]. These waveguides offer low loss and low dispersion guidance as a significant fraction of the modal fields is guided in

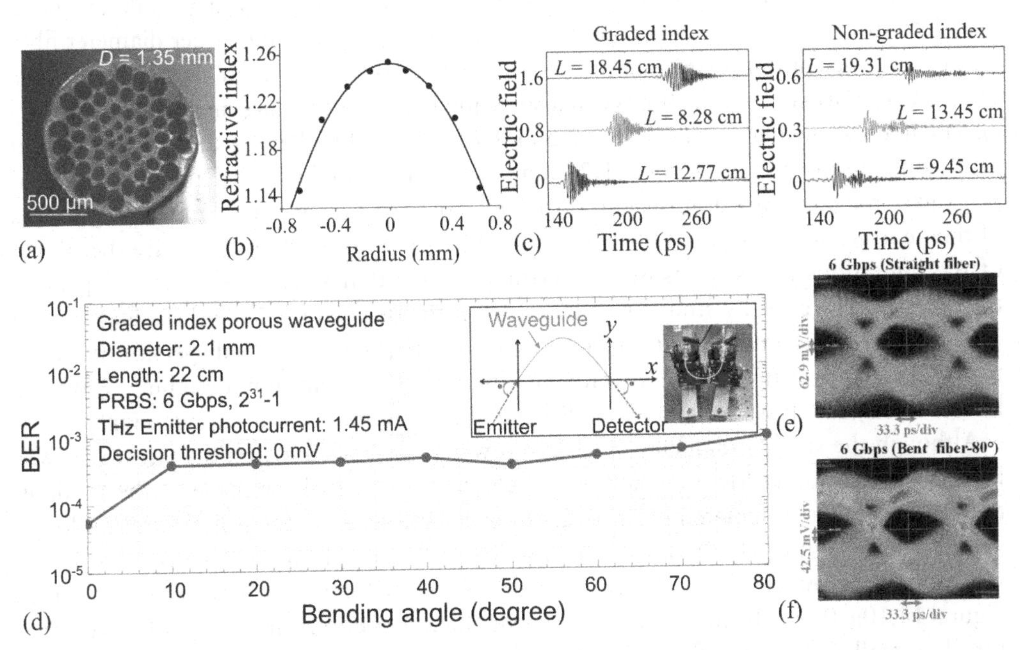

Figure 14.9 Graded index porous fiber. (a) Photograph of the graded-index fiber. (b) The refractive index profile along the cross-section of the porous fiber. (c) Comparison of the pulse propagation through different lengths of the graded-index fiber (left) and a non-graded index fiber of similar porosity (right) [93]. (d) Bit error rate measurements of the bent fiber for various bending angles at a carrier frequency of 140 GHz. © [2019] IEEE. Reprinted, with permission, from [17]. (e) Eye patterns for 6 Gbps in a straight fiber and (f) in a bent fiber with an angle of 80 degrees.

the low-loss air cladding surrounding the solid core [84,86]. However, handling the fiber is problematic, and several strategies have been used to isolate the waveguide from the environment, including suspending the solid core in a larger dielectric cladding [85], using a foam cladding [95] or a photonic crystal fiber geometry [96]. In such fibers, scattering from inhomogeneities along the fiber length such as diameter variation, micro/macro bending, and material density variation are the dominant loss mechanisms due to weak confinement in the fiber core [97]. These scattering losses can be somewhat mitigated by increasing the fiber diameter and realizing stronger confinement in the fiber core at the expense of increased material absorption losses. A good compromise can be found by choosing a low-loss polymer for the core material, and THz fiber links of several meters can be realized. Despite all these challenges, the rod-in-air/foam subwavelength fibers is a simple and reliable platform for enabling various short-range THz communication applications. Furthermore, such fibers can be used to design real-time signal processing components such as directional coupler, power dividers, and bandpass filter. These allow the building of complete transmission/signal processing subsystems using the same base technology [86,98,99]. An example of coupler is presented in Section 14.5.3.

In [100], Nallappan et al. characterized subwavelength polypropylene rod-in-air fibers of different diameters (0.57, 0.93, and 1.75 mm) in a THz communications system at a carrier frequency of 128 GHz. Bit error rate measurements (up to 6 Gbps and for distances up to 10 m) and corresponding eye patterns are shown in Figure 14.10a–e. The authors found that when using the fiber with smaller diameter, the propagating mode extended deep into the air, leading to low-transmission losses. In contrast, the larger diameter fiber had more propagation losses due to the confinement of the mode in the polypropylene core. The performance of the fibers can also be judged using eye diagrams (Figure 14.10c–e). The fiber of smaller diameters (0.57 and 0.93 mm) had higher eye amplitude (vertical axis) compared to the thick fiber (1.75 mm), which is related to the diameter-dependent propagation losses. However, synchronization errors are clearly visible in the eye patterns of the intermediate (0.93 mm) fiber (horizontal axis in Figure 14.10d) and are the source of the limitations in the measured bit error rates for that fiber. The authors explained this unexpected result by studying in more detail the dispersion of the three fibers. They concluded that the intermediate fiber showed larger group velocity dispersion than the two others. This example demonstrates the importance of carefully designing a fiber to have not only low losses but also low dispersion for communications applications.

Although the smaller diameter fibers had lower propagation losses, the larger one was less prone to bending losses, which was explained by the high presence of the mode in the solid core. The calculated bending losses in the absence of material absorption losses are shown in Figure 14.10f. These bending losses depended on the polarization due to the bending asymmetry. The electric field distributions for the bent fiber are shown in Figure 14.10g (for a bend radius of 3 cm in the x direction). It can be observed that for the smaller diameter fiber, the propagating mode completely leaks as radiation. Therefore, a good compromise must be found between transmission, dispersion, and bending losses, which depends on the material properties and the operation frequency. For practical applications, the rod-in-air can be encapsulated using a low-loss (<1 dB/m) and low refractive index (~1) foam as the cladding material. An example of such geometry is shown in Figure 14.10h. The diameter of the foam cladding must be chosen to accommodate ~90% of the power guided by the identical rod-in-foam waveguide with infinite cladding. Although in principle, the dielectric foams can contribute to additional propagation losses, the effect is negligible for short distances (several meters) and at low frequencies (below 200 GHz).

14.5 THz Waveguides as Communications Devices

In addition to the potential of guiding THz radiation over vast distances, waveguides can also be remarkable tools to manipulate THz beams in the context of communications. In this section, we cover some recent developments in THz waveguides used as devices for communications. For example, these devices can be placed at the output end of a THz waveguide to bridge between wired and wireless propagation.

14.5.1 Dispersion Compensation Waveguides

As we mentioned earlier, the dispersion is a significant limitation that restricts the transmission bandwidth. Even though THz waveguides can be engineered to possess low dispersion, its effect is significant, especially at large link distances. In those cases, dispersion compensation waveguides can be used to compress the broadened pulses and thereby improve the link performance. These waveguides are characterized by large dispersion values of opposite sign to the dispersion being compensated. To understand the action of

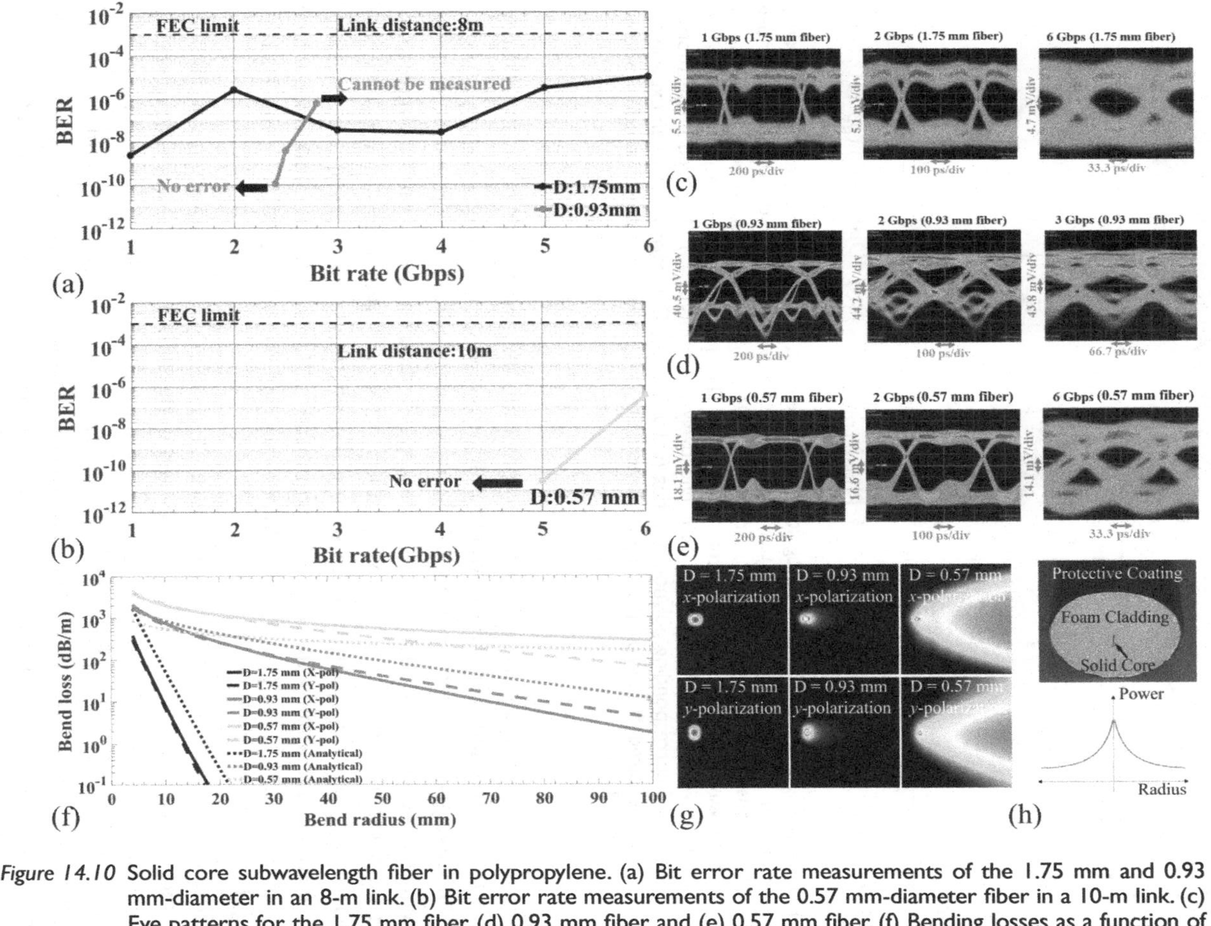

Figure 14.10 Solid core subwavelength fiber in polypropylene. (a) Bit error rate measurements of the 1.75 mm and 0.93 mm-diameter in an 8-m link. (b) Bit error rate measurements of the 0.57 mm-diameter fiber in a 10-m link. (c) Eye patterns for the 1.75 mm fiber, (d) 0.93 mm fiber and (e) 0.57 mm fiber. (f) Bending losses as a function of the bend radius for different polarizations. The solid curve corresponds to the *x*-polarized leaky mode and the dashed curve corresponds to the *y*-polarized leaky mode. The dotted lines are analytical estimations. (g) Electric field distribution of the leaky modes for different diameters of fiber and a bending radius of 3 cm. (h). Photograph of the rod-in-foam fiber that can accommodate ~90% of the guided power [100].

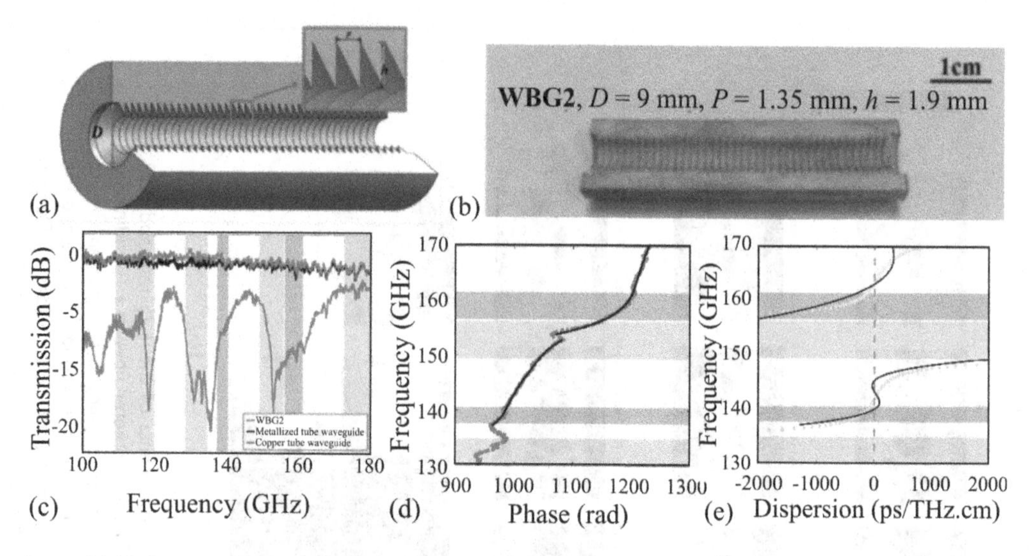

Figure 14.11 Dispersion compensation waveguide. (a) Schematic of the waveguide Bragg grating and (b) section of the 3D-printed metalized waveguide to show the internal structure. (c) Transmission measurements of the waveguide Bragg grating compared to a metalized tube waveguide and a copper tube waveguide. The gray regions correspond to spectral ranges where the Bragg waveguide is effectively single mode. (d) Measured phase (dotted line) and corresponding polynomial fitting (solid line). (e) Comparison between experimentally measured dispersion (black) and numerically computed dispersion of the fundamental mode. [37].

the compensating waveguide, consider the propagation in a waveguide of positive dispersion, where high frequencies travel slower than low frequencies. Adding a dispersion compensation waveguide at the output end of the waveguide induces the opposite effect: the high frequencies travel faster than the low frequencies, thus leading to pulse compression. Furthermore, this is done in an "analog" fashion before any digitalization, meaning that fewer errors are introduced in the data stream. Finally, the large dispersion in the compensating waveguide implies that it can be significantly shorter than the link fiber, which would, in turn, reduce the propagation losses incurred in the dispersion compensation waveguide.

The idea of dispersion management in waveguide systems is well known in optical fiber communications but has been much less explored in the THz range. An example of such a THz dispersion compensation waveguide was demonstrated by Ma et al. [37]. There, the authors fabricated dispersion compensation waveguides featuring negative dispersion of ~–100 $ps/(THz \cdot cm)$ at 0.14 THz. The waveguides were based on a metallic hollow-core geometry where periodic corrugations were introduced in the inner core to ensure high coupling efficiency in the lower order mode while opening a bandgap for the higher-order modes (Figure 14.11a). Then, by operating near the bandgap edge, strong negative dispersion was obtained.

The dispersion compensation waveguide was fabricated with a stereolithography technique and fully coated with a silver layer using the wet chemistry method detailed in [101] (Figure 14.11b). Optical characterization was performed with a continuous-wave THz spectroscopy system, which allowed to obtain phase and amplitude between 100 and 180 GHz with a spectral resolution of 0.1 GHz. Figure 14.11c shows the measured transmission spectra (waveguide referenced to an empty system). The amplitude measurements revealed two single-mode transmission bands centered around 140 and 160 GHz (gray regions in Figure 14.11c). From the phase measurements $\phi(\omega)$ (Figure 14.11d), the authors computed the group-velocity dispersion as $\frac{1}{L}\frac{\partial^2\phi}{\partial\omega^2}$ in the transmission windows (Figure 14.11e). They obtained negative dispersion ranging from –500 to $-100\,\mathrm{ps}/(\mathrm{THz}\cdot\mathrm{cm})$ in the 137–141 GHz range and –2000 to $-60\,\mathrm{ps}/(\mathrm{THz}\cdot\mathrm{cm})$ in the 156–162 GHz range.

14.5.2 Frequency-Division Multiplexing with Metallic Parallel-Plate Waveguides

It is necessary to encode information in multiple independent channels to increase the amount of transmitted data. The operation of mixing the independent channels in a communication link is known as multiplexing (mux), while demultiplexing (demux) is the opposite operation. Mux/demux can be performed in various ways, by taking advantage of light properties such as polarization, spatial mode, angular momentum, but most commonly, frequency. In [102], Karl et al. proposed to use a leaky-wave antenna based on a metal parallel-plate waveguide (PPWG) (Figure 14.12a,b). They introduced a narrow slot in the PPWG to allow radiation to leak out and couple into free-space modes with a frequency-dependent angle ϕ. Similarly, free-space radiation could couple into the PPWG's slot at the appropriate incident angle (Figure 14.12d). Consider the lowest-order transverse electric (TE_1) mode of the PPWG with the propagation constant given by [103]:

$$k_{\mathrm{PPWG}} = \frac{2\pi v}{c_0}\sqrt{1-\left(\frac{c_0}{2bv}\right)^2} \tag{14.15}$$

where b is the plate separation, and v is the frequency. Opening a slot in the PPWG allows this mode to couple to free-space modes with the phase-matching constraint $k_0 \cos\phi = k_{\mathrm{PPWG}}$, thus resulting in an angle-dependent emission frequency:

$$v(\phi) = \frac{c_0}{2b\sin\phi} \tag{14.16}$$

Considering an acceptance angle $\Delta\phi$, the corresponding spectral bandwidth is

$$\Delta v(\phi) = \left|\frac{dv}{d\phi}\right|\Delta\phi = \frac{c_0}{2b\sin\phi\tan\phi}\Delta\phi \tag{14.17}$$

This rather simple geometry was then used to achieve frequency-division multiplexing of real data flows, as shown by Ma et al. [104]. As a system demonstration, they showed real-time mux/demux of two independent channels with carrier frequencies of 264.7 and

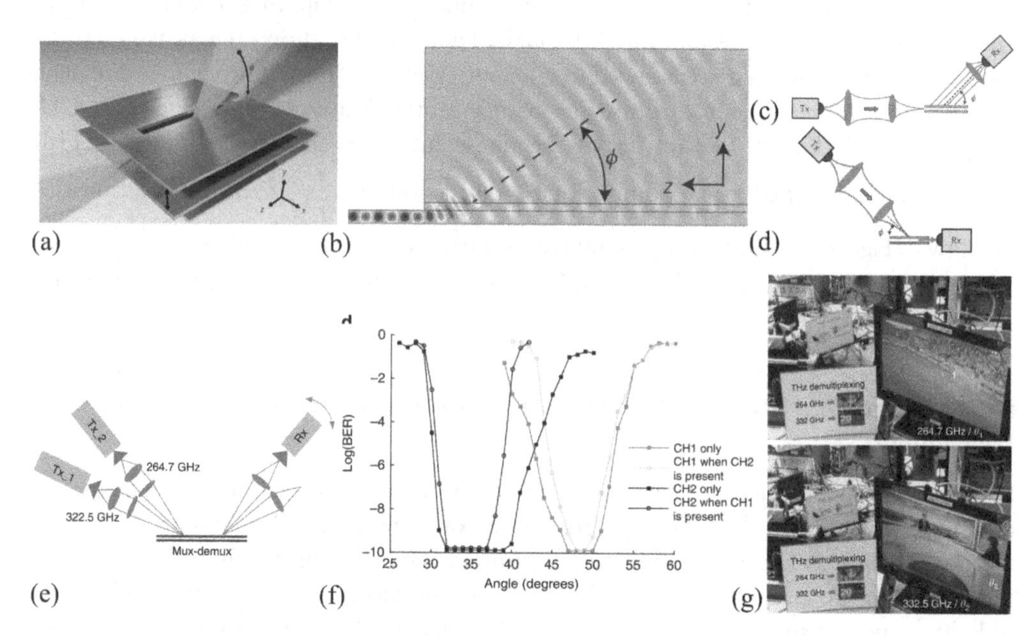

Figure 14.12 Leaky-wave antenna. (a) THz radiation propagating inside a metal parallel-plate waveguide is allowed to leak out by opening a narrow slot in one face of the waveguide. (b) Finite element simulation showing a side view of the wave emerging from the slot at an angle ϕ. (c) The same leaky-wave antenna can be used to couple to and (d) from free-space modes. Karl. N. J. et al., Frequency-division multiplexing in the terahertz range using a leaky-wave antenna, Nature Photonics, published 2015 [102] (e) Leaky-wave antenna used as a mux/demux of two spectral channels. (f) Bit error measurements as a function of the angle show error-free behavior when the detector is correctly positioned. (g) Real-time video data transmission of two different television broadcasts encoded in separate frequency bands. Ma J. et al., Frequency-division multiplexer and demultiplexer for terahertz wireless links. Nature Communications, published 2007 [104].

322.5 GHz (Figure 14.12e). These channels were coupled into the PPWG at different angles respecting Equation (14.16). Then, the two signals propagated simultaneously in the waveguide before leaking in free space. By moving the detector angularly, the authors were able to achieve an error-free (bit error rate < 10^{-10}) mux/demux for both channels (Figure 14.12f). They demonstrated this by demodulating real video data from two different televisions broadcasts. By simply rotating the detector, they were able to switch the received video from one channel to another (Figure 14.12g). Recently, Ghasempour et al. used the same leaky waveguide with broadband THz pulses to demonstrate single-shot link discovery for THz wireless networks [105].

14.5.3 Couplers, Splitters, and Add-Drop Multiplexers

To enable complex THz fiber architectures, it is necessary to develop waveguide components to route THz waves along specific paths. Examples of such couplers and splitters were demonstrated in [106]. Using 3D printing technology of polystyrene, Weidenbach et al. showed a Y-splitter and a multimode interference 1×3 splitter. They also demonstrated a variable waveguide coupler in which two parallel waveguides were brought close together

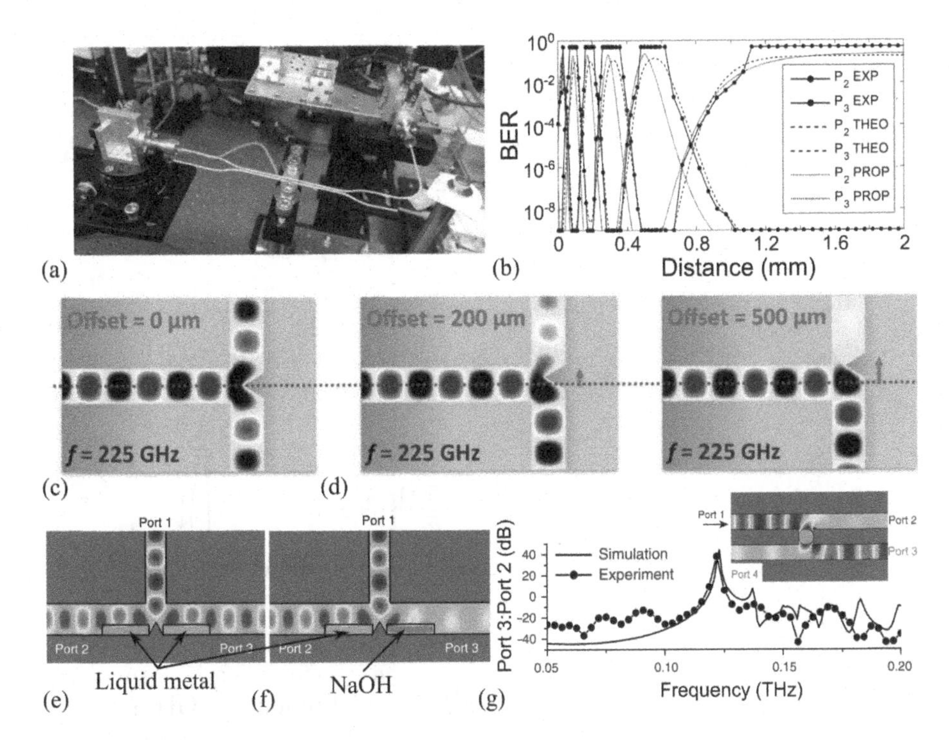

Figure 14.13 (a) 3D-printed coupler in the communications setup. The THz signal is launched on one of the inputs. (b) Bit error rate measurements as a function of the distance between the two parallel waveguides for the first output (P_2) and the second output (P_3). Reprinted by permission from Springer Nature Copyright 2017 [98]. (c) T-junction. When the triangular septum is positioned in the center, THz radiation is equally separated to both outputs. (d) By moving the septum, the amount of radiation outgoing one of the outputs can be tuned. Adapted from [107] under the terms of the Creative Commons CC BY License. With copyright permission (e) Electrically reconfigurable T-junction with liquid metal. When the liquid metal is placed, high output transmission is measured. (f) When the liquid metal is replaced by the NaOH solution, high losses reduce the output transmission. (g) Channel add-drop filter using metal liquid. The isolation (power ratio between the two outputs) is almost +40 dB at the resonant frequency of 123 GHz. Adapted from [108] under the terms of the Creative Commons CC BY License.

(Figure 14.13a). The energy could be transferred from one waveguide to the other by evanescent wave coupling by varying the distance between the two. In [98], this variable coupler was used in a 1 Gb/s communications system to demonstrate switching between two output waveguides. Bit error rate measurements (Figure 14.13b) confirmed that the distance between the two waveguides could be tuned to switch the output ports.

In [107] Reichel et al. designed a broadband T-junction variable power splitter based on the parallel-plate waveguide. The TE_1 mode was excited at the bottom of the "T" and propagated to the T-junction, where a triangular septum was positioned. When the septum was in the middle of the T-junction, the power was directed equally at each of the output ports, while minimizing back-reflections to the input port (Figure 14.13c). By laterally moving the septum, the authors were able to vary the power split on both output ports on the complete single-mode range of 150–300 GHz (Figure 14.13d). In [108], the idea was pushed one step further by using electrically actuated liquid metal (Galinstan,

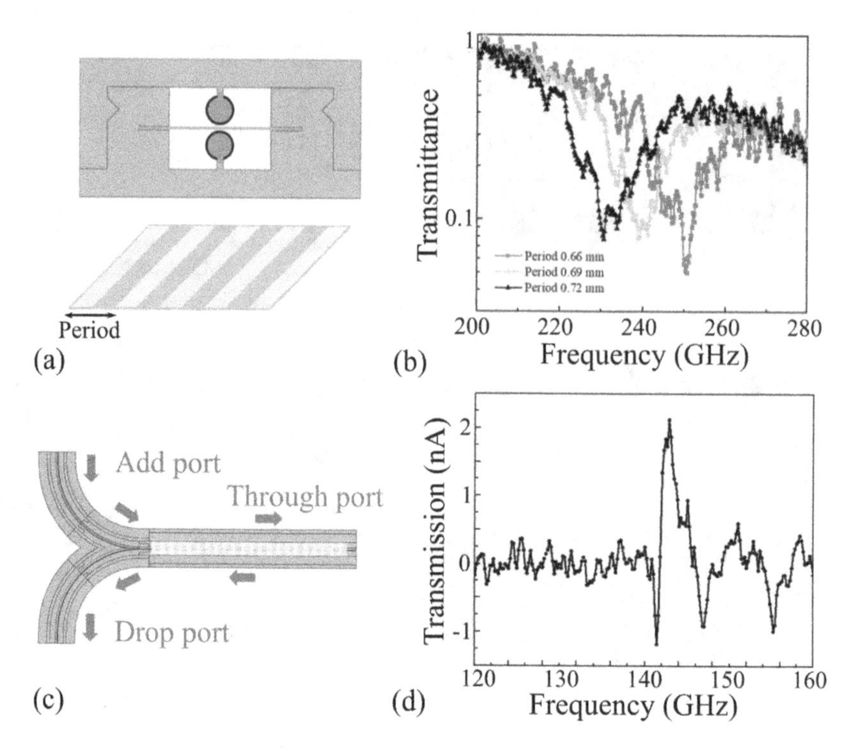

Figure 14.14 (a) Schematic of the grating based on a 3D-printed two-wire waveguide. (b) Transmittance spectra of the waveguide gratings featuring different periods. (c) Schematic diagram of the Y-shaped two-wire waveguide add-drop multiplexer with a Bragg grating. (d) Transmission at the drop port of the add-drop multiplexer defined as the difference between the measurement at the drop port with and without the Bragg grating [59].

gallium-indium-tin alloy) in the output ports. The liquid metal was introduced in rectangular glass tubes and connected a NaOH electrolyte solution. By applying a small ~4 V voltage, the liquid metal could move in and out of the channel. Therefore, the channel could either be made of a thin metal wall (with high output transmission, Figure 14.13e) or an electrolyte wall (with low output transmission, Figure 14.13f). A channel add-drop filter was designed using a similar concept in which an actuated liquid metal was placed between two PPWGs sharing the same inner wall. When the liquid metal was in the channel, the THz radiation from the upper waveguide could couple to the bottom through the capillary glass walls at a precise frequency (Figure 14.13g). Thus, a signal propagating at the resonant frequency in the top channel could be extracted in the bottom waveguide through this add-drop functionality.

Another example of add-drop multiplexer was shown by Cao et al. where Bragg gratings were used with the two-wire waveguide geometry [59]. The periodic structure (Bragg grating) was made using the toner-assisted metal foil transfer technique (also known as hot stamping [109]) to print metallic lines on a paper substrate. Then, the waveguide grating was obtained by inserting the patterned metallic film in the gap between the two wires of the 3D printed two-wire waveguide (Figure 14.14a). Figure 14.14b shows the transmittance spectra of three different waveguide gratings with different periods in the

200–280 GHz band. In this configuration, the fundamental TEM mode of the two-wire waveguide coupled into the radiation mode. By tuning the periodicity of the grating, the desired frequency channel could then be efficiently filtered. The Bragg resonance of the proposed waveguide grating was used to demonstrate a Y-shaped add-drop multiplexer around 140 GHz (Figure 14.14c). The signal was transmitted at the curved waveguide section (add port), followed by the waveguide with a paper-based metallic periodic pattern. The reflected signal at the Bragg wavelength was coupled into another curved waveguide section (drop port). The normalized transmission at the output port is shown in Figure 14.14d. As intended, the add-drop multiplexer reflects the Bragg frequency of 142.6 GHz.

14.6 Conclusion

In conclusion, we have presented the state-of-the-art of THz waveguides applied to communications. Most of the recent THz communications works were performed in a wireless context. Although THz radiation has several proven benefits in the context of wireless links, many challenges remain. Among these challenges, we note the high directionality of the THz beams, which requires precise alignment of the emitter and receiver, the increased risks of eavesdropping in free-space links, and the impact of environmental factors for robust communications. THz fiber links can provide additional flexibility and opportunities when designing future THz networks to overcome these limitations.

We started this chapter by providing an overview of the important parameters to control for efficient THz fiber links. First, fiber losses must be small. Dielectric materials such as polymers feature the lowest losses in the THz range, but dry gases such as air have virtually no absorption. Therefore, a common strategy used when designing low-loss THz waveguides is to increase the modal energy propagating in air. To improve the excitation efficiency, the waveguide dimensions must support a linearly polarized propagating mode similar to the input excitation mode. Dispersion in the waveguide is also a crucial parameter to reduce as it may lead to pulse broadening, adjacent bit overlapping as well as increased errors in the data stream. Finally, the waveguides must be relatively easy to manufacture. Fiber drawing towers similar to those used in drawing infrared optical fibers have been investigated. More recently, 3D printing approaches have been explored to build THz waveguides of more complex geometries.

We then turned to the study of metallic THz waveguides. In metallic single-wire waveguides, the mode propagates in the air surrounding a bare wire conductor. The mode bounded to the wire has low losses and low dispersion but suffers from low excitation efficiency due to its radial polarization. In the two-wire waveguide, a second metallic wire is placed parallel to the first one. The propagating mode is now confined in the inter-wire gap, immunizing the mode from environmental disturbances. The metallic parallel plate waveguide can support TEM and TE modes. The TEM modes have no cut-off frequency and losses that increase with frequency. In contrast, the TE modes have lower losses and can be designed to reduce the impact of the cut-off frequency.

Dielectric waveguides are the second class of THz waveguides, where the mode is supported by a low loss dielectric. Hollow-core dielectric waveguides are simple dielectric tubes that can propagate THz radiation through the antiresonant reflecting guiding mechanism (ARROW). Judiciously designed multiple alternate layers in the outer tube (photonic bandgap) allows for lower losses and more transmission bandwidths. In porous core dielectric waveguides, the presence of the mode is increased in subwavelength air inclusions. Graded index with low dispersion can then be designed by varying the pore size as a function of the radius. Solid core dielectric waveguides are simple dielectric fibers where the wave is propagating with a total internal reflection mechanism. To reduce the losses incurred while propagating in the dielectric core, the waveguide diameter can be reduced to subwavelength dimensions to increase the amount of energy in the air cladding.

Finally, the waveguides can be remarkable tools for communication applications. Dispersion compensation waveguides can be designed to counteract the effect of dispersion. Frequency-division multiplexing can be demonstrated by introducing a narrow slot in the metallic parallel plate waveguide and allowing THz radiation to leak in a predictable frequency-dependent angular distribution. Couplers, splitters, and T-junction to dynamically route the THz wave can be fabricated using the solid core waveguide and the metallic parallel plate waveguide. Add-drop multiplexing can be shown using the two-wire waveguide and a Bragg grating in a Y-coupler geometry.

Notes

[1] Nagatsuma, T., G. Ducournau, and C.C. Renaud, Advances in terahertz communications accelerated by photonics. *Nature Photonics*, 2016. **10**(6): 371–379.
[2] Dang, S., et al., What should 6G be? *Nature Electronics*, 2020. **3**(1): 20–29.
[3] Niu, Y., et al., A survey of millimeter wave communications (mmWave) for 5G: opportunities and challenges. *Wireless Networks*, 2015. **21**(8): 2657–2676.
[4] Zhang, J., et al., 6–100 GHz research progress and challenges from a channel perspective for fifth generation (5G) and future wireless communication. *Science China Information Sciences*, 2017. **60**(8): 1–8.
[5] Yang, Y., M. Mandehgar, and D.R. Grischkowsky, Understanding THz pulse propagation in the atmosphere. *IEEE Transactions on Terahertz Science and Technology*, 2012. **2**(4): 406–415.
[6] Yang, Y., M. Mandehgar, and D.R. Grischkowsky, THz-TDS characterization of the digital communication channels of the atmosphere and the enabled applications. *Journal of Infrared, Millimeter, and Terahertz Waves*, 2014. **36**(2): 97–129.
[7] Singh, R., and D. Sicker, An analytical model for efficient indoor THz access point deployment. arXiv e-prints, 2020. arXiv:2001.11167.
[8] Ma, J., et al., Security and eavesdropping in terahertz wireless links. *Nature*, 2018. **563**(7729): 89–93.
[9] Markov, A., H. Guerboukha, and M. Skorobogatiy, Hybrid metal wire–dielectric terahertz waveguides: challenges and opportunities [Invited]. *Journal of the Optical Society of America B*, 2014. **31**(11): 2587.
[10] Humbert, G., Optical fibers in terahertz domain, in *Handbook of Optical Fibers*, Springer, 2019. p. 1–49.
[11] Barh, A., et al., Specialty fibers for terahertz generation and transmission: A review. *IEEE Journal of Selected Topics in Quantum Electronics*, 2016. **22**(2): 365–379.
[12] Atakaramians, S., et al., Terahertz dielectric waveguides. *Advances in Optics and Photonics*, 2013. **5**(2): 169.
[13] Islam, M.S., et al., Terahertz optical fibers [Invited]. *Optics Express*, 2020. **28**(11): 16089–16117.
[14] Volkaerts, W., N. Van Thienen, and P. Reynaert, 10.2 An FSK plastic waveguide communication link in 40nm CMOS, in 2015 IEEE International Solid-State Circuits Conference – (ISSCC) Digest of Technical Papers. 2015. pp. 1–3.

[15] Voineau, F., et al., A 12 Gb/s 64QAM and OFDM compatible millimeter-wave communication link using a novel plastic waveguide design, in 2018 IEEE Radio and Wireless Symposium (RWS). 2018. pp. 250–252.
[16] Van Thienen, N., et al., An 18 Gbps polymer microwave fiber (PMF) communication link in 40 nm CMOS, in ESSCIRC Conference 2016: 42nd European Solid-State Circuits Conference. 2016. pp. 483–486.
[17] Nallappan, K., et al., Experimental demonstration of 5 Gbps data transmission using long subwavelength fiber at 140 GHz, in 2019 IEEE Radio and Wireless Symposium (RWS). 2019. pp. 1–4.
[18] Nallappan, K., et al., High bitrate data transmission using polypropylene fiber in terahertz frequency range, in 2019 International Workshop on Antenna Technology (iWAT). 2019. pp. 81–83.
[19] Priebe, S. and T. Kurner, Stochastic modeling of THz indoor radio channels. *IEEE Transactions on Wireless Communications*, 2013. **12**(9): 4445–4455.
[20] Petrov, V., et al., Terahertz band communications: Applications, research challenges, and standardization activities, in 2016 8th International Congress on Ultra Modern Telecommunications and Control Systems and Workshops (ICUMT). 2016. pp. 183–190.
[21] Mollahasani, S. and E. Onur, Evaluation of terahertz channel in data centers, in NOMS 2016–2016 IEEE/IFIP Network Operations and Management Symposium. 2016. pp. 727–730.
[22] Boujnah, N., S. Ghafoor, and A. Davy, Modeling and link quality assessment of THz network within data center, in 2019 European Conference on Networks and Communications (EuCNC). 2019. pp. 57–62.
[23] Zhang, C., et al., Breaking the blockage for big data transmission: gigabit road communication in autonomous vehicles. *IEEE Communications Magazine*, 2018. **56**(6): 152–157.
[24] Cacciapuoti, A.S., et al., Beyond 5G: THz-based medium access protocol for mobile heterogeneous networks. *IEEE Communications Magazine*, 2018. **56**(6): 110–115.
[25] Yu, X., et al., Direct terahertz communications with wireless and fiber links, in 2019 44th International Conference on Infrared, Millimeter, and Terahertz Waves (IRMMW-THz). 2019. pp. 1–2.
[26] Nallappan, K., et al., Live streaming of uncompressed HD and 4K videos using terahertz wireless links. *IEEE Access*, 2018. **6**: 58030–58042.
[27] Yang, Y., et al., Terahertz topological photonics for on-chip communication. *Nature Photonics*, 2020. **14**(7): 446–451.
[28] Gu, Q.J., THz interconnect: the last centimeter communication. *IEEE Communications Magazine*, 2015. **53**(4): 206–215.
[29] Alonso-del Pino, M., et al., Micromachining for advanced terahertz: interconnects and packaging techniques at terahertz frequencies. *IEEE Microwave Magazine*, 2020. **21**(1): 18–34.
[30] Li, J., et al., 3D printed hollow core terahertz Bragg waveguides with defect layers for surface sensing applications. *Optics Express*, 2017. **25**(4): 4126–4144.
[31] Guerboukha, H., K. Nallappan, and M. Skorobogatiy, Toward real-time terahertz imaging. *Advances in Optics and Photonics*, 2018. **10**(4): 843.
[32] Jin, Y.-S., G.-J. Kim, and S.-G. Jeon, Terahertz dielectric properties of polymers. *Journal of the Korean Physical Society*, 2006. **49**(2): 513–517.
[33] Siles, G.A., J.M. Riera, and P. Garcia-del-Pino, Atmospheric attenuation in wireless communication systems at millimeter and THz frequencies [Wireless Corner]. *IEEE Antennas and Propagation Magazine*, 2015. **57**(1): 48–61.
[34] Duvillaret, L., F. Garet, and J.L. Coutaz, A reliable method for extraction of material parameters in terahertz time-domain spectroscopy. *IEEE Journal of Selected Topics in Quantum Electronics*, 1996. **2**(3): 739–746.
[35] Dorney, T.D., R.G. Baraniuk, and D.M. Mittleman, Material parameter estimation with terahertz time-domain spectroscopy. *Journal of the Optical Society of America. A, Optics, Image Science, and Vision,* 2001. **18**(7): 1562–1571.

[36] Skorobogatiy, M., Hamiltonian formulation of Maxwell equations for the modes of anisotropic waveguides, in *Nanostructured and Subwavelength Waveguides*, John Wiley & Sons, 2012. p. 21–37.

[37] Ma, T., et al., Analog signal processing in the terahertz communication links using waveguide Bragg gratings: example of dispersion compensation. *Optics Express*, 2017. **25**(10): 11009–11026.

[38] Mendis, R., and D.M. Mittleman, An investigation of the lowest-order transverse-electric (TE_1) mode of the parallel-plate waveguide for THz pulse propagation. *Journal of the Optical Society of America B*, 2009. **26**(9): A6.

[39] Eugene, H., Modern optics: lasers and other topics, in *Optics*, Addison-Wesley, 2002. p. 581–648.

[40] Ung, B., et al., Polymer microstructured optical fibers for terahertz wave guiding. *Optics Express*, 2011. **19**(26): B848–861.

[41] Agrawal, G.P., Signal propagation in fibers, in *Lightwave Technology: Telecommunications System*. John Wiley & Sons. 2005. p. 63–106.

[42] Dupuis, A., et al., Fabrication and THz loss measurements of porous subwavelength fibers using a directional coupler method. *Optics Express*, 2009. **17**(10): 8012–8028.

[43] Skorobogatiy, M. and A. Dupuis, Ferroelectric all-polymer hollow Bragg fibers for terahertz guidance. *Applied Physics Letters*, 2007. **90**(11): 113514.

[44] Dupuis, A., et al., Spectral characterization of porous dielectric subwavelength THz fibers fabricated using a microstructured molding technique. *Optics Express*, 2010. **18**(13): 13813–13828.

[45] Atakaramians, S., et al., THz porous fibers: design, fabrication and experimental characterization. *Optics Express*, 2009. **17**(16): 14053–14062.

[46] Zhang, B., et al., Review of 3D printed millimeter-wave and terahertz passive devices. *International Journal of Antennas and Propagation*, 2017. **2017**: 1–10.

[47] D'Auria, M., et al., 3-d printed metal-pipe rectangular waveguides. *IEEE Transactions on Components, Packaging and Manufacturing Technology*, 2015. **5**(9): 1339–1349.

[48] Ma, T., et al., 3D printed hollow-core terahertz optical waveguides with hyperuniform disordered dielectric reflectors. *Advanced Optical Materials*, 2016. **4**(12): 2085–2094.

[49] Duangrit, N., et al., Terahertz dielectric property characterization of photopolymers for additive manufacturing. *IEEE Access*, 2019. **7**: 12339–12347.

[50] Guofu, X., et al., Fabrication of low loss and near zero dispersion suspended core polypropylene fibers for terahertz communications using infinity 3D printing technique. engrXiv, 2020.

[51] Wang, K. and D.M. Mittleman, Metal wires for terahertz wave guiding. *Nature*, 2004. **432**(7015): 376–379.

[52] Deibel, J.A., et al., Enhanced coupling of terahertz radiation to cylindrical wire waveguides. *Optics Express*, 2006. **14**(1): 279–290.

[53] Jeon, T.I., J.Q. Zhang, and D. Grischkowsky, THz Sommerfeld wave propagation on a single metal wire. *Applied Physics Letters*, 2005. **86**(16): 161904.

[54] Akalin, T., A. Treizebre, and B. Bocquet, Single-wire transmission lines at terahertz frequencies. *IEEE Transactions on Microwave Theory and Techniques*, 2006. **54**(6): 2762–2767.

[55] Galli, S., J. Liu, and G. Zhang, Bare metal wires as open waveguides, with applications to 5G, in 2018 IEEE International Conference on Communications (ICC). 2018. pp. 1–6.

[56] Mbonye, M., R. Mendis, and D.M. Mittleman, A terahertz two-wire waveguide with low bending loss. *Applied Physics Letters*, 2009. **95**(23): 233506.

[57] Markov, A., and M. Skorobogatiy, Two-wire terahertz fibers with porous dielectric support. *Optics Express*, 2013. **21**(10): 12728–12743.

[58] Shrestha, R., et al., A wire waveguide channel for terabit-per-second links. *Applied Physics Letters*, 2020. **116**(13): 131102.

[59] Cao, Y., et al., Additive manufacturing of highly reconfigurable plasmonic circuits for terahertz communications. *Optica*, 2020. **7**(9): 1112–1125.

[60] Mendis, R. and D. Grischkowsky, Undistorted guided-wave propagation of subpicosecond terahertz pulses. *Optics Letters*, 2001. **26**(11): 846–848.

[61] Mendis, R. and D.M. Mittleman, Comparison of the lowest-order transverse-electric (TE1) and transverse-magnetic (TEM) modes of the parallel-plate waveguide for terahertz pulse applications. *Optics Express*, 2009. **17**(17): 14839–14850.

[62] Mbonye, M., R. Mendis, and D.M. Mittleman, Inhibiting the TE1-mode diffraction losses in terahertz parallel-plate waveguides using concave plates. *Optics Express*, 2012. **20**(25): 27800–27809.

[63] Gerhard, M., M. Theuer, and R. Beigang, Coupling into tapered metal parallel plate waveguides using a focused terahertz beam. *Applied Physics Letters*, 2012. **101**(4): 041109.

[64] Mendis, R. and D.R. Grischkowsky, THz interconnect with low-loss and low-group velocity dispersion. *IEEE Microwave and Wireless Components Letters*, 2001. **11**(11): 444–446.

[65] Zhao, Y.G. and D.R. Grischkowsky, 2-D terahertz metallic photonic crystals in parallel-plate waveguides. IEEE Transactions on Microwave Theory and Techniques, 2007. **55**(4): 656–663.

[66] Astley, V., et al., Analysis of rectangular resonant cavities in terahertz parallel-plate waveguides. *Optics Letters*, 2011. **36**(8): 1452–1454.

[67] Harsha, S.S., N. Laman, and D. Grischkowsky, High-Q terahertz Bragg resonances within a metal parallel plate waveguide. *Applied Physics Letters*, 2009. **94**(9): 091118.

[68] Lee, E.S., et al., Terahertz band gaps induced by metal grooves inside parallel-plate waveguides. *Optics Express*, 2012. **20**(6): 6116–6123.

[69] Gingras, L., et al., Active phase control of terahertz pulses using a dynamic waveguide. *Optics Express*, 2018. **26**(11): 13876–13882.

[70] Gingras, L. and D.G. Cooke, Direct temporal shaping of terahertz light pulses. *Optica*, 2017. **4**(11): 1416–1420.

[71] Cooke, D.G. and P.U. Jepsen, Optical modulation of terahertz pulses in a parallel plate waveguide. *Optics Express*, 2008. **16**(19): 15123–15129.

[72] Mbonye, M., R. Mendis, and D.M. Mittleman, Measuring TE1 mode losses in terahertz parallel-plate waveguides. *Journal of Infrared, Millimeter, and Terahertz Waves*, 2013. **34**(7–8): 416–422.

[73] Cruz, A.L.S., C.M.B. Cordeiro, and M.A.R. Franco, 3D printed hollow-core terahertz fibers. *Fibers*, 2018. **6**(3): 43.

[74] Bao, H., et al., Dielectric tube waveguides with absorptive cladding for broadband, low-dispersion and low loss THz guiding. *Scientific Reports*, 2015. **5**: 7620.

[75] Lai, C.H., et al., Modal characteristics of antiresonant reflecting pipe waveguides for terahertz waveguiding. *Optics Express*, 2010. **18**(1): 309–322.

[76] Yang, S., et al., Novel pentagram THz hollow core anti-resonant fiber using a 3D printer. *Journal of Infrared Millimeter and Terahertz Waves*, 2019. **40**(7): 720–730.

[77] Yu, R.J., et al., Proposal for ultralow loss hollow-core plastic Bragg fiber with cobweb-structured cladding for terahertz waveguiding. *IEEE Photonics Technology Letters*, 2007. **19**(9–12): 910–912.

[78] Dupuis, A., et al., Transmission measurements of hollow-core THz Bragg fibers. *Journal of the Optical Society of America B-Optical Physics*, 2011. **28**(4): 896–907.

[79] Nielsen, K., et al., Bendable, low-loss Topas fibers for the terahertz frequency range. *Optics Express*, 2009. **17**(10): 8592–8601.

[80] Hassani, A., A. Dupuis, and M. Skorobogatiy, Low loss porous terahertz fibers containing multiple subwavelength holes. *Applied Physics Letters*, 2008. **92**(7): 071101.

[81] Atakaramians, S., et al., Porous fibers: a novel approach to low loss THz waveguides. *Optics Express*, 2008. **16**(12): 8845–8854.

[82] Hassani, A., A. Dupuis, and M. Skorobogatiy, Porous polymer fibers for low-loss Terahertz guiding. *Optics Express*, 2008. **16**(9): 6340–6351.

[83] Hasan, M.I., et al., Ultra-low material loss and dispersion flattened fiber for thz transmission. *IEEE Photonics Technology Letters*, 2014. **26**(23): 2372–2375.

[84] Chen, L.J., et al., Low-loss subwavelength plastic fiber for terahertz waveguiding. *Optics Letters*, 2006. **31**(3): 308–310.

[85] Roze, M., et al., Suspended core subwavelength fibers: towards practical designs for low-loss terahertz guidance. *Optics Express*, 2011. **19**(10): 9127–9138.
[86] Li, H.S., et al., Terahertz polarization-maintaining subwavelength dielectric waveguides. *Journal of Optics*, 2018. **20**(12): 125602.
[87] Van Thienen, N., W. Volkaerts, and P. Reynaert, A multi-gigabit CPFSK polymer microwave fiber communication link in 40 nm CMOS. *IEEE Journal of Solid-State Circuits*, 2016. **51**(8): 1952–1958.
[88] Argyros, A., Microstructures in polymer fibres for optical fibres, THz waveguides, and fibre-based metamaterials. *ISRN Optics*, 2013. **2013**: 1–22.
[89] Katyba, G.M., et al., Sapphire photonic crystal waveguides for terahertz sensing in aggressive environments. *Advanced Optical Materials*, 2018. **6**(22): 1800573.
[90] Lu, J.Y., et al., Terahertz air-core microstructure fiber. *Applied Physics Letters*, 2008. **92**(6): 064105.
[91] Guerboukha, H., et al., Planar porous components for low-loss terahertz optics. *Advanced Optical Materials*, 2019. **7**(15): 1900236.
[92] Nielsen, K., et al., Porous-core honeycomb bandgap THz fiber. *Optics Letters*, 2011. **36**(5): 666–668.
[93] Ma, T., et al., Graded index porous optical fibers - dispersion management in terahertz range. *Optics Express*, 2015. **23**(6): 7856–7869.
[94] Chen, L.-J., et al., Low-loss subwavelength plastic fiber for terahertz waveguiding. *Optics Letters*, 2006. **31**(3): 308–310.
[95] De Wit, M., Y. Zhang, and P. Reynaert, Analysis and design of a foam-cladded PMF link with phase tuning in 28-nm CMOS. *IEEE Journal of Solid-State Circuits*, 2019. **54**(7): 1960–1969.
[96] Han, H., et al., Terahertz pulse propagation in a plastic photonic crystal fiber. *Applied Physics Letters*, 2002. **80**(15): 2634–2636.
[97] Chen, H.W., et al., Investigation on spectral loss characteristics of subwavelength terahertz fibers. *Optics Letters*, 2007. **32**(9): 1017–1019.
[98] Ma, J., et al., Communications with THz waves: Switching data between two waveguides. *Journal of Infrared Millimeter and Terahertz Waves*, 2017. **38**(11): 1316–1320.
[99] Chen, H.W., et al., Subwavelength dielectric-fiber-based THz coupler. *Journal of Lightwave Technology*, 2009. **27**(11): 1489–1495.
[100] Nallappan, K., et al., Dispersion limited versus power limited terahertz transmission links using solid core subwavelength dielectric fibers. *Photonics Research*, 2020. **8**(11): 1757–1775.
[101] Jiang, S.Q., et al., Chemical silver plating on cotton and polyester fabrics and its application on fabric design. *Textile Research Journal*, 2016. **76**(1): 57–65.
[102] Karl, N.J., et al., Frequency-division multiplexing in the terahertz range using a leaky-wave antenna. *Nature Photonics*, 2015. **9**(11): 717–720.
[103] Mendis, R. and D.M. Mittleman, A 2-d artificial dielectric with $0<n<1$ for the terahertz region. *IEEE Transactions on Microwave Theory and Techniques*, 2010. **58**(7): 1993–1998.
[104] Ma, J., et al., Frequency-division multiplexer and demultiplexer for terahertz wireless links. *Nature Communications*, 2017. **8**(1): 729.
[105] Ghasempour, Y., et al., Single-shot link discovery for terahertz wireless networks. *Nature Communications*, 2020. **11**(1): 1–6.
[106] Weidenbach, M., et al., 3D printed dielectric rectangular waveguides, splitters and couplers for 120 GHz. *Optics Express*, 2016. **24**(25): 28968–28976.
[107] Reichel, K.S., R. Mendis, and D.M. Mittleman, A broadband terahertz waveguide t-junction variable power splitter. *Science Reports*, 2016. **6**: 28925.
[108] Reichel, K.S., et al., Electrically reconfigurable terahertz signal processing devices using liquid metal components. *Nature Communications*, 2018. **9**(1): 4202.
[109] Guerboukha, H., et al., Super-resolution orthogonal deterministic imaging technique for terahertz subwavelength microscopy. *ACS Photonics*, 2020. **7**(7): 1866–1875.

Chapter 15

NextGen Granular Resource Management in the THz Spectrum for Indoor and Outdoor Mobile Deployment

Rohit Singh and Douglas Sicker

Contents

15.1 Introduction

With each generation of wireless telecommunication, we have experienced a boost in the usage of mobile devices. The transition from 3G to 4G experienced an exponential growth in devices, which will likely become more explosive with the advent of 5G, resulting in extremely dense networks (EDN) in both indoor and outdoor environments. With the rise in popularity of the Internet of Everything (IoE), an average human can (or will) carry multiple devices demanding high data rate, low latency, and secure communication. In the future, the success of 5G will enable numerous industries or verticals, such as smart cities, smart homes, cellular drone communication, unmanned aerial vehicle (UAV), autonomous cars, machine-to-machine (M2M) communication, and EDNs [1,2]. Adding to this problem, the post-5G deployment might boost multiple non-communication-based industries and applications, which may positively result in an increased demand for devices with higher throughput and lower latency [3]. The 5G new radio (NR) standard is supposed to cater to use cases such as ultra-reliable low-latency communication (URLLC), enhanced mobile broadband (eMBB), and massive machine type communication (mMTC) [4]. These services will take a toll on the already heavily allocated spectrum.

15.1.1 Heading to a Chaotic Future

With the advent of 5G and multiple killer applications, there has been a surge in the demand for ultra-high throughput and ultra-low latency technologies. In the future, the networks will need to be tactile and self-adaptive to understand the *dense-heterogeneous* environment and the needs of users based on service type. It is expected that in the next 10 years, 5G will reach its maximum performance limit and might not be able to meet future demands. Thus, it is critical that we start to explore beyond 5G (B5G) or even 6G type technologies. 5G and even 6G will give rise to numerous verticals, such as smart home, industry automation, vehicle-to-infrastructure (V2X), eHealth, body area networks (BAN), UAVs, entertainment, education, and much more [1,2]. All these verticals will have heterogeneous devices, protocols, performance demands, and resource needs [2,5]. Moreover, it is expected that within the next decade, wireless systems might need to satisfy more than 1,000 devices per 100 m^2 [6,7]. Narrower antenna beams with beam tracking will be required to improve signal strength [3] to maintain such high system demands [8,9]. To complicate the situation further these devices might be mobile in nature with varying velocity,[1] resulting in multiple radio access technology (RAT) handovers, access point (AP) selection, frequent route optimization, time complexity, and signaling overhead [2,6,10]. Narrow antenna beams and high mobility of the users can cause frequent outages leading to high latency and low reliability [11,12]. These frequent resource handovers can also occur either due to changes in performance demands, or if a device changes its vertical, or if a device transits from indoor-to-outdoor or outdoor-to-indoor [13,14]. These dense mobile devices with heterogeneous needs in both indoor and outdoor environment will lead to a chaotic future if not properly managed [2–7]. Figure 15.1 hints at such a chaotic future.

Densification of users, where users have more than one device operating on heterogeneous demands and protocols, will make the future much more chaotic than we can imagine. Moreover, most of these devices will generate a huge amount of sensitive data, which, when mismanaged, can lead to network vulnerabilities, privacy risks, and impact national security [15–17]. Deploying heavy encryption and authentication tools will increase latency. Thus, most of the 5G systems rely on physical layer security (PLS) [18] or light authentication [19]. These methods might not solve all the existing security or privacy risks and might give rise to new risks in the process. A trade-off the operators and manufacturers are willing to tolerate for lower latency.

As shown in Figure 15.1, networks will not only be terrestrial, but also non-terrestrial in the form of UAVs, high-altitude platform systems (HAPS), and satellites. Considering a uniform distribution of both terrestrial and non-terrestrial APs to solve the user coverage problem will not work, since it will exponentially increase the capital expenditure (CAPEX) and operating expense (OPEX), which is already a bottleneck for 5G deployment in most of the countries [20]. Moreover, the need for too many distributed edge nodes [1], energy wastage [21], and backhauling [22] will also further complicate the process. It is estimated that in the next 10 years 5G will be saturated [23] and there will be a need for extra infrastructure and resources, which we are not ready for [23–25].

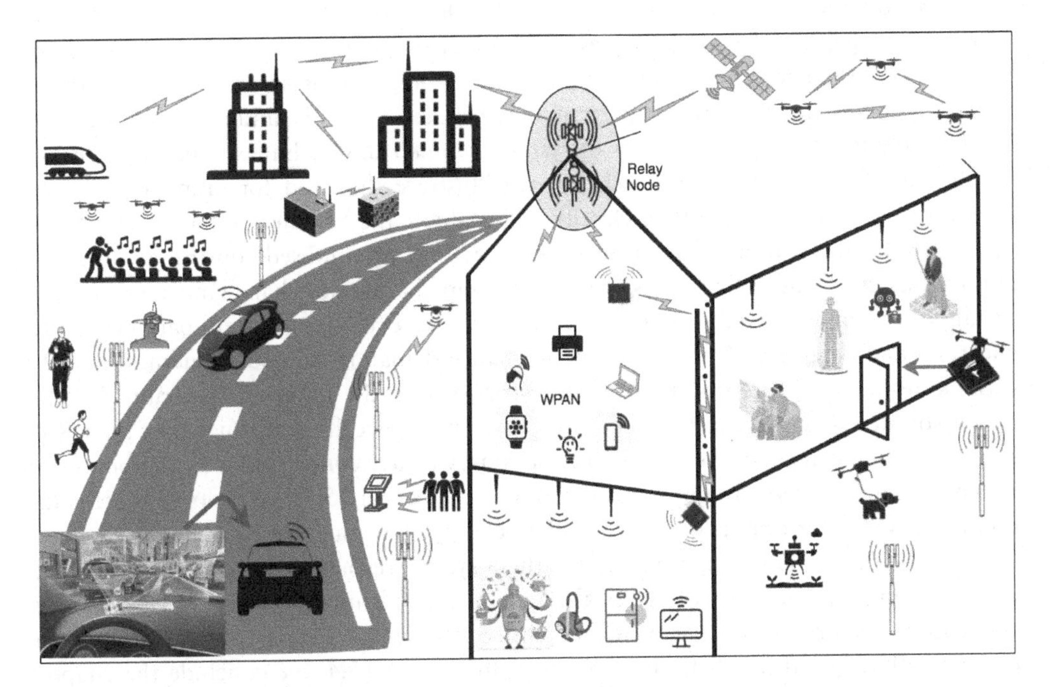

Figure 15.1 A future chaotic scenario with heterogeneous and dense deployment of UEs and APs in both indoor and outdoor environments.

15.1.2 THz Spectrum: Framing the Solution Space

For 5G, we are currently looking into more spectrum at Citizens Broadband Radio Service (CBRS) and mmWave bands, which will offer nearly 100 MHz and 1–2 GHz[2] of contiguous bandwidth respectively [3]. We acknowledge that this amount of spectrum is more than what we have ever used for International Mobile Telecommunication (IMT) services, but there are reasons to believe that this will not be enough [26]. We need to move beyond mmWave and explore more greenfield spectrum in the THz band (100 GHz to 10 THz) [3,7].

Exploiting frequency bands above 100 GHz opens multiple avenues for research and interesting deployment strategies for both communication and non-communication purposes [27]; however, it also comes along with multiple challenges [28–31]. Compared to the traditional lower frequency bands, THz or other very-high frequency bands suffer from high path loss, smaller coverage, sensitivity to mobility, and high equipment cost. Interestingly, it is these characteristics that could make THz communications friendly for indoor use and for some outdoor use cases. Although THz is aimed at static and nomadic services, it can be used for relatively constrained mobile users. Improvements in antenna design, small-scale mobility management, and detailed knowledge about the environment can further the use cases of the THz spectrum.

15.1.3 Our Contribution and Outline

THz is very sensitive to multiple system parameters and can negatively impact system performances. Keeping track of these parameters and opportunistically allocating resources can prove otherwise. To monitor this detailed knowledge or resources, we need novel radio resource management (RRM) strategies to meet higher system performance. RRM for 5G and 6G networks need to account for multiple factors, such as spectrum, system architecture, system parameters, traffic pattern, network heterogeneity, and backhaul [32–34]. For THz, there are a myriad number of *system parameters*, which needs special attention for the NextGen RRM. In Section 15.2, we illustrate the need for granular resource management in the THz and then classify and evaluate the system parameters into three classes: (a) fixed resources, (b) variable resources, and (c) imposed constraints. In every radio architecture, there are some system parameters that do not change much and can be pre-allocated, namely fixed resources. However, in the THz, some system parameters are uncertain and change with use cases, performance demands, and surroundings, namely variable resources. Moreover, regulatory restrictions on the usage of spectrum and the data associated with these parameters can negatively impact the performance of the RRM strategies. To manage these regulatory restrictions, we have considered it as another set of system parameters, namely imposed constraints. In Section 15.3, we propose three solution frameworks for opportunistically and efficiently allocating these granular resources. We propose three allocation strategies for these system resources, which is based on the user grouping: (a) mobility, (b) verticals, and (c) identification. In each of the schemes we propose resource allocation strategies, either through a resource trade-off triangle or dynamic allocation through ID broadcasting. In Section 15.4, we conclude the chapter with final remarks and future direction.

15.2 THz Resources

In this section, we illustrate the need for granular resource management in the THz spectrum and classify these resources into three categories, namely fixed resources, variable resources, and imposed constraints.

15.2.1 Need for Granular Resource Management

THz spectrum has a very high potential, but it also comes with many pre-conditions to be used in both indoors and outdoors. There are multiple economic and non-economic use cases which demand high-throughput and low-latency performance from the system, where using the THz spectrum is justified. THz spectrum can be used for application, such as virtual reality (VR), augmented reality (AR), mixed-reality (MR), uncompressed 4K streaming for ultra-high displays (UHD), holographic type communication (HTC), tactile and haptic services, backhauling, imaging, sensing, and spectroscopy [1,2,4,35]. However, all these applications have multiple challenges and special system performance demands. As shown in Figure 15.2, a single THz-AP will face multiple resource allocation challenges to cater to any of the above-mentioned applications. For example, a dense deployment of mobile users demanding high throughput and low latency services might compromise the efficiency of the whole system. Moreover, even while satisfying a smaller set of these users, multiple factors, such as environment awareness, health impacts,[3] security risks, and privacy risks, need to be monitored and managed [3].

Due to low penetration power, the THz-APs have limited coverage and requires line-of-sight (LOS) communication. Except for some lower THz bands, non-line-of-sight (NLOS)

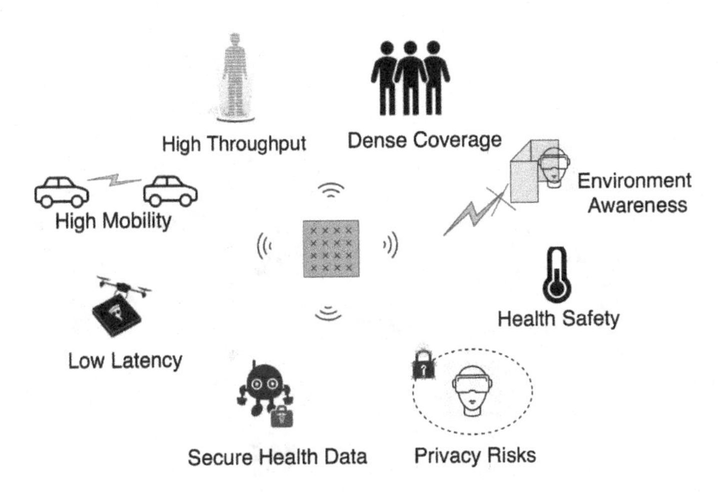

Figure 15.2 Resource allocation challenges for a single THz-AP.

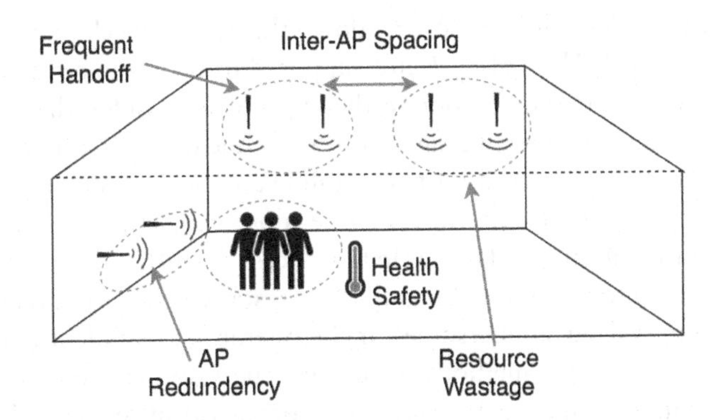

Figure 15.3 Resource Allocation Challenges for multiple Indoor THz-AP.

is not so effective, since it is sensitive to surfaces for reflection, refraction, and scattering [36–38]. Thus, a possible solution for the challenges shown in Figure 15.2 is to blanket the area with multiple THz-APs. However, deploying multiple THz-APs comes with its own set of challenges both for indoor and outdoor environments. In an indoor environment, AP parameters, such as number of THz-APs [39], number of antenna arrays per THz-AP [40], and THz-AP placement [39,41,42], can significantly impact the system efficiency. Poor deployment of these APs can lead to inefficient use of infrastructure, resource, spectrum, and energy [43], as shown in Figure 15.3.

One of the key reasons for using THz spectrum in an outdoor environment is to cater to applications such as smart cities, emergency services, ultra-high kiosks, aerial base stations (ABS), or wireless backhaul. In an outdoor environment, structures, such as building, vehicles, trees, and humans, can block THz signal. Thus, to improve coverage THz-APs will need to be placed wherever space is available, such us smart lamps, APs on buildings, and over-the-air through ABS, as shown in Figure 15.4. Moreover, THz-UEs and THz-APs may be present both in terrestrial and non-terrestrial [44–46] areas, which can impact the

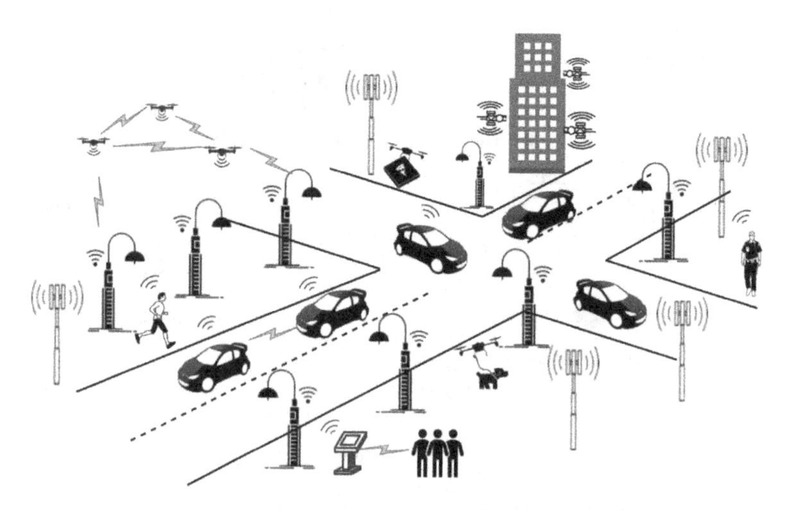

Figure 15.4 Resource Allocation Challenges for multiple Outdoor THz-AP.

city's aesthetic, a problem already slowing down the 5G small cell mmWave deployment in many US cities [47,48]. The THz-APs are supposed to be in the range of a femto or auto cell and might exacerbate the outdoor aesthetic problem. Besides, these THz-APs will also face multiple challenges concerning signal variation due to weather uncertainty [38,49–51] and too much infrastructure need for fronthauling and backhauling [46]. Outdoor deployment architecture for smart cities needs to be more end user focused and based on the specifications of the Internet of Things (IoT) devices [52–54].

In the future, monitoring and managing these resource allocation related challenges for the THz spectrum will be cumbersome for the operators, manufacturers, and regulators. There is an urgent need for methods to manage the resources for these dense heterogeneous devices. Multiple approaches have been proposed in the THz domain to improve the signal strength in a constrained environment [55–59]. However, most of the research has not provided a holistic roadmap to solve the challenges pertaining to granular RRM for higher frequencies [9,32–34]. Moreover, given these myriad number of devices and its multifarious parameters and restrictions, there is a need for a lightweight granular RRM modeling. Classification of resources will help us make better choices for resource allocation. Since factors, such as coverage range, penetration power, small-scale mobility [11,12], and antenna alignment, were not a critical challenge in the traditional RF bands, there was no urgency for classification of resources. However, as we move higher in the spectrum to mmWave and especially THz frequencies, the need for granular resource classification becomes prominent. In this work, we have incorporated the fixed resources, the uncertainties, and regulatory imposed constraints, to make resource management decisions at the THz. We have classified the system resources into three classes and tried to analyze their trade-offs in the following sub-sections.

15.2.2 Fixed Resources

Some system resources can be measured and planned before deployment and does not require dynamic adjustments and can be considered as fixed resources. The fixed resources for the THz are as follows:

(a) *Transmit Power and Hardware*: The scalability and deployment of THz systems is highly dependent on the NextGen nanofabrication and solid-state circuit technology. Challenges in higher transmit power, receiver sensitivity, antenna design, and THz signal generation can impact the end-use case. In the traditional RF spectrum, dynamic transmit power control (DTPC) has often been explored as a method to increase user coverage and reduce interference. However, DTPC is not that effective in the THz band due to its low penetration power and sensitivity to surrounding materials. Currently, it is technically and economically challenging to design a transmitter to transmit more than 2 W for the lower frequencies and ~1 mW for the upper frequencies [3,60]. Some close proximity applications might not require 2 W of power and increasing the power might result in harmful interference and health risks. Moreover, a device's transmit power might need to be regulated for health reasons based on the device proximity and duration of human use [3]. Antenna design will also have to improve, such that it can generate narrow antenna beams with minimal side-lobes, which can otherwise cause interference. Nevertheless, narrow beams and mobility (small and large scale) will lead to frequent beam misalignments and increased latency. The antennas will have to be smart to combat these unnecessary outages [58,59]. Moreover, oscillators, mixers, amplifiers, and signal generation will also have to be improved. Currently THz uses either an all-electronic or photonic using infrared laser-based system to generate THz signals [61,62]. Although the electronic system can achieve higher transmit power compared to the photonic system, frequency tuning is quite difficult using the electronics system. In most cases, the choice of hardware will be fixed, which will constrain the transmit power, operating frequencies, and antenna designs.

(b) *Operating Frequency*: Although the usable frequency windows in the THz is mostly dependent on the environmental conditions [3,63], factors, such as hardware and its application, will dominate the choice of operating frequency for most THz-APs. The THz band resonates at multiple frequencies due to the presence of oxygen and water molecules in the air, which can result in massive dips in throughput, as shown in Figure 15.5 [63,64]. The data rate R, shown in Figure 15.5, for the THz spectrum can be calculated using Shannon's capacity formula shown in Equation (15.1), where B is the bandwidth, which is 1 GHz, and SINR is the signal-to-interference-noise ratio [3,58]. Given the short range, limited penetration, and ample amount of spectrum, we can assume interference to be negligible, and just consider the signal-to-noise ratio (SNR) shown in Equation (15.2). SNR is directly dependent on the transmit power $P_t = 0dBm$, and the antenna gains for the transmitter and receiver pairs. If we assume a conical antenna beamwidth with perfect main lobe, then we can define $G_t = G_R = X / \delta^2$, where X is the antenna aperture dependent factor, which is 52, 525 for a uniformly illuminated circular aperture, and δ is the antenna beamwidth, where narrow antenna beams result in higher antenna gains. In case of mobile users these narrow antenna beams result in beam alignments and outages [12,58], in Figure 15.5 we have assumed a static scenario. While SNR is inversely proportional to the spreading loss $L_S = \left(\frac{4\pi d f_c}{c}\right)^2$, absorption loss $L_A = e^{\kappa(f_c,\rho,T)d}$, and the noise power spectral density $N_0 = 10\log_{10} N_f K T_K = -193.85\text{dB}/\text{Hz}$, where d is the Euclidian distance, f_c is the central frequency, c is the speed of light, κ is the medium absorptions coefficient dependent on f_c, the relative humidity ρ, and room temperature T (which can be calculated from the HIgh resolution TRANsmission molecular absorption (HITRAN) database), N_f is the noise figure of 10 dB, K is the Boltzmann constant of $1.3810e^{-23}$, and T_k is the temperature in Kelvin [39].

$$R = B\log_2(1 + SINR) \tag{15.1}$$

$$R = B\log_2\left(1 + \frac{P_t * G_t * G_R}{L_S * L_A * (N_0 * B)}\right) \tag{15.2}$$

From Figure 15.5, we can observe that low temperature and dry air are *favorable conditions* for using the THz spectrum, which is somewhat possible to maintain indoors through heating, ventilation, and air conditioning (HVAC) systems, but challenging for an outdoor environment. Although Figure 15.5 shows multiple usable frequency windows, even at *unfavorable conditions* of 100% humidity, most of it can only achieve a throughput of fewer than 1 Gbps. Narrow band-IoT (NB-IoT) devices, which require very less throughput, can be the right candidate even in these unfavorable conditions. Higher frequencies are suitable for manufacturing smaller devices, especially devices for nanonetworks and BAN, but will result in massive path loss. On the other hand, lower frequencies will be suitable for relatively larger mobile devices but can result in low throughput in unfavorable conditions. Moreover, the choice of the frequency range is also dependent on THz receivers. Cheap THz-IoT devices might have a wide antenna gain and compelling the manufacturers to make better receivers might stifle the growth of THz-IoT. Thus, the applications and manufacturers will dictate the choice of operating frequencies and will need to be allocated beforehand for reliability purposes.

(c) *Bandwidth*: Similar to the operating frequency, the usable bandwidth is also dependent on environmental conditions and applications [3,60]. Table 15.1 summarizes the

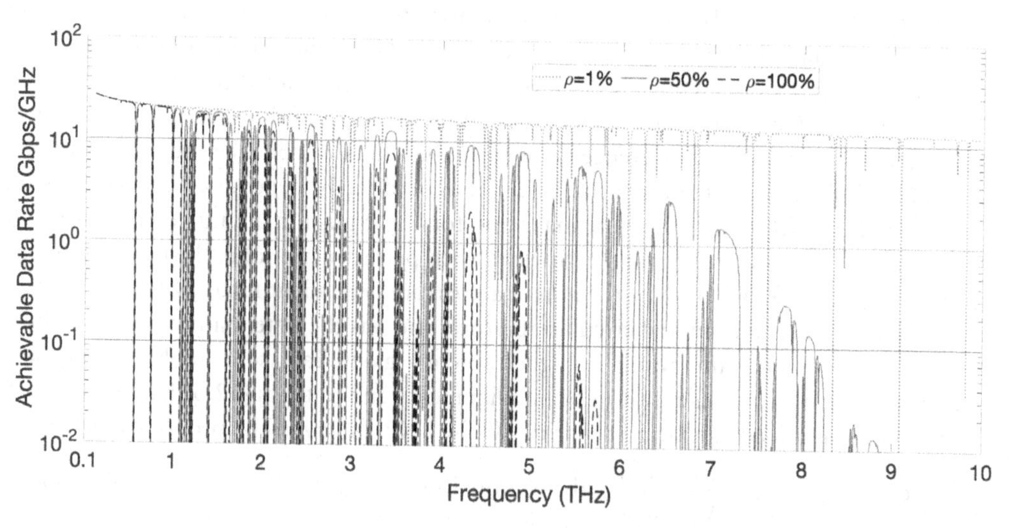

Figure 15.5 Achievable data rate at the THz Spectrum for fixed distance of 5m and varying air quality of dry air at 1% to very humid air at 100% relative humidity at a temperature of 25°C.

Table 15.1 Available contiguous bandwidths (in GHz) at the THz spectrum with a minimum achievable data rate of 1 Gbps, for varying distance and relative humidity

Distance	*RH*	*<1THz*	*1–2THz*	*2–3THz*	*3–4THz*	*4–6THz*	*6–8THz*	*8–10THz*
1 m	**50%**	450	170	80	130	190	310	310
	100%	440				120	300	220
5 m	**50%**	440	100	60	90	-	-	-
	100%	430	90	50	60	-	-	-
10 m	**50%**	430	90	-	-	-	-	-
	100%	270	80	-	-	-	-	-

Although there are multiple frequency windows, as shown in Figure 15.5, in this table we have shown the maximum contiguous bandwidth in that range.

available contiguous bandwidth for multiple chunks of THz bands and at varying environmental conditions. From Table 15.1, it is clear that most of the higher THz frequencies are not usable for high throughput applications in unfavorable conditions of high humidity and long distances. However, multiple applications with very low throughput requirements can operate all throughout the THz band. For example, secure communication, which uses wide bandwidth with low transmit power to hide the information, or NB-IoT, which require moderate throughput with very narrow bandwidth [5]. Due to massive contiguous bandwidth in the THz, there is no need for carrier aggregation, and an enormous number of NB-IoT devices can operate even in the smallest frequency window without the need for sophisticated multiple access methods. Thus, the choice of bandwidth and the operating frequency will be fixed in most cases.

(d) *AP Deployment*: Due to the limited coverage area and low penetration power of the THz spectrum, the number and location of THz-APs can be critical for the received signal strength. Thus, there is a need for an efficient AP deployment strategy. Similar deployment strategies were essential in the sub-6 GHz bands and will become a necessity in the higher frequency bands. A streamlined deployment strategy will need to factor in multiple aspects of the system, such as choice of operating frequency, source of backhaul, performance metric of an application or a vertical, blockages, user mobility and density, energy, and costs. Most of these parameters can be factored in by the operator or network administrator before the deployment and do not need dynamic monitoring and adjustment. Moreover, as we move higher in the frequency, the deployment of indoor and outdoor applications becomes more prominent. Although the use cases remain the same, the deployment strategies will change based on uncertainty or user specifications.

(e) *Modulation and Coding*: The THz channel is sensitive to free space loss, molecular absorption, noise, surrounding surfaces, and hardware limitation, which can negatively impact the signal strength. To mitigate these impacts, lightweight modulation and coding schemes can be utilized [55,56]. Efficient modulation schemes, such as on–off keying (OOK), pulse position modulation (PPM), quadrature phase-shift keying (QPSK), and quadrature amplitude modulation (QAM), will help maintain user throughput for fluctuating channels [65]. While efficient coding, such as Reed Solomon, low-density parity checks, and turbo coding, will help reduce the bit error rate [65]. However, the choice of modulation and coding schemes are highly dependent on the hardware, transmit power, bandwidth, and level of complexity. As explained above, most of these factors will be fixed by the operator or network administrator before deployment.

15.2.3 Varying Resources

Some system resources are uncertain and need to be tracked. Monitoring these resources and even adjusting them dynamically can help improve system performance and can be considered as variable resources. The variable resources for the THz are as follows:

(a) *Environmental Losses*: The THz channel can fluctuate due to uncertainties in multiple factors. Most of these uncertainties are due to environment losses, such as free-space, molecular absorption, sky noise, blackbody noise, scattering, and surface penetration [44,47]. In an outdoor setting dust, rain, fog, snow, and cloud can negatively degrade the already compromised THz channel [66,67]. Most of the analysis to understand the outdoor THz channels has been done through emulated weather chambers, which indicate the use of THz in outdoor is justified under specific weather conditions and can perform better than most of the free-space optics (FSO) systems. Still, obtaining a reliable connection is debated by researchers in this community. While we cannot prevent these losses, we can always work around them by building smart and opportunistic systems. Information about the aggregated losses will allow intelligent algorithms to better understand the channel and environment. Methods, such as AP handover, adjusting antenna gains, adaptive protocol, and negotiating user performance, can be used to ensure system efficiency.

(b) *Distance and Humidity Variance*: Variation in AP–UE distance and humidity can impact not only the signal strength, but also the effective number of usable frequency windows, as shown in Figure 15.6. As we move from a favorable condition of 1 m indoor with moderate relative humidity to an unfavorable condition of 10 m outdoor with high relative humidity, the usable frequency decreases. Although the AP deployment is fixed, as stated in the previous sub-section, small distance fluctuation for mobile devices can impact the signal strength. Multiple indoor models have been proposed in the THz to be distance adaptive [56–58]. Additionally, HVAC systems can be used for indoors or small outdoor enclosures to reduce the temperature and/or humidity and improve signal strength. On the contrary, in the case of free-space outdoor, the humidity information can be used to decide if a THz communication is justified at that time or not. Other vertical-specific applications can also benefit from this information. For example, the information can be helpful for UAV type communication, where the humidity decreases with an increase in altitude. Moreover, mobile applications and NB-IoT devices using frequency hopping can benefit from distance and humidity information. This information can also be transferred to MAC and transport layers to adaptively alter the frame lengths, frame duration, wait times, channel access protocols, and energy fluctuations.

(c) *Mobility*: Mobility management has always been a challenge for RRM. Due to a larger coverage area and higher penetration power of the lower RF bands, device orientation was not a big challenge. However, as we move higher in the spectrum, there is a

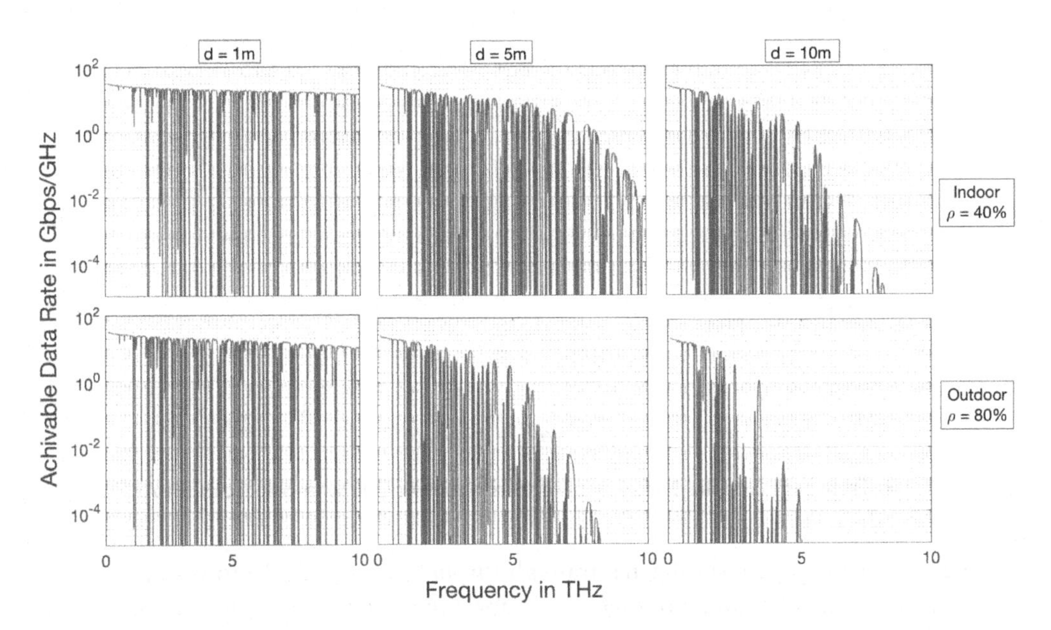

Figure 15.6 Usable THz frequency windows for varying relative humidity and distances for indoor and outdoor environments

need for using narrower antenna beamwidths to improve antenna gains, which makes these higher frequency bands sensitive to devise orientation. Even small-scale mobility such as yaw, pitch, and roll movement [58] of devices can cause significant outages. Beam misalignment for small-scale mobile users or handovers for highly mobile user need to be monitored for efficient resource management. Mobility in the THz band can also result in Doppler shifts [68], as shown in Figure 15.7 for the full THz band. Although these frequency shifts range a few MHz, which is small compared to the massive channel bandwidths, it can be a severe concern for manufacturers while making receivers. Moreover, for mobile NB-IoT devices using MHz or KHz range bandwidth can be critically impacted by this frequency shift. Although mobility is still an unexplored aspect of THz and a hard problem to solve, tracking mobility information can help predict or at least avoid outages. For example, the interruptions caused by the small-scale mobility of the devices or users can result in increasing the latency and reducing the reliability of the system. Based on the mobility type of the devices, use cases can be drawn where the THz spectrum can provide reliable communication and cases when the use of THz is not justified. Currently APs are static; however, in the future AP can be mobile and can be used to dynamically move around in an area to improve system performance [47]. In case of flying ad-hoc networks (FANETs) or ABS, mobility management for AP and the relative mobility between an AP–UE pair will become critical.

(d) *Alignment*: Multiple pieces of research have discussed the need for antenna beam alignment in the mmWave, THz, and FSO bands. Although AP deployment is a fixed resource, the antenna is free to change alignment both physically and logically based on the system needs. Perfect beam alignment for narrow antenna beams can result in

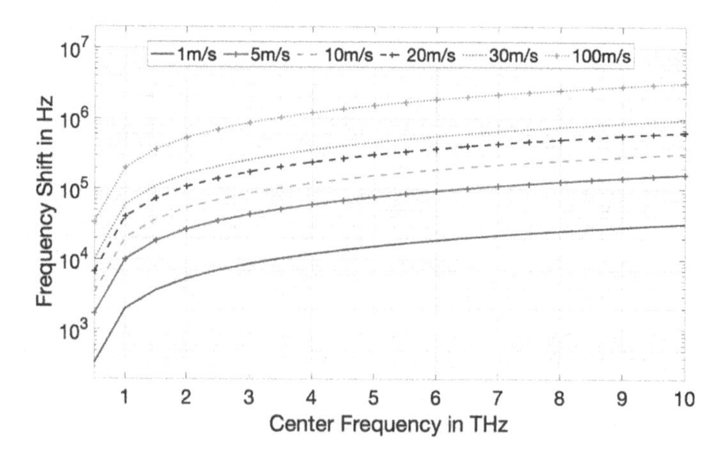

Figure 15.7 THz Frequency Doppler Shift due to relative mobility of UE with respect to the AP.

higher antenna gain, resulting in ultra-high-throughput [11,12,8]. However, uncertainties, such as mobility and blockages, can impact this alignment resulting in outages. To combat these outages, a shower of narrower antenna beams can be used to provide a momentary ultra-high-throughput connection with a user. Additionally, there is a relationship between antenna beamwidths (δ), misalignment, and antenna gain, which is shown in Figure 15.8. Every time a beam is misaligned between an AP–UE pair, both the devices go through a series of beam searching and beam training process, which is time-consuming and can result in severe delays to the system. However, for a realistic system, where the traffic is bursty, an AP–UE pair can tolerate a few degrees of beam misalignment, which might result in a slight drop in antenna gain [12]. By reducing the performance needs for an AP–UE pair, the AP can avoid a beam training process cycle and reduced system delays sufficiently. Although narrower beamwidths of $\delta = 1°$ or 5° can provide higher antenna gain, the freedom for beam misalignment is very narrow, as shown in Figure 15.8. On the other hand, beamwidths of $\delta = 20°$ or 15° will reduce antenna gain but can tolerate wider misalignments. Based on the mobility and location of the users, the AP can adaptively change the beamwidths. For example, narrow beams can be used for static and nomadic users, while wider beams can be used for relatively mobile users. Thus, alignment information, both horizontal and vertical, between an AP–UE antenna pair can be critical for resource allocation and optimization.

(e) *Density*: Most of the THz band cannot penetrate human skin, with some few exceptions at the lower THz bands. Thus, users can act as blockages for other users, which can be frequent in an ultra-dense mobile distribution. The system can use this information to predict the mobility pattern of other neighboring users and adaptively change resource allocation [42] or use surrounding reflective or intelligent surfaces to implement NLOS communication. Thus, information regarding the user density and their exact location can be used by the system to reduce latency.

(f) *MAC Protocols*: Designing efficient MAC layer protocols will be critical for all types of THz devices, whether it operates indoor or outdoor, static or mobile, or uses lower or higher THz frequency bands. The MAC layer is responsible for flow

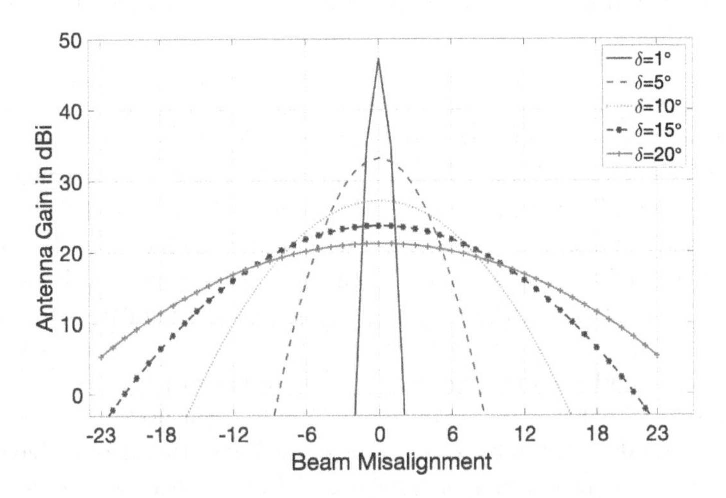

Figure 15.8 Change in antenna gain for varying user mobility and antenna beamwidth.

control, multiplexing, scheduling, beam management, capacity management, and interaction with the physical and upper TCP/IP layers. Thus, the MAC protocols and its parameters can be acute for system throughput and latency. For example, short interval frame gaps of few milliseconds for acknowledgment messages can result in severe delays in THz networks. Even scheduling between nodes and their beams, which is mostly based on time division techniques, can result in significant delays to the system if not properly managed. Thus, there is a need for an efficient MAC that can provide adaptive solutions [63]. For example, the MAC layer can be trained to change frame lengths and access protocols to adapt based on varying environmental uncertainties and performance metrics via information from the physical layer.

15.2.4 Imposed Constraints

Every radio device has to follow a set of policies or imposed constraints set by the regulators, which are generally constraints on the technical parameters or applicability of the devices. Generally, these imposed constraints are specific to a city, county, state, country, or a set of countries. The regulators enforce these constraints either through ex-ante approach, that is, preventive measure to avoid constraint violation, or ex-post approach, that is, punitive measures for constraint violation. Ex-ante approach has multiple disadvantages, which limit the opportunistic use of resource allocation and may stifle the growth of innovative devices and technology. Ex-post approach, which is a more dynamic way of monitoring constraint violation, may be ideal for THz systems. One can argue that the imposed constraints should be a part of fixed resources. However, with the ex-post approach, the constraint violations are subjected to multiple factors and do not fit into the definition of fixed or variable resources. There are agencies, working groups, and organizations that are working on setting global constraints for THz communication [69–72]. Imposed constraints are good, since it guarantees a minimum level of reliability to the system. However, too tight standardization of constraints can negatively affect the potential of THz and hold back multiple opportunities, or even stall the growth of THz. The imposed constraints for the THz are as follows:

(a) *License*: Although the choice of operating frequency in the THz is based on the environment uncertainties and hardware limitations, regulators and political agendas will also have an impact over it [31,71, 72]. Unlicensed spectrum has tremendous economic and technological benefits, as previously observed in the sub-6 GHz band. Most of the researchers and experts are inclined toward making the full THz band unlicensed. However, it might not be the case when the regulators finally decide to open THz bands for IMT use. Licensing of spectrum allows the regulators to reduce enforcement efforts and make sure that the spectrum is used for social benefit. Moreover, the operators and the manufacturers will also like to have licensed access to THz spectrum, so that they can guarantee reliability to their users and use their own proprietary methods. A plausible future for THz will be either deploy dynamic spectrum access (DSA), or have a mix of both licensed and unlicensed bands. In either case knowledge about the operating frequency and licensing will be necessary.
(b) *Interference*: The traditional RF bands can sustain communications over vast distances and can penetrate through most objects, which can cause harmful interference in multiple scenarios. Multiple constraints have been implemented by the regulators to mitigate adjacent-channel and co-channel interference. The drawbacks associated with the THz band, such as low penetration power and small coverage range, makes THz band *interference-friendly* [3]. For example, allocating relatively wider guard bands and the use of narrow direction beams can avoid adjacent-channel and co-channel interference. Nevertheless, THz devices might receive or transmit unintentional interference to other devices already operating in this band. Although the THz band is not allocated to IMT applications, it is currently being used by Earth Exploration Satellite Service (EESS) for passive communication only [31,73]. Using the THz spectrum indoor will not cause any form of interferences to these satellites, but using it outdoor might result in an aggregated interference. Restriction might apply for outdoor THz use, which can vary opportunistically. Furthermore, the THz devices might receive harmful interference from non-communication-related devices used for imaging, security scanners, and spectroscopy [30,74,75]. THz is also likely to be used by military and law enforcement agencies to perform secure communication, which makes them the primary users in the spectrum. Given all of these incumbent users and harmful interference-related challenges, information related to interference restriction will be beneficial for opportunistic allocation of resources.
(c) *Safety Standards*: Since THz belongs in the Infrared band, the THz radiation is often criticized to be hazardous for human health. First, THz is non-ionizing, so it poses less harm compared to higher frequency bands. Second, most of the THz devices are in their rudimentary stage, and there is not enough evidence to prove that THz might be harmful. However, standardization work on safety levels for THz devices to be used near human proximity is being developed, which will be varying based on environment, user density, AP deployment, and application [76–78]. Although the transmit power for most THz devices are extremely low compared to the RF devices and THz cannot even penetrate human skin, the narrow beams and beam exposure time might cause human tissue damage or injure the eye. Information on safety standards

will be critical for resource allocation since it might result in adjusting the antenna beamwidth, AP handovers, number of concurrent THz-AP transmissions, and channel occupation time (COT).

(d) *Security Issues*: Generally, to implement secure communication, a system uses heavy encryption and complex authentication schemes, which can cause system delays. The delays will exponentiate in case of ultra-dense or extremely dense networks. However, due to the ultra-low-latency demands of B5G and 6G applications, operators are shifting toward using physical layer security (PLS) and light encryption. These security methods might compromise the system and result in more security risks. THz devices can use methods, such as (i) large bandwidth coupled with low transmit power to spread the information, (ii) pencil beams with perfect beam alignment, or (iii) frequency hopping throughout the THz band, to prevent jamming and eavesdropping. However, these methods might not be feasible for cheap NB-IoT or eHealth devices that require highly secure communication with least complexity. The THz devices will face a trade-off between low latency and high security. Information about the level of security risk tolerance of a UE might impact multiple variable resources [15,79].

(e) *Privacy Issues*: Knowledge about a UE's precise location is critical for THz systems to perform beamforming, resulting in both higher throughput and better security. However, the location information can cause privacy risks for the users [16,17]. Moreover, other information, such as performance demands, bandwidth, and operating frequency, can be used to engineer out sensitive or behavioral information about the user. Thus, constraints on privacy will have to be regulated on what resource-related information can be shared, accordingly, impacting the resource allocation strategy and system efficiency. Nevertheless, the level of privacy is relative to a UE and can change with scenario or time. Therefore, users might want to share and update their level of privacy risk tolerance to the system, which can be used to serve these users better. In the case of resource deadlock, negotiations between high privacy and high performance might be the only solution. For example, UEs demanding high privacy will not share their real-time location, which might make it difficult for the algorithms to learn and provide high performance. One more concern is the use of crowdsourcing to gather resource-related information about a UE. Although a user might have a very low privacy risk tolerance level, the system might be able to crowdsource that same information from other sources. There are no current standards for privacy in RRM; however, with the popularity and success of the General Data Protection Regulation (GDPR) in the European Union (EU) and California Consumer Privacy Act (CCPA) in California, we might see regulators implementing these privacy-related imposed constraints.

(f) *Auxiliary Spectrum*: It will be highly optimistic and misleading to think that THz alone will be able to solve all the future needs related to B5G and 6G. On the contrary, THz will be used as a catalyst or a support system for other existing spectrum bands. There are multiple use cases and scenarios, where the use of THz is not justified, and switching to lower RF, sub-6 GHz, or mmWave bands will be beneficial. THz systems will need information about how much lower frequency spectrum is available as a fallback in case of unavailability of THz resources or a resource deadlock. This cross-spectrum information might be beneficial while allocating system resources or even designing cross-layer algorithms between RF, mmWave, and THz bands

15.3 THz Resource Management Schemes

In this section, we propose three different schemes to allocate the THz system resources, as illustrated in Section 15.2. Effective and opportunistic resource management can be done among similar users or users competing for similar resources. Although there are multiple ways to group resources [80–83], in this work, we have considered three different grouping factors – (a) mobility, (b) vertical, and (c) granular identification – and provided a strawman framework for each group.

15.3.1 Scheme A: Mobility

As emphasized before in the chapter and previous literature [11,12,58], the THz band is sensitive to mobility. Even small orientational changes in devices mounted over the head and hand can cause outages, leading to higher latency and lower reliability. Knowledge about device location, movement direction, and speed can be helpful while allocating resources. Therefore, the classification of applications based on mobility type can help us find which class of resources needs a tighter bound. To better understand and manage the demands for a mobile device, we classify it into four types of mobility: S: static, N: nomadic, CM: constrained mobility, and HM: high mobility, as shown in Table 15.2. Moreover, the performance and resource demands for these mobility types mentioned above will change based on the deployment, that is, indoor or outdoor. The list of applications shown in Table 15.2 is not exhaustive; however, this mobility classification will help understand the limitations and needs of mobile devices and, in the future, can fit in other use cases.

Now that we have classified these applications based on the mobility deployment type, we can figure out the resource trade-offs for each of these classes, through a resource triangle shown in Figure 15.9, which can help maximize system reliability. The three resource types explained in the previous section can be adjusted for each of these classes. Variable resources come with uncertainty, and we need to reduce their dependency by setting tighter bounds for fixed resources and imposed constraints. For example, in a constrained environment with lower uncertainty, such as indoors, tighter fixed resources can guarantee more reliability than depending on tighter imposed constraints. On the other hand, highly mobile application systems cannot be let free to work stochastically around with

Table 15.2 THz use cases classification based on mobility and deployment type

Mobility Type *Deployment Type*	*Static (S)*	*Nomadic (N)*	*Constrained Mobility (CM)*	*High Mobility (HM)*
Indoor	• UHD Displays • HTC • Data Center • Indoor Backhaul • On-Chip Comm.	• Hotspots • Laptops/Tablets • Robotic Surgery • Remote Diagnostics	• Walking & Streaming • Live Broadcast • Information Shower	• VR/AR/MR Headsets • Mobile Ad-Hoc Networks • Cm-Level Positioning
Outdoor	• Backhaul • Sensing • Remote Diagnostics	• Nomadic Vans • Ultra-High-Kiosks • Remote Diagnostics	• In-car virtual Systems • V2X • Drone Fleet Control • ABS	• VR/AR/MR Users • V2X

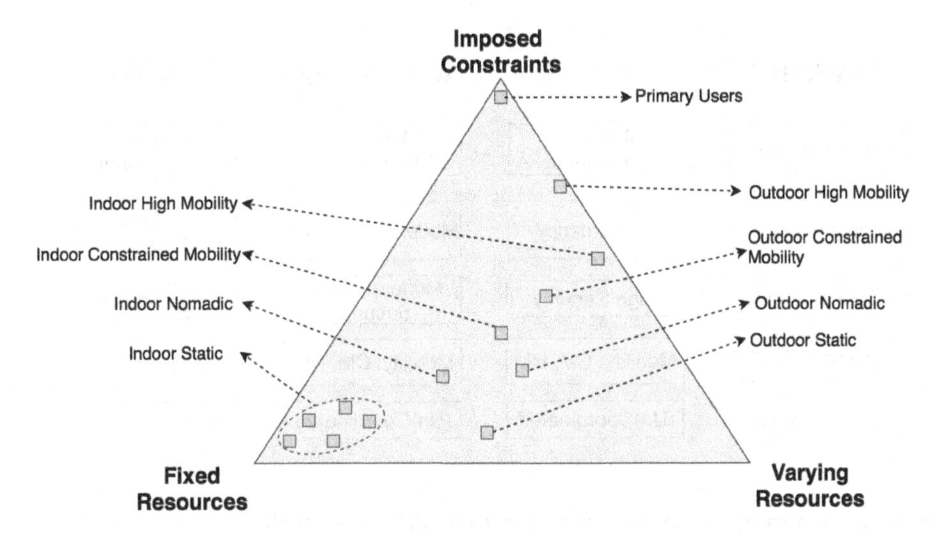

Figure 15.9 Resource triangle illustrating the resource dependencies and trade-off for mobile users.

the varying resources, especially for non-delay tolerant applications, which can lead to outages, delays, and resource wastage. Setting tighter imposed constraints will be a better solution for relatively mobile devices, both indoor and outdoor. A schematic illustration of this approach has been shown in Figure 15.9, for the classes introduced in Table 15.2.

By tighter bounds on fixed resources and imposed constraints, we can limit the uncertainty of the varying resources. For example, indoor static and indoor nomadic users do not have so much uncertainty in the system and can be just dependent on fixed resources for reliability. A similar argument holds for outdoor static and outdoor nomadic users, but the higher level of uncertainty in variable resources push these classes to the right corner of the triangle.

As the mobility increases from CM to HM, only depending on tighter fixed resources will not help and stronger imposed constraints need to be set. In the case of indoor CM or HM, users can also be catered with the deployment of a blanket of THz-APs. However, challenges, such as aesthetics, privacy, and health safety, will limit the densification. Most of the outdoor type applications have large uncertainties in varying resources, which might compel the regulators to impose stronger imposed constraints. However, too conservative imposed constraints might stifle the growth of outdoor THz. Maybe at a later stage with advanced machine learning algorithms and artificial intelligence systems might be able to predict these variabilities in the system, and the location in the triangle can be adjusted toward the fixed resources corner. While allocating resources, the system also needs to consider the primary user (PU), likely the EESS, military, or spectroscopy applications, and avoid interfering with them. We assume that there is a central decision making unit that can effectively place these applications correctly on the resource triangle shown in Figure 15.9.

15.3.2 Scheme B: Verticals

B5G and 6G will trigger multiple new verticals [2,5], and the performance and resource demands might be different for each of these verticals. Moreover, the resource management

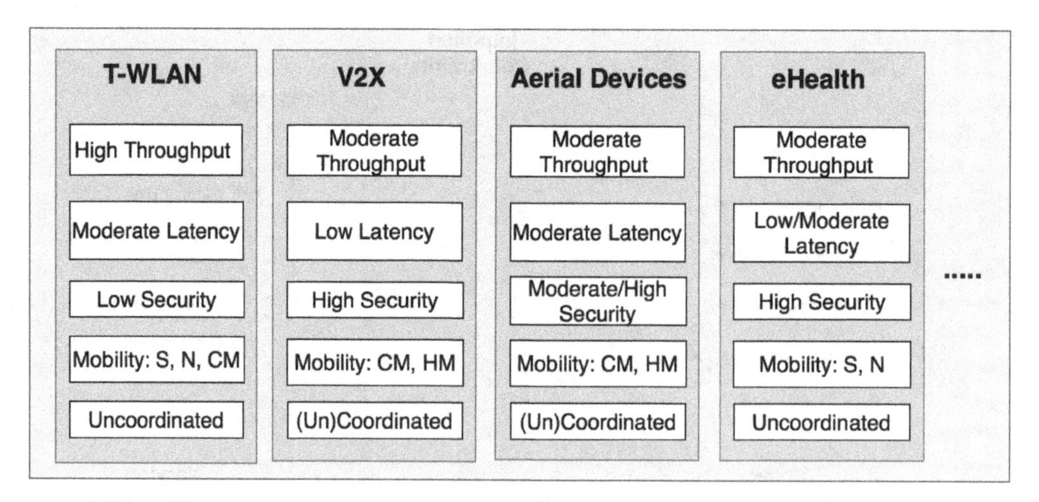

Figure 15.10 A comparative key performance index for different verticals.

and decision-making unit will also be different. For example, resources might be managed by a central service or an operator, and each vertical is provided with resources through network slicing, or the resources can be self-regulated by the vertical itself. Therefore, Scheme A, which was a general approach with a common decision maker, will not work when we consider resource allocation for verticals. For example, a wearable device can belong to both WLAN and eHealth verticals, but the performance and the resource needs within each vertical are completely different. There are multiple verticals, which will be born from B5G and 6G, but for simplicity, we consider four, namely, terabit wireless local area networks (T-WLAN), V2X, aerial devices, and eHealth. We list the key performance indices (KPI) for each of the verticals [2,5], which are essential for the system performance, in Figure 15.10.[4] Since the KPIs vary among verticals, the resource trade-offs will also change. Similar to Figure 15.10, the resource dependency for each of the verticals and the applications within the verticals will be different and have been illustrated in Figure 15.11.

The T-WLAN vertical will consist of indoor devices, such as laptops, tablets, wearables, UHD, HTC, real-time video, AR/VR/MR applications, data centers, centimeter-level positioning, and on-chip communication. These applications will likely demand high throughput, moderate latency, and with the least security compared to other verticals, as shown in Figure 15.10. Since most of these devices are indoor and operating within a constraint environment with minimum mobility, a tighter fixed resource can be used as an approach to ensure reliability to the system. Some applications within the verticals, which belong to a relatively more mobile class, are likely to be skewed to the variable resource corner of the triangle, shown in Figure 15.11. While the applications with sensitive information and demanding relatively higher reliability will be skewed to the imposed constraints corner of the resource triangle.

The V2X vertical will consist of outdoor applications, such as autonomous cars Society of Automotive Engineers (SAE) level 5 to 3,[5] vehicle platooning, autonomous trains, and nomadic vans or broadcast auxiliary service (BAS) [81, 84,85]. Although these applications have a moderate throughput demand, they require low latency and high security, as shown

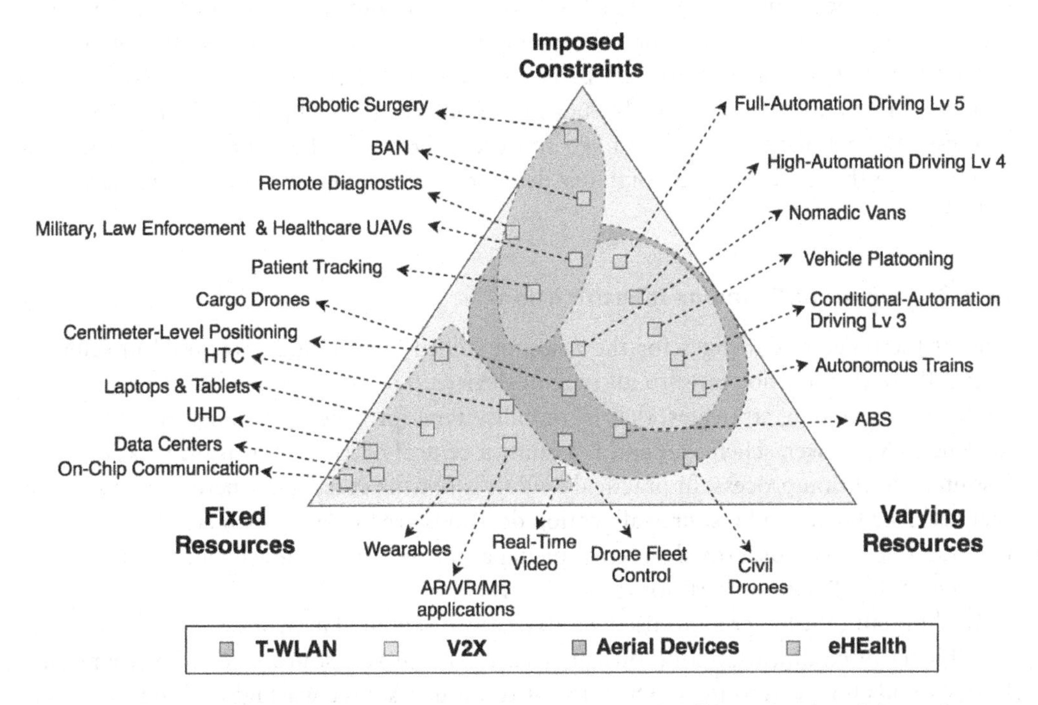

Figure 15.11 Resource triangle illustrating the dependency and trade-off for verticals and the applications within each vertical.

in Figure 15.10. Providing such top performances for relatively high mobile devices will be challenging, and tighter fixed resources will not solve the problem. Moreover, completely removing the dependency from variable resources will also be challenging. Nomadic vans or BAS experience less uncertainty compared to other V2X applications and can depend on the fixed resource for reliability. However, for other applications, setting tighter imposed constraints is the only viable solution, which can be a downside for the growth of THz in this vertical. Some applications can have relaxed imposed constraints and instead depend on intelligent algorithms to work around with the variable resources. Thus, applications, such as autonomous trains, conditional-automation driving, and vehicle platooning, which operate on a coordinated system, can have lighter imposed constraints compared to high-automation and full-automation driving, as shown in Figure 15.11.

Aerial devices, which is a booming vertical, have multiple applications ranging from military use, law enforcement, health services, cargo delivery, ABS, and data collection [84,85,86]. Similar to V2X, all of these are outdoor centric applications and will have similar resource trade-off as V2X. However, coordinated devices, such as drone fleet control and ABS can be operated using an intelligent algorithm and do not need tighter imposed constraints. Moreover, the civil drones, which are used for recreation and operate in an uncoordinated manner, also do not require tighter imposed constraints. On the contrary, applications, such as military, law enforcement, health services, cargo delivery, will depend on tighter imposed constraints for high reliability and highly secure communication, as shown in Figure 15.11.

eHealth applications are similar to T-WLAN vertical, but can be used both indoor and outdoor. Applications, such as patient tracking and remote diagnostics, do not require low latency, but robotic surgery and BAN demand the system to be tactile, as shown in Figure 15.10 [87,88,89]. The low latency and moderate throughput can be met for most of these applications using tighter fixed resources; however, the importance of the data handled and the need for high reliability demands tighter imposed constraints, as shown in Figure 15.11.

15.3.3 Scheme C: Granular Identification

One of the biggest challenges for the resource allocation methods proposed in schemes A and B are they do not cater to all type of devices. The schemes are based on user and application grouping strategies, that is, mobility type for Scheme A and vertical type for Scheme B. Moreover, schemes A and B assume a central or known decision-maker for the resource allocation process. In practical field, it is not the case, since there can be systems that are distributed, and resource allocation decisions are made through election of cluster nodes, ad-hoc methods, crowdsourcing, media access sensing, or greedy approaches. Both schemes A and B are not ideal for these scenarios.

Traditionally, the lower unlicensed bands or any distributed network can implement two possible resource allocation solutions: (a) least conservative resource allocation, that is, use loose bounds for the resources, which might result in resource wastage and reduce system efficiency; or (b) most conservative resource allocation, that is, use very tight bounds for the resources, which might result in reduced user coverage and system efficiency. In either case, there is a tendency that the system will face a "tragedy of the commons" problem, that is, individuals in a shared-resource system act independently for their self-interests, which results in a depletion or deadlock of resources for the system collectively [90,91]. THz systems are already very sensitive to uncertainties in resources, and an unmanaged distributed system will only make the system worse. RRM for distributed THz system will require more sophisticated, tactile, lightweight, and energy-efficient schemes. Therefore, we propose a third approach, scheme C, which is based on UEs sharing their performance needs and technical capabilities with their neighboring nodes. This information can then be used by the individual UE in the system to either predict or avoid a resource collision, and then make resource allocation strategies either in a coordinated or uncoordinated fashion. To share this information about the performance needs and technical capabilities, we propose granular identification, that is, uniquely identifying a device both logically and physically [91].

In this scheme, users can be from any vertical, demand any performance metric, and have any set of resources. The users are not pre-classified based on verticals, deployment environment, or mobility, like schemes A and B. On the other hand, the devices are free to work independently, like a normal distributed system, but along with an added extra knowledge about their neighbors and resources. The content of these granular ID structure can be based on a stacked architecture, illustrated in Figure 15.12. This stacked architecture is based on the similar resource classification strategies introduced earlier in the chapter. Devices through exchanging granular IDs can obtain information, already embedded in the ID structure, about each layer, that is, (a) decision maker, (b) vertical, (c) performance metric, and (d) available resources.

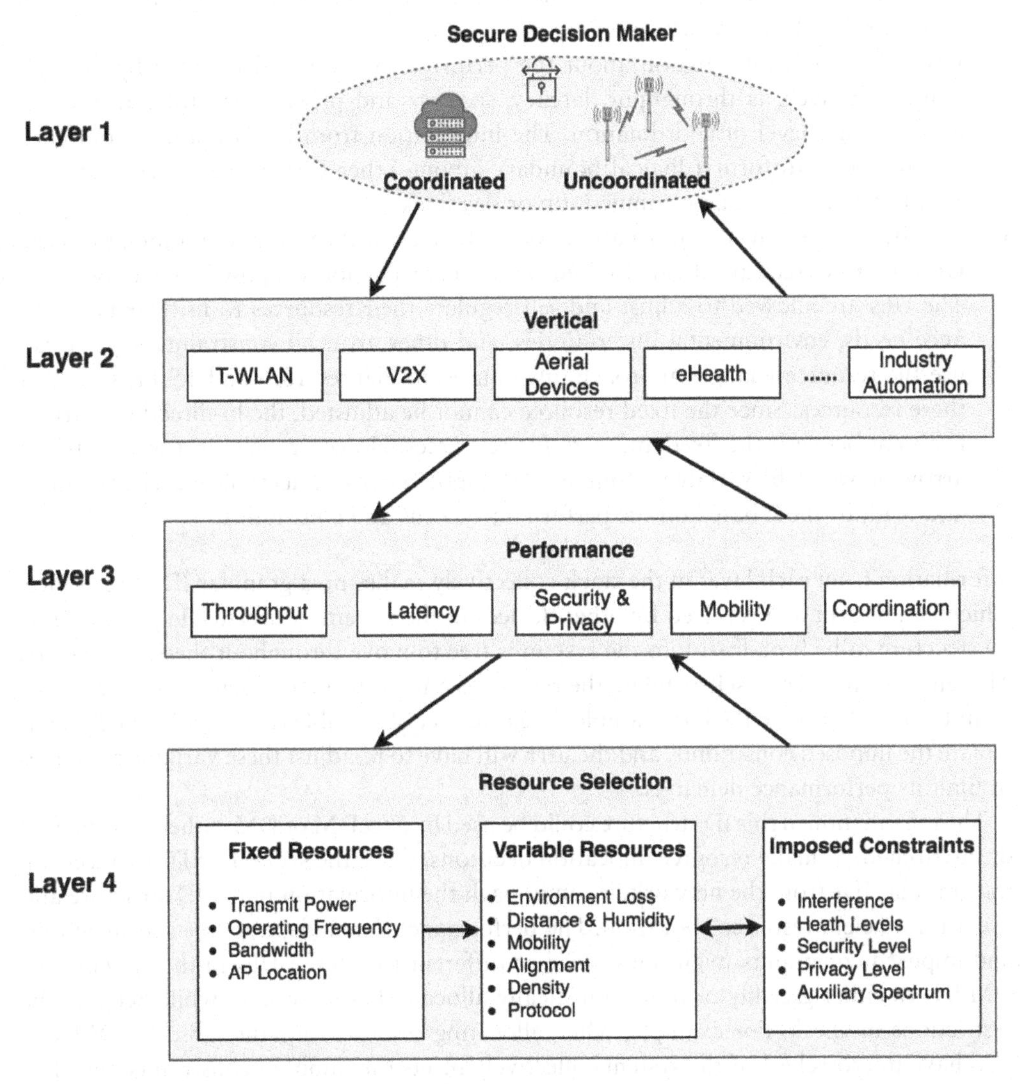

Figure 15.12 Granular ID management structure for granular resource management of coordinated and uncoordinated systems.

Classification of a device identification based on these four layers will allow the UEs to make informed resource allocation strategies. Each layer helps to narrow down a device to its exact resource needs. The four layers are as follows:

- Layer 1 embeds the information about the presence of the type of a decision maker (DM), coordinated or uncoordinated, and the security level of that DM. Information about the presence of a secure decision maker (secDM) will allow the UEs to decide on how granular information should be embedded further into the ID structure.
- Layer 2 embeds the information about a user's vertical, as discussed in Section 15.3.2. The vertical information will help the UEs to logically divide resources pertaining to

each vertical and its needs. This is somewhat similar to a network slicing approach but in a distributed framework.

- Layer 3 embeds information about the performance metrics demanded by an individual UE, such as throughput, latency, security and privacy risk tolerance levels, mobility, and level of coordination. The information from levels 2 and 3 combined will help UEs to form a logical boundary among other UEs in the system and may avoid any future resource competition or deadlocks.
- Finally, layer 4 embeds information about the technical resources pertaining to each UE. The resources are divided based on the classification proposed in Section 15.2. The UEs are allowed to adjust and self-regulate their resources to fit their performance needs, environmental uncertainties, and other artificial constraints. A user can use the resource triangle approach, introduced in Figures 15.9 and 15.11, to adjust these resources. Since the fixed resource cannot be adjusted, the bi-directional arrow is absent between the fixed and variable resources. However, there is a bi-directional arrow between the variable resources and imposed constraints since a user can adjust and readjust these based on the performance needs and constraints.

Information from each layer in the stack collectively makes up a granular ID for a device, which can be sent and received by other devices in the system. A user while preparing its ID structure to be broadcast into the system is free to move throughout these four layers. This enables the user to self-regulate the resources and performance demands, which they want to embed into the ID. For example, a selected set of variable resources by a UE might violate the imposed constraints, and the user will have to re-adjust these variable resources or limit its performance demands.

The information in this ID structure could be used by a secDM or DM, either coordinated or distributed, to make resource allocation decisions. Once the secDM or DM collects all the granular IDs from the network, it can unpack the information in the ID structure and use it for efficient resource allocation. The performance demands, resource specifications, and imposed constraints might be similar or different for all the UEs in the system. The DM has to either greedily or opportunistically allocate these resources while keeping the restrictions in check. For example, while allocating resources for the UEs, the DM will also have to crosscheck if the system collectively meets the imposed constraints, which is available in the ID structure of the UEs. Eventually a secDM or DM will come up with a resource triangle for the system, with all the users placed in this triangle, similar to the one showed in Figure 15.11. Please note that the efficiency of the allocation scheme and placement of the users on the resource triangle by a secDM or DM is limited based on the information present in the ID structure of the UEs. The ID structure sent back by the secDM or DM will be the set of suggested resources or performance metric, which might avoid a resource collision among users in the system. However, the users in a distributed network are not compelled to follow these resource allocation suggestions and can still act selfishly, which might result in resource scarcity.

In the case of a resource deadlock, the secDM or DM can also exchange the IDs with all or a set of UEs for negotiating the performance metric and resource allocation among devices. During the negotiation phase, the secDM or DM can decide to provide a lower performance metric compared to what the UE had requested, or re-adjust the variable resources based on the resource triangle of the UE. The DM, after selecting a set of variable resources for all the users in the system, will also have to crosscheck for violation in imposed constraints and might have to redo the variable resource assignment for all or some users. This negotiation might be helpful for static and nomadic users but can increase delays in the case of relatively higher mobile devices. Still, in case of the mobile devices, the ID structure can be used by the DM to train the MAC layer to change frame lengths and access protocols and adapt based on varying environmental uncertainties and performance metrics.

To summarize, using granular identification for distributed networks to improve coordination has been evaluated in [81]. It can be argued that using granular ID and resource allocation will result in increased packet load and delays. THz has a lot of bandwidth to transmit these granular ID structures, so packet load will not be a big concern. Decision making by secDM or DM, further negotiation, and implementing a security feature for the secDM, might increase network delays. However, in the future, with quantum computing, intelligent networks, and edge computing, these drawbacks can be overcome. Even if we remove the need for a secDM or DM from the system for making resource allocation strategies, just sharing the information through granular ID structure will help the UEs in a distributed THz network to avoid or predict any form of resource collision.

15.4 Conclusion

In this chapter, we proposed resource classification and allocation strategies for high-frequency networks, such as THz. We present a set of technical and policy approaches that can be used for modeling and deploying THz systems for both indoor and outdoor environments. One of the biggest challenges for the future THz network will be to perform efficient RRM for dense-heterogeneous networks. However, we show that with proper "knowledge" about the environment (humidity, temperature, turbulence, operating frequency, blockages, antenna type, UEs and APs location, and user profile) and smart algorithms, we can provide reliable ultra-high data rate for both indoor and outdoor devices. Granular knowledge will allow algorithms to provide reliable THz connection. Since gathering knowledge is hard, cumbersome, and raises security and privacy concerns, we preset resource classification and resource allocation models.

To understand these resources and their trade-offs, we classify them into fixed resources, variable resources, and imposed constraints. Although considering regulatory constraints (a.k.a, policies/ imposed constraints) as a part of a resource for RRM is not the general notion, these imposed constraints can act as limiting factors for the efficiency of future THz systems. Setting tighter imposed constraints assures reliability to the operators and users; however, it can also stifle the growth in manufacturing. To further understand the interdependencies and trade-off between these resources, we presented a resource triangle. Applications, either grouped based on mobility or verticals, can be strategically located in this resource triangle, illustrated in Figures 15.9 and 15.11. The location of these applications on the resource triangle is purely quantitative but can be evaluated to prove the same. The motivation for the resource triangles is to provide a framework for the operators, manufacturers, and regulators to start preparing the THz spectrum for the future B5G and 6G networks. For devices that cannot be grouped based on mobility and verticals, or are distributed in nature, they can still perform efficient RRM using granular identification. The THz devices using granular IDs and the resource information embedded within these IDs can either through a centralized, clustered, or distributed architecture, allocate or negotiate radio resources to improve system efficiency. With proper resource classification, understanding of resource dependencies through resource triangles, and efficient resource allocation schemes, THz can be a champion to the future mobile wireless communication.

Acknowledgments

This project was funded by CMU Portugal Program: CMU/ECE/0013/2017- THz Communication for Beyond 5G Ultra-fast Networks. We thank Dr. William Lehr, Dr. Kazi Mohammed Saidul Huq, and Dr. Tim Brown for discussions on this concept with the authors.

Notes

1 Throughout the chapter we use the following abbreviation to indicate the mobility type of a device: S: static, N: nomadic, CM: constrained mobility, and HM: high mobility.
2 FCC recently auctioned 425 MHz in 24 GHz band, and in the future plan to auction 1 GHz in the 47GHz band [3].
3 THz spectrum belongs in the infrared range and is thermal in nature, which, when exposed with high intensity and for a long duration, can heat up human skin [3].
4 Please note that there are no fixed benchmarks for the KPIs in each vertical, and are ranges with loose bounds. These KPIs are subjected to the operators, manufacturer, user, and regulator. Therefore, we consider three levels, high, moderate, and low, to differentiate between the KPI needs among the verticals.
5 The Society of Automotive Engineers (SAE) defines six levels of driving automation: Level 0 – no driving automation, Level 1 – driver assistance, Level 2 – partial automation, Level 3 – conditional automation, Level 4 – high automation, and Level 5 – full automation.

Notes

[1] International Telecommunication Union. 2018. "Network 2030 A Blueprint of Technology, Applications and Market Drivers Towards the Year 2030 and Beyond." Geneva. www.itu.int/en/ITU-T/focusgroups/net2030/Documents/White_Paper.pdf
[2] Giordani, M., Polese, M., Mezzavilla, M., et al. 2020. "Towards 6G Networks: Use Cases and Technologies." https://arxiv.org/pdf/1903.12216.pdf (accessed February 4, 2020)
[3] Singh, R., Lehr, W., Sicker, D., Huq, et al. 2019. "Beyond 5G: The Role of THz Spectrum." In Proceedings of TPRC 47: The 47th Research Conference on Communication, Information and Internet Policy. Washington D.C.
[4] International Telecommunication Union. 2017. "5G roadmap: Challenges and Opportunities Ahead." Information Session by ITU-R and ITU-T. www.itu.int/en/ITU-D/Conferences/GSR/Documents/GSR2017/IMT2020%20roadmap%20GSR17%20V1%202017-06-21.pdf
[5] Vannithamby, R., and Soong, A. 2020. *5G Verticals: Customizing Applications, Technologies and Deployment Techniques*. Hoboken, NJ: John Wiley & Sons.
[6] International Telecommunication Union. 2017. "Minimum requirements related to technical performance for IMT-2020 radio interface(s)." www.itu.int/md/R15-SG05-C-0040/en
[7] Han, C., Wu, Y., Chen, Z., et al., 2019. "Terahertz Communications (TeraCom): Challenges and Impact on 6G Wireless Systems." https://arxiv.org/abs/1912.06040
[8] Pal, R., Srinivas, K. V., Chaitanya, and A. K. 2018. "A Beam Selection Algorithm for Millimeter-Wave Multi-User MIMO Systems." *IEEE Communications Letters* 22(4): 852–855. doi:10.1109/LCOMM.2018.2803805

[9] Lin, C. and G Ye Li. 2015. "Adaptive beamforming with resource allocation for distance-aware multi-user indoor Terahertz communications." *IEEE Transaction on Communication.* 63(8): 2985–2995.

[10] Jain, A., Lopez-Aguilera, E., and Demirkol, I. 2019. "Are Mobility Management Solutions Ready for 5G?" https://arxiv.org/pdf/1902.02679.pdf

[11] Petrov, V., Moltchanov, D., Koucheryavy, Y., et al. 2018. "The Effect of Small-Scale Mobility On Terahertz Band Communications." In Proceedings of the 5th ACM International Conference on Nanoscale Computing and Communication (NANOCOM), Reykjavik, Iceland.

[12] Singh, R., Sicker, D., Huq, K. M. S., 2020. "MOTH – Mobility-induced Outages in THz: A Beyond 5G (B5G) Application." In Proceedings of IEEE Consumer Communications & Networking Conference (CCNC), Las Vegas, NV.

[13] Manan, W., Obeidat, H., Alabdullah, A., et al. 2018. "Indoor to Indoor and Indoor to Outdoor Millimeter Wave Propagation Channel Simulations at 26 GHz, 28 GHz and 60 GHz for 5G mobile networks." *The International Journal of Engineering and Science* 7(3): 8–18. https://doi.org/10.9790/1813-0703020818

[14] Zhong, Z., Zhao, J., and Li, C., 2019. "Outdoor-to-Indoor Channel Measurement and Coverage Analysis for 5G Typical Spectrums." *International Journal of Antennas and Propagation.* https://doi.org/10.1155/2019/3981678

[15] Fang, D., Qian, Y., and Hu, R. Q. 2018. "Security for 5G Mobile Wireless Networks." *IEEE Access* 6: 4850–4874.

[16] Liyanage, M., Salo, J., Braeken, A., et al. 2018. "5G Privacy: Scenarios and Solutions." IEEE 5G World Forum (5GWF), Silicon Valley, CA.

[17] Sağlam, E. T., and Bahtiyar, Ş. 2019. "A Survey: Security and Privacy in 5G Vehicular Networks." In Proceedings of 4th International Conference on Computer Science and Engineering (UBMK), Samsun, Turkey.

[18] Wu, Y., Khisti, A., Xiao, C., et al. 2018. "A Survey of Physical Layer Security Techniques for 5G Wireless Networks and Challenges Ahead." *IEEE Journal on Selected Areas in Communications* 36(4): 679–695.

[19] Knězevic, M., Nikov, and V., Rombouts, P. 2012. "Low-Latency Encryption – Is "Lightweight = Light + Wait"?" In Proceedings of Conference on Cryptographic Hardware and Embedded Systems (CHES), Leuven, Belgium.

[20] Oughton, E.J., and Frias, Z. 2018. "The Cost, Coverage and Rollout Implications Of 5G infrastructure in Britain." *Telecommunications Policy* 42(8): 636–652.

[21] Aykin, I., and Karasan, E., 2019. "An Activity Management Algorithm for Improving Energy Efficiency of Small Cell Base Stations in 5G Heterogeneous Networks." https://arxiv.org/abs/1901.10021

[22] Tripathi, N.D., and Reed, J. H. "5G Evolution: On the Path to 6G: Expanding the Frontiers of Wireless Communications," Rohde & Schwarz, Germany.

[23] Tariq, F., Khandaker, M., Wong, K.-K., et al., 2019. "A Speculative Study on 6G." https://arxiv.org/abs/1902.06700

[24] Dang, S., Amin, O., Shihada, B. *et al.* 2020. "What Should 6G Be?" *Nature Electronics* 3(1): 20–29. https://doi.org/10.1038/s41928-019-0355-6

[25] T. S. Rappaport, Xing, Y., Kanhere, O., et al. 2019. "Wireless Communications and Applications Above 100 GHz: Opportunities and Challenges for 6G and Beyond." *IEEE Access* 7: 78729–78757.

[26] International Telecommunication Union. 2015. "IMT Vision – Framework and Overall Objectives of the Future Development of IMT for 2020 and Beyond." Recommendation ITU-R M.2083-0. www.itu.int/dms_pubrec/itu-r/rec/m/R-REC-M.2083-0-201509-I!!PDF-E.pdf

[27] Dhillon, S. S., Vitiello, M. S., Linfield, E. H., et al. 2017. "The 2017 Terahertz Science and Technology Roadmap." *Journal of Physics D: Applied Physics* 50(4). doi:10.1088/1361-6463/50/4/043001

[28] Lemic, F., Abadal, S., Tavernier, W., et al. 2019. "Survey on Terahertz Nanocommunication and Networking: A Top-Down Perspective." https://arxiv.org/abs/1909.05703

[29] Elayan, H., Amin, O., Shihada, B., et al. 2020. "Terahertz Band: The Last Piece of RF Spectrum Puzzle for Communication Systems." *IEEE Open Journal of the Communications Society* 1: 1–32, 2020.

[30] Sarieddeen, H., Saeed, N., Al-Naffouri, T. Y., et al. 2019. "Next Generation Terahertz Communications: A Rendezvous of Sensing, Imaging and Localization." https://arxiv.org/pdf/1909.10462

[31] Tekbıyık, K., Ekti, A. R., Kurt, G. K., et al. 2019. "Terahertz Band Communication Systems: Challenges, Novelties and Standardization Efforts." *Physical Communication* 35:1–18.

[32] Li, Y., Pateromichelakis, E., Vucic, N., et al. 2017. "Radio Resource Management Considerations for 5G Millimeter Wave Backhaul and Access Networks." *IEEE Communications Magazine* 55(6): 86–92.

[33] Saddoud, A., Doghri, W., Charfi, E., et al. 2020. "5G Radio Resource Management Approach For Multi-Traffic IoT Communications." *Computer Networks* 166: 106936, issn 1389-1286, doi.org/10.1016/j.comnet.2019.106936, www.sciencedirect.com/science/article/pii/S1389128618303876

[34] Calabrese, F.D., Wang, L., Ghadimi, E., et al. 2016. "Learning Radio Resource Management in 5G Networks: Framework, Opportunities and Challenges." *IEEE Communications Magazine* 56(9): 138–145.

[35] Nallappan, K., Guerboukha, H., Nerguizian, C., et al. 2018. "Live Streaming of Uncompressed 4K Video Using Terahertz Wireless Links." In Proceeding of IEEE International Conference on Communications (ICC), Kansas City, MO.

[36] Piesiewicz, R., Jansen, C., Wietzke, S., et al. 2007. "Properties of Building and Plastic Materials in the THz Range." *International Journal of Infrared and Millimeter Waves* 28(5): 363–371. https://doi.org/10.1007/s10762-007-9217-9

[37] Ning, B., Chen, Z., Chen, W., et al., 2019. "Channel Estimation and Transmission for Intelligent Reflecting Surface Assisted THz Communications." https://arxiv.org/abs/1911.04719

[38] Piesiewicz, R., Jansen, C., Mittleman, D., et al., 2007. "Scattering Analysis for the Modeling of THz Communication Systems." *IEEE Transactions on Antennas and Propagation* 55(11): 3002–3009. DOI: 10.1109/TAP.2007.908559

[39] Singh, R., Sicker, D. 2020. "An Analytical Model for Efficient Indoor THz Access Point Deployment." In Proceedings of IEEE Wireless Communications and Networking Conference (WCNC). https://arxiv.org/abs/2001.11167

[40] Lin, C., and Li, G. Y. 2015. "Indoor Terahertz Communications: How Many Antenna Arrays Are Needed?" *IEEE Transactions on Wireless Communications* 14(6): 3097–3107. doi: 10.1109/TWC.2015.2401560

[41] Petrov, V., Kokkoniemi, J., Moltchanov, D., et al., 2018. "Last Meter Indoor Terahertz Wireless Access: Performance Insights and Implementation Roadmap." *IEEE Communications Magazine* 56(6): 158–165.

[42] Singh, R., and Sicker, D. 2020. "SHINE (Strategies for High-frequency INdoor Environments) with Efficient THz-AP Placement." In Proceedings of IEEE 91st Vehicular Technology Conference: VTC2020-Spring, Workshop on Terahertz Communication for Future Wireless Systems. https://arxiv.org/submit/3086645/view

[43] Lemic, F., Marquez-Barja, J., and Famaey, J. 2019. "Modeling and Reducing Idling Energy Consumption in Energy Harvesting Terahertz Nanonetworks." Proceedings of IEEE Global Communications Conference (GLOBECOM), Waikoloa, HI.

[44] Pandharipande, A., and Thijssen, P. 2019. "Connected Street Lighting Infrastructure for Smart City Applications." *IEEE Internet of Things Magazine* 2(2): 32–36.

[45] Huq, K. M. S., Busari, S. A., Rodriguez, J., et al. 2019. "Terahertz-Enabled Wireless System for Beyond-5G Ultra-Fast Networks: A Brief Survey." *IEEE Network* 33(4): 89–95.

[46] Singh, R., and Sicker, D. 2020. "Reliable THz Communications for Outdoor based Applications- Use Cases and Methods." Proceedings of IEEE Consumer Communications & Networking Conference (CCNC), Las Vegas, NV. https://arxiv.org/abs/1911.05330

[47] Stevens, I., 2019. "Imagining Future Cities: Aesthetic Design Guidelines for Small Cells." http://dx.doi.org/10.2139/ssrn.3427587

[48] Cramer, B. W. 2019. "Not Over My Backyard: The Regulatory Conflict between 5G Rollout and Environmental and Historic Preservation." http://dx.doi.org/10.2139/ssrn.3427211

[49] Ma, J., Adelberg, J., Shrestha, R., et al. 2018. "The Effect of Snow on a Terahertz Wireless Data Link." *Journal of Infrared Millimeter and Terahertz Waves* 39: 505–508.

[50] Federici, J.F., Ma, J. and Moeller, L., 2016. "Review of weather impact on outdoor terahertz wireless communication links." *Nano Communication Networks*, 10: 13–26.

[51] Amarasinghe, Y., Zhang, W., Zhang, R., Mittleman, D.M., and Ma, J., 2020. "Scattering of Terahertz Waves by Snow." *Journal of Infrared, Millimeter, and Terahertz Waves*, 41(2): 215–224.

[52] Bergés, M., and Samaras, C., 2019. "A Path Forward for Smart Cities and IoT Devices." *IEEE Internet of Things Magazine* 2(2):2–4.

[53] Davy, A., Pessoa, L., Renaud, C., et al., 2017. "Building an End User Focused THz Based Ultra High Bandwidth Wireless Access Network: The TERAPOD approach." In Proceedings of 9th International Congress on Ultra Modern Telecommunications and Control Systems and Workshops (ICUMT). doi: 10.1109/ICUMT.2017.8255205.

[54] Hammons, R., and Myers, J., 2019. "Architects of Our Future: Redefining Smart Cities to Be People-Centric and Socially Responsible." *IEEE Internet of Things Magazine* 2(2): 10–14.

[55] Jornet, J. M., and Akyildiz, I. F. 2011. "Channel Modeling and Capacity Analysis for Electromagnetic Wireless Nanonetworks in the Terahertz Band." *IEEE Transactions on Wireless Communications* 10(10): 3211–3221.

[56] Han, C., and Akyildiz, I. F. 2016. "Distance-aware Bandwidth-Adaptive Resource Allocation for Wireless Systems in the Terahertz Band." *IEEE Trans. Terahertz Science and Tech.* 6(4): 541–553.

[57] Akyildiz, I. F., Han, C., and Nie, S. 2018. "Combating the Distance Problem in the Millimeter Wave and Terahertz Frequency Bands." *IEEE Communications Magazine* 56(6): 102–108.

[58] Boulogeorgos, A.-A. A., Papasotiriou, E. N., and Alexiou, A., 2018. "A Distance and Bandwidth Dependent Adaptive Modulation Scheme for THz Communications." In Proceedings of 19th IEEE International Workshop on Signal Processing Advances in Wireless Communications (SPAWC). doi: 10.1109/SPAWC.2018.8445864

[58] Singh, R., and Sicker, D. 2019. "Parameter Modeling for Small-Scale Mobility in Indoor THz Communication." In Proceedings of IEEE Global Communications Conference (GLOBECOM). Waikoloa, HI.

[59] Priebe, S., and Kurner, T. 2013. "Stochastic Modeling of THz Indoor Radio Channels." *IEEE Transactions on Wireless Communications* 12(9): 4445–4455. doi:10.1109/TWC.2013.072313.121581

[60] Schneider, T., Wiatrek, A., Preussler, S., et al. 2012. "Link Budget Analysis for Terahertz Fixed Wireless Links." *IEEE Transactions on Terahertz Science and Technology* 2(2): 250–256. doi:10.1109/TTHZ.2011.2182118

[61] Nagatsuma, T., Ducournau, G. and Renaud, C.C., 2016. "Advances in terahertz communications accelerated by photonics." *Nature Photonics*, 10(6): 371–379.

[62] Ducournau, G., Szriftgiser, P., Pavanello, F., Peytavit, E., Zaknoune, M., Bacquet, D., Beck, A., Akalin, T. and Lampin, J.F., 2015. "THz communications using photonics and electronic devices: the race to data-rate." *Journal of Infrared, Millimeter, and Terahertz Waves*, 36(2): 198–220.

[63] International Telecommunication Union. 2013. "Recommendation ITU-R P.676–10: Attenuation by atmospheric gases- P Series Radiowave propagation." Tech. Rep. ITU-R P.676–10. www.itu.int/dms_pubrec/itu-r/rec/p/R-REC-P.676-10-201309-S!!PDF-E.pdf

[64] Jornet, J. M., Akyildiz, I. F., 2011. "Channel Modeling and Capacity Analysis for Electromagnetic Wireless Nanonetworks in the Terahertz Band." *IEEE Transactions on Wireless Communications* 10(10): 3211–3221. doi:10.1109/TWC.2011.081011.100545

[65] Ghafoor, S., Boujnah, N., Rehmani, M. H., et al. 2019. "MAC Protocols for Terahertz Communication: A Comprehensive Survey." https://arxiv.org/pdf/1904.11441.pdf

[66] Federici, J.F., Ma, J. and Moeller, L., 2016. "Review of weather impact on outdoor terahertz wireless communication links." *Nano Communication Networks*, 10: 13–26.

[67] Amarasinghe, Y., Zhang, W., Zhang, R., Mittleman, D.M., and Ma, J., 2020. "Scattering of Terahertz Waves by Snow." *Journal of Infrared, Millimeter, and Terahertz Waves*, 41(2): 215–224.
[68] Zhang, C., Wang, G., Jia, M., et al. 2019. "Doppler Shift Estimation for Millimeter-Wave Communication Systems on High-Speed Railways." *IEEE Access* 7: 40454–40462.
[69] Marcus, M. J. 2016. "WRC-19 Issues: Agenda Item 1.15 and the Use of 275–450 GHz." *IEEE Wireless Communications* 23(6): 2–3.
[70] International Telecommunication Union. 2019. "World Radiocommunication Conference 2019 (WRC-19) - Agenda and Relevant Resolutions." www.itu.int/pub/R-ACT-ARR.1
[71] Marcus, M.J., "US Spectrum Policy & the 95GHz Wall." https://ecfsapi.fcc.gov/file/60001569382.pdf
[72] Federal Communications Commission. 2018. "Spectrum Horizons- Rulemaking to Allow Unlicensed Operation in the 95–1,000 GHz Band." Tech. Rep. ET Docket No. 18–21. https://docs.fcc.gov/public/attachments/DOC-348982A1.pdf
[73] Kürner, T., and Priebe, S. 2014. "Towards THz Communications-Status in Research Standardization and Regulation." *Journal of Infrared Millimeter and Terahertz Waves* 35(1): 53–62.
[74] Federici, J. F., Schulkin, B., Huang, F., et al. 2005. "THz Imaging and Sensing For Security Applications – Explosives, Weapons and Drugs." *Semiconductor Science Technology.* 20: 266–280.
[75] Demers, J. 2015. "The THz Drone Project." https://hackaday.io/project/11634-the-thz-drone-project
[76] Michael, K. 2014. "Radiation Safety of handheld mobile phones." In Proceedings of the 2nd Pan African International Conference on Science, Computing and Telecommunications (PACT).
[77] Institute of Electrical and Electronics Engineers. 2005. "IEEE Standard for Safety Levels With Respect to Human Exposure to Radio Frequency Electromagnetic Fields, 3 kHz to 300 GHz." IEEE International Committee on Electromagnetic Safety, Tech.Rep. IEEE C95.1–2005.
[78] International Commission on Non-Ionizing Radiation Protection. 2013. "ICNIRP Guidelines On Limits of Exposure to Laser Radiation of Wavelengths Between 180 nm and 1,000 μm." Health Physics 105(3): 271–295. www.icnirp.org/en/frequencies/infrared/index.html
[79] Singh, R., Doug, Sicker. 2020. "THz Communications - a Boon and/or Bane for Security, Privacy, and National Security" In Proceedings of TPRC 48: The 48th Research Conference on Communication, Information and Internet Policy. Washington D.C.
[80] Zhang, Z., Xiao, Y., Ma, Z., et al. 2019. "6G Wireless Networks: Vision, Requirements, Architecture, and Key Technologies." *IEEE Vehicular Technology Magazine*, 14(3): 28–41.
[81] Mumtaz, S., Jornet, J.M., Aulin, J., Gerstacker, W.H., Dong, X. and Ai, B., 2017. "Terahertz Communication for Vehicular Networks." *IEEE Transactions on Vehicular Technology*, 66(7): 5617–5625.
[82] Boulogeorgos, A.A.A., Alexiou, A., Merkle, T., et al. 2018. "Terahertz Technologies to Deliver Optical Network Quality of Experience in Wireless Systems Beyond 5G." *IEEE Communications Magazine*, 56(6):144–151.
[83] Yang, P., Xiao, Y., Xiao, M. and Li, S., 2019. "6G Wireless Communications: Vision and Potential Techniques." *IEEE Network*, 33(4): 70–75.
[84] Yi, H., Guan, K., He, D., et al. 2019. "Characterization for the Vehicle-to-Infrastructure Channel in Urban and Highway Scenarios at the Terahertz Band." *IEEE Access* 7: 166984–166996.
[85] Domínguez, J.M.L., and Sanguino, T.J.M. 2019. "Review on V2X, I2X, and P2X Communications and Their Applications: A Comprehensive Analysis over Time." *Sensors* 19(12): 1–29. https://doi.org/10.3390/s19122756
[86] Sekander, S., Tabassum, H., and Hossain, E. 2018. "Multi-Tier Drone Architecture for 5G/B5G Cellular Networks: Challenges, Trends, and Prospects." *IEEE Communications Magazine* 56(3): 96–103.
[87] Guan, Z., and Kulkarni, T. 2019. "On the Effects of Mobility Uncertainties on Wireless Communications between Flying Drones in the mmWave/THz Bands." In Proceedings of IEEE

International Conference on Computer Communications Workshops (INFOCOM WKSHPS), Paris, France.

[88] Dressler, F., Fischer, S. 2015. "Connecting in-Body Nano Communication with Body Area Networks: Challenges and Opportunities of the Internet of Nano Things." *Nano Communication Network*. 6: 29–38.

[89] Farahani, B., Firouzi, F., Chang, V., et al. 2018. "Towards Fog-Driven IoT eHealth: Promises and Challenges of IoT in Medicine and Healthcare," *Future Generation Computer Systems* 78(2): 659–676. https://doi.org/10.1016/j.future.2017.04.036

[90] Mumtaz, S., Jamalipour, A., Gacanin, et al. 2019. "Licensed and Unlicensed Spectrum for Future 5G/B5G Wireless Networks." *IEEE Network*, 33(4), 6–8.

[91] Singh, R., Sicker, D. 2019. "Part 15++: An Enhanced ID-based Approach for Etiquette and Enforcement Management in Unlicensed Band." *IEEE Transactions on Cognitive Communications and Networking (TCCN)* 5(3): 754–767. doi:10.1109/TCCN.2019.2915662

Chapter 16

Smart Terahertz Wireless Communication Zones

Ravikant Saini and Deepak Mishra

Contents

16.1 Introduction

With the growing demand for required resources per person in the urban cities globally, there is an inherent challenge in planning efficient utilization of the available resources, be it either natural or human-made. The vision of improved well-being of the urban civilization is full of numerous challenges in transportation, housing, water, electricity, medical facility, safety, to name a few (Jin et al., 2014). Toward this end, automation has appeared as a promising option to outcast human errors and their repercussions. Thus, visionaries are looking forward to the possibility of smart cities that rely on modern technological infrastructure and will be able to meet the expectations of the next generation's smart citizens.

A smart city that has been envisioned as an autonomous infrastructure is based on smart devices that require frequent information exchange for implementing their functionality. Terahertz band is a promising option for supporting high-speed wideband connectivity between devices in a short coverage area. Due to high device density, the communication distances between devices are usually small. The communication zone can be considered as an information hot-spot that provides high-speed connectivity among devices in a small coverage area. Special communication zones can be envisaged to support a large number of IoT devices using the THz band. These terahertz-based high-speed communication zones (ThiSCoZ) appear as a promising solution to the last mile problem. In the following, after discussing IoT-based smart cities, the introduction of terahertz-based smart communication zones have been provided.

16.1.1 IoT-based Smart City

A smart city is a model city that utilizes information and communication technologies (ICT) to enable more efficient city services (Jin et al., 2014). A smart city is envisioned as a self-sustained solution that adapts itself to the demanding situations without any need of human interventions. Toward this end, advanced technological solutions for wireless communication and information processing are being looked for to support an autonomous system running on its own without any human control. Supported by the advancements in ICT, the Internet of Things (IoT) is a paradigm where devices of everyday life interact with each other as well as human beings and become an integral part of the Internet (Zanella et al., 2014). With the development of smart devices, supported by novel ICT solutions, IoT opens up the space for improved and modern societal behavior through the implementation of smart cities.

Internet of Things (IoT) is a paradigm shift from the traditional internet of computers to the internet of smart electronic devices that utilize their information exchange capabilities for improving the overall quality of human life (Stojkoska and Trivodaliev, 2017). The devices can be categorized as sensors – converting physical phenomena to an electrical signal; controllers – making decisions based on the sensed information input; and actuators – turning the decision (electrical signal) to physical activity. Each IoT device interfaces with the physical world in which they have been placed at the front end, and with the digital world where digital connection among devices is established for information exchange at the back-end. Thus, each IoT device behaves as a connector between the physical world and the digital world. On the front end, there are information-collecting units and units for information presentation. At the same time, at the back-end, there are computational devices that can run complex algorithms to make necessary decisions.

IoT-based networks are imagined as a realm of distributed smart computational devices that can adapt to the working conditions. These devices are spread all around, inside home as well as outside. They are an integral part of the overall infrastructure that lays the foundation of smart cities, including roads, offices, parks, hospitals, and markets. With the current technological advances in the field of chip fabrication technology, there is an increasing interest from electronic device manufacturers to propose smart devices that are capable of changing their working strategy based on the environmental conditions. Thus, going forward, there is an upsurge in the density of these smart devices.

16.1.2 Smart Communication Zones

The exponential increase in the number of active devices, the increasing number of information interactions, and the heterogeneity of data being shared among them are the key dimensions of the communication challenge faced by the ICT researchers envisioning IoT-based smart cities. The research communities are thinking beyond simple data connections between devices, and instead, they are envisioning information superhighways. Because of the very high device density, the involved communication distances are usually not very large. Thus, there is a need for special zones, catering to a cluster of nearby smart devices, which provides very high-speed data connectivity among the devices.

Terahertz (THz) communication offering data links in the range of terabits per second (Tbps) appears as a promising option for an indoor or wearable network where communication distances are usually limited to a few meters. Smart wireless communication zones are terahertz communication-based information hot spots that provide massive data rates to a cluster of smart devices within its limited coverage region (Petrov et al., 2016). Thus, researchers are considering the possibility of coming up with terahertz-based high-speed communication zones (THiSCoZ) as a promising option to cater to the need of data-hungry devices.

The infeasibility of providing high-speed connectivity at the user end is known as the last mile problem in the context of communication networks. High-speed data connectivity could be offered up to a last access node of the network only, but not to the end-user equipment. THiSCoZ is a promising solution to this ancient bottleneck in the communication industry. With the possibility of providing very high data rates in the range of Tbps, THiSCoZ opens a plethora of applications for the coming generation communication

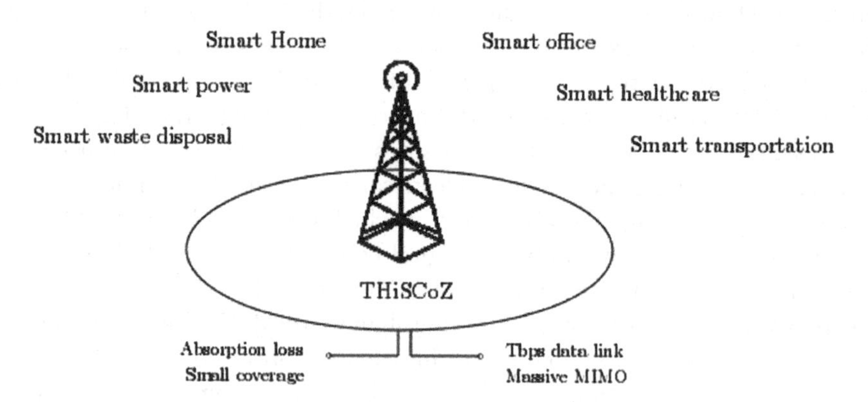

Figure 16.1 THz communication – a band of opportunities and challenges, with numerous application scenarios.

networks such as a smart house, smart office, smart hospital, smart parks, smart vehicles, smart transport, smart electricity, smart municipality, to name a few (Figure 16.1).

16.1.3 Terahertz Communication Opportunities Versus Challenges

Terahertz frequency band (0.1–10 THz) presents the possibility of an enormous bandwidth in the range of 10–100 GHz, which can provide data rates in the range of Tbps. This enormous data rate leads to the feasibility of seamless data transfer, the latency of the order of microseconds, and ultrafast downloads (Kawanishi, 2019). THz communication presents itself as a novel technological revolution that will leave a significant mark on the coming generation networks' ICT scenario. The utilization of high frequency leads to smaller wavelength and antenna size, which opens the feasibility of using massive MIMO techniques to improve spectral efficiency and directivity.

However, along with opportunities, there are some serious challenges associated with THz communication. The novel challenges associated with THz band include generation and detection of such high-frequency signals, designing of broadband antennas to facilitate ultra-wideband communication, and implementing an amplifier that could provide a flat response in this broad band of operation. THz communication has issues related to unfavorable propagation characteristics, including atmospheric absorption and spreading of the signal. The channel properties are very different in this region of operating frequency (Huq et al., 2019). To design an efficient communication system, channel properties must be studied, and proper channel models must be proposed. Modulation and coding techniques have to be designed according to the THz band's channel conditions to improve the communication system's spectral efficiency. Thus, there is a need for renewed interest in thorough investigations of the research challenges associated with THz communication. Even if the existing solutions can be evolved for utilization in the THz band, they may not be optimal as they had been designed for a lower frequency band having different characteristics than the THz band. Further, channel modeling, transceiver design, antenna design, and signal processing design are key research directions that require joint revolutionary development for the realization of THz band communication zones.

16.1.4 Outline

Since the base technology for ThisCoZ is terahertz communication, these communication zones face the same challenges as the terahertz band communication. With an idea to present a brief introduction to terahertz communication, this chapter attempts to provide first-hand background information and the research challenges associated with THz communication. First of all, a short survey on the state of the art is presented, highlighting the key challenges that the research community is engaged in. A detailed discussion is then presented on key issues related to physical layer communication techniques such as channel modeling, signal processing design for multiantenna techniques, optimization involved in efficient utilization of resources, and usage of nearby nodes as relays for information coverage enhancement. Finally, the chapter concludes with a short discussion on open research directions in the field of THz communication.

16.2 State-of-the-Art in Terahertz Wireless Communication

This section presents a brief survey of different research fronts being challenged by the research community worldwide. The fundamental techniques required for the terahertz band communication system's realization include the generation and detection of terahertz signal and the design and development of a communication system to cater to the THz band's challenges. In this regard, we first discuss the research developments in the direction of the transceiver's hardware design. Following this, a discussion on channel investigations includes modeling the THz band's channel and noise properties. Finally, modulation and coding techniques, along with MIMO techniques, which have been specially designed observing the challenges of the THz band, are discussed. These key research domains have been summarized in Figure 16.2.

16.2.1 Transceiver Hardware Design

This section begins with an introduction to issues in the design of the signal generator and detector at THz frequency. Then, various semiconductor-based sources, considered by researchers, have been discussed. Finally, the section concludes with a discussion on the possibility of utilization of graphene for generation as well as detection of THz signal.

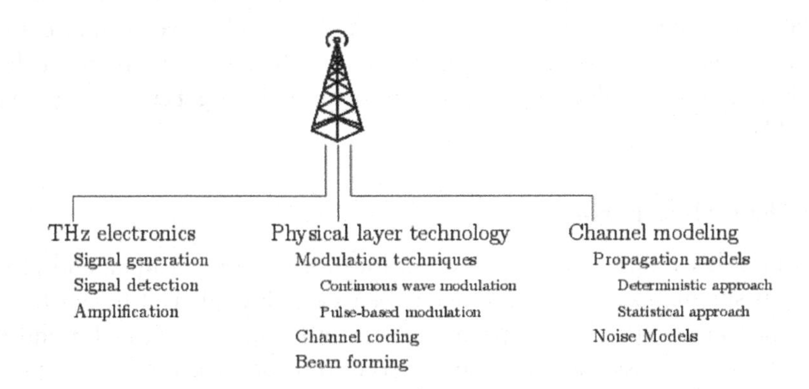

Figure 16.2 Key Research Domains in THz communications.

Due to the location of the THz band between microwave (300 MHz to 300 GHz) and infrared frequency (300 GHz to 430 THz), the THz band is called as THz gap because of the lack of maturity in the signal generation and detection techniques (Akyildiz et al., 2014). Thus, electronic and photonic devices can be used for the generation of THz signals. However, the conceptual issue is that, because of the high frequency of operation, the polarity changes quickly. Because of the fast change in polarity, the direction of electron movement changes very fast. Since the electrons are not able to travel a minimum required distance, there is no significant current. On the photonic side, it is difficult to generate photons in the THz range due to the vigorous movement of electrons between energy levels and the difficulty of controlling the required discrete energy level jump.

Since the transceiver has to face high path loss due to molecular absorption, the design should consider high power, high sensitivity, and low noise figure (Tekbyk et al., 2019). Silicon–Germanium (SiGe) based transistors have been investigated for utilization up to a frequency range of 798 GHz (Chakraborty et al., 2014). Although silicon-based devices can be improved for usage up to 1 THz, they cannot be efficiently used beyond 500 GHz due to inadequate transistor breakdown voltage. Due to high path loss, there is a need for high gain–wideband power amplifiers. Gallium nitride (GaN) based high electron mobility transistor (HEMT) has been recently investigated for utilization in millimeter-wave and THz applications (Tang et al., 2015).

Graphene has recently attracted the research community's attention due to its favorable properties, such as electrical and thermal conductivity. It shows plasmonic effects that allow signal propagation at THz range in the form of surface plasmon polariton (SPP) wave (Bird et al., 2017). While studying III–V semiconductor-based HEMT, which has been enhanced through graphene, it is observed that the application of a voltage between drain and source of HEMT causes electron acceleration, resulting in the formation of SPP waves on the graphene gate (Jornet and Akyildiz, 2014b).

Due focus is required for the implementation of antennas as well. Because of the THz frequency range, the antenna size is small. Thus, diversity, as well as directivity gain, can be obtained by using multiple-input multiple-output (MIMO). More specifically, ultramassive MIMO with a 1024*1024 system has been investigated for improving the communication distance (Akyildiz and Jornet, 2016). The graphene-based antenna has been considered for the possibility of high directivity, which is a very important feature required to take care of high path loss. Compared to copper, graphene provides relatively better performance in terms of small antenna size and high directivity (Dash and Patnaik, 2018). Plasmonic graphene-based antenna, which allows the formation of SPP waves, can be considered for nano-scale devices. The frequency of operation of SPP waves can be tuned by material doping as graphene's conductivity varies with chemical doping, Fermi energy, and electron mobility (Abadal et al., 2017).

16.2.2 Channel Properties

Channel models capturing the effects of various losses like spreading and blocking and noise effects summarize the changes that a wave goes through as it travels from a source to its destination. Since channel properties are different for this high-frequency band, appropriate channel loss and noise models need to be proposed before designing an efficient communication system. Since most use-cases include short-range communication, reflection, and penetration from the nearby walls, ceilings and objects are to be taken into account. Further, at this frequency band, almost all objects behave as scatterers for THz waves.

Usually, two methods are employed for channel model investigations: deterministic methods and statistical methods (Han and Chen, 2018). Even though the deterministic model, such as the ray tracing method, is specific to the study environment, it is highly accurate. These methods need detailed geometrical information about the environment and are computationally intensive. Further, the model's complexity increases with the increase in the size of the investigated environment, making them more suitable for indoor environments due to controlled variations (Bile Peng et al., 2016). Their effectiveness gets reduced further because the results can hardly be utilized in a modified scenario despite minimal changes.

Observing the complexity involved in deterministic models, statistical models based on the empirical channel studies have been considered (Chen et al., 2019). Due to higher frequency bandwidth, the multipath components can be easily resolved in the time domain. Therefore, rather than using narrow-band based models like Rayleigh, or Rician models, tap delay line based models have been proposed. In this direction, statistical investigations are performed for different multipath components to achieve a tap delay line model, which requires study of the direction of departure, direction of arrival, time of arrival, and complex amplitude. Considering the possibility of combining arriving multipath components as a cluster, a cluster-based statistical channel model has been proposed in Akdeniz et al. (2014). However, these models are limited because the information regarding channel correlation function and power delay profile is difficult to obtain.

During the electromagnetic transmission of the THz wave through the medium, the medium molecules shift from one energy level to another level due to atmospheric absorption. This phenomenon is referred to as molecular absorption and presents a loss of transmitted signal because a part of signal energy gets converted into kinetic energy of molecules. Along with the background thermal additive white Gaussian noise, which is termed as sky noise, there is another component called atmospheric absorption noise. Molecular absorption noise is that random component which results from the emission of energy absorbed by medium molecules while returning to their original states. Since this noise component is caused by the transmission of the signal in the medium, it is termed as self-induced noise (Kokkoniemi et al., 2016).

16.2.3 Physical Layer Techniques

For efficient utilization of available communication bandwidth, the physical layer's modulation and coding techniques should be designed to match the THz channel's characteristics. Modulation helps in tuning the transmitter characteristics to that of channel behavior, and coding attempts to overcome the effect of loss due to channel impairments. Due to smaller antennas, a large number of antennas can be utilized under the MIMO scheme, which can help obtain diversity and directivity.

Simple continuous-wave modulation techniques such as on–off keying (OOK) have been considered for on-chip wireless communications between inter-connects (Laha et al., 2015). Because of the variation in the transmission window, the channel shows distance-dependent behavior, so distance-aware modulation schemes should be considered. For short-range communication, a pulse-based ON–OFF modulation scheme has been discussed in Jornet and Akyildiz (2014a). Further, it is worthwhile to investigate if the complete channel should be used as a single broadband channel or a collection of parallel channels in multi-carrier communication. Toward this end, the creation of parallel channels for sharing the total bandwidth and utilization of M-ary QAM has been investigated (Han and Akyildiz, 2014). Although this scheme proposes to offer data rates in the range of Tbps, there is high complexity in the overall transmission system.

Conventional channel codes attempt to provide error-free communication. However, the requirements in the THz environment are different. Here, the coding schemes should consider the decoding complexity, decoding power, and decoding time. This new requirement stems from the fact that due to Tbps links, the involved data payload is huge, and the power limitations are imposed because of the constrained nodes involved in IoT-based communication. Instead of attempting error correction, error prevention can be a suitable strategy (Jornet, 2014). Further, similar to the distance-based modulation scheme, there could be a distance-dependent coding scheme that varies according to the channel conditions (Moshir and Singh, 2016).

Multiple antennas can be considered at the transmitter to overcome the short-distance limitation of THz communication and help improve the communication distance (Lin and Li, 2015). However, they can be employed for increasing data rates as well. Due to very small wavelengths, the feasibility of using multiple nano-antennas leads to massive MIMO usage as another opportunity that can be used to serve multiple users (Zakrajsek et al., 2017a). However, the usage of massive MIMO for data rate enhancement does not appear as a suitable option for nano-networks due to the involved computational complexity and resource requirements.

16.3 Terahertz Wireless Channel Modeling and Characterization

Since the channel properties are different at the terahertz range, suitable channel models have to be investigated. Analytical studies look for a simpler channel model, while more advanced and sophisticated channel models are needed for studies leading to the realization of practical communication networks. Channel properties are different for the line of sight (LOS), and non-line of sight (NLOS) communication because signal suffers losses due to blockage and atmospheric absorption. In this section, we present some of the basic properties of THz channel and then briefly discuss some of the suitable channel models investigated in the literature.

16.3.1 Basic Channel Properties

This section presents the mathematical models for the key attributes of the THz channel. The transmitted signal suffers from spreading loss and atmospheric absorption loss. The channel observes sky noise and absorption noise.

16.3.1.1 Spreading Loss

The path loss for an electromagnetic wave in the THz frequency domain, traveling in the air, is composed of spreading loss and absorption loss. Spreading loss caters to the loss of signal due to the expansion of the wave as it propagates in the medium, that is, air. The loss is measured through free-space path loss which can be described as (Jornet and Akyildiz, 2011)

$$A_{\text{spread}}(f,r) = \left(\frac{4\pi fr}{c}\right)^2$$

where f is the frequency of the electromagnetic wave, c is the speed of light, and r is the path length. Due to significantly high spreading loss in the THz band, the communication distances are usually small.

16.3.1.2 Absorption Loss

In the frequency range of THz, electromagnetic transmission through wireless medium results in molecular absorption. Usually, air molecules have random motions, and the absorption of electromagnetic energy either increases their kinetic energy or move them to higher energy states. Both of these results in loss of communication power termed as absorption loss. Total absorption loss that a wave of frequency f suffers while traveling distance r can be expressed as a function of transmittance of the medium (Jornet and Akyildiz, 2011)

$$A_{\text{abs}}(f,r) = \frac{1}{\tau(f,r)} = e^{k(f)r}$$

where $\tau(f,r)$ is the fraction of electromagnetic radiation that can pass through the medium. The absorption coefficient $k(f)$ depends on the particular mixture of molecules found along the channel. $k(f)$ is obtained as a summation of individual absorption coefficients $k^{i,g}(f)$ for all the isotopologues of each gas in the terahertz band. The absorption coefficient of an isotopologue i of gas g, which is a molecule that differs only in isotopic composition, for a molecular volumetric density $Q^{i,g}$ at pressure p and temperature T is given as

$$k^{i,g}(f) = \frac{p}{p_o}\frac{T_{STP}}{T}Q^{i,g}\sigma^{i,g}(f)$$

where p_o and T_{STP} are the standard pressure temperature values, and $\sigma^{i,g}$ is the absorption cross-section for the isotopologue i of gas g. For a given gas, molecular density $Q^{i,g}$ is obtained using the ideal gas law. Absorption cross-section $\sigma^{i,g}$ is obtained using line intensity, which identifies the absorption strength of each molecule and spectral line shape, depending on the resonant frequency of each isotopologue.

16.3.1.3 Sky Noise

Sky noise is caused by the temperature of the signal absorbing atmosphere behaving like a black body radiator. Thereby, sky noise can be considered as background noise. This phenomenon is measured in terms of emissivity of the channel ϵ, which is defined as (Jornet and Akyildiz, 2011)

$$\epsilon(f,r) = 1 - \tau(f,r)$$

where $\tau(f,r)$ is the transmissivity of the medium at frequency f for a path length r. The equivalent noise temperature of the atmosphere is obtained as

$$T_{sky}(f,r) = T_0\epsilon(f,r)$$

where T_o T_0 is the reference temperature of the atmosphere. Assuming no other losses, the sky noise can be stated as

$$N_{sky}(T_0,f,r) = k_B T_0 B F \epsilon(f,r)$$

where k_B k_B is the Boltzmann constant, B is the transmission bandwidth of the system, and F is sky noise acceptance factor of the receiver.

While discussing the general behavior of sky noise, it can be alternatively described using the Planck's function $B(T,f) = \frac{2h\pi f^3}{c^2}\left(e^{\frac{hf}{k_B T}} - 1\right)^{-1}$ as (Kokkoniemi et al., 2016)

$$N_{sky}(T_0,f,r) = B(T_0,f)\epsilon(f,r) \approx B(T_0,f)$$

where h is the Planck's constant. Observing that the antenna temperature is approximately constant in THz band, the atmosphere radiates according to Planck's law. Sky noise using the Planck's function is the background noise, while molecular absorption noise is an additional component over it. Considering the ideal antenna temperature, the power spectral density of sky noise can be further approximated as

$$N_{ps}(T_0,f) = \frac{c^2}{4\pi f^2} B(T_0,f)$$

16.3.1.4 Molecular Absorption Noise

While passing through the medium, electromagnetic waves excite molecules to higher states, which leads to random noise signals due to emission by these molecules while coming back to their original states. This factor affects the normal propagation of energy through the medium. Assuming that all the energy absorbed from the transmitted signal gets converted into molecular absorption noise, it can be expressed as (Kokkoniemi et al., 2016)

$$N_{pm}(r,f) = \frac{P_{Tx}(f)}{4\pi r^2}\epsilon(f,r)$$

where P_{Tx} is power spectral density of transmitted signal, and $4\pi r^2$ takes care of spreading loss.

The total noise power spectral density at the receiver side is calculated as the summation of sky noise and molecular absorption noise as $N(f,r) = N_{ps}(r,f) + N_{pm}(r,f)$. The equivalent noise power, for a transmission bandwidth B can be calculated as

$$P_n(f,r) = \int_B \mathcal{N}(f,r)\mathrm{d}f$$

16.3.2 Multi-ray Channel Model

Ray tracing method has been utilized to identify the THz channel model using the propagation properties of the line of sight (LOS), reflected, scattered, and diffracted electromagnetic waves (Han et al., 2015). The line of sight component faces the spreading loss and the absorption, as described in Section 16.3.1.1. In the THz band, any surface with its roughness of the order of the communication wavelength range behaves as scatterers. Thus, a surface could be smooth for lower frequency signals but can scatter THz signals. Next, we discuss the behavior of different multipath components, such as reflected, scattered, and diffracted waves (Figure 16.3).

16.3.2.1 Reflected Wave

Denoting the reflection coefficient as R, the distance between the transmitter and reflector as r_1, and between reflector and receiver as r_2, the transfer function of the reflected wave can be expressed as

$$H_{\mathrm{Ref}}(f) = \left(\frac{c}{4\pi f(r_1 + r_2)}\right) e^{-j2\pi f \tau_{\mathrm{Ref}} - \frac{1}{2}k(f)(r_1 + r_2)} * R(f)$$

where τ_{Ref} is arrival time of the reflected wave which can be given as $\tau_{\mathrm{Ref}} = \tau_{\mathrm{LoS}} + (r_1 + r_2 - r)/c$ and $\tau_{\mathrm{LoS}} = r/c$.

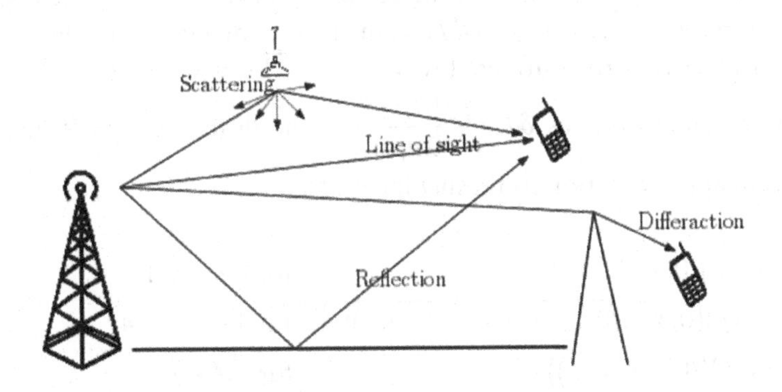

Figure 16.3 **Principles of wave propagation in wireless communication.**

Using Kirchhoff scattering theory, the reflection coefficient of a rough surface can be stated as

$$R(f) = \gamma_{\mathrm{TE}}(f)\rho(f)$$

where $\gamma_{\mathrm{TE}}(f)$ is the Fresnel reflection coefficient for TE-polarized wave for a smooth surface. $\rho(f)$ is the Rayleigh roughness factor of the surface, having surface height standard deviation as σ, which can be characterized as

$$\rho(f) = e^{\left(\frac{-8\pi^2\sigma^2 cos^2(\theta_i)}{c^2}\right)}$$

where $\theta_i = \frac{1}{2}cos^{-1}\left(\frac{r_1^2 + r_2^2 - r^2}{2r_1r_2}\right)$ is the angle of incidence, obtained after geometrical simplifications. For a TE-polarized wave, the Fresnel reflection coefficient over a smooth surface can be given as

$$\gamma_{\mathrm{TE}}(f) \approx -exp\left(-\frac{2\cos(\theta_i)}{\sqrt{n_i^2 - 1}}\right)$$

where n_i is the refractive index of the medium.

16.3.2.2 Diffracted Wave

Due to the shorter wavelength (THz frequency), the diffraction effect around the edges can be ignored in an indoor environment (Han et al., 2015). For non-line of sight (NLOS) cases, diffraction can be characterized by the uniform geometrical theory of diffraction. The diffraction channel transfer function can be stated as

$$H_{\mathrm{Dif}}(f) = \left(\frac{c}{4\pi f(d_1 + d_2)}\right)e^{-j2\pi f\tau_{\mathrm{Dif}} - \frac{1}{2}k(f)(d_1+d_2)} * L(f)$$

where d_1 is the distance from the transmitter to the diffractor, d_2 is the distance between diffractor and receiver, $\tau_{\mathrm{Dif}} = \tau_{\mathrm{Los}} + \delta d / c$ is the time of arrival of the diffracted wave. The path difference between the diffracted wave and the LOS wave δd can be obtained using geometrical simplifications as $\delta d = \frac{h_d^2(d_1 + d_2)}{2d_1d_2}$. The diffraction coefficient $L(f)$ can be obtained as an approximation to Fresnel integral as

$$L(f) = \begin{cases} \mu_1(f)\left(0.5e^{-0.95v(f)}\right), & \text{for } 0 < v \leq 1 \\ \mu_2(f)\left(0.4 - \sqrt{0.12 - (0.38 - 0.1v(f))^2}\right), & \text{for } 1 < v \leq 2.4 \\ \mu_3(f)\left(0.225 / v(f)\right), & \text{for } 2.4 < v \end{cases}$$

where $v(f) = \sqrt{\frac{2f\delta d}{c}}$, and μ_1, μ_2, and μ_3 are frequency-dependent parameters that are suitably chosen to match the empirical data.

16.3.2.3 Scattered Wave

In the THz frequency band, the surface roughness is of the order of the wavelengths of the electromagnetic wave, which causes the scattering phenomenon to be highly predominant. The transfer function of a scattered wave can be given as (Han et al., 2015)

$$H_{\text{Sca}}(f) = \left(\frac{c}{4\pi f (s_1 + s_2)}\right) e^{-j2\pi f \tau_{\text{Sca}} - \frac{1}{2}k(f)(s_1+s_2)} * S(f)$$

where $\tau_{\text{Sca}} = \tau_{\text{LoS}} + (s_1 + s_2 - r)/c$ is the time of arrival of the scattered wave when the scatterer is at s_1 distance from the transmitter, and the receiver is s_2 distance from the scatterer. For large angles of incidence and scattering, the scattering coefficient of rough surfaces can be approximated as (Ragheb and Hancock, 2007)

$$S(f) = -\exp\left(\frac{-2\cos(\theta_1)}{\sqrt{n_t^2 - 1}}\right) \cdot \sqrt{\frac{1}{1 + g + \frac{g^2}{2} + \frac{g^3}{6}}} \cdot \sqrt{\rho_o^2 + \frac{\pi \cos(\theta_1)}{100}\left(g e^{-v_s} + \frac{g^2}{4} e^{-\frac{v_s}{2}}\right)}$$

where θ_1 is the zenith angle of the incident wave. For a detailed discussion on evaluation of other parameters like ρ_o, g, v_s etc. kindly refer (Ragheb and Hancock, 2007).

16.3.3 LOS and NLOS Characterization

Developing on the fundamental properties of the channel, detailed LoS, and NLOS channel model suitable for frequency range in 0.1–10 THz has been presented (Moldovan et al., 2014). To identify molecular absorption loss, fraction of water in the air is obtained using the relative humidity (RH). RH expressed as a percentage ratio of dry mass mixing ratio with saturation mixing ratio as $RH = \frac{q_{\text{wv}}}{q_{\text{sat}}}$ can be obtained using

$$q_{\text{wv}} = 0.622 \frac{p_{\text{wv}}}{p - p_{\text{wv}}}, \quad \text{and} \quad q_{\text{sat}} = 0.622 \frac{p_{\text{sat}}}{p - p_{\text{sat}}}$$

where p_{wv} and p_{sat}, are actual water vapor pressure and saturation vapor pressure. p_{sat} can be obtained using Magnus formula

$$p_{\text{sat}} = K_1 exp\left(\frac{K_2 T}{K_3 + T}\right)$$

where $K_1 = 610.94P_a$, $K_2 = 17.625$, and $K_3 = 23.04$ are obtained by measurements.

In the indoor environment, the LOS may be blocked, and NLoS plays a vital role in wave propagation. The characteristics of reflections are obtained as a function of the scattering coefficient, which is obtained as a sum of specular spike component of the reflection and specular lobe component, which relates to the scattered field because of the surface and is distributed around the specular spike (Moldovan et al., 2014).

16.4 Signal Processing Design and Multiantenna Techniques for Spectral Efficiency Enhancement

Modulation and coding techniques have to be weaved around the channel properties to improve spectral efficiency. Since the data rates are very high, in the range of Tbps, a loss in communication may lead to burst errors. Thus, suitable burst error correction schemes are to be investigated. With the possibility of placing a large number of antennas, massive MIMO is a promising option for improving spectral efficiency. Suitable signal processing schemes such as beamforming needs more in-depth investigations.

16.4.1 Modulation Schemes

Even though the existing modulation schemes can be utilized in the THz band, they may be highly ineffective because the THz channel shows the frequency and distance-dependent behavior, which requires specifically designed modulation techniques.

16.4.1.1 Pulse-based Modulation

For short-range communication, a femtosecond pulse-based asymmetric on–off keying modulation spread in time (TS-OOK) scheme was investigated (Jornet and Akyildiz, 2014a). To keep the transmission and reception simple, a logical 1 is represented by a pulse, while the absence of the pulse represents a logical 0. Each burst of transmission begins with a preamble to differentiate between logical 0 and the silence period of inactivity between communicating devices. The time between consecutive transmissions is fixed such that devices need not sense the channel all the time, and this also removes strict synchronization requirements. Since the time between successive transmissions is more than the pulse duration, there are significant silence zones. This offers the flexibility of multiple access between communicating devices with very little possibility of collision and that too without a central controller's need.

TSOOK has an inherent limitation: if data of any two users collide during one slot, it is bound to collide at all slots. By varying the time slot durations of users, and thereby changing their rate of transmissions, an improvement in terms of rate division TSOOK (RD-TSOOK) has been proposed (Pujol et al., 2011). These two modulation schemes have been compared in Figure 16.4. Figure 16.4(a) describes the TSOOK scheme with transmitter-1 and transmitter-2 transmitting their respective data, (1101) and (1110), using the same slot duration. In case there is a collision at one slot, then data of both the users will collide at all the slots. Figure 16.4(b) describes RD-TSOOK scheme, where usage of different slot duration results in better collision avoidance.

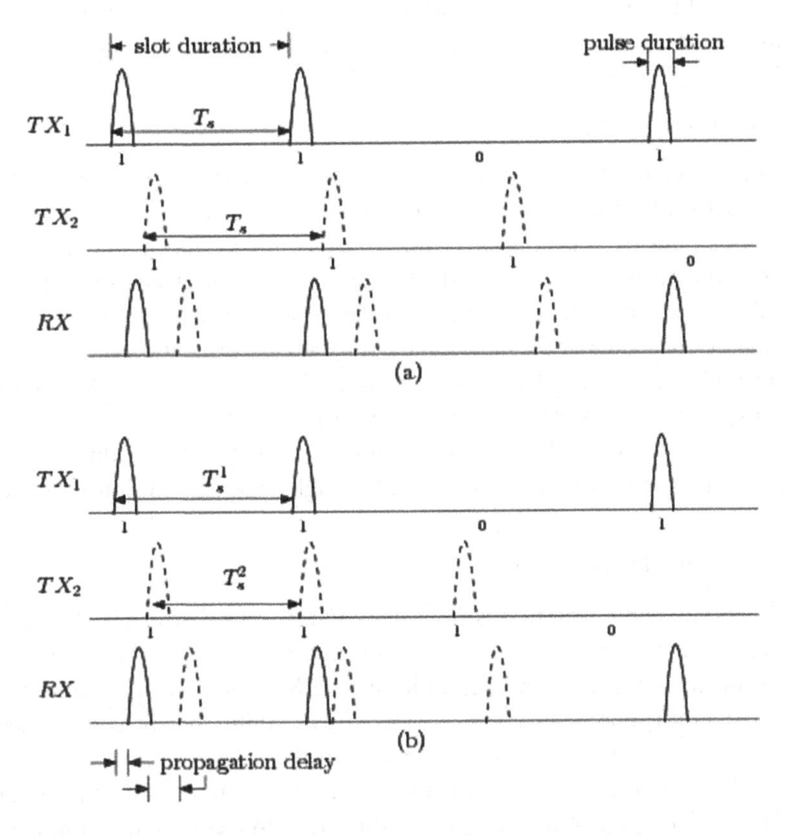

Figure 16.4 Examples of modulation techniques utilized in terahertz communication.

16.4.1.2 Distance-Aware Multi-Carrier Modulation

To utilize the distance and frequency-dependent behavior of THz channel, a distance aware multi-carrier (DAMC) modulation scheme is proposed in Han and Akyildiz (2014). DAMC is realized in a centralized control unit, which identifies the communication windows based on the communication distance. Each transmission window is further divided into smaller sub-windows used for multi-carrier communication, where each sub-carrier is modulated using MQAM modulation. The modulation index M is suitably chosen to satisfy the bit error rate requirements at the receiver. Orthogonal frequency division multiple access (OFDMA), another potential multi-carrier technique, is not suitable in THz band due to strict inherent synchronization requirements and unavailability of digital processors at such high rates. In the proposed DAMC scheme, there is a trade-off on the number of subcarriers. More number of subcarriers reduces the sub-window, which reduces the transmission complexity. However, a large number of carriers increases the overall complexity of the complete modulation scheme.

16.4.2 Coding Schemes

Channel coding can improve data rate by detecting and correcting any transmission errors due to channel effects. However, detection would cost in terms of re-transmission, and correction would cost in terms of decoding time and complexity, which is an inherent limitation on a network of resource-limited IoT devices.

16.4.2.1 Error Preventing Codes

Decoding time is a crucial parameter because, at the rate of Tbps, it appears far better to prevent errors rather than doing error correction. Toward this end, an error preventing coding strategy has been proposed in Jornet (2014). It has been analytically shown that the effect of the absorption noise and interference can be reduced by varying the weight of the code (i.e., the number of 1's in the codeword) without compromising on the information rate performance of the code. Thus, through the utilization of low-weight channel codes, the effect of noise can be minimized. With the lower code weight resulting in larger code words, there is a simple doubt if it is worth to send more bits for the same amount of information. It has been shown that the information rate after coding gets improved by this scheme, and there exists an optimal code weight that can minimize the noise effects.

16.4.2.2 Minimum Energy Coding

Minimum energy coding (MEC) is a strategy where codewords are chosen to minimize the average codeword power. While retaining the inherent benefits of minimum energy codes in transmission power saving, an improved MEC satisfying a minimum Hamming distance criteria for more robust error-correcting capability has been proposed (Kocaoglu and Akan, 2012). After identifying the minimum required code word weight for a given Hamming distance, the error performance of larger Hamming distance codes has been investigated. It has been observed that reliable error-free communication can be ensured by keeping a sufficiently larger Hamming distance provided source set cardinality is less than the inverse of symbol error probability.

16.4.2.3 Minimum Energy Source Coding

Based on the same concept, minimum energy source coding that attempts to reduce the number of 1's compared to 0's in a source symbol has been presented for nano-networks. The algorithm uses a dictionary that maps fixed-size input symbols to output symbols such that frequent symbols are attached with a code having a lesser number of 1's (Zainuddin et al., 2014).

16.4.3 MIMO Schemes

The THz band provides a unique opportunity in terms of massive MIMO as a large number of antennas can be placed inside a small footprint. Utilizing the MIMO beamforming technique, a narrow beam can be directed toward a distant user to overcome the issues with channel propagation like absorption loss.

16.4.3.1 Adaptive Beamforming

To tune to the distance and frequency-dependent behavior of the THz band of communication, a distance-aware multi-carrier transmission scheme has been proposed Lin and Li

(2015). A hybrid approach utilizing analog as well as digital beamforming has been submitted. Analog beamforming has been considered in the radiofrequency domain for user grouping and interference cancellation, while digital beamforming has been considered at baseband for dynamically selected sub-arrays. The scheme proposes adaptive power allocation and low complexity antenna sub-array selection policy that can serve users located at different distances from the transmitter.

Recently, the concept of ultramassive multi-carrier multiple-input multiple-output (UMMC MIMO) communication using ultra-dense frequency-tunable plasmonic nano-antenna arrays has been proposed and its performance has been analytically evaluated in Zakrajsek et al. (2017b). Utilizing the THz plasmonic nano-antenna array's abilities, in which each element of the array can be tuned independently at the transmitter and the receiver, an optimization framework has been proposed that caters to the spreading loss and absorption loss of the THz channel.

16.4.3.2 Multiplexing

It has been observed that molecules absorbing electromagnetic waves re-emit them in the environment at the same frequency. This re-emission behaving as the delayed version of the original signal has been highly correlated with the primary signal. It can be considered for reaping the benefits of multiplexing even on an LOS path (Hoseini et al., 2017). It has been observed that the gain from multiplexing can override that from beamforming. After deriving a beam domain channel model suitable for massive MIMO communication in the THz band, a beam division multiple access scheme is proposed that uses per beam synchronization. It has been observed that the effective delay and Doppler spread of wideband channels are reduced compared to conventional synchronization approaches (You et al., 2017).

16.5 Optimal Cooperation over Shorter Links for Enhancing the QoE for High Data Rate Applications

THiSCoZ is a high activity zone, which is limited in coverage. This would lead to a relatively more number of cells in a given area. With a very high density of small communication zones, energy-efficient optimization of resources is an important research direction toward the vision of green communication. Further, medium access control strategy and relaying strategies can be employed for better resource management.

16.5.1 Medium Access Control

Medium access control (MAC) in THz has to be designed from a different perspective because the transmissions are distance and frequency-dependent and also are of short time duration. The later feature helps in reducing the probability of collision. Thus, an efficient MAC needs to be investigated. A physical layer aware MAC protocol utilizing pulse-based communication, and using a low-weight channel coding scheme, maximizing the probability of successful decoding of transmission has been proposed in Jornet et al. (2012). A detailed classification of MAC protocols for the THz band, emphasizing the need for efficient MAC design and a comparison of the existing approaches, has been presented in Han et al. (2019).

16.5.2 Cooperation among Nodes

Since communication distances are limited, there is a possibility of cooperation among the IoT nodes by forwarding neighboring node data. Thus, it may be possible to identify some of the capable nodes as cluster heads, which may help communicate the information of nodes in their neighborhood. The outage capacity of a cluster-based forwarding scheme has been investigated in a body area network in Afsana et al. (2015). An energy-efficient forwarding scheme for a cooperative nano communication system has been studied in Afsana et al. (2018) for a wireless nano-network consisting of hybrid clusters.

16.5.3 Relaying

The terahertz band of communication offers high data rates, but the channel characteristics are such that the communication distances are quite small. This leads to the requirement of the utilization of relays for coverage enhancement. Considering the bit error rate as the performance metric, both decode and forward (DF) as well as amplify and forward (AF) relays have been studied in a nano-network system model, and BER expressions have been derived (Rong et al., 2017). Having observed the utility of relays, the next design issue is optimal resource allocation, which includes power allocation at source and relay, and relay placement for efficient utilization of given resources. Toward this end, a mathematical framework has been investigated in Xia and Jornet (2017) to optimize the relaying distance to maximize the system throughput considering the effect of the channel, antenna, and network layer, that is, cross-layer optimization.

16.6 Concluding Remarks and Future Research Directions

This chapter presented a comprehensive reading in a step by step manner. After providing a brief survey on state of the art, discussions on the key aspects like channel modeling, signal processing, and communication techniques, and cooperation among nodes have been presented (refer Figure 16.5). Even though THiSCoZ offers fast data rates in special communication zones with short coverage distance, there are many issues to be addressed before it comes to reality, which has been presented in the following.

16.6.1 Deeper Channel Investigations

While improving the channel model, there is a need to study reflection and scattering of THz waves over various materials found in an indoor channel environment. Further, small changes in the vicinity may harshly affect the channel properties, making the channel conditions pretty random and unpredictable. Thus, the channel model should be adaptive to accommodate randomness at micro as well as macro level. Investigations on time-varying behavior of channel require research attention, as outdoor channel environments may not necessarily be static. To realize a network of wearable devices, short-distance channels over different body organs with skin tissues' dependent signal adsorptions need special investigations. The channel properties may be highly random due to the movement of various organs of the body, making it a challenging task. A real-time simulator can help investigate multiple technological advancements in the lab environment, which will lead to more extensive THz technology applications in real life.

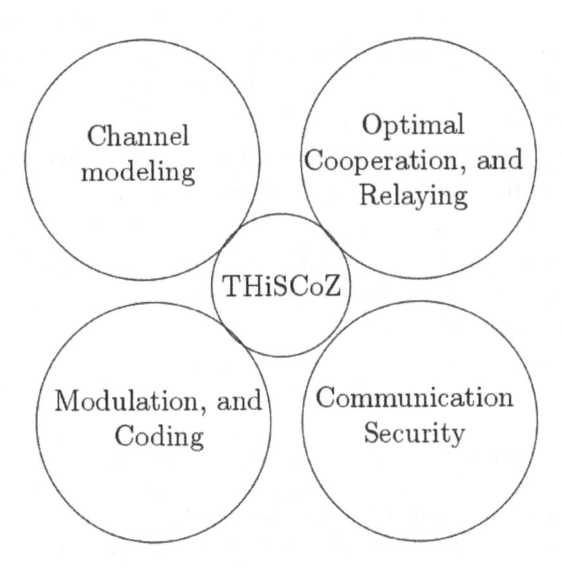

Figure 16.5 Research directions in terahertz communication.

16.6.2 Communication Techniques

Because of the inherent presence of frequency windows, the femtosecond pulses used for TS-OOK communication are broadened, leading to severe communication challenges that need detailed investigation. The interference thereby introduced may further hamper the rate of transmission. With massive MIMO, inherent difficulties of synchronization and channel estimations appear, both of which require a thorough understanding of the system properties such as channel characteristics, before realizing practical communication zones.

16.6.3 Cooperation and Relaying

While implementing cooperation among nodes in a dense network through cluster heads, cluster size selection, head selection, number of clusters, routing protocol, and amount of off-loading are some of the critical questions before congestion comes into the picture. While considering relays, optimal resource allocation between source and relay, relay selection, and relay placement are design considerations that need to be investigated from the energy efficiency point of view. Handovers between THiSCoZ is another critical challenge due to the involvement of very high data rates. Due to the smaller coverage area, there exists a high possibility of crossing many zones during a call by a mobile device. In such a scenario, handover latency, be it either soft or hard, is another exciting research direction to keep the connection alive. Moreover, novel green-optimized medium access control protocols for maximizing the achievable throughput in both dense and sparse heterogeneous ad-hoc nano-networks need investigations.

16.6.4 Security Concerns

IoT, being a network of devices that are either collecting, processing, or receiving information based on a different layer of a system architecture that they are placed in, is prone to security concerns. Privacy issues arise at all layers of IoT architecture, namely, perception layer, network layer, and application layer. Maintaining the critical aspects of security such as confidentiality, integrity, availability, and trust requires a layer-wise thorough investigation based on the severity of the attack's impact, which largely depends on the criticality of the information. Building up a common trust platform where devices from various vendors could share their data reliably with each other is a significant focus area before the idea of THiSCoZ comes to reality. Investigation on the impact of false information injection by a malicious node is required, as it could lead to manipulation in the system's behavior. Similarly, ensuring identity verification is a crucial challenge as impersonation could lead to information leakage.

References

Abadal, S., S. E. Hosseininejad, A. Cabellos-Aparicio, and E. Alarcn (2017, July). Graphene-based Terahertz antennas for area-constrained applications. In 2017 40th International Conference on Telecommunications and Signal Processing (TSP), pp. 817–820.

Afsana, F., M. Asif-Ur-Rahman, M. R. Ahmed, M. Mahmud, and M. S. Kaiser (2018). An energy conserving routing scheme for wireless body sensor nanonetwork communication. *IEEE Access* 6, 9186–9200.

Afsana, F., S. A. Mamun, M. S. Kaiser, and M. R. Ahmed (2015, December). Outage capacity analysis of cluster-based forwarding scheme for body area network using nano-electromagnetic communication. In 2015 2nd International Conference on Electrical Information and Communication Technologies (EICT), pp. 383–388.

Akdeniz, M. R., Y. Liu, M. K. Samimi, S. Sun, S. Rangan, T. S. Rappaport, and E. Erkip (2014, June). Millimeter wave channel modeling and cellular capacity evaluation. *IEEE Journal on Selected Areas in Communications* 32(6), 1164–1179.

Akyildiz, I. F., J. M. Jornet, and C. Han (2014, August). Teranets: ultra-broadband communication networks in the Terahertz band. *IEEE Wireless Communications* 21(4), 130–135.

Akyildiz, I. F., and J. M. Jornet (2016). Realizing ultra-massive MIMO (1024×1024) communication in the (0.06–10) Terahertz band. Nano Com. Net., pp. 46–54.

Bile Peng, S. Rey, and T. Krner (2016, April). Channel characteristics study for future indoor millimetre and submillimeter wireless communications. In 2016 10th European Conference on Antennas and Propagation (EuCAP), pp. 1–5.

Bird, J. P., J. M. Jornet, E. Einarsson, and G. R. Aizin (2017). Prospects for the application of two-dimensional materials to Terahertz-band communications. In Proceedings of the 4th ACM International Conference on Nanoscale Computing and Communication, NanoCom '17, New York, pp. 28:1–28:2. ACM.

Chakraborty, P. S., A. S. Cardoso, B. R. Wier, A. P. Omprakash, J. D. Cressler, M. Kaynak, and B. Tillack (2014, February). A 0.8 THz f_{max} SIGE HBT operating at 4.3 k. *IEEE Electron Device Letters* 35(2), 151–153.

Chen, Z., X. Ma, B. Zhang, Y. Zhang, Z. Niu, N. Kuang, W. Chen, L. Li, and S. Li (2019, February). A survey on Terahertz communications. *China Communications* 16(2), 1–35.

Dash, S., and A. Patnaik (2018). Material selection for THz antennas. *Microwave and Optical Technology Letters* 60(5), 1183–1187.

Han, C. and I. F. Akyildiz (2014, June). Distance-aware multi-carrier (damc) modulation in Terahertz band communication. In 2014 IEEE International Conference on Communications (ICC), pp. 5461–5467.

Han, C., A. O. Bicen, and I. F. Akyildiz (2015, May). Multi-ray channel modeling and wideband characterization for wireless communications in the Terahertz band. *IEEE Transactions on Wireless Communications* 14(5), 2402–2412.

Han, C., and Y. Chen (2018). Propagation modeling for wireless communications in the Terahertz band. *IEEE Communications Magazine* 56(6), 96–101.

Han, C., X. Zhang, and X. Wang (2019). On medium access control schemes for wireless networks in the millimeter-wave and Terahertz bands. *Nano Communication Networks* 19, 67–80.

Hoseini, S. A., M. Ding, and M. Hassan (2017, December). Massive MIMO performance comparison of beamforming and multiplexing in the Terahertz band. In 2017 IEEE Globecom Workshops (GC Wkshps), pp. 1–6.

Huq, K. M. S., S. A. Busari, J. Rodriguez, V. Frascolla, W. Bazzi, and D. C. Sicker (2019). Terahertzenabled wireless system for beyond-5g ultra-fast networks: A brief survey. *IEEE Network* 33(4), 89–95.

Jin, J., J. Gubbi, S. Marusic, and M. Palaniswami (2014). An information framework for creating a smart city through internet of Things. *IEEE Internet of Things Journal* 1(2), 112–121.

Jornet, J. M. (2014). Low-weight error-prevention codes for electromagnetic nanonetworks in the Terahertz band. *Nano Communication Networks* 5(1), 35–44.

Jornet, J. M. and I. F. Akyildiz (2011, October). Channel modeling and capacity analysis for electromagnetic wireless nanonetworks in the Terahertz band. *IEEE Transactions on Wireless Communications* 10(10), 3211–3221.

Jornet, J. M. and I. F. Akyildiz (2014a, May). Femtosecond-long pulse-based modulation for terahertz band communication in nanonetworks. *IEEE Transactions on Communications* 62(5), 1742–1754.

Jornet, J. M. and I. F. Akyildiz (2014b, April). Graphene-based plasmonic nano-transceiver for Terahertz band communication. In The 8th European Conference on Antennas and Propagation (EuCAP 2014), pp. 492–496.

Jornet, J. M., J. C. Pujol, and J. S. Pareta (2012). Phlame: A physical layer aware mac protocol for electromagnetic nanonetworks in the terahertz band. *Nano Communication Networks* 3(1), 74–81.

Kawanishi, T. (2019). THz and photonic seamless communications. *Journal of Lightwave Technology* 37(7), 1671–1679.

Kocaoglu, M. and O. B. Akan (2012, March). Minimum energy coding for wireless nanosensor networks. In 2012 Proceedings IEEE INFOCOM, pp. 2826–2830.

Kokkoniemi, J., J. Lehtomki, and M. Juntti (2016). A discussion on molecular absorption noise in the terahertz band. *Nano Communication Networks* 8, 35–45.

Laha, S., S. Kaya, D. W. Matolak, W. Rayess, D. DiTomaso and A. Kodi (Feb., 2015). A new frontier in ultralow power wireless links: Network-on-chip and chip-to-chip interconnects. *IEEE Transactions on Computer-Aided Design of Integrated Circuits and Systems* 34(2), 86–198.

Lin, C. and G. Y. Li (2015, August). Adaptive beamforming with resource allocation for distance-aware multi-user indoor terahertz communications. *IEEE Transactions on Communications* 63(8), 2985–2995.

Moldovan, A., M. A. Ruder, I. F. Akyildiz, and W. H. Gerstacker (2014, December). LOS and nLOS channel modeling for terahertz wireless communication with scattered rays. In 2014 IEEE Globecom Workshops (GC Wkshps), pp. 388–392.

Moshir, F. and S. Singh (2016). Modulation and rate adaptation algorithms for Terahertz channels. *Nano Communication Networks* 10, 38–50.

Petrov, V., D. Moltchanov, and Y. Koucheryavy (2016). Applicability assessment of Terahertz information showers for next-generation wireless networks. In 2016 IEEE International Conference on Communications (ICC), pp. 1–7.

Pujol, J. C., J. M. Jornet and J. S. Pareta (2011). PHLAME: A physical layer aware MAC protocol for electromagnetic nanonetworks. In 2011 IEEE Conference on Computer Communications Workshops (INFOCOM WKSHPS), pp. 431–436.

Ragheb, H., and E. R. Hancock (2007). The modified Beckmann-Kirchhoff scattering theory for rough surface analysis. *Pattern Recognition* 40, 2004–2020.

Rong, Z., M. S. Leeson, and M. D. Higgins (2017). Relay-assisted nano-scale communication in the THz band. *Micro Nano Letters* 12(6), 373–376.

Stojkoska], B. L. R. and K. V. Trivodaliev (2017). A review of Internet of Things for smart home: Challenges and solutions. *Journal of Cleaner Production* 140, 1454–1464.

Tang, Y., K. Shinohara, D. Regan, A. Corrion, D. Brown, J. Wong, A. Schmitz, H. Fung, S. Kim, and M. Micovic (2015, June). Ultrahigh-speed gan high-electron-mobility transistors with fT =fmax of 454/444 GHz. *IEEE Electron Device Letters* 36(6), 549–551.

Tekbyk, K., A. R. Ekti, G. K. Kurt, and A. Grin (2019). Terahertz band communication systems: Challenges, novelties and standardization efforts. *Physical Communication* 35, 100700.

Xia, Q. and J. M. Jornet (2017, October). Cross-layer analysis of optimal relaying strategies for Terahertzband communication networks. In 2017 IEEE 13th International Conference on Wireless and Mobile Computing, Networking and Communications (WiMob), pp. 1–8.

You, L., X. Gao, G. Y. Li, X. Xia, and N. Ma (2017, July). Bdma for millimeter-wave/Terahertz massive mimo transmission with per-beam synchronization. *IEEE Journal on Selected Areas in Communications* 35(7), 1550–1563.

Zainuddin, M. A., E. Dedu, and J. Bourgeois (2014, Dec). Nanonetwork minimum energy coding. In 2014 IEEE 11th Intl Conf on Ubiquitous Intelligence and Computing and 2014 IEEE 11th Intl Conf on Autonomic and Trusted Computing and 2014 IEEE 14th Intl Conf on Scalable Computing and Communications and Its Associated Workshops, pp. 96–103.

Zakrajsek, L. M., D. A. Pados, and J. M. Jornet (2017a). Design and performance analysis of ultramassive multi-carrier multiple input multiple output communications in the Terahertz band. In N. K. Dhar and A. K. Dutta (Eds.), *Image Sensing Technologies: Materials, Devices, Systems, and Applications IV*, Volume 10209, pp. 26–36. International Society for Optics and Photonics: SPIE.

Zakrajsek, L. M., D. A. Pados, and J. M. Jornet (2017b). Design and performance analysis of ultramassive multi-carrier multiple input multiple output communications in the Terahertz band. In N. K. Dhar and A. K. Dutta (Eds.), *Image Sensing Technologies: Materials, Devices, Systems, and Applications IV*, Volume 10209, pp. 26–36. International Society for Optics and Photonics: SPIE.

Zanella, A., N. Bui, A. Castellani, L. Vangelista, and M. Zorzi (2014). Internet of Things for smart cities. *IEEE Internet of Things Journal* 1(1), 22–32.

Chapter 17

Integration Frameworks for THz Wireless Technologies in Data Centre Networks

Sean Ahearne

Contents

17.1 Introduction

This chapter looks to investigate and detail the possible methods of integration of THz wireless links as part of a traditional data centre network. In order to achieve this integration, a number of questions need to be answered on how THz links can interact with all other devices and links typically found in a data centre network. Software-defined networking (SDN) plays a prominent role in modern data centres, meaning integration of THz links using SDN principles is required.[1] Achieving this integration enables a new form of communication technology to be introduced to SDN data centres, with the vast majority of previous networks only utilizing fixed point-to-point wired links.[2]

17.1.1 The Need for Terahertz Technology

Wireless network technologies are seldom seen in modern data centres. The primary reason for this is the inability of current wireless technologies to meet data centre network performance requirements. Both very high data rates and very high reliability are key performance metrics for good data centre network performance. Bit rates of 25 Gigabits per seconds (25 Gbps) for a single link with bit error rates (BER) of 1×10^{-9} or greater are expected in a modern data centre environment.[3] Currently these requirements can only be met using wired communication technologies such as copper or optical fibre links. The latest standard of the wireless network technology Wi-Fi (802.11ax) in comparison achieves a theoretical data rate of 9.5 Gbps, which is less than what is required, with a BER of 1×10^{-5} being several orders of magnitude below requirements also.[4] It is also well known that Wi-Fi networks links rarely achieve their peak theoretical performance due to several environmental variables such as interference and other nearby devices sharing the same wireless channel.[5]

A new type of device capable of wireless communication is currently in development. This device is known as a uni-travelling carrier photodiode (UTC-PD).[6] Where Wi-Fi networks typically operate in the 2.4 and 5 GHz frequency bands, UTC-PDs can transmit data at wireless frequencies of 300 GHz and beyond. The increased frequency and bandwidth available to these devices allows them to achieve performance levels which may be physically impossible for Wi-Fi due to spectrum saturation. Laboratory tests of these devices show data rates of 100 Gbps[7] and a BER of 1×10^{-9} at 60 Gbps.[8] This performance is much higher than Wi-Fi and these THz wireless devices are thus potentially capable of satisfying the requirements of data centre networks. This presents the potential for adoption of these new types of wireless links into a data centre environment. The introduction of a wireless network technology into an environment which previously relied on wired links has the potential to bring broad changes to future data centre network's physical design and logical architecture. In order for this to occur, a number of challenges must be overcome to integrate these THz UTC-PD devices with data centre network devices.

The first step towards this integration is a method to enable data centre network devices (such as a network switch) to monitor and control THz wireless devices. An interface between the THz device and a network switch needs to be defined. This interface needs to be compatible with the electronics controlling the THz device, and the operating system (OS) running on the network switch. Once this interface is defined, a method to abstract this interface such that it can be controlled using SDN principles needs to be investigated. A number of SDN protocols exist that are capable of this abstraction, provided extensions are made to the protocol to support THz links. This chapter will detail the implementation of these extensions with three SDN protocols: OpenFlow,[9] P4,[10] and NETCONF.[11]

Once the architecture and abstraction of THz link parameters is defined, designing an architecture of SDN virtual network functions (VNFs) to utilize THz links is required. VNFs are a virtual and software-defined implementation of network features frequently used and utilized in hardware in traditional enterprise networks such as routers, load balancers, and firewalls.[12] In a similar fashion to the previous requirement, these VNFs require an extension in order to be able to control and manage THz wireless links. As these functions are software-defined, this enables the extensions to be implemented relatively quickly and without any possible hardware incompatibility issues.[13] If the required THz parameters necessary to perform the required function have been abstracted using an SDN protocol the VNF can quickly access and control the THz link through the protocol's application programming interface (API).[14]

Finally, the creation of a network architecture capable of controlling THz links using an SDN network protocol is also required. This entails the definition of a hardware system capable of processing and controlling the various physical layer (PHY) parameters necessary for THz links to perform effectively. This system needs to be accessible using an SDN networking protocol through an API. Once the SDN protocol has access to these parameters, an autonomous SDN controller can use these parameters to operate VNFs provided the VNF has been programmed to process the now available THz wireless parameters during operation. These requirements will be detailed and defined in the following section of this chapter.

17.1.2 Background

A modern data centre architecture revolves around a leaf-spine fabric topology.[15] This topology is typically used in a data centre as it is scalable and reliable. It is also flexible in the sense that it can usually achieve a high average data throughput for most application workloads. The physical network itself, however, is not flexible with the point-to-point optical links in the network rarely being modified once installed. Thus a leaf-spine topology is considered a good general approach to data centre network architecture but will often not achieve maximum performance for a given network application. With the advent of SDN, an abstraction was created that enabled the forwarding plane of network devices to be programmable, and more flexible than traditional networking. Applications could now take advantage of features such as network function virtualization (NFV), which could program the forwarding plane of their network to achieve more optimal performance.[9] The point-to-point nature of optical links still presents an optimization problem as this physical constraint means that the leaf-spine topology remains the architecture of choice. For applications that require ultra-low latency for example, this topology can potentially have a large negative performance impact on a large-scale network. Racks of servers in the data centre could be potentially physically close to each other but logically distant in the network. For each switch, a packet is transmitted through in the network that switch has to process that packet in its forwarding plane. This takes a certain amount of time with switching latencies ranging from 5 to 125 microseconds per packet.[16] The likelihood of a network packet being delayed due to other network traffic also increases with each switch it must pass through. Therefore for ultra-low latency applications reducing the number of switches a packet must pass through in the network is key to maintaining a high network application performance level. Such a change in the network topology is not possible in current wired datacentre networks. With the implementation of THz wireless links, however, flexible and service-driven network topologies are possible. THz wireless links can support point-to-multipoint connections and are capable of features such as beam steering to different nodes. These features enable one to create a network topology more efficiently architected to the application workload that will operate within it. The nature of THz wireless technology also allows it to utilize other advanced network features being deployed in 5G wireless networks but not yet seen in datacentre networks

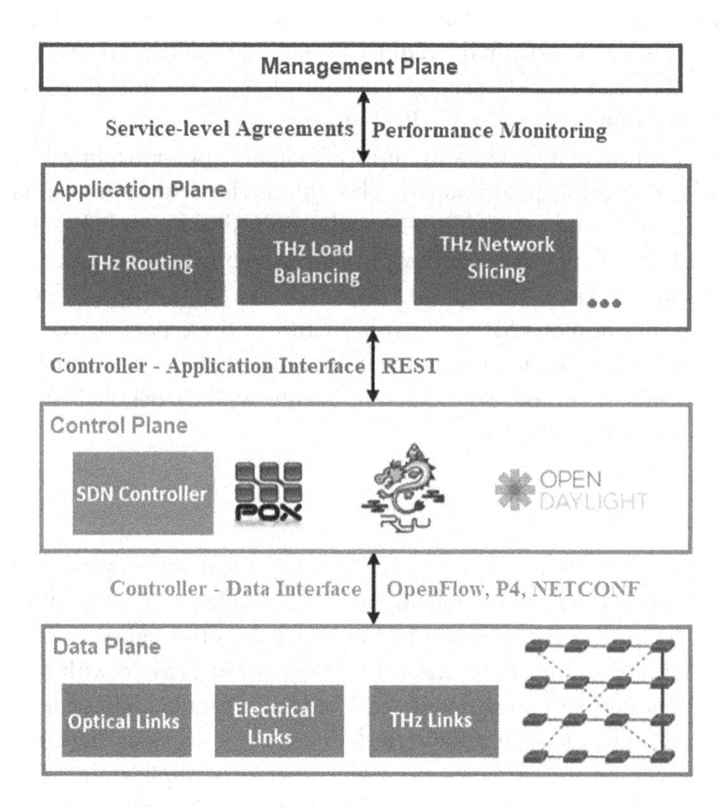

Figure 17.1 SDN network architecture.

such as network slicing.[17] This feature among others may play a key role in the future of THz networks and their application to not only datacentre networks but also mobile-edge and transport networks.

SDN is a key technology implemented in modern data centre networks enabling separation of a modern network's control plane from the packet processing data plane. This separation allows network functions to be virtualized into software, with data plane devices to be controlled and managed inter-operably using a common API. The general architecture of SDN can be seen in Figure 17.1. Data plane devices such as network switches enable connectivity between all other devices on the network using a variety of physical connections. In this example, the addition of THz links to the network is shown as a dotted line alongside traditional copper and optical connections in black.

To achieve SDN control of THz links and advanced THz VN's, the data plane devices must be capable of passing information about the physical operating parameters of the THz links to an SDN controller via an API. Once the controller has access to the THz parameter information VNFs can be created for using that information to enable network-level support for advanced THz features. Once these THz VNFs have been standardized, they can be integrated with network management plane software, which can maintain a heterogeneous network of copper, optical and THz links using its available VNFs. The operation of these VNFs and SDN controllers will be defined by service-level agreements, which define what network performance requirements are desired by the users of this network. This in turn will impact how the SDN controller autonomously manages its network of links in order to achieve the desired requirements.

The integration of THz links with SDN will pave the way for adoption of THz technology in data centre environments. The ability for THz links to be autonomously configured in real time with future advanced features such as multiple input multiple output (MIMO), beamsteering and switching, and time division multiple access (TDMA) could present a drastic change in the network architecture of future data centre deployments.[18] The increased flexibility offered by THz links over their copper or optical counterparts could enable future THz data centre designs and topologies to utilize all possible available network bandwidth provided by the underlying infrastructure, something previously impossible with wired links. This combined with the ability to modify available bandwidth between devices within the infrastructure in real time based on user requirements presents a strong argument for continued research and development for wireless data centre links.

17.1.3 Comparison with Related Works

The idea of using THz devices in a data centre network environment is still relatively new given the recent development of THz devices capable of high-speed communication at the time of this publication. Much of the current research with THz devices and data centre environments revolves around modelling and characterization of THz signals and their behaviour within a typical data centre.[19],[20],[21],[22] These PHY and medium-access control (MAC) layer research efforts are an important part of THz device development for data centre use and will be incorporated into future THz devices as they progress towards commercialization. Research to develop network protocols for THz devices at the network layer (NET) is comparatively sparse, with most of the research in this area focused on NET integration of THz devices in 5G and telecom communication infrastructure as opposed to data centre networks.[23],[24] The TERAPOD research project is specifically focused on the development of THz devices for the data centre, with some initial focus and research on development of NET protocols for integrating THz devices into an SDN data centre.[25],[26]

17.1.4 Assumptions

In order for development of NET protocols to continue for THz links a number of assumptions must be made. In Section 17.2, the development of application-specific integrated circuit (ASIC) used to control THz devices is described followed by the definition of SDN protocols based on this architecture. In Section 17.3, it is assumed that this architecture has been already created for the purposes of implementing THz network functions. As ASIC development for THz devices will not occur until much later in the THz development and commercialization process, the THz network functions have been designed using simulated values the SDN controller would receive from the ASIC.

17.2 Integration Frameworks for THz Wireless Links in a Software-Defined Network

17.2.1 A Hardware Control System for THz Wireless Links

The first step towards full integration of THz wireless links as part of SDN is development of an electronic system that can monitor and control the various physical-layer parameters

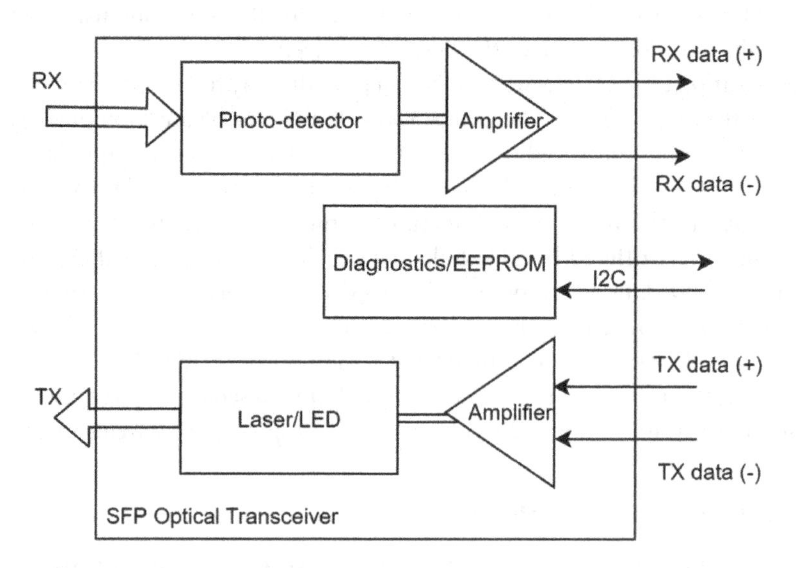

Figure 17.2 Internal configuration of an optical SFP module.

necessary for THz wireless communication. For a THz link to be useful in data communication, a number of physical parameters such as wireless frequency, power, noise, and more need to be known. These types of parameters must be monitored by various sensors placed throughout the THz system in order to ensure the hardware is operating within specification. As the development and commercialization of THz devices continues, the creation of an ASIC solely designed for the management of a THz wireless device will be developed. This type of ASIC is often developed using a simple field-programmable gate array (FPGA) with basic functions and memory developed to prevent damage to underlying THz hardware and maintain stable operation.[27] In data centre networks, this type of ASIC design already exists to assist in the operation of laser-based optical links. The small form-factor pluggable (SFP) standard is a widely used method in data centres to insert various types of communication equipment into network devices. These SFP modules primarily operate using copper or optical cables with various levels of bit rate and range available depending on user requirements. In a laser-based optical SFP module, an ASIC exists within the packaged module that automatically monitors and maintains operation of the laser according to specifications set by its manufacturer. Accompanying this ASIC is a block of electrically erasable programmable read-only memory (EEPROM).[28] This memory is used by the ASIC to store values recorded by connected hardware sensors such as the lasers temperature, power consumption, wavelength and more. An example of the configuration of a laser SFP module can be seen in Figure 17.2 with the ASIC and EEPROM memory located in the 'Diagnostics' block.

While this ASIC can provide robust and stable operation of the hardware components inside its SFP module, it would be beneficial for the sensor data stored in EEPROM to be accessible to the host device the module is plugged in to. This would allow the host device to also monitor hardware performance and notify a network administrator with a warning if an SFP module is overheating or has encountered some other type of error. To facilitate communication between the host device and SFP module, an electrical communication bus is required. The inter-integrated circuit (I^2C) bus is the accepted standard to

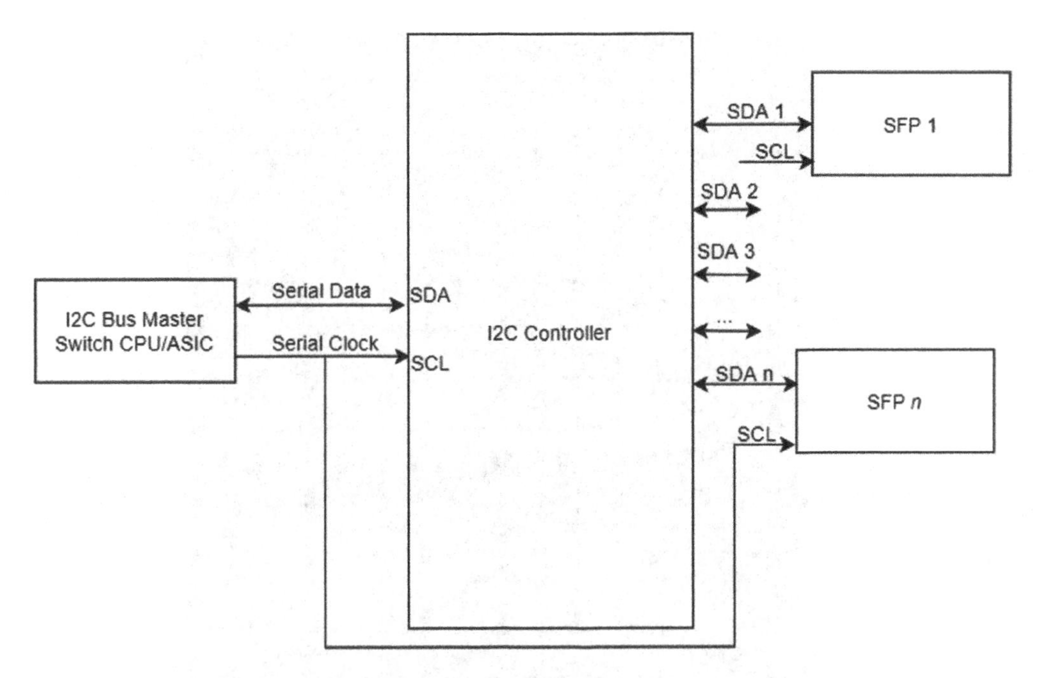

Figure 17.3 Overview on an I²C controller, typically found on a network switch.

enable this communication.[29] This bus provides a two-wire (serial clock and serial data) limited electrical interface between the host device and the modules EEPROM.

As data centre network switches typically contain several SFP ports, an I²C controller is often used to simplify access to each SFP modules EEPROM from the perspective of the host device. An example of the operation of this controller can be seen in Figure 17.3, where the I²C bus master is the central processing unit (CPU) of a network switch in this instance.

In order for a network switch to correctly read and interpret the sensor data contained in an SFP's EEPROM, the switches' operating system (OS) must contain the appropriate software to support interaction with an I²C bus. The majority of modern data centre switches run an OS based on a Linux kernel with driver support for I²C communication available.[30] Software can be written utilizing this driver to access and read the EEPROM memory of an SFP module. Values can also be written in certain circumstances, allowing the ASIC to execute a function based on a memory write event from the host device. The parameters and values stored in an SFP EEPROM block are standardized. A multi-source agreement (MSA) exists between all major manufacturers of SFP modules to store certain information and sensor values in specific memory locations. The SFF-8472 specification details the use of two addressable EEPROM memory blocks accessible via the I²C data bus.[31] One of these blocks is known as the information block containing details about the SFP modules manufacturer, serial number and other product information. The second diagnostic memory block contains sensor information about the optical hardware within the module. These memory blocks are arranged in a series of registers. Each register contains two hexadecimal digits the product of which represents a decimal value range from 0 to 255 (1 byte). Both memory blocks contain 256 registers split into 128 block

```
[+] INFO
03 04 07 00 00 00 00 00 00 00 00 06 67 00 50 FF
00 00 00 00 4A 44 53 55 20 20 20 20 20 20 20 20
20 20 20 20 00 00 01 9C 57 52 54 2D 53 46 50 50
54 30 31 35 53 43 20 20 30 30 30 30 06 15 00 1B
06 5A 0A 04 46 45 34 38 35 35 37 38 30 31 30 34
20 20 20 20 31 34 31 31 32 36 20 20 68 F0 05 4F
31 A9 4A 44 53 55 4E 41 20 20 20 20 20 20 20 20
53 46 50 50 54 30 31 35 53 43 FF FF FF FF FF FF
00 00 00 00 00 00 00 00 00 00 00 00 00 00 00 00
00 00 00 00 00 00 00 00 00 00 00 00 00 00 00 00
00 00 00 00 00 00 00 00 00 00 00 00 00 00 00 00
00 00 00 00 00 00 00 00 00 00 00 00 00 00 00 00
00 00 00 00 00 00 00 00 00 00 00 00 00 00 00 00
00 00 00 00 00 00 00 00 00 00 00 00 00 00 00 00
00 00 00 00 00 00 00 00 00 00 00 00 00 00 00 00
00 00 00 00 00 00 00 00 00 00 00 00 00 00 00 00

[+] DIAG
49 00 F8 00 46 00 FB 00 8D CC 74 04 87 5A 7A 75
D6 D8 1D 4C B9 8C 30 D4 4D F0 13 93 3D E8 18 A5
0F 8D 00 0C 09 CF 00 13 00 00 00 00 00 00 00 00
00 00 00 00 00 00 00 00 00 00 00 00 00 00 00 00
00 00 00 00 00 00 00 00 00 00 00 00 01 00 00 00
01 00 00 00 01 00 00 00 01 00 00 00 00 00 00 DF
24 3A 7C 5D 4A 43 27 10 32 7D 00 00 00 00 00 00
00 80 00 00 00 80 00 00 00 00 00 00 00 00 00 00
00 00 00 00 00 00 00 00 00 00 00 00 00 00 00 00
00 00 00 00 00 00 00 00 00 00 00 00 00 00 00 00
00 00 00 00 00 00 00 00 00 00 00 00 00 00 00 00
00 00 00 00 00 00 00 00 00 00 00 00 00 00 00 00
00 00 00 00 00 00 00 00 00 00 00 00 00 00 00 00
00 00 00 00 00 00 00 00 00 00 00 00 00 00 00 00
00 00 00 00 00 00 00 00 00 00 00 00 00 00 00 00
00 00 00 00 00 00 00 00 00 00 00 00 00 00 00 00
```

Figure 17.4 Contents of an SFP modules EEPROM memory blocks.

'pages'. The SFF specification details what information is stored in each register location with some values represented as a product or function of two or more register values. To correctly interpret the information stored in these registers, a piece of software running on the switch's OS must read the memory blocks using the I²C bus and decode the values it receives based on the SFF specification. When writing a value to EEPROM if input is given in decimal form or another format, it must be re-calculated and re-encoded to the correct hexadecimal format required. A 'raw' memory dump of both memory blocks can be seen in Figure 17.4. This particular memory dump being from a 1550 nm SFP optical transceiver module. Note the large number of unused '00' registers, suggesting available space for future specification expansion.

The SFF-8472 specification has been updated numerous times since its inception with the current revision version number being 12.3. Separate to this a number of other SFF specifications have been created to expand the functionality of optical SFP modules, in significant enough ways to warrant a new specification. One of these specifications is SFF-8690, which standardizes the method to interact with and control optical SFP modules with tuneable laser wavelengths.[32] This specification achieves the implementation of

A2h Address	Size	Name	Description
Bytes 132 (MSB) & 133 (LSB)	2 bytes	LFL1	Lasers First Frequency (THz)
Bytes 134 (MSB) & 135 (LSB)	2 bytes	LFL2	Lasers First Frequency (GHz*10), in units of 0.1 GHz
Bytes 136 (MSB) & 137 (LSB)	2 bytes	LFH1	Lasers Last Frequency (THz)
Bytes 138 (MSB) & 139 (LSB)	2 bytes	LFH2	Lasers Last Frequency (GHz*10), in units of 0.1 GHz
Bytes 140 (MSB) & 141 (LSB)	2 bytes	LGrid	Laser's minimum supported grid spacing (GHz*10), i.e., in units of 0.1 GHz NOTE: LGrid can be a positive or negative number.

Figure 17.5 Excerpt from SFF-8690 showing the definition of registers for controlling wavelength.[28]

this feature through the use of memory paging. On a tuneable SFP module, the diagnostic EEPROM block contains a third block of 128 memory registers. This third memory block is accessed by writing '02' to the 127th register of the first block which triggers the ASIC to change the information contained in the second-half of the EERPROM block to page 2, referring to the new register definitions created in SFF-8690 for tuneable SFPs. Having the new memory page accessible using this method allows these modules to be backwards compatible with existing SFP network devices though without tuning functionality. It may be possible for this functionality to be added to existing network devices with a software update, though some devices may not feature the EEPROM write functionality required. An excerpt from the SFF-8690 specification can be seen in Figure 17.5 showing some of the newly defined registers and what their values represent. Note that the manufacturer is given multiple options for which type of physical parameter measurements and units they wish to use and store. Different unit types can be stored at different register locations with flag registers defined to identify which particular units and mode the manufacturers ASIC operates with. As there are a number of different possible units and modes of operation of tuneable optical modules all these different possibilities must be accounted for in the software of any network device and OS that wishes to support them.

With the hardware-level operation of optical SFP modules now known, the applicability of this system to THz wireless links is evident. Development of THz wireless links based on the principles of the SFP specifications is a logical way of implementing THz device compatibility with standard data centre network equipment where the SFP standard is already ubiquitous. Like SFF-8690, we propose a similar extension, which enables a new memory page containing all physical parameters related to the monitoring and operation of THz wireless links. Once THz devices reach maturity an ASIC can be developed to maintain operating stability of the THz device. This ASIC will also perform all the new THz-related functions available with it being capable of executing these functions when a user writes a new value to one of its EEPROM registers. An example of some of the possible register definitions can be seen in Table 17.1. Note that these values are for reference only to define a prototype for what an official specification may look like. The decision on what information to store and in what registers and format will be standardized when THz devices reach a higher technology readiness level (TRL). Even after an official specification is defined such specifications can still be revised to add new THz functions in the future. Expanding on some of the example definitions provided in the table, the WF1 variable refers to the centre frequency of the THz link. This variable is two bytes in size,

Table 17.1 Example of some possible register definitions for THz wireless links

A2h Address	*Size*	*Name*	*Description*
Bytes 132 (MSB) and 133 (LSB)	2 bytes	WF1	Wireless link frequency (GHz), in units of 0.1 GHz
Bytes 134 (MSB) and 135 (LSB)	2 bytes	CW1	Wireless channel width (GHz), units of 0.01 GHz
Bytes 136 (MSB) and 139 (LSB)	4 bytes	TX1	Wireless TX power (W), units of 10^{-9} W (nW)
Bytes 140 (MSB) and 143 (LSB)	4 bytes	RX1	Wireless RX power (W), units of 10^{-9} W (nW)
Bytes 144 (MSB) and 145 (LSB)	2 bytes	DR1	TX Bitrate (Gbps), units of 0.1 Gbps
Bytes 146 (MSB) and 147 (LSB)	2 bytes	DR2	RX Bitrate (Gbps), units of 0.1 Gbps
Byte 148	1 byte	BER1	RX Bit Error Rate (BER), measured as 1×10^{-X}
Byte 149	1 byte	MCS1	TX Modulation scheme, measured as MCS index

represent a maximum possible decimal value range from 0 to 65,535. With units of 0.1 GHz this gives a maximum possible range of 0.1 to 6553.5 GHz. While this is excessive given the current state of the art of THz devices it leaves a very wide frequency range available for future THz devices without requiring a new revision. It must be considered that once a memory location has been defined it cannot be changed in future revisions for backwards compatibility reasons. This means only registers which were previously unused can be defined in future revisions. The CW1 represents channel width with a range of 0.01 to 655.35 GHz. If reduction in memory space is required, CW1 could be reduced to a single byte value with a range of 0.01 to 2.25 GHz, 0.1 to 22.5 GHz or 1 to 255GHz depending on requirements and device characteristics. TX power is a 4-byte value ranging from a power output of 1 nW up to 4.29W (2^{32}). Bitrate shares the same range as WF1 representing 100 Mbps to 6.5 Tbps. Bit error rate (BER) is a single byte value with a range from 1×10^{-1} to 1×10^{-255}. Other methods to detect and control the error rate of the active THz link can be defined and can be automatically controlled by the PHY, MAC or NET layers depending on user demands. The modulation scheme used can similarly be automatically decided by the ASIC based on other THz device factors, or it can be specifically chosen by the MAC or NET layers by writing the desired value to the register. MCS index in this context refers to an index used in the 802.11 Wi-Fi specifications related to modulation formats (QPSK, QAM-16 QAM-256 etc.). Alternatively, a number of the previously mentioned variables can be defined such that they are compliant with an associated IEEE specification such as IEEE 802.15.3d-2017.[33] Defining variables in this fashion ensures they are compliant with the specification while also potentially reducing the memory space required for some variables such as frequency and channel width due these variables now having a fixed range according to the specification.

In the event that more than 128 bytes of memory are required to store all the parameters necessary for THz operation extra memory 'pages' can be added as required as the majority of the 255 available pages are currently unassigned. Certain parameters can also be derived using a more complex formula instead of a simple product.

These example values represent the structure which can be defined to integrate THz wireless links with the SFP specification. A specification extension designed in this way would allow for backward compatibility with existing SFP network devices. Functionality would be limited for existing devices due to the lack of software functionality to inspect for and decode the new THz pages. The THz hardware can still operate and establish a THz link in this case however, as the embedded ASIC can sweep and search for other THz links in the area and establish a point-to-point link if another device is found. For more advanced features such as THz beamsteering or point to multipoint communication the ability of the SFP host device to communicate with the ASIC is required. This can be achieved on standard data centre switch by updating its OS with new software code to read and write to the new THz EEPROM page(s). This would enable the switch and hence the network administrator or SDN controller to control and configure the THz link as they desire.

While user control of THz links has been established at this point integration of THz links with SDN principles has not yet been achieved. At this point, controlling the THz link can only be done from within the switch itself either manually by a user or automatically via the OS. In order for the links to be considered part of SDN they need to be controllable and configurable from an autonomous SDN controller. This means that the THz EEPROM values need to be abstracted such that they are accessible and able to be modified by a remote device. The use on an SDN process, protocol, and API are necessary to achieve this type of functionality. There are a number of SDN protocols available and the implementation of new features such as THz link monitoring and control is highly varied and non-trivial for each protocol. In many cases once a specification or revision is created for an SDN protocol it is not backwards compatible with legacy devices. The reason for this is that many SDN control protocols are highly integrated into the data plane meaning support for a specific version of that protocol is defined in the hardware of a network switch and its data packet processor. This often means that any revision of an SDN protocol that requires an increase in memory space is unlikely to be supported by an existing packet forwarding processor as that processor was fabricated without the required memory space. This is not necessarily true for all SDN protocols, however. An investigation on how to abstract these THz parameters with three of the major SDN protocols follows.

17.2.2 OpenFlow Implementation

The OpenFlow protocol is the original and most widely known protocol in SDN.[34] This pioneering protocol introduced the concept of a network with a centralized control plane. This control plane forms the basis of the 'brain' of the network controlling network functions such as routing, firewalls, and load balancing.[35] This control plane is fully software-based and can be run on any commodity server. This control plane connects to an OpenFlow-supporting network switch using the OpenFlow API. This switch no longer operates using its own internal control plane leaving only the packet processor (known as the data plane) available to be configured.[36] OpenFlow-supporting switches contain a data path architecture that processes incoming packets through a series of flow tables containing flow rules. Each incoming packet progresses through the available flow tables until it successfully matches an existing flow rule. Each rule in a flow table has an associated action to perform when a packet is matched to it such as sending the packet out a specific port, sending the packet to an SDN controller for inspection, dropping the packet, and more.

A diagram of the standard OpenFlow architecture can be seen in Figure 17.6. An SDN controller connects to the OpenFlow Agent service, which runs on the switch OS and installs new flow rules and actions into its packet processor based on what network functions the user has chosen to implement on the network. Thus network

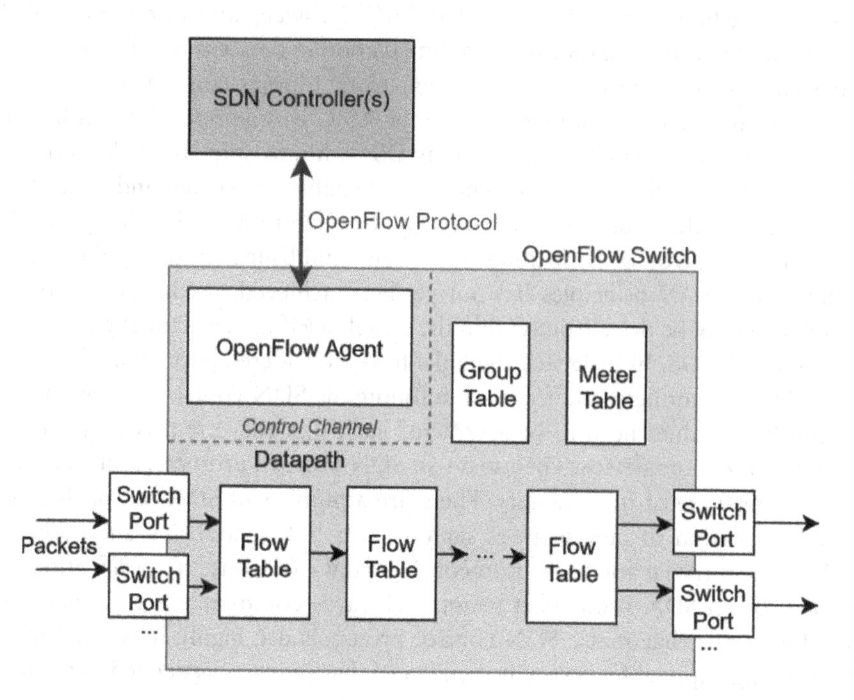

Figure 17.6 Overview of OpenFlow architecture.

functions such as routing and load balancing can now be achieved on a network by installing the correct set of flow rules and actions on each device in the network. This eliminates the need for separate hardware devices to perform these actions as is required in traditional networking. It also removes the need for complex and proprietary control planes in each network device, which enables heterogeneous network infrastructures with inter-operable OpenFlow network devices available from any supporting manufacturer. This is in contrast to traditional networking, where many control plane management systems were proprietary and specific to each manufacturer. The advantages presented by SDN functions and commodification of network devices has led SDN and the OpenFlow protocol to become widely adopted in data centre environments and beyond.

The OpenFlow protocol has seen many major revisions since its inception adding new and improved features for each successive revision. One major revision of the OpenFlow protocol is version 1.4, which added support for tunable optical SFP modules based on the previously mentioned SFF-8690 specification.[37] To enable this functionality in OpenFlow, the addition of new memory blocks (also known as structs) was required. These structs play a key role in the operation of the packet processor as it uses them to store and update many variables used in real time as packets are processed through the flow table pipeline. One of these structs can be seen in Figure 17.7 specifying a number of variables which can be used to store information for the optical properties of tuneable SFP modules. It is noticeable that the size in bytes and names of these variables are not exactly the same as the SFF-8690 specification, thus signifying that a software translation of these variables to the specification occurs elsewhere in the process.

```
/* THz port description property. */
struct ofp_port_desc_prop_thz {
    uint16_t        type; /* OFPPDPT_THZ */
    uint16_t        length; /* Length in bytes of this property. */
    uint8_t         pad[4]; /* Align to 64 bits. */

    uint32_t supported; /* Features supported by the port. */
    uint32_t tx_freq; /* THz TX Frequency */
    uint32_t tx_rate; /* THz TX bitrate */
    uint32_t tx_chan_freq; /* TX THz Channel Width Frequency */
    uint32_t rx_freq; /* RX Frequency */
    uint32_t rx_rate; /* RX bitrate */
    uint32_t rx_chan_freq; /* RX Channel Width Frequency */
    uint16_t tx_pwr; /* TX power */
    uint16_t rx_pwr; /* RX power */
};
OFP_ASSERT(sizeof(struct ofp_port_desc_prop_thz) == 40);
```

Figure 17.7 The addition of a struct to control optical wavelength in OpenFlow 1.4.[48]

As the addition of these new structs in version 1.4 required significant changes to both the memory and hardware functionality of OpenFlow-supporting devices it resulted in the version being incompatible with all previous versions of OpenFlow. This means that a network ASIC manufactured to the 1.3 specification is not capable of being updated to 1.4. This disadvantage has to be taken into consideration when designing and implementing OpenFlow support of THz links. The implementation of THz links in OpenFlow follows the same design principles as the implementation of support for SFF-8690. A new subsequent protocol version can be defined and a series of memory structs can be created that mimics the design of the EEPROM parameters for THz links. An example of this THz struct can be seen in Figure 17.8. Based on the OpenFlow 1.4 specification, these variables will also be independent of the associated THz SFF specification and will be transformed to the required values by a software function. As previously mentioned, this implementation will be incompatible with devices supporting previous version of OpenFlow due to memory and hardware differences. A THz SFP module could still transmit and receive data on a legacy OpenFlow switch but visibility and control of the THz parameters and features would not be available to the SDN controller. Nevertheless, developing support of THz links with the OpenFlow protocol should be considered as it remains the most ubiquitous SDN protocol with a large development community surrounding it. It is also the simplest method of achieving full data-plane integration of THz links using a single protocol. Its fixed hardware nature and high production volumes allow for rapid prototyping with lower development costs and likely a shorter time to a production-ready release in comparison to other SDN protocols.

17.2.3 P4 Implementation

P4 is a programming language specifically designed for use with data plane devices. P4 aims to change the way traditional packet forwarding is designed and implemented. The original specification was created in 2014 with an updated specification released in 2016.[38] P4 follows the same principle for forwarding packets as OpenFlow as it also

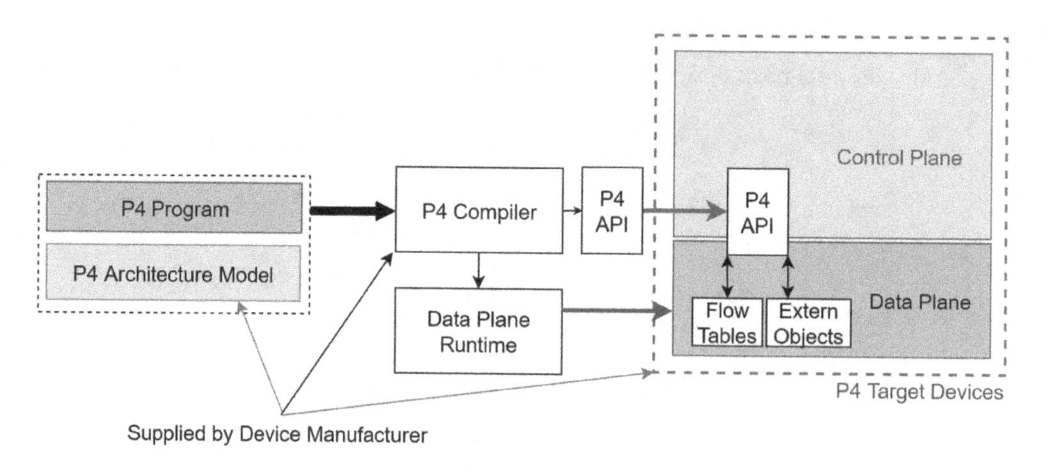

Figure 17.8 Example possible OpenFlow struct for THz parameters.

contains a match-action table structure. Unlike OpenFlow, however, the process by which a packet is processed through the pipeline is now programmable instead of fixed in nature as it is OpenFlow. P4 achieves this by having programmable parsing logic.[39] In a traditional network packets sent though the network are constructed following the standard open systems interconnect (OSI) model.[40] Packets in this model contain strictly defined headers with each header identifying the protocol it is operating on and other relevant values related to the use of that protocol. A traditional packet processor contains a fixed-function pipeline where each function is designed to parse and process a specific header and protocol. OpenFlow devices also use fixed parsing logic. By using programmable parsing logic, P4 network devices are protocol independent meaning its packet processing capabilities are not limited by the OSI model or currently defined network protocols. A network built entirely of P4 devices could enable a fully software-defined data plane where packet headers can be created, parsed, and processed using a definition entirely developed by the owner of the network. This definition can also be re-configured on any device at any time after deployment enabling new network protocols to be created and implemented on existing devices. In comparison the creation of a new packet processor would be required in OpenFlow to support the new protocol. This software-defined parser also enables the P4 architecture to be compiled and implemented across multiple different device architectures while sharing the same functionality. The devices for which P4 code can be compiled for are known as targets and can include standard CPUs, FPGAs, systems on a chip (SoCs) and ASICs.[41],[42] Programmable actions are also possible with P4, which gives the user a high amount of flexibility in what to do with a packet compared to the fixed number of actions possible using OpenFlow. P4 also contains a feature known as externs, which are values or parameters than can be used to store information outside of the packet processing pipeline thus enabling stateful processing. The P4 architecture model can be seen in Figure 17.9.

A user creates their P4 program containing all their desired network functions in the form of parsing and action code. This code is created based on a manufacturer-created architecture model, which specifies the externs available on the target device and the arithmetic logic operations it is capable of performing. This program is then compiled by a P4

```
extern Register<T> {
    Register(bit<32> size);
    T read(bit<32> index);
    void write(bit<32> index, T value);
}
```

The type T has to be specified when instantiating a set of registers, by specializing the Register type:

```
Register<bit<32>>(128) registerBank;
```

Figure 17.9 P4 architecture model.

```
Register<bit<8>>(128) THzBank;

extern Register<thz_freq> {
    Register(bit<16> size);
    thz_freq read(bit<16> index);
    void write(bit<16> index, thz_freq value);
}
//index = 4 (132 - 128)
```

Figure 17.10 Example of how to access registers as extern values in a P4 program.[41]

compiler, which is also supplied by the manufacturer of the target device. The compiler loads the defined parsing and action logic into the data plane of the device and specifies the desired API to allow the control plane SDN controller to interact with the device. The SDN controller can use this API to install new rules in the target devices match-action tables and read from and write to the available external "extern" objects. The capabilities of P4 make it a very powerful language for data centre networks with the improved flexibility it brings considered to outweigh the disadvantage of increased complexity in comparison to OpenFlow.

In order for THz links to be configurable by an SDN controller using P4 a number of conditions must be met. First, the data plane device must contain an I^2C bus in order to access the SFP ASIC. Second, control logic must exist within the P4 packet processor to allow access to the THz SFP EEPROM in the form of addressable extern values. The ability for registers to be defined can be seen in Figure 17.10 showing how to initialize a bank of registers and how to read and write to a specific register. This is a standard feature defined in the P4-16 specification.

The ability to define a bank of registers means that definition of a register bank that matches the proposed THz SFF specification is possible. Example P4 code to create a THz parameter register bank and associated read and write functions can be seen in Figure 17.11. There are a few things to note about this example. The first is the definition of the memory bank follows that of the EEPROM with a bank of 128 registers (2nd page of the diagnosis block), which are 8 bits (1 byte) in size. Defining the bank in this way would allow for a direct mapping of the EEPROM registers to the P4 memory bank, thus

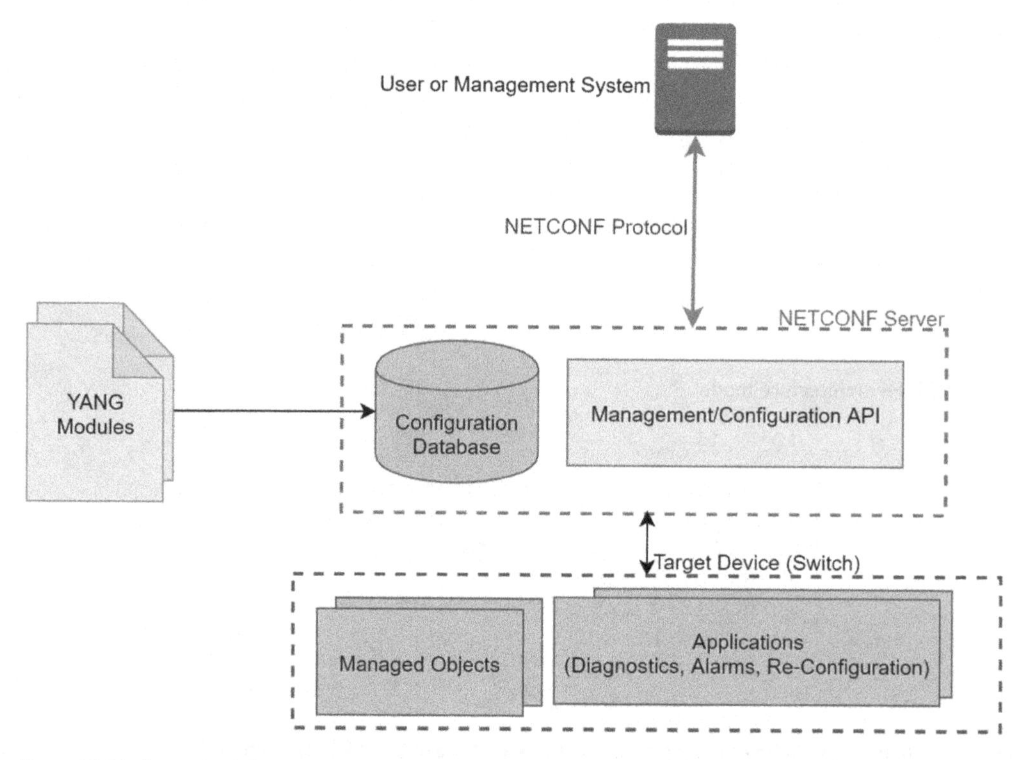

Figure 17.11 Example definition of P4 THz extern value.

requiring no translation function to be necessary. It can be seen in the register definition for THz frequency, however, that it requires a size of 16 bits as given in the SFF specification. There are a number of solutions to this. The first is to simply modify the code to split the register definition into two separate 8-bit sections and re-combine them to get the correct value and performing the inverse for a write operation. The next solution is for the device manufacturer to define a function in their architecture model, which automatically recognizes when a defined register exceeds the size of a single register in the memory bank. When the read or write function is called it will automatically re-factor the register for the correct size of the bank based on the index value adding 1 to the index value for every 8 bits. An indirect memory bank mapping is also possible as seen in OpenFlow where a memory bank with 32-bit registers is defined. In this case a function to translate the values to the SFF specification will be required in the manufacturer's architecture model. The index value of 4 in this example refers to the first register used to define THz frequency in the example SFF specification. Byte 132 is 4th byte in the new EEPROM page.

A THz memory bank should be available for each SFP port contained on the switch. Once the extern mapping is completed match-actions can be created that read or write to the extern registers. Programmable actions can be created, which can access and modify THz link parameters in real time on a packet by packet basis if required. This could be useful for implementing THz features such as high-speed beamsteering in the future. These extern values will also be accessible to the SDN controller via the chosen control plane API where a network of P4 THz switches can be monitored by the controller and network functions can be implemented. Given the programmable nature of P4 it is also possible for new THz features added in a future SFF specification to be programmed into already existing P4 devices, provided it has the correct architecture model to allow for changes in register definitions. The software-defined nature of controllers also allows them to be updated to support new THz features as they become available.

17.2.4 NETCONF

The NETwork CONFiguration (NETCONF) protocol pre-dates the era of SDN and previously existed as a network protocol which could be used to remotely change the configuration of supporting network devices. A remote network device such as a user or SDN controller can connect to a network device such as a switch or router using the NETCONF protocol and read or update the devices current configuration. To achieve this a NETCONF server must be ran as a service contained within the network devices OS. This NETCONF service maintains a data store of yet another next generation (YANG) modules. These modules are used to define specific parameters that are available to be configured on the network device. When a user or controller wants to change or read the configuration of a device, it connects to the NETCONF server and sends or requests data in the same structure and format of that YANG module. An example diagram of the NETCONF architecture can be seen in Figure 17.12 where a modern SDN controller replaces the user and management systems previously used. The ease of implementation of the NETCONF protocol has enabled its use in optical networks on devices such as reconfigurable optical add-drop multiplexers (ROADMs), a device which previously had limited configurability and flexibility.[43]

A NETCONF server is designed to contain an API that enables OS-level applications to access and update parameters stored within its YANG-defined datastore. It also allows applications to contain a call-back function, which executes when the parameter it is monitoring in the data store is updated from a remote device. To achieve SDN control of THz links with the NETCONF protocol, a NETCONF server and data handling service needs to be installed on the OS of the network device (in this example a switch). 'Sysrepo' is an application which provides this service as a complete solution for network devices operating on a Linux-based OS.[44] An architecture diagram for this solution can be seen

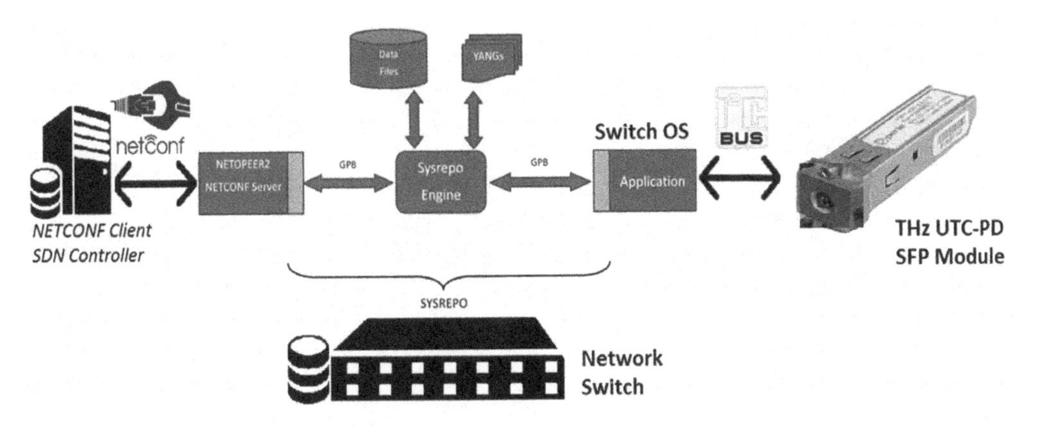

Figure 17.12 NETCONF system architecture.

in Figure 17.13 showing the process of accessing and configuring a THz SFP module from the SDN controller from left to right.

The first step for integrating control of THz links with the Sysrepo engine is development of an application which can access and modify the THz EEPROM registers using the I²C bus. This can be achieved using the C programming language as it contains native driver headers needed to utilize the I²C bus on the switch and is also the API used between

```
module sysrepo-thz-system
{
  yang-version 1;
  namespace "urn:ietf:params:xml:ns:yang:sysrepo-thz-system";
  prefix sysrepo-thz;
  description "Sysrepo THz Wireless SFP YANG Module";
  container thz
  {
    description "Container for THz parameters";
    leaf port
    {
      type uint8;
      description "Which SFP port on the switch will be accessed";
    }
    leaf frequency
    {
      type uint16;
      description "THz link frequency value";
    }
    leaf width
    {
      type uint16;
      description "THz Channel Width";
    }
    leaf txpower
    {
      type uint32;
      description "THz Transmit Power";
    }
    leaf txbitrate
    {
      type uint32;
      description "THz Current Bitrate";
    }
    leaf ber
    {
      type uint8;
      description "Current Bit Error Rate";
    }
    leaf modulation
    {
      type uint8;
      description "Current Modulation Scheme";
    }
  }
}
```

Figure 17.13 NETCONF architecture to implement THz links.

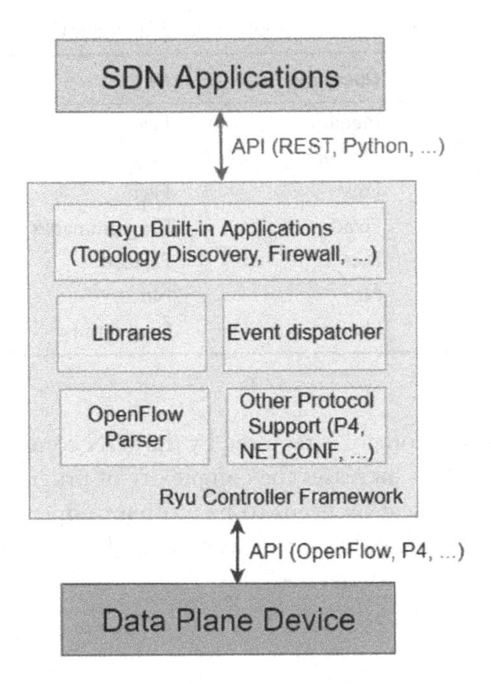

Figure 17.14 Example YANG model containing THz link parameters.

Sysrepo and the host OS. Once application control of the THz device is verified a YANG module needs to be created that is based on the EEPROM register definitions in the example SFF specification. An example of this YANG module can be seen in Figure 17.14, which defines the type and size of the parameters utilized by the THz module.[45]

Once this YANG module is installed in the Sysrepo data store, the C application can be modified to connect to the Sysrepo service, and independent functions can be defined for each leaf parameter value found within the YANG module. The C application can periodically update the data store with new information taken from the THz EEPROM or can wait for the data store to be updated remotely, signifying a desired configuration change and writing the new value in the data store to the THz EEPROM. The function performed by the C application can be tailored to user requirements and use cases. Similar to previous protocols a C translation function can exist in the application to convert between the YANG values and the EEPROM values. Once the YANG module is installed and the C application is connected to Sysrepo, the THz SFP module is available to be configured from an SDN controller. The controller can connect to the NETCONF server running on the switch and using the THz YANG module to read or write to parameters within that module to control the THz link. A major advantage to this implementation compared to OpenFlow or P4 is that it is backwards-compatible with existing network devices. Unlike those protocols where a new specification or architecture model is required the NETCONF system is fully software-based and as such is implementable on any device using a Linux OS kernel and containing an I^2C bus at any time. Like P4 it can also be updated to enable new THz features on any device as they become available as it only requires an updated YANG module and C code. One major disadvantage to this implementation is the lack of interoperability with packet forwarding processors. This means traditional networking or

Table 17.2 Comparison of Requirements for implementing SDN control of THz links

Requirements	*OpenFlow*	*P4*	*NETCONF*
Complexity	Medium	High	Low
Development time	Medium	High	Low
Upgradeability	Low	High	High
Data plane control	Fixed	Programmable	None
THz control speed	Fast	Fastest	Slow
Backwards compatible	No	No	Yes
Real-time features	Limited	Supported	None

the use of a second SDN protocol is required by the SDN controller to achieve full control of all network devices. This increases the complexity of programming VNFs for the SDN controller as now the application needs to be written to operate with two different SDN APIs at once. This trade-off could be seen as acceptable, however, to enable SDN control of THz links on legacy network devices.

17.2.5 Summary

In summary, Section 17.2 has defined how to implement SDN control of THz links in data centre networks based on an extension of the SFF specification for SFP modules. Abstraction of this specification to be configurable via an SDN controller was also described for three common SDN API's OpenFlow, P4, and NETCONF. As shown in the description of these APIs, there are a number of advantages and disadvantages to the implementation of each, thus the protocol chosen to be implemented may depend on the requirements of the THz link user and their requirements. Table 17.2 shows a summary comparison of some important requirements that should be considered when implementing SDN control of THz links. Real-time features is an important metric for THz wireless links and describes functions that require a very high processing speed when a configuration change is made. THz beamsteering can be considered a feature, which requires this in order to implement advanced wireless network techniques such as time division multiple access (TDMA). Such features are not possible with the NETCONF protocol as the control plane is not capable of changing the configuration at the rate required. While possible with the OpenFlow protocol its fixed-function data plane limits the capabilities of real time features. This makes P4 the protocol of choice for development of advanced THz network features and should be considered as the long-term choice for future THz SDN development.

17.3 THz Network Function Virtualization

Once abstraction of the parameters to control THz links at the network layer is achieved (as described in the previous section), virtual network functions must be created which use these parameters in order to effectively manage THz wireless links within a software-defined network. The first step in this process is to choose an SDN controller to test implementation of THz VNFs. An SDN controller known as Ryu was chosen for this task. It

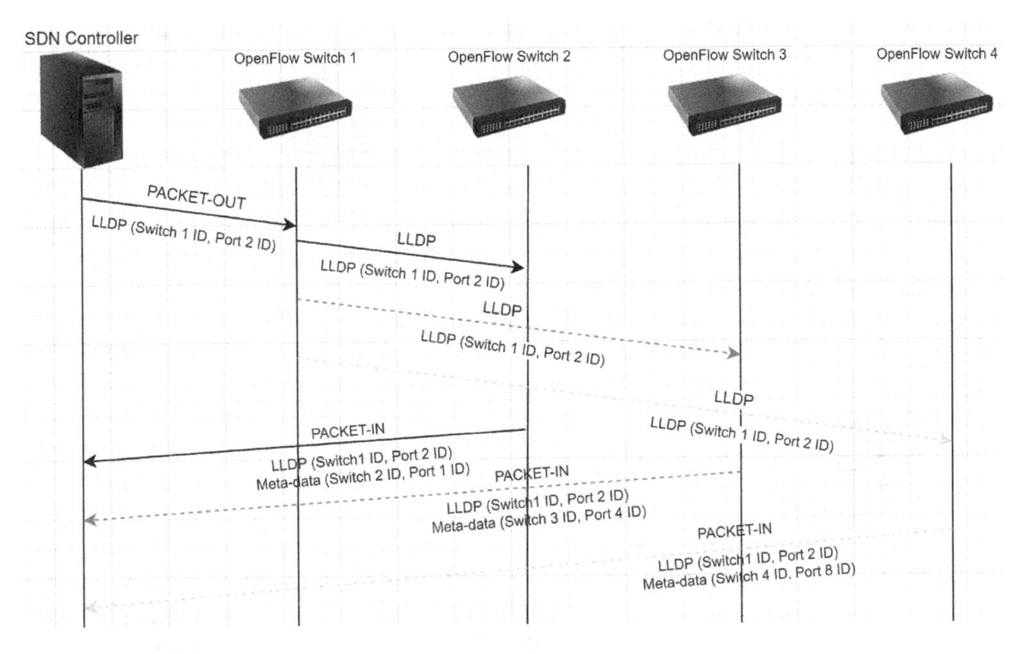

Figure 17.15 Ryu SDN Controller Architecture.[49]

is an SDN controller based on the Python programming language and contains several functions and features useful for SDN development.[47] Ryu's software architecture can be seen in Figure 17.15 showing the data plane network switches at the bottom connecting to Ryu via OpenFlow or another protocol. Ryu contains a number of built-in apps to provide basic NET functions such as topology discovery, firewalls and more. A user could choose to expand the built-in apps in Ryu to support THz links or develop an independent VNF which interfaces with the controller, denoted by the SDN applications in this diagram.

17.3.1 Topology Discovery

One potentially major change to the way SDN will operate with THz wireless links is in the method of topology discovery. Traditionally a wired point-to-point link only has one possible destination. The link layer discovery protocol (LLDP) exploits this principal to determine what devices are connected to each other on the networks data plane.[46] An SDN controller sends a command to a connected switch to perform LLDP on a specific port. This port sends an LLDP packet out of its interface, which is received by a port on another switch. When this switch detects it received an LLDP packet, it responds with its own unique identifier and the port on which it received the packet contained in a metadata format. With future THz links capable of beamsteering, point-to-multipoint communication between devices becomes possible. In order for LLDP to support this new form of communication, a number of changes have to be made. The topology discovery function contained within the SDN controller must be modified in order to support the concept of point-to-multipoint communication. This can be achieved by programming the controller and topology discovery function to maintain metadata tables, which can contain more than a single instance of an LLDP respondent. This table can later be used to determine

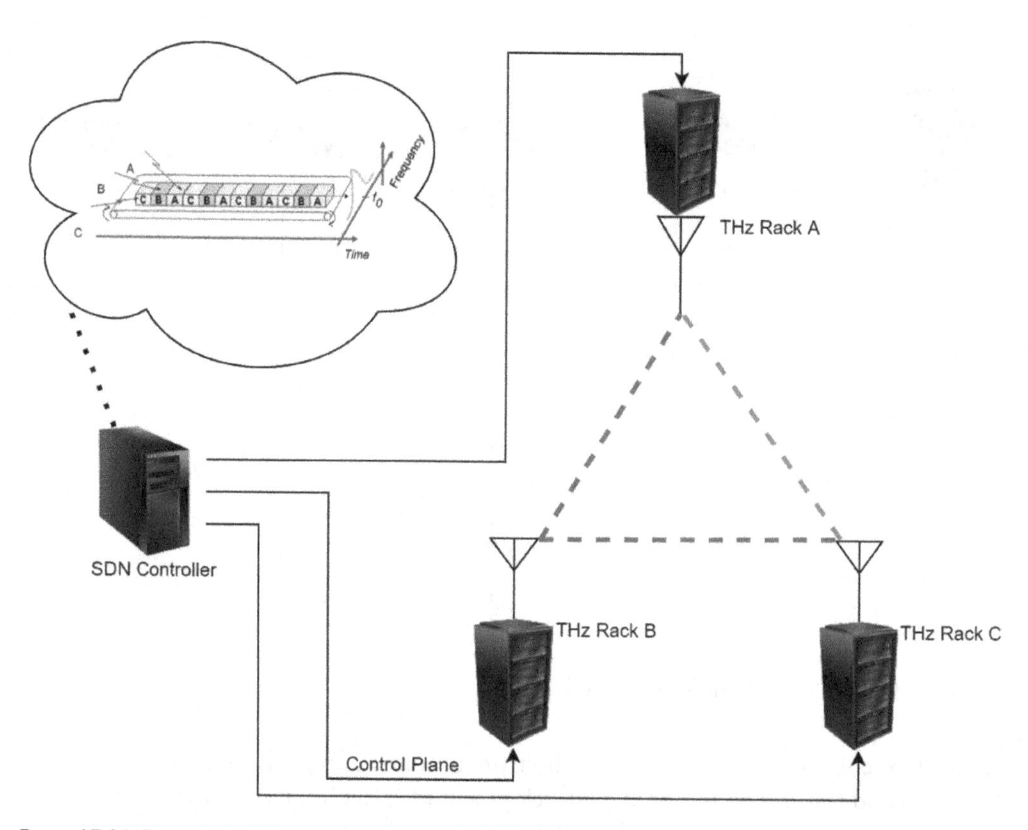

Figure 17.16 Example LLDP sequence diagram for THz links.

what THz-enabled switches are within range to establish a connection with each other and if more than one THz link is available between a set of switches.

In order to detect other THz links in range, the THz device will need to perform a detection process to scan for other possible THz links. This detection process can be standardized as an ASIC function in the THz SFF specification or it can be implemented as a MAC function embedded in the switches packet processor. When an LLDP packet is received, this triggers the THz detection function, which sweeps and sends a further LLDP packet to all THz receivers within its range. The receiving switches recognize this LLDP packet and respond to the controller on the control plane network specifying what switch they are and what port they received the LLDP packet on. A sequence diagram for the operation of the new THz LLDP function can be seen in Figure 17.16 where a traditional point-to-point LLDP sequence can be seen in black. The expanded operation of the new point-to-multipoint THz LLDP can be seen in green and red, showing the THz detection process and switch metadata response to the controller.

17.3.2 Real-Time Features: Time Division Multiple Access

As mentioned in the previous section, the use of technologies such as phased array antennas to enable THz beamsteering provides an opportunity to implement advanced wireless networking features previously unused in a data centre environment.[47] In a large data

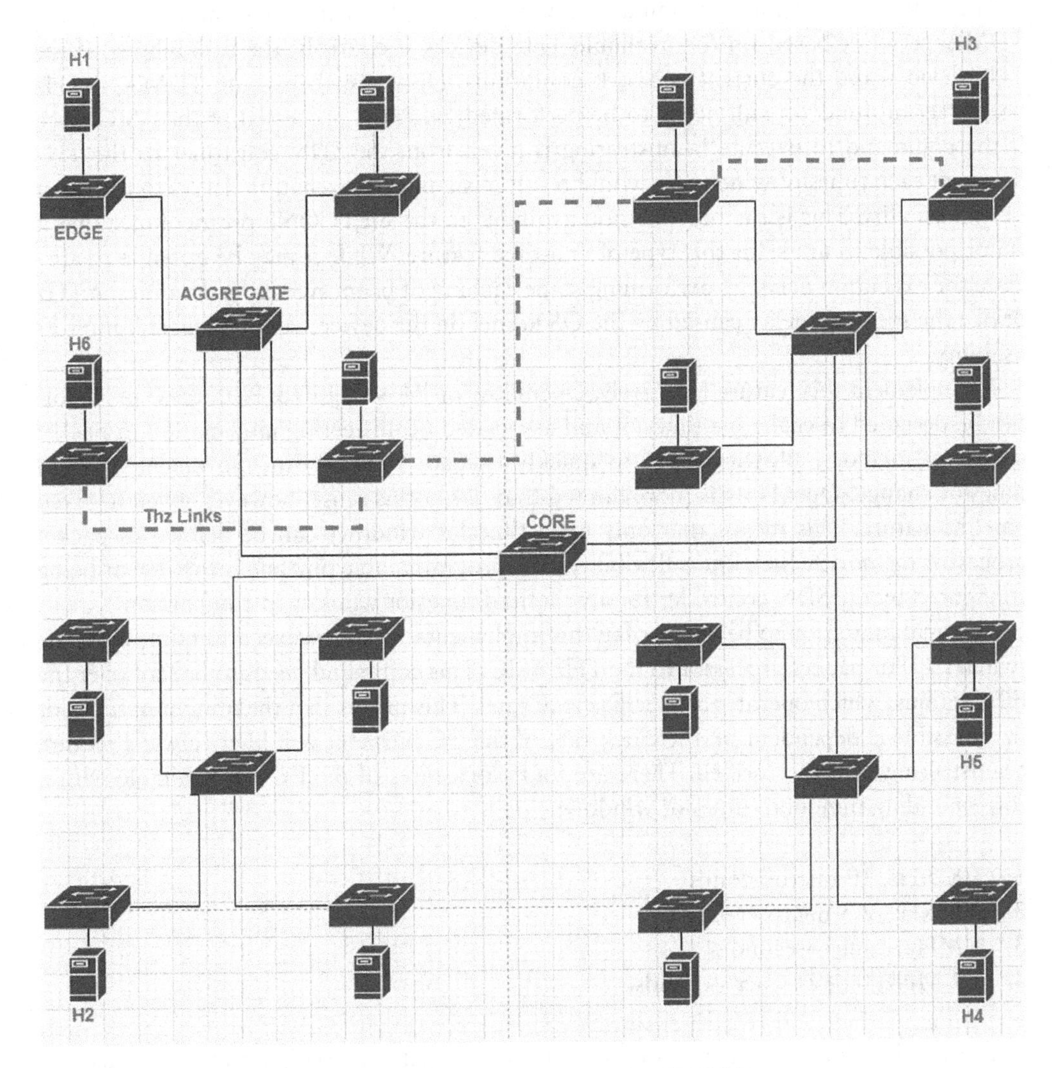

Figure 17.17 Visualization of THz TDMA being implemented as an SDN VNF.

centre network with these beamsteerable THz devices techniques such as TDMA can be implemented in order to improve wireless network's flexibility and performance. TDMA allows multiple wireless devices to access a shared medium using timeslots to specify when each device on the network can transmit data. This can greatly improve performance of wireless networks by reducing interference and collisions. It also enables point-to-multipoint communication where high-speed THz beam switching can be used to maintain an active network connection between a single source THz device and multiple receiver devices. Through the use of SDN the TDMA configuration of devices can be updated in real time, thus greatly improving the flexibility offered particularly in comparison to traditional wired point-to-point networks in most datacentres. Figure 17.17 gives a visual representation of a possible TDMA implementation in a datacentre between 3 THz racks in communication range with each other. The coloured and dashed lines represent which connection is active during each timeslot.

The SDN controller in this implementation is capable of defining and configuring a TDMA network based on the available THz devices it controls, the capabilities of the THz devices and the current network conditions. However, to execute TDMA, a THz beamsteering function still needs to be performed, which is processed at the ASIC level. If the command to execute beamsteering is given from the SDN controller to the THz device at each timeslot window, it would result in significant execution delays to the point of unfeasibility. This is the only method available to the NETCONF protocol meaning it is not possible to use it for this type of real-time feature. While it may be possible to alleviate the execution delay by performing some caching of beam switching data on the THz device the software delay caused by the OS kernel on the device would likely still limit its usability.

To implement this feature effectively, OpenFlow or P4 is required. Both these protocols can be designed to cache the timeslot and associated beamsteering information in packet processor memory, allowing them to quickly execute a THz beam movement on each timeslot change. OpenFlow is more limited than P4 in this sense, however, given its fixed pipeline nature. This means that only fixed timeslot windows can be defined in packet processor memory, which limits flexibility. P4's programmable pipeline is capable of being configured by the SDN controller for any desired timeslot window in comparison.

One final limitation to be aware of in this implementation is the execution time from the switch CPU or packet processor to the THz ASIC. This command needs to be sent over the I^2C data bus, which operates at specific clock rates. This means that the minimum timeslot size possible is dependent on the clock rate of the I^2C bus as it will likely always require beamsteering to be performed. There are four clock rates of the I^2C bus available with a corresponding minimum timeslot window:

1. 100 kHz, 10 microseconds
2. 400 kHz, 2.5 microseconds
3. 1 MHz, 1 microsecond
4. 3.2 MHz, 0.3125 microseconds

Note that this does not include any overhead required for executing the beamsteering function or time taken to physically perform beam steering. These overheads need to be included for effective operation. For optimal performance, a 3.2MHz I^2C controller should be used with these devices as data centre workloads can often be sensitive to latency. Thus the minimum timeslot size is preferred. This can depend on the overhead, however, as smaller window sizes reduce data throughput due to more of the timeslot being used to perform beamsteering. The SDN controller can configure what length of timeslot to use to favour throughput or latency depending on the services operating on its network.

17.3.3 THz Routing, Load Balancing and Fail-over

For effective utilization of THz links, common network functions need to be updated to interact with the new THz parameters. While an SDN controller can now successfully access and modify the THz controlling registers located in a THz SFP module, the usefulness of this ability is limited until it is incorporated into an SDN controllers built-in or external network functions. Once incorporated into a VNF, control of the THz link

```
def find_path_cost(self,path):
    path_cost = []
    for i in range(len(path) - 1):
        port1 = self.neigh[path[i]][path[i + 1]]
        bandwidth_between_two_nodes = self.bw[path[i]][port1]
        if (self.thz[path[i]][port1]):
            path_cost.append(THZ_POWER / bandwidth_between_two_nodes)
        path_cost.append(bandwidth_between_two_nodes)
    return sum(path_cost)
```

Figure 17.18 Network topology for THz NFV testing.

```
"Host: h6"
[286]  5.00-6.00   sec  1.01 GBytes  8.64 Gbits/sec  4218   826 KBytes
[286]  6.00-7.00   sec  1.31 GBytes  11.2 Gbits/sec  2366   423 KBytes
[286]  7.00-8.00   sec  1.04 GBytes  8.89 Gbits/sec  1814  1.36 MBytes
[286]  8.00-9.00   sec  1.22 GBytes  10.5 Gbits/sec  8775   229 KBytes
[286]  9.00-10.00  sec  1.09 GBytes  9.33 Gbits/sec  1086   482 KBytes
[286] 10.00-11.00  sec  1.26 GBytes  10.8 Gbits/sec  2012   197 KBytes
[286] 11.00-12.00  sec  1018 MBytes  8.53 Gbits/sec  1419   635 KBytes
[286] 12.00-13.00  sec  1.38 GBytes  11.8 Gbits/sec  667    857 KBytes
[286] 13.00-14.00  sec   839 MBytes  7.03 Gbits/sec  9202  1.61 MBytes
[286] 14.00-15.00  sec   139 MBytes  1.16 Gbits/sec     2  5.42 MBytes
[286] 15.00-16.00  sec  88.8 MBytes   745 Mbits/sec     5  6.08 MBytes
[286] 16.00-17.00  sec   145 MBytes  1.22 Gbits/sec    96  2.85 MBytes
[286] 17.00-18.00  sec   130 MBytes  1.09 Gbits/sec     4  5.97 MBytes
[286] 18.00-19.00  sec   105 MBytes   881 Mbits/sec  1357  4.64 MBytes
[286] 19.00-20.00  sec  77.5 MBytes   650 Mbits/sec     5  5.74 MBytes
[286] 20.00-21.00  sec   108 MBytes   902 Mbits/sec     5  6.74 MBytes
```

Figure 17.19 Excerpt of Ryu THz Routing algorithm.

can be automated instead of requiring manual input. To test this a network, a testing environment needs to be created. This network environment was created in the network emulation tool Mininet mimicking a collapsed tree model typically found in a data centre environment. The network topology tested can be seen in Figure 17.18 with 16 hosts and 21 switches. An SDN controller (not pictured) is controlling all switches via the OpenFlow protocol. A number of THz links exist as seen with the dotted red lines between hosts H6 and H3 which will be the basis for testing of network functionality of new THz NFVs.

The first step in improving utilization of THz links is the development of a routing algorithm, which considers various THz parameters when calculating its routing decisions. In this example, all THz links in the demonstration are set to a bandwidth of 10 Gbps while all wired links are set to a bandwidth of 1 Gbps. The reason for this is to design the THz routing algorithm to initially favour routing traffic between H3 and H6 over the THz links. This routing algorithm is quite simple and is based on creating a path cost value for various potential routes between H3 and H6. The switch begins the route calculation with the switch connected to H3 and calculates the cost of the available paths based on the available link bandwidth. The available bandwidth cost is measured based on current throughput load of the link versus its maximum bandwidth. In this case, a new parameter is added when calculating the cost of a route path: the THz transmission power. If the

Time	Source	Destination	Protocol	Info	Delta Time
0.469839815	fe:45:dc...	CayeeCom_...	LLDP	TTL = 4919 Syste...	0.454461193
0.779967703	10.0.0.6	10.0.0.3	TCP	[TCP Spurious Re...	0.310127888
0.004055260	10.0.0.6	10.0.0.3	TCP	[TCP Previous se...	0.003415025
0.015378622	10.0.0.3	10.0.0.6	TCP	[TCP ACKed unsee...	0.001243542
0.010547433	10.0.0.6	10.0.0.3	TCP	58236 → 5201 [AC...	0.000745424
0.000613535	10.0.0.6	10.0.0.3	TCP	[TCP Previous se...	0.000593535
0.780531362	10.0.0.6	10.0.0.3	TCP	[TCP Previous se...	0.000548579
0.001165694	10.0.0.3	10.0.0.6	TCP	5201 → 58236 [AC...	0.000538799
0.001134234	10.0.0.3	10.0.0.6	TCP	5201 → 58236 [AC...	0.000508209
0.781027424	10.0.0.3	10.0.0.6	TCP	5201 → 58236 [AC...	0.000482992

Figure 17.20 TCP test showing re-routing of THz traffic through loss of throughput.

algorithm detects that the connection between the current hop and the next is a THz link, it will adjust the cost associated with path based on the current THz transmission power of the device. An excerpt from this algorithm can be seen in Figure 17.19.

As previously mentioned, the initial state of the routing algorithm should favour using the 10 Gbps THz links. During link testing, however, the THz transmission power is set to slowly decrease over time. This should eventually cause the routing algorithm to switch the data flow between H3 and H6 over to the 1 Gbps wired link when the path cost for the THz link exceeds that of the wired link. In this example, an available bandwidth of 1 Gbps between nodes is given a cost of 10 while a 10 Gbps link is given a cost of 1. A THz transmission power of 1 micro-watt (μW) is given a cost of 1 with a transmission power of 0.1 μW given a proportional cost of 10. This routing algorithm runs continuously within the SDN controller as a background process. This means as the transmission power of the THz device weakens over time the SDN routing algorithm should automatically switch the flow of traffic between H3 and H6 from the THz link to the wired link once the cost of the THz link exceeds the cost of the wired link. If the controller performs this action successfully it signifies the algorithm is capable of enhanced routing, load balancing and fail-over capabilities. While this example algorithm is simple in nature it provides a proof of concept for the successful implementation of future and further advanced THz VNFs. The results of the test can be seen in Figure 17.20. A TCP is sending the maximum amount of data possible between H3 and H6. During this test THz transmission power is slowly decreased until the route cost for the THz link exceeds that of the wired link at t=14 seconds. At this point the controller successfully re-routes the traffic to be sent over the wired link at 1 Gbps.

It is important for network reliability purposes that this path change and fail-over procedure occurs with minimal delay. Figure 17.21 contains packet capture data, which shows the service interruption delay caused by the path change. In this example, network connectivity was restored in 310 ms, as noted by the maximum TCP delta time. This time represents the value of time in seconds, which passed between this packet being received and the previous one in the TCP sequence. In this example the TCP transmission was able to resume after the service interruption.

Time	Source	Destination	Protocol	Info	Delta Time ^
0.469839815	fe:45:dc...	CayeeCom ...	LLDP	TTL = 4919 Syste...	0.454461193
0.779967703	10.0.0.6	10.0.0.3	TCP	[TCP Spurious Re...	0.310127888
0.004055260	10.0.0.6	10.0.0.3	TCP	[TCP Previous se...	0.003415025
0.015378622	10.0.0.3	10.0.0.6	TCP	[TCP ACKed unsee...	0.001243542
0.010547433	10.0.0.6	10.0.0.3	TCP	58236 → 5201 [AC...	0.000745424
0.000613535	10.0.0.6	10.0.0.3	TCP	[TCP Previous se...	0.000593535
0.780531362	10.0.0.6	10.0.0.3	TCP	[TCP Previous se...	0.000548579
0.001165694	10.0.0.3	10.0.0.6	TCP	5201 → 58236 [AC...	0.000538799
0.001134234	10.0.0.3	10.0.0.6	TCP	5201 → 58236 [AC...	0.000508209
0.781027424	10.0.0.3	10.0.0.6	TCP	5201 → 58236 [AC...	0.000482992

Figure 17.21 Latency test result for THz fail-over function.

17.4 Conclusions

Based on the investigations performed and results gathered we can conclude that full integration of terahertz-based wireless devices with software-defined datacentre networks is possible. A number of methods have been found to implement SDN control of THz links and their introduction into a datacentre environment could cause a dramatic change in the architecture and behaviour of networks that use them. With the bitrates offered by THz devices continuing to increase their improved flexibility combined with their convenience and potentially reduced physical footprint make them a very lucrative technology in space-saving datacentre network designs. THz technology features such as beamsteering have the potential to enable radically different network architectures in the future with the future potential features and performance of this technology in infrastructure networks exceeding that of both wired and other wireless technologies.

17.4.1 Future Research Directions and Challenges

Given the information discovered throughout the course of this chapter, further research into the development of an SDN protocol capable of controlling THz links should specifically focus on the P4 protocol. The performance advantages presented by the P4 protocol in comparison to other SDN protocols particularly in the enablement of real-time THz network features make it a primary candidate for future research into its implementation and increasing the number of THz network functions and features it provides.

There are several challenges that need to be overcome in future THz network research. The development of an ASIC capable of controlling a THz device which is also compatible with existing network switches and SDN protocols is not a trivial task. Development of such an ASIC is required for THz devices to become commercially feasible thus is an important area of research and a development challenge that must be overcome for future THz devices. Standardization is also a future area and challenge for THz device research. While standards are available for THz frequency bands for communication, there is currently no standard specification or method to interface THz devices with the existing network equipment, either in terms of physical interfacing or in terms of standard network protocols. This chapter envisions the use of SFP and P4 as the standard methods for interfacing THz devices with data centre network equipment, but officiating such a standard is challenge and is the subject of future research efforts.

References

[1] Kreutz, Diego, Fernando MV Ramos, Paulo Esteves Verissimo, Christian Esteve Rothenberg, Siamak Azodolmolky, and Steve Uhlig. "Software-defined networking: A comprehensive survey." *Proceedings of the IEEE* 103, no. 1 (2014): 14–76.

[2] McKeown, Nick. "Keynote talk: Software-defined networking." In *Proc. of IEEE INFOCOM*, vol. 9. 2009.

[3] Mazzini, M., M. Traverso, M. Webster, C. Muzio, S. Anderson, P. Sun, D. Siadat et al. "25GBaud PAM-4 error free transmission over both single mode fiber and multimode fiber in a QSFP form factor based on silicon photonics." In *Optical Fiber Communication Conference*, pp. Th5B-3. Optical Society of America, 2015.

[4] Sharon, Oran, and Yaron Alpert. "Single User MAC Level Throughput Comparison: IEEE 802.11 ax vs. IEEE 802.11 ac." *Wireless Sensor Network* 9, no. 5 (2017): 166–177.

[5] Deng, Der-Jiunn, Shao-Yu Lien, Jorden Lee, and Kwang-Cheng Chen. "On quality-of-service provisioning in IEEE 802.11 ax WLANs." *IEEE Access* 4 (2016): 6086–6104.

[6] Nagatsuma, Tadao, Guillaume Ducournau, and Cyril C. Renaud. "Advances in terahertz communications accelerated by photonics." *Nature Photonics* 10.6 (2016): 371.

[7] Chinni, V. K., et al. "Single-channel 100 Gbit/s transmission using III–V UTC-PDs for future IEEE 802.15. 3d wireless links in the 300 GHz band." *Electronics Letters* 54.10 (2018): 638–640.

[8] Yu, Xianbin, Rameez Asif, Molly Piels, Darko Zibar, Michael Galili, Toshio Morioka, Peter U. Jepsen, and Leif K. Oxenløwe. "400-GHz wireless transmission of 60-Gb/s Nyquist-QPSK signals using UTC-PD and heterodyne mixer." *IEEE Transactions on Terahertz Science and Technology* 6, no. 6 (2016): 765–770.

[9] McKeown, Nick, Tom Anderson, Hari Balakrishnan, Guru Parulkar, Larry Peterson, Jennifer Rexford, Scott Shenker, and Jonathan Turner. "OpenFlow: enabling innovation in campus networks." *ACM SIGCOMM Computer Communication Review* 38, no. 2 (2008): 69–74.

[10] Bosshart, Pat, Dan Daly, Glen Gibb, Martin Izzard, Nick McKeown, Jennifer Rexford, Cole Schlesinger et al. "P4: Programming protocol-independent packet processors." *ACM SIGCOMM Computer Communication Review* 44, no. 3 (2014): 87–95.

[11] Enns, Rob, Martin Bjorklund, and Juergen Schoenwaelder. NETCONF Configuration Protocol. RFC 4741, December 2006.

[12] Mijumbi, Rashid, Joan Serrat, Juan-Luis Gorricho, Niels Bouten, Filip De Turck, and Raouf Boutaba. "Network function virtualization: State-of-the-art and research challenges." *IEEE Communications Surveys & Tutorials* 18, no. 1 (2015): 236–262.

[13] Han, Bo, Vijay Gopalakrishnan, Lusheng Ji, and Seungjoon Lee. "Network function virtualization: Challenges and opportunities for innovations." *IEEE Communications Magazine* 53, no. 2 (2015): 90–97.

[14] Shin, Myung-Ki, Ki-Hyuk Nam, and Hyoung-Jun Kim. "Software-defined networking (SDN): A reference architecture and open APIs." In 2012 International Conference on ICT Convergence (ICTC), pp. 360–361. IEEE, 2012.

[15] Alizadeh, Mohammad, and Tom Edsall. "On the data path performance of leaf-spine datacenter fabrics." In 2013 IEEE 21st Annual Symposium on High-Performance Interconnects, pp. 71–74. IEEE, 2013.

[16] Fiber Optic Network Products. 2018. *Understanding Network Latency In Ethernet Switches*. [online] Available at: www.fiberopticshare.com/network-latency-in-ethernet-switches.html.

[17] Ordonez-Lucena, Jose, Pablo Ameigeiras, Diego Lopez, Juan J. Ramos-Munoz, Javier Lorca, and Jesus Folgueira. "Network slicing for 5G with SDN/NFV: Concepts, architectures, and challenges." *IEEE Communications Magazine* 55, no. 5 (2017): 80–87.

[18] Nelson, Randolph, and Leonard Kleinrock. "Spatial TDMA: A collision-free multihop channel access protocol." *IEEE Transactions on communications* 33, no. 9 (1985): 934–944.
[19] Cheng, Chia-Lin, Seun Sangodoyin, and Alenka Zajić. "THz cluster-based modeling and propagation characterization in a data center environment." *IEEE Access* 8 (2020): 56544–56558.
[20] Boujnah, Noureddine, Saim Ghafoor, and Alan Davy. "Modeling and link quality assessment of THz network within data center." In 2019 European Conference on Networks and Communications (EuCNC), pp. 57–62. IEEE, 2019.
[21] Cheng, Chia-Lin, Seun Sangodoyin, and Alenka Zajić. "THz MIMO channel characterization for wireless data center-like environment." In 2019 IEEE International Symposium on Antennas and Propagation and USNC-URSI Radio Science Meeting, pp. 2145–2146. IEEE, 2019.
[22] Cheng, Chia-Lin, Seun Sangodoyin, and Alenka Zajić. "Terahertz MIMO fading analysis and Doppler modeling in a data center environment." In 2020 14th European Conference on Antennas and Propagation (EuCAP), pp. 1–5. IEEE, 2020.
[23] Cacciapuoti, Angela Sara, Ramanathan Subramanian, Kaushik Roy Chowdhury, and Marcello Caleffi. "Software-defined network controlled switching between millimeter wave and terahertz small cells." *arXiv preprint arXiv:1702.02775* (2017).
[24] Akyildiz, Ian F., Pu Wang, and Shih-Chun Lin. "SoftAir: A software defined networking architecture for 5G wireless systems." *Computer Networks* 85 (2015): 1–18.
[25] Davy, Alan, Luis Pessoa, Cyril Renaud, Edward Wasige, Mira Naftaly, Thomas Kürner, Glenn George, Oleg Cojocari, Niamh O'Mahony, and Marco AG Porcel. "Building an end user focused THz based ultra-high bandwidth wireless access network: The TERAPOD approach." In 2017 9th International Congress on Ultra Modern Telecommunications and Control Systems and Workshops (ICUMT), pp. 454–459. IEEE, 2017.
[26] Ahearne, Sean, Niamh O'Mahony, Noureddine Boujnah, Saim Ghafoor, Alan Davy, Luis Gonzalez Guerrero, and Cyril Renaud. "Integrating THz Wireless Communication Links in a Data Centre Network." In 2019 IEEE 2nd 5G World Forum (5GWF), pp. 393–398. IEEE, 2019.
[27] Brown, Stephen D., Robert J. Francis, Jonathan Rose, and Zvonko G. Vranesic. *Field-Programmable Gate Arrays*. Vol. 180. Springer Science, 2012.
[28] Chan, T. Y., K. K. Young, and Chenming Hu. "A true single-transistor oxide-nitride-oxide EEPROM device." *IEEE Electron Device Letters* 8, no. 3 (1987): 93–95.
[29] Semiconductors, Philips. "The I2C-bus specification." *Philips Semiconductors* 9397, no. 750 (2000): 00954.
[30] LI, Xiang-bing, and Kou-gen ZHENG. "Design and implementation of linux I2C bus driver [J]." *Computer Engineering and Design* 1 (2005): 013.
[31] SFF Committee. "Sff-8472 specification for diagnostic monitoring interface for optical transceivers." https://members.snia.org/document/dl/25916 PDF (2012).
[32] SFF Committee. "Sff-8690 specification for Tunable SFP+ Memory Map for ITU Frequencies." https://members.snia.org/document/dl/25977 PDF (2013).
[33] IEEE Standard for High Data Rate Wireless Multi-Media Networks – Amendment 1: High-Rate Close Proximity Point-to-Point Communications," in IEEE Std 802.15.3e-2017 (Amendment to IEEE Std 802.15.3-2016), vol., no., pp.1–178, 7 June 2017, doi: 10.1109/IEEESTD.2017.7942281.
[34] Lara, Adrian, Anisha Kolasani, and Byrav Ramamurthy. "Network innovation using openflow: A survey." *IEEE Communications Surveys & Tutorials* 16, no. 1 (2013): 493–512.
[35] Shalimov, Alexander, Dmitry Zuikov, Daria Zimarina, Vasily Pashkov, and Ruslan Smeliansky. "Advanced study of SDN/OpenFlow controllers." In Proceedings of the 9th Central & Eastern European Software Engineering Conference in Russia, pp. 1–6, 2013.
[36] Farrington, Nathan, Erik Rubow, and Amin Vahdat. "Data center switch architecture in the age of merchant silicon." In 2009 17th IEEE Symposium on High Performance Interconnects, pp. 93–102. IEEE, 2009.
[37] Yamashita, Shinji, Akiko Yamada, Keiichi Nakatsugawa, Toshio Soumiya, Masatake Miyabe, and Toru Katagiri. "Extension of OpenFlow protocol to support optical transport network,

and its implementation." In 2015 IEEE Conference on Standards for Communications and Networking (CSCN), pp. 263–268. IEEE, 2015.

[38] Bosshart, Pat, Dan Daly, Glen Gibb, Martin Izzard, Nick McKeown, Jennifer Rexford, Cole Schlesinger et al. "P4: Programming protocol-independent packet processors." *ACM SIGCOMM Computer Communication Review* 44, no. 3 (2014): 87–95.

[39] Sivaraman, Anirudh, Changhoon Kim, Ramkumar Krishnamoorthy, Advait Dixit, and Mihai Budiu. "Dc. p4: Programming the forwarding plane of a data-center switch." In Proceedings of the 1st ACM SIGCOMM Symposium on Software Defined Networking Research, pp. 1–8. 2015.

[40] Briscoe, Neil. "Understanding the OSI 7-layer model." *PC Network Advisor* 120, no. 2 (2000): 13–16.

[41] Ibanez, Stephen, Gordon Brebner, Nick McKeown, and Noa Zilberman. "The P4→ NetFPGA workflow for line-rate packet processing." In Proceedings of the 2019 ACM/SIGDA International Symposium on Field-Programmable Gate Arrays, pp. 1–9, 2019.

[42] Wang, Han, Robert Soulé, Huynh Tu Dang, Ki Suh Lee, Vishal Shrivastav, Nate Foster, and Hakim Weatherspoon. "P4fpga: A rapid prototyping framework for P4." In Proceedings of the Symposium on SDN Research, pp. 122–135. 2017.

[43] Collings, Brandon. "New devices enabling software-defined optical networks." *IEEE Communications Magazine* 51, no. 3 (2013): 66–71.

[44] Sysrepo – YANG-based configuration and operational state data store for Unix/Linux applications. Project homepage. 2020. [Online]. Available: www.sysrepo.org/

[45] Scott, M., and M. Bjorklund. "YANG module for NETCONF monitoring." RFC 6022 (Proposed Standard), Internet Engineering Task Force (2010).

[46] Attar, Vahida Z., and Piyush Chandwadkar. "Network discovery protocol LLDP and LLDP-MED." *International Journal of Computer Applications* 1, no. 9 (2010): 93–97.

[47] Tousi, Yahya, and Ehsan Afshari. "14.6 A scalable THz 2D phased array with+ 17dBm of EIRP at 338GHz in 65nm bulk CMOS." In 2014 IEEE International Solid-State Circuits Conference Digest of Technical Papers (ISSCC), pp. 258–259. IEEE, 2014.

[48] Specification, OpenFlow Switch. "Oct. 14, 2013, version 1.4. 0 (Wire Protocol 0x05)." The Open Networking Foundation.

[49] Ryu – Component-based Software Defined Networking Framework. Project Homepage. 2020. [Online]. Available: https://ryu-sdn.org/

Index